ISBN 13: 978-1-6027-7305-9
ISBN 10:   1-6027-7305-X

9 1421 15 14
4500454501

SAXON® GEOMETRY

Student Edition

HOUGHTON MIFFLIN HARCOURT
Supplemental Publishers

www.SaxonPublishers.com
800-531-5015

Table of Contents

Section 1: Lessons 1–10, Investigation 1

DISTRIBUTED STRANDS

Geometry Foundations		Quadrilaterals	
Logic and Reasoning		Right Triangles and Trigonometry	
Construction		Circles	
Coordinate Geometry		Solids	
Triangles: Congruence and Similarity		Transformations	
Polygons			

Section 2: Lessons 11–20, Investigation 2

Section 3: Lessons 21–30, Investigation 3

DISTRIBUTED STRANDS

Geometry Foundations

Logic and Reasoning

Construction

Coordinate Geometry

Triangles: Congruence and Similarity

Polygons

Quadrilaterals

Right Triangles and Trigonometry

Circles

Solids

Transformations

Section 4: Lessons 31–40, Investigation 4

Section 5: Lessons 41–50, Investigation 5

DISTRIBUTED STRANDS

Geometry Foundations

Logic and Reasoning

Construction

Coordinate Geometry

Triangles: Congruence and Similarity

Polygons

Quadrilaterals

Right Triangles and Trigonometry

Circles

Solids

Transformations

Section 6: Lessons 51–60, Investigation 6

Section 7: Lessons 61–70, Investigation 7

DISTRIBUTED STRANDS

▢ Geometry Foundations	■ Quadrilaterals
▢ Logic and Reasoning	▢ Right Triangles and Trigonometry
▢ Construction	■ Circles
■ Coordinate Geometry	▢ Solids
■ Triangles: Congruence and Similarity	▢ Transformations
▢ Polygons	

Section 8: Lessons 71–80, Investigation 8

Section 9: Lessons 81–90, Investigation 9

DISTRIBUTED STRANDS

- Geometry Foundations
- Logic and Reasoning
- Construction
- Coordinate Geometry
- Triangles: Congruence and Similarity
- Polygons
- Quadrilaterals
- Right Triangles and Trigonometry
- Circles
- Solids
- Transformations

Section 10: Lessons 91–100, Investigation 10

Section 11: Lessons 101–110, Investigation 11

DISTRIBUTED STRANDS

▢ Geometry Foundations		▣ Quadrilaterals	
▢ Logic and Reasoning		▢ Right Triangles and Trigonometry	
▢ Construction		▣ Circles	
▣ Coordinate Geometry		▢ Solids	
▣ Triangles: Congruence and Similarity		▢ Transformations	
▢ Polygons			

Section 12: Lessons 111–120, Investigation 12

Points, Lines, and Planes

Warm Up

Start off each lesson by practicing prerequisite skills and math vocabulary that will make you more successful with today's new concept.

1. Vocabulary The _____ plane contains the *x*-axis and the *y*-axis.
(SB 13)

2. Kira needs to buy a piece of pipe that is 40% longer than the 7-inch piece
(SB 4) she already has. What length of pipe does Kira need?

3. Simplify $\sqrt[4]{81}$.
(SB 6)

4. Evaluate $\dfrac{4(n + 6)}{2n}$ for $n = 2$.
(SB 14)

New Concepts

In geometry, a **definition** of a term is a statement that defines a mathematical object. Definitions usually reference other mathematical terms. A basic mathematical term that is not defined using other mathematical terms is called an **undefined term**. In geometry, points, lines, and planes are undefined terms that are the building blocks used for defining other terms.

A **point** names a location and has no size. It is represented by a dot and labeled using a capital letter, such as *P*.

A **line** is a straight path that has no thickness and extends forever. There are an infinite number of points on a line. A line is named using either a lowercase letter or any two points on the line. Two possible names for the line shown in the diagram are $\overleftrightarrow{AB}$ and line *x*.

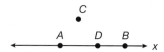

Any set of points that lie on the same line are called **collinear** points. In the diagram, *A*, *B*, and *D* are collinear. If points do not lie on the same line, they are **noncollinear**. Points *A*, *B*, and *C* are noncollinear.

Hint

A ruler can be used to determine if points are collinear or noncollinear. A ruler can always connect two points, so two points are always collinear. Three points are only collinear if you can use the ruler to draw a line passing through all three of them.

Online Connection
www.SaxonMathResources.com

Example 1 | **Identifying Lines and Collinear Points**

a. Give two different names for the line.

SOLUTION
Two possible names for the line are line *y* and $\overleftrightarrow{CD}$. The order of the points does not matter, so $\overleftrightarrow{DC}$ would also be correct.

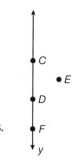

b. Name three collinear points and three noncollinear points.

SOLUTION
Points *C*, *D*, and *F* are collinear. Points *C*, *D*, and *E* are noncollinear.

A **plane** is a flat surface that has no thickness and extends forever. A plane is named using either an uppercase letter or three noncollinear points that lie in the plane. The plane in the diagram below could be called plane $\mathcal{P}$ or plane *ABC*.

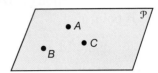

Lines or points that are in the same plane are said to be **coplanar**. If there is no plane that contains the lines or points, then they are **noncoplanar**. Space is the set of all points. Therefore, space includes all lines and all planes.

Example 2 Identifying Planes

What are two different names for this plane?

SOLUTION
Two possible names for the plane are plane *FGH* or plane $\mathcal{M}$.

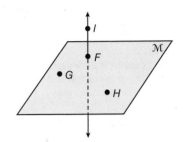

Example 3 Identifying Coplanar Lines

a. Identify the coplanar and noncoplanar lines in the diagram.

SOLUTION
Lines *m* and *n* are coplanar. Line ℓ is noncoplanar with lines *m* and *n*.

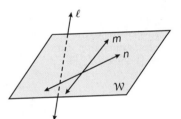

b. Identify the coplanar and noncoplanar lines in the diagram.

SOLUTION
Lines *r* and *s* are coplanar. Line *t* is noncoplanar with lines *r* and *s*.

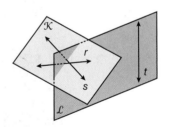

Each day brings you a **New Concept** where a new topic is introduced and explained through thorough **Examples** — using a variety of methods and real-world applications.

You will be reviewing and building on this concept throughout the year to gain a solid understanding and ensure mastery on the test.

Math Reasoning

Model Can two planes have no intersections at all? What common objects illustrate what this might look like?

An **intersection** is the point or set of points in which two figures meet. When two lines intersect, their intersection is a single point. When two planes intersect, their intersection is a single line. If a line lies in a plane, then their intersection is the line itself. If the line does not lie in the plane, then their intersection is a single point.

Lines *q* and *m* intersect at point *Q*. Plane $\mathcal{R}$ intersects plane $\mathcal{P}$ at line *m*. The intersection of plane $\mathcal{R}$ and line *m* is line *m*. Line *q* intersects planes $\mathcal{P}$ and $\mathcal{R}$ at point *Q*.

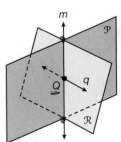

Example 4 Intersecting Lines and Planes

a. What is the intersection of $\overleftrightarrow{AB}$ and $\overleftrightarrow{CD}$?

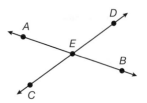

SOLUTION
The intersection of $\overleftrightarrow{AB}$ and $\overleftrightarrow{CD}$ is point E.

b. What is the intersection of $\overleftrightarrow{PQ}$ and $\overleftrightarrow{RS}$? What is the intersection of planes $\mathcal{M}$ and $\mathcal{L}$?

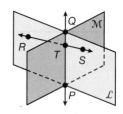

SOLUTION
The intersection of $\overleftrightarrow{PQ}$ and $\overleftrightarrow{RS}$ is point T. The intersection of the planes $\mathcal{M}$ and $\mathcal{L}$ is $\overleftrightarrow{PQ}$.

Lesson Practice

Identify each of the following from the diagram.

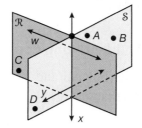

    a. All of the lines.
(Ex 1)
    b. A pair of collinear points.
(Ex 1)
    c. All of the planes.
(Ex 2)
    d. Three coplanar points.
(Ex 2)
    e. Two coplanar lines.
(Ex 3)
    f. A pair of noncoplanar lines.
(Ex 3)

Use the diagram to answer each question.

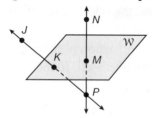

    g. What is the intersection of $\overleftrightarrow{JK}$ and $\overleftrightarrow{NM}$?
(Ex 4)
    h. What is the intersection of $\overleftrightarrow{JK}$ and plane $\mathcal{W}$?
(Ex 4)
    What is the intersection of $\overleftrightarrow{NP}$ and plane $\mathcal{W}$?

1. In the diagram, which set of three points are collinear?
₍₁₎ Which point cannot be included in a collinear set of three points?

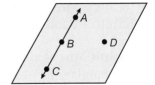

2. Can two points be noncollinear?
₍₁₎

3. Can two noncoplanar lines intersect?
₍₁₎

4. Name three undefined basic figures of geometry.
₍₁₎

5. What term describes two lines that have a point in common?
₍₁₎

6. Generalize Can three points be noncoplanar? Explain.
₍₁₎

7. Name the coplanar points shown on plane $\mathcal{M}$.
₍₁₎

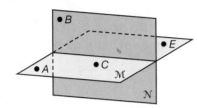

> The *italic numbers* refer to the lesson(s) in which the major concept of that particular problem is introduced. You can refer to the examples or practice in that lesson, if you need additional help.

8. Write The floor, ceiling, and walls of a room are all parts of planes. How many
₍₁₎ planes intersect the plane of the floor in your classroom? What geometric figures are formed where the planes intersect? Explain.

Use the diagram to answer problems 9–11.

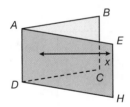

9. What is the intersection between the planes?
₍₁₎

10. Justify Are $\overleftrightarrow{AD}$, $\overleftrightarrow{CD}$, and $\overleftrightarrow{CH}$ coplanar? Explain.
₍₁₎

11. State the intersection point of $\overleftrightarrow{BC}$ and line x.
₍₁₎

12. Multiple Choice Which statement is true?
₍₁₎   **A** Any two lines are coplanar.   **B** Any three lines are coplanar.
   C Any two intersecting lines are coplanar.   **D** Any two perpendicular lines are noncoplanar.

13. How many different points can a line contain?
₍₁₎

14. How many different lines can a plane contain?
₍₁₎

15. Multiple Choice Which statement is true?
(1)

    A The intersection of two lines forms a one-dimensional figure.

    B The intersection of two planes forms a two-dimensional figure.

    C The intersection of two planes forms a one-dimensional figure.

    D The intersection of two lines forms a two-dimensional figure.

In the **Practice,** you will review today's new concept as well as math you learned in earlier lessons. By practicing problems from many lessons every day, you will begin to see how math concepts relate and connect to each other and to the real world.

Also, because you practice the same topic in a variety of ways over several lessons, you will have "time to learn" the concept and will have multiple opportunities to show that you understand.

16. Error Analysis Miri used two points to name a plane. What mistake did Miri make in naming the plane?
(1)

17. Evaluate: $5 - (7 + 8) \div 5 + (-2)^3$
(SB 1)

18. Name the property of addition shown by this equation:
(SB 2) $(+3) + (-4) = (-4) + (+3)$

19. Error Analysis Jacob said that -3 is a rational number. Aaron said that -3
(SB 3) is an irrational number. Who is correct? Explain.

20. Justify Is $0.\overline{3}$ irrational? Why or why not?
(SB 3)

21. (**Baseball**) A ball player is at bat 55 times and hits the ball 33 times. What ratio of
(SB 41) his times at bat does he not hit the ball?

22. Evaluate: $(-3)^3 - \left(\frac{1}{3}\right)^{-3}$
(SB 1)

23. Simplify: $2\sqrt{12} + 6\sqrt{27}$
(SB 6)

24. (**Construction**) A concrete pad has dimensions 9 feet by 9 feet by 4 inches. How
(SB 9) many cubic yards of concrete does it contain?

25. (**Meteorology**) Determine the mean and median values for the weekly rainfall
(SB 11) data.

Day	Monday	Tuesday	Wednesday	Thursday	Friday	Saturday	Sunday
Rainfall (mm)	0.5	2	0	4	2.5	5	7

26. (**Physics**) On average, the acceleration due to gravity is 9.807 m/s^2. A science
(SB 10) student measured it as 9.760 m/s^2. To the nearest hundredth of a percent, what was the student's percent error?

27. Coordinate Geometry Plot these points on the coordinate plane: $(2, 1)$, $(-1, -1)$,
(SB 13) $(0, 0)$, and $(3, -1)$.

28. Algebra Evaluate the expression $xy^{-2} + \frac{x}{y}$, where $x = -2$, $y = \frac{1}{2}$.
(SB 14)

The mixed set of Practice is just like the mixed format of your state test. You'll be practicing for the "big" test every day!

29. Algebra Transform the formula $I = Prt$ to solve for r.
(SB 16)

30. Algebra State the slope of the line $3x + 4y - 15 = 0$.
(SB 19)

Segments

Warm Up

1. **Vocabulary** Points that lie on the same line are called _____ points.
(1)

2. Solve for x: $5x + 6 = 2x - 5$
(SB 15)

3. Simplify: $5(2x - 6) + 3x - 7$
(SB 1)

New Concepts

A **line segment** is a part of a line consisting of two **endpoints** and all points between them.

Math Language

Point C is **between** points A and B if A, B, and C are collinear and $AC + CB = AB$.

The diagram above depicts a line segment with endpoints A and B. A segment is named by its two endpoints in either order with a straight segment drawn over them. This segment could be called either $\overline{AB}$ or $\overline{BA}$.

Two geometric objects that have the same size and shape are **congruent**. **Congruent segments** have the same length.

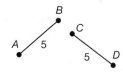

Caution

When comparing segments or other geometric figures, congruence statements are used. When the length of segments are being compared, or any other measurements that can be expressed as numbers, an equal sign is used.

In this figure, $\overline{AB}$ and $\overline{CD}$ are congruent. As shown on the diagram, they both have a **length** of 5 units.

A **congruence statement** shows that two segments are congruent. The symbol $\cong$ is read "is congruent to." The congruence statement for the segments above is $\overline{AB} \cong \overline{CD}$.

In a diagram, congruent segments are shown with **tick marks**. The diagram below shows congruent segments indicated by tick marks.

Hint

For more on the Reflexive, Symmetric, and Transitive Properties of Equality, see the Skills Bank at the back of this textbook.

The following properties apply to all congruent segments.

Reflexive Property of Congruence	$\overline{AB} \cong \overline{AB}$
Symmetric Property of Congruence	If $\overline{AB} \cong \overline{CD}$, then $\overline{CD} \cong \overline{AB}$.
Transitive Property of Congruence	If $\overline{AB} \cong \overline{CD}$ and $\overline{CD} \cong \overline{EF}$, then $\overline{AB} \cong \overline{EF}$.

Example 1 Using Properties of Equality and Congruence

Identify the property that justifies each statement.

a. $\overline{WX} \cong \overline{YZ}$, so $\overline{YZ} \cong \overline{WX}$

SOLUTION
Symmetric Property of Congruence

b. $\overline{PQ} \cong \overline{RS}$ and $\overline{RS} \cong \overline{TU}$, so $\overline{PQ} \cong \overline{TU}$

SOLUTION
Transitive Property of Congruence

c. $\overline{GH} \cong \overline{GH}$

SOLUTION
Reflexive Property of Congruence

Reading Math

The length of $\overline{AB}$ is denoted AB.

A ruler can be used to measure the lengths of segments. The points on a ruler correspond with the points on a line segment. This concept is presented in the Ruler Postulate. A **postulate** is a statement that is accepted as true without proof.

Postulate 1: Ruler Postulate
The points on a line can be paired in a one-to-one correspondence with the real numbers such that: **1.** Any two given points can have coordinates 0 and 1. **2.** The distance between two points is the absolute value of the difference of their coordinates.

Distance is the measure of the segment connecting two points. The distance between two points can be represented by those two points with no segment symbol. For example, AB means "the distance between A and B."

Distance is always positive, so absolute values are used to calculate distances.

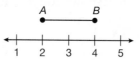

Online Connection
www.SaxonMathResources.com

$|\text{Point } A - \text{Point } B| = |2 - 4| = |-2| = 2$

The distance from point A to point B is 2.

Example 2 Finding Distance on a Number Line

Find each distance.

a. AB

SOLUTION

$AB = |6 - 3|$
$\quad = |3|$
$\quad = 3$

b. BC

SOLUTION

$BC = |3 - (-5)|$
$\quad = |8|$
$\quad = 8$

c. CD

SOLUTION

$CD = |(-5) - (-1)|$
$\quad = |-4|$
$\quad = 4$

d. AC

SOLUTION

$AC = |6 - (-5)|$
$\quad = |11|$
$\quad = 11$

Hint

Postulates are statements that are accepted as true without proof. See the Postulates and Theorems section in the back of this book for a complete list of postulates in this program.

In the example above, notice that $AC = AB + BC$. This is not a coincidence.

Postulate 2: Segment Addition Postulate
If B is between A and C, then $AB + BC = AC$.

Example 3 Using the Segment Addition Postulate

a. Point S lies on $\overline{RT}$ between R and T. $RS = 12$ and $RT = 31$. Find ST.

SOLUTION

$RT = RS + ST$ Segment Addition Postulate
$31 = 12 + ST$ Substitute.
$19 = ST$ Subtract 12 from both sides.

b. Find AC in terms of x.

$$2x - 7 \qquad 3x + 9$$
$$A \quad\quad B \qquad\qquad C$$

SOLUTION

$AC = AB + BC$ Segment Addition Postulate
$AC = (2x - 7) + (3x + 9)$ Substitute.
$AC = 5x + 2$ Simplify.

The **midpoint** of a segment is the point that divides the segment into two congruent parts. If M is the midpoint of $\overline{AB}$, then $AM = MB$.

Midpoint

Example 4   Application: Hiking

A hiker is traveling up a mountain towards the summit. The distance from the base of the mountain to the summit is 2.5 miles, as shown. How far will she have traveled when she reaches the midpoint (Y) of the hike?

SOLUTION

$XZ = XY + YZ$	Segment Addition Property
$XY = YZ$	Definition of midpoint
$XZ = XY + XY$	Substitute *XY* for *YZ*.
$2.5 = 2(XY)$	Substitute.
$1.25 = XY$	Divide both sides by 2.

The hiker will have traveled 1.25 miles when she reaches the midpoint.

Lesson Practice

a. Identify the property that justifies the statement,
(Ex 1) $\overline{KL} \cong \overline{MN}$, so $\overline{MN} \cong \overline{KL}$.

b. Find the distance between the points A and B.
(Ex 2)

c. Find AC in terms of x.
(Ex 3)

$$x^2 - x + 3 \quad\quad 2x + 7$$

d. The drive from Seattle to San Francisco is 811 miles. How many miles
(Ex 4) is the midpoint from either city?

Practice   Distributed and Integrated

1. Measure the line segments at right to determine
(2)   which segments are congruent.

2. Estimate $\sqrt{32}$ to the nearest tenth.
(SB 6)

3. What property is illustrated by this statement?
(SB 2)
   If a and b are real numbers, then a + b = b + a.

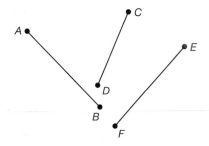

4. Multiple Choice What is the least number of points that can determine a plane?
(1)
   A 1   **B** 2

   C 3   **D** 4

Calculate the length of each segment using the diagram below.

5. $\overline{AD}$
(2)

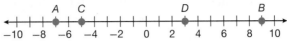

6. $\overline{BC}$
(2)

7. $\overline{DA}$
(2)

8. $\overline{AC}$
(2)

9. Data Analysis The numbers in this set are the math test scores for a class. Find the
(SB 11) mode.

$$\{62, 53, 74, 55, 66, 72, 80, 83, 83, 86, 92, 93, 40, 51, 61, 71\}$$

10. Evaluate $3m^2 - 5m + 17$ for $m = -3$.
(SB 14)

11. In what quadrant is point S if its coordinates are $(2, -4)$?
(SB 13)

12. Find the median of the set $\{2, 2, 3, 5, 6, 7, 7, 8, 9\}$.
(SB 11)

13. Algebra A, B, and C are collinear, $AB = 5x - 19$, and $BC = 3x + 4$. Find an
(2) expression for AC if B is between A and C.

14. (City Planning) The city is planning to install streetlights and wants five lights
(2) along a walkway of 60 yards. If there is a light at the beginning and at the end of
the walkway and the lights are evenly spaced, what is the distance between each
light?

15. Error Analysis Sunil stated that three points determine a unique plane. Explain
(1) Sunil's error and and give a corrected statement.

16. Algebra Suppose $AB = 3x$, $BC = 2y + 16$, $AC = 60$, and B is the midpoint of AC.
(2) Find the values of x and y.

17. Analyze Points D, E, and F are collinear with E between D and F. $DE = 15$,
(2) $EF = x + 17$, and $DF = 3x - 10$. Find EF and DF.

18. Write Describe how equality and congruence are used to describe two line
(1, 2) segments and their lengths.

19. Are points A, B, and C collinear, coplanar, both, or neither?
(1)

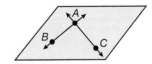

20. How many points are required to determine a line?
(1)

21. Multiple Choice Which of these statements is false?
(1)
   A Two distinct points determine a line.

   B Three noncollinear points determine a plane.

   C Two noncoplanar lines determine a space.

   D Four noncoplanar points determine a space.

22. Lines MN and PQ intersect at point E. Name two sets of three collinear points.
$_{(1)}$

23. Factor: $2x^2 - 16x - 66$
$_{(SB\ 18)}$

24. Expand: $(x - 4)(x + 7)$
$_{(SB\ 18)}$

25. $\boxed{\textbf{Carpentry}}$ A piece of wood that is 8 feet long needs to be 7 feet 4 inches long.
$_{(SB\ 9)}$ How much has to be cut off? Express your answer in feet.

26. Simplify: $\dfrac{-36x^{-4}y^5}{12x^2y^{-3}}$. Express your answer with positive exponents.
$_{(SB\ 6)}$

27. $\boxed{\textbf{Sales Tax}}$ If sales tax is 6%, estimate the sales tax on an item that retails for
$_{(SB\ 4)}$ $48.99.

28. Estimate the sum to the nearest whole number.
$_{(SB\ 5)}$ $1.8 + 2.345 + 0.65 + 13.56$

29. Multi-Step There are 1000 liters in a cubic meter and approximately 3.85 liters
$_{(SB\ 9)}$ in one gallon. A swimming pool measures 8 meters wide by 4 meters long by 1.5 meters deep. Approximately how many gallons of water can the pool hold?

30. $\boxed{\textbf{Physics}}$ In a free fall, acceleration due to gravity is approximately 9.8 m/s^2.
$_{(SB\ 9)}$ Convert this rate into ft/s^2.

Angles

1. **Vocabulary** Two figures that have the same size and shape are called _____ figures.
   _(2)_

2. Simplify: $\dfrac{1}{2}\left(\dfrac{2^2}{2} - 2\right)$
   _(SB 1)_

3. Convert $\dfrac{22}{7}$ into a decimal number. Is it a terminating decimal number,
   _(SB 3)_ repeating decimal number, or non-terminating and non-repeating decimal number?

New Concepts   A **ray** is a part of a line that starts at an endpoint and extends infinitely in one direction.

A ray is named by its endpoint and any other point on the ray. For example, the ray in the diagram is called $\overrightarrow{AB}$, which is read "ray AB."

Two rays that have a common endpoint and form a line are called **opposite rays**.

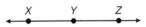

Rays $\overrightarrow{YX}$ and $\overrightarrow{YZ}$ are opposite rays.

An **angle** is a figure formed by two rays with a common endpoint. The common endpoint is the angle's **vertex**. The rays are the **sides** of the angle. The sides of this angle are $\overrightarrow{BA}$ and $\overrightarrow{BC}$. The vertex is B.

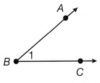

An angle can be named in several different ways: by its vertex, by a point on each ray and the vertex, or by a number. For example, the angle in the diagram could be called $\angle B$, $\angle ABC$, $\angle CBA$, or $\angle 1$.

The exterior of an angle is the set of all points outside the angle. The interior of an angle is the set of all points between the sides of an angle.

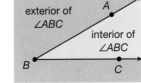

Caution

An angle can be named by its vertex only if it is clear that there is only one angle at the vertex.

Example 1 Naming Angles and Rays

a. Name three rays in the diagram.

SOLUTION
$\overrightarrow{SP}$, $\overrightarrow{SQ}$, and $\overrightarrow{SR}$

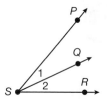

b. Name three angles in the diagram.

SOLUTION
$\angle PSQ$ or $\angle 1$, $\angle QSR$ or $\angle 2$, and $\angle PSR$.

c. Could $\angle PSQ$ also be referred to as $\angle S$?

SOLUTION
No, there are three different angles with S as a vertex.

A **protractor** is a tool used to measure angles. Unlike segments, angles are measured in **degrees**. One degree is a unit of angle measure that is equal to $\frac{1}{360}$ of a circle.

Postulate 3: Protractor Postulate

Given a point X on $\overleftrightarrow{PR}$, consider rays $\overrightarrow{XP}$ and $\overrightarrow{XR}$, as well as all the other rays that can be drawn with X as an endpoint, on one side of $\overleftrightarrow{PR}$. These rays can be paired with the real numbers from 0 to 180 such that:

1. $\overrightarrow{XP}$ is paired with 0, and $\overrightarrow{XR}$ is paired with 180.

2. If $\overrightarrow{XA}$ is paired with a number c and $\overrightarrow{XB}$ is paired with a number d then $m\angle AXB = |c - d|$.

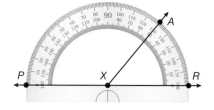

Angles are classified according to their angle measure.

An **acute angle** measures greater than 0° and less than 90°.

An **obtuse angle** measures greater than 90° and less than 180°.

A **right angle** measures exactly 90°. A box drawn at the vertex of an angle shows that it is a right angle, as shown in the diagram.

A **straight angle** measures exactly 180°.

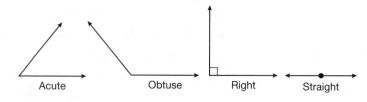

Example 2 **Measuring and Classifying Angles**

a. Use a protractor to measure ∠ABC, then classify the angle.

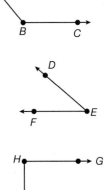

SOLUTION ∠ABC measures 130°, so it is an obtuse angle.

b. Use a protractor to measure ∠DEF, then classify the angle.

SOLUTION ∠DEF measures 40°, so it is an acute angle.

c. Use a protractor to measure ∠GHI, then classify the angle.

SOLUTION ∠GHI measures 90°, so it is a right angle.

Angles can be added in the same way that segments are added.

Postulate 4: The Angle Addition Postulate

If point D is in the interior of ∠ABC, then

m∠ABD + m∠DBC = m∠ABC.

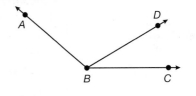

Example 3 **Using the Angle Addition Postulate**

The measure of ∠RST = 22° and m∠TSU = 69°. Find m∠RSU. Classify the angle.

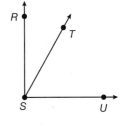

SOLUTION

m∠RST + m∠TSU = m∠RSU   Angle Addition Postulate

22° + 69° = m∠RSU   Substitute.

91° = m∠RSU   Simplify.

∠RSU is an obtuse angle.

To **bisect** a figure is to divide it into two congruent parts. An **angle bisector** is a ray that divides an angle into two **congruent angles**. Congruent angles have the same measure. They are marked with **arc marks**, as shown in the diagram.

∠XYW ≅ ∠WYZ and ∠ZYS ≅ ∠SYT.

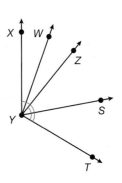

Example 4 Using Angle Bisectors and Congruence Marks

The measure of $\angle ABC = 44°$. $\overrightarrow{BC}$ bisects $\angle ABD$. The measure of $\angle EBF = 23°$. Find the measure of $\angle CBE$.

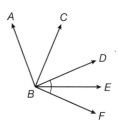

SOLUTION Since $\overrightarrow{BC}$ bisects $\angle ABD$, it divides $\angle ABD$ into two congruent angles. So, $\angle ABC \cong \angle CBD$ and $m\angle ABC = m\angle CBD$. Since $m\angle ABC = 44°$, $m\angle CBD = 44°$.

Using the congruence marks in the diagram, $\angle DBE \cong \angle EBF$, so $m\angle DBE = m\angle EBF$. Since $m\angle EBF = 23°$, $m\angle DBE = 23°$.

$$
\begin{aligned}
m\angle CBE &= m\angle CBD + m\angle DBE &&\text{Angle Addition Postulate} \\
&= 44° + 23° &&\text{Substitute.} \\
&= 67° &&\text{Add.}
\end{aligned}
$$

The measure of $m\angle CBE$ is 67°.

Example 5 Application: Interpreting Statistics

Hint

For more about displaying data using circle graphs, see the Skills Bank at the back of this textbook.

Louis runs a restaurant. He knows that he has about 900 customers a day. The circle graph in the diagram shows what percentage of his customers fall into the given age brackets. He wants to know exactly how many of his customers are between ages 15 and 20. Use a protractor to measure the angle and find the number of Louis's customers that fall into the 15–20 age bracket.

SOLUTION

Measure the angle of the sector that represents 15–20-year-old customers. The sector has an angle measure of 120°. Since an entire circle is 360°, this is $\frac{120}{360} = \frac{1}{3}$ of the circle. One third of Louis's customers is $\left(\frac{1}{3}\right)(900) = 300$ customers.

Customers by Age

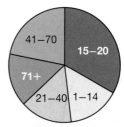

Lesson Practice

a. Name three rays and three angles in the diagram.
(Ex 1)

b. Classify $\angle ABC$ and use a protractor to find the measure of it.
(Ex 2)

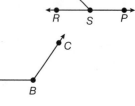

c. Determine $m\angle AEB$ if $m\angle AED = 120°$.
(Ex 3)

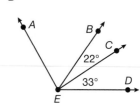

d. The measure of $\angle WXY = 32°$. $\overrightarrow{XY}$ bisects $\angle WXZ$.
(Ex 4) The measure of $\angle UXV = 35°$. Find the measure of $\angle YXU$.

e. A survey shows that 10% of students in a class
(Ex 5) did not eat lunch. What would be the degree measure of an angle indicating these students on a circle graph?

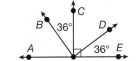

Practice Distributed and Integrated

Use the diagram to classify and find the measure of each angle.

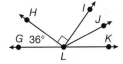

1. $\angle AFC$
(3)

2. $\angle CFD$
(3)

3. $\angle BFD$
(3)

4. $\angle AFD$
(3)

5. Multiple Choice $\overrightarrow{LJ}$ bisects $\angle ILK$. What is the measure of $\angle GLJ$?
(3)
   A 27°   **B** 117°
   C 54°   **D** 153°

6. What is the measure of a straight angle?
(3)

7. Find AB.
(2)

8. Which of the following triplets are equivalent?
(SB 4)
   A $\frac{1}{2}$, 50%, 5:10   **B** $\frac{20}{20}$, 100%, 7:7

   C $\frac{30}{36}$, 83.$\overline{3}$%, 5:6   **D** All of the above

9. What is the fewest number of points that can determine a line?
(1)

10. How many significant digits are there in 10,001?
(SB 7)

11. Error Analysis Jackson said, "Congruent segments are the same as equal segments."
(2) Explain how he has misunderstood congruence and equality.

12. Algebra Ray BD bisects $\angle ABC$. If $m\angle ABD = (x^2 + x + 12)°$ and
(3) $m\angle DBC = (x^2 + 3x + 4)°$, find x and $m\angle ABC$.

13. (**Landscaping**) A homeowner wants to put up a fence around her square yard.
(2) She has 8 posts, which must be equally spaced. If the perimeter of her yard is 16 meters, how far apart must the posts be?

14. **Verify** If $\overline{MN} \cong \overline{NH}$ and $\overline{NH} \cong \overline{ES}$, then $\overline{MN} \cong \overline{ES}$. What property justifies this?
(2)

15. Points A, B, and C are collinear. Point B is between points A and C. $AB = 12$,
(2) $AC = 7x + 5$ and $BC = 4x - 1$. Find x.

16. Find the midpoint of the segment connecting points with coordinates -142 and
(2) 53 on a number line.

17. **Write** If four points are collinear, are they also coplanar? Explain.
(1)

18. The ratio of boys to girls at the camp is 14:17. If there are 186 children at the
(SB 4) camp altogether, how many are boys?

19. **Algebra** Factor: $10x^2 + x - 21$
(SB 18)

20. The mean of these six numbers is 20.38. What is the missing number?
(SB 11)
25.14, 17.22, _____, 23.04, 20.21, 21.27

21. (Surveying) A surveyor measured a building lot. She recorded the corner points
(SB 13) of the lot as $M(-80, 210)$, $N(-80, -120)$, $P(130, 210)$, and $Q(130, -120)$. The
coordinates represent the distances, in feet, from the axes of a coordinate grid.
Calculate the area of the lot.

22. (Farming) A farmer wishes to plant a field with flax. According to his budget, the
(SB 22) total cost of planting and harvesting cannot exceed $12,000. The planting and
harvesting costs are $0.015/square yard for flax. What is the maximum area the
farmer could plant with flax while keeping within the budget?

23. How does the graph of $f(x)$ compare to the graph of $f(x) + 3$?
(SB 17)

24. Solve the equation $\sqrt{6g} = (j + y)$ for g.
(SB 16)

25. A stone is dropped from a cliff into a river. The function $h(t) = 82 - 4.9t^2$ gives
(SB 22) the height h of the stone in meters t seconds after it is dropped. What is the height
of the stone, to the nearest tenth of a meter, after 1.7 seconds?

26. (Solar System) Sirius, the brightest visible star in the sky, is about 8.6 light years
(SB 7) away from Earth. One light year is 9.46×10^{15} kilometers. Find the distance
between Earth and Sirius in kilometers. Express your answer in scientific notation
with 3 significant figures.

27. The instructions on a juice concentrate container read "Five parts water,
(SB 4) 1 part concentrate." How much juice can you make if you have 500 mL
of concentrate?

28. To what sets of numbers does the number -8 belong?
(SB 3)

29. Is the square root of any prime number a real number?
(SB 3)

30. (Chemistry) Liquid nitrogen is extremely cold. The freezing point for nitrogen
(SB 1) is approximately $-346°F$ and the boiling point for nitrogen is approximately
$-320°F$. Use this data to find the range.

Congruent Segments and Angles

Construction Lab 1 *(Use with Lesson 3)*

You can use a compass and a straightedge to construct figures. In Lessons 2 and 3, you learned about congruent segments and angles. The first figure you will construct using tools of geometry is a segment congruent to a given segment.

(1.) Begin with a given segment. Label the endpoints of the segment A and B. Set the compass to have radius AB.

(2.) Draw point C not on the segment and sweep an arc from that point using the compass setting from step 1. Because congruency does not depend on orientation, any of the points on the arc may serve as the other endpoint of the second segment.

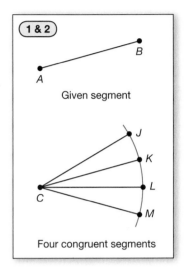

You can use the same tools you used to construct a segment to construct an angle congruent to a given angle.

(1.) To construct congruent angles, begin by drawing an angle of any measure and label it $\angle TSP$.

(2.) Draw $\overrightarrow{GH}$. This ray will be one side of the congruent angle.

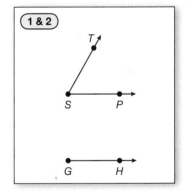

(3.) Use a compass to draw an arc across $\angle TSP$ centered at S. Be sure that the arc intersects both $\overrightarrow{ST}$ and $\overrightarrow{SP}$.

(4.) Label the points of intersection A and B.

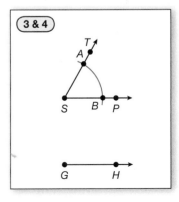

5. Use the same compass setting to draw an arc centered at *G*.

6. Label the point of intersection of this arc and $\overrightarrow{GH}$ as *C*.

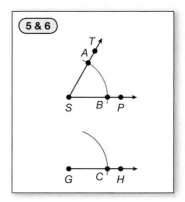
5 & 6

7. Adjust the compass setting to the distance between points *A* and *B*. Use this setting to draw an arc centered at *C* which intersects the first arc you drew on $\overrightarrow{GH}$.

8. Label the point of intersection of the two arcs as point *D*.

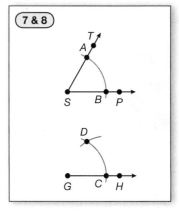
7 & 8

9. Draw $\overrightarrow{GD}$.

$\angle DGH \cong \angle TSP$

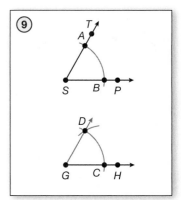
9

Math Reasoning

Verify Use a protractor to verify that the given angle and the angle you constructed are congruent.

Lab Practice

Practice constructing congruent segments and angles in pairs. Sketch a segment and an angle, then trade sketches with a partner. Next, using your compass, construct a segment and angle congruent to the ones given.

Postulates and Theorems About Points, Lines, and Planes

Warm Up

1. **Vocabulary** The _____ states that $a = a$ for any value of a.
   (2)

2. Classify the angle at right.
   (3)

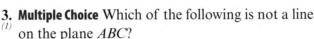

3. **Multiple Choice** Which of the following is not a line
   (1) on the plane *ABC*?

   A $\overleftrightarrow{AB}$ **B** $\overleftrightarrow{CD}$

   C $\overleftrightarrow{BC}$ **D** $\overleftrightarrow{GH}$

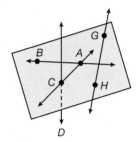

New Concepts

The postulates below, all of which help define the properties of points, lines, and planes, are essential to proving and understanding many of the most important theorems of geometry.

Postulate 5
Through any two points there is exactly one line.

Postulate 5 simply states that, given a point *A*, and a point *B*, there is only one line that can be drawn through both points. The line $\overleftrightarrow{AB}$ is therefore unique.

Hint

Unlike postulates, theorems must be proved to be accepted as true. See the Postulates and Theorems section at the back of this textbook for a complete list of theorems and their proof locations.

Theorem 4-1
If two lines intersect, then they intersect at exactly one point.

Postulate 5 can be used to prove the theorem above, which states that when two lines intersect, they must intersect at only one point.

Example 1 Using Postulates and Theorems

Does this diagram show two distinct lines through points *A* and *B*? Use Postulate 5 and Theorem 4-1 in your answer.

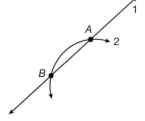

SOLUTION
Postulate 5 says that through any two points, there can be only one line. In the diagram, two "lines" pass through points *A* and *B*. However, the curve in this diagram is not straight and therefore is not an example of a line. Theorem 4-1 provides further evidence that the curve is not a line, since we know that two lines must intersect at only one point, whereas the line and the curve in this diagram intersect in two points.

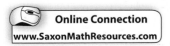
Online Connection
www.SaxonMathResources.com

Postulate 6
Through any three noncollinear points there exists exactly one plane.

If three points are collinear, then there are an infinite number of planes that can be drawn through them.

Theorem 4-2
If there is a line and a point not on the line, then exactly one plane contains them.

Theorem 4-3
If two lines intersect, then there exists exactly one plane that contains them. 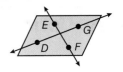

Math Reasoning

Model For three points to define a plane, they must be noncollinear. Draw three collinear points. Then draw two planes, both of which contain all three points.

Postulate 6, Theorem 4-2, and Theorem 4-3 give conditions for determining exactly one plane. Any of the following are sufficient: three noncollinear points, a line and a point not on the line, or two intersecting lines.

Example 2 **Identifying Points and Lines in Planes**

Name the following:

a. four points

b. two lines

c. two planes

SOLUTION

a. Points A, B, C and D

b. $\overleftrightarrow{AB}$ and $\overleftrightarrow{CD}$

c. Planes $\mathcal{M}$ and $\mathcal{N}$

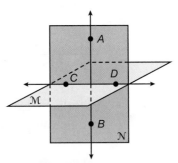

Just as two lines intersect in exactly one point, two planes intersect in exactly one line.

Postulate 7

If two planes intersect, then their intersection is a line. Planes $\mathcal{M}$ and $\mathcal{N}$ intersect at $\overleftrightarrow{AB}$.

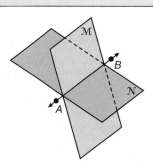

Math Reasoning

Generalize Is it possible to name three noncoplanar points in the diagram? Is it ever possible to name three noncoplanar points? Is it possible to name four noncoplanar points in the diagram?

Example 3 **Identifying Intersections of Lines and Planes**

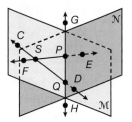

a. Identify the intersection of planes $\mathcal{M}$ and $\mathcal{N}$.

SOLUTION

The intersection of two planes is a line. Though there are several lines in the diagram, only one of them lies in both plane $\mathcal{M}$ and plane $\mathcal{N}$. The intersection is $\overleftrightarrow{GH}$.

b. Identify all points of intersection of lines on plane $\mathcal{M}$.

SOLUTION

Lines CD, FE, and GH all intersect in plane $\mathcal{M}$. The points of intersection are points P, Q, and S.

Any two points in a plane are contained by a line on the same plane.

Postulate 8
If two points lie on a plane, then the line containing the points lies in the plane.

Postulate 9 gives the minimum number of points needed to define exactly one line, one plane, and space.

Postulate 9
A line contains at least 2 points. A plane contains at least 3 noncollinear points. Space contains at least 4 noncoplanar points.

Example 4 **Application: Carpentry**

Often small tables have only three legs. They have the advantage of not tipping if placed on an uneven surface. Use postulates given in this lesson to describe why this is true and why a three-legged table will wobble less than a four-legged table.

SOLUTION The legs of a three-legged table make a single plane, as shown in Postulate 7. The legs are noncollinear points. Even if they are uneven, the table will be stable. If the table has four legs and one of the legs is higher than the other three, the table will tip. Four points can be noncoplanar, so the table can tip between them.

Lesson Practice

a. Can another plane be drawn that contains
(Ex 1) points *A*, *B*, and *C*? Justify your answer using a postulate.

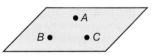

b. Identify a line in plane M. What is the
(Ex 2, Ex 3) intersection of $\overleftrightarrow{CD}$ and $\overleftrightarrow{AB}$? What is the intersection of planes M and N? Are points *E* and *F* coplanar?

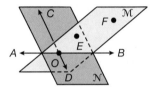

c. Often camping supplies are made with three
(Ex 4) legs. For example, chairs, stoves, and small tables often have three legs. Explain why they might be made this way and use postulates to explain your reasons.

Practice Distributed and Integrated

* **1. Multiple Choice** Which of the following is a set of noncollinear points?
(4)

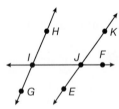

A *H, I, G*   **B** *I, J, F*
C *H, I, K*   **D** *K, J, E*

* **2. Write** Describe the difference between collinear points and coplanar lines.
(4)

* **3.** Find the theorem or postulate that justifies the following statement.
(4)
If two lines are on the same plane, they will only intersect at one point.

24   *Saxon* Geometry

Draw each of the following.

* **4.** Two planes that intersect at a line.
(4)

* **5.** Two lines that intersect on a plane at point *O*.
(4)

* **6.** Is this statement always true, never true, or sometimes true?
(4)
If two lines intersect then they are both contained in a single plane.

Use the figure to answer problems 7–9.

* **7.** Name two rays.
(3)

* **8.** What is the measure of ∠*ABC*?
(3)

* **9.** What name is given to this type of angle?
(3)

***10.** The base of a crane at a construction site makes an angle of 30° with the ground.
(3) As it lifts its load to clear an obstruction, the base makes an additional angle of 13°. What is the base's angle with the ground after lifting the load?

***11.** If m∠*ABC* = 62° and $\overline{DB}$ bisects ∠*ABC*, what is the measure of ∠*ABD*?
(3)

***12.** **Write** Describe how to locate the quarter points of a line segment, $\overline{AB}$.
(2)

***13.** **Algebra** Find the expression to represent *AC* if *AB* = 2*x* + 5, *BC* = 5*x* − 16 and *A*,
(2) *B* and *C* are collinear. *B* is between *A* and *C*.

 14. **Probability** Handel flips a fair coin seven times and it comes up heads seven times.
(SB 12) What is the probability his next flip will be heads? Why?

***15.** (**Traffic**) There are three traffic lights on a street. If the second traffic light is
(2) 75 yards from the first, and all three traffic lights span 120 yards, how far is the second traffic light from the third one?

***16.** **Multiple Choice** Which statement is true? Make a drawing to show why each other
(1) statement is not always true.
   A Any two lines are coplanar.
   B Any two intersecting lines are coplanar.
   C Any three intersecting lines are coplanar.
   D Any two perpendicular lines are noncoplanar.

***17.** How many points are between any two points on a line?
(1)

***18.** What word describes two coplanar lines that never meet?
(1)

 19. What is the parent function of $\dfrac{15 - 6x^2}{2}$?
(SB 17)

 20. **Analyze** Graph *y* = |*x* − 3|. Is it a function?
(SB 17)

 21. **Algebra** Solve for *j*. $\dfrac{4j - 12x^2}{3} = j$
(SB 16)

 **22.** **Algebra** Solve for m. $-4m + 2 < 26$
(SB 15)

*23. Determine AC.
(2)

A ⎯⎯ B ⎯⎯ C
-4 -2 0 2 4 6

24. Jaime measures the volume of a container to be 1.15 liters. If its actual volume is
(SB 10) 1.25 liters, what was Jaime's percent error?

25. Convert 18 centimeters to inches. Round to the nearest tenth.
(SB 9)

26. How many inches equal one-half of a mile?
(SB 9)

27. Express 23,000,000 in scientific notation.
(SB 7)

28. Evaluate 3^5.
(SB 6)

29. Round 34,016.45 to the nearest ten.
(SB 5)

30. What property of arithmetic is shown by $(a)(0) = 0$?
(SB 2)

More Theorems About Lines and Planes

Warm Up

1. **Vocabulary** Lines that lie in the same plane are _____.
   (1)
2. What geometric figure is formed when two planes intersect?
   (4)
3. **Multiple Choice** Which of these is not an undefined term?
   (1)
   A line   **B** plane
   C point   **D** perpendicular

4. How many points are needed to define a line?
   (4)
5. How many noncollinear points are needed to define a plane?
   (4)

New Concepts

Lines and planes are classified by whether or not they intersect and how they intersect. When lines intersect to form a right angle, they are called **perpendicular** lines. The symbol to show that two lines are perpendicular is ⊥. In the diagram, $\overline{AE}$ and $\overline{EH}$ are perpendicular, or simply: $\overline{AE} \perp \overline{EH}$.

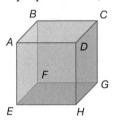

Math Language

Planes, segments, and rays can also be perpendicular to one another if they intersect at 90° angles.

Coplanar lines that do not intersect are called **parallel lines**. The symbol to show that two figures are parallel is ∥. In the diagram, $\overline{AB} \parallel \overline{DC}$. Planes that do not intersect are **parallel planes**. In the diagram, the plane *ABC* is parallel to the plane *EFG*.

To indicate that lines are perpendicular on a diagram, it is only necessary to indicate that two segments intersect at a right angle, as $\overline{WX}$ and $\overline{WZ}$ do in the diagram of the square *WXYZ*. To indicate that lines are parallel in a diagram, arrowheads are drawn on them. In the diagram, the corresponding arrowheads indicate that $\overline{WX} \parallel \overline{ZY}$ and $\overline{WZ} \parallel \overline{XY}$.

If lines are not in the same plane and do not intersect, they are **skew lines**. In the cube shown above, the lines that contain $\overline{DH}$ and $\overline{EF}$ are skew.

Theorem 5-1 can be used to identify parallel lines.

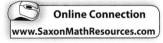

Online Connection
www.SaxonMathResources.com

Theorem 5-1
If two parallel planes are cut by a third plane, then the lines of intersection are parallel.

Example 1 **Identifying Parallel Lines**

In the figure, planes S and T are parallel. Identify two pairs of parallel lines.

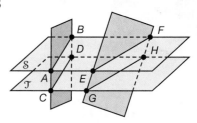

SOLUTION

Since planes S and T are parallel, their intersections with any plane that crosses them form a pair of parallel lines. Therefore, $\overleftrightarrow{AB} \parallel \overleftrightarrow{CD}$ and $\overleftrightarrow{EF} \parallel \overleftrightarrow{GH}$.

Theorems 5-2 and 5-3 can be used to determine if coplanar lines are parallel or perpendicular to each other.

Theorem 5-2
If two lines in a plane are perpendicular to the same line, then they are parallel to each other.

Theorem 5-3
In a plane, if a line is perpendicular to one of two parallel lines, then it is perpendicular to the other one.

In the diagram, for example, if $p \parallel q$ and $r \perp q$, then r is also perpendicular to p by Theorem 5-3.

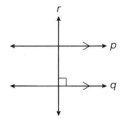

Lines are considered perpendicular when their intersection creates a single right angle, but all perpendicular lines actually create four right angles. Theorems 5-4, 5-5, and 5-6 describe the angles formed by perpendicular lines.

Math Reasoning

Formulate Recall that a circle is 360°. How could you use this information to show that four congruent adjacent angles must all be right angles?

Theorem 5-4
If two lines are perpendicular, then they form congruent adjacent angles.

Theorem 5-5
If two lines form congruent adjacent angles, then they are perpendicular.

Theorem 5-6
All right angles are congruent.

Example 2   Classifying Pairs of Lines

In the figure, $\overleftrightarrow{AB} \parallel \overleftrightarrow{DC}$, $\overleftrightarrow{KL} \perp \overleftrightarrow{DC}$, and $\overleftrightarrow{PQ} \perp \overleftrightarrow{DC}$.

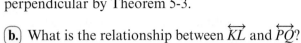

a. What is the relationship between $\overleftrightarrow{KL}$ and $\overleftrightarrow{AB}$?

SOLUTION

Since $\overleftrightarrow{AB}$ and $\overleftrightarrow{DC}$ are parallel, and $\overleftrightarrow{KL}$ and $\overleftrightarrow{DC}$ are perpendicular, $\overleftrightarrow{KL}$ and $\overleftrightarrow{AB}$ are perpendicular by Theorem 5-3.

b. What is the relationship between $\overleftrightarrow{KL}$ and $\overleftrightarrow{PQ}$?

SOLUTION

Since both $\overleftrightarrow{KL}$ and $\overleftrightarrow{PQ}$ are perpendicular to $\overleftrightarrow{DC}$, $\overleftrightarrow{KL}$ and $\overleftrightarrow{PQ}$ are parallel to each other by Theorem 5-2.

c. What is the measure of $\angle 1$? What is the measure of $\angle 2$?

SOLUTION

Since $\overleftrightarrow{KL} \perp \overleftrightarrow{DC}$, they form pairs of congruent adjacent angles. Therefore, $\angle 1$ and $\angle 2$ are congruent, and $\angle 1$ is congruent to $\angle 3$, which is known to be 90°. By the definition of congruency, $\angle 1$ and $\angle 2$ must also measure 90°.

For any given line, there are an infinite number of other lines that are parallel to it. However, there is only one line that is parallel to another through a given point, as stated in the Parallel Postulate.

Postulate 10: The Parallel Postulate
Through a point not on a line, there exists exactly one line through the point that is parallel to the line.

Example 3   Using the Parallel Postulate

Draw as many lines as possible that are parallel to $\overleftrightarrow{DE}$, through a point K that is not on $\overleftrightarrow{DE}$.

SOLUTION

The Parallel Postulate indicates that there is only one line that can be drawn through a point not on a line that is parallel to the given line.

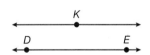

Hint

The distance between two parallel lines is the same at every point.

One final property of parallel lines is the Transitive Property.

Theorem 5-7: Transitive Property of Parallel Lines
If two lines are parallel to the same line, then they are parallel to one other.

Example 4 **Application: Power Lines**

Felix is repairing power lines and he needs to ensure that the power lines are parallel. After taking some measurements, he determined that the upper power line and the phone line are parallel and the lower power line and the phone line are parallel. Does Felix have enough information to conclude that the upper and lower power lines are parallel?

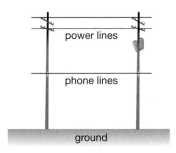

SOLUTION

Yes. The Transitive Property of Parallel Lines can be applied. Since each power line is parallel to the phone line, the power lines are parallel to each other.

Lesson Practice

a. If the horizontal planes are parallel, identify
(Ex 1) the lines that are parallel.

b. In this figure, $\overleftrightarrow{CD} \parallel \overleftrightarrow{GH}$, and $\overleftrightarrow{CD} \perp \overleftrightarrow{AB}$. What
(Ex 2) is the relationship between $\overleftrightarrow{AB}$ and $\overleftrightarrow{GH}$?

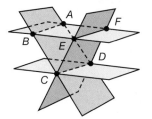

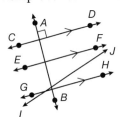

c. If $\overleftrightarrow{RS} \perp \overleftrightarrow{XY}$ and $\overleftrightarrow{VW} \perp \overleftrightarrow{XY}$, what is the
(Ex 2) relationship between $\overleftrightarrow{RS}$ and $\overleftrightarrow{VW}$?

d. What is the relationship among the angles at
(Ex 2) the points of intersection of $\overleftrightarrow{RS}$, $\overleftrightarrow{TU}$, and $\overleftrightarrow{VW}$
with $\overleftrightarrow{XY}$?

e. If line $\overleftrightarrow{XY}$ goes through point M, and $\overleftrightarrow{XY} \parallel \overleftrightarrow{CD}$,
(Ex 3) how do you know that $\overleftrightarrow{JK}$ is not parallel to $\overleftrightarrow{CD}$?

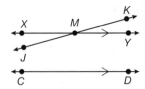

f. Carlotta is assembling the floorboards for her deck. How can she be
(Ex 4) sure that all of the boards are parallel to one another?

1. Draw a diagram showing three planes that intersect at a common line.
(1)

* 2. Error Analysis Damon states that at least four collinear points are needed to determine a plane. Is Damon correct? Explain.
(4)

3. Write Why is a plane not defined by the three given points?
(4)

4. Predict The Pythagorean Theorem, $a^2 + b^2 = c^2$, relates the lengths of the legs of right triangle, a and b, to the length of the hypotenuse, c. There are cases where all three numbers are positive integers. One example is $3^2 + 4^2 = 5^2$. Find one other example. Are there an infinite number of positive integers that satisfy the Pythagorean Theorem?
(SB 3)

5. What is the measure of the angle between two opposite rays?
(3)

* 6. Justify Can two lines be drawn from a point not on a line that will meet the line at a right angle? at a 45° angle? Explain each answer.
(5)

7. There are 100 centimeters in 1 meter. How many square centimeters are there in 1 square meter? Explain how you found your answer.
(SB 9)

8. Algebra Simplify $\dfrac{2\,(2 + 4)}{6} - |-2|$.
(SB 1)

* 9. Consider $\overline{AB}$ in this figure. Name a segment that is skew, one that is parallel, and one that is perpendicular to $\overline{AB}$.
(5)

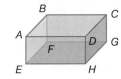

10. Which property is used to state, "If $5 = x$, then $x = 5$"?
(2)

11. Plot the points $A\,(0, 0)$, $B\,(4, 0)$, $C\,(4, 4)$ and $D\,(0, 4)$ on a coordinate grid. Describe the shape that results when the points are connected in alphabetical order.
(SB 13)

*12. (**Air Traffic Controller**) An air traffic controller requested that a pilot taxi on the runway from point A to point B by traveling in a straight line. How many different options does the pilot have to get from point A to point B? How do you know?
(4)

13. For the function $f(x) = 3x^2 - 2x + 5$, find $f(1)$, $f(-2)$ and $f(a)$.
(SB 17)

14. Multiple Choice Which point has the coordinates $(3, -3)$ on the graph shown?
(SB 13)
   A W **B** Y
   C X **D** Z

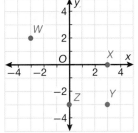

15. (**Photography**) It is very important that a camera be held as still as possible when using certain photographic techniques. Explain why professional photographers use tripods to hold their cameras steady instead of a stand with four or more legs. Name a postulate or theorem in your explanation.
(4)

*16. If $\overleftrightarrow{XY}$ is parallel to $\overleftrightarrow{LM}$, what is the relationship between $\overleftrightarrow{XY}$ and $\overleftrightarrow{UT}$?
(5)

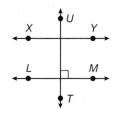

17. **Food Preparation** The instructions on a container of juice concentrate read,
(SB 4) "Seven parts water per 1 part juice concentrate." How much juice can this container make if it holds 250 milliliters of juice concentrate?

18. Are the points (1, 1), (5, 5), and (7, −4) collinear or coplanar? Explain.
(1)

19. Transform this formula to solve for b: $a^3 + b^3 + c^3 = d^3$.
(SB 16)

20. What is the probability, as a fraction, of rolling an even number on a six-sided
(SB 12) number cube?

21. **Justify** How can you use a ruler to determine if a point M is the midpoint of a
(2) segment AB?

22. **Tipping** In some places, it is customary to give a tip of 15% at restaurants. Does
(SB 4) it make a difference whether the 15% is paid on the bill before taxes are added or after taxes are added? Explain.

23. **Write** How many lines connect points A and B? Explain.
(4)

24. **Generalize** What are the properties of all ordered pairs in Quadrant II of the
(SB 13) coordinate plane?

25. Are angles A and B in the triangles shown congruent? Explain.
(3)

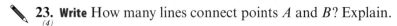

26. **Algebra** Transform the formula for the area of a triangle to solve for the height, h.
(SB 16)

*27. In most rooms, the floor is perpendicular to the wall and the wall is perpendicular
(5) to the ceiling. What can you say about the relationship between the floor and ceiling of most rooms if those two statements are true?

28. **Generalize** Give an example of two acute angles whose measures could be added
(3) together to form another acute angle. Give an example of two acute angles whose measures could be added together to form an obtuse angle.

29. **Music** Musical notes can be identified by their frequencies. The frequency
(SB 22) doubles when going up an octave. Middle C is denoted as C_4 and has a frequency of 261.63 hertz. What is the frequency of the note C_5, which is one octave above C_4?

*30. Does the diagram at right show $\overleftrightarrow{AB}$ perpendicular to $\overleftrightarrow{DE}$? Explain.
(5)

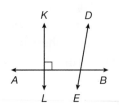

Perpendicular Line through a Point on a Line

Construction Lab 2 (Use with Lesson 5)

In Lesson 5, you learned about perpendicular lines. This construction lab shows you how to construct a perpendicular line through a given point on a line.

1. Draw $\overleftrightarrow{LM}$ and choose point P on the line as the point at which to construct a perpendicular line.

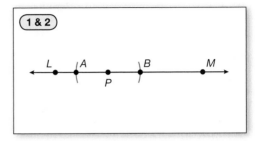

2. Sweep two arcs centered at P through $\overleftrightarrow{LM}$. Label the two points of intersection with $\overleftrightarrow{LM}$ as A and B, respectively.

3. Sweep two arcs, one centered at A and one centered at B. Be sure to choose a radius large enough so that the arcs intersect. Label one point of intersection of these two arcs T.

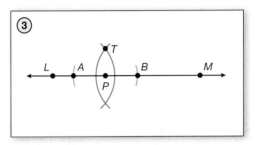

4. Draw $\overleftrightarrow{TP}$.

   Line TP is perpendicular to $\overleftrightarrow{LM}$ at P.

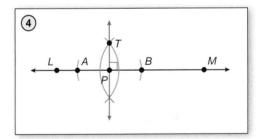

Hint

Instead of sweeping arcs to find points A and B, you can construct a whole circle centered on point P. See the diagram at right.

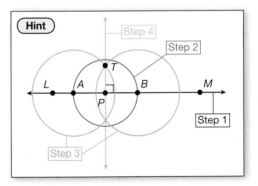

Lab Practice

a. Draw $\overline{AB}$ that is 3 inches long. Construct a segment perpendicular to $\overline{AB}$ through point P such that $\overline{AP}$ is 1.75 inches.

b. Draw $\overline{MN}$ that is 6 inches long. Construct a segment perpendicular to $\overline{MN}$ through point P such that $\overline{NP}$ is 2 inches.

Identifying Pairs of Angles

Warm Up

1. **Vocabulary** Two angles that have the same measure are
(3)
said to be _____.

2. Solve the equation $x + 135 = 180$.
(SB 15)

3. Match each term below with the correct definition.
(1)
- collinear   **A** the point or set of points common to different figures
- space   **B** on the same line
- coplanar   **C** the set of all points
- intersection   **D** in the same plane

4. Identify one acute angle and one obtuse angle
(3)
in the diagram.

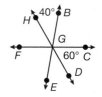

New Concepts

A pair of angles can sometimes be classified by their combined measure.
These pairs are known as complementary and supplementary angles.

Math Language

The sum of an angle and its **complement** is 90°.
The sum of an angle and its **supplement** is 180°.

Complementary Angles	Supplementary Angles
Two angles are **complementary angles** if the sum of their measures is 90°.	Two angles are **supplementary angles** if the sum of their measures is 180°.
$m\angle ABC + m\angle CBD = 90°$, so $\angle ABC$ is complementary to $\angle CBD$.	$m\angle PQR + m\angle RQS = 180°$, so $\angle PQR$ is supplementary to $\angle RQS$.

Example 1   Finding Complements and Supplements

a. Find the angles complementary to $\angle KLM$ if
$m\angle KLN = 90°$.

SOLUTION

From the diagram, $m\angle JLK + m\angle KLM = 90°$
and $m\angle KLM + m\angle MLN = 90°$.

So $\angle JLK$ and $\angle MLN$ are complementary to
$\angle KLM$.

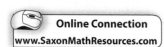

b. Find the angles supplementary to ∠*DGF*.

SOLUTION

From the diagram, m∠*EGD* + m∠*DGF* = 180°
and m∠*DGF* + m∠*FGC* = 180°.
So, ∠*EGD* and ∠*FGC* are supplementary to ∠*DGF*.

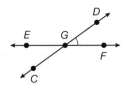

Theorem 6-1: Congruent Complements Theorem

If two angles are complementary to the same angle or to congruent angles, then they are congruent.

Theorem 6-2: Congruent Supplements Theorem

If two angles are supplementary to the same angle or to congruent angles, then they are congruent.

According to Theorem 6-1, if ∠1 is complementary to ∠2 and ∠3 is also complementary to ∠2, then ∠1 ≅ ∠3. Likewise, according to Theorem 6-2, if ∠1 is supplementary to ∠2 and ∠3 is also supplementary to ∠2, then ∠1 ≅ ∠3.

Example 2 **Solving with Complements and Supplements**

Find the measures of the angles labeled *x* and *y*.

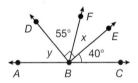

SOLUTION

To find *x*, notice that ∠*DBF* and ∠*FBE* are complementary.

m∠*DBF* + m∠*FBE* = 90°	Definition of complementary angles
55° + *x* = 90°	Substitute.
55° + *x* − 55° = 90° − 55°	Substract 55° from each side.
x = 35°	Simplify.

To find *y*, notice that ∠*ABD* and ∠*DBC* are supplementary.

m∠*ABD* + m∠*DBC* = 180°	Definition of supplementary angles
y + 55° + 35° + 40° = 180°	Substitute.
y + 130° = 180°	Simplify.
y + 130° − 130° = 180° − 130°	Subtract 130° from each side.
y = 50°	Simplify.

Hint

Notice that m∠*DBC* is equal to the sum of the measures of three angles: m∠*DBF* + m∠*FBE* + m∠*EBC*.

Complementary angles and supplementary angles are related to each other by their angle measures. Angles are also related to each other by their positions relative to each other in the same plane.

Two angles in the same plane that share a vertex and a side, but share no interior points are **adjacent angles**. In the diagram, $\angle TSL$ is adjacent to $\angle LSM$, and $\angle RST$ is adjacent to $\angle TSL$.

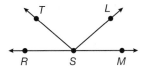

Adjacent angles whose non-common sides are opposite rays are a **linear pair**. In the diagram, $\angle RST$ and $\angle TSM$ are a linear pair. Recall that the measure of a straight line is 180°. Since a linear pair composes a straight line, linear pairs are supplementary.

Theorem 6-3: Linear Pair Theorem
If two angles form a linear pair, then they are supplementary.

Example 3 **Identifying Angle Pairs**

Identify two sets of adjacent angles and one linear pair.

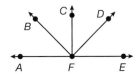

SOLUTION
There are many adjacent angles in the diagram. Two possible sets are $\angle AFB$ and $\angle BFC$, and $\angle AFC$ and $\angle CFE$.
There are also several linear pairs shown. One is $\angle AFD$ and $\angle DFE$.

Nonadjacent angles formed by two intersecting lines are **vertical angles**. Vertical angles share the same vertex and have no common sides.

Theorem 6-4: Vertical Angle Theorem
If two angles are vertical angles, then they are congruent.

Math Reasoning

Generalize Are the statements below true or false?

a. *If two angles are not adjacent, then they are vertical angles.*

b. *If two angles form a linear pair, then they are adjacent angles.*

Example 4 **Solving with Vertical Angles**

Determine the values of x and y.

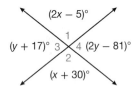

SOLUTION
Since $\angle 1$ and $\angle 2$ are vertical angles, they are congruent. The same is true of $\angle 3$ and $\angle 4$. Therefore, $m\angle 1 = m\angle 2$ and $m\angle 3 = m\angle 4$.

$$m\angle 1 = m\angle 2$$
$$2x - 5 = x + 30$$
$$2x - 5 + 5 = x + 30 + 5$$
$$2x = x + 35$$
$$2x - x = x + 35 - x$$
$$x = 35$$

$$m\angle 3 = m\angle 4$$
$$y + 17 = 2y - 81$$
$$y + 17 - 17 = 2y - 81 - 17$$
$$y = 2y - 98$$
$$y - 2y = 2y - 98 - 2y$$
$$-y = -98$$
$$\frac{-y}{-1} = \frac{-98}{-1}$$
$$y = 98$$

Example 5 **Application: Bridge Supports**

The diagram shows the part of a bridge where it contacts a vertical cliff, so that the bridge and the cliff are perpendicular. The angle between the surface of the road and the line extended from the bridge's support measures 50°. It is important that the bridge's support be set at the correct angle to hold the weight of the bridge. What is the angle x that the support makes with the cliff?

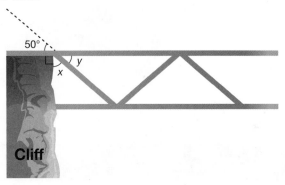

Math Reasoning

Formulate How could you have solved this problem by looking at a linear pair of angles instead of a vertical pair of angles?

SOLUTION

The angle that measures 50° and the angle labeled y are vertical angles. The angles labeled x and y are complementary angles.

$$x + y = 90°$$
$$x + 50° = 90°$$
$$x + 50° - 50° = 90° - 50°$$
$$x = 40°$$

The angle between the support and the cliff measures 40°.

Lesson Practice

Refer to the diagram to find the complementary and supplementary angles.

a. Which angle is complementary to $\angle PKM$?
(Ex 1)

b. Which angle is supplementary to $\angle JKL$?
(Ex 1)

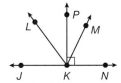

c. Find the value of x.
(Ex 2)

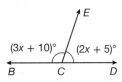

d. Identify three pairs of adjacent angles and two linear pairs.
(Ex 3)

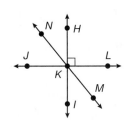

e. Determine the value of x.
(Ex 4)

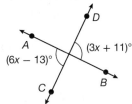

f. A rectangular roof is made of four triangular sections and is
(Ex 5) supported at its peak by a diamond-shaped support and four wooden
beams, as shown in the diagram. What is the value of x? What are the
angle measures of $\angle 1$ and $\angle 2$, respectively?

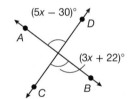

Support Beams

Practice   Distributed and Integrated

* 1. An angle measuring $32°$ has a complement that measures $(2x - 16)°$. What is the
(6) value of x?

* 2. An angle measures $51°$. What is the measure of its supplementary angle?
(6)

3. **Algebra** Compute $|3 \times 2 - 8| \times 2$.
(SB 1)

4. Four noncoplanar points define space, but could they be used to form two
(4) intersecting planes? Explain why or why not.

* 5. Find the value of x in the diagram at right.
(6)

6. **Write** Stanley says, "If a point is not on a line, then the line and the point
(4) exist on exactly one plane." If the point *is* on the line, how would his
statement need to be changed?

7. **Mapping** On a grid of their community, Silvio lives at the point $(7, 3)$ and
(2) his friend Andy lives at the point $(-11, 3)$. If they decide to meet halfway
between their houses, at what point on the grid will they meet?

8. Evaluate the expression $3xy + 5x + 2y$ for $x = 2$ and $y = -1$.
(SB 14)

* 9. What angle is equal to its complementary angle?
(6)

10. What is the probability, expressed as a percentage, of choosing a red marble if a
(SB 12) bag of marbles contains 20 red, 15 green, 25 orange, and 40 blue marbles?

11. **Error Analysis** A student says that two points define a line, three points define a
(1) plane, and four points define space. Are the student's definitions correct? Explain.

12. Find the midpoint of the values 4 and 8 on the x-axis and the midpoint of the
(2) values 1 and 7 on the y-axis.

13. If D is in the interior of $\angle ABC$, which measures $74°$, and $m\angle ABD = 32°$, what is
(3) $m\angle DBC$?

14. (**Competition**) In 2007, the World Rock-Paper-Scissors Championship was held in
$_{(SB\ 12)}$ Toronto, Ontario, Canada. What is the probability that any two competitors will
both have the same combination in a game of rock-paper-scissors, expressed as a
fraction?

*15. Simone has been given the challenge of drawing two coplanar lines that are
$_{(5)}$ perpendicular to the same line, but the lines must not be parallel to each other.
Will she succeed? Explain why or why not.

*16. Write Two planes are parallel to each other and are both intersected by a third
$_{(5)}$ plane. Describe the property of each intersection and how these intersections
are related.

*17. Evaluate the expression $-x^y + 3xy - \dfrac{x}{y}$ for $x = 4$ and $y = 2$.
$_{(SB\ 14)}$

18. Multiple Choice Which is *not* a correct way to name the angle in the diagram?
$_{(3)}$ **A** $\angle 1$ **B** $\angle B$
  C $\angle ABC$ **D** $\angle A$

19. Write Give an example that requires estimation.
$_{(SB\ 5)}$

20. (**Handicap Access**) Use a protractor to determine the angle
$_{(3)}$ measure of the wheelchair ramp shown in the diagram.

21. Use the Ruler Postulate to find the distance between
$_{(2)}$ the values -7 and 11 on a number line.

*22. Analyze If $\overleftrightarrow{AB}$ and $\overleftrightarrow{CD}$ are both parallel to $\overleftrightarrow{EF}$, what conclusion
$_{(5)}$ can be made regarding the relationship between $\overleftrightarrow{AB}$ and $\overleftrightarrow{CD}$?

23. Find 20% of 654.
$_{(SB\ 4)}$

24. Justify Does subtraction exhibit the property of closure over the set of real
$_{(SB\ 2)}$ numbers? Is subtraction commutative? If not, give an example to demonstrate.

*25. (**Landscaping**) A tree is secured with a support wire that makes a 40° angle with the
$_{(6)}$ ground. The angle between the tree and the wire is complementary to the angle
between the wire and the ground. What is the measure of the angle between the
tree and the wire?

26. If $\overleftrightarrow{AB}$ and $\overleftrightarrow{CD}$ intersect at point K, how many planes are needed to contain the two
$_{(4)}$ lines?

*27. List all of the lines that are parallel to $\overleftrightarrow{CD}$ in the diagram at right.
$_{(5)}$

28. (**Comparison Shopping**) A price tag reads \$1.59/kg for apples, and \$0.82/lb
$_{(SB\ 22)}$ for oranges. Which fruit is cheaper by weight? Show your steps.

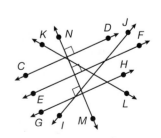

*29. If $m\angle DEF = (3x + 5)°$ and $m\angle KLM = (x + 31)°$, and both angles are
$_{(6)}$ complementary to $\angle PQR$, what are the measures of the angles?

30. Algebra Evaluate the function $f(x) = \dfrac{5}{x + 2}$ for $x = \dfrac{2}{3}$.
$_{(SB\ 17)}$

Perpendicular Bisectors and Angle Bisectors

Construction Lab 3 *(Use with Lesson 6)*

Math Language

A **locus** is a set of points that satisfies a list of given conditions.

A **perpendicular bisector** is a line, segment, or ray that intersects a segment at its midpoint, forming 90° angles. Another way to define a perpendicular bisector is as a locus of points that are equidistant from the endpoints of a segment.

(1.) To construct a perpendicular bisector of a segment, begin with a segment, $\overline{AB}$.

(2.) Set your compass to a setting wider than half the length of the segment and place one end on A.

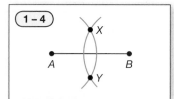

(3.) Use your compass to draw an arc through the segment. Be sure to draw the arc both above and below the segment.

(4.) Repeat this process with the same compass setting, starting at point B.

(5.) Label the points where the two arcs intersect as points X and Y.

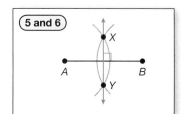

(6.) Draw the line through points X and Y. Line XY is the perpendicular bisector of $\overline{AB}$.

There are two important theorems about perpendicular bisectors that you will use throughout this program.

Theorem 6-5
If a point lies on the perpendicular bisector of a segment, then the point is equidistant from the endpoints of the segment.

Theorem 6-6
If a point is equidistant from the endpoints of a segment, then the point lies on the perpendicular bisector of the segment.

An **angle bisector** is a line, segment, or ray that divides an angle into two congruent adjacent angles.

(1.) To construct the bisector of an angle, begin with an angle, ∠QRS.

(2.) Using a compass, draw an arc centered at R that intersects both sides of the angle. Label these points of intersection J and K.

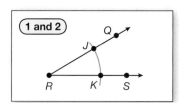

3. Using the same compass setting, draw an arc centered at *J* as shown.

4. Repeat this process with the same compass setting to draw an arc centered at *K*. Label the intersection of the two arcs *M*.

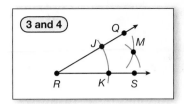

5. Draw $\overrightarrow{RM}$, the angle bisector of ∠*QRS*.

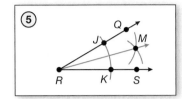

There are also two important theorems about angle bisectors that you will use throughout this program.

Theorem 6-7
If a point lies on the bisector of an angle, then the point is equidistant from the sides of the angle.

Theorem 6-8
If a point is equidistant from the sides of an angle, then the point lies on the bisector of the angle.

Lab Practice

a. Use a ruler to draw a 4 inch segment, then use construction techniques to draw a perpendicular bisector.

b. Use a straightedge to draw a segment of any length, then use construction techniques to bisect it. Use a ruler to verify that your bisector divides the segment into two congruent parts.

c. Use a protractor to draw a 108° angle, then use construction techniques to bisect it.

d. Use a straightedge to draw an angle of any measure, then use construction techniques to bisect it. Use a protractor to verify that your bisector divides the angle into two congruent angles.

Using Inductive Reasoning

1. **Vocabulary** A statement that has been proved true is a _____.
   (6)

2. Which statement about postulates and/or theorems is false?
   (4)
   A A postulate is a proven statement.
   B A postulate is assumed to be true.
   C A theorem is a proven statement.
   D Both **B** and **C** are false.

3. Which statement is true?
   (5)
   A If two lines are perpendicular, they make a 90° angle.
   B If two lines are parallel, they make a 90° angle.
   C Both **A** and **B** are correct.
   D Neither **A** nor **B** is correct.

New Concepts

Inductive reasoning is the process of reasoning that a rule or statement is true because several specific cases are true. Inductive reasoning can be used to formulate a conjecture about something. A **conjecture** is a statement that is believed to be true. If a conjecture can be proven, it becomes a theorem.

> **Math Reasoning**
>
> **Analyze** Tad saw a brown bear and made the conjecture, "All bears are brown." Did Tad use inductive reasoning appropriately? Explain.

Example 1 Formulating a Conjecture

Look at the progression of the pattern and formulate a conjecture regarding the number of blocks there will be in the fifth arrangement of this series.

1st   2nd   3rd

SOLUTION
Each successive step of the series adds a row of blocks to the bottom of the figure that is one block longer than the previous row of blocks. For example, the second arrangement has a bottom row that is 2 blocks long, so to make the third one, a row of 3 blocks is added to the bottom.

To continue, the fourth step would have a row of 4 blocks added for a total of 10, and the fifth would have 5 more blocks added for a total of 15. The statement, "There will be 15 blocks in the fifth step of this pattern," is a conjecture.

Instead of formulating a conjecture by looking at data, it may be necessary to test a conjecture using given data. If even one example can be found that does not support the conjecture, then the conjecture must be incorrect. An example that does not support the conjecture is called a **counterexample**. You will learn more about counterexamples in Lessons 10 and 14.

Online Connection
www.SaxonMathResources.com

Example 2 Testing a Conjecture

Caution

Testing a conjecture is not the same as proving it. Testing a conjecture can disprove it, or it can prove it for certain values, but it cannot prove the conjecture for all possible values.

a. Michelle made the conjecture, "The expressions $6n + 1$ and $6n - 1$ will always result in two prime numbers." Show that this conjecture is true for $n = 1$, 2, and 3, but not true for $n = 4$.

SOLUTION

For $n = 1$: $6(1) + 1 = 7$ and $6(1) - 1 = 5$; both are prime.
For $n = 2$: $6(2) + 1 = 13$ and $6(2) - 1 = 11$; both are prime.
For $n = 3$: $6(3) + 1 = 19$ and $6(3) - 1 = 17$; both are prime.
But for $n = 4$: $6(4) + 1 = 25$ and $6(4) - 1 = 23$; 25 is not prime.

Since we have found one case in which the conjecture is incorrect, we can conclude that the conjecture is false.

b. Maria looks at the diagram below and conjectures that the number of triangles in the figure is given by the expression $2n + 1$. Is this conjecture true for the four steps of the pattern shown below?

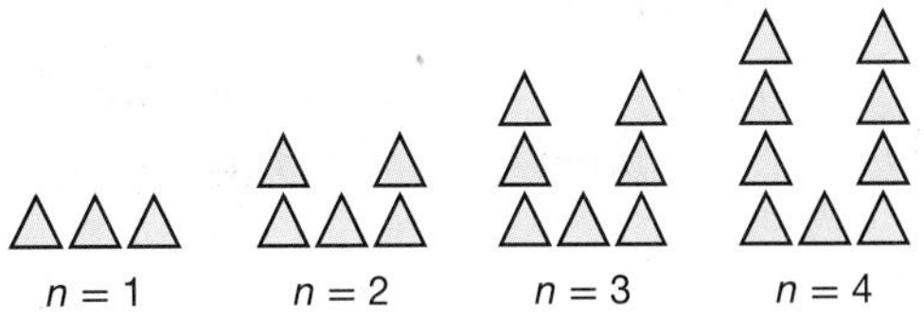

SOLUTION

Yes, the conjecture is true for all 4 steps.

For $n = 1$, $2(1) + 1 = 3$
For $n = 2$, $2(2) + 1 = 5$
For $n = 3$, $2(3) + 1 = 7$
For $n = 4$, $2(4) + 1 = 9$

Even though conjectures are not proven, scientists often use conjectures to describe real-world phenomena. These conjectures are carefully studied and tested, but it is often difficult to prove them formally.

Example 3 Application: Research

A researcher studying crows for several years made the observation that every crow she studied was black. Her research assistant made this conjecture: "All crows are black." How can this conjecture be tested? Can it be proved?

SOLUTION

The conjecture can be tested by observing as many crows as possible. If even one crow is found that is not black, then the conjecture is disproved. The only way to prove this conjecture is to observe every crow. If every crow can be studied, and they are all black, then the conjecture is true. However, it is impossible to study every crow that exists, so the conjecture cannot be proved.

a. Formulate a conjecture about how the next value in this pattern would be found: 1, 1, 2, 3, 5, 8, 13, 21, 34, 55, …
(Ex 1)

b. Test the conjecture that every even integer 4 through 14, can be written as the sum of two prime numbers.
(Ex 2)

c. How might you disprove the conjecture below?
(Ex 3)

Apples, pears, lemons, and peaches all grow on trees, therefore all fruits grow on trees.

Practice   Distributed and Integrated

1. **Justify** In the following description, are rational and irrational numbers
(SB 3) defined adequately? Explain.

Rational numbers are defined to be all terminating or repeating decimals, and irrational numbers are defined to be all nonterminating, nonrepeating decimals.

* 2. **Analyze** Leilani suggests that if the midpoint of a segment connecting 6 and 8 on
(2) the *x*-axis is 7, and the midpoint of a segment connecting 2 and 10 on the *y*-axis is 6, then the midpoint of a segment connecting the points (6, 2) and (8, 10) must be (7, 6). Is this a valid conclusion? Explain.

3. What is the value of *x* in the diagram?
(6)

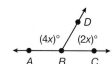

* 4. **Model** Draw an example of skew lines using a geometric figure. Clearly show the
(5) skew lines.

* 5. **Generalize** Formulate a conjecture about how the next step in this pattern would be
(7) found: 4, 16, 36, 64, 100, 144, …

6. (Air Traffic Control) Two airplanes have the same latitudinal and longitudinal
(4) coordinates, but the air traffic controller is not concerned. Why would she not be concerned?

7. **Write** A number is rounded to 9000. Is it possible to tell exactly how many
(SB 7) significant figures the number had before rounding? Explain.

8. Find the value of *x* in the diagram.
(6)

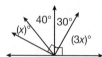

9. **Justify** What is the fewest number of points that can define a plane? Use a
(4) postulate or theorem to justify your answer.

*10. Justify When Raul studied objects that were released from a position of rest, all
(7) of the objects fell to the ground. He conjectured that any object released from
a position of rest will immediately fall to the ground. Find an example that
disproves Raul's conjecture.

*11. Is the following conjecture true for $x = 5$, $x = 10$, and $x = 15$?
(7)

For any x, *the sum of the whole numbers 1 through* x *is equal to* $\frac{1}{2}$x(x + 1).

12. (Summer Employment) Devondra estimated that she would make $4000 this summer
(SB 10) at her part-time job. She actually made $3450. What was her percent error, to the
nearest percentage?

13. Write Explain what is meant when a term in mathematics is said to be *undefined*.
(1)

14. Algebra Find the value of x in the equation $\frac{2x - 5}{3} = \frac{x + 10}{4}$.
(SB 15)

15. (Landscaping) A rectangular field that is 119 square yards in area is to be covered
(SB 22) with sod. It is known that the field is 3 yards longer than twice the width. Find the
dimensions of the field.

16. (Travel) Marcella lives at (4, 22) on a city grid, and Arnault lives at (−3, −4) on
(SB 13) the same grid. One grid line represents one mile. If Arnault's mom drives him to
Marcella's house by driving on a road that is parallel to the x-axis, then turns and
drives along a road that is parallel to the y-axis, how many miles does Arnault's
mom drive?

*17. Write Can a conjecture be proved or disproved by studying examples or cases of
(7) the conjecture? Explain.

18. Algebra Ray BD is the angle bisector of $\angle ABC$. What is the value of x?
(3)

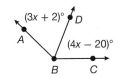

*19. Sketch an example of two lines that are coplanar.
(4)

20. How many significant figures does the number 0.0100 have?
(SB 7)

*21. Draw a line parallel to and a line perpendicular to $\overleftrightarrow{AB}$ from each of the
(5) points K and L. Discuss the relationship between these drawn lines.

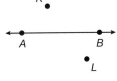

*22. What type of reasoning is used in the statement below? Provide a
(7) counterexample to the conjecture if possible.

Every car I have ever owned ran on gas, therefore all cars run on gas.

23. Multi-Step The formula for converting kilometers to miles is $m = 1.609k$, where m is
(SB 16) miles and k is kilometers. Transform this formula to find a formula for converting
miles to kilometers. Use this formula to determine the number of kilometers in
539 miles and in 7380 miles, rounded to the nearest kilometer.

24. Justify Can two acute adjacent angles form a linear pair? Explain why or why not.
(6)

25. **Temperature** The conversion for temperature from degrees Celsius to degrees
(SB 16) Fahrenheit is given by the formula $F = \frac{9}{5}C + 32$. Transform this formula to find
a formula to convert degrees Fahrenheit to degrees Celsius.

26. **Multiple Choice** Which statement is true?
(4)
   A Two distinct coplanar lines can intersect at two points.
   B Two distinct coplanar lines can intersect at one point or never intersect if they
     are parallel.
   C Two distinct coplanar lines can intersect at one point or intersect everywhere.
   D Two distinct coplanar lines intersect at every point if they are parallel.

27. The sum of the measures of three congruent angles is 180°. What would be the
(6) complementary angle to each of the angles?

*28. **Justify** Is it possible for three planes to intersect at a point? Explain why or why
(1) not, and if it is possible, sketch an example showing how.

29. If B is the midpoint of $\overline{AC}$, find the value of x.
(2)

$$\underset{A}{\bullet} \quad \overset{2x-2}{\underset{B}{\bullet}} \quad \overset{x+11}{\underset{C}{\bullet}}$$

30. The measure of $\angle XYZ$ is 110°. What would the resulting angles' measures be if
(3) the angle bisector to $\angle XYZ$ were drawn?

Using Formulas in Geometry

1. **Vocabulary** _____ lines are coplanar lines that do not intersect.
   (5) (*parallel, perpendicular, skew*)

2. If the length of $\overline{AC}$ is 12, and $\overline{AC}$ is bisected at point B, what are the
   (2) lengths of $\overline{AB}$ and $\overline{BC}$?

3. **Multiple Choice** Which of the following is not a theorem?
   (6) **A** Right Angle Congruence   **B** Congruent Complements
   Theorem   Theorem
   C Congruent Supplements Theorem   **D** Reflexive Property of Equality

New Concepts

A formula is a mathematical relationship expressed with symbols. Some formulas have already been encountered in algebra.

A familiar formula is the formula for perimeter. The perimeter is the sum of the side lengths of a closed geometric figure. It is often thought of as the distance around a figure.

Math Reasoning

Write List some other formulas used in other math classes, such as in algebra. How might these formulas be helpful in geometry?

There is a special formula to find the perimeter of a rectangle, where P is the perimeter, ℓ is the length of the rectangular base, and w is the width, or height, of the rectangle.

$$P = 2\ell + 2w$$

Example 1   **Finding Perimeter of a Figure**

a. Find the perimeter of the triangle.

SOLUTION Add the lengths of the sides together.
$8 + 8 + 8 = 24$
The perimeter of the triangle is 24 inches.

b. Find the perimeter of the rectangle.

SOLUTION Use the formula for the perimeter of a rectangle.
$P = 2\ell + 2w$ Perimeter formula
$P = 2(12) + 2(8)$ Substitute.
$P = 40$ in.   Simplify.

c. If a regular pentagon has a side length of 8 inches, what is its perimeter?

SOLUTION There are five sides in a pentagon and each side of a regular pentagon has the same measure. Therefore, the perimeter is $5 \times 8 = 40$ inches.

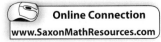
Online Connection
www.SaxonMathResources.com

Math Reasoning

Formulate Draw a diagonal from one corner of a rectangle to the other. What shapes does the diagonal create? Explain how this relates to the formula for area of a triangle.

The **area** of a figure is the size of the region bounded by the figure.

The area of a rectangle is found by the following formula, where ℓ is the length of the figure's base and w is the length of the figure's height:

$A = \ell w$

The area of a triangle is found by the following formula:

$A = \dfrac{1}{2}bh$

The area of a figure is always expressed in square units.

Example 2   **Using the Area Formula for a Rectangle**

(a.) Find the area of the rectangle.

SOLUTION

$A = \ell w$	Area formula
$A = (14)(3)$	Substitute.
$A = 42 \text{ cm}^2$	Simplify.

(b.) Find the length of the rectangle.

SOLUTION

$A = \ell w$	Area formula
$108 = \ell(9)$	Substitute.
$12 \text{ in.} = \ell$	Divide both sides by 9.

Theorem 8-1: Pythagorean Theorem

The sum of the square of the lengths of the legs, a and b, of a a right triangle is equal to the square of the length of the hypotenuse c and is written $a^2 + b^2 = c^2$.

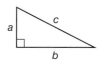

Example 3   **Using the Pythagorean Theorem**

(a.) Find the length of the hypotenuse.

12 cm

5 cm

SOLUTION

$a^2 + b^2 = c^2$	Pythagorean Theorem
$12^2 + 5^2 = c^2$	Substitute.
$144 + 25 = c^2$	Simplify.
$\sqrt{169} = \sqrt{c^2}$	Square root of both sides
$13 \text{ cm} = c$	Simplify.

b. Find the area of the triangle.

SOLUTION

Use the Pythagorean Theorem to find the length of b.

$$a^2 + b^2 = c^2 \qquad \text{Pythagorean Theorem}$$
$$3^2 + b^2 = 5^2 \qquad \text{Substitute.}$$
$$9 + b^2 = 25 \qquad \text{Simplify.}$$
$$9 + b^2 - 9 = 25 - 9 \qquad \text{Subtract 9 from both sides.}$$
$$b^2 = 16 \qquad \text{Simplify.}$$
$$\sqrt{b^2} = \sqrt{16} \qquad \text{Square root of both sides}$$
$$b = 4 \text{ ft} \qquad \text{Simplify.}$$

Then calculate the area of the triangle.

$$A = \frac{1}{2}bh \qquad \text{Formula for area of a triangle}$$
$$A = \frac{1}{2}(4)(3) \qquad \text{Substitute.}$$
$$A = 6 \text{ ft}^2 \qquad \text{Simplify.}$$

Example 4 Application: Measuring Temperature

Different countries use different units to measure the temperature. Much of the world uses degrees Celsius, but a few countries use degrees Fahrenheit. For scientists and travelers, converting between Celsius and Fahrenheit is an important skill.

To convert to Celsius from Fahrenheit, use the formula:
$$C = \frac{5}{9}(F - 32).$$

a. If it is 77°F, what is the temperature in degrees Celsius?

SOLUTION

$$C = \frac{5}{9}(F - 32) \qquad \text{Conversion formula}$$
$$C = \frac{5}{9}(77 - 32) \qquad \text{Substitute.}$$
$$C = 25 \qquad \text{Simplify.}$$

b. If it is 10°C, what is the temperature in degrees Fahrenheit?

SOLUTION

$$C = \frac{5}{9}(F - 32) \qquad \text{Conversion formula}$$
$$10 = \frac{5}{9}(F - 32) \qquad \text{Substitute.}$$
$$10 \times \frac{9}{5} = \frac{5}{9}(F - 32) \times \frac{9}{5} \qquad \text{Multiply by the reciprocal of } \frac{5}{9}, \frac{9}{5}.$$
$$18 = F - 32 \qquad \text{Simplify.}$$
$$18 + 32 = F - 32 + 32 \qquad \text{Add 32 to both sides}$$
$$50 = F \qquad \text{Simplify.}$$

Caution

The Pythagorean Theorem only applies to right triangles.

A right angle is denoted with a small square in the corner that has a measure of 90°.

a. Find the perimeter of a triangle with congruent side lengths all equal
(Ex 1) to 6.5 meters.

b. Find the perimeter of the rectangle.
(Ex 1)

123 cm

145 cm

c. Find the perimeter of a six-sided figure with side lengths that are all
(Ex 1) equal to 16 inches.

d. Find the area of the rectangle.
(Ex 2)

21.2 ft

14.5 ft

e. Find the base of a rectangle with an area of 12 cm^2 and a length of
(Ex 2) 6 centimeters.

f. Use the Pythagorean Theorem to find b.
(Ex 3)

8 m   10 m

b

g. Use the Pythagorean Theorem to find the area of a triangle with a
(Ex 3) hypotenuse of 17 millimeters and a side length of 15 millimeters.

h. If it is 0° Fahrenheit, what is the temperature in degrees Celsius?
(Ex 4) Round to the nearest tenth.

i. If it is 100° Celsius, what is the temperature in degrees
(Ex 4) Fahrenheit?

Practice **Distributed and Integrated**

1. Complete the following conjecture.
(7) *The product of an even and an odd number is _____.*

2. (**Plants**) A bean sprout grows 3 inches in its first week, 2 inches in its second week,
(7) and 1.333... inches in its third week. If the sprout's growth follows this pattern,
how much will the bean sprout have grown in its fourth week?

3. What are three undefined terms of geometry?
(1)

4. **Coordinate Geometry** State the quadrant in which each point is located.
(SB 13)
   a. $(-2, -2)$ **b.** $(-2, 2)$ **c.** $(2, -2)$

5. If there are twelve turtles in the pet store, what is the probability that a turtle
_(SB 12) chosen at random will weigh more than the median turtle weight?

6. Algebra The area of a rectangle is 54 cm². The side lengths are $2x + 1$ and $x + 2$.
₍₈₎ What is the measure of each side?

7. Analyze Can a line be intersected at the same point with two different perpendicular
₍₅₎ lines? Explain why or why not.

8. How far does a jogger travel to reach the midpoint between a jogging track and
₍₂₎ a parking lot that are 0.25 miles apart?

9. Find the next item in the pattern. 2, 4, 8, 16, 32, …
₍₇₎

10. Multiple Choice Which pair of angles are congruent?
₍₆₎
  A $\angle 1$ and $\angle 8$ **B** $\angle 2$ and $\angle 3$
  C $\angle 1$ and $\angle 2$ **D** $\angle 5$ and $\angle 6$

11. Multiple Choice Which pair of angles are a linear pair?
₍₆₎
  A $\angle 5$ and $\angle 6$ **B** $\angle 6$ and $\angle 7$
  C $\angle 1$ and $\angle 4$ **D** $\angle 4$ and $\angle 6$

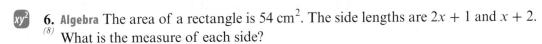

12. Name all the pairs of supplementary angles formed by line m.
₍₆₎

13. Write Karl wrote that any three points define a plane. Is his statement true? If not,
₍₄₎ rewrite his statement so that it is true.

14. Verify A rectangle has side lengths of 30 meters and 36 meters. If the side lengths
₍₈₎ of the rectangle are doubled, verify that the perimeter also doubles.

15. Error Analysis A student has solved the equation for Pythagorean Theorem in the
₍₈₎ following way. It is given that $a = 5$ and $b = 12$.
$$a^2 + b^2 = c^2$$
$$(a + b)^2 = c^2$$
$$(5 + 12)^2 = c^2$$
$$17^2 = c^2$$
$$\sqrt{289} = \sqrt{c^2}$$
$$17 = c$$
Where did the student make an error? What is the actual answer?

16. Identify the intersection between planes $\mathcal{P}$ and $\mathcal{Q}$.
₍₄₎

17. Name the two rays that make $\angle LNA$.
₍₃₎

18. What figure is formed at the intersection of two lines?
₍₁₎

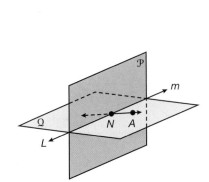

19. Multiple Choice Which of the following terms cannot describe
₍₅₎ lines?
  A perpendicular   **B** congruent
  C skew   **D** parallel

20. Evaluate the expression $\dfrac{x + (3x)^2 + x}{2}$ for $x = \sqrt{2}$. Express the answer in
_(SB 14) simplified radical form.

21. (Music) Musical notes are identified by their frequencies, which are recorded in
(SB 22) hertz (Hz). Middle C is denoted as C_4 and has a frequency of 261.63 Hz. What is
the frequency of the C_3 note, which is one octave below C_4? (Frequency doubles
when going up an octave.)

22. Find the perimeter of a triangle if each side length is 136 millimeters.
(8)

23. If $f(x) = 3x + x^2$, what does $f(2)$ equal?
(SB 17)

24. Statistics What is the mode of the following set? {A, B, M, M, H, H, J, P}
(SB 11)

25. What is the angle between two opposite rays?
(3)

26. (Biology) In a lab experiment, there is a culture that contains 25 bacteria at 2:00.
(7) At 2:15, there are 50 bacteria. At 3:15, there are 800 bacteria. Make a conjecture
about the rate at which the bacteria increases.

27. By what property is the statement $AB = AB$ justified?
(2)

28. (Banking) King is baking a recipe that serves 6 and calls for 1.5 cups of flour. He
(SB 22) expects 8 guests and decides to increase the recipe so that it can serve 8. How
much flour will he need to make the larger recipe?

29. (Building) A housing developer wants to build a fence around the backyard of a
(8) house. If the backyard forms a square and one side is bounded by the house, what
formula could the developer use to calculate the amount of fence required?

30. Verify Show that the Commutative and Associative Properties of Addition are true
(SB 2) when adding the linear functions $f(x) = 3x + 2$, $g(x) = -2x + 3$, and $h(x) = 0$.
The first line of each is given below.

Associative Property of Addition   Commutative Property of Addition
$[f(x) + g(x)] + h(x) = f(x) + [g(x) + h(x)]$ $f(x) + g(x) = g(x) + f(x)$

Finding Length: Distance Formula

1. **Vocabulary** The distance of a number from zero on the number line is the
(SB 1)
_____ of that number.

2. In what quadrant is the point $(5, -3)$ on the coordinate plane?
(SB 13)

3. Evaluate the expression $\dfrac{|5x - 8|}{2}$ for $x = 3$.
(SB 14)

New Concepts

Often, the length of a segment can be measured using a ruler. At other times, it may be necessary to find length by looking at a number line or a coordinate plane.

To find the distance between two points on a number line, take the absolute value of the difference between those points' coordinates.

$$d = |a_2 - a_1|$$

Hint

An absolute value is used to determine distance and length because it is impossible for something to have a negative length, or for the distance between two points to be negative.

The distance between points a_1 and a_2 is $|4 - 1| = 3$.

Example 1 **Distance Between Two Points on a Line**

Find the distance between the points on the number line.

SOLUTION
Use the formula:

$d = |a_2 - a_1|$
$\quad = |4 - (-3)|$
$\quad = |7|$
$\quad = 7$

On a coordinate plane, the distance between two points can be found using the distance formula.

Distance Formula
In a coordinate plane, the distance d between two points (x_1, y_1) and (x_2, y_2) is given by the formula: $$d = \sqrt{(x_2 - x_1)^2 + (y_2 - y_1)^2}$$

Online Connection
www.SaxonMathResources.com

Example 2 Using the Distance Formula

Find the distance between the two points.

SOLUTION
First, choose one point's coordinates to be (x_1, y_1). The other point will be (x_2, y_2). It does not matter which point is chosen.

Let $(1, 2)$ be (x_1, y_1) and $(4, 6)$ be (x_2, y_2).

Substitute into the distance formula.

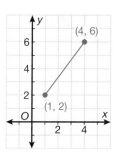

$$d = \sqrt{(x_2 - x_1)^2 + (y_2 - y_1)^2}$$
$$d = \sqrt{(4 - 1)^2 + (6 - 2)^2}$$
$$d = \sqrt{3^2 + 4^2}$$
$$d = \sqrt{25}$$
$$d = 5$$

It does not matter which ordered pair is chosen to be (x_1, y_1). It is important, however, that x_1 and y_1 come from the same ordered pair.

When two points share the same x-value or y-value, the distance formula can be simplified as shown in the next example.

Math Reasoning

Verify Since the diagram shows a horizontal line, an easy way to verify the result of the distance formula is to simply count the number of unit squares the line crosses. Would the result be any different if you flipped the x and y coordinates of these points?

Example 3 Distance Between Points That Share One Coordinate

Find the distance between the two points.

SOLUTION
Since $y_1 = y_2$, we can substitute and simplify.

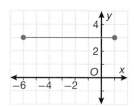

$$d = \sqrt{(x_2 - x_1)^2 + (y_2 - y_1)^2}$$
$$d = \sqrt{(x_2 - x_1)^2 + (y_2 - y_2)^2}$$
$$d = \sqrt{(x_2 - x_1)^2 + 0^2}$$
$$d = \sqrt{(x_2 - x_1)^2}$$

The square root and the square cancel, so with the two identical y-values, the distance formula becomes:
$$d = |x_2 - x_1|$$

An absolute value is used because squaring and then taking the square root of a number always results in a positive number. The resulting formula is identical to the one used to find distance on a number line.

$$d = |x_2 - x_1|$$
$$d = |-6 - 1|$$
$$d = |-7|$$
$$d = 7$$

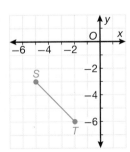

Example 4 Application: Navigation

Use the following map for each question. The distance is measured from the dot on each building.

a. What is the distance from John's house to the school if each unit on the coordinate plane represents 100 meters? Round to the nearest hundredth.

SOLUTION

John's house is at (4, 9) and the school is at (2, 5).

$$d = \sqrt{(x_2 - x_1)^2 + (y_2 - y_1)^2}$$
$$d = \sqrt{(4 - 2)^2 + (9 - 5)^2}$$
$$d = \sqrt{2^2 + 4^2}$$
$$d = \sqrt{20}$$
$$d \approx 4.4721$$

Since each unit represents 100 meters, multiply the answer by 100. The distance from John's house to the school is about 447.21 meters.

b. What is the distance from Sandra's house to the store?

SOLUTION

Sandra's house and the store have the same y-coordinate, so the formula for distance on a number line can be used.

$$d = |x_2 - x_1|$$
$$d = |7 - 2|$$
$$d = 5$$

Since each unit represents 100 meters, the distance from Sandra's house to the store is 500 m.

Math Reasoning

Analyze Can you find a particular route on the map where the distance you would have to walk between two buildings is not accurately expressed by applying the distance formula? Why does the distance formula not work in these situations?

Lesson Practice

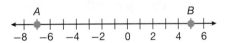

a. Find AB.
(Ex 1)

b. What is the distance between points S and T? Round to the nearest hundredth.
(Ex 2)

c. Find the distance between the points (2, 3) and (2, −4).
(Ex 3)

d. The peak of a mountain is located at the coordinate (120, 0). The hiker starts at the bottom of the trail at coordinate (0, 125). If each unit on the coordinate plane represents 10 meters, how far will the hiker walk if he gets to the peak? Round to the nearest tenth.
(Ex 4)

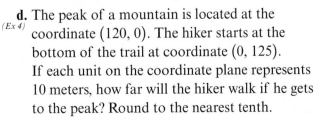

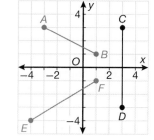

1. What is the approximate distance between points A and B in the
(9) coordinate grid at right?

2. Verify What is the approximate distance between the points E and F?
(9) Is the answer the same if you find the distance between the points F
and E?

3. Factor $(9x^2 - 18x - 7)$.
(SB 18)

4. Coordinate Geometry Calculate the distance between the points $(9, 14)$
(9) and $(-5, 13)$ to the nearest hundredth.

5. (Clocks) What is the angle between the hands of a clock when it is exactly 4 o'clock?
(3)

6. If an angle is obtuse, must its supplement be acute, obtuse, right, or none?
(6)

7. Error Analysis A student calculated the formula for the perimeter (P) of a regular
(SB 22) dodecagon, where n is the side length.
$P = 10n$
Is the student correct? Explain.

8. Evaluate this expression: $\dfrac{2(2+4)}{6} - |-2|$.
(SB 1)

9. Is the following statement sometimes, always, or never true?
(4)
Two planes intersect at a point.

10. Make a conjecture about the pattern at right and find the next line in the pattern.
(7)

11. (Maps) On a map, the Tropic of Capricorn is parallel to the equator, and the
(5) Tropic of Cancer is also parallel to the equator. Do the two tropics ever meet?
How do you know?

12. Write Describe the difference between a postulate and a theorem.
(4)

13. Algebra Find the value of x in the diagram at right.
(6)

14. What two objects intersect at a line? Explain.
(4)

15. Analyze Solve the Pythagorean Theorem for c. How is it like the
(8) distance formula? Explain your reasoning.

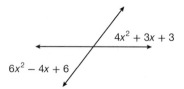

16. (Fencing) If Sven has a rectangular yard with the corners given in the
(9) diagram, to the nearest hundredth meter, how much fencing will he
need to divide the yard into two triangles? Each unit on the grid
represents one meter.

17. There are six bananas, two oranges, and seven apples in a bag. At
(SB 12) random, pieces of fruit are chosen from the bag. What is the probability
that an apple is chosen first and an orange is chosen second?

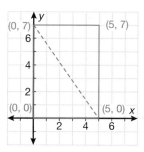

18. What is the measure of ∠BXC?
(3)

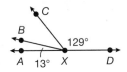

19. Multiple Choice Which of the following is not a formula?
(8)
   A distance formula
   B perimeter formula
   C Pythagorean Theorem
   D protractor formula

20. Find the product of $(2x - 2)$ and $(x + 1)$.
(SB 18)

21. What property justifies this statement?
(2)
   If $\overline{AB} \cong \overline{XY}$ *and* $\overline{CD} \cong \overline{XY}$, *then* $\overline{AB} \cong \overline{CD}$.

22. Explain Can two noncoplanar lines intersect? Explain.
(1)

23. ⬡**Shopping** If a carton of eggs costs $2.05 with tax included, how many cartons
(SB 22) can you buy with $10? Explain how you estimated the answer.

24. Multiple Choice Which of the following is not a pair of angles?
(6)
   A supplementary
   B adjacent
   C right
   D vertical

25. Algebra If B is the midpoint of $\overline{AC}$, and $AB = \frac{5}{2}x + 12$ and $AC = 12x - 4$, what is
(2) the length of $\overline{BC}$?

26. What is the domain of the function $f(x) = \sqrt{x}$?
(SB 17)

27. How does the distance formula simplify when finding the distance
(9) between two points which are directly above each other on a coordinate grid?

28. Solve $-3w + 2 < 4w + 16$ for w.
(SB 15)

29. A line is perpendicular to another line. Classify the supplementary angles it
(5) creates. Are they congruent?

30. What is the difference between a conjecture and a theorem?
(7)

Using Conditional Statements

1. **Vocabulary** _____ reasoning is the process of reasoning in which a
(7) rule or statement is considered true because specific cases are true.

2. Melissa notices that all the flowers in her yard bloom during the spring.
(4) Based on this observation, she says, "All flowers bloom in the spring." Is
this an example of a postulate, a theorem, or a conjecture?

3. Solve the equation $2x + 7 = 4$.
(SB 15)

New Concepts

A **conditional statement** is a statement in the form, "If *p*, then *q*," where *p* is
the hypothesis and *q* is the conclusion. For example:

If it is morning, then the sun is in the east.

The **hypothesis** of a conditional statement is the part of the statement that is
between the words *if* and *then*. In the statement above, the hypothesis, *p*, is
"it is morning." The **conclusion** of a conditional statement is the part of the
statement that follows the word *then*. In the statement above, the conclusion,
q, is "the sun is in the east."

Example 1 Identifying the Hypothesis and Conclusion

Identify the hypothesis and conclusion of each conditional statement.

a. *If 2x + 1 = 5, then x = 2.*

SOLUTION
Hypothesis: $2x + 1 = 5$
Conclusion: $x = 2$

b. *If a plant is growing, then it needs water.*

SOLUTION
Hypothesis: *A plant is growing*.
Conclusion: *It needs water*.

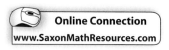

Online Connection
www.SaxonMathResources.com

Some conditional statements are true and some are false. This is called the
truth value of a conditional statement. A statement is only false when the
hypothesis is true and the conclusion is false. For example:

*If a rectangle has a width of 5 feet and a height of 4 feet, then its area is
30 square feet.*

The hypothesis is true, but the conclusion of this statement is false. Since the
hypothesis is true but the conclusion is false, the statement's truth value is
false.

Math Reasoning

Connect Using the
formula you learned
in Lesson 9, write the
sample statement so it
is true.

If a conditional statement's hypothesis is false, then the statement could still be true. For example, consider the statement, "If Ai wins the lottery, he will take a vacation." The hypothesis is false if Ai does not win the lottery, but the statement is still true, because the statement only applies if Ai does win the lottery.

Example 2 Evaluating the Truth Value of a Conditional Statement

Determine whether each statement is true or false. If it is false, explain your reasoning.

(a.) *If an angle is obtuse, it measures 120°.*

SOLUTION
The hypothesis of this statement is true, but the conclusion is false. An obtuse angle can measure anything greater than 90° and less than 180°. Any obtuse angle that is not 120° could be used to contradict this statement. Therefore, the statement is false.

(b.) *If two parallel lines intersect, then they form acute angles.*

SOLUTION
The hypothesis of this statement is false because parallel lines are defined as lines that never intersect. When the hypothesis of a conditional statement is false, the conditional statement as a whole has a truth value of "true." The statement cannot be said to be false unless a situation exists where the hypothesis is true.

Math Reasoning

Analyze What is the result of taking the converse of a converse statement?

The **converse** of a statement is the statement formed by exchanging the hypothesis and conclusion. The converse of a statement "if p, then q" has the form "if q, then p." Consider the following conditional statement.

If it is morning, then the sun is in the east.

The converse of this statement is,

If the sun is in the east, then it is morning.

Even if a conditional statement is true, the converse of that statement is not necessarily true. For example:

If an animal is a duck, then it can fly.

The converse of this statement is,

If an animal can fly, then it is a duck.

This statement is not true. There are many animals that can fly that are not ducks.

Example **3** Stating Converses

Write the converse of each statement and determine whether the converse is true.

a. *If an animal is a dog, then it has four legs.*

SOLUTION
Converse: *If an animal has four legs, then it is a dog.*
The converse is not true. We can prove it is not true by finding an untrue example. For example, a cat is also an animal with four legs, but it is not a dog.

b. *If x = 4, then 3x + 7 = 19.*

SOLUTION
Converse: *If 3x + 7 = 19, then x = 4.*
The converse is also true.

Example **4** Application: Biology

Write the converse of each conditional statement. Use the Venn diagram to determine if the converse is true.

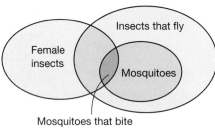

a. *If an insect is a mosquito, then it can fly.*

SOLUTION
Converse: *If an insect can fly, then it is a mosquito.*
There are many flying insects besides mosquitoes, so the converse is false.

b. *If a mosquito bites, then it is female.*

SOLUTION
Converse: *If a mosquito is female, then it bites.*
We see from the diagram that the entire region indicating "female mosquitoes" overlaps completely with "mosquitoes that bite," so the converse of this statement is also true.

Hint

For more about Venn diagrams, see the Skills Bank at the back of this textbook.

Lesson Practice

Identify the hypothesis and conclusion of each statement.
(Ex 1)
   a. *If x = 4 and y = 2, then 2x + 3y = 14.*

   b. *If an apple is a golden delicious apple, then it is yellow in color.*

   c. Determine whether the statement is true or false.
(Ex 2)
   If two points are collinear, then they are also coplanar.

Find the converse of each statement and determine whether it is true.
(Ex 3)
   d. *If $x^2 = 9$, then x = 3 or −3.*

   e. *If it is Thanksgiving Day, then it is Thursday.*

   f. *If a cardinal is a male, then it is bright red.*

1. **(Driving)** Three cities lie on a straight highway. Dawson City is between
   (2) Orangebourgh and Danteville. If the distance from Dawson City to Danteville is
   19 miles and the distance from Orangebourgh to Danteville is 32 miles, what is the
   distance from Orangebourgh to Dawson City? Name a theorem or postulate that
   justifies your answer.

* 2. **(Signs)** What is the perimeter of a regular, triangular yield sign with a side length
   (8) of 15 inches?

3. **Justify** If two lines are parallel, are they contained in a plane?
   (4) Why or why not?

* 4. Use the Venn diagram to write a conditional statement.
   (10)

5. Factor: $x^2 - 4x - 21$.
   (SB 18)

6. Can the expression $\dfrac{\left| x^2 - 4x + 3 \right|}{2}$ ever be a negative number?
   (SB 1) Explain.

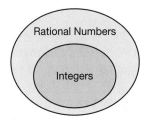

*7. **Multiple Choice** The point $(4, 7)$ is an endpoint of a line segment. The segment
   (9) is 5 units long. Which point is a possible second endpoint of the line?
   A $(1, 4)$ **B** $(1, 11)$
   C $(-1, 6)$ **D** $(0, 3)$

8. **Write** What is the difference between the Ruler Postulate and the Protractor
   (3) Postulate?

* 9. **Write** A certain conditional statement's conclusion is always true when its
   (10) hypothesis is true, so the conditional statement is always true. Can you determine
   for certain that the statement's converse will always be true? Explain and give an
   example to support your answer.

*10. **(Life Span)** If an animal is a loggerhead sea turtle, then its expected life span is
   (10) approximately 70 years. Is this a conditional statement? If it is not, rewrite it as a
   conditional statement. If so, write its converse.

11. An experiment designed to find the speed of sound shows that it is 768 miles per
    (SB 10) hour. If the speed of sound is actually 770 miles per hour, what is the experiment's
    percent error, to the nearest hundredth?

12. Transform the formula $E = mc^2$ to solve for m.
    (SB 16)

13. **Model** Copy the figure at right and draw a line through the point that
    (5) is parallel to $\overleftrightarrow{AB}$. Try to draw a different line through the point that is also
    parallel to $\overleftrightarrow{AB}$. Is it possible to draw a second line? Why or why not?

*14. Find the distance between the two points $(2, 3)$ and $(-4, 1)$ to the nearest
    (9) hundredth.

*15. The length of a line segment is 7. Its endpoints are $(1, 3)$ and $(k, 3)$. Solve for k.
$_{(9)}$ Is there more than one solution? Explain.

16. Model Is it possible to draw two intersecting lines where the vertical angles are not
$_{(6)}$ equal? Why or why not?

17. Two streets intersect and are not perpendicular. Two different corners of the
$_{(6)}$ streets are at angles that have equal measures. What is another term to describe
the angle pair?

18. Is it possible to compare two quantities that are recorded in different units of
$_{(SB\,9)}$ measure? Explain how.

19. Give an example of a situation in which scientific notation is commonly used.
$_{(SB\,7)}$

20. Justify What property justifies the statement $x = x$?
$_{(2)}$

*21. (**Gardening**) A gardener wants to split a rectangular bed of flowers diagonally to
$_{(8)}$ make two separate triangular beds. What is the area of one of the right triangles
the gardener will form if the diagonal is 13 feet and one side is 12 feet?

22. Find the next number in this pattern: 3, 9, 27, 81, …
$_{(7)}$

23. Evaluate the expression mc^2 for $m = 1$ and $c = 3 \times 10^8$. Express your answer in
$_{(SB\,7)}$ scientific notation.

*24. Algebra Complete the conditional statement.
$_{(10)}$
 If $2x + 7 = 13$, then $x =$ _____.

*25. Error Analysis Kareem wrote the following statement.
$_{(10)}$
 If the measures of three angles combined equal 180°, then the angles are all acute.

 Use an example to disprove Kareem's conditional statement, then write a true
 conditional statement.

26. Error Analysis A student multiplied the binomial below. Explain where the student
$_{(SB\,18)}$ made an error. Multiply the binomial correctly.
 $(x + 2)(x - 3) = x^2 + x - 6$

27. List all the sets of numbers to which -2.8 belongs.
$_{(SB\,3)}$

28. Analyze Find the value of the expression $|3 + (-4) + 6|$. Now find the value of the
$_{(SB\,1)}$ expression $|3| + |-4| + |6|$. Why are the answers different?

29. Multiple Choice Which pair of terms is equivalent?
$_{(6)}$
 A supplementary angles and vertical   **B** vertical angles and complementary
 angles   angles
 C linear pair and adjacent, supplementary   **D** linear pair and adjacent angles
 angles

*30. Draw a Venn diagram to show the relationship in the conditional statement.
$_{(10)}$
 If a number is a natural number, then it is a whole number.

Transversals and Angle Relationships

A **transversal** is a line that intersects two or more coplanar lines at different points.

1. Identify the transversal.

2. How many angles are formed by a transversal crossing two lines?

When two lines are intersected by a transversal, the angles formed are classified according to four types of angle pairs. The example column shows two angles that fit each classification.

Classification	Example
A pair of **corresponding angles** is any pair of angles that lie on the same side of the transversal and on the same sides of the other two lines.	
A pair of **alternate interior angles** is any pair of nonadjacent angles that lie on opposite sides of the transversal and between the other two lines.	
A pair of **alternate exterior angles** is any pair of angles that lie on opposite sides of the transversal and outside the other two lines.	
The **same-side interior angles**, also called the **consecutive interior angles**, are a pair of angles that lie on the same side of the transversal and between the other two lines.	

Give one example of each type of angle pair.

3. corresponding angles

4. alternate interior angles

5. alternate exterior angles

6. same-side interior angles

7. **Generalize** When a transversal intersecting two lines is moved, what happens to the measures of the two angles in a linear pair?

Identify the following in the town map.

8. A street that is a transversal and the streets that it intersects.

9. Two businesses on street corners that represent an alternate exterior angle pair.

10. The angle pair of the street corners with a shopping mall and parking lot.

A transversal may also intersect two parallel lines.

Multi-Step Use the diagram to answer the following questions.

11. What type of angle pair is ∠1 and ∠2?

12. Using a protractor, measure ∠1 and ∠2.

13. What conjecture can you make regarding the measure of a pair of corresponding angles formed when a transversal intersects parallel lines?

When a transversal intersects parallel lines, the angle pairs that are formed are either supplementary or congruent. Postulate 11 and the theorems below indicate which pairs are congruent and which pairs are supplementary.

> **Hint**
>
> All of the postulates and theorems presented here refer only to transversals that intersect a pair of parallel lines.

Postulate 11: Corresponding Angles Postulate
If two parallel lines are cut by a transversal, then the corresponding angles are congruent.

Theorem 10-1: Alternate Interior Angles Theorem
If two parallel lines are cut by a transversal, then the alternate interior angles are congruent.

Theorem 10-2: Alternate Exterior Angles Theorem
If two parallel lines are cut by a transversal, then the alternate exterior angles are congruent.

Use the diagram to answer these questions.

14. If m∠1 = 50°, what is m∠2?

15. **Write** If you know m∠4, is it possible to know m∠2? Explain.

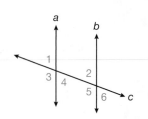

Theorem 10-3: Same-Side Interior Angles Theorem
If two parallel lines are cut by a transversal, then the same-side interior angles are supplementary.

16. If $\angle ABC$ and $\angle DEF$ are a pair of same-side interior angles, what is m$\angle ABC$ when m$\angle DEF = 75°$?

17. Multi-Step Lines ℓ and m are intersected by transversal n.
  a. What angle pair is represented by the expressions?
  b. Find b if $\ell \parallel m$.
  c. What is the measure of $\angle 1$?

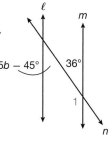

Investigation Practice

What types of angle pairs are the following?

  a. $\angle 5$ and $\angle 4$

  b. $\angle 6$ and $\angle 7$

  c. $\angle 1$ and $\angle 3$

  d. $\angle 1$ and $\angle 8$

  e. $\angle 3$ and $\angle 6$

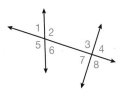

Multi-Step Use the diagram to identify the following.

  f. A pair of same-side interior angles with transversal b.

  g. A pair of corresponding angles with transversal a.

  h. A pair of alternate interior angles with transversal c.

  i. Identify the transversal such that $\angle 4$ and $\angle 10$ are a pair of alternate exterior angles.

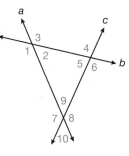

A sign on a hill has posts that are parallel.

  j. Identify the components of the diagram that represent a transversal and the two lines it intersects.

  k. What is m$\angle 1$ if m$\angle 2 = 135°$? Justify your answer by naming the theorem used.

 l. Algebra Determine m$\angle LMP$ and m$\angle ONQ$ when $\overleftrightarrow{MP} \parallel \overleftrightarrow{NQ}$.

  m. Analyze When a transversal intersects two parallel lines, are all the acute angles congruent? Draw a sketch to demonstrate your answer.

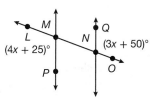

Finding Midpoints

1. **Vocabulary** On a coordinate plane, the ordered pair (4, 4) are
(SB 13) the _____ of a point.

2. What is the distance between points *C* and *D* on the graph?
(9)

3. **Multiple Choice** Which number is halfway between −3 and 5 on the number
(2) line?

A 0   **B** 0.5
C 1   **D** 4

New Concepts For two points on a number line *A* and *B*, the midpoint of $\overline{AB}$ is the point
that is **equidistant** from both *A* and *B*. For point *C* to be equidistant from *A*
and *B* means that the distance from *A* to *C* is the same as the distance from
B to *C*.

Midpoint on a Number Line
The midpoint *C* of $\overline{AB}$ has a coordinate that is the average of the coordinates of *A* and *B*: $$C = \frac{A + B}{2}$$

Math Reasoning

Formulate Describe how the midpoint formula can be inferred from the formula for midpoints on a number line.

The midpoint of $\overline{AB}$ on a coordinate plane is the point *M* on $\overline{AB}$ that is
equidistant from *A* and *B*. To find the midpoint of a segment on a coordinate
plane, use the midpoint formula given below.

Midpoint on a Coordinate Plane
The midpoint *M* of $\overline{AB}$ with endpoints $A(x_1, y_1)$ and $B(x_2, y_2)$, has coordinates that are given by the formula: $$M\left(\frac{x_1 + x_2}{2}, \frac{y_1 + y_2}{2}\right)$$

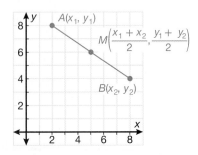

Online Connection
www.SaxonMathResources.com

Example 1 **Finding the Midpoints**

a. What is the coordinate of the midpoint of $\overline{AB}$?

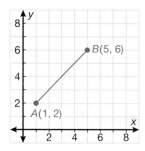

SOLUTION The midpoint is the coordinate on the number line that is the average of the coordinates of the points:

$$C = \frac{2 + 6}{2}$$

$$C = 4$$

b. Determine the midpoint M of $\overline{AB}$ connecting $(1, 2)$ and $(5, 6)$.

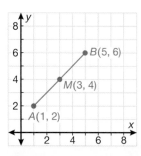

SOLUTION
Substitute $(1, 2)$ for (x_1, y_1) and $(5, 6)$ for (x_2, y_2).

$$M\left(\frac{x_1 + x_2}{2}, \frac{y_1 + y_2}{2}\right)$$

$$M\left(\frac{1 + 5}{2}, \frac{2 + 6}{2}\right)$$

$$M(3, 4)$$

To check, plot the point $(3, 4)$. It should lie on $\overline{AB}$.

Also, the distance formula can be used to verify that $(3, 4)$ is equidistant from A and B:

$$MA = \sqrt{(3 - 1)^2 + (4 - 2)^2} \qquad MB = \sqrt{(3 - 5)^2 + (4 - 6)^2}$$

$$MA = \sqrt{4 + 4} \qquad MB = \sqrt{4 + 4}$$

$$MA = \sqrt{2(4)} \qquad MB = \sqrt{2(4)}$$

$$MA = 2\sqrt{2} \qquad MB = 2\sqrt{2} \checkmark$$

Math Reasoning

Connect Determine the mean of each pair of numbers: 0 and 4, 1 and 3, and 1.5 and 2.5. How are the concepts of mean and midpoint related?

Math Reasoning

Estimate Before solving Example 2 , look at each side of the triangle and estimate where you think the midpoints might be. This is a useful way to check your answer. How close were your estimates to the actual values?

Example 2 Finding Midpoints of Sides

Determine the midpoint of each side of △MNP.

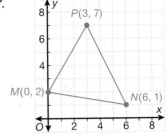

SOLUTION Use the midpoint formula to find A, the midpoint of $\overline{MN}$.

$$A\left(\frac{x_1 + x_2}{2}, \frac{y_1 + y_2}{2}\right)$$

$$A\left(\frac{0 + 6}{2}, \frac{2 + 1}{2}\right)$$

$A(3, 1.5)$

Similarly, the midpoints B of $\overline{NP}$ and C of $\overline{MP}$ have coordinates:

$$B\left(\frac{x_1 + x_2}{2}, \frac{y_1 + y_2}{2}\right) \qquad C\left(\frac{x_1 + x_2}{2}, \frac{y_1 + y_2}{2}\right)$$

$$B\left(\frac{3 + 6}{2}, \frac{7 + 1}{2}\right) \qquad C\left(\frac{0 + 3}{2}, \frac{2 + 7}{2}\right)$$

$B(4.5, 4)$ $\qquad\qquad C(1.5, 4.5)$

Example 3 Application: Navigation

A fishing boat dropped its anchor equidistant from Cape Spirit and Endeavor Rock Lighthouse, on the segment joining the two locations. Find the coordinates of the boat.

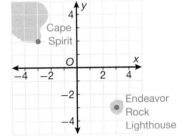

SOLUTION Let point T represent the location of the boat. Point T is the midpoint of the segment with endpoints (−3, 2) and (3, −3).

$$T = \left(\frac{x_1 + x_2}{2}, \frac{y_1 + y_2}{2}\right)$$

$$T = \left(\frac{-3 + 3}{2}, \frac{2 + (-3)}{2}\right)$$

$$T = (0, -0.5)$$

The diagram shows the location of the boat at (0, −0.5).

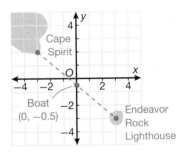

Lesson Practice

a. On the number line below, what is the midpoint of $\overline{AB}$?
(Ex 1)

```
        A     B
 +--●--+--●--+--+→
 0  2  4  6
```

b. Determine the coordinates of the midpoint M for $\overline{AB}$ connecting
(Ex 2) A(5, 1) and B(3, 7).

c. Determine the midpoint of the segment connecting (−3, 2) and
(Ex 2) (4, 2).

d. Determine the coordinates of the midpoint of
(Ex 3) each side of △*JKL*.

e. **Navigation** The map below shows the
(Ex 4) locations of three buoys. Find the midpoint
of the line segments connecting each pair
of buoys. State the coordinates of the
midpoints.

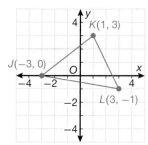

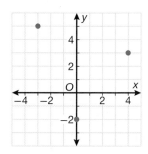

Practice Distributed and Integrated

1. What is the midpoint of the segment connecting 2 and 9 on a number line?
(11)

2. Determine the midpoint *M* of $\overline{AB}$ connecting *A*(3, 2) and *B*(7, 4).
(11)

3. Determine the midpoint *M* of $\overline{PQ}$ connecting *P*(−3, 2) and *Q*(2, −2).
(11)

4. **Algebra** Use the number line to determine the value of *a*, where the midpoint of *a*
(11) and 5 is 2.

$$\overset{\longleftrightarrow}{\underset{-4 \quad -2 \quad 0 \quad 2 \quad 4 \quad 6}{\vdash\!\!+\!\!+\!\!+\!\!+\!\!+\!\!\bullet\!\!+\!\!+\!\!\bullet\!\!+\!\!+}}$$

5. **Error Analysis** Fred claims that the midpoint of $\overline{AB}$ connecting *A*(−2, 5) and *B*(2, 0)
(11) is (0, 5). What mistake did he make?

Refer to the figure to answer each question.

6. **Drafting** Hoyt used engineering software to draft a sketch of a metal
(Inv 1) plate etched with grooves that is needed to complete a highly secure lock.
Segments *CD* and *AB* are parallel, and $\overline{EF}$ intersects each at 17° from
vertical. What is the measure of ∠*AEF*?

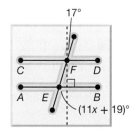

7. **Drafting** What is the value of *x*?
(Inv 1)

8. **Verify** What theorem or postulate can be used to justify the statement
(Inv 1) ∠1 ≅ ∠7?

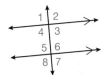

**Determine whether each conditional statement is true. If the statement is false,
give a counterexample.**

9. If it is winter, then the month is January.
(10)

10. If 2*x* + 2*y* = 6, then *x* = 2 and *y* = 1.
(10)

11. Find the distance between points (3, 4) and (−9, 7) to the nearest hundredth.
(9)

12. Find the distance between points (−1, 4) and (5, −6) to the nearest
(9) hundredth.

13. **(Skiing)** A 180-yard ski slope is 120 yards high. What is the
(8) horizontal distance of the hill to the nearest yard?

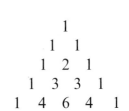

14. Find the perimeter of a right triangle with two legs which each
(8) measure 6 units. Round to the nearest hundredth.

15. Find the side length of a square with a perimeter of
(8) 458 centimeters.

16. **Generalize** Fill in the next three rows of the following pattern:
(7)

17. Use inductive reasoning to determine the next term:
(7) 1, 3, 9, 27, 81, 243, _____

```
          1
        1   1
      1   2   1
    1   3   3   1
  1   4   6   4   1
```

18. **Write** Morgan has been following college football for many years and makes the
(7) statement that her team will win if she wears her lucky sweater. Comment on the
validity of her conjecture.

19. Find the complement of a 14° angle.
(6)

20. Find the supplement of an 85° angle.
(6)

21. Two coplanar lines are parallel when a third line intersects these two lines at 90°.
(5) Can it also be said that two planes are parallel if a third plane intersects these
two planes at 90°?

22. In this figure, what is the value of k?
(5)

23. What is the greatest number of planes determined by
(4) three points? What is the least?

$(18k)°$

24. Is this statement always, sometimes, or never true? *Two lines intersect at one point.*
(4)

25. **Multi-Step** If $\overleftrightarrow{SU}$ lies in the right angle $\angle RST$ and $\angle RSU$ is one-fourth the measure
(3) of $\angle UST$, what is m$\angle UST$?

26. **(Clocks)** What is the angle of a clock's hands when it is 2:30?
(3)

27. **Write** Explain the difference between congruence and equality. Use an example.
(2)

28. Given that P, Q, and R are collinear and Q is between P and R, what postulate
(2) can be used to justify this addition? $PQ + QR = PR$

29. **Error Analysis** Joy made the statement, "Any three points are noncollinear." Explain
(1) where Joy made an error and suggest a new, true statement.

30. **Multiple Choice** Which statement is true?
(1)
   A Any two planes are coplanar.   **B** Any two points are noncollinear.
   C Any two points are coplanar.   **D** Any two lines are coplanar.

Proving Lines Parallel

1. **Vocabulary** In this diagram, line ℓ
(Inv 1) is a(n) _____ to lines p and r.

2. What type of angles are $\angle 1$ and $\angle 2$?
(Inv 1)

3. **Multiple Choice** What type of angles are $\angle 3$ and $\angle 4$?
(Inv 1)

   A Alternate interior angles   **B** Alternate exterior angles

   C Corresponding angles   **D** Same-side interior angles

New Concepts

In Investigation 1, you learned Postulate 11: if two parallel lines are cut by a transversal, the corresponding angles formed are congruent. The converse of Postulate 11 is also true, and can be used to show that two lines are parallel.

Postulate 12: Converse of the Corresponding Angles Postulate

If two lines are cut by a transversal and the corresponding angles are congruent, then the lines are parallel.

If $\angle 1 \cong \angle 2$, then $m \parallel n$.

Hint

The postulate and theorems in Investigation 1 refer to two parallel lines cut by a transversal to prove that certain angles are congruent or supplementary. The postulate and theorems in this lesson work conversely. That is, they use known angle relationships to prove that lines are parallel.

Example 1 **Proving Parallelism: Corresponding Angles**

Prove that lines m and n in this diagram are parallel.

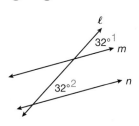

SOLUTION

Angles 1 and 2 both measure 32°, so by the definition of congruent angles, $\angle 1$ and $\angle 2$ are congruent. Since $\angle 1$ and $\angle 2$ are corresponding congruent angles, lines m and n are parallel by Postulate 12.

Theorem 12-1: Converse of the Alternate Interior Angles Theorem

If two lines are cut by a transversal and the alternate interior angles are congruent, then the lines are parallel.

If $\angle 1 \cong \angle 2$, then $m \parallel n$.

Online Connection
www.SaxonMathResources.com

Example 2 **Proving Parallelism: Alternate Interior Angles**

Prove that lines *j* and *k* in this figure are parallel.

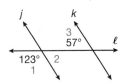

SOLUTION

Angles 1 and 2 form a linear pair, which means they are supplementary.

$$m\angle 1 + m\angle 2 = 180°$$
$$123° + m\angle 2 = 180°$$
$$m\angle 2 = 57°$$

Since $m\angle 2 = m\angle 3$, $\angle 2 \cong \angle 3$. Angles 2 and 3 are congruent alternate interior angles, so by Theorem 12-1, lines *j* and *k* are parallel.

Math Language

Two angles are **supplementary** if the sum of their measures equals 180°.

Theorem 12-2: Converse of the Alternate Exterior Angles Theorem
If two lines are cut by a transversal and the alternate exterior angles are congruent, then the lines are parallel.   If $\angle 1 \cong \angle 2$, then $m \parallel n$.

Example 3 **Proving Parallelism: Alternate Exterior Angles**

a. Identify both pairs of alternate exterior angles in this figure.

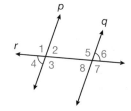

SOLUTION

$\angle 1$ and $\angle 7$ are alternate exterior angles.

$\angle 4$ and $\angle 6$ are also alternate exterior angles.

b. Prove that lines *p* and *q* are parallel.

SOLUTION

The angle congruency marks show that the alternate exterior angles, $\angle 4$ and $\angle 6$, are congruent. Therefore lines *p* and *q* are parallel by the Converse of the Alternate Exterior Angles Theorem (Theorem 12-2).

Theorem 12-3: Converse of the Same-Side Interior Angles Theorem
If two lines are cut by a transversal and the same-side interior angles are supplementary, then the lines are parallel.   If $m\angle 1 + m\angle 2 = 180°$, then $m \parallel n$.   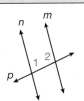

Example 4 Proving Parallelism: Same-Side Interior Angles

a. Identify both pairs of same-side interior angles in this figure.

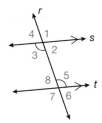

SOLUTION

Angle 2 and ∠5 are same-side interior angles. Angle 3 and ∠8 are also same-side interior angles.

b. Use the Converse of the Same-Side Interior Angles Theorem (Theorem 12-3) to prove that lines *s* and *t* are parallel.

SOLUTION

It is shown in the drawing that ∠3 ≅ ∠5. Angle 5 and ∠8 are supplementary since they form a straight line. Therefore, by substitution, ∠3 and ∠8 are supplementary. Since ∠3 and ∠8 are also same-side interior angles, Theorem 12-3 proves that lines *s* and *t* are parallel.

Example 5 Application: City Planning

In San Francisco, California, Columbus Avenue crosses Stockton, Powell, Mason, and Taylor Streets as shown on the map. Columbus Avenue makes a 40° angle with each of these four streets.

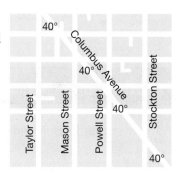

a. What geometric term best describes Columbus Avenue?

SOLUTION

Columbus Avenue is a transversal.

b. Prove that Powell, Mason, and Taylor streets are all parallel to each other.

SOLUTION

The two 40° angles at the intersections of Columbus and Mason, and Columbus and Powell are congruent by definition. They are also corresponding angles. By Postulate 12, Mason and Powell are parallel. Taylor is parallel to Mason and Powell for the same reason. Since two lines that are parallel to the same line are also parallel to each other (Theorem 5-7), all three streets are parallel to one another.

Lesson Practice

a. Prove that lines *a* and *b* in this figure are parallel.
(Ex 1)

b. Prove that lines *u* and *v* in this figure are parallel.
(Ex 2)

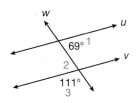

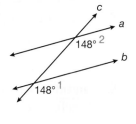

Use the diagram to answer problems c through f.
(Ex 3, 4)

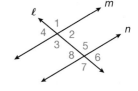

c. Identify both pairs of alternate exterior angles in this figure.

d. Given that $\angle 1 \cong \angle 7$, prove that lines m and n are parallel.

e. Identify both pairs of same-side interior angles in this figure.

f. Given that $\angle 2 \cong \angle 6$, use Theorem 12-3 to prove that lines m and n are parallel.

Use the diagram to answer problems g and h.
(Ex 5)

g. (**City Planning**) Speer Boulevard crosses Fox Street, Elati Street, and Delaware Street. Give the geometric term for Speer Boulevard.

h. (**City Planning**) Prove that Fox, Elati, and Delaware streets are all parallel to one another.

Practice   Distributed and Integrated

* 1. Prove that lines m and n are parallel.
(12)

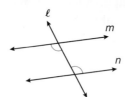

2. **Justify** What type of reasoning is being used in forming the following statement?
(7)

The numbers 5, 10, 15, 20, and 25 all end in either a 5 or a 0, and can all be divided by 5. Therefore, all numbers ending in 5 or 0 can be divided by 5.

* 3. a. Use the Converse of the Corresponding Angles Postulate to prove that lines x and y are parallel.
(12)
   b. Identify the same-side interior angles in the figure.

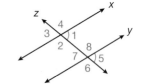

* 4. **Write** A pair of same-side interior angles formed by a transversal across two lines measure $(3x + 10)°$ and $(2x)°$. Explain how to find a value of x that would indicate that two lines are parallel.
(12)

5. What is the midpoint of the segment connecting 4 and 8 on the number line?
(11)

6. (**Carpentry**) Karen drilled holes in a 12-inch wooden ruler at 1 inch, 6 inches, and 11 inches along its length. She wants to drill two more holes at the midpoints between the holes she already has. Where should the new holes go?
(11)

7. Write One endpoint of a segment lies on the origin. If the second endpoint is
(9) known to lie on the line $x = 4$, and the segment is a units long, explain how you
could find the coordinates of the other endpoint.

Refer to the figure to find the requested points.

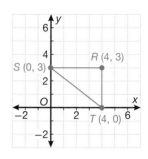

8. What is the midpoint of $\overline{SR}$? of $\overline{RT}$?
(11)

9. What is the midpoint of $\overline{ST}$?
(11)

**Determine whether each conditional statement is true. If the statement
is false, give an example why it is false.**

10. If you have two points, then they can be connected by exactly one line.
(10)

11. (Track and Field) If a runner wins the gold medal, then that runner was the fastest in
(10) the race.

12. Algebra Two angles are supplementary. One measures $(4x + 2)°$ and the other
(6) $(6x + 8)°$. What is the value of x?

13. (Building) A builder wants to build a diagonal beam across a decorative window.
(8) If the window is 3 feet by 5 feet, then to the nearest hundredth, how long is the
diagonal?

14. What is the midpoint of the segment connecting the points $(-3, -4)$ and $(4, 3)$?
(11)

15. Is this statement always, sometimes, or never true?
(4)
Two planes intersect in exactly one line.

16. Find the side length of a square with an area of 182.25 m^2.
(8)

17. (Tiling) A square ceramic tile is 18 inches in length. If a room is 12 feet by 18 feet,
(8) how many tiles are needed?

18. The parallel planes *PQT* and *SRU* are cut by the plane *PQR*. Name
(5) one pair of parallel lines in the figure.

*19. Prove that lines *a* and *b* are parallel.
(12)

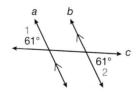

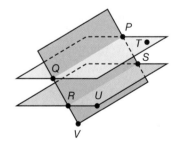

20. (Landscaping) A support wire to a newly planted tree makes a 32° angle with the
(6) ground. Determine the angle made by the wire at the point of contact to the tree
by finding the complement of 32°.

21. Multiple Choice Which statement disproves the following conditional statement?
(10) *If $y = 2x^2 - 5$, then y is positive.*

A $x = 1$ **B** $x = 2$

C $x = 3$ **D** $x = 100$

22. Is $\overline{AB}$ parallel to $\overline{DF}$? Explain.
₍₅₎

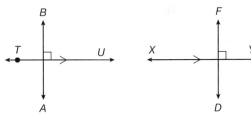

23. Multiple Choice Use inductive reasoning to determine the
₍₇₎ next item in the sequence.

A   **B**

C   **D**

24. Write How many different ways can coplanar lines intersect? Explain each
₍₅₎ situation.

25. Multi-Step Find the approximate area of the rectangle if the
₍₈₎ perimeter is 84 inches.

$3x - 7$

$5x - 5$

26. Error Analysis Sarita said that if there are two lines, they are both
₍₄₎ contained in the same plane. Explain a situation where she could be
wrong.

27. (Clocks) What is the angle made by the hands of a clock when it is 8 o'clock?
₍₃₎

28. Multiple Choice Which of the following is not a property of congruence?
₍₂₎ **A** Mirror Property   **B** Transitive Property
  C Symmetric Property   **D** Reflexive Property

29. Point C lies between A and B on $\overline{AB}$. If $CB = 18$ and $AB = 42$, what is AC?
₍₂₎

30. How many possible points of intersection can three different coplanar
₍₁₎ lines have?

Parallel Line through a Point

Construction Lab 4 (Use with Lesson 12)

In Lesson 12, you practiced proving lines parallel by showing that certain angle pairs are congruent. This lab will show you how to construct a line parallel to a given line through a point using corresponding angles.

(1.) Begin by drawing a line and a point not on the line, labeled *P*. Label any two points on the line *A* and *B*.

(2.) Draw a line through *P* such that it intersects $\overleftrightarrow{AB}$ at *A*.

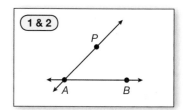

(3.) With the compass centered at *A*, draw an arc across both rays of ∠*PAB*. Label the intersection points *C* and *D* as shown.

(4.) Using the same compass setting, center the compass at *P*. Draw an arc across $\overleftrightarrow{AP}$ such that it intersects $\overleftrightarrow{AP}$ at point *E*, as shown. Be sure the arc extends downward below *P*. Point *P* should lie between points *A* and *E*.

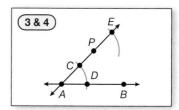

(5.) Set the compass width to the distance between *C* and *D*.

(6.) Center the compass at *E* and draw an arc that intersects the arc drawn in step 4. Label the intersection of the two arcs as point *F*.

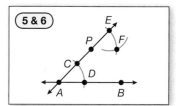

(7.) Draw a line through points *P* and *F*.

Line *PF* is parallel to $\overleftrightarrow{AB}$.

$$\overleftrightarrow{PF} \parallel \overleftrightarrow{AB}$$

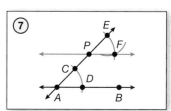

Hint

This is the same procedure you learned in Construction Lab 2 for constructing an angle congruent to a given angle.

This construction is a direct application of the Converse of the Corresponding Angles Postulate.

Lab Practice

Use a straightedge to draw a line and a point not on the line on a blank sheet of paper. Trade with a partner and construct a line parallel to the given line through the given point. Use a protractor to verify that corresponding angles are equal.

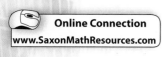
Online Connection
www.SaxonMathResources.com

Introduction to Triangles

Warm Up

Use the diagram to answer problems 1 to 3.

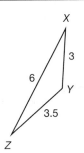

1. Vocabulary In △XYZ, ∠Y is a(n) _____ angle.
(3) (*obtuse, acute, right, straight*)

2. Determine the perimeter of △XYZ.
(8)

3. Multiple Choice Classify the angles X and Z in △XYZ.
(3)
 A acute, right   **B** acute, acute
 C obtuse, obtuse   **D** congruent

New Concepts

A **triangle** is a three-sided polygon. A triangle can be classified by its angles or by its sides. The table below shows three ways to classify a triangle according to its angles.

Math Reasoning

Model An obtuse triangle has exactly one obtuse angle. Try to draw a triangle with two obtuse angles. What do you notice?

Acute Triangle	Obtuse Triangle	Right Triangle
Any triangle that has three acute angles is an **acute triangle**.	Any triangle that has one obtuse angle is an **obtuse triangle**.	Any triangle that has one right angle is a **right triangle**.

A special kind of acute triangle is an **equiangular triangle**, which has three congruent angles.

Example **1**   **Classifying Triangles by Angles**

a. In the diagram, which triangle is obtuse?

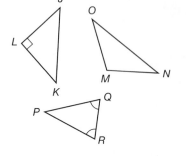

SOLUTION
△MNO is obtuse because it has one obtuse angle M.

b. Which triangle is a right triangle?

SOLUTION
△JLK is a right triangle because ∠L is a right angle.

c. Are any of the triangles equiangular?

SOLUTION
No. △JLK and △MNO are not acute, so they cannot be equiangular. △PQR is acute (because all its angles are acute), but is not equiangular because ∠P is not congruent to the other two angles.

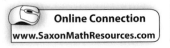

Triangles may also be classified by the lengths of their sides. The table below summarizes three ways to classify a triangle by its sides.

Equilateral Triangle	Isosceles Triangle	Scalene Triangle
Any triangle that has three congruent sides is an **equilateral triangle**.	Any triangle with at least two congruent sides is an **isosceles triangle**.	Any triangle that does not have any congruent sides is a **scalene triangle**.

Example 2 Classifying Triangles by Sides

a. In the diagram, which triangle is scalene?

SOLUTION
△*GHJ* is scalene, because none of its sides are congruent.

b. Which triangle is equilateral?

SOLUTION
△*ABC* is equilateral, because all three sides are congruent.

c. Are any of the triangles isosceles but not equilateral?

SOLUTION
Yes. △*ABC* and △*DEF* are both isosceles, because at least two sides are congruent. △*DEF* is not equilateral because its third side is not congruent to the other two.

A **vertex of a triangle** is one of the points where two sides of the triangle intersect. A **base of a triangle** can be any one of the triangle's sides. The **height of a triangle** is the perpendicular segment from a vertex to the line containing the opposite side. The length of that segment is also called the height.

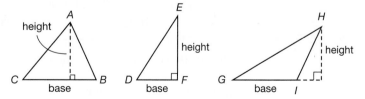

In △*GHI*, the perpendicular segment from *H* does not intersect the base. The base is extended so a perpendicular segment can be drawn to show the height. To find the area of a triangle, both the base and the height must be known.

> ### Area of a Triangle
>
> The area of a triangle is given by the formula below, where b is the length of the triangle's base and h is the height.
>
> $$A = \frac{1}{2}b \times h$$

The diagram shows $\triangle ABC$ enclosed in rectangle $ABDE$. Notice that $\triangle AFC$ and $\triangle CEA$ have the same base and height, so areas A_1 and A_2 are equal. Similarly, $A_3 = A_4$. The area of rectangle $ABDE$ is $b \times h$. Therefore,

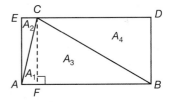

$$\text{Area of } ABDE = A_1 + A_2 + A_3 + A_4$$
$$b \times h = A_1 + A_1 + A_3 + A_3$$
$$b \times h = 2(A_1 + A_3)$$
$$\frac{1}{2}b \times h = A_1 + A_3$$
$$\frac{1}{2}b \times h = \text{Area of } \triangle ABC$$

Example 3 Finding Perimeter and Area of a Triangle

(a.) Determine the perimeter of $\triangle RST$.

SOLUTION
$$P = TR + RS + ST$$
$$= 40.26 + 31.5 + 12.5$$
$$= 84.26$$
The perimeter is 84.26 cm.

(b.) Determine the area of $\triangle RST$.

SOLUTION
$$A = \frac{1}{2}bh$$
$$= \frac{1}{2}(31.5)(10)$$
$$= 157.5$$
The area is 157.5 square centimeters.

Example 4 Application: Farming

A triangular plot of land has a northwestern boundary measuring 64.6 yards, a southern boundary measuring 138.0 yards, and a northeastern boundary measuring 114.1 yards. The perpendicular distance from the southern boundary to the northern corner of the plot is 53.0 yards.

(a.) How much fencing is required to surround the plot?

SOLUTION The perimeter is
$$P = 64.6 + 138.0 + 114.1$$
$$= 316.7$$
316.7 yards of fencing are required.

Math Language

The **perimeter** of a closed figure is the sum of its side lengths. So, to find the perimeter of a triangle, add the lengths of its three sides together.

Hint

To aid in solving this example, draw a picture of the plot of land being described.

b. It takes 100 pounds of barley seed to seed 2400 square yards of land. How much seed is needed for the whole plot, to the nearest pound?

SOLUTION The area of the plot is

$$A = \frac{1}{2}b \times h$$

$$= \frac{1}{2}(138.0)(53.0)$$

$$= 3657 \text{ yd}^2$$

100 pound of barley covers 2400 square yards

Use a proportion:

$$\frac{100}{2400} = \frac{x}{3657}$$

$$(3657)(100) = (2400)(x)$$

$$365{,}700 = 2400x$$

$$\frac{365{,}700}{2400} = x$$

$$152.375 = x$$

To the nearest pound, 152 pounds of seed is needed for the whole plot.

Lesson Practice

Use the diagram to answer problems a through d.

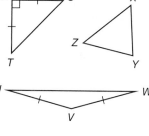

a. Which triangle is obtuse?
(Ex 1)

b. Which triangle is acute?
(Ex 1)

c. Which triangle is a right isosceles triangle?
(Ex 2)

d. Are any of the triangles scalene?
(Ex 2)

e. A right isosceles triangle has legs measuring 13.2 centimeters and a hypotenuse measuring 18.7 centimeters. What is its perimeter?
(Ex 3)

f. What is the area of the triangle in part e?
(Ex 3)

g. (**Surveying**) A golf course is planned for the plot of land shown on the right. How much boundary fencing is required to surround the plot?
(Ex 4)

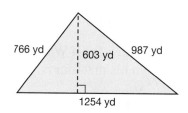

h. (**Surveying**) Grass sod is needed for 95% of the area of the golf course shown above. What area of sod is needed, rounded to the nearest square yard?
(Ex 4)

1. Identify a pair of angles in the diagram that can be used to prove lines parallel
(12) using the Converse of the Same-Side Interior Angles Theorem.

2. **Write** Write a conditional statement that has a false converse.
(10)

3. Classify $\triangle ABC$ as acute, right, or obtuse.
(13)

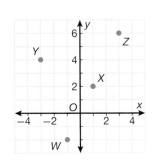

4. **Analyze** Is it possible to draw three points that are
(4) noncoplanar? What about drawing four points?

5. **Algebra** Test the conjecture that the sum of the first n
(7) odd whole numbers is given by n^2 using $n = 3$, $n = 5$,
and $n = 7$.

Use the diagram for the next two questions.
Round to the nearest hundredth.

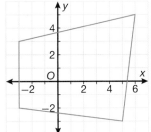

6. What is the length of $\overline{YZ}$?
(9)

7. What is the length of $\overline{WX}$?
(9)

8. (**Surveying**) A hedge surrounds the field shown on the grid. There are
(11) openings at the midpoints of the northern and southern sides of the
hedge for a hiking trail. Determine the coordinates of the openings
in the hedge.

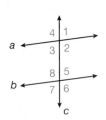

9. **Write** Shanice hears that the weather forecast predicts rain tomorrow.
(7) She thinks, "Since they have been wrong every day this week, it will
not rain." Is Shanice's conjecture valid? Explain.

10. Is $\overleftrightarrow{AB}$ perpendicular to $\overleftrightarrow{DE}$? Explain.
(5)

11. Suppose $\angle 3$ and $\angle 8$ are supplementary. Which postulate
(12) or theorem should be used to prove that lines a and b
are parallel?

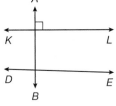

12. **Multiple Choice** What is the greatest number of intersection
(1) points that four coplanar lines can have?
   A 0   **B** 2
   C 4   **D** 6

13. Classify $\triangle DEF$ by its sides. Is $\triangle DEF$ also equilateral?
(13)

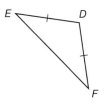

14. **Error Analysis** Alejandro and Jessica were asked to find the next number in the
(7) sequence 2, 4, ___. Alejandro makes the conjecture that the next number will be
6, while Jessica makes the conjecture that the number will be 8. Who is correct?

15. **Multiple Choice** Which point is the midpoint of the
(11) segment connecting -4 and 11?
   A 1   **B** 3
   C 3.5   **D** 5

16. Write Explain why an obtuse angle cannot be congruent to an acute angle.
(3)

17. (**Passports**) Write a conditional statement from the following statement.
(10) *People born in the United States can have an American passport.*

18. (**Agriculture**) A farmer is building a rectangular pen for his chickens with an area of
(8) 40 square feet. If one side measures 8 feet, what is the length of the other side?

19. Draw a pair of parallel lines and label them *n* and *m*. Then draw a transversal that
(Inv 1) is not perpendicular to either line and label it *k*. Mark all congruent acute angles
on the figure.

20. (**Architecture**) A certain tower has eight triangular faces. Four of these faces have
(13) approximate base lengths of 200 feet and heights of 1280 feet. Estimate the
number of panes of glass needed for one of these faces, if each pane of glass has
an area of approximately 45 ft².

21. Write the converse of the following conditional statement.
(10) *If a bird is a flamingo, then it is pink.*

22. Justify Is ∠*ABD* congruent to ∠*CBE*? Justify your answer.
(6)

23. Is the following statement always, sometimes, or never true?
(4) *Three noncoplanar points are contained in a single space.*

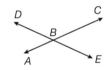

24. a. Of the three triangles shown, which triangle is a right triangle?
(13) **b.** Which triangle is obtuse?
c. Which triangle is equiangular?

25. a. Find the midpoint of the segment connecting −5
(11) and 0.
b. Draw a number line containing −5 and 0. Label
the midpoint.

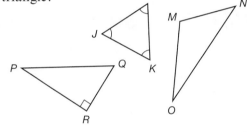

26. What is the perimeter of the pentagon shown?
(8)

27. Algebra Suppose *E* is the midpoint of $\overline{CD}$,
(2) *CD* = 2*x* + 7, and *DE* = 4*x* − 13. What is *CE*?

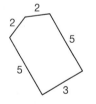

28. Justify Three planes, *BCD*, *ABE*, and *ACD* are arranged so that they
(5) intersect in pairs, forming the plane arrangement shown at right. Is the
information given sufficient to determine that $\overline{AE} \parallel \overline{CD}$? Explain.

29. (**Civil Engineering**) Each diagonal girder on this bridge
(12) forms a transversal with the upper and lower
horizontal girders.

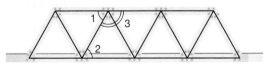

a. Suppose the angles 1 and 2 are congruent. Prove
that the upper and lower girders are parallel.
b. Describe another way you could prove that the upper and lower girders
are parallel.

30. If the measure of an acute angle is (12*x* + 30)°, what is the range of values for *x*?
(3)

Disproving Conjectures with Counterexamples

Warm Up

1. **Vocabulary** A statement that is believed to be true but has not been
(7) proved is a _____.

2. **Multiple Choice** What is the hypothesis of the conditional statement below?
(10) *If it rains today, the road will be slick.*

   A It rains today.
   B The road will be slick.
   C It does not rain today.
   D The road will not be slick.

3. **Multiple Choice** What is the conclusion of the conditional statement below?
(10) *If a triangle has one acute angle and one obtuse angle, then the remaining angle is acute.*

   A A triangle has one acute angle and one obtuse angle.
   B The remaining angle is acute.
   C A triangle has one obtuse angle.
   D A triangle has one acute angle.

New Concepts

Consider the simple conjecture given below.

If two lines are both intersected by a transversal, then they are parallel.

This conjecture is false: two lines do not have to be parallel to be intersected by a transversal. A simple way to prove that this statement is not true is to use a counterexample.

A **counterexample** is an example that proves a conjecture or statement false. For example, the diagram shows a pair of lines that are not parallel, but they are intersected by a transversal. It disproves the statement given above because it gives a specific example where the statement is *not* true. To construct a counterexample, find a situation where the hypothesis of the statement is true but the conclusion is false.

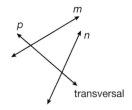

> **Math Reasoning**
>
> **Write** Give a counterexample to prove the following conjecture false: *if a fruit is an apple, then it is green.* Write a false conjecture about an object you see or use every day, and give a counterexample to show it is false.

| Example 1 | Finding a Counterexample to a Geometric Conjecture |

Use the conjecture to answer **a** and **b**.

If a triangle is isosceles, then it is acute.

a. What is the hypothesis of the conjecture? What is its conclusion?

SOLUTION

Hypothesis: *The triangle is isosceles.*
Conclusion: *The triangle is acute.*

(b.) Find a counterexample to the conjecture.

SOLUTION
A counterexample would be an example of a triangle for which the hypothesis is true, but the conclusion is false; that is, a triangle that is isosceles but not acute. Consider this right triangle, ABC. Since $\overline{BC}$ and $\overline{AB}$ are congruent, $\triangle ABC$ is isosceles.

Since $\angle B$ is a right angle, $\triangle ABC$ is not an acute triangle.

Therefore, $\triangle ABC$ is a counterexample to the conjecture.

Math Reasoning

Model Is there another kind of triangle that could be a counterexample to this statement?

Not all conjectures are geometric. Counterexamples can be used to disprove algebraic conjectures or any other kind of conjecture.

Example 2 **Finding a Counterexample to an Algebraic Conjecture**

(a.) Find a counterexample to the conjecture.
Every quadratic equation has either no solution or two solutions.

SOLUTION
You probably remember that sometimes quadratic equations can have only one solution. Such an equation would have to have only one x-intercept. A simple example is the parent function for quadratic equations.

$0 = x^2$

This equation can be solved by graphing, using the quadratic formula, or factoring. The answer is $x = 0$. Since this is a quadratic equation that has only one solution, the statement is proven false by this counterexample.

(b.) Find a counterexample to the conjecture.
If $5x - 10 = 15$, then $2x + y > 9$.

SOLUTION
First, solve the hypothesis of this statement. We find that for the hypothesis to be true, $x = 5$. Then substitute $x = 5$ into the conclusion to solve for y.

$2x + y > 9$
$2(5) + y > 9$
$y > -1$

So for the conclusion to be true, y must be greater than -1. A counterexample to the statement is any value of y that is less than -1. Only one counterexample is needed, so a possible answer is $y = -2$.

Example 3 **Application: Astronomy**

Use the data in the table to prove the conjecture false.

If a planet orbits our Sun, its orbital period (year) is proportional to its distance from the Sun.

Planet	Orbital Period (days)	Distance from Sun (million miles)
Earth	365	93
Mars	687	142
Saturn	10,760	888

SOLUTION The hypothesis of the conjecture, *the planet orbits our Sun*, is true for all three planets in the table. If the conclusion were true, the ratio $\frac{\text{orbital period}}{\text{distance from Sun}}$ should be the same for all three planets. Extend the table by calculating this proportion for each planet:

Planet	Orbital Period (days)	Distance from Sun (million miles)	Proportion
Earth	365	93.0	$\frac{365}{93.0} \approx 3.92$
Mars	687	142	$\frac{687}{142} \approx 4.84$
Saturn	10,760	888	$\frac{10,760}{888} \approx 12.12$

By looking at the fourth column of the table, it is clear that the proportion is not always the same. Any two of these planets provide a counterexample that proves the statement false.

Lesson Practice

Use the conjecture below to answer a and b.
(Ex 1)
If line a is perpendicular to line b and to line c, then lines b and c are perpendicular.

    a. What is the hypothesis of the conjecture? What is its conclusion?

    b. Find a counterexample to the conjecture.

Use the conjecture below to answer c and d.
(Ex 2)
If $x^2 = 9$, then x = 3.

    c. What is the hypothesis of the conjecture? What is its conclusion?

    d. Find a counterexample to the conjecture.

    e. The masses of two sedimentary rocks are 327 grams and 568 grams,
(Ex 3) respectively. Their volumes are 275 cm^3 and 501 cm^3, respectively. Explain how this data disproves the conjecture below.
    If a rock is sedimentary, then its mass is proportional to its volume.

1. This table shows the base lengths and areas of some triangles.
(14)
   a. Does the table give a counterexample to this conjecture? Explain.
(12)   *If the base length of a triangle is less than 3 centimeters, then its area is less than 10 cm².*
   b. Draw a triangle, not listed in the table, that is a counterexample to the conjecture.

Base Length (cm)	Area (cm²)
5.2	15.3
2.9	9.7
1.8	5.6
10.3	20.6
0.5	2.4

2. Classify the following statement as sometimes true, always true, or
(1) never true. *Three points can be noncoplanar.*

3. **Analyze** How could you find the perimeter of an isosceles right triangle if you
(8) already know the area? Is it possible to find the leg lengths?

4. List two adjacent angles and a linear pair from the diagram.
(6)

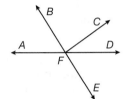

5. **Algebra** If a regular triangle has a perimeter of $12x + 3$, what is an
(8) expression for the length of one of its sides?

6. Line p is a transversal that intersects parallel lines m and n. If two of the
(Inv 1) alternate exterior angles measure $(2x - 10)°$ and $(-3x + 20)°$, what is the measure of each exterior angle?

7. Consider this conjecture. *If a shape is a pentagon, then all its interior angles*
(14) *are obtuse.*
   a. What is the hypothesis of the conjecture? What is its conclusion?
   b. Find a counterexample to the conjecture.

8. **Multiple Choice** Which point is the midpoint of -13 and 5?
(11)   **A** -3 **B** -4
       C -6 **D** -8

9. An angle in a triangle is labeled as $\angle ABC$. Explain why $\overleftrightarrow{AB}$ and $\overleftrightarrow{BC}$ can be used to
(5) define a plane.

10. a. In this figure, which triangle is acute?
(13)   b. Which triangle is obtuse?
       c. Which triangle is equiangular?

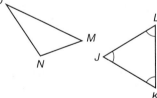

11. **Write** Do complementary angles need to be adjacent? Explain.
(6)

12. Draw and label two opposite rays with a common endpoint, A.
(4)

13. Identify the hypothesis and conclusion of this statement. *If one dozen eggs cost*
(10) *$2.49, then two dozen eggs cost $4.98.*

14. **Error Analysis** A student calculated the distance between $G(12, 4)$ and $H(12, 2)$ as
(9) approximately 12.81. Where did the student make the error and what should the answer be?

15. Multi-Step A star-shaped logo for a shop consists of three isosceles triangles, each
(13) with a base length of 12.5 millimeters and a height of 24.5 millimeters, fitted
around an equilateral triangle with a base length of 12.5 millimeters and a height
of 10.8 millimeters. Find the approximate sum of the four triangles' areas.

16. State the converse of the statement below.
(Inv 1) *If two lines are cut by a transversal and the alternate interior angles are congruent,
then the lines are parallel.*

Write a conditional statement from each sentence.

17. (Football) The team scores three points if the kicker makes a field goal.
(10)

18. A number squared is a positive number.
(10)

19. Find a counterexample to this conjecture. *If $(1 - a)(1 - b) = 0$, then $a = b = 1$.*
(14)

20. (Packaging) This figure shows the triangular shaped label for a brand of cheese
(13) wedges. Classify the triangle by its sides and angles.

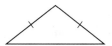

21. If a pair of parallel lines are cut by a transversal, which pairs of angles will be
(Inv 1) supplementary?

22. Analyze If you are given the first endpoint of a segment and the length of the
(9) segment, how many possible locations are there for the second endpoint of the
segment? Explain.

23. (Visual Arts) Seth is creating an abstract design with parallel lines, shown here. He
(12) adds angle markings as a reminder that the angles are congruent. How do you
know that $a \parallel b$?

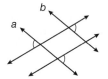

24. Write When can a conjecture be considered a theorem?
(7)

25. Determine the midpoint of each side of $\triangle KLM$ with coordinates $K(-2, -5)$,
(11) $L(-7, 4)$, and $M(0, -1)$.

26. Describe all of the possible types of intersections of three planes.
(5)

27. Use inductive reasoning to find the next term: $-4, -1, 2, \underline{\hspace{1cm}}$
(7)

28. Find a counterexample to this conjecture. *If a triangle is isosceles, then it is equilateral.*
(14)

29. (Painting) A man wants to paint the walls of a room. If the floor is a rectangle
(8) that is 10 feet by 12 feet and the ceiling is 10 feet high, what is the area he will
have to paint?

30. (Geography) Jacob lives in Philadelphia and his friend Antonio
(11) lives in Baltimore. The two plan to meet in the town that is closest to the
midpoint of these two cities.
    a. What are the coordinates of the midpoint of Philadelphia and
    Baltimore on the grid?
    b. What is the name and the coordinates of the town where Jacob and
    Antonio will meet?

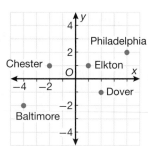

Introduction to Polygons

Warm Up

1. **Vocabulary** A and B are the _____ of $\overline{AB}$.
   (2)

2. How many endpoints does a ray have?
   (3)

3. Classify each of the triangles in the diagram by both sides and angles.
   (13)

New Concepts

A **polygon** is a closed plane figure formed by three or more segments. Each segment intersects exactly two other segments only at their endpoints. No two segments with a common endpoint are collinear.

The segments that form a polygon are called its **sides**. A **vertex of a polygon** is the intersection of two of its sides.

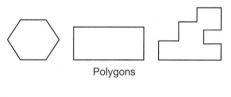

Polygons

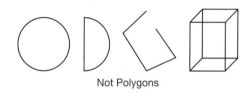

Not Polygons

> **Hint**
>
> Equilateral and equiangular polygons have the same traits as equilateral and equiangular triangles, as introduced in lesson 13.

An **equiangular polygon** is a polygon in which all angles are congruent. An **equilateral polygon** is a polygon in which all sides are congruent. If a polygon is both equiangular and equilateral, then it is called a **regular polygon**. If a polygon is not equiangular and equilateral, then it is called an **irregular polygon**.

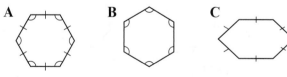

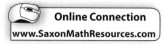
Online Connection
www.SaxonMathResources.com

In the diagram, polygons A and B are equiangular. Polygons A and C are equilateral. Since polygon A is both equiangular and equilateral, it is a regular polygon. Polygons B, C and D are all irregular.

Polygons are named by the number of sides they have. The chart below shows some common polygons and their names.

Name	Sides	Regular Polygon	Irregular Polygon
Triangle	3		
Quadrilateral	4		
Pentagon	5		
Hexagon	6		
Heptagon	7		
Octagon	8		
Nonagon	9		
Decagon	10		
Hendecagon	11		
Dodecagon	12		

Example 1 **Classifying Polygons**

Classify each polygon. Determine whether it is equiangular, equilateral, regular, irregular, or more than one of these.

A B C D

SOLUTION

Polygon **A** has 5 sides, so it is a pentagon. It is equiangular.

Polygon **B** has 7 sides, so it is a heptagon. It is equilateral and irregular.

Polygon **C** is a dodecagon. It is irregular.

Polygon **D** is a quadrilateral. It is equilateral and equiangular, so it is regular.

A **diagonal of a polygon** is a segment that connects two nonconsecutive vertices of a polygon. For example, pentagon *ABCDE* has two diagonals, $\overline{AC}$ and $\overline{AD}$, from vertex *A*. Three other diagonals could be drawn: $\overline{BD}$, $\overline{BE}$, and $\overline{CE}$.

Diagonals can help determine whether a polygon is concave or convex. In a **convex polygon**, every diagonal of the polygon lies inside it, except for the endpoints. In a **concave polygon**, at least one diagonal can be drawn so that part of the diagonal contains points in the exterior of the polygon.

Convex polygon   Concave polygon

If two polygons have the same size and shape, they are **congruent polygons**.

Example 2   Identifying Polygon Properties

a. Find a diagonal that contains points in the exterior of polygon *ABCD*.

SOLUTION
Diagonal $\overline{BD}$ lies outside polygon *ABCD*, except for its endpoints.

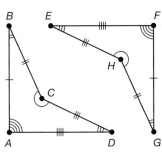

b. Determine whether polygon *EFGH* is convex or concave. Explain.

SOLUTION
Diagonal $\overline{EG}$ contains points in the exterior of polygon *EFGH*. Therefore, polygon *EFGH* is concave.

c. Are polygons *ABCD* and *FGHE* congruent? Justify your answer.

SOLUTION
Write a congruency statement for all corresponding sides and angles. Angle pairs $\angle A \cong \angle F$, $\angle B \cong \angle G$, $\angle C \cong \angle H$, and $\angle D \cong \angle E$. Sides $\overline{AB} \cong \overline{FG}$, $\overline{BC} \cong \overline{GH}$, $\overline{CD} \cong \overline{HE}$, and $\overline{DA} \cong \overline{EF}$. Therefore, $ABCD \cong FGHE$.

At each vertex of a polygon, there are two special angles. An **interior angle of a polygon** is an angle formed by two sides of a polygon with a common vertex. An **exterior angle of a polygon** is an angle formed by one side of a polygon and the extension of an adjacent side. In the diagram, $\angle CDA$ is an interior angle and $\angle ADE$ is an exterior angle.

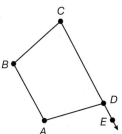

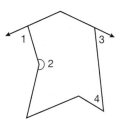

Math Reasoning

Generalize How many interior angles does a convex *n*-gon have? How many exterior angles does a convex *n*-gon have?

Example 3 Identifying Interior and Exterior Angles of Polygons

For each numbered angle in the polygon, determine whether it is an interior angle or an exterior angle.

SOLUTION

Angles 2 and 4 are interior. Angles 1 and 3 are exterior.

Example 4 Application: Tile Patterns

This floor tile pattern uses polygonal tiles that fit together exactly.

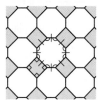

a. Name the two types of polygons used in the pattern. Are they regular or irregular? Explain.

SOLUTION

Square and octagon; both types are regular, because they have all sides and all angles congruent, respectively.

b. Pick any pair of unshaded polygons. Are they congruent? Are they convex or concave? Explain.

SOLUTION

All pairs of unshaded polygons are congruent, because corresponding sides and angles are congruent. Each unshaded polygon is convex, because none of the polygon's diagonals contain points in its exterior.

Lesson Practice

a. Name each polygon. Determine whether it is equiangular, equilateral, regular, irregular, or more than one of these.
(Ex 1)

A   B   C   D

 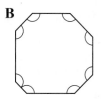 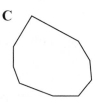 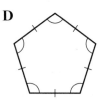

b. Find a diagonal in polygon *GHJKL* that contains points in the exterior of the polygon.
(Ex 2)

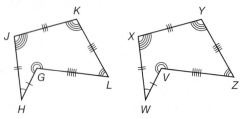

c. Determine whether polygon *VWXYZ* is convex or concave. Explain.
(Ex 2)

d. Are polygons *GHJKL* and *VWXYZ* congruent? Justify your answer.
(Ex 2)

e. For each numbered angle in the polygon, determine whether it is an interior angle or an exterior angle.
(Ex 3)

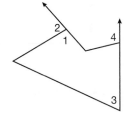

f. Name the type of polygon used in this pattern. Are the polygons regular or irregular? Explain.
(Ex 4)

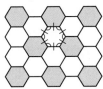

g. Pick any pair of polygons in this pattern. Are they congruent? Are they convex or concave? Explain.
(Ex 4)

Practice   Distributed and Integrated

1. Write Explain why the pair of numbers 3 and 3 is a counterexample to this
(14) statement. *If the sum of two numbers is even, then both numbers are even.*

2. (**Agriculture**) A farm is being divided so that each section of land has equal access
(5) to the canal running through the property for watering crops. If the road on the opposite side of the property runs parallel to the canal, explain how this can be done.

3. (**Wallpaper**) A family wants to install wallpaper around the bottom half of a room.
(8) If the room has 12-foot tall ceilings and each wall is 14 feet long, calculate the area the wallpaper will cover.

4. In polygon *JKLMN*, name each angle and identify it as an interior
(15) or exterior angle.

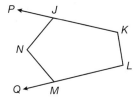

Write a conditional statement from each sentence.

5. The absolute value of a number is a nonnegative number.
(10)

6. A bilingual person speaks two languages.
(10)

7. Determine the midpoint *M* of $\overline{XY}$ connecting *X*(0, 2) and *Y*(6, 1).
(11)

8. Model Find a counterexample to this conjecture.
(14) *If two lines intersect, then any third coplanar line intersects both of them.*

9. Use inductive reasoning to determine the pattern in the following sequence:
(7) 2, 3, 5, 9, 17, 33, 65

10. Identify the property that justifies this statement:
(2) *a = b and b = c, so a = c.*

11. Name all the pairs of angles that are congruent when a transversal cuts a pair of parallel lines.
(Inv 1)

12. Find the length of the segment connecting (1.3, 4.1) and (2.8, 6.1).
(9)

13. **Verify** Rectangle *PQRS* is divided into two triangles by diagonal $\overline{QS}$.
(13)
  a. Determine the area of rectangle *PQRS*.
  b. Given that $\triangle PQS$ and $\triangle QRS$ have equal areas, use your answer to part **a** to determine the area of $\triangle PQS$.
  c. Verify that the formula for the area of a triangle gives the same answer as part **b** for the area of $\triangle PQS$.

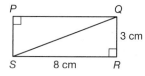

14. Classify this polygon. Is it equiangular? Is it equilateral? Is it regular?
(15)

15. (**Home Renovation**) Tasha is using two wooden rails to construct a pair of stair
(12) rails along the walls of a staircase. To make sure that the rails are parallel, she measures the acute angle each rail makes with the vertical edge of the wall at the base of the stairs.
  a. What type of angles are these?
  b. Tasha measures each angle to be 42°. Explain how she can make sure that the rails are parallel.

16. **Verify** Confirm that this quadrilateral is a counterexample to the conjecture.
(14)
  If a quadrilateral has two pairs of congruent sides, then both pairs of opposite sides are parallel.

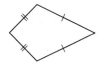

17. If two parallel lines are cut by a transversal and one pair of same-side interior
(Inv 1) angles has angles that measure $(10x + 90)°$ and $(4x + 6)°$, what is the measure of each angle?

18. **Write** In this figure, the transversal line ℓ intersects lines m and n. Write a
(12) paragraph explaining how you know that m and n are parallel.

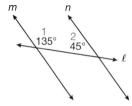

19. This figure shows a polygon with one vertex and two sides missing.
(15)
  a. Copy the figure and add a point *G* that makes *ABCDEFG* concave.
  b. Make a second copy of the figure and add a point *H* that makes *ABCDEFH* convex.

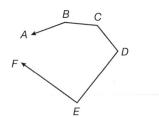

20. Name every pair of corresponding angles in the diagram.
<small>(Inv 1)</small>

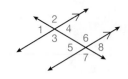

21. (Building) A builder wants to add a diagonal beam to add support to
<small>(8)</small> a structure. The beam needs to extend across a height of 15 feet and a distance of 28 feet. What will the length of the beam be to the nearest hundredth of a foot?

22. Determine the midpoint of each side of △*ABC*.
<small>(11)</small>

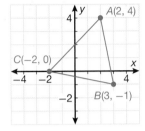

23. **Algebra** The height of a triangle is 12.7 centimeters and its area
<small>(13)</small> is 31.75 square centimeters. Use the Triangle Area Formula to determine its base length.

24. **Multiple Choice** Which statement is always true?
<small>(4)</small>
  A Two planes intersect in a straight line.
  B Two lines are contained in exactly one plane.
  C Two lines can intersect at two points.
  D Any four points can be contained in exactly one plane.

25. **Multiple Choice** If ∠1 and ∠2 are congruent, which of these should be used to
<small>(12)</small> prove that lines *p* and *q* are parallel?
  A Converse of the Alternate Exterior Angles Theorem
  B Converse of the Alternate Interior Angles Theorem
  C Converse of the Corresponding Angles Postulate
  D Converse of the Same-Side Interior Angles Theorem

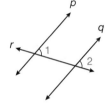

26. **Predict** Use inductive reasoning to find the next term in this sequence. Explain
<small>(7)</small> the rule for the pattern.

  2, 3, 5, 9, 17, …

27. **Multi-Step** Find the perimeter of a square if its area is 289 square centimeters.
<small>(8)</small>

28. If two parallel lines are intersected by a transversal, what is the sum of the
<small>(Inv. 1)</small> measures of all four interior angles that are formed?

29. Find the perimeter of a regular hexagon with side lengths of 6.8 inches
<small>(8)</small>

30. If the points *A*, *B*, *C*, *D*, and *E* are connected, is polygon *ABCDE*
<small>(15)</small> convex or concave?

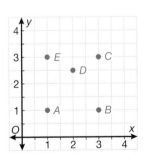

Finding Slopes and Equations of Lines

Warm Up

1. **Vocabulary** The pair of numbers $(3, -1)$ is referred to as a(n)
   _____. (*linear pair, ordered pair*)
   (SB 13)

2. Given the equation $x + 2y = 8$, find the value of y when $x = 3$.
   (SB 14)

3. Which of these equations has $x = -2$ as its solution?
   (SB 15)
   A $x - 2 = 0$ **B** $-2 + x = 0$
   C $2 + x = 0$ **D** $2 - x = 0$

New Concepts

A **linear equation** is an equation whose graph is a line. Some examples are:

$$y = 3x - 1 \qquad 2x + 5y = 7$$

$$10 = 2x \qquad \frac{x}{4} + \frac{y}{13} = 1$$

The variables in linear equations never have exponents other than 1. Linear equations connect algebra (equations in x and y) to geometry (lines in a coordinate plane).

Suppose (x_1, y_1) and (x_2, y_2) are points P and Q on a line in a coordinate plane. The **rise** from P to Q is the *vertical* change between P and Q, and equals $y_2 - y_1$. The **run** from P to Q is the *horizontal* change between P and Q, and equals $x_2 - x_1$.

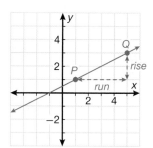

Math Reasoning

Analyze If one point on a line is in Quadrant I and another point is in Quadrant III, what can be determined about the slope of the line?

Example 1   **Finding the Slope of a Line**

Determine the slope of this line.

SOLUTION
Use the points $(1, 5)$ and $(3, 9)$ to calculate the slope. The rise is 4 units and the run is 2 units.

The slope is $\dfrac{\text{rise}}{\text{run}} = \dfrac{4}{2} = 2$.

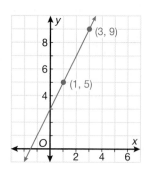

Math Language

Slope is a **rate of change**, which is a ratio that compares the amount of change in a dependent variable to the amount of change in an independent variable.

The **slope** of a line is defined as the ratio of the vertical change (rise) between two points on a line to the horizontal change (run).

$$\text{slope} = \frac{\text{rise}}{\text{run}} = \frac{y_2 - y_1}{x_2 - x_1}$$

There are two special cases. For a horizontal line, the rise is always zero, so the slope is $\frac{0}{\text{run}} = 0$. For a vertical line, the run is zero, so the slope is undefined because division by zero is undefined.

The **slope-intercept form** of a linear equation is a way of writing a linear equation using the slope (*m*) and the *y*-intercept (*b*) of the line. This way of writing the equation has the form $y = mx + b$.

Example 2 Writing the Equation of a Line

a. Use this graph of a line to write its equation.

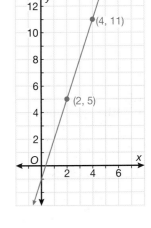

SOLUTION

First, determine the slope *m* using the points (2, 5) and (4, 11).

$$m = \frac{y_2 - y_1}{x_2 - x_1}$$
$$= \frac{11 - 5}{4 - 2}$$
$$= 3$$

Read the *y*-intercept *b* directly from the graph.

$b = -1$

Substite for *m* and *b* in the slope-intercept form.

$$y = mx + b$$
$$y = 3x - 1$$

Math Reasoning

Analyze How can the slope-intercept form be used to find the equation of this line, instead of the slope formula?

b. Write the equation of the line that has slope $\frac{2}{3}$ and passes through (−2, 4).

SOLUTION

Since $m = \frac{2}{3}$, substitute it into the slope formula using the given point (−2, 4) and a general point.

$$m = \frac{y_2 - y_1}{x_2 - x_1}$$
$$\frac{2}{3} = \frac{y - 4}{x - (-2)}$$

Rearrange this equation to remove the denominators.

$$\frac{2}{3} = \frac{y - 4}{x - (-2)}$$

$2(x - (-2)) = 3(y - 4)$	Cross-multiply.
$2x + 4 = 3y - 12$	Simplify.
$2x = 3y - 16$	Subtract 4 from each side.

Finally, solve for *y* to put this equation into slope-intercept form.

$2x = 3y - 16$	
$2x - 3y = -16$	Subtract 3*y* from each side.
$-3y = -16 - 2x$	Subtract 2*x* from each side.
$y = \frac{2}{3}x + \frac{16}{3}$	Divide by −3.

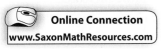

Online Connection
www.SaxonMathResources.com

Example 3   Graphing a Linear Equation

a. Graph the line that has the equation
$y = -5x + 3$.

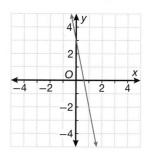

SOLUTION
The equation is in slope-intercept form. Since
the y-intercept is 3, the line passes through 3 on
the y-axis. The slope is -5. For each run of 1
unit, the line has a rise of -5, so it drops 5 units.
Use this fact to plot points starting at $(0, 3)$.

b. Graph the line that has the equation $2y - 4x = 7$.

SOLUTION
The equation is not in slope-intercept form.
Convert it to slope-intercept form.

$$2y - 4x = 7 \qquad \text{Add } 4x \text{ to each side.}$$
$$2y = 7 + 4x \qquad \text{Divide each side by 2.}$$
$$y = 2x + 3.5 \qquad \text{Simplify.}$$

The slope of the line is 2 and the y-intercept
is 3.5. The graph is shown in the diagram.

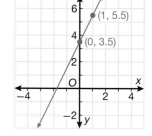

Math Reasoning

Write How can the x-
and y-intercepts of
$2y - 4x = 7$ be used to
graph the equation?

Example 4   Application: Meteorology

Kim believes that there is a linear
relationship between the average
July temperature in the city of
Brightdale in a particular year, and
the number of days of sunshine
Brightdale enjoys that year. She
defines x to be the average July
temperature (daily high, in Fahrenheit),
and y to be the number of days of sunshine.

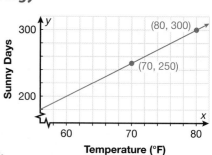

Kim's model is shown on this graph.

a. Determine the slope of the graph. What does the slope represent?

SOLUTION
Use the formula to find the slope.

$$m = \frac{y_2 - y_1}{x_2 - x_1}$$
$$m = \frac{300 - 250}{80 - 70}$$
$$m = 5$$

A slope of 5 means that for every 1°F increase, the number of days of
sunshine increases by 5. The slope represents the rate at which the number
of days of sunshine increases as the average temperature increases.

b. Write an equation for Kim's model.

SOLUTION

Find the y-intercept using the slope and a point on the line.

$y = mx + b$	
$250 = 5(70) + b$	Substitute.
$250 = 350 + b$	Simplify.
$250 - 350 = 350 + b - 350$	Subtract 350 from each side.
$b = -100$	Simplify.

Kim's model has the equation $y = 5x - 100$.

c. Use the equation to predict the average July temperature if there are 280 days of sunshine.

SOLUTION

Substitute 280 for y in your equation from part **b**:

$280 = 5x - 100$	Substitute.
$280 + 100 = 5x - 100 + 100$	Add 100 to each side.
$380 = 5x$	Simplify.
$\dfrac{380}{5} = \dfrac{5x}{5}$	Divide each side by 5.
$76 = x$	Simplify.

The average July temperature should be 76°F.

Math Reasoning

Analyze Through what quadrants could Kim's graph reasonably cross? Why?

Lesson Practice

a. Determine the slope of this line.
(Ex 1)

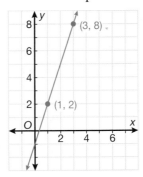

b. Use the slope formula to determine the slope of the line passing
(Ex 1) through (1, 2) and (3, 4).

c. Use the graph to write an equation of the line.
(Ex 2)

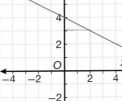

d. Write the equation of the line that has slope $\frac{3}{2}$ and passes through
(Ex 2) (4, −1).

e. Write the equation of the line that passes through
(Ex 2) (0, −2) and (5, 2).

f. Graph the line with the equation $y = 3 - \frac{1}{3}x$.
(Ex 3)

g. Graph the line with the equation $3x + y = 6$.
(Ex 3)

Practice   Distributed and Integrated

1. **Packaging** This figure shows the cardboard which will be folded into a
(15) package.

    a. Determine whether the shape is equiangular and whether it is equilateral.

    b. Determine whether the shape is regular or irregular.

    c. Determine whether the shape is concave or convex, and explain how you
    know.

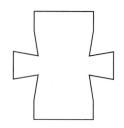

2. Determine the slope of the line passing through (3, 1) and (5, 5).
(16)

3. **Verify** Which postulate or theorem can you use to determine that x and y in this
(12) figure are parallel? Explain.

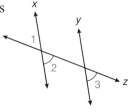

4. a. Determine the slope of this line.
(16) **b.** Write an equation for the line.

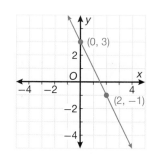

5. **Write** Write a conditional statement that has a false converse.
(10)

6. **Tiling** The area of a single tile is exactly 1 square foot. If a developer wants to
(8) put tile in a room that measures 17 feet by 14 feet and each tile costs $1.29, how
much will it cost to tile the room?

7. Graph the line with equation $y + 2x = 0$.
(16)

8. Two parallel lines cut by a transversal form corresponding angles that measure
(Inv 1) $(2x)°$ and $(4x − 60)°$. What is the measure of each angle?

9. **Multi-Step** If the midpoint of the segment connecting $(−8, x)$ and $(x, 4x)$ is (y, y),
(11) what is the value of y?

10. **Space Exploration** NASA plans to launch Terrestrial Planet Finder 1,
(11) a telescope array that uses five spacecraft flying in formation to
search for Earth-like planets in other solar systems. Suppose the
locations of the top-left and bottom-right telescopes in the picture
are $(−100, 40)$ and $(40, −60)$. Given that the central combiner
telescope lies exactly halfway between these two, what are the
coordinates of the central combiner telescope?

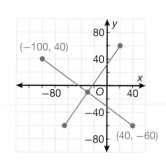

11. (**Farming**) A plow has a length of 6 feet. If the plow has dimensions
(8) as shown, how far off the ground are the plow handles, to the
nearest hundredth of a foot?

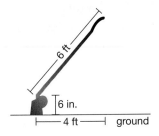

12. Multi-Step If $\overrightarrow{BD}$ intersects straight angle $\angle ABC$ and $\angle ABD$ is one-
(3) eighth the size of $\angle DBC$, what is m$\angle DBC$?

13. Algebra The base of a triangle measures 4.5 inches and its area is
(13) 24.75 in². Use the Triangle Area Formula to determine its height.

14. Error Analysis Mohammed reads a theorem that says there is exactly one line
(5) through a point not on a given line that is perpendicular to that line, and extends
this statement to read that a point on the line would also have one perpendicular
line through it. Comment on the validity of his conclusion in two dimensions and
in three dimensions.

15. Copy polygon *UVWXYZ* and draw in all diagonals. Classify polygon *UVWXYZ* as
(15) convex or concave.

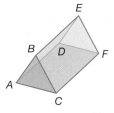

16. Find the area of a right triangle with a hypotenuse of 36 inches and
(8) a leg of 15 inches. Round to the nearest hundredth.

17. In the right triangular prism shown, list pairs of lines that appear to be parallel,
(5) perpendicular, and skew.

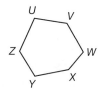

18. Verify Points *A*, *B*, *C*, and *D* all lie on the same line. Are the points also
(4) coplanar? Which postulate or theorem can be used to justify your response?

19. Formulate Show that the conjecture "$2n - 1$ is a prime number" is true for $n = 3$
(7) and $n = 4$.

20. Multiple Choice Which angle measure is supplementary to 80°?
(6) **A** 100°   **B** 10°
     C 80°   **D** 110°

21. $\angle ABD$ and $\angle CBE$ are right angles. Are $\angle ABC$ and $\angle DBE$ congruent? Explain.
(6)

22. Does the line $y = -1.2x + 3$ slope up or down when you move left to right?
(16) How do you know?

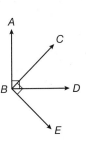

23. Analyze Is it possible to draw two planes that intersect at a point? Explain.
(4)

24. **Write** In this figure, the transversal line r intersects lines q and p.
(12) Write a paragraph explaining why q and p are parallel.

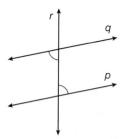

25. Use inductive reasoning to determine the next term in the series:
(7) 11, 16, 15, 20, 19, 24, 23, _____

26. What are the values of x and y in the diagram, given that $m \parallel n$?
(Inv 1)

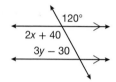

27. **Write** State the hypothesis and the conclusion of this conjecture. *If the product of*
(14) *two numbers is at least 4, then both numbers are at least 2.* Is the conjecture true?
Explain how you know.

28. Determine the slope of this line.
(16)

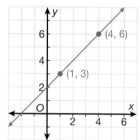

29. **Model** Use a straightedge and a protractor to draw a scalene, an isosceles, and
(13) an equilateral triangle. Mark each congruent angle with arc marks and each
congruent side with tick marks.

30. **Multiple Choice** Which is a counterexample to the following conjecture?
(14) *If two numbers are not positive, then their product is positive.*
  A 2 and −3   **B** −1 and 0
  C −4 and −1   **D** 2 and 7

More Conditional Statements

Warm Up

1. **Vocabulary** In the statement below, "A triangle is equilateral," is the
(10) _____ of the statement.
If a triangle is equilateral, then it is isosceles.

2. Give the converse of the conditional statement.
(10) *If I drink enough water, I am not dehydrated.*

3. For a polygon, which statement is true?
(15) **A** If a polygon is irregular, then it is not equilateral.
B If a polygon is irregular, then it is not equiangular.
C If a polygon is not equilateral, then it is irregular.
D If a polygon is equiangular, then it is equilateral.

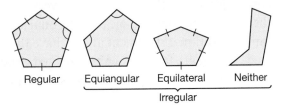

Regular   Equiangular   Equilateral   Neither

Irregular

New Concepts

In Lesson 10, you learned that a conditional statement has the form, "If p, then q." It is formed from two other statements: the hypothesis, p, and the conclusion, q. If you switch the statements, the result is the converse of the conditional statement, "If q, then p."

Math Language

The **converse** of a statement is formed by switching the conclusion and the hypothesis. The converse of a statement is sometimes true but can also be false.

Example 1 **Analyzing the Truth Value and Converse of Conditional Statements**

Consider the conditional statement, "If Sylvester walks to work, then it is a Wednesday."

a. State the hypothesis and conclusion of this statement, and write its converse.

SOLUTION
Hypothesis: *Sylvester walks to work.*
Conclusion: *It is a Wednesday.*
Converse: *If it is a Wednesday, then Sylvester walks to work.*

b. If the original statement is true, is the converse true?

SOLUTION
The converse is not necessarily true. Though we know that Sylvester only walks to work on Wednesdays, we do not know that he walks to work every Wednesday. He might drive to work on some Wednesdays. Therefore, the converse is false.

The **negation** of a statement is the opposite of that statement. The negation of a statement p is "not p," and is written as $\sim p$. For example, the negation of "a pentagon is regular" is "a pentagon is not regular."

Example 2 Examining the Negation of Conditional Statements

Identify the hypothesis and the conclusion in the statement below. Then, write the negation of each.
If a pentagon is regular, then it is equiangular.

SOLUTION
The hypothesis is: *A pentagon is regular.*
The conclusion is: *It is equiangular.*
The negation of the hypothesis is: *A pentagon is not regular.*
The negation of the conclusion is: *It is not equiangular.*

The **inverse** of a conditional statement is formed when its hypothesis and conclusion are both negated. The inverse of "If p, then q" is "If $\sim p$, then $\sim q$." The converse of a conditional statement and the inverse of the same conditional statement always have the same truth value: either both are true or both are false. When two related conditional statements have the same truth value, they are called **logically equivalent statements.**

Math Reasoning

Verify Find the inverse of the statement in Example 1. Is the inverse true or false? Do the inverse and the converse have the same truth value?

Example 3 Examining the Inverse of Conditional Statements

Write the inverse of the statement below. Is the statement true? Is the inverse of the statement true?

For two lines that are cut by a transversal, if alternate interior angles are congruent, then the lines are parallel.

SOLUTION
The inverse of the statement is, "For two lines that are cut by a transversal, if alternate interior angles are not congruent, then the lines are not parallel."

Since the converse and the inverse of a statement have the same truth value, the converse can be used to determine the truth value of the inverse.

The original statement is the converse of Theorem 10-1. Since the converse of the statement is known to be true, the inverse of the statement is also true.

The **contrapositive** of a conditional statement is formed by both exchanging and negating its hypothesis and conclusion. The contrapositive of, "If p, then q," is, "if $\sim q$, then $\sim p$." A conditional statement and its contrapositive are logically equivalent statements: either both are true or both are false.

We summarize the different types of conditional statements in this table.

	Form
Statement	If p, then q
Converse	If q, then p
Inverse	If $\sim p$, then $\sim q$
Contrapositive	If $\sim q$, then $\sim p$

Example 4 **Examining the Contrapositive of Conditional Statements**

a. Determine the contrapositive of the statement.
If Mai finishes school at 1 p.m., then it is a Thursday.

SOLUTION
In this statement, p is "Mai finishes school at 1 p.m." and q is "it is a Thursday." Therefore, the contrapositive statement "If $\sim q$, then $\sim p$" is "If it is not a Thursday, then Mai does not finish school at 1 p.m."

b. Determine the contrapositive of the solution to part **a**. What do you notice?

SOLUTION
The new p and q are "it is not a Thursday" and "Mai does not finish school at 1 p.m." Therefore, its contrapositive is "If $\sim$(Mai does not finish school at 1 p.m.), then $\sim$(it is not a Thursday)," which is the same as "If Mai finishes school at 1 p.m., then it is a Thursday." This is the original statement.

Math Reasoning

Verify A conditional statement and its contrapositive always have the same truth value. If the conditional statement in this example is true, is the contrapositive true? If the statement is false, is the contrapositive false?

Lesson Practice

a. State the hypothesis and conclusion of this statement and its converse.
(Ex 1) *"If a polygon is regular, then it is convex"*

b. If the statement in problem **a** is true, is the converse true?
(Ex 1)

c. Identify the hypothesis and the conclusion in the statement below.
(Ex 2) Then write the negation of each.
If Durrell buys juice, then he buys pretzels.

d. Write the inverse of the statement below. Is the statement true? Is the
(Ex 3) inverse of the statement true?
For two lines that are cut by a transversal, if the lines are parallel, then the same-side interior angles are congruent.

e. Determine the contrapositive of the statement.
(Ex 4) *If two angles are complementary, then the sum of their measures is 90°.*

f. Determine the *converse* of the solution to problem **e**. What do you
(Ex 4) notice?

1. **Music** Chris is planning to make an electric guitar in this shape.
(13)
    a. What area of material will he need for the face of the triangular body, to the nearest square inch?

    b. Chris wants to edge the body in luminescent material. If the body is $\frac{3}{4}$ inch. thick, what area of material will he need for the edging? (Round to the nearest square inch.)

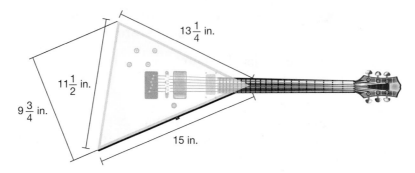

2. **Economics** A cost function relates the total cost (C) of production to the number
(14) of items (x) produced per day. Acme Industries has a cost function $C_1 = 300x + 500$, while Amalgamated Widgets' cost function is $C_2 = 420x + 250$. Using these facts, find a counterexample to this conjecture.
For any value of x, Acme Industries has lower costs than Amalgamated Widgets.

3. If a right triangle has a base of 6 inches and a height of 12 inches, what is the
(8) length of the hypotenuse to the nearest inch?

4. **Formulate** Points $G(7, 1)$ and $H(-1, 4)$ lie on the coordinate plane.
(11)
    a. Think of the x-coordinates 7 and -1 as points on a number line. What is the midpoint of the segment connecting them? What is the midpoint of the segment connecting y-coordinates 1 and 4? What is the midpoint of $\overline{GH}$?

    b. Use part **a** to write a statement of the form: $\left(\frac{-+-}{2}\right), \left(\frac{-+-}{2}\right) = (\_, \_)$

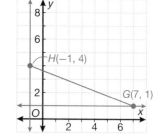

5. Write the inverse of this statement. *If Noah has pasta for lunch, then it is*
(17) *Tuesday.*

6. Classify the statement as sometimes true, always true, or never true. Explain why.
(1)
If two lines are noncoplanar, then they are skew.

7. This conditional statement is not in if-then form. Rewrite it in if-then form, then
(10) identify the hypothesis and the conclusion.

All PE students wear blue shorts to class.

8. Multiple Choice In this figure, ∠5 and which other angle
(12) can prove the lines parallel using the Alternate Exterior
Angles Converse?

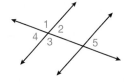

A 1   **B** 2
C 3   **D** 4

9. Write In this figure, the transversal line z intersects lines x and y.
(12) Write a paragraph explaining how to show that x and y are parallel.

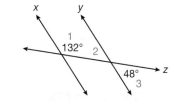

10. Multiple Choice The slope of this graph is:
(16)

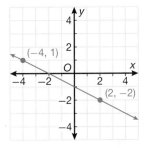

A 0.5   **B** 2
C −0.5   **D** −2

11. Algebra Examine the statement, "If $x^2 > 4$, then $x > 2$." Is the statement true or
(17) false? Write its converse. Is the converse true or false?

12. Write Explain why the midpoint of a and $-a$ on a number line is 0, for
(11) any value of a.

13. (Physics) This graph shows the distance a train must travel in miles
(16) against the length of time of its journey in minutes. Determine the
slope of the line. What does it represent?

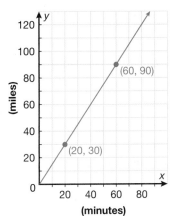

14. Identify the property that justifies the statement:
(2) $WX = YZ$, so $YZ = WX$.

15. Draw an acute, a right, and an obtuse triangle to test the conjecture
(7) made by Marc that the side which is opposite the angle with the
greatest measure in a triangle is the longest side in the triangle.

16. Prove that lines r and q in this figure are parallel.
(12)

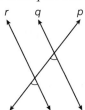

17. Copy and complete this table showing the side lengths of several triangles.
(13)

Side Lengths	Isosceles or Scalene?	Equilateral?
3, 4, 5		
7, 13, 7		
39, 39, 39		

18. State the contrapositive of this statement.
₍₁₇₎

If a bird is black, then it is not a swan.

Does an individual white swan prove the contrapositive statement? If all the swans you have ever seen are white, does that prove the statement? Does it provide evidence to support the statement? Explain your reasoning.

19. Verify Confirm that a triangle with side lengths of 5, 4, and 3 units is a
₍₁₄₎ counterexample to this statement.

If a triangle has no congruent sides, then it is obtuse or acute.

20. Multiple Choice Many rooms have a baseboard that lines the room where the walls
₍₄₎ and floor intersect. Which of the following best describes this situation?
A Two unique points make a line.
B Two noncollinear points define a plane.
C Two noncoplanar lines make a plane.
D Two distinct planes intersect in a line.

21. Identify the hypothesis and conclusion of this statement.
₍₁₀₎

A drink is a soda if it has bubbles.

22. (**Civics**) An opinion poll shows support for Sally Smith at 60% in her congressional
₍₁₆₎ race. Assuming the poll is accurate, plot a line to represent how many votes Smith would get against different voter turnouts. Use scales from 0 to 300,000 in steps of 50,000.

23. Find a counterexample to this conjecture. *If three lines in a plane are all non-*
₍₁₄₎ *parallel, then they divide the plane into seven regions.*

24. Write the converse of this statement. *If a polygon is a triangle, then the sum of its*
₍₁₇₎ *angle measures is 180°.*

xy^2 25. Algebra Find a counterexample to this conjecture. *If xy < 1, then x < 1 and y < 1.*
₍₁₄₎

26. Multi-Step Find the length of the given segments shown and
₍₉₎ determine if they are congruent.

xy^2 27. Analyze A heptagon has sides with positive integer lengths. If the heptagon
₍₁₅₎ is regular, what can you conclude about its perimeter? Explain.

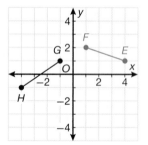

28. Find the area of a right triangle with the hypotenuse of 17 centimeters
₍₈₎ and a leg of 10 centimeters. Round to the nearest hundredth.

29. Multi-Step Find the perimeter of a right triangle with both legs equal, if the
₍₈₎ area is 32 square meters. Round to the nearest hundredth.

30. Draw an irregular convex hexagon. Draw and label an exterior angle and the
₍₁₅₎ associated interior angle. What is the relationship between these two angles?

Triangle Theorems

Warm Up

1. **Vocabulary** An angle that measures less than 90° is a(n) _____ angle.
(3)

2. Classify all three angles of a right triangle.
(13)

3. What kind of triangle is formed by two sides and one diagonal of a
(15) square?

New Concepts

Exploration Developing the Triangle Angle Sum Theorem

In this exploration, you will use patty paper to discover the relationship between the measures of the interior angles of a triangle.

1. On a sheet of notebook paper, trace and label △ABC in Theorem 18-1.

2. On a piece of patty paper, draw a line and label a point on the line P.

3. Place the patty paper on top of △ABC. Align the papers so that $\overline{AB}$ is on the line you drew and P and B coincide. Trace ∠B. Rotate the triangle and trace ∠C adjacent to ∠B. Rotate the triangle once more and trace ∠A adjacent to ∠C. The diagram shows your final step.

4. What do you notice about the three angles of the triangle you traced?

5. Draw a new triangle and repeat the activity using the new triangle. What is the result?

6. Write an equation describing the relationship you found between the three angles of △ABC.

The angles of a triangle have special relationships. The most basic relationship is given by the Triangle Angle Sum Theorem.

Theorem 18-1: Triangle Angle Sum Theorem
The sum of the measures of the angles of a triangle is equal to 180°. $$m\angle A + m\angle B + m\angle C = 180°$$

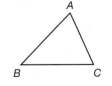

Online Connection
www.SaxonMathResources.com

Example 1 Using the Triangle Angle Sum Theorem

In the right triangle $\triangle ABC$, m$\angle B = 35°$ and the right angle is at vertex A. Find the measure of $\angle C$.

SOLUTION
From the Triangle Angle Sum Theorem:

$$m\angle A + m\angle B + m\angle C = 180°$$
$$90° + 35° + m\angle C = 180° \qquad \text{Substitute.}$$
$$m\angle C = 55° \qquad \text{Solve.}$$

Math Reasoning

Justify Explain in words how Corollary 18-1-2 follows from Theorem 18-1.

A **corollary** to a theorem is a statement that follows directly from that theorem. The Triangle Angle Sum Theorem has several useful corollaries.

Triangle Angle Sum Theorem Corollaries
Corollary 18-1-1: If two angles of one triangle are congruent to two angles of another triangle, then the third angles are congruent.
Corollary 18-1-2: The acute angles of a right triangle are complementary.
Corollary 18-1-3: The measure of each angle of an equiangular triangle is 60°.
Corollary 18-1-4: A triangle can have at most one right or one obtuse angle.

Each of these corollaries is a direct result of the Triangle Angle Sum Theorem. For example, Corollary 18-1-3 is true because every triangle has three angles and by Theorem 18-1, they add up to 180°. For all three angles to be congruent, they must each measure $\frac{180°}{3} = 60°$.

Example 2 Finding Angle Measures in Right Triangles

a. Find the measure of $\angle D$ in $\triangle DEF$.

SOLUTION
By Corollary 18-1-2, $\angle D$ and $\angle E$ are complementary.

$$m\angle D + m\angle E = 90°$$
$$m\angle D + 58° = 90° \qquad \text{Substitute for m}\angle E.$$
$$m\angle D = 32° \qquad \text{Subtract 58° from each side.}$$

b. In right $\triangle KLM$, $\angle K \cong \angle L$. Determine m$\angle K$.

SOLUTION
By Corollary 18-1-4, $\angle K$ and $\angle L$ cannot both be right angles. Therefore they are the acute angles of $\triangle KLM$. Since they are congruent, $\angle K$ can be substituted for $\angle L$.

$$m\angle K + m\angle L = 90°$$
$$m\angle K + m\angle K = 90° \qquad \text{Substitute for m}\angle L.$$
$$2m\angle K = 90° \qquad \text{Simplify.}$$
$$m\angle K = 45° \qquad \text{Divide both sides by 2.}$$

In any polygon, a **remote interior angle** is an interior angle that is not adjacent to a given exterior angle. In triangles, every exterior angle has a special relationship to its two remote interior angles.

Theorem 18-2: Exterior Angle Theorem

The measure of each exterior angle of a triangle is equal to the sum of the measures of its two remote interior angles.

$$m\angle DCA = m\angle A + m\angle B$$

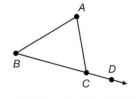

Example 3 Using the Exterior Angle Theorem

a. For $\triangle XYZ$, determine the measure of $\angle WYZ$.

SOLUTION
The remote interior angles are at X and Z. Therefore,
$$m\angle X + m\angle Z = m\angle WYZ$$
$$70° + 50° = m\angle WYZ$$
$$120° = m\angle WYZ$$

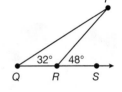

b. Determine the measure of $\angle P$ in $\triangle PQR$.

SOLUTION
Apply the Exterior Angle Theorem with the exterior angle at vertex R:
$$m\angle P + m\angle Q = m\angle PRS$$
$$m\angle P + 32° = 48° \qquad \text{Substitute.}$$
$$m\angle P = 16° \qquad \text{Subtract } 32° \text{ from each side.}$$

Example 4 Application: Civil Engineering

A bridge uses cables to support its 2000 foot span. Use the data in the image to determine the measure of the angle at the apex of the marked cable structure.

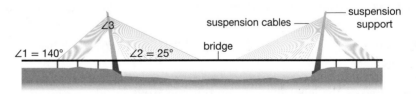

SOLUTION
Apply the Exterior Angle Theorem.
$$m\angle 2 + m\angle 3 = m\angle 1$$
$$25° + m\angle 3 = 140° \qquad \text{Substitute.}$$
$$m\angle 3 = 115° \qquad \text{Subtract } 25° \text{ from each side.}$$
The angle at the apex measures $115°$.

Use this figure to answer a and b.

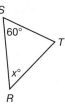

a. If $x = 50$, determine the measure of $\angle T$.
(Ex 1)

b. Determine m$\angle T$ if $x = 60$.
(Ex 1)

c. In right triangle $\triangle PQR$, the measure of one acute angle is 20°. What is the measure of the other acute angle in $\triangle PQR$?

d. In $\triangle ABC$, determine the measure of $\angle DAB$.
(Ex 3)

e. In $\triangle JKL$, $\angle K$ measures 60° and the exterior angle at vertex L measures 100°. Make a sketch of $\triangle JKL$ showing the given interior and exterior angle measures.
(Ex 3)

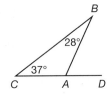

f. Determine the measure of $\angle J$ in $\triangle JKL$ in problem **e**.
(Ex 3)

g. (**Civil Engineering**) A planned glass pyramid structure has four triangular faces. The angles at the base of each face are congruent. Each of the angles at the apex of the pyramid measures 68°. What are the measures of the congruent base angles?
(Ex 3)

Practice **Distributed and Integrated**

1. Use $\triangle XYZ$ to answer problems **a** and **b**.
(8)
    a. In $\triangle XYZ$, what is the length of side $\overline{YZ}$? Round to the nearest eighth of an inch.
    b. Determine the perimeter of $\triangle XYZ$.

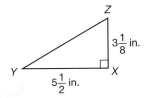

 2. **Algebra** Two congruent angles are complementary to a
(6) 22° angle. These angles are defined by the expressions $3x + 5$ and $7y - 2$, respectively. What are the values of x and y?

3. (**Geology**) Consider this statement. *If a rock is metamorphic, then it is crystalline.*
(17)
    a. Write the inverse of this statement.
    b. Assuming that the original statement is true, can you conclude that its inverse is true?

4. **Verify** Two of the labeled points on this coordinate grid have a third
(11) labeled point as their midpoint. Identify the two points and their midpoint, and verify your answer.

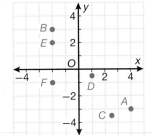

5. Find an example to disprove the converse of the following statement.
(10)
    If $x = -4$, then $x^2 = 16$.

6. **Multiple Choice** If a line has the length of 5 units and it has one endpoint
(9) of $(3, -2)$, what is a possible second endpoint?
    A $(0, 1)$ **B** $(-1, 2)$
    C $(-1, 1)$ **D** $(6, -6)$

7. Is $\overleftrightarrow{EF}$ parallel to $\overleftrightarrow{AB}$? Explain.
(5)

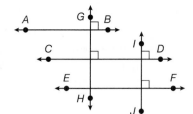

8. Algebra A rectangle has a perimeter of 40 inches. If the longer side is $(4x + 8)$ inches and the shorter side is $(3x - 2)$ inches, what are the dimensions of the rectangle?
(8)

9. What is the measure of angle 1 in the diagram?
(Inv 1)

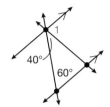

10. Geography Mia, a traveling sales representative, has to stop at five customers' locations in one day: A, B, C, D, and E. She begins and ends at her hotel, H. Mia figures that she will drive the shortest total distance if she makes the stops in the sequence A, C, D, E, B. Classify the polygon that her route, *HACDEB* forms, including any features which are congruent, and whether it is convex or concave.
(15)

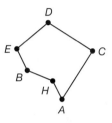

11. If two parallel lines are cut by a transversal that is perpendicular to both lines, which pairs of angles will be supplementary?
(Inv 1)

12. Analyze Use the Alternate Interior Angles Theorem to explain why corresponding angles are congruent.
(Inv 1)

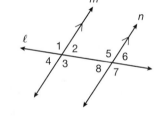

13. Justify Use inductive reasoning to create the next item in the sequence, and explain your reasoning.
(7)

14. Two angles of a triangle measure 30° and 45°. What is the measure of the third angle?
(18)

15. Algebra Transform the equation $x - 3y = 6$ to put it in slope-intercept form.
(16)

16. Determine the measure of the exterior angle at C in $\triangle ABC$.
(18)

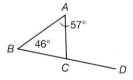

17. Verify Consider this statement. "If a polygon is irregular, then it is concave."
(17)
  a. Write the converse of the statement.
  b. Write the inverse of the original statement.
  c. Verify that the inverse statement and converse statement have the same truth value.

18. All exterior angles of a certain triangle are obtuse. Classify the triangle by its angles and explain how you know.
(18)

19. (Computer Science) An "OR" logic gate has two inputs and one
(14) output, related by the table shown.

   a. Find a counterexample to this conjecture.
   If the output of an OR logic gate is 1, input B is 1.

   b. Violet made the conjecture: *If neither input of an OR gate is 1,*
   the output is 0.
   Does the table contain a counterexample to her conjecture?

In		Out
A	*B*	
0	0	0
0	1	1
1	0	1
1	1	1

20. Write the inverse of this statement, "If a triangle has all three
(17) sides congruent, then it is obtuse." Which is true: the statement, its inverse, both,
or neither?

21. Write Write a conditional statement that has a true converse.
(10)

22. Find an expression for the perimeter of a regular hexagon if the sides are $2x - 6$.
(8)

23. (Optics) This optical diagram shows the paths of light rays
(12) in a reflecting telescope. Identify the two angles you need
to prove congruent in order to prove that rays *a* and *b* are
parallel. Which postulate or theorem would you use,
and why?

24. Multiple Choice Polygon *JKLMN* is a(n):
(15)
   A irregular pentagon   **B** equiangular nonagon
   C irregular hexagon   **D** regular pentagon

25. Multi-Step How many true conditional statements can you
(10) write using the following three statements?

   p: ∠1 and ∠2 are complementary
   q: m∠1 + m∠2 = 90°
   r: ∠1 and ∠2 are acute

26. Error Analysis Ricardo has drawn this graph for the line with
(16) equation $x - 2y = -4$. Identify two errors Ricardo has made.

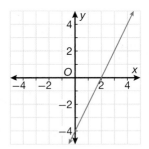

27. Write Explain why the distance formula is not needed to
(9) determine if three points lie on the same line.

28. In △*DEF*, the exterior angle at *D* measures 58° and m∠*E* = 36°.
(18) Determine m∠*F*.

29. Classify △*DEF* by its angles.
(13)

30. Two distinct lines which are not contained in plane *H* intersect plane *H*. At how
(4) many points could the lines intersect the plane?

Introduction to Quadrilaterals

Warm Up

1. *(15)* **Vocabulary** A polygon with all angles and all sides congruent is _____.

2. *(15)* **Vocabulary** A four-sided polygon is called a _____.

3. *(15)* Identify the figure at right as a square, a rectangle, or neither.

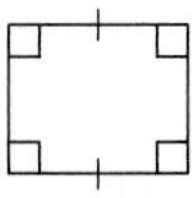

4. *(15)* The interior angles of a square are:

    A Right angles   **B** Congruent to each other

    C Congruent to the   **D** All of the above
        corresponding exterior angles

New Concepts

A **quadrilateral** is a polygon with four sides. Quadrilaterals are classified according to the number of congruent and parallel sides they have.

Quadrilateral	Properties	Example
Parallelogram	Both pairs of opposite sides are parallel.	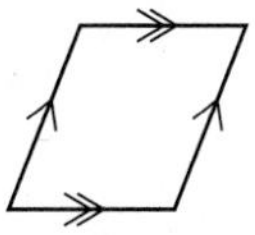
Kite	Exactly two pairs of consecutive sides are congruent.	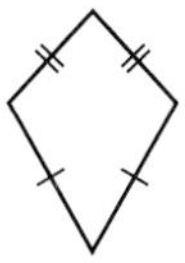
Trapezoid	Exactly one pair of opposite sides are parallel.	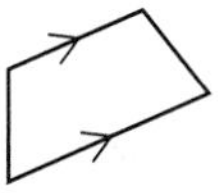
Trapezium	No sides are parallel.	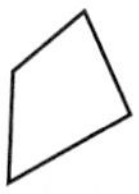

> **Reading Math**
>
> Sometimes symbols are used to name quadrilaterals. For example, ▭*PQRS* means "rectangle *PQRS*" and ▱*WXYZ* means "parallelogram *WXYZ*."

In addition to the quadrilaterals listed above, there are three types of parallelograms. Parallelograms are classified based on whether or not their sides are congruent and whether or not they have right angles.

> **Hint**
>
> Though parallelograms can often be given several names, always try to find the most specific name. For example, a quadrilateral with four right angles could be called a parallelogram, but it is more specific to call it a rectangle.

Parallelogram	Properties	Example
Rectangle	A parallelogram with four right angles	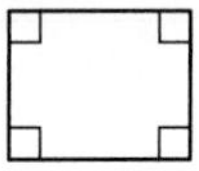
Rhombus	A parallelogram with four congruent sides	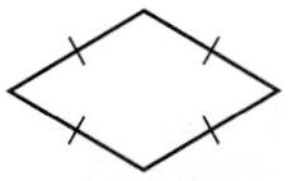
Square	A parallelogram with four right angles and four congruent sides	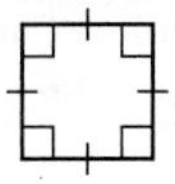

Some quadrilaterals can be named in several ways. For example, a square is also a rectangle, a rhombus, and a parallelogram; a kite is also a trapezium.

Example 1 **Classifying Quadrilaterals**

Classify each quadrilateral. Give multiple names if possible.

Math Reasoning

Model Could a kite or a trapezoid ever be a parallelogram? If so, draw an example.

SOLUTION

In quadrilateral *ABCD*, sides $\overline{AB}$ and $\overline{CD}$ are parallel. Also, $\overline{AD} \parallel \overline{BC}$. Therefore, *ABCD* is a parallelogram.

In quadrilateral *EFGH*, $\overline{EF} \cong \overline{FG}$ and $\overline{EH} \cong \overline{GH}$. Since both of these pairs of sides are consecutive, *EFGH* is a kite. Since no sides are parallel, *EFGH* is also a trapezium.

In quadrilateral *KLMN*, $\overline{KL} \parallel \overline{MN}$. *KLMN* is a trapezoid.

Quadrilateral *PQRS* has four right angles, so it is a rectangle.

Example 2 **Sketching Quadrilaterals**

Sketch each quadrilateral based on its description.

a. In quadrilateral *ABCD*, each side measures 3 feet.

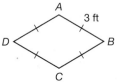

SOLUTION

1. Draw four sides of equal length. (The sides need not be perpendicular.)
2. Label the vertices *A*, *B*, *C*, and *D* in order.
3. Mark one side "3 ft".
4. Use tick marks on all four sides to show that they are congruent.

b. In quadrilateral *WXYZ*, each angle measures 90°.

SOLUTION

1. All four angles are right angles. Draw a rectangle.
2. Label the vertices *W*, *X*, *Y*, and *Z* in order.
3. Mark each angle with the symbol for a right angle.

c. In quadrilateral *RSTU*, $\overline{ST} \parallel \overline{RU}$.

SOLUTION

1. Draw one pair of parallel sides.
2. Mark these two sides with arrows to show they are parallel.
3. Draw the other two sides connecting the first two sides.
4. Label the vertices so that the parallel sides are $\overline{ST}$ and $\overline{RU}$.

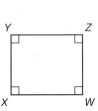

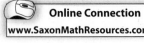
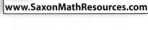

If a rectangle has a base of length b and height h, then its area is $A = bh$. Since all four sides of a square are congruent, its base and height are identical. Therefore, a square's area is given by squaring the length of any side, s, of the square: $A = s^2$.

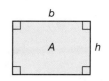

Example 3 Finding Perimeters and Areas of Rectangles and Squares

a. Determine the perimeter and area of this rectangle.

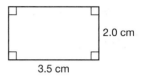

2.0 cm

3.5 cm

SOLUTION

The length of the rectangle is 3.5 centimeters and its width is 2.0 centimeters.

The perimeter is the sum of the side lengths:

$P = 3.5 + 2.0 + 3.5 + 2.0$
$P = 2(3.5) + 2(2.0)$
$P = 11.0$

The area is:

$A = bh$
$A = (3.5)(2.0)$
$A = 7.0$

The rectangle's perimeter is 11.0 cm and its area equals 7.0 cm^2.

b. Determine the perimeter and area of this square.

SOLUTION

The square has side lengths of $5\frac{1}{2}$ inches.

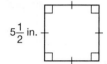

$5\frac{1}{2}$ in.

$P = 5\frac{1}{2} + 5\frac{1}{2} + 5\frac{1}{2} + 5\frac{1}{2}$ $A = s^2$
$P = 4(5.5)$ $A = (5.5)^2$
$P = 22$ $A = 30.25$

The perimeter of the square is 22 inches and its area equals 30.25 in².

Example 4 Sports

Each side of a baseball diamond measures 30 yards. Each of its corners is a right angle.

a. What kind of quadrilateral is a baseball diamond? Give as many different names for it as possible.

SOLUTION

Since the sides are congruent, each measuring 30 yards, and each corner forms a right angle, the most specific name for the diamond is a square. So, it follows that it can also be called a rhombus, rectangle, and parallelogram.

b. What distance must a batter run for a homerun?

SOLUTION

The distance is the perimeter of the diamond:

$P = 30 + 30 + 30 + 30$

$P = 120$

The batter must run 120 yards.

Math Reasoning

Model If one base of a rectangle is slid over, so it no longer had right angles, what quadrilateral results? Does this new quadrilateral have the same area or a different area than the rectangle?

a. Classify this quadrilateral. Give multiple names if possible.
(Ex 1)

b. In quadrilateral *PQRS*, $\overline{PQ} \parallel \overline{RS}$ and $\overline{PS} \parallel \overline{QR}$. Also, $\overline{PQ}$
(Ex 2) is approximately twice as long as $\overline{QR}$. Sketch *PQRS*.

c. **Sports** An Olympic swimming pool is a rectangle
(Ex 3, Ex 4) measuring 25 meters by 50 meters. Find its perimeter and area.

d. Determine the perimeter and area of this square.
(Ex 3)

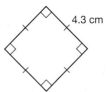

4.3 cm

Practice Distributed and Integrated

* 1. In quadrilateral *STUV*, $\overline{ST}$ and $\overline{TU}$ both measure 2.5 centimeters, and
(19) *SV* = *UV* = 5.0 centimeters. Sketch *STUV*.

2. Multiple Choice Choose the best classification of △*PQR* by its angles.
(13)
  A right   **B** acute and equiangular
  C acute   **D** obtuse

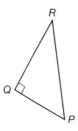

* 3. Write Write the negation of the Converse of Alternate Exterior Angles Theorem.
(17)

4. Algebra A triangle has a perimeter measuring 133.1 centimeters, and two of its
(13) sides measure 23.8 centimeters and 49.3 centimeters, respectively. The height from the third side to the opposite vertex is 19.0 centimeters. Determine the triangle's area.

5. Determine the slope of this line.
(16)

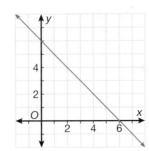

6. Generalize Use inductive reasoning to determine the missing value: 11, 13, 17, 19,
(7) 23, _____, 31, 37

7. A pair of lines is cut by a transversal, forming a pair of alternate interior angles
(Inv 1) that measure 85° and $(3x - 17)°$. For what value of x are the lines parallel?

8. What is the intersection of a line, a plane containing the line, and a point on
(4) the line?

9. Find a diagonal in polygon *PQRSTUVW* that lies partly in the exterior of the
(15) polygon. Classify polygon *PQRSTUVW* as convex or concave.

*10. Two angles of a triangle measure 68° and 43°, respectively. What is the measure of
(18) the third angle?

*11. Classify this quadrilateral.
(19)

12. Verify Confirm that a triangle with side lengths 5, 5,
(14) and 3 units is a counterexample to the statement,
"If a triangle has two congruent sides, then it is obtuse."

13. Identify the hypothesis and conclusion of this conditional statement.
(10) *If you have the blood type O, then you are the universal donor.*

14. (**Air Traffic Control**) On her screen, Rita notices that two planes are flying paths
(5) that are parallel to each other. She knows that if they continue flying straight,
their paths will never cross. What theorem or postulate can be used to justify
Rita's conclusion?

15. Determine the midpoint of the points $P(-3, 0)$ and $Q(-7, -3)$.
(11)

*16. Write Write the inverse of the statement.
(17)
If an animal is warm-blooded, then it is a mammal.

17. Verify Let M be the midpoint of $A(3, 7)$ and $B(5, 1)$ on the coordinate plane. Verify
(11) that M lies on $\overline{AB}$ and that $AM = BM$.

18. Find an expression for the area of a rectangle if one side is twice as long as the
(8) other side.

*19. Larissa is redecorating her room. The ceiling will be white and she plans to paint
(19) the walls blue. The diagram shows the room's dimensions.
   a. What is the area to be painted white?
   b. To the nearest square foot, what is the area to be painted blue?
   c. The actual area to be painted blue does not include the 6-by-2 foot door
      and does not include one 3-by-4 foot window. Find the actual area to be
      painted.

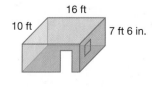

*20. (14) **Engineering** This table shows the gas mileage of a car at various cruising speeds. Do the data give a counterexample to the conjecture, "The product of cruising speed and gas mileage is constant"? Explain.

Cruising Speed (mi/h)	Gas Mileage (mi/gal)
37	35
56	30
75	22

*21. (18) In triangles *JKL* and *MNO*, ∠*J* is congruent to ∠*M*, and ∠*K* is congruent to ∠*N*. What can you say about ∠*L* and ∠*O*?

*22. (19) Classify this quadrilateral.

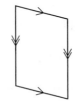

23. (11) What is the midpoint of (−4, 1) and (4, −1)?

24. (7) **Write** In a biological experiment looking at the migratory patterns of geese, all of the geese that were observed actually flew to a warmer climate during the cold winter months. Is the conjecture, "All geese fly to warmer climates for the winter," valid? Why or why not?

25. (9) **Multi-Step** Find the length of the given segments and determine if they are congruent.

26. (2) Identify the property that justifies the statement.
If $\overline{AB} \cong \overline{CD}$, then $\overline{CD} \cong \overline{AB}$.

27. (Inv 1) **Analyze** Use the Same-Side Interior Angles Theorem to explain why corresponding angles are congruent. Assume that *m* is parallel to *n*.

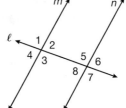

28. (16) **Write** an equation for the line passing through (3, 0) and (5, −1).

*29. (18) **Geography** Three streets in Chicago, Illinois—N. Wabash St., N. Rush St., and E. Chestnut St.—make a city block in the shape of a right triangle. The acute angle at the north end of the block measures 28°. What is the exterior angle measure at N. Rush St. and E. Chestnut St.?

30. (6) **Error Analysis** Victoria states that angles that measure 50° and 130°, respectively, are complementary, but her friend Shen disagrees, stating that they are not complementary, but instead are supplementary. Who is correct?

Interpreting Truth Tables

1. **Vocabulary** The statement "If q, then p," is the _____ of the
$^{(10)}$ statement "If p, then q."

Consider this statement to answer problems 2 and 3.
$^{(10)}$ *If two numbers are both positive, then their product is positive.*

2. Write the converse of the statement.

3. Which is true: the statement, its converse, both, or neither?

New Concepts A conditional statement has the form, "If p, then q." Recall that several other statements can be constructed from the conditional statement.

• the converse statement, "If q, then p"

• the inverse statement, "If $\sim p$, then $\sim q$"

• the contrapositive statement, "If $\sim q$, then $\sim p$"

The combination of a conditional statement and its converse is called a **biconditional statement**. A biconditional statement is true only when both the original statement and its converse are true. The biconditional of "If p, then q" and "if q, then p" can be written as "p if and only if q."

Reading Math

If p is the statement, "An angle is acute," then the **negation** $\sim p$ ("not p") is the statement, "An angle is not acute."

Example 1 Analyzing Conditional Statements

a. State the converse of this statement: *If* $x^2 \le 4$, *then* x ≤ 2.

SOLUTION Switch the hypothesis and conclusion: *If* x ≤ 2, *then* $x^2 \le 4$.

b. Determine if the statement and converse from part **a** are true.

SOLUTION Suppose the hypothesis is true: $x^2 \le 4$. Solving for x shows that x must be 2 or -2. Values greater than -2 and less than 2 also satisfy this equation. In fact, all the possible solutions are less than or equal to 2. Therefore, the statement is true.

For the converse, you can find a counterexample. For example, if $x = -3$, the conclusion is not true.

$$x^2 \le 4$$
$$(-3)^2 \le 4$$
$$9 \le 4$$

So, the converse is not true.

c. Write the biconditional of the statement. "If $x^2 \le 4$, then $x \le 2$." Is it true? Explain why or why not.

SOLUTION The biconditional is "$x^2 \le 4$ if and only if $x \le 2$."

For the biconditional to be true, both the statement and its converse must be true. In this case, the converse is false, so the biconditional is not true.

A **truth table** is a table that lists all possible combinations of truth values for a hypothesis, a conclusion, and the conditional statement or statements they form. Truth tables are useful tools because they show all the true/false possibilities at a glance:

Hypothesis: p	Conclusion: q	Statement: If p, then q
T	T	T
T	F	F
F	T	T
F	F	T

Math Reasoning

Analyze Why is a conditional statement true even when both its hypothesis and conclusion is false?

The highlighted row is the only combination for which the statement is *not* true. For example, to prove that the statement, "If a quadrilateral is equiangular, then it is a rhombus," is false, you need an example of an equiangular quadrilateral (p is true) that is not a rhombus (q is false).

Example 2 **Using a Truth Table**

(a.) Complete a truth table for the statement in Example 1.

SOLUTION There are four possibilities for the hypothesis and conclusion: both are true, only the hypothesis is true, only the conclusion is true, or both are false. Use these to create the rows of the truth table.

Hypothesis: $x^2 \leq 4$	Conclusion: $x \leq 2$	Statement: If $x^2 \leq 4$, then $x \leq 2$
T	T	T
T	F	F
F	T	T
F	F	T

Notice that the statement is only false when the hypothesis is true but the conclusion is false (the highlighted row of the table). It is impossible for the hypothesis $x^2 \leq 4$, to be true and the conclusion $x \leq 2$ to be false. Therefore, the statement is always true.

(b.) Add columns to your truth table for the statement's converse and its biconditional. Complete the table for these two statements.

SOLUTION

Hypothesis: $x^2 \leq 4$	Conclusion: $x \leq 2$	Statement: If $x^2 \leq 4$, then $x \leq 2$	Converse: If $x \leq 2$, then $x^2 \leq 4$	Biconditional: $x^2 \leq 4$ if and only if $x \leq 2$
T	T	T	T	T
T	F	F	T	F
F	T	T	F	F
F	F	T	T	T

Math Reasoning

Analyze Is a biconditional statement a conjunction or disjunction? Explain.

A **compound statement** combines two statements using *and* or *or*. It is similar to a conditional statement, except that *p* and *q* are related by "and" or "or" rather than by "if" and "then".

A compound statement that uses *and* is called a **conjunction**. Conjunctions have the form "*p* and *q*." For example, statement *p* stands for "I had bacon for breakfast," and *q* stands for "I had eggs for breakfast." If you have bacon and eggs for your breakfast, the conjunction, "I had bacon and eggs for breakfast," is true because *p* and *q* are both true. But if you have bacon and toast, the conjunction is false, because *p* is true but *q* is not.

A compound statement that uses *or* is called a **disjunction**. Disjunctions have the form "*p* or *q*." For example, suppose a lunch menu offers the choice of "soup or salad" as an appetizer. This is considered a disjunction because you can choose one or the other.

Example 3 Analyzing Compound Statements

A clothing store accepts cash or credit cards but not personal checks. It gives discounts on all cash purchases. Consider the statements, "a customer makes a credit-card purchase," and, "a customer gets a discount."

a. What is the conjunction of these statements? Use a truth table to assess its truth value.

SOLUTION
The conjunction is: "a customer makes a credit-card purchase *and* gets a discount."

Statement: *p*	Statement: *q*	Conjunction: *p* and *q*
T	T	T
T	F	F
F	T	F
F	F	F

For the conjunction to be true, the store would have to give discounts (*q* is true) on credit-card purchases (*p* is true). Discounts are only given on cash purchases though, so the conjunction is false.

b. What is the disjunction of these statements? Is it true or false?

SOLUTION
The disjunction is, "A customer makes a credit-card purchase *or* gets a discount." Extend your truth table:

Statement: *p*	Statement: *q*	Conjunction: *p* and *q*	Disjunction: *p* or *q*
T	T	T	T
T	F	F	T
F	T	F	T
F	F	F	F

For the disjunction to be false, the sale would have be a cash purchase (*p* is false) with no discount (*q* is true). Since this does not happen, the disjunction is true.

Math Reasoning

Justify Why is a disjunction true in more cases than a conjunction?

Example 4 **Application: Astronomy**

When stars run out of fuel, they either become black holes or degenerate stars. Consider the statements, "A star will become a degenerate star," and "A star will become a black hole." Form the conjunction and disjunction of these statements. Is the conjunction true? Is the disjunction true? Explain.

SOLUTION

Conjunction: *A star will become a degenerate star and a black hole.*

Disjunction: *A star will become a degenerate star or a black hole.*

The conjunction is false because a star cannot be both a degenerate star and a black hole. The disjunction is true, because all stars eventually become either degenerate stars or black holes.

Lesson Practice

For a–c, consider the statement, "If a quadrilateral is equiangular, then it is a rhombus."

a. State the converse of the statement.
(Ex 1)

b. Determine whether the statement is true. Also determine whether its
(Ex 1) converse is true.

c. Write the biconditional of the statement. Is it true? Explain.
(Ex 1)

Use the description of the restaurant to answer d–g.

(Restaurants) The chef's special at a five-star restaurant offers its customers a complimentary appetizer based on their choice of entrée. If customers order a filet mignon entrée, they receive leek soup for the appetizer. If customers order grilled salmon for the entrée, they receive baby spinach salad for the appetizer.

d. Use truth tables to represent the given statements. Interpret the tables
(Ex 2) for the statements.

e. Add columns to your truth tables to address the statements converses
(Ex 2) and biconditionals. Interpret the tables for these statements.

For f and g, consider the statements, "A customer orders a filet mignon entrée," and, "A customer receives a baby spinach salad appetizer."

f. What is the conjunction of these statements? Use a truth table to assess
(Ex 3, Ex 4) its truth value.

g. What is the disjunction of these statements? Is it true or false?
(Ex 3, Ex 4)

Practice Distributed and Integrated

1. Graph the line with equation $y = 0.5x - 1.5$.
(16)

2. Classify $\triangle DEF$ by its sides.
(13)

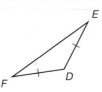

* 3. Find the converse of the statement, "If two angles are congruent, then their
(20) measures are equal," and write a biconditional statement equivalent to them.

* 4. (**Architecture**) Each floor of an office building has 30 rectangular windows
(19) measuring 57 inches by 38 inches. The building has 23 floors.
   a. Determine the total area of glass in all the windows, in square inches.
   b. Express this area in square feet.

5. Draw a Venn diagram to represent the following conditional statement.
(10) *A number is an integer if it is a whole number.*

6. (**Aim**) Julio is playing a game where he rolls a ball from one corner of the room to
(4) the opposite corner, without touching the walls. In how many different directions
could Julio roll the ball to reach its target? How do you know?

7. Find the converse of the following statement.
(10) *If you are an only child, then you do not have a brother.*

8. Multiple Choice Which is a counterexample to the conjecture, "If a quadrilateral is
(14, 19) equiangular, then it is equilateral"?
   A A rectangle that is not a rhombus   **B** A rhombus that is not rectangular
   C Any square   **D** Any parallelogram

9. (**Geography**) Write the converse of the following statement.
(17) *If a state lies to the east of longitude 125°W, then it is one of the lower 48 states.*

10. Write Can an obtuse and an acute angle make a straight angle? Can an obtuse and
(3) acute angle make a right angle? Explain.

11. Write Explain why testing a conjecture is not a generally accepted method of proof.
(7)

12. Analyze Given two lines m and n with a transversal ℓ, consider these statements:
(12) p: *Alternate interior angles formed by ℓ are congruent.*
   q: *Alternate exterior angles formed by ℓ are congruent.*
   r: *Lines m and n are parallel.*
   a. Prove that if statement p is true, then statement q is true.
   b. Use part **a** and the Converse of the Alternate Exterior Angles Theorem to
      prove the Converse of the Alternate Interior Angles Theorem.

13. Determine the midpoint of each side of square *ABCD*.
(11)

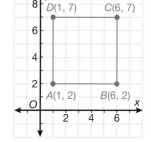

14. Write What can be done to disprove a conjecture? Explain.
(14)

*15. Error Analysis A student calculates the area and perimeter of a rectangle with
(19) length 7.5 cm and width 3 cm as follows:

$$A = 3(7.5) \qquad \text{and} \qquad P = 3 + 7.5$$
$$= 22.5 \text{ cm}^2 \qquad\qquad\qquad = 10.5 \text{ cm}$$

Explain the student's error.

*16. a. State the converse of the statement, "If a quadrilateral is a square, then it has
(20)   four congruent sides." Write the statement and its converse as a biconditional
    statement.
   b. Determine whether the biconditional is true. Use a truth table to explain why or
    why not.

*17. **Games and Puzzles** This tangram set includes two quadrilaterals. Classify them.
(19)

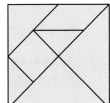

*18. An exterior angle of a triangle is a right angle. What can you say about the two
(18) remote interior angles? Explain.

19. **Algebra** What value of *x* allows you to apply the Converse
(12) of the Corresponding Angles Theorem to this figure?
If *x* has this value, how are lines *m* and *n* related?

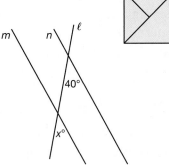

20. Determine the slope of the line passing through
(16) (3, 6) and (6, 0).

21. **Write** Can the sum of the angles in a triangle be considered a
(6) group of supplementary angles? Explain.

Use the diagram at right for the next two questions. Round to the nearest hundredth.

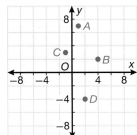

22. What is the length of $\overline{AB}$?
(9)

23. What is the length of $\overline{CD}$?
(9)

*24. Either Soo-Lin takes cheese sandwiches to school or she takes a salad,
(20) but not both or neither. On days when she takes cheese sandwiches, she
takes milk. Sometimes she takes milk even if she does not take cheese
sandwiches. Consider the statements "Soo-Lin takes cheese sandwiches,"
"Soo-Lin takes a salad," and "Soo-Lin takes milk." Write two disjunctions
that are true, each using two of these statements.

25. Is this conditional statement true? If so, explain why. If not, give a
(15) counterexample.
If a polygon is regular, then it is equiangular.

26. **Astronomy** This star chart shows the constellation Triangulum. Given
(18) that there is an almost perfect right angle at β ("beta"), what can you say
about the angles at the other two vertices?

*27. a. State the converse of the statement, "If the month is December, then
(10) it has 31 days."
  b. Determine whether the statement is true, and whether its converse
  is true.

28. **Woodworking** Scarlet measures two sides of a board of wood as 14 inches and 7
(8) inches. If she wants to find out the perimeter of the board, what formula should
she use? What is the perimeter of the board?

29. Determine the midpoint of each side of triangle *KLM* with coordinates $K(-3, 8)$,
(11) $L(1, 4)$, and $M(5, 2)$.

30. If a pair of parallel lines is cut by a transversal, and the alternate exterior angles
(Inv 1) measure $(160 - 2x)°$ and $(5x + 55)°$, what is the measure of each angle?

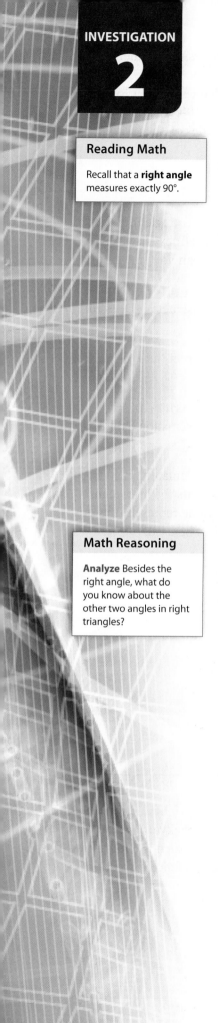

Proving the Pythagorean Theorem

A carpenter has built a rectangular frame with dimensions of 3 feet by 4 feet. To help secure this frame and ensure that the corners are at 90° angles, she places a brace across the diagonal between two opposite corners.

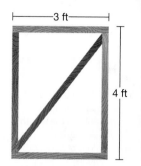

1. What type of triangle is formed when the diagonal brace is positioned across the frame?

Recall from Lesson 13 that a triangle with one right angle is classified as a right triangle. In right triangles, the side of the triangle that is opposite the right angle is called the **hypotenuse**. The other two sides of the triangle that form the right angle are called the **legs of a right triangle**.

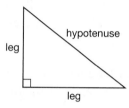

The carpenter can determine the length of the brace if she knows the lengths of the triangle's legs by using the Pythagorean Theorem. The Pythagorean Theorem states that for a right triangle, the sum of the squares of the lengths of the legs is equal to the square of the length of the hypotenuse.

The concept of the Pythagorean Theorem can be demonstrated using a model. In the diagram below, each side of the triangle is also a side of a square with each side congruent to that side of the triangle. All three squares are illustrated as blocks to make it easier to calculate the area of each one.

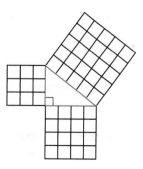

In the figure, the length of the sides of the triangle are 3, 4, and 5 units and the area of each square is 9, 16, and 25 square units, respectively.

2. Draw a right triangle with side lengths of 6 centimeters, 8 centimeters, and 10 centimeters, using a protractor to ensure that one angle is exactly 90°. Now make and cut out three squares; one with 6-centimeter long sides, one with 8-centimeter long sides, and one with 10-centimeter long sides. Place each square next to the corresponding sides of the triangle.

Which side is the hypotenuse in your diagram?

3. Which square has the largest area?

4. Try to fit the squares you made for the two legs of the triangle into the square for the hypotenuse. Do the two smaller squares fit exactly into the third square if you cut up the smaller squares? What can you say about the relationship between the areas of the squares?

5. Calculate the area of each of the squares that you constructed. Does this support your answer for part 4?

6. **Write** How do these results demonstrate the relationship in the Pythagorean Theorem?

Algebraically, the Pythagorean Theorem can be expressed as $a^2 + b^2 = c^2$ where a and b represent the lengths of the legs and c represents the length of the hypotenuse.

The length of the brace that the carpenter needs can be calculated algebraically. If the brace in the carpenter's frame represents the hypotenuse of a right triangle, c, and the length and width of the frame represent the legs of a right triangle, a and b, the equation $a^2 + b^2 = c^2$ can be used.

$$a^2 + b^2 = c^2$$
$$3^2 + 4^2 = c^2$$
$$25 = c^2$$
$$\sqrt{25} = \sqrt{c^2}$$
$$5 = c$$

Therefore, a 5-foot long brace is needed.

7. Determine c for this triangle.

8. A right triangle has a hypotenuse that measures 20 centimeters and one leg that is 16 centimeters. What is the length of the other leg?

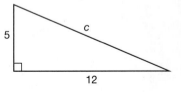

9. (**Landscaping Design**) Jillian has a corner garden in her backyard and wants to place decorative edging along the front edge. The garden is 6 feet along the back fence and 8 feet along the side fence. The two fences meet at a right angle. How much edging will Jillian need?

a. If the length of one leg of a right triangle increases while the length of the other leg is constant, what happens to the length of the hypotenuse in a right triangle? the measure of the right angle? the measure of the acute angles?

b. Find the length of side x in the triangle.

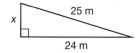

c. A right triangle has a hypotenuse of 15 inches and one leg that measures 12 inches. What is the length of the third side?

A builder is framing a house and wants to make sure that the walls will be at 90° angles to each other. The diagram shows the length measurements of two walls. Use this figure to answer the following questions.

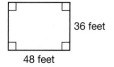

d. How could the Pythagorean Theorem be used to determine if the walls meet at a right angle?

e. If the builder strings a rope across the diagonal of the floor, what length would the rope be if the walls were at right angles to each other?

Laws of Detachment and Syllogism

Warm Up

1. Vocabulary The statement below is called a(n) _____ statement.
(10) (*conditional*, *inductive*)

If it is almost 8 p.m., then my favorite television show is about to start.

For each statement below, indicate if each conclusion is true or false.

2. *If all the students on a school bus are athletes from a local school, then all*
(10) *the school's athletes are on the school bus.*

3. *If x − 6 = 9, then x = 15.*
(17)

4. *All odd numbers are prime numbers.*
(17)

5. *All prime numbers greater than 2 are odd numbers.*
(17)

New Concepts

Deductive reasoning is the process of using logic to draw conclusions from given facts, definitions, and properties.

When two related statements are true, deductive reasoning can be used to make a conclusion. For example:

The bakery makes fresh bread every morning. It is morning.
Therefore, the bakery is making fresh bread.

Example 1 Using Deductive Reasoning

Use deductive reasoning to form a "Therefore" concluding statement from the given statements.

a. *All human beings need to breathe. Marla is a human being.*

SOLUTION
Therefore, Marla needs to breathe.

b. *All the chess team members won their opening match in the last tournament. Jeffery is on the chess team.*

SOLUTION
Therefore, Jeffery won his opening match in the last tournament.

c. *All the women in the royal family were wearing hats at the ball. Melissa is in the royal family.*

SOLUTION
Therefore, Melissa was wearing a hat at the ball.

Math Reasoning

Analyze If the second statement in this example was instead *"Marla needs to breathe,"* could you conclude that she is a human being?

Law of Detachment
For two statements p and q, when "If p, then q" is a true statement and p is true, then q is true.

Online Connection
www.SaxonMathResources.com

The Law of Detachment is a form of deductive reasoning that can be used to draw valid concluding statements. When the given facts are true, then correct logic can lead to a valid conclusion.

For example:
If it is Monday, then Marc will go to work.
Today is Monday.
Therefore, Marc will go to work today.

In the statements above, *p* represents the phrase "*it is Monday*," and *q* represents the phrase "*Marc will go to work.*"

Example 2   **Using the Law of Detachment**

For the following statements, use the Law of Detachment to write a valid concluding statement. Assume each conditional statement is true.

(a.) *When it is cold outside, I wear my warm jacket. It is cold outside today.*
SOLUTION
Therefore, I will wear my warm jacket today.

(b.) *If an angle is acute, then it cannot be obtuse. Angle D is acute.*
SOLUTION
Therefore, angle D cannot be obtuse.

(c.) *If a number is even, then it can be divided by 2. The number 104 is even.*
SOLUTION
Therefore, 104 can be divided by 2.

Law of Syllogism
When "If *p*, then *q*" and "If *q*, then *r*" are true statements, then "If *p*, then *r*" is a true statement.

The Law of Syllogism is another form of deductive reasoning. In this case, a third conditional statement is based on two conditional statements in which the conclusion of one is the hypothesis of the other.

This law poses that an intermediate truth is a valid progression from the original statement to a valid conclusion. For example:

If there is a power outage, then the freezer does not work.
   AND
If the freezer does not work, then the ice cream will eventually melt.
   THEN
If there is a power outage, then the ice cream will eventually melt.

Example 3 **Using the Law of Syllogism**

Use the Law of Syllogism to write a third conditional statement based on the statements below.

If Annika jumps higher than 5 feet 3 inches in this event, then she will win first place.

If Annika wins first place, then she will receive a medal.

SOLUTION

"Annika jumps higher than 5 feet 3 inches in this event" is *p*. *"She will win first place"* is *q*. *"She will receive a medal"* is *r*. Using the Law of Syllogism, write "if *p*, then *r*."

If Annika jumps higher than 5 feet 3 inches in this event, then she will receive a medal.

Math Reasoning

Write Write about one instance in the past few days where you have used the Law of Detachment or the Law of Syllogism to draw a conclusion about something in your life.

Example 4 **Using the Laws of Detachment and Syllogism**

For each of the given statement sets, draw a valid conclusion. Identify which law is used to reach the conclusion. Assume each conditional statement is true.

(a.) *If Maria wants to see a movie, then she goes to the theater. If Maria goes to the theater, then she buys popcorn.*

SOLUTION

If Maria wants to see a movie, then she buys popcorn. The Law of Syllogism is used. The first statement is of the form "If *p*, then *q*." The second statement is of the form "If *q*, then *r*." The conclusion follows, "If *p*, then *r*."

(b.) *If it is raining, then I will take an umbrella to school. Today, it is raining.*

SOLUTION

Therefore, today I will bring an umbrella to school. The Law of Detachment is used. The first statement is of the form "If *p*, then *q*." The second statement is of the form "*p* is true," which leads to the conclusion, "then *q*."

(c.) *All bibbles are bobbles. All bobbles play bubbles.*

SOLUTION

Recall that conditional statements are not always in 'if-then' form, but they can be rewritten that way. In 'if-then' form, the given statements read as follows:

If something is a bibble, then it is also a bobble. If something is a bobble, then it plays bubbles.

If something is a bibble, then it plays bubbles. The Law of Syllogism is used. The first statement is of the form "If *p*, then *q*." The second statement is of the form "If *q*, then *r*." The conclusion follows, "If *p*, then *r*."

Use deductive reasoning to form a concluding statement from the given information.
(Ex 1)

   a. *All the girls on the swim team are left-handed. Lorissa is on the swim team.*

   b. *When it is below 32°F for at least a week, the pond freezes. It has been below 32°F for a week.*

   c. *When every answer on a math test is correct, a student will get a perfect score on the test. Michael got every answer correct on the last math test.*

   d. *When employees work more than 40 hours in a week, they get paid overtime. Dominiqua worked 43 hours this week.*

   e. Use the Law of Detachment to write a valid conclusion to the
(Ex 2) statements below.
   If the gift I bought for my cousin is a toy truck, then it has four wheels. The gift I bought for my cousin is a toy truck.

   f. Write a third conditional statement using the Law of Syllogism:
(Ex 3) *If Nafeesa enrolls in an elective, then she will enroll in Orchestra. If Nafeesa enrolls in Orchestra, then she will play the violin this semester.*

   g. What conclusion can be drawn from the following set of statements?
(Ex 4) *If I oversleep tomorrow morning, then I will miss my bus. If I miss my bus, then I will be late for my appointment.*

   h. Which law was used to reach the conclusion in problem **g**?
(Ex 4)

Use detachment or syllogism to draw a valid conclusion to the following statements. Identify which law was used in reaching the conclusion.
(Ex 4)

   i. *If a gumble is hungry, it craves gloop. If a gumble craves gloop, he must hunt for gloop.*

   j. *If a vehicle is a unicycle, then it has only one wheel. This vehicle is a unicycle.*

Practice Distributed and Integrated

* **1.** **Analyze** What conclusion can be drawn from these statements?
 (21)
   If the quality of the granite is high, then our company will purchase it.
   The quality of the granite is high.

2. What is the length of a rectangle with a perimeter of 300 feet and a width of 32 feet?
(8)

3. **Error Analysis** Tim states that the slope between the points (1, 1) and (5, 5) is (4, 4).
(16) Yasmini states that the slope is $\frac{4}{4} = 1$. Who is correct? Explain.

4. Probability Write the converse of this statement and state whether the converse is
(20) always true. If both the statement and its converse are always true, write it as a
biconditional statement.

If it is certain that a sock I pull out of the drawer will be red, then all the socks in the drawer are red.

5. Verify Genji says that angle A measures $108°$. Is she correct? What theorem can be
(Inv 1) used to justify her answer?

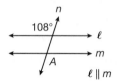

* 6. (Landscaping) A landscape designer needs to fence in the area indicated in the
(19) diagram. How much fencing will be needed for the whole area?

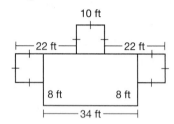

7. Find the distance between the points $A(4, 5)$ and $B(7, 1)$.
(9)

* 8. Multiple Choice A defense lawyer in court states in her cross examination of a
(20) witness, "You were either at the scene of the accident or you were not. Which is
it?" What type of statement is she using?

   A a compound statement   **B** a disjunction

   C both **A** and **B** are correct   **D** neither **A** nor **B** is correct

9. (Safety) It is recommended that the ratio of $a{:}b$ in the figure at right
(Inv 2) be 4:1 to prevent the ladder from shifting. According to this ratio, how
far from the base of the wall should you place the foot of a 16-foot ladder?
Round to the nearest inch.

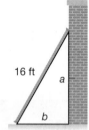

10. Write Disprove the statement by writing a counterexample.
(14)

All numbers that are either one more or one less than a multiple of 6 are prime.

*11. Use the Triangle Angle Sum Theorem to find the missing angles in the triangle.
(18)

Write the converse of each given statement, then determine whether the converse is true.

12. Analyze *If $3x + 4 = 10$, then $x = 2$.*
(17)

13. Analyze *If a number is prime, then it is odd.*
(17)

14. Classify the following quadrilaterals.
(19)

   a.   **b.**

15. Find the midpoint of the points $(-2, 5)$ and $(6, 9)$.
(11)

*16. Use the Law of Detachment to draw a valid concluding statement from
(21)　the following:

If the bread has just come out of the oven, then it is fresh. The bread has just come out of the oven.

17. Disprove the statement, "All irregular polygons are concave polygons," by drawing
(14)　a counterexample.

18. Justify Are the lines m and n parallel? Explain.
(12)

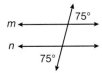

19. Find the missing angles in the figure on the right.
(Inv 1)

20. Algebra Find the slope of the line segment that joins points E and F.
(16)

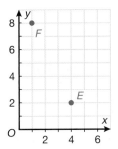

*21. Convert the following statements into a compound statement with a disjunction.
(20)
　　a. *Some numbers between 20 and 30 are prime numbers. The other numbers between 20 and 30 are composite numbers.*
　　b. *The lunch special can come with a soup. The lunch special can come with a salad.*

22. Is this polygon concave or convex? Explain.
(15)

*23. Write a third conditional statement using the Law of Syllogism:
(21)
　　If the season is rainy, then the crops will flourish.
　　If the crops flourish, then there will be plenty of corn for the livestock.

*24. Write a statement that can be deduced from the following statements.
(21)
　　All the children in the Dument family are the tallest in their classes at school.
　　Lucy is a child in the Dument family.

25. **Formulate** Find the equation of the line that passes through points *A* and *B* on
(16) the graph.

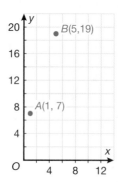

26. **Sewing** Thomas is sewing a tent with a triangular flap for a door. If the height of
(13) the door is 4 feet and the door is $4\frac{1}{2}$ feet wide, how many square feet of fabric will
he need to make the door?

27. **Justify** Are the lines *m* and *n* parallel? Explain.
(12)

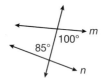

*28. Convert the following statements into compound statements with a conjunction.
(20) **a.** *Today it is cold. Today it is snowing.*
b. *Later this week is the big game at our school. Anita is playing in the big game.*

29. **Parking** A construction firm is designing a large parking lot. The lines that mark
(Inv 1) the widths of the spaces will be parallel. For optimum parking safety and visibility
when backing out, the measure of ∠1 should be half the measure of ∠2. What are
the measures of both angles?

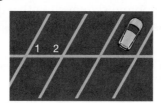

30. Find the measure of each angle in this figure. Lines *AB* and *CD* are parallel.
(Inv 1)

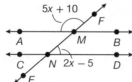

Finding Areas of Quadrilaterals

Warm Up

1. **Vocabulary** A segment from a vertex that forms a right angle with
(13) the line that contains the base is called the _____ of the triangle.
(**base**, **height**, **hypotenuse**)

2. Identify each type of triangle.
(13)
    a.   **b.**

    c.

3. Identify each type of quadrilateral.
(19)
    a.   **b.**

    c.

4. Evaluate the expression, $2x^2 + 3xy - y^2$, for $x = -2$ and $y = 1$.
(SB 14)

New Concepts

Recall from Lesson 13 that the area of a triangle can be determined using the
formula $A = \frac{1}{2}bh$ where b represents the base, h represents the height, and A
is in square units.

If a diagonal is drawn in a parallelogram, two identical triangles are formed.

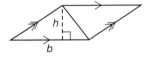

> **Math Language**
>
> In a parallelogram, the
> **height** is a segment
> perpendicular to both
> bases. Its length is used
> to calculate the area of
> the parallelogram.

The area of the parallelogram is twice the area of one of the triangles, so an
expression to find the area of a parallelogram could be written $A = 2\left(\frac{1}{2}bh\right)$.
Simplifying this expression results in the formula below, which can be used to
find the area of a parallelogram.

Area of a Parallelogram
To find the area of a parallelogram (A), use this formula, where b is the length of the base, and h is the height. 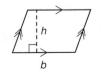 $$A = bh$$

> **Online Connection**
> www.SaxonMathResources.com

Since rectangles, rhombuses, and squares are all types of parallelograms, the areas of these shapes can also be found using this formula.

Example 1 Finding Areas of Parallelograms

Find the area of each parallelogram.

(a.)

SOLUTION
$A = bh$
$= (22 \text{ in.})(12 \text{ in.})$
$= 264 \text{ in.}^2$

(b.)
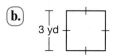

SOLUTION
$A = bh$
$= (3 \text{ yd})(3 \text{ yd})$
$= 9 \text{ yd}^2$

(c.)
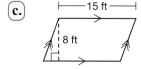

SOLUTION
$A = bh$
$= (15 \text{ ft})(8 \text{ ft})$
$= 120 \text{ ft}^2$

Hint

Always label the answer to an area problem with the correct unit. Area is measured in square units.

Recall that a trapezoid is a quadrilateral with exactly one pair of parallel sides. The height of a trapezoid is a segment that is perpendicular to both parallel sides of the trapezoid. The parallel sides of a trapezoid are known as the bases, b_1 and b_2.

Two congruent trapezoids can be connected to form a parallelogram with side lengths equal to $b_1 + b_2$, illustrated in the following diagram.

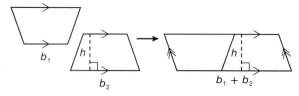

The area of the parallelogram is found by multiplying the length of its base by its height. Since the base of the parallelogram is $b_1 + b_2$, the area is $A = (b_1 + b_2)h$. Since this is the area of two trapezoids put together, the area of one trapezoid will be half this amount. The following formula can be used to find the area of any trapezoid.

Area of a Trapezoid

To find the area of a trapezoid (A), use the
following formula, where b_1 is the length of one
base, b_2 is the length of the other base of the
trapezoid, and h is the trapezoid's height.

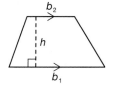

$$A = \tfrac{1}{2}(b_1 + b_2)h$$

Example 2 **Finding Areas of Trapezoids**

Find the area of each trapezoid.

a.

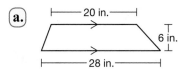

b.

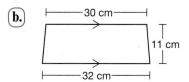

SOLUTION

$$A = \frac{1}{2}(b_1 + b_2)h$$

$$= \frac{1}{2}(20 \text{ in.} + 28 \text{ in.})(6 \text{ in.})$$

$$= 144 \text{ in}^2$$

SOLUTION

$$A = \frac{1}{2}(b_1 + b_2)h$$

$$= \frac{1}{2}(30 \text{ cm} + 32 \text{ cm})(11 \text{ cm})$$

$$= 341 \text{ cm}^2$$

Recall that a rhombus is a type of parallelogram with four congruent sides.
There are a few different methods that can be used to find the area of a
rhombus. One method is to apply the formula developed for a parallelogram,
$A = bh$. A second method uses the diagonals of the rhombus, which are
labeled d_1 and d_2.

The area of a rhombus is equal to one-half the product of its diagonals. The
formula below can be used to find the area of any rhombus.

Verify

A rhombus's diagonals
are perpendicular and
bisect each other. For a
rhombus with diagonals
x and y, show that the
area of the four triangles
in the rhombus's interior
is equal to the area
found by using the
formula given here.

Area of a Rhombus

To find the area of a rhombus (A), use the following
formula, where d_1 is the length of one diagonal,
and d_2 id the length of the other diagonal of the
rhombus.

$$A = \frac{1}{2}d_1 d_2$$

Example 3 — Finding Areas of Rhombuses

Find the area of each rhombus.

a.

9.0 cm

9.5 cm

b.

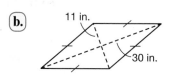

11 in.

30 in.

SOLUTION

$A = bh$

$= 9.5 \text{ cm} \times 9.0 \text{ cm}$

$= 85.5 \text{ cm}^2$

SOLUTION

$A = \frac{1}{2}d_1d_2$

$= \frac{1}{2}(11 \text{ in.} \times 30 \text{ in.})$

$= \frac{1}{2}(330 \text{ in}^2)$

$= 165 \text{ in}^2$

Example 4 — Application: Carpeting

Two areas of a day care need to be carpeted. The play area is shaped like a trapezoid, and the supplies area is shaped like a rectangle. Use the diagram of these two areas to determine the total area that needs to be carpeted.

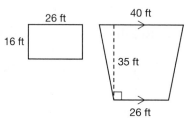

26 ft

16 ft

40 ft

35 ft

26 ft

> **Caution**
>
> This problem asks for the total area that is to be carpeted. To find this, you will have to combine the areas of both figures.

SOLUTION

For the rectangular supplies area,

$A = lw$

$= 26 \times 16$

$= 416 \text{ ft}^2$

For the trapezoidal play area,

$A = \frac{1}{2}(b_1 + b_2)h$

$= \frac{1}{2}(40 \text{ ft} + 26 \text{ ft})(35 \text{ ft})$

$= \frac{1}{2}(66 \text{ ft})(35 \text{ ft})$

$= 1155 \text{ ft}^2$

The total area can be found by adding these two areas:

$(416 + 1155) \text{ ft}^2 = 1571 \text{ ft}^2$

Therefore, a total area of 1571 square feet needs to be carpeted.

Find the area of each parallelogram.
(Ex 1)

a.

14 yd

8 yd

b.

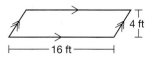

4 ft

16 ft

c. Find the area of a trapezoid with parallel sides measuring
(Ex 2) 14 centimeters and 21 centimeters and a height of 13 centimeters.

d. Find the area of this figure.
(Ex 2)

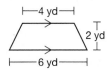

4 yd

2 yd

6 yd

e. Find the area of a rhombus that has diagonal lengths of 8 inches and
(Ex 3) 11 inches.

f. Find d_2 of a rhombus if d_1 is 6 meters and it has an area of 12 square
(Ex 3) meters.

g. **Carpeting** Dakota needs to put new flooring in the cafeteria of a high
(Ex 4) school. The cafeteria is shaped like a diamond with four congruent
sides. What measurements should Dakota take to find the total area of
the cafeteria?

Practice Distributed and Integrated

* 1. Find the area of a rectangle with a diagonal measuring 25 inches and one side that
(22) measures 20 inches.

2. Analyze Are the following pairs of polygons congruent? Explain.
(15)
a.

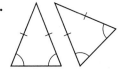

b.

3. Multiple Choice The exact conversion from Celsius to Fahrenheit is $°F = \frac{9}{5}(°C) + 32$. The actual
(8) temperature at which the two temperature scales have the same numerical value is
A −30 degrees   B 30 degrees
C −40 degrees   D 40 degrees

4. Use deductive reasoning to write a valid conclusion to this statement.
(21)
All games played at the baseball tournament were decided by one run. Allentown
played Holdenville at the baseball tournament.

* 5. Find the area of the following quadrilaterals.
<small>(22)</small>

a.
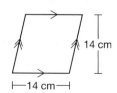
14 cm

14 cm

b.
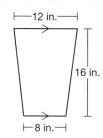
12 in.

16 in.

8 in.

6. **Justify** Write the converse to each conditional statement. State if each converse is true.
<small>(10)</small>
 a. *If x + 11 = 5, then x = −6.*
 b. *If it rains, then I bring my umbrella to work.*

7. **Justify** What is the value of *x*? Explain how you found it.
<small>(18)</small>

31°

x°

8. State whether the following statements are true or false.
<small>(6)</small>
 a. Linear pairs of angles are supplementary.
 b. Complementary angle pairs can be linear.
 c. The adjacent angles formed by two intersecting lines
   are called vertical angles.
 d. If two angles are in the same plane and share a vertex and a side, but no
   interior points, then the two angles are adjacent angles.

xy² *** 9.** **Algebra** Determine the measure of each angle.
<small>(18)</small>

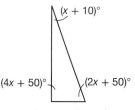
(*x* + 10)°

(4*x* + 50)°

(2*x* + 50)°

*10. Use the Law of Detachment to write
<small>(21)</small> a concluding statement.

 *If today is Thursday, then it is Mrs. Wu's turn
 to drive the children to their piano lesson. Today
 is Thursday.*

11. ⟮**Urban Planning**⟯ On a scale map of a state, each side of a square represents 5 miles.
<small>(9)</small> The water works company needs to connect two purification stations that are
 located at the points (3, 11) and (14, 26) on the grid. About how many miles apart
 are these two stations?

*12. Write each set of statements as a biconditional statement.
<small>(20)</small>
 a. Two angles are congruent if they have the same measure. Two angles that have
   the same measure are congruent.
 b. A triangle is equilateral if it has three lines of symmetry. A triangle that has
   three lines of symmetry is equilateral.

13. The following statement is an example of a _____. (*conjunction, disjunction*)
<small>(20)</small>
 A set of two statements can often be joined as a conjunction or a disjunction.

*14. Find the areas of the following shapes.
<small>(22)</small>
 a.
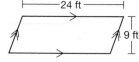
24 ft

9 ft

 b. *d₁* = 16 yd
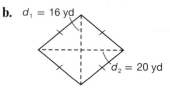
d₂ = 20 yd

15. Find the equation of the line that:
⁽¹⁶⁾
   a. passes through the point (0, 5) with a slope of −2.

   b. passes through the point (4, 11) with a slope of $\frac{1}{2}$.

16. Classify each triangle according to its sides.
⁽¹³⁾
   a.   **b.**

17. (**Yachting**) A race course for a yachting regatta is in the form of a right triangle. If
⁽¹³⁾ the two legs of the triangle are 6 miles and 8 miles long, how much distance does the race cover?

*18. **Error Analysis** Veronica thought the equation of the line that passes through the
⁽¹⁶⁾ points (1, 5) and (5, 17) is $y = 2x + 3$, but then she found out that her answer was incorrect. What has she done wrong and what is the correct equation?

19. **Analyze** Are the following pairs of lines parallel? Explain.
⁽¹²⁾
   a.   **b.**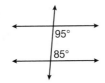

20. **Write** Draw an equiangular octagon with sides that are each 2 centimeters long.
⁽¹⁵⁾ Is it regular? Explain.

*21. (**Landscaping**) The following diagram represents a patio that is
⁽¹⁹⁾ to be covered in stones. Determine the total area of the patio.

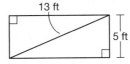

22. (**Urban Planning**) Three transformer stations exist on a grid at positions *A*, *B*, and
⁽¹¹⁾ *C*. The power company needs to locate the three points that are exactly halfway between each of the three stations to determine where new stations must be built. If the existing stations are located at grid points *A*(5, 9), *B*(4, −4), and *C*(8, 17), find the locations of the new stations.

23. Graph the line $y = 3x + 1$ using a table of values.
⁽¹⁶⁾

24. Identify the hypothesis and the conclusion in each conditional statement.
⁽¹⁰⁾
   a. *If today is Wednesday, then Jasmine needs to take out the trash.*

   b. *If* x − 3 = 5, *then* x = 8.

25. (**Emergency Response**) Two fire stations are located at grid points *A*(4, 8) and
⁽⁹⁾ *B*(17, 29). The closest fire station should respond to any emergency call. Which station should respond to a fire located at the grid point (9, 21)?

*26. Use the Law of Syllogism to write a third conditional statement.
(21)

If Mark's pen runs out of ink, then he must get a new pen to finish his assignment.
If Mark must get a new pen to finish his assignment, then he must go to the store.

27. An approximate method for converting from Celsius to Fahreneit is to double the
(8) Celsius temperature and add 30. Use this method to convert 25° C and −30° C to
Fahrenheit.

*28. (Landscaping) A landscape architect needs to place sod on the area within the shape
(22) shown below. How many square feet of sod are needed to cover the area?

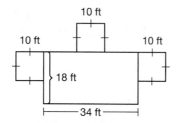

29. Find the measure of each acute angle and each obtuse angle in the diagram.
(Inv 1) Line *a* is parallel to line *b*.

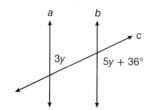

30. **Predict** Draw a line *l*, and a point *P* not on *l*. Then, construct a line parallel to *l*
(5) through point *P*. Can you construct a different line that is parallel to *l* through the
same point *P*? Write a conjecture about the number of lines parallel to *l* through
point *P*.

Introduction to Circles

Warm Up

1. **Vocabulary** The sum of the side lengths of a closed polygon is called
(8)   the _____.

2. Evaluate $2x + 5y$ for $x = 3$ and $y = -2$.
(SB 14)

3. For the following statements, use the Law of Detachment to write a valid
(21)   concluding statement.

If I forget to close the front door, then the dog will get out. I forgot to close the front door.

Evaluate each expression. Round to the nearest hundredth.
(SB 6)

4. $2(15)^2$

5. $3.27(6.5)^2$

New Concepts

A **circle** is the set of points in a plane that are a fixed distance from a given point. This point is called the **center** of the circle. To name a circle, use the ⊙ symbol and the center point. For example, ⊙A is read, "circle A."

All the points within the circle are called the **interior** of the circle. Any segment whose endpoints are the center of the circle and a point on the circle is called a **radius**. Any segment with both endpoints on the circle that passes through the center is called a **diameter**. The length of a diameter is always twice the length of a radius.

Two circles are congruent if they have congruent radii.

Math Reasoning

Write Name a few objects you see regularly that are circles. Do they have easily identifiable centers, radii, and diameters?

┌─ **Example 1** **Naming Parts of a Circle**

Identify a diameter, a radius, and the center of the circle at right.

SOLUTION
$\overline{AB}$ is a diameter, $\overline{AC}$ and $\overline{BC}$ are both radii, and the center of the circle is point C.

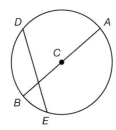

As seen in Lesson 8, the sum of the side lengths of a closed figure is called the perimeter. The perimeter of a circle is the distance around the circle, and is called the **circumference**.

Circumference of a Circle
To find the circumference (C) of a circle, use the formula below, where r is the circle's radius. $C = 2\pi r$

Pi, represented by the symbol π, is an irrational number that is defined as the ratio of the circumference of a circle to its diameter. There are several ways to approximate the value of pi. Some common approximations are 3.14 and the fraction $\frac{22}{7}$. Most problems in this book will tell you which approximation to use, or will instruct you to leave your answer in terms of π. If no approximation is specified, use the π key on your calculator for a more precise answer.

Note that, given the relationship between radius and diameter, the formula for the circumference of a circle can also be expressed as $C = \pi d$, where d is the diameter.

Example 2 **Finding Circumference**

Find the circumference of the circle to the nearest hundredth of an inch. Use 3.14 for π.

SOLUTION
The radius of the circle is 14.00 inches.

$C = 2\pi r$

$\approx 2(3.14)(14.00)$

≈ 87.92

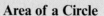

14.00 in.

Therefore, the circumference is approximately 87.92 inches.

Area of a Circle
To find the area (A) of a circle, use the formula below, where r is the circle's radius. $$A = \pi r^2$$ 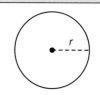

Another way to think of the area of a circle is the area that is swept out by rotating a radius of the circle one full rotation.

Example 3 **Finding Area**

Find the area of each circle to the nearest hundredth of a square unit. Use 3.14 for π.

2 m

SOLUTION
The radius of the circle is 2 meters.

$A = \pi r^2$

$\approx (3.14)(2)^2$

≈ 12.56

Therefore, the area is approximately 12.56 m^2.

b.

├── 26 in. ──┤

Math Reasoning

Verify Check that the formula $A = \dfrac{\pi d^2}{4}$ can be used to solve Example 3b, instead of finding the radius first. Why does this formula work?

SOLUTION

Divide the diameter by 2 to determine the radius measurement.

$$r = \frac{26}{2}$$
$$= 13$$

The radius of 13 inches can then be substituted into the formula.

$$A = \pi r^2$$
$$\approx (3.14)(13)^2$$
$$\approx 530.66$$

Therefore, the area is approximately 530.66 in².

Example 4 **Application: Urban Design and Planning**

A dog park is being constructed with a circular fence surrounding the park. The fence has a radius that is 50 yards long. Use 3.14 for π.

a. What is the distance around the fence to the nearest yard?

SOLUTION

To find the total distance around the fence, the circumference must be calculated.

$$C = 2\pi r$$
$$\approx 2(3.14)(50)$$
$$\approx 314$$

Therefore, the total distance around the fence is approximately 314 yards.

b. Approximately how many square yards of sod would be needed to completely cover the area inside the fence with grass?

SOLUTION

The area must be calculated in order to determine the amount of sod required.

$$A = \pi r^2$$
$$\approx (3.14)(50)^2$$
$$\approx 7850$$

Therefore, the total area to be covered with sod is approximately 7850 yd².

Lesson Practice

    a. Draw $\odot P$ with a radius, a diameter, and the center labeled.
(Ex 1)

    b. Find the circumference of a circle with a radius of 0.5 meters.
(Ex 2)   Use 3.14 for π and round to the nearest hundredth of a meter.

c. Find the area of a circle with a radius of 31 centimeters. Use 3.14 for π
(Ex 3) and round to the nearest hundredth of a square centimeter.

d. Find the area of a circle with a diameter of 1 yard. Use 3.14 for π and
(Ex 3) round to the nearest hundredth of a square yard.

e. The lid to a sewer access opening is 35 inches in diameter. If you roll
(Ex 4) it in a straight line along the ground for three rotations, how much
distance would it cover? Use $\frac{22}{7}$ for π.

Practice Distributed and Integrated

* 1. To the nearest kilometer, how far would you go if you traveled along the
(23) circumference of the equator, assuming the equator is a circle with a radius of
6378 kilometers? Use 3.14 for π.

2. Write Explain what it means for a mathematical statement to be a conjecture and give
(7) an example.

3. Algebra Find the hypotenuse of a right triangle with legs measuring $3x$ and $4x$.
(Inv 2)

* 4. ⬡Handball A semicircular area of the gym floor has been marked off with
(23) tape to make room for a handball game. If the radius of the semicircle is
16 feet, about how much tape is needed? Use 3.14 for π and round to the
nearest tenth of a foot.

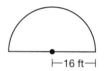

5. Multi Step Find the distance between the point $(5, 7)$ and the midpoint of the
(9) segment with endpoints $(1, 4)$ and $(11, 6)$.

6. Find the area of a trapezoid that is 12 centimeters tall, and whose bases are
(22) 12 centimeters and 10 centimeters long.

7. Give three possible sets of dimensions for a right triangle with an area of
(13) 36 square inches.

8. Can the number of sides of a polygon be different than its number of
(15) vertices?

9. Coordinate Geometry Find the midpoint of each side of
(11) triangle *DEF*.

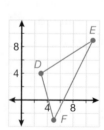

10. Draw a regular convex octagon.
(15)

*11. Write Use the Law of Detachment to draw a conclusion to the following
(21) statements.

If I have the correct combination, then I can open the safe.
I have the correct combination.

12. **Algebra** Find the hypotenuse of a right triangle with congruent leg lengths that are both b.
(Inv 2)

*13. Name the conditional statement and the converse that are true if the following
(20) biconditional statement is true.

The rooms in the arena will get a new coat of paint if and only if new funding from the local council is approved.

14. **Multiple Choice** Which of these choices *cannot* describe a polygon?
(15)   **A** more than 12 sides   **B** a curved side
      C closed   **D** concave

15. State the measure of the angle that is supplementary to each angle.
(6)   **a.** 75°   **b.** 126°

*16. Draw two different rectangles, each with a perimeter of 100 centimeters, and
(19) calculate the area of each.

*17. (Carpentry) A carpenter is going to use a protractor to determine if
(12) the sides of the door she just installed are parallel to the corresponding sides of the door frame. To do this, she attaches a brace to the door, as shown, and measures the indicated angles. How can she use the values of these two angles to determine if the sides of the door are aligned correctly?

18. **Multiple Choice** In this set of statements using syllogism, one of the
(21) statements is missing. Which one of the given statements below would correctly complete this syllogism?

If I plant a seedling in my yard, then a tree will grow.
   AND

——————————————————————————————————————

   THEN
If I plant a seedling in my yard, then my house will be worth more.

   A *If I don't plant a seedling in my yard, then my house will be worth less.*
   B *If a tree grows in my yard, then my house will be worth more.*
   C *If my house is worth more, then a tree will grow in my yard.*
   D *If my house is worth more, then I will plant a seedling in my yard.*

19. **Error Analysis** Dustin reasons that the missing angle in this triangle is 53°.
(18) Is he correct? If not, what is the measure of the missing angle?

*20. (Sewing) Rhonda is making a circular pillow with a diameter of 18 inches, and
(23) she needs to sew trim around the edge. To the nearest tenth of an inch, how much trim would Rhonda need to sew around the edge of the pillow? Use 3.14 for π.

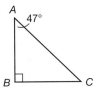

*21. **Write** Write a true conditional statement whose inverse is also true.
(17)

22. (Painting) If one can of spray paint can cover 32 square feet, how many cans of
(13) spray paint are needed to paint a wall in the shape of an isosceles triangle with a base of 18 feet and a height of 12 feet?

23. (**Surveying**) Civil engineers are drafting plans for a road with a slope that cannot
(16) exceed a value of 0.08. If the total rise of the road from the bottom to the top
point needs to be 24 meters, how long must the horizontal distance of the road be
to maintain the 0.08 slope?

24. Disprove the conditional statement by giving a counterexample.
(14)
*If a figure is formed by connecting five points with line segments, then the figure
is a polygon.*

*25. Two circles on the ground at the baseball field are to be painted red for the
(23) upcoming playoffs. If the diameter of each circle is 6 feet, what is the total area
to be painted? Use 3.14 for π and round to the nearest tenth of a square foot.

26. The coordinates of the vertices of a right triangle are $R(3, 4)$, $S(9, 4)$, and $T(3, 11)$.
(13) If each unit on the grid represents one foot, determine the approximate perimeter
and exact area of the resulting triangle.

27. Ana knows that $\angle B$ measures 72°. She says that if lines l and m
(Inv 1) are parallel, then $\angle G$ also measures 72°. What theorem supports
Ana's claim?

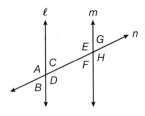

*28. A right triangle has side lengths of 15, 36, and 39 centimeters. What is
(13) the area of the triangle?

29. **Algebra** Find the area of a trapezoid with a height of $\frac{1}{4}x$ and base
(22) lengths of $3x$ and $5x$.

30. **Justify** Compare the areas of these two trapezoids. Do the figures have the same
(19) area? Explain your reasoning.

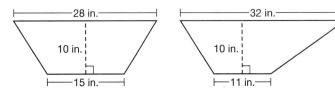

Algebraic Proofs

1. **Vocabulary** The statement below is an example of the Law of _____.
   (21) (**Detachment, Syllogism**)
   If 1 + 1 = 2 and 2 = a, then 1 + 1 = a.

2. State the opposite operation to each of the following:
   (SB 2)
   a. addition
   b. division
   c. finding the square root

3. **Multiple Choice** Which value of x is valid for the equation
   (SB 14) $3x - 2 = 2x - 3$?

   A -1 **B** -2
   C 1 **D** 2

4. Write the converse and the inverse of the statement below.
   (17) *If all the sides of a parallelogram are congruent, then it is a rhombus.*

New Concepts

A **proof** is an argument that uses logic to show that a conclusion is true. Since you have already learned to solve an algebraic equation, you have already performed a proof. An algebraic proof uses properties of equality to solve an equation. These properties are listed in the table below.

> **Hint**
>
> See the Skills Bank for a list of other properties of arithmetic that can be used as justifications in an algebraic proof.

Properties of Equality

Property	Example
Addition Property of Equality	If $a = b$, then $a + c = b + c$.
Subtraction Property of Equality	If $a = b$, then $a - c = b - c$.
Multiplication Property of Equality	If $a = b$, then $ac = bc$.
Division Property of Equality	If $a = b$ and $c \neq 0$, then $\frac{a}{c} = \frac{b}{c}$
Symmetric Property of Equality	If $a = b$, then $b = a$.
Reflexive Property of Equality	$a = a$
Transitive Property of Equality	If $a = b$ and $b = c$, then $a = c$.
Substitution Property of Equality	If $a = b$, then b can be substituted for a in any expression.

> **Online Connection**
> www.SaxonMathResources.com

> **Hint**
>
> Think of each justification in a proof as the mathematical answer to the question, "what did you do in this step?"

An algebraic proof shows step-by-step how a problem is solved. Each step has to be justified with one of the properties above, or by a property of arithmetic. For example, if a step required adding a number to both sides of an equation, it would be justified by the Addition Property of Equality.

Whenever a step requires that you perform basic mathematical operations on a single side of the equation (like addition, subtraction, multiplication, or division), the step is justified by the term, "Simplify."

Most proofs begin by presenting the facts of a problem. The first line restates what you have already been told, and is justified as "Given." When you are given an equation to solve for an algebraic proof, the original equation is the "Given" statement.

Example 1 Writing an Algebraic Proof

(a.) Solve this equation. Provide a justification for each step.

$$2(x + 1) = x + 9$$

SOLUTION

$2(x + 1) = x + 9$	Given
$2x + 2 = x + 9$	Distributive Property
$2x + 2 - 2 = x + 9 - 2$	Subtraction Property of Equality
$2x = x + 7$	Simplify.
$2x - x = x + 7 - x$	Subtraction Property of Equality
$x = 7$	Simplify.

(b.) Solve this equation. Provide a justification for each step.

$$\frac{3x - 1}{5} = \frac{2x + 3}{3}$$

SOLUTION

$\frac{3x - 1}{5} = \frac{2x + 3}{3}$	Given
$15\left(\frac{3x - 1}{5}\right) = 15\left(\frac{2x + 3}{3}\right)$	Multiplication Property of Equality
$\frac{15}{5}(3x - 1) = \frac{15}{3}(2x + 3)$	Associative Property of Multiplication
$9x - 3 = 10x + 15$	Distributive Property
$9x - 3 + 3 = 10x + 15 + 3$	Addition Property of Equality
$9x = 10x + 18$	Simplify.
$9x - 10x = 10x + 18 - 10x$	Subtraction Property of Equality
$-x = 18$	Simplify.
$\frac{-x}{-1} = \frac{18}{-1}$	Division Property of Equality
$x = -18$	Simplify.

Example 2 Verifying Algebraic Reasoning

The steps of the algebraic proof for solving the equation $2(a + 1) = -6$ are given below in the correct order. However, the justifications for each step are out of order. Determine the correct order for the justifications.

$2(a + 1) = -6$	Simplify.
$2a + 2 = -6$	Distributive Property
$2a + 2 - 2 = -6 - 2$	Division Property of Equality
$2a = -8$	Given
$\frac{2a}{2} = \frac{-8}{2}$	Subtraction Property of Equality
$a = -4$	Simplify.

Math Reasoning

Verify Explain how the solution to example 1a can be checked.

Math Reasoning

Justify Edwin performed the same algebraic proof as Example 2, but his started with the Division Property of Equality. Is Edwin also correct? Explain.

SOLUTION

The correct order of the proof is:

$2(a + 1) = -6$	Given
$2a + 2 = -6$	Distributive Property
$2a + 2 - 2 = -6 - 2$	Subtraction Property of Equality
$2a = -8$	Simplify.
$\dfrac{2a}{2} = \dfrac{-8}{2}$	Division Property of Equality
$a = -4$	Simplify.

Example 3 Application: Finding Dimensions

The area of a rectangular patio is 28 square feet. The patio's length is $(3x + 1)$ feet and the patio's width is $2x$ feet. Find the dimensions of the patio. Provide a justification for each step.

SOLUTION

The formula for the area of a rectangle is $A = lw$, so

$A = 28, l = (3x + 1), w = 2x$	Given
$A = lw$	Area formula for a rectangle
$28 = (3x + 1)(2x)$	Substitution Property of Equality
$28 = 6x^2 + 2x$	Distributive Property
$6x^2 + 2x = 28$	Symmetric Property of Equality
$\dfrac{6x^2 + 2x}{2} = \dfrac{28}{2}$	Division Property of Equality
$3x^2 + x = 14$	Simplify.
$3x^2 + x - 14 = 14 - 14$	Subtraction Property of Equality
$3x^2 + x - 14 = 0$	Simplify.
$(3x + 7)(x - 2) = 0$	Factor.

There are two solutions to this factorization, $3x + 7$, and $x - 2$. However, the solution to $3x + 7$ is negative. It does not make sense for a side of the rectangle to have a negative length, so that solution is thrown out.

Therefore,

$x - 2 = 0$	Given
$x - 2 + 2 = 0 + 2$	Addition Property of Equality
$x = 2$	Simplify.

Now, substitute $x = 2$ into the expressions for length and width of the rectangle to find the dimensions.

$$\begin{aligned} \text{length} \quad &= 3x + 1 \\ &= 3(2) + 1 \\ &= 7 \end{aligned}$$

$$\begin{aligned} \text{width} \quad &= 2x \\ &= 2(2) \\ &= 4 \end{aligned}$$

Therefore, the patio is 7 feet long and 4 feet wide.

Math Reasoning

Verify Explain how you could verify that the solution to Example 3 is valid.

a. Solve the equation $x + 5 = 4x + 2$. Provide a justification for
(Ex 1) each step.

b. Solve the equation $\dfrac{4x + 5}{3} = \dfrac{5x + 7}{4}$. Provide a justification for
(Ex 1) each step.

c. The steps of the proof below are given in the correct order. However,
(Ex 2) the justifications for each step are out of order. Determine the correct
order of the justifications.

$\dfrac{2}{3}x + 6 = 4 - 2x$	Given
$2x + 18 = 12 - 6x$	Subtraction Property of Equality
$2x = -6 - 6x$	Multiplication Property of Equality
$8x = -6$	Addition Property of Equality
$x = -\dfrac{2}{3}$	Division Property of Equality

d. The area of the rectangular floor of a shed is 40 yd^2. The length of the
(Ex 3) shed is $(x + 2)$ yd and the width is $(x - 1)$ yd. Find the dimensions of
the shed. Provide a justification for each step.

Practice   Distributed and Integrated

1. (**Monitors**) Computer monitors are usually measured in terms of the lengths of
(Inv 2) their diagonals. If a 20-inch monitor has a horizontal/vertical aspect ratio of 4:3,
what are the horizontal and vertical dimensions of the monitor?

2. Write Explain how to check if two shelves are parallel by dropping a string and
(12) measuring the angles between the string and the shelves.

3. Write the converse of the theorem, "If two parallel lines are cut by a transversal,
(Inv 1) then the same-side interior angles are supplementary." Is the converse true?

* 4. Error Analysis Minh and Soon Yu are asked to solve the expression $2(x + 1) = 12$.
(24) Here are their two solutions. Is Minh's solution correct? Is Soon Yu's solution
correct? Explain.

Minh's solution:		Soon Yu's solution:	
$2(x + 1) = 12$	Given	$2(x + 1) = 12$	Given
$2x + 2 = 12$	Distributive Property	$x + 1 = 6$	Division Property of
$2x = 10$	Subtraction Property		Equality
	of Equality	$x = 5$	Subtraction Property of
$x = 5$	Division Property of		Equality
	Equality		

5. Write the converse of the following statement. *If the window is open, then it is not closed.* Is the
(17) converse true?

* 6. Multiple Choice If $x = 3$ and $3 = y$, then $x = y$. This is an example of which property
(24) of equality?

    A Symmetric   **B** Reflexive   **C** Transitive   **D** Comparative

7. Multi Step What is the width of a rectangular driveway with an area of 2592 square
(8) feet and a length of 27 yards?

8. Generalize What is the relationship between the interior angle and its adjacent
(18) exterior angle at any vertex of any polygon?

9. (Art) Tony wants to paint a triangular pattern on a rectangle which
(13) is 10 feet long and 4 feet wide. He divides the rectangle into four
triangles by stretching wire across the diagonals. Determine the area
of triangles A and B.

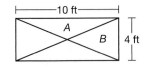

10. Draw a convex hexagon with all sides congruent and a concave
(15) hexagon with all sides congruent.

*11. **Write** Write the converse of the following statement and state whether it is true.
(20) If the converse is true, write the two statements as a biconditional.

*If all of the students in the school achieve over 70% on the state exams, then the
school will receive a new computer lab from the district.*

12. Justify If the diagonals of a rectangle form two equilateral triangles, classify the
(13) other two triangles by their angles. How do you know?

*13. Outline the solution and justification for each step in solving the equation
(24) $2x - 1 = 5$.

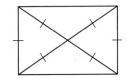

14. Multiple Choice Suzanne suggests that multiples of 3 never end in a 4, but her friend
(14) Maxine disproves her by stating that 24 is a multiple of 3. What has Maxine just
used to disprove Suzanne's suggestion?
A a conjecture   **B** inductive reasoning
C a theorem   **D** a counterexample

xy^2 *15. **Algebra** Miguel painted an equal number of squares with sides measuring
(22) 6 centimeters and 12 centimeters, respectively. How many of each type
of square did he paint if the squares cover 4500 cm^2?

16. Graph the equation $2x + 3y - 15 = 0$ using the slope and y-intercept.
(16)

17. (Landscaping) A patio in the shape of a trapezoid is to be covered in flagstone.
(22) The parallel sides of the trapezoid measure 20 feet and 32 feet respectively,
and are separated by a perpendicular distance of 14 feet. How many square
feet need to be covered?

*18. A figure skater is practicing her figure-eights by repeatedly skating around two
(23) circles that join to form a figure "8." If the diameter of the circles is 30 feet each,
how much total distance, to the nearest foot, will she cover in 5 full trips around
the figure?

19. A right triangle has a side length of 6 inches and a hypotenuse of 10 inches. What
(8) is the length of the third side?

20. Write the inverse of the following statement. Then state whether the inverse is true.
(17)
If it is raining, then I will use my umbrella.

*21. A right triangle has side lengths of 48 meters, 20 meters, and 52 meters. What is
(13)   the area of this triangle?

22. Name an interior and an exterior angle of the figure at right.
(15)

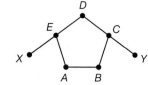

23. The midpoint of (4, 5) and a second point is (7, 1). What is the second
(11)   point?

24. (**Balancing**) A rectangular box sitting on a ramp must not be inclined more
(6)   than 30° from vertical or it will slide down the ramp. What is the
relationship between $\angle ABD$ and $\angle CBE$? How do you know? What
is the greatest value $\angle EBC$ can have before the box slides?

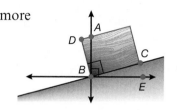

*25. Jack is drawing circles on a playground for a game. If he draws a
(23)   circle with a radius 3 feet, what is the area of the circle, to the
nearest square foot?

26. (**Fitness Membership**) At a fitness club, the equation for the cost of a membership is
(16)   given by $C = 30m + 200$, where C is dollars and m is months. In this application
of a linear equation, what is the slope and what does it represent? How much does
it cost to be a member of this gym for 5 months?

27. Error Analysis Explain why the following inductive reasoning is not a valid proof.
(7)
*All the birds I have seen in this park have brown feathers. All the birds in this park must
have brown feathers.*

*28. For any given angle, if the angle is obtuse, then it cannot be acute. A given angle
(21)   is found to be obtuse. What conclusion can be made from this and which law was
used to reach this conclusion?

29. Write the equation of a line with slope $\frac{1}{2}$ that passes through (3, 4).
(16)

*30. Verify To solve the linear equation $3(x + 5) = 9$, Cynthia begins her solution
(24)   like this:
$$3(x + 5) = 9$$
$$3(x + 5) - 5 = 9 - 5$$
$$3x = 4$$
Is Cynthia proceeding correctly? If not, what mistake has she made?

Triangle Congruence: SSS

Warm Up

1. **Vocabulary** A triangle with no congruent sides is a(n) _____ triangle.
   (13) (*equilateral, isosceles, scalene*)

2. Solve for x: $\frac{3.5}{x} = \frac{17.5}{40}$.
   (SB 4)

3. Two lines intersect as shown. Which angle is equal to $\angle 2$?
   (6)

4. What is the length of side x in this right triangle?
   (Inv 2)

   12 in.   13 in.

   x

5. What is the measure of $\angle A$?
   (18)

   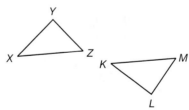

   45°

   80° A

New Concepts

Corresponding sides and **corresponding angles of polygons** are those that are in the same position in two different polygons with the same number of sides. These corresponding parts are indicated by the names of the polygons.

When naming congruent polygons, it is important that the order of the points, or vertices, in the names correspond.

> **Math Reasoning**
>
> **Model** Is it possible to make two different quadrilaterals with four sides of the same length? Make a sketch to illustrate your answer.

Example 1 **Identifying Corresponding Parts**

Identify the corresponding angles and sides for $\triangle XYZ$ and $\triangle KLM$.

Y

X   Z   K   M

L

SOLUTION

The names of the triangles show that $\angle X$ corresponds to $\angle K$, $\angle Y$ corresponds to $\angle L$, and $\angle Z$ corresponds to $\angle M$. Since we know which angles correspond with one another, the sides enclosed by those angles must also correspond. Side $\overline{XY}$ corresponds to side $\overline{KL}$, side $\overline{XZ}$ corresponds to side $\overline{KM}$, and side $\overline{YZ}$ corresponds to side $\overline{LM}$.

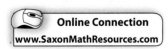

Online Connection
www.SaxonMathResources.com

Triangles are said to be **congruent triangles** when all of their corresponding sides and angles are congruent. One way to determine if two triangles are congruent is to use the Side-Side-Side Congruence Postulate.

Postulate 13: Side-Side-Side (SSS) Congruence Postulate
If three sides of one triangle are congruent to three sides of another triangle, then the triangles are congruent.

Exploration · **Exploring the SSS Postulate**

In this exploration, you will work with a partner to find out if three congruent line segments can form two different triangles.

1. Your teacher will give three segments of different lengths to each student. Make sure that your segments are the same lengths as your partner's.

2. Assemble a triangle using the segments provided.

3. Compare the triangle that you assembled with your partner's triangle. Are they the same or different? If they are different, in what way(s) are they different?

4. Working with your partner, try to assemble two triangles that are different than the ones originally formed. What do you notice about the triangles you assemble?

Side-side-side congruence indicates that if all the sides of a triangle are of a fixed length, the triangle can have only one size and shape. This is called **triangle rigidity**.

If triangles $\triangle ABC$ and $\triangle DEF$ are congruent, their relationship can be shown by the congruence statement $\triangle ABC \cong \triangle DEF$.

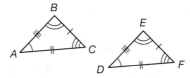

Example 2 **Naming Congruent Triangles**

Write a congruence statement for the two triangles below.

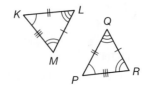

Math Reasoning

Verify In Example 2, is the congruence statement $\triangle LMK \cong \triangle QRP$ true? Explain.

SOLUTION

In these two triangles, M corresponds to R, K corresponds to P, and L corresponds to Q. Therefore, $\triangle MKL \cong \triangle RPQ$.

CPCTC
An abbreviation for the phrase "Corresponding **P**arts of **C**ongruent **T**riangles are **C**ongruent."

When two triangles are congruent, CPCTC states that the corresponding angles and sides of those triangles will also be congruent. For example, if △*ABC* ≅ △*DEF*, then by CPCTC, all of the following congruence statements can be written.

Congruent Angles	Congruent Sides
$\angle A \cong \angle D$	$\overline{AB} \cong \overline{DE}$
$\angle B \cong \angle E$	$\overline{BC} \cong \overline{EF}$
$\angle C \cong \angle F$	$\overline{AC} \cong \overline{DF}$

Example 3 **Writing Congruence Statements**

Identify the congruent sides and angles of the two triangles below and write six congruence statements.

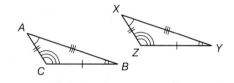

SOLUTION

In these congruent triangles, *A* corresponds to *X*, *B* corresponds to *Y*, and *C* corresponds to *Z*. Therefore,

Congruent Angles	Congruent Sides
$\angle A \cong \angle X$	$\overline{AB} \cong \overline{XY}$
$\angle B \cong \angle Y$	$\overline{BC} \cong \overline{YZ}$
$\angle C \cong \angle Z$	$\overline{AC} \cong \overline{XZ}$

Example 4 **Application: Making a Kite**

Regina is making her own kite. It is made of two perpendicular pieces of wood, to which she will attach a plastic kite shape. The kite shape is made of two congruent triangles as shown in the picture below. Regina has already found the measures that two of the angles need to be so that the kite can fit on the wooden frame. These measures are: m∠*DAB* = 40° and m∠*DCB* = 80°. What should the measure of ∠*B* be?

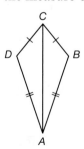

SOLUTION

Since the two triangles are congruent, their corresponding parts are congruent. This means that ∠*BAC* ≅ ∠*DAC* and, since together they measure 40°, each one must measure 20°. Similarly, each of the angles that make up ∠*DCB* must be 40°. Using the Triangle Angle Sum Theorem, △*ABC*'s angles must add up to 180°, so m∠*B* = 180° − 40° − 20° = 120°. So m∠*B* = 120°.

a. Identify the corresponding angles and sides.
(Ex 1)

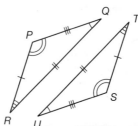

b. Write a congruence statement for the two triangles below.
(Ex 2)

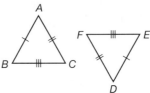

c. Identify the congruent sides and angles of the two triangles below and write six congruence statements.
(Ex 3)

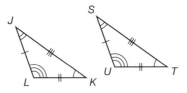

d. (**Kites**) Imagine you are making a kite, as in Example 4, with two
(Ex 4) congruent triangles that make up the kite shape. You know that one obtuse angle of the kite shape is 110°. What is the measure of the other obtuse angle? What will be the total measure of the kite's other two angles?

Practice Distributed and Integrated

1. (**Art**) A painting covers a wall as shown. What is the perimeter of the
(8) painting, to the nearest tenth of a meter?

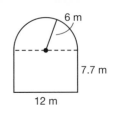

* 2. An area of 400 square meters is to be roped off for a farmers' market that is
(23) to be in the shape of a circle. What will be the diameter of this circle, to the nearest tenth of a meter?

* 3. Give the congruence statement for the triangles shown.
(25)

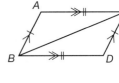

4. Classify the quadrilateral shown here.
(19)

* 5. (**Baseball**) The bases on a baseball diamond form a
(19) square that is 90 feet on each side. What is the minimum distance that would be covered when running around all four bases?

6. (Railroads) Railway tracks consist of two parallel steel rails. If an inspector places
(Inv 1) a straight board across the rails and determines that the measures of the alternate
interior angles differ by several degrees, is there anything wrong? Explain.

* 7. Model Draw two congruent triangles that correspond to the following statements:
(25) $\angle E \cong \angle N$, $\angle F \cong \angle L$, $\angle G \cong \angle M$, $\overline{EF} \cong \overline{NL}$, $\overline{EG} \cong \overline{NM}$, and $\overline{FG} \cong \overline{LM}$.

8. Find a counterexample to prove that the following conjecture is false.
(14) *For every natural number n, $4n > n^2$.*

**Use the grid to determine how far each friend lives from the other. Each
grid unit is 0.1 mile. Round to the nearest hundredth of a mile.**

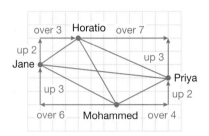

9. How far apart do Jane and Horatio live?
(8)

10. How far apart do Horatio and Priya live?
(8)

*11. Error Analysis Polly determines that for $\triangle ABC$ and $\triangle XYZ$, $\angle A \cong \angle X$, $\angle B \cong \angle Y$,
(25) and $\angle C \cong \angle Z$. She also concludes that $\overline{AB} \cong \overline{XY}$, $\overline{BC} \cong \overline{XZ}$, and $\overline{AC} \cong \overline{YZ}$. Are her
conclusions correct? Explain.

*12. Algebra Triangles *ABC* and *DEF* are congruent. If $AB = 2x + 10$ and
(24) $DE = 4x - 20$, find the value of *x* and include justifications for each step in
the solution.

*13. Multiple Choice Use the Law of Detachment to identify a valid concluding
(21) statement.

*If we get more than 12 inches of snow in the next hour, then the roads will be closed.
The weather network predicts that there will be 15 inches of snow in the next hour.*

A *Based on the weather network, the roads will be open in the next hour.*
B *The snow should stop after 15 inches of snow has fallen.*
C *Based on the weather network, the roads will be closed in the next hour.*
D *It will continue to snow, even after 12 inches of snow has fallen.*

14. Classify the quadrilateral.
(19)

15. Write What can be deduced from the following statements?
(21)

*The oldest person in each class in the school is wearing a blue shirt.
Kelly is the oldest person in one of the classes in the school.*

16. Is the statement below a biconditional statement? If so, are both conditional
(20) statements true?

An angle is a right angle if and only if it measures 90°.

17. Solve for *x* in the equation $\frac{x + 5}{3} = 3$. Justify each step.
(24)

Determine whether each figure is a polygon. If it is, name it.

18.
(15)
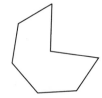

19.
(15)

20. **Justify** Is the following statement true or false? If true, provide a theorem to
(19) support it. If false, provide a counterexample.

If the sides of a quadrilateral have equal length, then it is a square.

*21. In the triangles shown, $\angle C \cong \angle N$ and $\overline{AC} \cong \overline{LN}$. If $\triangle ABC \cong \triangle LMN$,
(25) provide four more congruence statements using CPCTC.

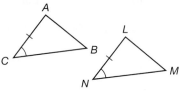

*22. **Algebra** Find the value of x for two supplementary angles if the larger
(24) angle is 20° more than three times the smaller angle. Justify each step.

23. (**Urban Planning**) Find the slope between two grid points marked on a road:
(16) $A(-3, 4)$ and $B(6, -4)$. Is the road going uphill or downhill from A to B? Explain.

24. Identify whether the statement is conditional or biconditional.
(20)

A phone rings if and only if someone is calling.

25. Find a counterexample to prove that the conjecture is false.
(14) $\frac{1}{n}$ *is a rational number for every whole number.*

26. (**Construction**) Workers are stringing caution tape around a circular parking lot that
(23) has just been repaired. To the nearest foot, how much tape is needed if the parking
lot has a diameter of 40 feet?

*27. **Formulate** What conclusion can be drawn from these statements? How do you
(21) know?

If the team wins the remainder of the games, then they will make the playoffs.
The team won the remainder of their games.

28. **Write** What is the contrapositive of the following statement?
(17)

If a bee is the queen bee, then it does not leave the hive.

29. **Model** Line ℓ_1 and ℓ_2 are parallel lines. Point P lies on ℓ_1, and point M lies on
(Inv 1) ℓ_2. Draw a line through P and M. Label one pair of alternate interior angles, and
label one pair of corresponding angles.

30. Find m$\angle 1$ and m$\angle 2$ if m$\angle 1 = 2x^2 + 2x + 3$ and
(18) m$\angle 2 = 3x^2 - 5x + 3$.

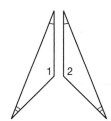

Central Angles and Arc Measure

Warm Up

1. **Vocabulary** A _____ is a line segment with the center of a circle as one
   (23) endpoint and any point on the circle as the other endpoint.

2. **Multiple Choice** Which of these formulas can be used to find the area of a
   (23) half circle?

   A $2\pi r$
   C πd

   B πr
   D $\dfrac{\pi r^2}{2}$

3. Find the value of x.
   (6)

4. What percent of a full circle
   (SB 4) is a rotation of 45°?

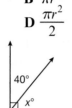

New Concepts An **arc** is a part of a circle consisting of two points on the circle, called endpoints, and all the points on the circle between them. When two arcs on a circle share exactly one endpoint, they are called **adjacent arcs**. An angle whose vertex is at the center of a circle is called a **central angle**. The diagram below shows a central angle that intercepts an arc on the circle.

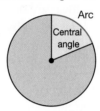

Math Reasoning

Generalize What is the relationship between the arc measures of a minor arc and the major arc that share endpoints?

Arcs of a Circle		
A **minor arc** is an arc that is smaller than half a circle.	A **major arc** is an arc that is larger than half a circle.	A **semicircle** is an arc equal to half a circle.
The **measure of a minor arc** is the same as the measure of its central angle. The measure of a minor arc must be greater than 0° and less than 180°.	The **measure of a major arc** is the difference of 360° and the measure of the associated minor arc. The measure of a major arc must be greater than 180° and less than 360°.	The measure of a semicircle is 180°. Like major arcs, semicircles can be named with the two endpoints of the semi-circle and a point on the circle between the endpoints.
All minor arcs are named using the two endpoints of the arc.	All major arcs are named using the two endpoints of the arc and a point on the circle between the endpoints.	

Example 1 Identifying Arcs and Angles

Identify a central angle, minor arc, major arc, and semicircle in $\odot P$.

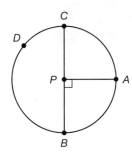

SOLUTION
Two central angles are pictured: $\angle APB$ and $\angle APC$. Each central angle forms a minor arc: $\overarc{AB}$ and $\overarc{AC}$. There are also two major arcs: $\overarc{ABC}$ and $\overarc{ACB}$. Finally, there are two semicircles: $\overarc{BAC}$ and $\overarc{BDC}$.

Example 2 Finding Arc Measures

What is $m\overarc{AB}$?

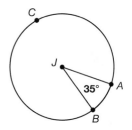

SOLUTION
The central angle's measure is 35°, so the measure of the arc is also 35°.

Two arcs that are in the same circle or in congruent circles and that have the same measure are called **congruent arcs.**

Math Reasoning

Predict In Example 2, what would be the measure of the major arc $\overarc{ACB}$?

Example 3 Congruent Arcs

The measure of $\overarc{DE}$ is given by the expression $3x + 10$, and the measure of $\overarc{HJ}$ is given by the expression $5x - 40$. It is given that $\overarc{DE} \cong \overarc{HJ}$. Determine the value of x and the measure of each arc.

SOLUTION
Since the two arcs are congruent, the expressions for their measures must be equal.

Therefore,

$$
\begin{aligned}
3x + 10 &= 5x - 40 \\
3x + 10 - 10 &= 5x - 40 - 10 \\
3x &= 5x - 50 \\
3x - 5x &= 5x - 50 - 5x \\
-2x &= -50 \\
\frac{-2x}{-2} &= \frac{-50}{-2} \\
x &= 25
\end{aligned}
$$

Therefore, $m\overarc{DE} = 3(25) + 10$
$= 75 + 10$
$= 85°$

Since the arcs are congruent, $m\overarc{HJ} = 85°$.

Postulate 14: Arc Addition Postulate
The measure of an arc formed by two adjacent arcs is the sum of the measures of the two arcs.

The Arc Addition Postulate says that if two arcs are adjacent, the sum of their measures is equal to the measure of the larger arc that they form. In the diagram, $m\overset{\frown}{QR} + m\overset{\frown}{RS} = m\overset{\frown}{QS}$.

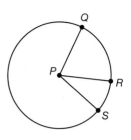

Example 4 **Using the Arc Addition Postulate**

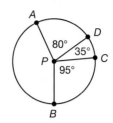

(a.) Use the Arc Addition Postulate to write an expression that represents $m\overset{\frown}{AC}$.

SOLUTION

While there are several arcs indicated, the only two that are adjacent and make up the same arc as $\overset{\frown}{AC}$ are $\overset{\frown}{AD}$ and $\overset{\frown}{DC}$. Therefore, $m\overset{\frown}{AD} + m\overset{\frown}{DC} = m\overset{\frown}{AC}$.

(b.) Find $m\overset{\frown}{AC}$.

SOLUTION

Since $m\overset{\frown}{AC} = m\overset{\frown}{AD} + m\overset{\frown}{DC}$

$\qquad\qquad = 80° + 35°$

$\qquad\qquad = 115°$

Therefore, $m\overset{\frown}{AC}$ is 115°.

Math Reasoning

Analyze For Example 4, Rahel suggested $m\overset{\frown}{AB} + m\overset{\frown}{BC} = m\overset{\frown}{AC}$. Is she correct? Explain.

Example 5 **Application: Surveillance Cameras**

A surveillance camera has a viewing angle of 42°. How many surveillance cameras would be needed to cover a semicircle of a room, with minimal overlap of the area to be viewed? How much of an overlap would these cameras produce?

SOLUTION

The Arc Addition Postulate can be used to determine the number of cameras to be used. A circle is 360°, so a semicircle is 180°. To find the number of cameras needed, divide 180° by the viewing angle of a camera: $\frac{180°}{42°} = 4.3$. Four cameras would not quite cover the area, so 5 cameras are needed. To find the overlap, multiply the number of cameras by the viewing angle and subtract 180°: $(42° \cdot 5) - 180° = 30°$.

Hint

The cameras' overlap will be any number of degrees over 180° that they cover when their viewing angles are added together.

Lesson Practice

a. Draw a diagram of a circle, identifying a central angle, a minor arc, and a major arc.

(Ex 1)

b. Identify the measure of the minor arc.
(Ex 2)

c. The measure of $\overset{\frown}{JK}$ is given by the expression
(Ex 3) $2x - 15$, and the measure of $\overset{\frown}{LM}$ is given by
the expression $x + 30$. It is given that
$\overset{\frown}{JK} \cong \overset{\frown}{LM}$. Determine the value of x and the
measure of each arc.

d. Use the Arc Addition Postulate to write an
(Ex 4) expression that represents $m\overset{\frown}{AB}$.

e. Find $m\overset{\frown}{DEG}$.
(Ex 4)

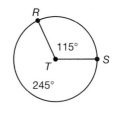

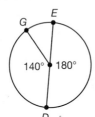

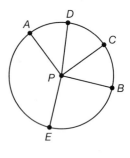

f. **Outdoor Lighting** A lamp projects a beam of light over a 100° arc.
(Ex 5) How many lamps facing outward from the center of a circle would
be needed to form a full circle of light at the center of a park? What
would be the overlap of these beams?

Practice Distributed and Integrated

1. Find the slope of the line containing points $C(7, 3)$ and $D(12, -2)$.
(16)

* 2. **Analyze** What conclusion can be drawn from the following statements?
(21)

   All plants require water to develop. A cactus is a plant.

3. **Algebra** Triangle ABC has side lengths of $AB = x^2 - x + 5$, $BC = 5x - 4$, and $AC = 3x^2 - 2x - 1$.
(13) If $\triangle ABC$ is isosceles where $AB = BC$, find the lengths of all sides.

* 4. **Geography** The Kentland Impact Crater in Indiana is in the shape of a circle
(23) whose diameter is approximately 13 kilometers. To the nearest kilometer, how far
would Farah walk if she were to travel the full distance around the outside of the
crater?

5. Construct a truth table for not P and not Q.
(20)

* 6. What is the measure of a central angle of a circle if the associated minor arc has
(26) an angle measure of 65°?

7. **Multiple Choice** Which of the following classifications allows you to know the
(13) measures of the triangle's angles with no additional information? What are the
angle measures of such a triangle?
   A equilateral   **B** equiangular
   C isosceles   **D** both **A** and **B**

8. **Geography** Use *A* and *B* to determine the truth value of the statement, "*A* and not *B*".
(20)
    A: New York City is the most populous city in the United States.
    B: New York City is near the Pacific Ocean.

9. **Algebra** The supplement of an angle is four times its complement. Find the
(6)   measure of the angle.

10. Find a counterexample to the conjecture $n^3 < 3n$.
(14)

11. **Algebra** If the slope of $\overline{XY}$ is 0, and $X(2, 3)$ and $Y(-1, k)$, find k.
(16)

12. Angles *A* and *B* are a pair of alternate interior angles on a pair of parallel lines
(Inv 1)   intersected by a transversal, and $\angle A$ and $\angle C$ are a pair of same-side interior
    angles on the same figure. If m$\angle B = 30°$, what is m$\angle C$?

*13. Which side of $\triangle PRG$ corresponds to $\overline{XT}$, if $\triangle XTW \cong \triangle PRG$?
(25)

14. **Error Analysis** A student wrote the following conjecture. Find a counterexample and
(14)   explain the error.

    If x is a positive even number and y is any integer, then the product xy will be
    positive.

*15. Write the congruence statement for the given triangles.
(25)

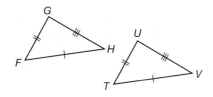

16. **Write** What is the inverse of the following statement?
(17)

    If a building has more than four floors, then it has an elevator.

17. If lines *f* and *g* are parallel, find *x*.
(Inv 1)

*18. **Multiple Choice** If there are a total of 6 non-overlapping adjacent
(26)   minor arcs in a circle, the Arc Addition Postulate gives
    A a total arc measure of 90°.
    B a total arc measure of 180°.
    C a total arc measure of 360°.
    D a total that is the sum of the
        measures of the adjacent arcs.

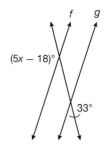

19. Is this figure a polygon? If so, name it. If not,
(15)   explain why not.

20. What is the area of this triangle?
(13)

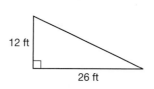

21. Janie has just solved an equation for x and found $4 = x$. She writes her answer as $x = 4$.
(24) What property of equality is she using to do this?

22. (Phone Plans) To calculate monthly charges, a cell phone provider uses the equation
(16) $C = 0.25m + 60$, with C being dollars and m representing minutes used beyond the ones provided with the plan. In this linear equation, what is the value of the vertical intercept and what does it represent?

23. *Algebra* Two congruent arcs have central angle measures that are $3x + 10$ and $50 - 2x$.
(26) Find x and the measure of the arcs.

*24. Write six congruence statements for $\triangle ABC \cong \triangle WXY$.
(25)

25. **Write** Write the contrapositive of the following statement.
(17)
If a polygon is a triangle, then it has three sides.

*26. What property of equality is used as a justification in solving
(24) for x in the problem below?
$$x + 5 = 11$$
$$x + 5 - 5 = 11 - 5$$
$$x = 6$$

27. (Geography) Roderick calculated the distance from New York City to Montreal
(Inv 2) using indirect measurement. On his map, New York City is 188 miles from Boston. Boston is 253 miles from Montreal, Canada. Assuming the angle from New York City to Boston to Montreal is 90°, what is the distance between New York City and Montreal, to the nearest mile?

*28. A major arc has a measure of 227°. What is the corresponding minor arc
(26) measure?

29. **Write** The Fibonacci Sequence is a famous pattern which begins: 1, 1, 2, 3, 5,
(7) 8, 13, 21, ... Describe the pattern and give the next four terms.

30. What are four different names for this quadrilateral?
(19)

Two-Column Proofs

Warm Up

1. **Vocabulary** A process that uses logic to show that a conclusion is true is
(24) called a _____.

2. Solving the equation $3x = 6$ yields the answer $x = 2$. Which property of
(24) equality was used to solve for x?

3. In solving $6 = 3x$, Sofia obtains an answer of $2 = x$. She writes it as
(24) $x = 2$. Which property of equality allows her to write the solution
this way?

4. **Multiple Choice** In a right triangle, one of the acute angles measures 43°.
(13) What is the measure of the second acute angle?
   A 37°   **B** 57°
   C 43°   **D** 47°

New Concepts

In a proof, deductive reasoning is used to develop a logical argument from
given information to prove a conclusion. Proofs in geometry must be done
step by step, and each step must have a justification. These justifications
can include the given information, definitions, postulates, theorems, and
properties, as seen in the two-column proofs in this lesson.

Hint

Review Lesson 24 for
some of the justifications
that can be used
in a proof. You may
wish to make a list of
justifications for your
reference. See the Skills
Bank for more properties
that can be used as
reasons in a proof.

> **Example 1** **Justifying Statements in a Two-Column Proof, Part 1**
>
> Fill in the justifying statements to support
> the proof of Theorem 4-2: If there is a line and
> a point not on the line, then exactly one plane
> contains them.
>
>
>
> **Given:** Point C is not on $\overleftrightarrow{AB}$.
> **Prove:** Exactly one plane contains $\overleftrightarrow{AB}$ and C.

Statements	Reasons
1. Point C is noncollinear with $\overleftrightarrow{AB}$.	1.
2. Exactly one plane contains points A, B, and C.	2.
3. Exactly one plane contains $\overleftrightarrow{AB}$ and C.	3.

SOLUTION

The missing justifying statements are:

1. Given

2. Through any three noncollinear points there exists exactly one plane.
   (Postulate 6)

3. If two points lie in a plane, then the line containing the points lies in the
   plane. (Postulate 8)

Example 2 **Justifying Statements in a Two-Column Proof, Part 2**

Prove Theorem 6-1: If two angles are complementary to the same angle, then they are congruent.

Given: ∠1 is complementary to ∠2.
         ∠3 is complementary to ∠2.

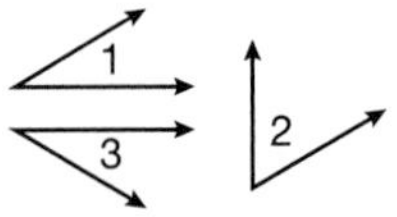

Prove: ∠1 ≅ ∠3

Statements	Reasons
1. ∠1 is complementary to ∠2. ∠3 is complementary to ∠2.	1. Given
2. m∠1 + m∠2 = 90° m∠3 + m∠2 = 90°	2.
3. m∠1 + m∠2 = m∠3 + m∠2	3.
4. m∠1 + m∠2 − m∠2 = m∠3 + m∠2 − m∠2	4.
5. m∠1 = m∠3	5.
6. ∠1 ≅ ∠3	6.

Hint

A proof does not always have to begin by restating the given, but doing so may help you understand what the next step should be.

SOLUTION

The missing justifying statements are:

2. Definition of complementary angles

3. Substitution Property of Equality

4. Subtraction Property of Equality

5. Simplify

6. Definition of congruent angles

Two-column proofs have a format that is composed of five parts.

Five Parts of a Two-Column Proof
1. Given statement(s): The information that is provided.
2. Prove statement: The statement indicating what is to be proved.
3. Diagram: A sketch that summarizes the provided information. Sometimes you will need to draw the sketch yourself based on given information.
4. Statements: The specific steps that are written in the left-hand column.
5. Reasons: Postulates, theorems, definitions, or properties written in the right-hand column, which justify each statement.

Example 3   Writing a Two-Column Proof, Part 1

Prove Theorem 6-4: If two angles are vertical angles, then they are congruent. (Vertical Angles Theorem)

Given: $\overleftrightarrow{AB}$ and $\overleftrightarrow{DE}$ intersect at point C
Prove: $\angle ACD \cong \angle BCE$

SOLUTION

Proof:

Statements	Reasons
1. $\overleftrightarrow{AB}$ and $\overleftrightarrow{DE}$ intersect at point C	1. Given
2. $m\angle BCD + m\angle ACD = 180°$	2. Linear Pair Theorem
3. $m\angle BCD + m\angle BCE = 180°$	3. Linear Pair Theorem
4. $m\angle BCD + m\angle ACD = m\angle BCD + m\angle BCE$	4. Transitive Property of Equality
5. $m\angle ACD = m\angle BCE$	5. Subtraction Property of Equality
6. $\angle ACD \cong \angle BCE$	6. Definition of congruent angles

Math Reasoning

Verify Prove that the other pair of vertical angles in Example 3 are also equal.

Example 4   Writing a Two-Column Proof, Part 2

Prove Theorem 5-3: If a transversal is perpendicular to one of two parallel lines, then it is perpendicular to the other one.

Given: $\overleftrightarrow{AD} \parallel \overleftrightarrow{BC}$
$\overleftrightarrow{EB} \perp \overleftrightarrow{AD}$
Prove: $\overleftrightarrow{EB} \perp \overleftrightarrow{BC}$

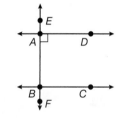

SOLUTION

Proof:

Statements	Reasons
1. $\overleftrightarrow{AD} \parallel \overleftrightarrow{BC}$	1. Given
2. $\angle EAD \cong \angle ABC$	2. Postulate 11: Corresponding angles are congruent.
3. $m\angle EAD = 90°$	3. Definition of perpendicular
4. $m\angle ABC = m\angle EAD$	4. Definition of congruent angles
5. $m\angle ABC = 90°$	5. Transitive Property of Equality
6. $\overline{EB} \perp \overline{BC}$	6. Definition of perpendicular

a. If a triangle is obtuse, what can you conclude about the measures of its
(Ex 1) two non-obtuse angles? Justify your answer.

b. Fill in the reasons of the proof of Theorem 5-5: If
(Ex 2) two lines form congruent adjacent angles, then they
are perpendicular.

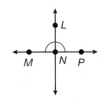

Given: $\angle LNM \cong \angle LNP$
Prove: $LN \perp MP$

Statements	Reasons
1. $\angle LNM \cong \angle LNP$	1.
2. $m\angle LNM = m\angle LNP$	2.
3. $m\angle MNP = 180°$	3.
4. $m\angle LNM + m\angle LNP = m\angle MNP$	4.
5. $2m\angle LNM = 180°$	5.
6. $m\angle LNM = 90°$	6.
7. $LN \perp MP$	7.

c. Given $\triangle ABC$ with exterior angle $\angle ACD$, write a
(Ex 3) two-column proof to prove the Exterior Angle
Theorem.

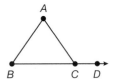

Given: $\angle ACD$ is an exterior angle of $\triangle ABC$.
Prove: $m\angle ACD = m\angle CAB + m\angle ABC$

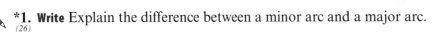

Practice **Distributed and Integrated**

*1. **Write** Explain the difference between a minor arc and a major arc.
(26)

2. Is the statement, "A square is a rhombus," sometimes, always, or never
(19) true?

3. **Error Analysis** Parveer calculated the area of the given triangle as follows.
(13) Is he correct? Explain.

$A = \frac{1}{2}bh$
$\quad = \frac{1}{2}(5)(6)$
$\quad = 15$

4. Prove Theorem 10-2: If two parallel lines are cut by a transversal, then
(27) the alternate exterior angles are congruent.

Given: $m \parallel n$, transversal
Prove: $\angle 1 \cong \angle 2$

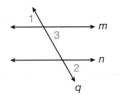

* 5. **Landscaping** A sprinkler system used for watering a field sprays water in a circular
(23) pattern with a radius of 20 feet. To the nearest square foot, what area is covered
by this sprinkler?

* 6. **Verify** If M is the midpoint of $\overline{AB}$, write a two-column proof to show that
(27) $AM = \frac{1}{2} AB$.

7. **Coordinate Geometry** Find the coordinates of the midpoints of $\overline{AB}$ and $\overline{CD}$ given the
(11) following coordinates: $A(8, 1)$, $B(14, 7)$, $C(7, 11)$, and $D(5, 5)$.

8. **Write** Use the Law of Syllogism to write a valid conclusion.
(21)
If all the math classes are full, then Raymond must enroll in an elective.
If Raymond must enroll in an elective, then he will enroll in theater.

9. **Furniture Design** The seating surface of a stool is a woolen circle with a diameter
(23) of 14 inches. To the nearest tenth of a square inch, how much material is used for
the top surface of the stool?

10. **Woodworking** If a rectangular table with an area of 35 square feet has a width
(8) that is 2 feet less than the length, what are the dimensions of the table?

11. Write the disjunction of the statements "Priyanka takes the bus," and "Priyanka goes to work
(20) early."

12. Prove Theorem 5-6: All right angles are congruent.
(27)
Given: Angle 1 and angle 2 are right angles.
Prove: $\angle 1 \cong \angle 2$

13. Adrienne claims that for transversals of two lines, all the acute angles generated
(Inv 1) by the transversal are congruent and all the obtuse angles are congruent. Is she
correct? Explain.

14. **Road Repair** In the diagram, a parallelogram-shaped area
(22) of a parking lot is to be repaved. How many square meters are
to be repaved?

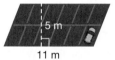

*15. Prove Theorem 5-7: If two lines are parallel to the same line,
(27) then they are parallel to each other.
Given: $\overleftrightarrow{AB} \parallel \overleftrightarrow{CD}$ and $\overleftrightarrow{EF} \parallel \overleftrightarrow{CD}$
Prove: $\overleftrightarrow{AB} \parallel \overleftrightarrow{EF}$

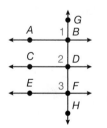

16. Draw a quadrilateral with no two sides parallel. Classify this
(19) shape.

17. Prove Theorem 12-1: If two lines are cut by a transversal and the
(12) alternate interior angles are congruent, then the lines are parallel.
Given: $\angle 1 \cong \angle 2$
Prove: $m \parallel n$

18. What is one-fourth of the circumference of a circle with a diameter
(23) of 10 meters?

19. **Construction** A construction worker is framing a new roof and needs to make a
(25)   triangular structure congruent to one already on the roof. If she only has tools
for linear measurement, what measurements should she take to ensure that the
triangles are congruent?

*20. **Justify** Write an algebraic proof to solve $4x = 8 - 2x$ for x.
(24)

21. **Algebra** Solve for y, if the slope between points U and V is $-\frac{2}{3}$, with points
(16) $U(-2, -3)$ and $V(-5, y)$.

*22. Which angle corresponds to $\angle R$, if $\triangle XTW \cong \triangle PRG$?
(25)

23. **Multiple Choice** Which triangle classification pertains to the lengths of the sides of a
(13)   triangle?

    A scalene   **B** equiangular

    C acute   **D** right

24. **Algebra** Quadrilateral $PQRS$ is a rhombus. If $PQ = 4x + 7$ and $QR = 8x - 5$, what
(19)   is the length of $\overline{QR}$?

*25. Two congruent arcs have measures $3(x + 4)°$ and $4(x - 5)°$. What is the value
(26)   of x?

26. Construct and label a diagram based on the following information:
(Inv 1)

    Transversal k intersects lines m and n to produce corresponding angles
$\angle 1$ and $\angle 2$ that are both acute, alternate interior angles $\angle 3$ and $\angle 4$ that
are both obtuse, and alternate exterior angles $\angle 1$ and $\angle 5$ that are both acute.

27. Jameer states that the difference between any two even integers is positive.
(14)   Give a counterexample to disprove this conjecture.

*28. Which property of equality supports the statement if $x = k$ and $k = y$,
(24)   then $x = y$?

*29. **Track and Field** The area used for shot-put events is a sector of a circle with a
(26)   central angle of 45°. What would be the arc measure (x) 50 meters away from the
center of the area?

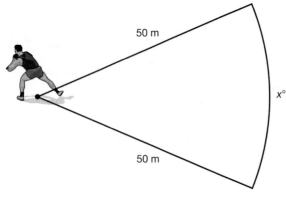

50 m

50 m

$x°$

30. Classify $\triangle DEF$ by its angles and sides.
(13)

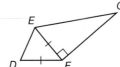

Triangle Congruence: SAS

Warm Up

1. **Vocabulary** Two triangles that have the same shape and size are said to
   (25)   be _____.

2. What angle is vertical to ∠2?
   (6)

3. **Multiple Choice** What is the measure of the
   (6)   complement of a 25° angle?

   A 75°   **B** 155°

   C 65°   **D** 25°

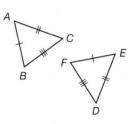

4. Explain why "△ABC ≅ △DEF" is a false statement.
   (25)   Write a true congruence statement for the triangles.

New Concepts

Sides and angles of polygons have special relationships to each other. The angle formed by two adjacent sides of a polygon is called an **included angle.** The common side of two consecutive angles of a polygon is called an **included side.**

Example 1 **Identifying Included Angles and Sides**

What is the included side of ∠A and ∠B? What is the included angle of $\overline{BC}$ and $\overline{CD}$?

SOLUTION

Angles A and B share the side $\overline{AB}$, so $\overline{AB}$ is the included side. The angle between $\overline{BC}$ and $\overline{CD}$ is ∠C, so ∠C is the included angle.

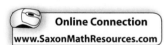
Online Connection
www.SaxonMathResources.com

Caution

In the SAS Postulate, the included angle is the angle that is formed by the two congruent sides. Remember that in the SAS Postulate, the *A* is between the two *S*'s, showing that the angle is between the two sides.

You learned in Lesson 25 how to prove triangles congruent by using the SSS Triangle Congruence Postulate. This lesson presents another way to prove that triangles are congruent. It uses two sides and the included angle.

Side-Angle-Side (SAS) Triangle Congruence Postulate
If two sides and the included angle of one triangle are congruent to two sides and the included angle of another triangle, then the triangles are congruent by side-angle-side congruence.

Example 2 **Using the SAS Postulate to Determine Congruency**

Determine whether the pair of triangles is congruent by the SAS Postulate.

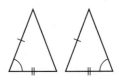

SOLUTION

The two indicated triangles are not necessarily congruent, even though they have two congruent sides and one congruent angle. In the second triangle, the angle that is congruent is not the included angle of the two congruent sides.

Math Reasoning

Analyze The two triangles that are made by drawing a diagonal in a square are congruent by the SSS Postulate. Could the SAS Postulate also be used?

Example 3 **Finding Missing Angle Measures**

Find the value of x that makes the triangles congruent.

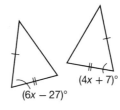

$(4x + 7)°$

$(6x - 27)°$

SOLUTION

For the two triangles to be congruent, the measures of the included angles must be equal. Therefore,

$$6x - 27 = 4x + 7$$
$$6x = 4x + 34$$
$$2x = 34$$
$$x = 17$$

The triangle congruence postulates and theorems will be used as justifications in proofs. An example is given below.

Example 4 **Using the SAS Postulate in a Proof**

Triangles make an "X" design on this barn door. Use the SAS Postulate to write a two-column proof.

Given: $\overline{AB} \cong \overline{DC}$

Prove: $\triangle ABD \cong \triangle DCA$

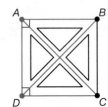

Hint

Recall that in a proof, every algebraic statement must be justified using properties of equality. See Lesson 24 and the Skills Bank for lists of these properties.

SOLUTION

1. $\overline{AB} \cong \overline{DC}$ 1. Given
   $\angle ADC$ and $\angle DAB$ are right angles
2. $m\angle DAB = m\angle ADC$ 2. All right angles are congruent.
3. $\overline{AD} \cong \overline{AD}$ 3. Reflexive Property of Congruence
4. $\triangle ABD \cong \triangle DCA$ 4. SAS Postulate

Example 5 **Application: Design**

An artist is designing patterned wallpaper made of congruent triangles. He starts by drawing △*ABC*, shown below. He wants to design a mirror image of △*ABC*, shown as △*EDC* below. How can he make sure that this new triangle is congruent to △*ABC* using the SAS pattern of triangle congruence?

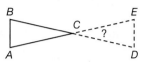

Math Reasoning

Verify The two triangles that are formed by drawing a diagonal through a parallelogram can be proven congruent by the SSS Postulate. Could the SAS Postulate also be used? Explain.

SOLUTION

To ensure that the two triangles are congruent, he should first measure $\overline{BC}$ and $\overline{AC}$. He can then extend both segments at points *E* and *D*, respectively, such that *C* is the midpoint of both $\overline{AE}$ and $\overline{BD}$. Since ∠*BCA* and ∠*ECD* are vertical angles, they are congruent, and the triangles must also be congruent by the SAS Postulate.

Lesson Practice

a. Determine whether the pair of triangles is congruent
(Ex 1) by the SAS Postulate.

b. Find the value of *x* that makes the triangles
(Ex 2) congruent.

c. Use the SAS Postulate to prove
(Ex 3) △*WXY* ≅ △*WZY* if $\overline{WZ} ≅ \overline{WX}$ and ∠*ZWY* ≅ ∠*XWY*.

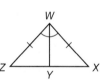

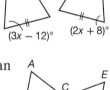

$(3x - 12)°$ $(2x + 8)°$

d. A land dispute over two triangular parcels of land can
(Ex 4) be ended by showing that the parcels have the same shape and size. In the figure, two sides of each parcel of land were measured and it was found that point *C* is the midpoint of both $\overline{AD}$ and $\overline{BE}$. End the dispute by proving △*ABC* ≅ △*DEC*.

Practice   Distributed and Integrated

1. Henry wants to hang a picture frame on to the wall but does not have a leveling
(Inv 1) tool. However, the striped wallpaper on the wall has parallel lines that form 45-degree angles to the floor and the ceiling. How can Henry use a protractor to hang the picture frame so that it is not tilted?

2. **Justify** Is the following conclusion valid? Why or why not?
(21)
Some skates that fit well are expensive. For my feet to be comfortable, I need skates that fit well. As a result, it is expensive to keep my feet comfortable.

xy² * **3. Algebra** Two congruent arcs have measures $(x^2 - 4x - 1)°$ and $(x^2 - 3x + 10)°$.
(26) What is the value of x?

* **4.** If two angles form a linear pair, then they are supplementary.
(27)
Given: $\angle 1$ and $\angle 2$ are a linear pair.
Prove: $m\angle 1 + m\angle 2 = 180°$

5. (**City Planning**) A rectangular city block measures 105 meters by 88 meters. The
(8) city council wants to divide the block into two triangles of equal area. What is the
length of the hypotenuse of each triangular area?

6. A minor arc has a measure of 122°. What is the measure of the corresponding
(26) major arc?

7. Find the measure of $\angle P$ in the figure shown.
(18)

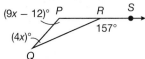

8. Generalize The numbers in the geometric sequence $1, \frac{1}{4}, \frac{1}{9}, \frac{1}{16}, \ldots$ continue to get
(7) smaller in value. Make a general formula for the sequence. State whether a term of
the sequence will ever have a value of 0.

* **9. Error Analysis** Is the following proof for congruent triangles correct? Explain.
(28)
   1. $\overline{AC} \cong \overline{DE}$ 1. Given

     $\angle ABC \cong \angle EDF$
     $\overline{BC} \cong \overline{DF}$

   2. $\triangle ABC \cong \triangle EFD$ 2. SAS Postulate

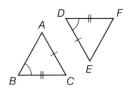

10. Multiple Choice What are the coordinates of the midpoint of a segment that connects
(11) the points $(1, 1)$ and $(-3, 5)$?

  A $(2, 3)$ **B** $(-1, -3)$
  C $(-1, 3)$ **D** $(4, 6)$

* **11.** Prove Theorem 5-2: If two lines in a plane are perpendicular to the
(27) same line, then they are parallel to each other.

  Given: $\overleftrightarrow{ST} \perp \overrightarrow{PQ}$, $\overrightarrow{RU} \perp \overrightarrow{PQ}$
  Prove: $\overleftrightarrow{ST} \parallel \overleftrightarrow{RU}$

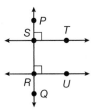

12. Write What is the inverse of this statement?
(17)

If John is at least 5'10" tall, then he is taller than Herschel.

* **13. Write** When can the SAS Postulate be used?
(28)

14. (**Art**) Paolo was hired to do an abstract mural on the side of a building. His mural
(23) includes three circles. The first circle has a diameter of 34 feet, the second has
a diameter of 30 feet, and the third has a diameter of 26 feet. What is the total
circumference of the three circles, to the nearest foot?

* **15.** Explain this statement: In $\odot P$, with two points anywhere on the circle, K and L,
(23) $\overline{KP} \cong \overline{LP}$.

xy² *16. **Algebra** Figure *DEFG* is a parallelogram. If m∠*D* = 8*x* + 17 and m∠*E* = 5*x*² − 2,
(19) what are the measures of the other two angles?

17. Write a congruence statement for the triangles and give a justification.
(25)

18. Write the equation of line *AB* in slope-intercept form if *A* is at $(-2, -3)$
(16) and *B* is at $(2, -1)$.

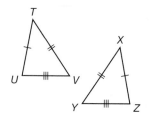

19. Given these statements, is the conclusion valid?
(21)

   Wilhelm will be in Paris on Friday if he leaves New York by Wednesday.
   Wilhelm was not in Paris on Friday.
   Wilhelm did not leave New York by Wednesday.

20. **Write** If △*KLM* ≅ △*TUV* by the SSS Postulate, is ∠*K* ≅ ∠*U*? Explain.
(25)

21. A rhombus has diagonals that measure 22 inches and 14 inches. What is the area
(22) of the rhombus?

22. (**U.S. History**) Use the Law of Detachment to make a valid conclusion based on
(21) these statements.

   If you live in Arkansas, then you live in the 25th state to enter the Union.
   Clark lives in Arkansas.

23. (**Farming**) A crop disease has started at a central point in a field. The disease is
(23) spreading in a circular pattern whose radius increases at a rate of 2 meters per
day. At this rate, how large an area will be affected in 3 days, to the nearest square
meter?

*24. **Write** If two triangles are proven congruent by the SAS Postulate, what can be said
(28) about the third sides of the two triangles?

*25. **Model** Graph the line *y* = 2*x* + 4.
(16)

*26. A properly-tied bow tie can be modeled as two congruent
(28) triangles. If $\overline{BA} \cong \overline{BD}$ and $\overline{BE} \cong \overline{BC}$, prove △*ABE* ≅ △*DBC*.

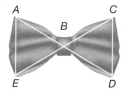

27. (**Road Signs**) A triangular yield sign is to be rust-proofed
(13) with a layer of plastic. Its base is 18 inches and its height
is 24 inches. How many square feet of plastic are needed
to cover both the front and the back of the sign?

xy² 28. **Algebra** What is the hypotenuse of a triangle whose legs are *a* and $\sqrt{3}a$?
(Inv 2)

29. Find m∠*W* and m∠*J*.
(18)

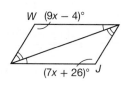

30. Find the area of the trapezoid.
(22)

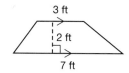

Congruent Triangles

Construction Lab 5 *(Use with Lesson 28)*

In Lesson 28, you learned some triangle congruence patterns. This lab shows you how to use a compass and straightedge to construct a triangle congruent to a given triangle using the SAS Postulate.

1. Sketch a triangle and label it *ABC*.

2. Using the skills you learned in Construction Lab 1, construct $\overline{DE}$ congruent to $\overline{AB}$.

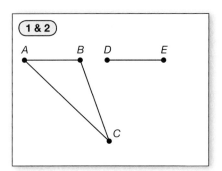

Hint

Refer to Construction Lab 1 for a reminder on how to construct congruent angles and segments. To copy this triangle, you will need to use both techniques at once.

3. Again, just like in Construction Lab 1, construct ∠*DEF* congruent to ∠*ABC*. Use your compass to place point *F* on the ray so that $\overline{EF}$ is congruent to $\overline{BC}$.

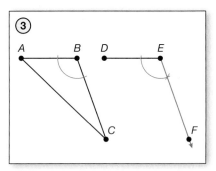

4. Draw segment $\overline{FD}$ to complete △*DEF*.

You constructed the figures so that:

$\overline{DE} \cong \overline{AB}$

∠*DEF* ≅ ∠*ABC*

$\overline{EF} \cong \overline{BC}$

Therefore, by the SAS Postulate, △*DEF* ≅ △*ABC*.

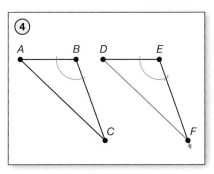

Lab Practice

Sketch a triangle using a straightedge, then trade with a partner and construct triangles congruent to each other's sketches. Verify that the triangles are congruent by measuring all the sides and angles using a ruler and protractor.

Using the Pythagorean Theorem

Warm Up

1. **Vocabulary** The side that is opposite the right angle in a right triangle is
(Inv 2) the _____.

2. Classify the triangle according to its sides and
(13) angles.

3. **Multiple Choice** Simplify: $\sqrt{3^2 + 4^2}$
(SB 6)
   A 7 **B** 5
   C 49 **D** 25

4. Simplify: $\sqrt{640}$
(SB 6)

New Concepts

When the side lengths of a right triangle are nonzero whole numbers that satisfy the Pythagorean Theorem, they form a **Pythagorean triple**.

Pythagorean Triples
A Pythagorean triple is a set of three nonzero whole numbers a, b, and c such that: $$a^2 + b^2 = c^2.$$

Math Reasoning

Verify Use your calculator to verify that multiples of these two common Pythagorean triples are also Pythagorean triples.

Two of the most well-known sets of Pythagorean triples are (3, 4, 5) and (5, 12, 13). An easy way to find Pythagorean triples is to multiply one of these two sets by a whole number. For example, multiplying the first set by 2 yields (6, 8, 10), which is also a Pythagorean triple.

Example **1** **Finding Pythagorean Triples**

Find the unknown length in the triangle.
Do the side lengths form a Pythagorean triple?

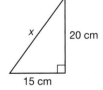

SOLUTION The length of the hypotenuse is the unknown, $c = x$. The lengths of the legs are 15 centimeters and 20 centimeters, which are represented by a and b. Note that it does not matter which value is a and which is b.

$$a^2 + b^2 = c^2$$
$$(15)^2 + (20)^2 = x^2$$
$$225 + 400 = x^2$$
$$625 = x^2$$
$$25 = x^2$$

Therefore, the length of the hypotenuse is 25 centimeters. The set (15, 20, 25), which gives the side lengths of this triangle, is the Pythagorean triple (3, 4, 5) multiplied by 5.

Math Reasoning

Justify When solving $x^2 = 625$, remember that $x = 25$ or $x = -25$. However in this type of problem, the negative value of the square root is ignored. Why?

Example 2 Using Pythagorean Triples To Find the Legs

Find the unknown length in the triangle.
Do the side lengths form a Pythagorean triple?

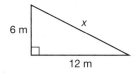

SOLUTION The hypotenuse is known, $c = 26$ inches.
The value of 24 inches is substituted for either a or b,
with x representing the other leg.

$$a^2 + b^2 = c^2$$
$$(24)^2 + x^2 = (26)^2$$
$$576 + x^2 - 576 = 676 - 576$$
$$x^2 = 100$$
$$x = 10$$

Therefore, the length of the other leg is 10 inches.
Since the side lengths are nonzero whole numbers that satisfy the equation
$a^2 + b^2 = c^2$, they form a Pythagorean triple. The set which gives the side
lengths of this triangle $(10, 24, 26)$, is the Pythagorean triple $(5, 12, 13)$
multiplied by 2.

However, not all right triangles are composed of side lengths that are
nonzero whole numbers. In such cases, one or more side lengths may be
written as a **radical expression**. A radical expression is any expression
that contains a root. Typically, a radical expression should be reduced to
simplified radical form.

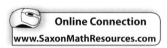
Online Connection
www.SaxonMathResources.com

Example 3 Simplifying Radicals

a. Find the value of x. Give your answer in
simplified radical form.

SOLUTION The length of the hypotenuse is the
unknown, x, and the legs are 6 meters and 12 meters
in length.

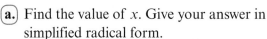

$$a^2 + b^2 = c^2$$
$$6^2 + 12^2 = x^2$$
$$36 + 144 = x^2$$
$$180 = x^2$$
$$\sqrt{180} = x$$

To write the answer in simplified radical form, you must factor out all
perfect square factors of the number under the radical sign. The largest
perfect square that is a factor of 180 is 36, so 180 is factored out as 36×5.

$$\sqrt{36 \times 5} = x$$
$$6\sqrt{5} = x$$

Therefore, the length of the hypotenuse is $6\sqrt{5}$ m.

Hint

When substituting
numbers into
$a^2 + b^2 = c^2$, always
substitute the
largest number (the
hypotenuse) for c. The
other two numbers
can be substituted for
a and b.

Math Reasoning

Analyze List the first
ten perfect squares.
These are examples of
factors that you look for
to express a radical in
simplified form.

b. Find the value of x. Give your answer in simplified radical form.

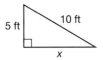

SOLUTION The hypotenuse is 10 feet, one of the legs is 5 feet, and the other leg is x ft.

$$a^2 + b^2 = c^2$$
$$x^2 + 5^2 = 10^2$$
$$x^2 + 25 = 100$$
$$x^2 = 75$$
$$x = 5\sqrt{3}$$

Therefore, the length of the second leg is $5\sqrt{3}$ feet.

Example 4 Application: TV Aspect Ratios

The aspect ratio of a TV screen is the ratio of the width to the height of the image. Find the height and the width of a 42-inch TV screen with an aspect ratio of 4:3 to the nearest tenth of an inch. The length 42 inches refers to the diagonal distance across the screen.

SOLUTION

Understand

The problem asks for the height and width of a TV screen. There are two important pieces of information: the ratio of the TV's width to its height is 4:3, and the diagonal of the TV is 42 inches long.

Plan

The diagonal, width, and height of the screen form a right triangle. Therefore, the Pythagorean Theorem can be used.

Although both a and b are unknown, it is known that the ratio of their lengths is 4:3. The width can be represented by $4x$ and the height by $3x$. Substitute $a = 4x$, $b = 3x$, and $c = 42$ to find the value of x. Then, use the value of x to determine the width and the height.

Solve

$$a^2 + b^2 = c^2$$
$$(4x)^2 + (3x)^2 = 42^2$$
$$25x^2 = 42^2$$
$$x^2 = \frac{42^2}{25}$$
$$x = \frac{42}{5}$$
$$x = 8.4$$

Width: $= 4x$ Height: $= 3x$
$\quad\quad\quad = 4(8.4)$ $\quad\quad\quad\ = 3(8.4)$
$\quad\quad\quad = 33.6$ inches $\quad\quad\quad\ = 25.2$ inches

Check

Check your answer by substituting the calculated height and width into the equation $a^2 + b^2 = c^2$ and solving for c. It should simplify to 42 inches.

$$a^2 + b^2 = c^2$$
$$(33.6)^2 + (25.2)^2 = c^2$$
$$42 = c$$

> **Hint**
>
> Taking the square root of a fraction is equivalent to taking the square root of both the denominator and the numerator separately. In other words:
>
> $$\sqrt{\frac{x}{y}} = \frac{\sqrt{x}}{\sqrt{y}}$$

a. Find the hypotenuse of the triangle. Do the side lengths form a Pythagorean triple?
(Ex 1)

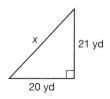

b. Find the value of p in the triangle at right. Do the side lengths form a Pythagorean triple?
(Ex 2)

c. Find the value of s in the triangle at right. Give your answer in simplified radical form.
(Ex 3)

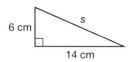

d. Find the value of y in the triangle at right. Give your answer in simplified radical form.
(Ex 3)

e. A ratio of a TV's width to its height is 16:9. If its width is 32 inches, what is the length of its diagonal?
(Ex 4)

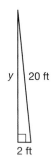

Practice Distributed and Integrated

 * 1. a. Algebra In the figure shown, if $x = 5$, find m$\angle 1$ and m$\angle 2$ where
(12) m$\angle 1 = (x^2 + 3x - 5)°$ and m$\angle 2 = (x^3 - 2x^2 - 2x - 10)°$.

b. Is $\overleftrightarrow{DE}$ parallel to $\overleftrightarrow{FG}$?

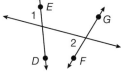

2. (**Air Traffic Controller**) An air traffic controller uses radar to scan the airspace
(23) above an area near the airport. A controller's radar covers a circular area of 250,000 square kilometers. What is the radius of the circle, to the nearest kilometer?

3. Construct a truth table for "s and not t," where s and t are logical statements.
(20)

* 4. Write Are the two triangles congruent? Explain.
(25)

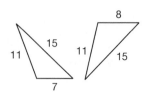

* 5. Write When can SAS congruency be used to conclude two triangles are congruent?
(28)

6. In $\triangle DEF$, m$\angle D = 71°$ and the measure of exterior angle F is 107°.
(18) Find m$\angle E$.

7. (**Road Signs**) While driving, Anil saw the following signs. Which of these is not a regular polygon?
₍₁₅₎

A

B

C OAK ST

D

8. What conclusion can be drawn from the following statements using the Law of Syllogism?
₍₂₁₎

My family room is a place where I like to study.
The places where I like to study are quiet.

* 9. Write the congruency statement for the triangles shown below.
₍₂₈₎

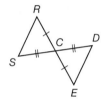

10. **Justify** The congruence statement for the two triangles is written as
₍₂₅₎ $\triangle DEF \cong \triangle PQR$. Find the values of x and y.

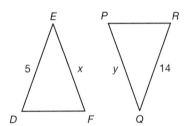

11. **Multiple Choice** Which of the following conjectures is false?
₍₁₄₎
A If x is odd, then $2x$ is even.
B If x is prime, then $x + 1$ is not prime.
C The product of two odd numbers is odd.
D Division of an integer by a prime number gives a rational number.

12. (**Geography**) On a grid map, Albuquerque is located at $(-10, 4)$ and Oklahoma
₍₁₁₎ City is at $(4, 2)$. If Amarillo is the midpoint between the two cities, where is it located?

*13. If a point lies on the perpendicular bisector of a segment, then the point is
₍₂₇₎ equidistant from the endpoints of the segment.

Given: A lies on the perpendicular bisector of $\overline{BC}$.
Prove: $AB = AC$

14. A small car has a turning radius of 22 feet. If the car makes a complete 360° turn
₍₂₃₎ at this radius, what is the area of the circle made to the nearest square foot?

15. Find m∠Q.
(18)

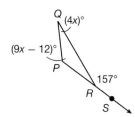

*16. Find the unknown length of the side in the triangle.
(29)

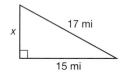

*17. If an arc of 120° were to be divided into three equal arcs, what would be the
(26) measure of each arc?

18. Prove Theorem 10-3: If two parallel lines are cut by a transversal, then the
(Inv 1) same-side interior angles are supplementary.
Given: *m* ∥ *n*, transversal *q*
Prove: ∠1 is supplementary to ∠2.

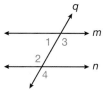

19. (**Astronomy**) The moon orbits the Earth with an average radius of
(23) 380,000 kilometers. Assuming the orbit is circular, what is the distance
the moon travels in one orbit, to the nearest kilometer?

20. Graph the line $y = -3x - 6$.
(16)

21. Classify △*EFG* by its angles and sides.
(13)

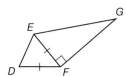

22. Prove Theorem 5-1: If two parallel planes are cut by a third plane, then the lines
(5) of intersection are parallel.
Given: Planes *CDE* and *ABF* are parallel. Plane *ABC* is a transversal intersecting
both planes.
Prove: $\overleftrightarrow{AB} \parallel \overleftrightarrow{CD}$

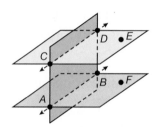

23. (**Landscaping**) A school's soccer field is rectangular with dimensions 110 yards
(22) by 60 yards. If the field is temporarily fenced and covered with a protective tarp
during the off-season, how many yards of fence are needed? How many square
yards of tarp are needed?

24. **Analyze** Use the statements to determine the truth value for "*A* and not *B*."
(20)

> *A: Thanksgiving is a holiday.*
> *B: Thanksgiving is celebrated in January.*

*25. **Error Analysis** Oshwinder found the solution 13.9 feet when solving for the length of
(29) one leg of a right triangle with a hypotenuse of 13 feet and a side length of 5 feet.
Is he correct? Explain.

*26. These two triangles are congruent, $\triangle KLM \cong \triangle TVU$. Find the values of *x* and *y*.
(28)

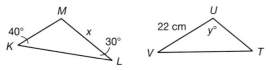

27. **Error Analysis** A student wrote the following as a conjecture. Is he correct? If not,
(14) provide a counterexample.

> *Any sequence starting with $\frac{1}{2}$ which keeps increasing will eventually include a term greater than 1.*

*28. Give two Pythagorean triples that are related to the triple (7, 24, 25).
(29)

29. Is the statement, "A parallelogram is a rectangle," always, sometimes,
(19) or never true?

*30. **Carpentry** A shelf bracket is needed to secure the shelf in the diagram. Use the
(29) Pythagorean Theorem to find the length of the support.

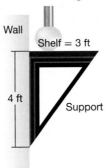

Triangle Congruence: ASA and AAS

Warm Up

1. **Vocabulary** The angle between two adjacent sides of a polygon is called
   (28)
   a(n) _____ angle. (*exterior, vertical, included*)

2. What is the value of x in the triangle?
   (29)

3. **Multiple Choice** Which of these is *not* used
   (25, 28) to prove that two triangles are congruent?

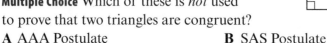

   A AAA Postulate   **B** SAS Postulate
   C SSS Postulate   **D** none of the above

New Concepts

There are many ways to prove triangle congruence. In Lesson 25, you learned that three congruent sides between two triangles prove triangle congruence (SSS Congruence Postulate). In Lesson 28, you learned that congruence of two pairs of corresponding sides and the corresponding included angles of two triangles also proves triangle congruence (SAS Congruence Postulate). In this lesson, you will learn two more triangle congruence patterns—the Angle-Side-Angle Postulate and the Angle-Angle-Side Theorem.

> **Hint**
>
> Recall that two congruent triangles have 3 congruent sides and 3 congruent angles. Although there are 6 congruent parts, only 3 properly chosen ones are needed to prove triangle congruence.

Postulate 16: Angle-Side-Angle (ASA) Congruence Postulate
If two angles and the included side of one triangle are congruent to two angles and the included side of another triangle, then the triangles are congruent.

Example 1 **Using the ASA Postulate**

Use ASA congruence to determine the measure of the sides of $\triangle DEF$.

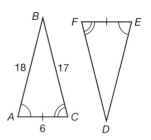

SOLUTION
In the two triangles, it is given that $\angle C \cong \angle F$, $\overline{AC} \cong \overline{EF}$, and $\angle A \cong \angle E$.

Since two angles and the included side of $\triangle BAC$ are congruent to two angles and the included side of $\triangle DEF$, by the ASA Postulate, $\triangle BAC \cong \triangle DEF$.

Therefore,

since $\overline{AC} \cong \overline{EF}$, $EF = 6$;

since $\overline{CB} \cong \overline{FD}$, $FD = 17$;

and since $\overline{AB} \cong \overline{ED}$, $ED = 18$.

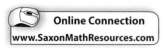
Online Connection
www.SaxonMathResources.com

Caution

Be sure that the congruent side is an included side of the two congruent angles when using ASA congruence.

Example 2 Using the ASA Postulate in a Proof

Prove that $\triangle SWT \cong \triangle UVT$, given that T is the midpoint of $\overline{WV}$ and $\overline{VU} \parallel \overline{WS}$.

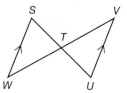

SOLUTION

1. T is the midpoint of $\overline{WV}$	1. Given
2. $\overline{WT} \cong \overline{VT}$	2. Definition of midpoint
3. $\angle SWT \cong \angle TVU$	3. If two parallel lines are cut by a transversal, then alternate interior angles are congruent.
4. $\angle WTS \cong \angle VTU$	4. Vertical angles are congruent.
5. $\triangle SWT \cong \triangle UVT$	5. ASA Congruence Postulate

Another way to prove that two triangles are congruent involves two angles and a non-included side.

Theorem 30-1: Angle-Angle-Side (AAS) Triangle Congruence Theorem

If two angles and a nonincluded side of one triangle are congruent to two angles and the corresponding nonincluded side of another triangle, then the triangles are congruent.

Example 3 Using the AAS Congruence Theorem

Given that $\overline{DE} \cong \overline{LK}$, find the area of each triangle shown below.

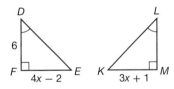

SOLUTION

It is given that $\overline{DE} \cong \overline{LK}$. It is also given in the illustration that $\angle D \cong \angle L$, and since all right angles are congruent, $\angle F \cong \angle M$. Since two angles and a non-included side of $\triangle DEF$ are congruent to the corresponding angles and non-included side of $\triangle LKM$, $\triangle DEF \cong \triangle LKM$ by the AAS Congruence Theorem.

Therefore, solve for x using CPCTC:

$$EF = KM$$
$$4x - 2 = 3x + 1$$
$$x = 3$$

Since the value of x is 3, the measure of $\overline{EF}$ is $4 \cdot 3 - 2$, or 10. So, $\overline{EF}$ and $\overline{KM}$ each have a length of 10. Since $\overline{DF} \cong \overline{LM}$, they both have a length of 6. The area of each triangle is $\frac{1}{2}bh = \frac{1}{2}(10)(6) = 30$.

Therefore, the area of each triangle is 30 square units.

Example 4   Using the AAS Theorem in a Proof

Given: $\overline{BD}$ bisects $\angle ADC$ and $\angle A \cong \angle C$.
Prove: $\triangle ABD \cong \triangle CBD$

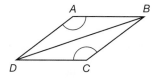

SOLUTION

Statements	Reasons
1. $\angle A \cong \angle C$	1. Given
2. $\angle ADB \cong \angle CDB$	2. Definition of angle bisector
3. $\overline{DB} \cong \overline{DB}$	3. Reflexive Property of Congruence
4. $\triangle ABD \cong \triangle CBD$	4. AAS Theorem

Math Reasoning

Justify Write a congruence statement for each of the parts of $\triangle ABD$ and $\triangle CBD$ that are given as congruent. What other congruence statements can you write once you have proven that the two triangles are congruent?

Example 5   Application: Bridges

A diagram of a portion of the truss system of a new bridge is shown below. Prove $\triangle ABC \cong \triangle DCB$.

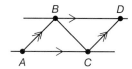

SOLUTION

Statements	Reasons
1. $\overline{BD} \parallel \overline{AC}$ $\overline{AB} \parallel \overline{CD}$	1. Given
2. $\angle DBC \cong \angle ACB$	2. If parallel lines are cut by a transversal, then alternate interior angles are congruent (Theorem 10-1).
3. $\angle ABC \cong \angle DCB$	3. Theorem 10-1
4. $\overline{BC} \cong \overline{BC}$	4. Reflexive Property of Congruence
5. $\triangle ABC \cong \triangle DCB$	5. ASA Theorem

Lesson Practice

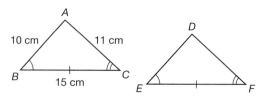

a. State the postulate that can be used to prove the triangles congruent,
(Ex 1) and state the measure of the sides of $\triangle DEF$.

b. Prove that $\triangle ABC \cong \triangle DEC$, given that $\overline{AB} \cong \overline{DE}$ and $\overline{AB} \parallel \overline{DE}$.
(Ex 2)

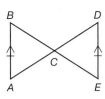

c. If the two triangles are congruent by the AAS Theorem, what is the area of each triangle?
(Ex 3)

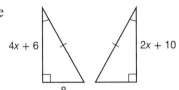

d. Prove that $\triangle ADC \cong \triangle BDC$.
(Ex 4)

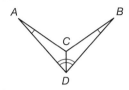

e. A standard-sized envelope is a $9\frac{1}{2}$-inch by 4-inch rectangle. The envelope is folded and glued from a sheet of paper shaped like the figure shown. Prove that if $JKNM$ is a rectangle, then $\overline{JI} \cong \overline{ML}$.
(Ex 5)

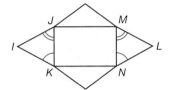

Practice   Distributed and Integrated

1. Construct a truth table for the biconditional statement, "(if x, then y) and (if y, then x)."
(20)

2. (**Mapping**) After walking 0.5 kilometers north and 0.4 kilometers east of her house, Juanita is one-third the distance to her grandfather's house. If the coordinates of Juanita's house are (1, 2), what are the coordinates of her grandfather's house?
(9)

3. (**Landscaping**) A newly planted tree will be secured by wires, each attached to the ground 1.5 meters from the base of the tree. If each wire is 4 meters long, how high up the tree should each wire be attached, to the nearest tenth of a meter?
(29)

4. Justify Can the SAS Postulate be used to prove these triangles congruent? Explain.
(28)

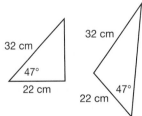

* 5. Prove that the two triangles are congruent.
(30)

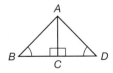

6. Algebra Solve the equation $\frac{2x-5}{5} = 3$, and justify each step.
(24)

7. Justify Given: $PQRS$ is a parallelogram.
(27) **Prove:** $\angle 1 \cong \angle 5$, $\angle 2 \cong \angle 4$ and $\angle 6 \cong \angle 3$.

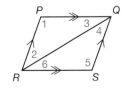

8. (**Safety**) A lid is needed to cover a well that has a 14-foot circumference. What is
(23) the area of the lid that is needed to cover the well, to the nearest tenth of a square
foot?

* 9. If the triangles are congruent, what is the perimeter of △PRQ?
(30)

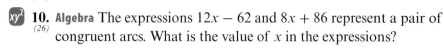

10. **Algebra** The expressions $12x - 62$ and $8x + 86$ represent a pair of
(26) congruent arcs. What is the value of x in the expressions?

*11. Determine the value of x.
(28)

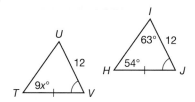

*12. Given that △ABC ≅ △DEF, write the six congruence
(25) statements for the triangles.

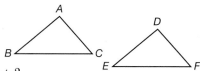

13. **Muiltiple Choice** What is the relationship between these two statements?
(17)

If Ming rides the bus to school, then she will not ride with Debra.
If Ming rides with Debra, then she will not ride the bus to school.

A They are converses.   **B** They are contrapositives.
C They are inverses.   **D** none of the above

*14. **Write** What theorem or postulate can be used to
(30) prove that the two triangles are congruent? Explain.

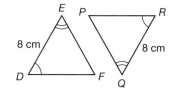

15. **Justify** Use *a, b,* and *c* to determine the truth value of "*a* or (*b* and *c*)".
(20)

a: New York City is the most populous city in the United States.
b: Florida is the most northwestern of all states.
c: The first president of the United States was George Washington.

16. Find a counterexample for the following conjecture.
(14)

The difference between x and y is a positive, even integer.

17. **Algebra** Find the length of each leg of an isosceles right triangle with a hypotenuse
(29) of 24 inches, to the nearest inch.

18. **Analyze** Draw a valid conclusion from the following statements.
(21)

Whoever dropped the bowl made a mess.
The clerk dropped the bowl.

*19. A minor arc has a measure of 135°. If the corresponding major arc is divided into
(26) 5 equal parts, what is the measure of each part?

*20. **Surveying** A road is to be built on an incline to reach a new bridge. The surveyor
(29) measures a horizontal distance of 660 yards and a vertical distance of 60 yards for
the incline. How long is the actual road surface, to the nearest yard?

21. **Multi-Step** In $\triangle GHI$, $m\angle G = 33°$ and the measure of an exterior angle at I is 141°.
(18) Find $m\angle H$.

22. The diagonals of a rhombus are 14 centimeters and 22 centimeters long,
(22) respectively. If the rhombus is cut apart and reassembled into a rectangle with
a width of 5 centimeters, what would be the length of the rectangle?

*23. **Error Analysis** Jamal says that $\overline{AB} \cong \overline{DE}$ is needed to prove that
(30) $\triangle ABC \cong \triangle DEC$ by the AAS Theorem. Identify and correct
Jamal's mistake.

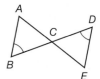

*24. Prove Theorem 6-7: If a point lies on the bisector of an angle,
(27) then the point is equidistant from the sides of the angle.
Given: K lies on the angle bisector of $\angle HIJ$.
Prove: $HK = JK$

25. **Multi-Step** Find the slope and the y-intercept of the line $C(1, 1)$ and $D(-2, 3)$.
(16)

26. A circular lid has an area of 42 square inches. If the lid rolls 8 times, what distance
(23) will it cover, to the nearest inch?

*27. **Model** How many triangles could you construct if you were given three different
(25) line segments? Explain.

28. The measure of one acute angle in a right triangle is y. What is the measure of the
(18) other acute angle?

29. **Stereo Equipment** The front of a speaker box is shaped like a trapezoid with a
(22) height of 2 feet, a lower base of 4 feet, and an upper base of 3 feet. How many
square feet of mesh is needed to cover the front of the speaker?

30. What is the relationship between the following two statements?
(17)
If a rectangle is regular, then it is a square.
If a rectangle is not regular, then it is not a square.

Exploring Angles of Polygons

In Lesson 15, you learned to identify interior and exterior angles in polygons. In this investigation, you will explore ways of determining angle measures in regular polygons.

Interior Angles in Regular Polygons

1. Using a protractor, measure an interior angle of each of the polygons and record your measurements in the second row of the table below.

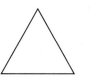

2. **Analyze** Compare the interior angle measures for each polygon. What patterns do you notice? How does the data support your observations?

Regular Polygon	Triangle	Quadrilateral	Pentagon	Hexagon
Interior Angle Measure				
Sum of Interior Angle Measures				

3. Since the polygons are regular, how can you calculate the sum of the interior angles for each polygon?

4. Record the sums of the interior angle measures in the table and compare these sums for the different polygons. What do you notice?

Rather than directly measuring interior angles, there is a way to calculate their measures for regular polygons. By drawing all possible diagonals from one vertex, a polygon can be divided into triangles.

The sum of the interior angle measures for all the triangles is the sum of the interior angle measures for the polygon. For any convex polygon, the number of triangles formed is two fewer than the number of sides. The formula for the sum of the interior angles of a polygon is the number of triangles it can be divided into multiplied by 180°.

Formula for the Sum of the Interior Angles of a Polygon
To find the sum of the interior angles of a polygon, use the formula below, where n is the number of sides of the polygon. $(n - 2)180°$

For example, if you draw a diagonal on a quadrilateral there are 2 triangles formed. The sum of the interior angles of the polygon is equal to 2 times 180°, or 360°.

To calculate the angle measure of a regular quadrilateral:

$$\text{measure of each interior angle in the square} = \frac{360°}{4}$$

Thus, the measure of each interior angle is 90°. Since each interior angle measure in a regular polygon is the same, the following formula can be used for any regular polygon.

Formula for Interior Angle Measure of a Regular Polygon
To find the measure of each interior angle, use the formula below, where n is the number of sides of the polygon. $$\frac{(n-2)180°}{n}$$ 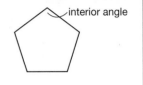

5. **Verify** Using the regular polygons for which you measured the interior angles:

   a. Calculate the interior angle measure for each using the formula above.

   b. Compare your measurements with the values in part **a**.

Exterior Angles in Regular Polygons

6. Trace each regular polygon from the previous page, then draw an exterior angle on each. Using a protractor, measure the exterior angle of each of the polygons and record your measurements in the table below.

7. Since the polygons are regular, what conjecture can you make about the exterior angle measures at each vertex for each polygon?

Regular Polygon	Triangle	Quadrilateral	Pentagon	Hexagon
Exterior Angle Measurements				
Sum of Exterior Angle Measurements				

8. **Analyze** Compare the exterior angle measures for each polygon. What do you notice? Does the data support your conjecture?

9. **Generalize** Add the exterior angle measures for each polygon, and record the results in the table. What do you notice?

This illustrates the fact that for all convex polygons, the sum of the exterior angles is 360°. For a regular polygon, each exterior angle measure is equivalent. Therefore, the following formula can be used to calculate the exterior angle measure in regular polygons.

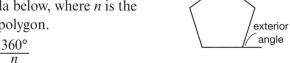

Formula for Exterior Angle Measure of a Regular Polygon

To find the exterior angle measures of a regular polygon, use the formula below, where n is the number of sides in the polygon.

$$\frac{360°}{n}$$

exterior angle

Math Reasoning

Generalize What are some examples of regular polygons that you see every day? Can you identify the interior angles? The exterior angles? How do they compare to the regular polygons you studied in this investigation?

10. **Write** Using the regular polygons for which you measured the exterior angles:

    a. Calculate the exterior angle measure for each using the formula above.

    b. Compare the results you obtained by direct measurement with your values in part **a**.

Central Angles in Regular Polygons

The **center of a regular polygon** is the point that is equidistant from each of the polygon's vertices. A **central angle of a regular polygon** has its vertex at the center of the polygon and its sides pass through consecutive vertices.

$\angle APB$, $\angle BPC$, $\angle CPD$, and $\angle DPA$ are the four central angles in polygon $ABCD$.

11. For each of the polygons at the beginning of the investigation, sketch all of the central angles and measure each angle formed using a protractor. Record your results in the table.

12. Since the polygons are regular, what conjecture can you make about the central angles in each polygon?

Regular Polygon	Triangle	Quadrilateral	Pentagon	Hexagon
Central Angle Measurements				
Sum of Central Angle Measurements				

13. **Generalize** Compare the central angle measures for each polygon. What do you notice? Does the data support your conjecture?

14. Add the central angle measurements for each of the polygons and record the results in the table. What is the sum of the central angles for each polygon and what do you notice?

15. How many central angles are in each polygon?

16. What conjecture can you write regarding the number of sides in a regular polygon and the number of central angles?

For each polygon, the sum of the central angles is 360° and the measure of each central angle is found using the number of sides in the polygon. The following formula can be used to calculate central angle measure of a regular polygon.

Formula for Central Angle Measure of a Regular Polygon

To find the measure of each central angle of a regular polygon, use the formula below, where n is the number of sides in the polygon.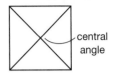

$$\frac{360°}{n}$$

17. Using the regular polygons for which you measured the central angles:
    a. Calculate the central angle measure for each using the formula above.

    b. Compare the results you obtained by direct measurement with your values in part **a**.

Investigation Practice

a. As the number of sides increases, what happens to the measure of each interior angle of a regular polygon? What happens to the measure of each exterior angle? The measure of each central angle?

b. What is the interior angle measure of a regular octagon?

c. What is the interior angle measure of a regular 20-gon?

d. Determine the measure of each exterior angle for a regular decagon.

e. Determine the measure of each central angle for a regular 30-gon.

f. Write Could you use the same formula to determine the interior angle measures of an irregular polygon as you did for regular polygons? Explain.

g. Write Could you use the same formula to determine the central angle measures of an irregular polygon as you did for regular polygons? Explain.

Flowchart and Paragraph Proofs

1. **Vocabulary** The process of using logic to draw conclusions is called
(21)
_____ reasoning.

2. **Multiple Choice** Which set of numbers is a Pythagorean triple?
(29)
A (1, 2, 3)
B (1, 1, $\sqrt{2}$)
C (1, 1, 1)
D (3, 4, 5)

3. Write a conclusion based on these statements. What law was used to make
(21) this conclusion?
If the milk has gone bad, I will go to the store. The milk has gone bad.

New Concepts

A **flowchart proof** is a style of proof that uses boxes and arrows to show the
structure of the proof.

A flowchart proof should be read from left to right or from top to bottom.
Each part of the proof appears in a box, while the justification for each
step is written under the box. The arrows show the progression of the
proof's steps.

| Example 1 | Interpreting a Flowchart Proof |

Use the given flowchart proof to write a two-column proof.
Given: ∠1 and ∠3 are congruent.
        ∠1 and ∠2 are supplementary.
Prove: ∠2 and ∠3 are supplementary.

Hint

A flowchart proof has all
the same components
as a 2-column proof,
including a diagram. If
you are having trouble
writing or interpreting
a flowchart proof, try
writing it as a 2-column
proof first.

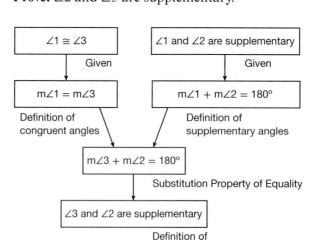

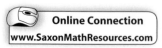
Online Connection
www.SaxonMathResources.com

SOLUTION

Write the steps and justifications of the proof as a 2-column proof.

Statements	Reasons
1. ∠1 and ∠3 are congruent.	1. Given
2. ∠1 and ∠2 are supplementary.	2. Given
3. m∠1 = m∠3	3. Definition of congruent angles
4. m∠1 + m∠2 = 180°	4. Definition of supplementary angles
5. m∠3 + m∠2 = 180°	5. Substitution Property of Equality
6. ∠3 and ∠2 are supplementary.	6. Definition of supplementary angles

Flowchart proofs are useful when a proof has two different threads that could be performed at the same time, rather than in sequence with one another. Whenever a proof does not proceed linearly from one step to another, a flowchart proof should be considered.

Math Language

Line *a* is an **auxiliary line**. Auxiliary lines are drawn only to aid in a proof.

Example 2   Writing a Flowchart Proof

Prove the Triangle Angle Sum Theorem: The sum of the interior angles of a triangle is 180°.

Given: △*ABC*
Prove: m∠1 + m∠2 + m∠3 = 180°

SOLUTION

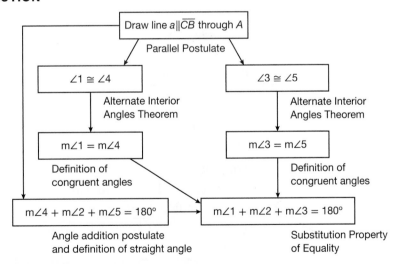

A **paragraph proof** is a style of proof in which statements and reasons are presented in paragraph form.

In a paragraph proof, every step of the proof must be explained by a sentence in the paragraph. Each sentence contains a statement and a justification.

Example 3 Reading a Paragraph Proof

Use the given paragraph proof to write a two-column proof.
Given: ∠1 and ∠2 are complementary.
Prove: ∠3 and ∠4 are complementary.

∠1 and ∠2 are complementary, so m∠1 + m∠2 = 90° by the definition of complementary angles. Angle 1 is congruent to ∠4, and ∠2 is congruent to ∠3, by the Vertical Angles Theorem. So m∠1 = m∠4, and m∠2 = m∠3. By substitution, m∠4 + m∠3 = 90°. Therefore, ∠3 and ∠4 are complementary by the definition of complementary angles.

SOLUTION
Put the steps in the paragraph proof into a two-column proof:

Statements	Reasons
1. ∠1 and ∠2 are complementary.	1. Given
2. m∠1 + m∠2 = 90°	2. Definition of complementary angles
3. ∠1 and ∠4 are congruent. ∠2 and ∠3 are congruent.	3. Vertical Angles Theorem
4. m∠1 = m∠4 and m∠2 = m∠3	4. Definition of congruent angles
5. m∠3 + m∠4 = 90°	5. Substitution Property of Equality
6. ∠3 and ∠4 are complementary.	6. Definition of complementary angles

A paragraph proof is good for short proofs where each step follows logically from the one before. Paragraph proofs are usually more compact than two-column proofs.

Caution

When writing a paragraph proof, make sure that every statement is accompanied by a justification. If necessary, make a 2-column proof of the most important steps as a plan.

Example 4 Writing a Paragraph Proof

Prove Theorem 10-1: If two parallel lines are cut by a transversal, then alternate interior angles are congruent.
Given: Lines p and q are parallel.
Prove: ∠2 ≅ ∠3

SOLUTION
It is given that lines p and q are parallel. It is known that ∠1 ≅ ∠3, by the Corresponding Angles Postulate (Postulate 11). By the Vertical Angles Theorem, ∠2 ≅ ∠1, so by the Transitive Property of Congruence, ∠2 ≅ ∠3.

Lesson Practice

a. Write a two-column proof from the given flowchart proof.
(Ex 1)
Given: ∠1 and ∠2 are complementary.
∠1 and ∠3 are congruent.
Prove: ∠2 and ∠3 are complementary.

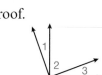

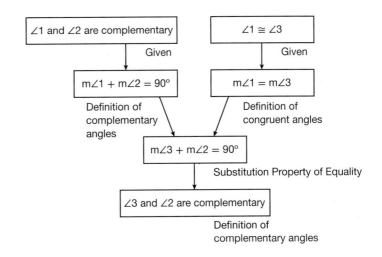

b. Prove the Vertical Angles Theorem using a flowchart proof.
(Ex 2) **Given:** *a* and *b* are intersecting lines.
Prove: ∠1 and ∠3 are congruent.

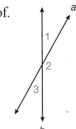

c. Write a two-column proof from the given
(Ex 3) paragraph proof.
Given: ∠1 and ∠4 are complementary.
Prove: ∠2 and ∠3 are complementary.

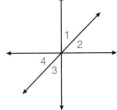

∠1 and ∠4 are complementary, so
m∠1 + m∠4 = 90° by the definition of
complementary angles. Angle ∠1 ≅ ∠3, and
∠2 ≅ ∠4, by the Vertical Angles Theorem.
Therefore, m∠1 = m∠3 and m∠2 = ∠4 by the definition of congruent
angles, and m∠2 + m∠3 = 90° by substitution. Therefore, ∠2 and ∠3
are complementary by the definition of complementary angles.

d. Prove Theorem 5-4: If two lines are
(Ex 4) perpendicular, then they form congruent
adjacent angles.
Given: Lines *s* and *t* are perpendicular.
Prove: Angles 1 and 2 are congruent
adjacent angles.

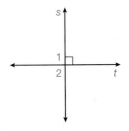

Practice   Distributed and Integrated

* 1. (**Baseball**) Vernon is standing on the pitcher's mound of a baseball field. He turns
(Inv 3) from looking toward home plate to first base. If the pitcher's mound is at the
center of the regular polygon made by the bases, how many degrees did
Vernon rotate?

2. **Analyze** In the diagram given, prove that $\overline{AD} \cong \overline{BC}$,
(28) using a two-column proof.

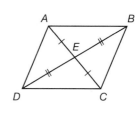

3. Determine m$\angle Y$ in $\triangle XYZ$.
(18)

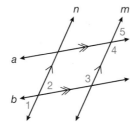

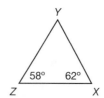 **4.** **Algebra** Erin drew a parallelogram with an area of 18 square centimeters
(22) and a base of 6 centimeters. What was the height?

5. If m$\overset{\frown}{AB} = 70°$, what is the measure of the associated major arc?
(26)

* 6. Write a paragraph proof.
(31) **Given:** $a \parallel b$, $n \parallel m$
Prove: $\angle 1 \cong \angle 5$.

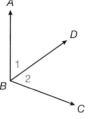

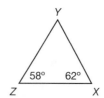 **7.** **Algebra** $\triangle ABC$ and $\triangle DEF$ are congruent triangles with m$\angle A = 90°$, $AB = 7$, and
(25) $DF = 24$. Based on the given information, what is the length of BC?

* 8. (**Home Repair**) A contractor arrives at a house with a ladder that is 4 meters long. If
(29) the closest he can safely place the ladder to the house is 1.5 meters, will the top of
the ladder reach the edge of the roof, which is 3.8 meters above the ground?

* 9. (**Design**) Whitney is making a stop sign, a regular octagon. What should be the
(Inv 3) measure of each angle of her sign?

10. (**Kites**) Mohinder wants to make a rhombus-shaped kite with an area of
(22) 224 square inches and a frame using one diagonal crosspiece that is 14 inches
long. What is the length of the other diagonal crosspiece?

*11. Write a flowchart proof.
(31) **Given:** $\overrightarrow{BD}$ bisects $\angle ABC$ and $\angle 1$ is congruent to $\angle 3$.
Prove: Angle 2 is congruent to $\angle 3$.

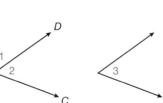

12. **Justify** If $\triangle ABC \cong \triangle LMN$ and $\triangle LMN \cong \triangle XYZ$, write the
(25) congruence statements between $\triangle ABC$ and $\triangle XYZ$. What
property justifies these conclusions?

13. (**Landscaping**) Prove that if the area of a rectangular patio $ABCD$ is 25 square units
(27) and one of the lengths is 5 units, then $ABCD$ is a square.

14. Write the converse of the following statement.
(17)
If two numbers have midpoint 0 on a number line, then they are opposites.

Is the original statement true? Is the converse?

*15. **Multiple Choice** The length of a leg of a right triangle with a hypotenuse of
(29) 20 centimeters and another leg of 12 centimeters is _____.

    A 6 centimeters   **B** 4 centimeters

    C 16 centimeters   **D** 10 centimeters

16. Determine the equation of each line graphed on this grid.
(16)

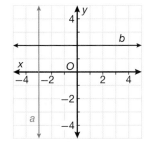

17. ⟨**Meteorology**⟩ When cumulus clouds appear, there is a good chance of rain
(21) later in the day. When it rains, the air cools and descends, creating wind.
What can students conclude from these statements if they see cumulus
clouds forming?

*18. **Error Analysis** Suresh and Gayle are having a debate. Suresh believes that
(30) since each angle in an equilateral triangle measures 60°, all equilateral
triangles are congruent. Gayle disagrees since she can draw two equilateral
triangles that are different sizes. Who is correct? Explain.

19. **Algebra** Find a counterexample to the conjecture.
(14)

    A linear equation in one unknown has exactly one solution.

*20. Using the diagram given, prove that $\overline{AE} \cong \overline{CE}$.
(30)

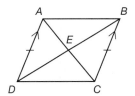

*21. Write a paragraph proof.
(31)   **Given:** $\overline{AD}$ bisects $\overline{CB}$ and $\overline{AD} \perp \overline{CB}$.
  Prove: $\triangle ACD \cong \triangle ABD$.

22. City Hall is located at (1, 1) on a coordinate grid of a city. The police station is
(24) located at point $(x - 3, 4x + 1)$. If the police station is 5 units away from City
Hall and x is positive, what are the numeric coordinates of the police station?
Provide a justification for each step.

23. Write an equation for the line passing through $(-3, 2)$ that has slope 4.
(16)

*24. Write a flowchart proof.
(31)   **Given:** $\overline{LP}$ is congruent to $\overline{MN}$ and $\overline{LO}$ is congruent to $\overline{MO}$.
  Prove: $\overline{OP}$ is congruent to $\overline{ON}$.

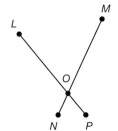

*25. Generalize If $\triangle ABC$ and $\triangle DEF$ are right triangles, $\angle B$ and $\angle E$ are right angles, AC
(28) $= DF$, and $AB = DE$, is it possible to conclude that $\triangle ABC \cong \triangle DEF$?
If yes, use the SAS Theorem to prove it. If no, explain why.

26. a. A right triangle has a base length of 1.5 meters and a height of 1.2 meters.
(13)   Determine its area.

b. The other side of the same triangle has a length of approximately 1.9 meters.
Determine the triangle's perimeter.

*27. a. Generalize Using symbolic logic, write the disjunction of "not p" and "q."
(20) **b.** Write the negation of the statement, "p or q" as a compound statement
involving p, q, and/or their negations.

c. Write the negation of your answer to part **a** as a compound statement involving
p, q, and/or their negations.

28. In the figure below, name each angle marked with a dot and identify it as an
(15) interior or exterior angle.

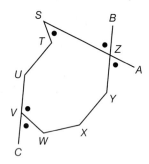

29. Verify For a circle with a diameter of 13 centimeters, Sal calculated the
(23) circumference to be 81.64 centimeters. Is Sal's calculation correct?

*30. Algebra If in $\triangle ABC$, $AB = (3x + 11)$, $m\angle ABC = 45°$, $m\angle BCA = 75°$, and $BC = 15$,
(30) and in $\triangle DEF$, $DE = (7x - 9)$, $m\angle DEF = 45°$, $m\angle EFD = 75°$, and $EF = 15$, what is the
value of x?

Altitudes and Medians of Triangles

1. **Vocabulary** A triangle with three congruent angles is called
(13) a(n) _____ triangle.

2. Is this statement true or false?
(13) *The height of a triangle is always a segment in the interior of the triangle.*

3. What is the midpoint of the points $(-3, 4)$ and $(2, 8)$?
(11)

4. What is the midpoint of the points $(4, 2)$ and $(-1, 3)$?
(11)

New Concepts

The **median of a triangle** is a segment whose endpoints are a vertex of the triangle and the midpoint of the opposite side.

Every triangle has one median for each vertex, or three medians total. The point of concurrency of these three medians also has a special name. The **centroid of a triangle** is the point of concurrency of the three medians of a triangle. This point is also called the center of gravity. The three medians of a triangle are always concurrent lines.

centroid
intersection of the medians

Math Language

A **point of concurrency** is a point where three or more lines intersect. When three or more lines intersect at one point, they are called **concurrent lines**.

Caution

Be careful not to flip the equalities given by the Centroid Theorem. Remember that the centroid is always closer to the side than to the vertex.

Theorem 32-1: Centroid Theorem

The centroid of a triangle is located $\frac{2}{3}$ the distance from each vertex to the midpoint of the opposite side.

In $\triangle ABC$:

$$CP = \frac{2}{3}CZ \qquad BP = \frac{2}{3}BY \qquad AP = \frac{2}{3}AX$$

Example 1 **Using the Centroid to Find Segment Lengths**

In $\triangle LMN$, $LA = 12$ and $OC = 3.1$. Find LO.

SOLUTION

$LO = \frac{2}{3}LA$ Centroid Theorem

$LO = \frac{2}{3}(12)$ Substitution Property of Equality

$LO = 8$ Simplify.

Find the length of $\overline{NC}$.

SOLUTION

$NO = \frac{2}{3}NC$ Centroid Theorem

$NO + OC = NC$ Segment Addition Postulate

$\frac{2}{3}NC + 3.1 = NC$ Substitution Property of Equality

$3.1 = \frac{1}{3}NC$ Subtraction Property of Equality

$9.3 = NC$ Multiplication Property of Equality

Online Connection
www.SaxonMathResources.com

Example 2 Finding a Centroid on the Coordinate Plane

Find the centroid of $\triangle DEF$ with vertices at $D(-3, 5)$, $E(-2, 1)$, and $F(-7, 3)$.

SOLUTION

Graph the triangle. Start by finding the midpoint of each segment of the triangle.

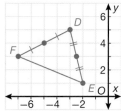

The midpoint of $\overline{DF}$ is $(-5, 4)$ and the midpoint of $\overline{DE}$ is $(-2.5, 3)$.

Since all three medians meet at the same point, the intersection of any two will give the location of the centroid.

Use the two points on each median to find an equation for the medians. The slope of the median that extends from E to $(-5, 4)$ is -1.

$$y - y_1 = m(x - x_1) \qquad \text{Point-slope formula}$$
$$y - 4 = -1(x + 5) \qquad \text{Substitute.}$$
$$y = -x - 1 \qquad \text{Simplify.}$$

The equation of the median from F to $(-2.5, 3)$ is $y = 3$. Solve these two equations as a system.

$$y = -x - 1 \qquad\qquad y = 3$$
$$3 = -x - 1$$
$$x = -4$$

We already know the y-coordinate is 3, so the centroid is located at $(-4, 3)$.

Hint

Try to pick a median with a simple equation to make it easier to find the centroid. In this example, one of the medians is horizontal, so the equation, $y = 3$, is easy to find.

The **altitude of a triangle** is a perpendicular segment from a vertex to the line containing the opposite side. The **orthocenter of a triangle** is the point of concurrency of the three altitudes of a triangle.

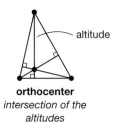

orthocenter
intersection of the altitudes

Example 3 Locating the Orthocenter of a Triangle

Draw an acute triangle, a right triangle, and an obtuse triangle. Sketch the altitudes of each triangle and find their orthocenters. Is the orthocenter always in the interior of a triangle?

SOLUTION

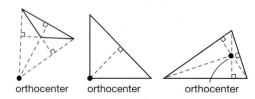

orthocenter   orthocenter   orthocenter

The diagram shows an obtuse, right, and acute triangle. From the diagram, the orthocenter of the obtuse triangle is in the exterior of the triangle. So orthocenters are not always in the interior of triangles.

Example 4 Application: Balancing Objects

The centroid of a triangle is the point where the triangle can be balanced on a point. Suppose a student wants to perform a balancing act with triangles as part of a talent show. Each of the triangles are the same size, and the three medians are 4.2 inches, 7.2 inches, and 8.1 inches long, respectively. What is the distance from each vertex to the centroid of each of these triangles?

SOLUTION

Use the Centroid Theorem. The distance from each vertex to the centroid is two-thirds the length of the median.

$$4.2 \times \frac{2}{3} = 2.8$$

$$7.2 \times \frac{2}{3} = 4.8$$

$$8.1 \times \frac{2}{3} = 5.4$$

The balancing point, or center of gravity, for each triangle is 2.8 inches, 4.8 inches, and 5.4 inches away from their respective vertices.

Hint

Remember that the centroid is also known as the center of gravity. This is due to the fact that a triangle suspended or held up by the centroid will be perfectly level.

Lesson Practice

a. In $\triangle ABC$, $AD = 5$ and $EO = 4.2$.
(Ex 1) Use the Centroid Theorem to find the lengths of $\overline{OD}$ and $\overline{BE}$ to the nearest hundredth.

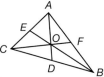

b. Use the coordinate plane to find the centroid of
(Ex 2) $\triangle JKL$ with vertices $J(-9, 1)$, $K(-1, 5)$, and $L(-5, 9)$.

c. Where is the orthocenter of a right triangle located?
(Ex 3)

d. Clara is hanging triangles from a mobile. She needs to find the centroid
(Ex 4) of each triangle for the mobile to hang correctly. If the triangles have medians of 3.6 inches, 6.9 inches, and 4.5 inches respectively, how far is the centroid from each vertex?

Practice Distributed and Integrated

1. Formulate An arc measure of 180° creates a semicircle. Use this fact to derive a
(26) formula for the area of a semicircle.

2. Three sides of convex quadrilateral $ABCD$ are congruent, but the fourth is not.
(19) What possible shape(s) could $ABCD$ be?

3. Algebra A trapezoid has an area of 22, a height of 4, and bases of $x + 2$ and
(22) $3x - 3$. Solve for x.

4. Make a conjecture about the next item in the list and explain the basis of
(7) your conjecture.

Triangle, square, pentagon, hexagon, heptagon…

* 5. A median of a triangle connects a vertex and the _____ of the opposite side.
(32)

6. (**Marketing**) In the diagram, a corporate logo is designed in the shape shown
(15) at right.

Arrow
Delivery

    a. Name the polygon.
    b. Determine whether the polygon is equiangular, equilateral, regular,
       irregular, or more than one of these.
    c. Determine whether the polygon is concave or convex. Explain.

7. Consider the conjecture.
(14)

*If △MNO is an acute triangle, D lies on $\overline{MN}$, and E lies on $\overline{MO}$, then △MDE is an
acute triangle.*

    a. What is the hypothesis of the conjecture? What is its conclusion?
    b. Find a counterexample to the conjecture.

8. In the diagram given, $\overline{RU}$ and $\overline{ST}$ are diameters of the circle and O is the
(27) center of the circle. Prove that m∠SRO = m∠UTO.

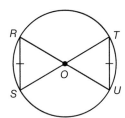

Use the diagram to answer the next two questions.

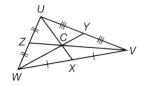

* 9. If the length of $\overline{ZV}$ is 72, determine the length of $\overline{VC}$.
(32)

*10. If the length of $\overline{CX}$ is 18, determine the length of $\overline{UC}$.
(32)

11. (**Designing**) Spencer is making two triangular flags, △ABC and △DEF, for the
(28) upcoming football game. Given that if △ABC is a right triangle at ∠B and △DEF
is a right triangle at ∠E with $\overline{AB} \cong \overline{DE}$ and $\overline{BC} \cong \overline{EF}$, which congruence postulate
would prove △$ABC \cong$ △DEF?

*12. Using the diagram shown, prove $KM = LM$.
(30)

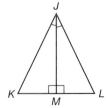

13. (**Art**) Using a technique known as vanishing point to give a three-dimensional
(22) perspective to the painting, an artist drew a rectangular table as a trapezoid with a
height of 3 inches and with upper and lower bases of 3 and 4 inches, respectively.
What is the area of the painting of the table?

*14. Write a paragraph proof.
(31) **Given:** $\overline{AB} \cong \overline{BC}$ and $\overline{BC} \cong \overline{DE}$
Prove: $\overline{AB} \cong \overline{DE}$

15. Prove that if all angles in $\triangle PQR$ are equal, then each angle is 60°.
(27)

16. Error Analysis In calculating the length of a hypotenuse, Beatrice and Amelie both
(29) got an answer of $\sqrt{360}$. However, when asked to provide their answer in simplified radical form, Beatrice's answer was $36\sqrt{10}$ and Amelie's was $6\sqrt{10}$. Who is incorrect and why?

*17. Write a paragraph proof.
(31) **Given:** Angle 1 and $\angle 2$ are complementary.
        Angle 3 and $\angle 4$ are complementary.
        $\angle 2 \cong \angle 3$
Prove: $\angle 1 \cong \angle 4$

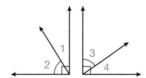

 18. Algebra Two vertical angles measure $(4x - 4)°$ and $(3x + 7)°$. What is the measure
(24) of each angle? Provide a justification for each step.

19. Consider the following conjecture.
(14)
If the product of three numbers is positive, then all three numbers are positive.
a. What is the hypothesis of the conjecture? What is its conclusion?
b. Find a counterexample to the conjecture.

20. Use the Law of Detachment to fill in the missing statement in this conjecture.
(21)
_____

M is the midpoint of $\overline{WX}$.
Therefore, $WM = MX$.

21. Verify A triangle has an exterior angle measuring 98°. One remote interior angle
(18) measures 49° and the other remote angle is congruent to it. Verify the Exterior Angle Theorem for this exterior angle.

22. (Machinery) Two gears are interlocked. One has a radius of 10 centimeters and for
(23) each complete rotation, it rotates the second gear 0.77 of a full turn.
a. What is the relationship between the circumferences of the two gears?
b. What is the second gear's radius, to the nearest centimeter?

23. Prove that if the heights of two parallelograms, h_1 and h_2, lie on the same
(27) parallel lines and have the same base, b, then the parallelograms are equal in area.

*24. Multiple Choice If the length of a median of a triangle equals 4.5, what is the
(32) distance from the centroid to the opposite side?
  A 4.5   **B** 3
  C 2.25   **D** 1.5

25. **Analyze** In the diagram given, prove that $\triangle ABC \cong \triangle BDC$ using the SAS
$_{(28)}$ Triangle Congruence Theorem. *Hint: Show* $m\angle BAC = m\angle DBC$.

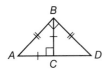

26. **Generalize** Complete the table by writing the contrapositive of each statement.
$_{(17)}$

Statement	Contrapositive
If p, then q	If $\sim q$, then $\sim p$
If $\sim p$, then q	
If p, then $\sim q$	
If $\sim p$, then $\sim q$	

*27. (**Design**) Ian is making a kite he has labeled *VWXY*. He has already made
$_{(30)}$ $m\angle VWY = m\angle XWY$. What pair of angles can he make equal to ensure
$\triangle VWY \cong \triangle XWY$?

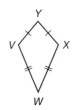

28. $\triangle HIJ \cong \triangle LKN$ and $m\angle H = 50°$, $m\angle K = 100°$, and $HI = 5$ centimeters.
$_{(25)}$ What is $m\angle J$?

*29. Write a flowchart proof. Refer to the figure shown.
$_{(31)}$ **Given:** $m\angle BAC = m\angle EAF$, $m\angle CAD = m\angle DAE$
Prove: $m\angle BAD = m\angle DAF$

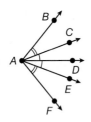

30. **Generalize** What is the negation of the statement, "$\sim p$ or q?" Write a truth table to
$_{(20)}$ help you.

Converse of the Pythagorean Theorem

Warm Up

1. **Vocabulary** A statement formed by exchanging the hypothesis and
(10) conclusion of a conditional statement is called the _____.

2. Find the length of $\overline{PQ}$.
(Inv 2)

3. Write the converse of the conditional statement. Is the converse true?
(10) *If two angles are vertical angles, then they are congruent.*

New Concepts

In Investigation 2, the Pythagorean Theorem is presented. The Pythagorean Theorem states that for any right triangle, the sum of the squares of the lengths of its two shortest sides is equal to the square of the length of its longest side.

As an equation, the Pythagorean Theorem reads: $a^2 + b^2 = c^2$.

Each of the shorter sides is called a leg, and the longest side is called the hypotenuse.

Example 1 Applying the Pythagorean Theorem

Find the value of x. Write the answer in simplified radical form.

a.

SOLUTION

$a^2 + b^2 = c^2$	Pythagorean Theorem
$5^2 + 5^2 = x^2$	Substitute.
$50 = x^2$	Simplify.
$x = \sqrt{50} = 5\sqrt{2}$	Solve for x.

b.

SOLUTION

$a^2 + b^2 = c^2$	Pythagorean Theorem
$6^2 + x^2 = \left(5\sqrt{13}\right)^2$	Substitute.
$36 + x^2 = 325$	Simplify.
$x = 17$	Solve for x.

The converse of the Pythagorean Theorem is true and can be used to determine whether or not a triangle is a right triangle.

Theorem 33-1: Converse of the Pythagorean Theorem
If the sum of the squares of the two shorter sides of a triangle is equal to the square of the longest side of the triangle, then the triangle is a right triangle.

Example 2 Proving the Converse of the Pythagorean Theorem

Given: $a^2 + b^2 = c^2$
Prove: $\triangle STU$ is a right triangle.

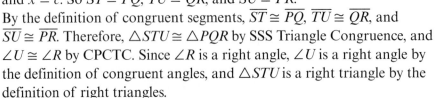

Hint

In order to do this proof, draw another right triangle with the same side lengths as $\triangle STU$. Sometimes proofs require that a figure be added to the given diagram.

SOLUTION
Draw right triangle $\triangle PQR$ with leg lengths identical to $\triangle STU$ and a third side of length x. In $\triangle STU$, it is given that $a^2 + b^2 = c^2$. In $\triangle PQR$, $a^2 + b^2 = x^2$ by the Pythagorean Theorem. Since $a^2 + b^2 = c^2$ and $a^2 + b^2 = x^2$, it follows by substitution that $x^2 = c^2$. Take the positive square root of both sides, and $x = c$. So $ST = PQ$, $TU = QR$, and $SU = PR$. By the definition of congruent segments, $\overline{ST} \cong \overline{PQ}$, $\overline{TU} \cong \overline{QR}$, and $\overline{SU} \cong \overline{PR}$. Therefore, $\triangle STU \cong \triangle PQR$ by SSS Triangle Congruence, and $\angle U \cong \angle R$ by CPCTC. Since $\angle R$ is a right angle, $\angle U$ is a right angle by the definition of congruent angles, and $\triangle STU$ is a right triangle by the definition of right triangles.

Example 3 Applying the Converse of the Pythagorean Theorem

Determine whether each triangle is a right triangle.

a.

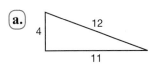

SOLUTION
Use the Pythagorean Theorem.

$a^2 + b^2 = c^2$ Pythagorean Theorem
$4^2 + 11^2 \stackrel{?}{=} 12^2$ Substitute.
$137 \neq 144$ Add and compare.

This triangle is not a right triangle by the Converse of the Pythagorean Theorem.

b.

SOLUTION

Use the Pythagorean Theorem.

$$a^2 + b^2 = c^2 \qquad \text{Pythagorean Theorem}$$
$$5^2 + 12^2 \stackrel{?}{=} 13^2 \qquad \text{Substitute.}$$
$$169 = 169 \qquad \text{Add and compare.}$$

This triangle is a right triangle by the Converse of the Pythagorean Theorem.

The Pythagorean Theorem can also be used to determine if a triangle is an acute triangle or an obtuse triangle by using inequalities. This is called the Pythagorean Inequality Theorem.

Theorem 33-2: Pythagorean Inequality Theorem
In a triangle, let a and b be the lengths of the two shorter sides and let c be the length of the longest side. If $a^2 + b^2 < c^2$, then the triangle is obtuse. If $a^2 + b^2 > c^2$, then the triangle is acute.

Example 4 **Classifying Triangles Using the Pythagorean Inequality Theorem**

Determine whether $\triangle JKL$ is an obtuse, acute, or a right triangle.

a.

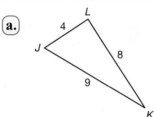

SOLUTION

The longest side is $\overline{JK}$, so it will represent c, with the other two sides representing a and b.

$$a^2 + b^2 = c^2$$
$$4^2 + 8^2 \stackrel{?}{=} 9^2$$
$$80 < 81$$

Since $a^2 + b^2 < c^2$, the triangle is an obtuse triangle.

b.

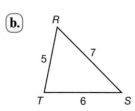

SOLUTION

$$a^2 + b^2 = c^2$$
$$6^2 + 5^2 \stackrel{?}{=} 7^2$$
$$61 > 49$$

Since $a^2 + b^2 > c^2$, the triangle is an acute triangle.

Example 5 **Application: Building a Wheelchair Ramp**

The local library needs a ramp for wheelchair access. If the city says that the ramp should be built with a slope of $\frac{1}{4}$, and must reach a platform that is 2 yards high, what must be the length of the ramp? Round to the nearest hundredth.

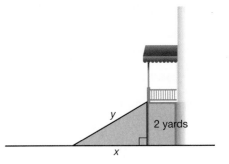

SOLUTION

Slope is in the form $\frac{\text{rise}}{\text{run}}$. The ramp rises 2 yards. Let x be its run.

$$\frac{\text{rise}}{\text{run}} = \frac{1}{4} = \frac{2}{x}$$ Substitute a rise of 2 and a run of x.

$$x = 8$$ Cross multiply.

Therefore, the run of the ramp x is equal to 8 yards.

Now use the Pythagorean Theorem to solve for y.

$$x^2 + 2^2 = y^2$$ Pythagorean Theorem
$$8^2 + 2^2 = y^2$$ Substitute.
$$64 + 4 = y^2$$ Simplify.
$$y \approx 8.25$$ Solve for y.

The ramp needs to be about 8.25 yards long.

Lesson Practice

a. Use the Pythagorean Theorem to solve for
(Ex 1) the missing side length. Give your answer in simplified radical form.

b. Is $\triangle ABC$ a right triangle?
(Ex 3)

c. Is $\triangle PQR$ obtuse, acute, or right?
(Ex 4)

d. Is $\triangle STU$ obtuse, acute, or right?
(Ex 4)

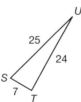

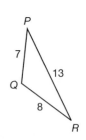

e. A ladder leans against a wall. Its top
(Ex 5) reaches a window that is 6 feet above the ground and the bottom of the ladder is 7.5 feet from the base of the wall, measured along the ground. Do the ladder, the wall, and the ground form a right triangle? Explain.

1. The two triangles at right are congruent. Write the six congruency
(25) statements for them.

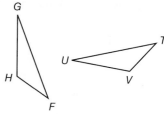

2. In △*ABC*, m∠*BAC* = 57°, *AB* = 10, and m∠*ABC* = 52°. In △*DEF*,
(30) m∠*EDF* = 57°, *DE* = 10, and m∠*DFE* = 71°. Is △*ABC* ≅ △*DEF*?
Explain.

* 3. Multi-Step What is the value for *x* if the triangle is a right triangle? Write an
(33) inequality to show the smallest value for *x* that makes the triangle an obtuse
triangle. Write your answer in simplified radical form.

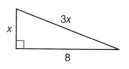

4. Write the converse of the following false statement.
(17)
    If a triangle is isosceles, then it is obtuse.

* 5. Write a paragraph proof.
(31) **Given:** ∠*STV* ≅ ∠*TVU*, ∠*STU* ≅ ∠*UVS*
Prove: ∠*SVT* ≅ ∠*UTV*

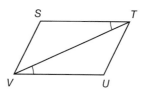

6. The angles of △*ABC* are m∠*BAC* = 9*x* − 77°, m∠*ABC* = 5*x* + 17°, and
(28) m∠*ACB* = 6*x* − 33°. If m∠*DEF* = 90°, $\overline{AB}$ ≅ $\overline{DE}$, and $\overline{BC}$ ≅ $\overline{EF}$, is
△*ABC* ≅ △*DEF*?

7. Find an expression for the area of a parallelogram with a height of 2*y* and a base
(22) of 3*x*.

* 8. (Architecture) The Leaning Tower of Pisa in Italy would be 185 feet tall if
(33) it were standing vertically.
   a. If you are standing 80 feet from the base of the tower, and the tower is
leaning away from you, what type of triangle would be formed? What is
the minimum length of *y*?
   b. If you are standing 125 feet from the base of the tower, and the tower
is leaning toward you, what type of triangle would be formed? What
is the maximum length of *x*?

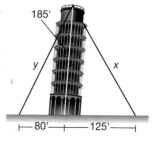

9. Verify Write the inverse of the statement, "If a triangle does not have an obtuse
(17) angle, then it is acute." Which is true: the statement, its inverse, both, or neither?
Explain.

*10. Write a flowchart proof.
(31) **Given:** $\overline{EF}$ ≅ $\overline{HI}$, *H* is the midpoint of $\overline{GI}$
Prove: $\overline{EF}$ ≅ $\overline{GH}$

11. a. Generalize How many different radii does a circle have?
(23) **b.** How many different measures do the radii of a circle have?

*12. Algebra Write an inequality to show the largest value for *x* that makes the
(33) triangle obtuse if the longest side length is 11.

13. **Write** Explain how the slope-intercept form of a line can be used to sketch
(16) its graph, without creating a table of values.

14. (**Geology**) A volcano is active if it has erupted recently and is expected to do so
(20) again soon. It is dormant if it is not active, but is expected to erupt again at some
future time. It is extinct if it is not ever expected to erupt again.
   a. Write the conjunction of the statements, "A volcano is active," and "A volcano
   is not ever expected to erupt again." Is this conjunction true? Explain.
   b. Write the disjunction of the same two statements. Is the disjunction true?
   Why or why not?

15. (**Gardening**) Ashanti has a 40-square foot garden in the shape of a trapezoid. If
(24) the diagram shows the garden's dimensions, what are the lengths of the two bases?
Provide a justification for each step.

*16. **Multiple Choice** The altitudes are used to find which of the following?
(32) **A** orthocenter   **B** median
   C area   **D** centroid

17. In the figure, name each angle marked with a dot and identify
(15) it as an interior or exterior angle.

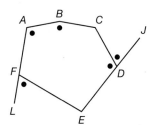

18. **Algebra** A convex polygon has interior angles $(x + 4)°$, $(2x + 5)°$, $(3x + 6)°$,
(Inv 3) $(4x + 7)°$, and $(5x + 8)°$. What is the value of x?

*19. (**Furniture**) The height of a triangular cabinet is 8 feet. There are four
(29, 33) shelves which are evenly spaced. What is the distance from the top of the
cabinet to each of the shelves along the hypotenuse, to the nearest inch?

20. **Error Analysis** The statements, "If a, then b; if c, then b; and a" are all true
(21) statements. A student incorrectly concluded, "c is a true statement." What
mistake did the student make? What conclusion should have been made?

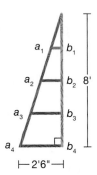

21. What is the measure of each exterior angle in a regular dodecahedron?
(Inv 3)

22. Justify An exterior angle of a triangle measures 136°. One remote interior angle
(18) measures 56°. Determine the other two interior angle measures, justifying each step in your reasoning.

 23. Algebra In the diagram, m$\widehat{TU}$ = (5x + 10)°, m$\widehat{UV}$ = (3x + 12)° and
(26) m$\widehat{TV}$ = (6x + 50)°. What is the value of x?

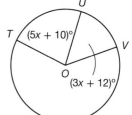

24. Write Describe how to write the equation of a line, given that its slope is
(16) −1 and that the point (4, 5) lies on it.

Use the diagram to answer the following questions.

*25. Given that AP = 123, determine the length of $\overline{AD}$.
(32)

*26. Given that BD = 63, determine the length of $\overline{DQ}$.
(32)

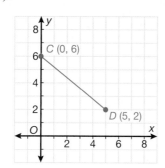

27. (**Gardening**) Julian installed a total of 10 yards of fence which
(29) runs along the perimeter of his square garden. He decides to run a watering hose between two opposite corners. What length of hose will he need, to the nearest tenth of a yard?

28. If m$\widehat{AB}$ = 23° and m$\widehat{BC}$ = 33°, what is the measure of $\widehat{AC}$ if $\widehat{AB}$
(26) and $\widehat{BC}$ are non-overlapping, adjacent arcs?

29. Determine the midpoint M of the line segment $\overline{CD}$ with endpoints
(11) C(0, 6) and D(5, 2).

30. What is the circumference of a circle with a radius measure of 7x? Express
(23) your answer in terms of π.

Properties of Parallelograms

Warm Up

1. **Vocabulary** The _____ of a line divides a segment into two
(2) congruent segments.

2. Find the equation of line $\overline{ST}$.
(16)

3. What is the name for an equiangular and
(19) equilateral quadrilateral?

A parallelogram   **B** rhombus

C square   **D** trapezium

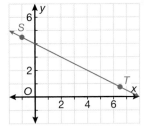

New Concepts

Squares, rhombuses, and rectangles are all types of parallelograms. All of them share some basic properties of parallelograms.

Exploration **Exploring Diagonals of Parallelograms**

1. Draw a parallelogram $ABCD$. Trace it into a piece of patty paper and label the new figure $QRST$. Notice that $ABCD \cong QRST$.

2. Lay $QRST$ over $ABCD$ so that $\overline{ST}$ overlays $\overline{AB}$. What do you notice about their lengths? What do you suppose is the relationship between $\overline{AB}$ and $\overline{CD}$? What does this suggest about $\overline{AD}$ and $\overline{BC}$?

3. Lay $QRST$ over $ABCD$ so that $\angle S$ overlays $\angle A$. What do you notice about their measures? What do you suppose is the relationship between $\angle A$ and $\angle C$? What does this suggest about $\angle B$ and $\angle D$?

4. Draw diagonals $\overline{AC}$ and $\overline{BD}$. Fold $ABCD$ so that A overlays C, making a crease which represents points that are equidistant from A and C. Unfold the paper and fold it again so that B overlays D, making another crease. What do you notice about the creases? What can you conclude about the diagonals?

> **Math Reasoning**
>
> **Generalize** In what kind of parallelogram are adjacent sides congruent? ... adjacent angles?

Properties of Parallelograms

1. If a quadrilateral is a parallelogram, then its opposite angles are congruent.

$\angle J \cong \angle L$ $\angle M \cong \angle K$

2. If a quadrilateral is a parallelogram, then its consecutive angles are supplementary.

m$\angle S$ + m$\angle T$ = 180°   m$\angle T$ + m$\angle U$ = 180°
m$\angle U$ + m$\angle V$ = 180°   m$\angle V$ + m$\angle S$ = 180°

3. If a quadrilateral is a parallelogram, then its opposite sides are congruent.

$\overline{AD} \cong \overline{BC}$ $\overline{AB} \cong \overline{DC}$

4. If a quadrilateral is a parallelogram, then its diagonals bisect each other.

$NE = EP$ $QE = EO$

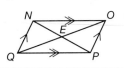

Example 1 Proving Opposite Angles of a Parallelogram Are Congruent

Given: $WXYZ$ is a parallelogram.
Prove: $m\angle W = m\angle Y$, $m\angle X = m\angle Z$

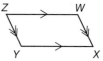

SOLUTION

Statements	Reasons
1. $WXYZ$ is a parallelogram	1. Given
2. $\overline{WX} \parallel \overline{ZY}$ and $\overline{WZ} \parallel \overline{XY}$	2. Definition of parallelograms
3. $m\angle W + m\angle X = 180°$ $m\angle Z + m\angle Y = 180°$	3. Same-side interior angles are supplementary
4. $m\angle W + m\angle Z = 180°$ $m\angle X + m\angle Y = 180°$	4. Same-side interior angles are supplementary
5. $m\angle X + m\angle Y = m\angle W + m\angle X$	5. Substitution Property of Equality
6. $m\angle Y = m\angle W$	6. Subtraction Property of Equality
7. $m\angle Z + m\angle Y = m\angle X + m\angle Y$	7. Substitution Property of Equality
8. $m\angle Z = m\angle X$	8. Subtraction Property of Equality

Example 2 Proving Consecutive Angles of a Parallelogram Are Supplementary

Given: $WXYZ$ is a parallelogram.
Prove: $\angle W$ and $\angle X$, $\angle Z$ and $\angle Y$, $\angle W$ and $\angle Z$, and $\angle X$ and $\angle Y$ are supplementary.

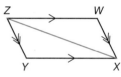

SOLUTION

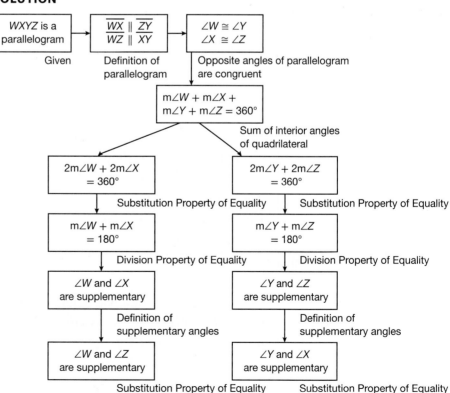

Example 3 **Proving Opposite Sides of a Parallelogram Congruent**

Given: $WXYZ$ is a parallelogram.
Prove: $\overline{WZ} \cong \overline{XY}, \overline{WX} \cong \overline{ZY}$

SOLUTION
$WXYZ$ is a parallelogram, so $\overline{WZ} \parallel \overline{XY}$ and $\overline{WX} \parallel \overline{ZY}$. The diagonal $\overline{ZX}$ is congruent to itself by the Reflexive Property. Since $\overline{ZX}$ is a transversal for both sets of parallel lines, $\angle WXZ \cong \angle XZY$ and $\angle ZXY \cong \angle WZX$. Since two angles and the included side of $\triangle WZX$ are congruent to two angles and their included side in $\triangle YXZ$, $\triangle WZX \cong \triangle YXZ$ by ASA Triangle Congruence. By CPCTC, $\overline{WZ} \cong \overline{XY}$ and $\overline{WX} \cong \overline{ZY}$.

Math Reasoning

Model Draw a parallelogram, a rhombus, a square, and a rectangle. Which of these parallelograms, if any, appear to have congruent diagonals?

Example 4 **Finding Unknown Measures of a Parallelogram**

$PQRS$ is a parallelogram.

a. Find the value of x.

SOLUTION
Opposite sides of parallelograms are congruent. Therefore, $PQ = SR$.
$3x + 10 = 7x - 8$
$x = 4.5$

b. Find the value of y.

SOLUTION
Diagonals of a parallelogram bisect each other. Therefore, $PT = TR$.
$\frac{1}{2}y + 10 = 2y - 2$
$y = 8$

Hint

When finding unknown values in a parallelogram, first think about each of the four properties of parallelograms and decide which one could apply to the problem.

Example 5 **Application: Farming**

In Kansas, the Highway Department builds gravel roads to evenly divide parcels of land. If a farmer has a plot of land shaped like a parallelogram with a diagonal of 1 mile, as shown, calculate the values of x and y to the nearest hundredth.

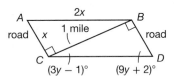

SOLUTION
Use the Pythagorean Theorem to solve for x.
$x^2 + 1^2 = (2x)^2$
$x^2 + 1^2 = 4x^2$
$0.58 \approx x$

Consecutive angles of a parallelogram are supplementary. Therefore, $(m\angle ACB + m\angle BCD) + m\angle CDB = 180°$.
$90 + (3y - 1) + (9y + 2) = 180$
$91 + 12y = 180$
$12y = 89$
$y \approx 7.42$

a. Find the value of y in parallelogram *ABCD*.
(Ex 1, 4)

b. Find the values of x and z in parallelogram *RSTU*.
(Ex 2, 4)

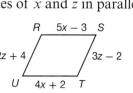

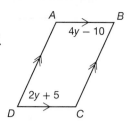

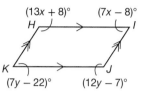

c. Find the values of x and y in parallelogram *HIJK*.
(Ex 3, 4)

d. Prove Property 4 of parallelograms: if a quadrilateral is a parallelogram, then its diagonals bisect each other. Justify your reasoning in a paragraph proof, and draw an example.
(Ex 3, 4)

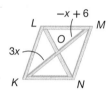

e. An old fence gate, shown here, is starting to lean. Find each measure.
(Ex 5)
   1. *KM*
   2. *KO*

Practice Distributed and Integrated

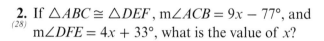

* 1. **Multi-Step** Figure *KLMN* is a parallelogram. Find the length of the diagonals.
(34)

2. If $\triangle ABC \cong \triangle DEF$, m$\angle ACB = 9x - 77°$, and m$\angle DFE = 4x + 33°$, what is the value of x?
(28)

3. **Analyze** Classify this quadrilateral, and explain how you know.
(19)

4. Draw a valid conclusion from these conditional statements:
(21)

If Molly does her homework, then she learns the material.
If she learns the material, then she will do well on the test.
If she does well on the test, then she will get a good grade in the course.

5. **Justify** In the diagram, $AC = BC$, and $DC = BC$. Prove that m$\angle BCD = 2x + 2y$.
(27)

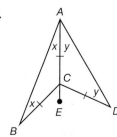

* 6. (**Construction**) In order to place a square toolshed in his backyard, Landon measures
(22) an area with a diagonal length of 12 feet. What area of the yard will the shed occupy?
Hint: Use the Pythagorean Theorem to find the length of one side of the shed.

7. Multiple Choice Consider the following statements.
(17)

> *If Martha wears jeans, then it is the weekend.*
> *If it is the weekend, then Martha wears jeans.*

The second statement is the _____ of the first.
A inverse **B** negation
C contrapositive **D** converse

8. Write a flowchart proof.
(31) **Given:** ∠2 and ∠3 are supplementary.
Prove: ∠1 and ∠3 are supplementary.

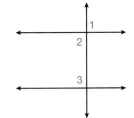

9. What is the sum of all the interior, exterior, and central angles in a convex hexagon?
(Inv 3)

10. Write Explain why the measure of an exterior angle of a triangle is equal to the
(18) sum of its two remote interior angles.

*11. Briefly state the four properties of parallelograms.
(34)

12. What is the area of a rectangle with a length of 4 and a diagonal length of 5?
(22)

13. Error Analysis For the contrapositive of the statement, "If a bus is green, then it
(17) is a downtown bus," Keisha has written, "If a bus is not green, then it is not a
downtown bus." Identify any errors Keisha has made.

14. Multiple Choice Which of these statements describes these triangles?
(30) **A** $\triangle VWU \cong \triangle ZYX$ **B** $\triangle XYZ \cong \triangle WVU$
C $\triangle WUV \cong \triangle ZXY$ **D** $\triangle YZX \cong \triangle UWV$

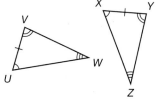

15. (**Irrigation**) In order to water plants efficiently, many farmers use a long
(23) metal pole which sweeps over a circular area, spraying water on the
crops below. If the radius of one such device is 150 feet, what is the
area of the circle that gets watered, to the nearest ten square feet?

*16. Multiple Choice Which of the following sets of numbers represents the leg lengths of
(33) an acute triangle?
A 7, 4, 8 **B** 4, $2\sqrt{2}$, 5 **C** 9, 7, $2\sqrt{3}$ **D** 14, 15, 4

17. (**Chemistry**) Linear alkanes are a class of organic molecules with only hydrogen (H)
(16) and carbon (C) atoms. This graph plots the number of carbon atoms of some
linear alkanes on the *x*-axis and the number of hydrogen atoms on the *y*-axis.
What is the equation of the line connecting them?

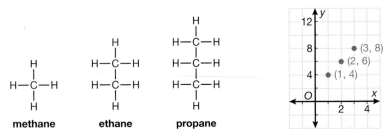

methane   ethane   propane

18. Write Quadrilateral *WXYZ* has two pairs of parallel sides. Describe the steps
⁽¹⁹⁾ needed to sketch *WXYZ*, and then sketch it.

*19. *HIJK* is a parallelogram. Find the measure of each
⁽³⁴⁾ angle.

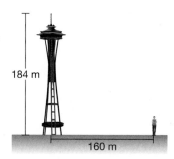

20. Name the included angle between the sides $\overline{RS}$ and $\overline{TR}$ of
⁽²⁸⁾ $\triangle RST$.

*21. (**Architecture**) The Seattle Space Needle is 184 meters tall. If
^(29, 33) a person is standing 160 meters away from the base, how far is
the person from the top, to the nearest meter?

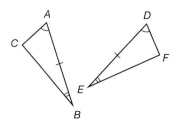

*22. Generalize Are there any proofs that can be done as a flowchart
⁽³¹⁾ proof but cannot be proved with a paragraph proof? If so, give an
example. If not, explain why.

23. Algebra In $\triangle ABC$, $m\angle BAC = 67°$, $AB = (5x - 7)$, and
⁽³⁰⁾ $m\angle ABC = 23°$. In $\triangle DEF$, $m\angle EDF = 67°$, $EF = (2x + 4)$,
$DE = (3x + 1)$, and $m\angle DEF = 23°$. What is the length of $\overline{EF}$?

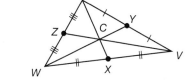

24. Algebra A right triangle has a hypotenuse of 15 inches, with
⁽²⁹⁾ one leg twice the length of the other leg. What is length of
each leg, to the nearest tenth of an inch?

25. Justify Solve the equation $3(x - 2) = 4(x + 1)$ with a justification for each step.
⁽²⁴⁾

*26. Three vertices of parallelogram *ABCD* are $A(-1, 5)$, $B(4, 5)$ and $C(3, 2)$. Find the
⁽³⁴⁾ coordinates of *D*.

Use the diagram to find the following measures.

27. Given that $CX = 18$, determine the length of $\overline{UX}$.
⁽³²⁾

28. Given that $ZV = 72$, determine the length of $\overline{ZC}$.
⁽³²⁾

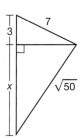

29. Algebra Consider the following statement.
⁽²⁰⁾

If two positive numbers are less than 1, then their product is positive and less than 1.

a. Rewrite the statement using inequalities.

b. State the converse and biconditional of your statement from part **a**. Is the
biconditional true? Explain why or why not.

*30. Algebra Determine the value of *x*. Write your answer in simplified
⁽³³⁾ radical form.

Finding Arc Lengths and Areas of Sectors

Warm Up

1. **Vocabulary** A segment that has both endpoints on a circle and passes
   (23) through the center of that circle is the _____ of the circle.

2. Find the area and circumference of a circle with a 5-centimeter radius, to
   (23) the nearest hundredth.

3. What is the radius of a circle with an area of 40 square units, to the
   (23) nearest hundredth?

4. What is the area of a semicircle with a diameter of 10 inches, to the
   (23) nearest hundredth?

New Concepts Recall from Lesson 26 that minor and major arcs of a circle are parts of the
circle's circumference. The length of these arc segments can be determined if
the circle's radius and the arc's degree measure are known. The **arc length** of
a circle is the distance along an arc measured in linear units.

Math Reasoning

Formulate What parts
of the arc length formula
do you recognize? What
does $\left(\frac{m^\circ}{360^\circ}\right)$ represent?

Arc Length
To find the length of an arc, use this formula, where m is the degree measure of the arc. $$L = 2\pi r\left(\frac{m^\circ}{360^\circ}\right)$$ 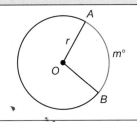

Example 1 **Finding Arc Length**

Find each arc length. Give your answer in terms of π.

a. Find the length of $\overparen{XY}$.

SOLUTION

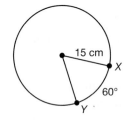

$L = 2\pi r\left(\frac{m^\circ}{360^\circ}\right)$

$L = 2\pi(15)\left(\frac{60^\circ}{360^\circ}\right)$

$L = 5\pi$ cm

b. Find the length of an arc with a measure of 75° in a circle
with a radius of 4 feet.

SOLUTION

$L = 2\pi r\left(\frac{m^\circ}{360^\circ}\right)$

$L = 2\pi(4)\left(\frac{75^\circ}{360^\circ}\right)$

$L = \frac{5}{3}\pi$ ft

Online Connection
www.SaxonMathResources.com

Every arc on a circle encompasses a portion of the circle's interior. The region inside a circle bounded by two radii of the circle and their intercepted arc is known as a **sector of a circle**. Finding the area of a sector is similar to finding arc length.

Area of a Sector
To find the area of a sector (A), use the following formula, where r is the circle's radius and m is the central angle measure: $$A = \pi r^2 \left(\frac{m°}{360°}\right)$$ 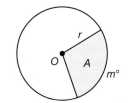

Math Reasoning

Formulate What part of the formula for area of a sector do you recognize?

Example 2 Finding the Area of a Sector

Find the area of each sector. Give your answer in terms of π.

(a.) Find the area of sector XOY.

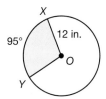

SOLUTION

$A = \pi r^2 \left(\frac{m°}{360°}\right)$

$A = \pi (12)^2 \left(\frac{95°}{360°}\right)$

$A = 38\pi \text{ in}^2$

(b.) Find the area of a sector with an arc that measures 174° in a circle with a radius of 13 meters.

$A = \pi r^2 \left(\frac{m°}{360°}\right)$

$A = \pi (13)^2 \frac{174°}{360°}$

$A = \frac{4901}{60} \pi \text{ m}^2$

Example 3 Solving for Unknown Radius

Find the radius of the circle to the nearest hundredth of a meter.

SOLUTION

Substitute the known measures into the formula for the area of a sector, then solve for r.

$A = \pi r^2 \left(\frac{m°}{360°}\right)$

$100 = \pi r^2 \left(\frac{246°}{360°}\right)$

$r \approx 6.83 \text{ m}$

Example 4 Solving for Unknown Central Angle

Find the central angle measure of $\overset{\frown}{RS}$ to the nearest hundredth of a degree, if the length of the arc is 12 centimeters.

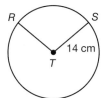

SOLUTION

$$L = 2\pi r\left(\frac{m°}{360°}\right)$$

$$12 = 2\pi(14)\left(\frac{m°}{360°}\right)$$

$$m° \approx 49.11°$$

Example 5 Application: Farming

A spray irrigation system has a radius of 150 feet. If it rotates through a 175° central angle, what is the area that the system covers? Round your answer to the nearest square foot.

SOLUTION

$$A = \pi r^2\left(\frac{m°}{360°}\right)$$

$$A = \pi(150)^2\left(\frac{175°}{360°}\right)$$

$$A \approx 34,361 \text{ ft}^2$$

Hint

When you are finding the area of a sector, do not forget to express the answer in square units.

Lesson Practice

a. Find the length of an arc with a measure of 125° in a circle and
(Ex 1) 12-mile radius. Round to the nearest hundredth of a mile.

b. Find the area of the sector to the nearest hundredth square inch.
(Ex 2)

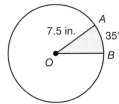

c. Find the radius to the nearest hundredth of a centimeter.
(Ex 3)

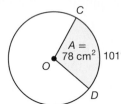

d. If a farmer wants his irrigation system to cover an area of 2 square
(Ex 5) miles, and his sprinkler rotates through 50°, what is the diameter of his circular field to the nearest hundredth of a mile?

✎ * **1.** **Write** Explain how the formula for arc length is similar to the formula for the area
(35)   of a circle.

* **2.** **Model** Find the centroid of △HIJ with vertices at $H(-6, 10)$, $I(-4, 2)$, and
(32) $J(-14, 6)$.

3. Given that △$BCD \cong$ △EFG, write the six congruence statements
(25)   for the triangles.

4. **Analyze** What conclusion can be drawn from these statements?
(21)
   *If my computer is not working properly, then I will restart it.
   My computer is not working properly.*

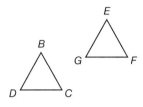

* **5.** Three vertices of parallelogram $HIJK$ are $H(0, 10)$, $I(2, 10)$, and $K(-2, 0)$. Find
(34)   the coordinates of J.

6. ⬡ **Visual Arts** An artist is making a motif design using the two triangles
(18)   shown. The measurements must be exact in order to make the
   repeating pattern of the motif. What are the measurements of
   x, y, and z?

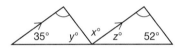

7. **Multiple Choice** A minor arc has a measure of $(5x + 20)°$. Its corresponding major
(26)   arc has a measure of $(8x - 50)°$. What is the correct value of x?

   A 23.3   **B** 30
   C 170   **D** 190

8. **a.** State the converse of the following statement,
(20)   *If a triangle is obtuse, then it has exactly two acute angles.*

   b. Determine whether the statement is true, and whether its
      converse is true.

* **9.** $QRST$ is a parallelogram. Find the length of the diagonals $\overline{TR}$ and $\overline{QS}$.
(34)

10. ⬡ **Carpentry** Margot is making a coffee table with a top in the shape of
(Inv 3)   a regular hexagon. What measure should she make each angle of the
   table top?

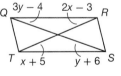

xy^2 **11.** **Algebra** Solve the equation $x + 3 = \frac{4x + 5}{2}$, and justify each step.
(24)

12. Find the unknown length in the triangle. Are these side lengths a
(29)   Pythagorean triple?

13. Prove Theorem 26-1: In the same or congruent circles,
(26)   congruent arcs have congruent central angles.
   Given: $\overset{\frown}{JK} \cong \overset{\frown}{HI}$
   Prove: $\angle 1 \cong \angle 2$

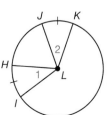

14. The side lengths of a triangle are $4\sqrt{2}$, $4\sqrt{3}$, and 9. Is it
(33)   a right triangle?

15. (Calendars) A page of a calendar is 10 inches by $12\frac{1}{4}$ inches. The page is divided
$^{(22)}$ into 5 rows of 7 congruent boxes. What is the area of each of these boxes?

16. Write a two-column proof.
$^{(27)}$ **Given:** $\angle KLM$ and $\angle NML$ are right angles and $\angle 2 \cong \angle 3$.
Prove: $\angle 1 \cong \angle 4$

*17. **Error Analysis** Two students tried to find the area of the shaded region
$^{(35)}$ in the circle shown. Which solution is incorrect? Explain where the
error was made.

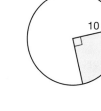

Marc

$A = 2\pi r\left(\frac{m°}{360°}\right)$

$A = 2\pi(10)\left(\frac{90°}{360°}\right)$

$A = 5\pi$

$A \approx 15.71$

Stephanie

$A = \pi r^2\left(\frac{m°}{360°}\right)$

$A = \pi(10)^2\left(\frac{90°}{360°}\right)$

$A = 25\pi$

$A \approx 78.54$

18. What conclusion can be drawn from these statements?
$^{(21)}$

If I lose my driver's license, then I need to go get a new one.
If I need to go get a new driver's license, then I will have to go to the DMV.

*19. (Games) D.J. is designing a spinner for a game. He wants the spinner to have
$^{(35)}$ a $\frac{3}{8}$ probability of landing on a particular sector. What size arc should he assign to
this sector?

*20. These triangles are congruent. What is the perimeter of $\triangle LMN$?
$^{(30)}$

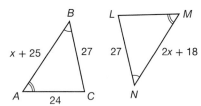

21. State the disjunction of the statements, "A triangle is acute," and
$^{(20)}$ "A triangle has exactly two acute angles." Is the disjunction true or false?

*22. **Multi-Step** The major arc of a circle measures 240° and has a length of
$^{(35)}$ 18 centimeters. What is the length of the minor arc?

23. **Justify** Alejandro made the conjecture, "For every integer x, $x^2 + 2x - 1$
$^{(14)}$ is divisible by 2." Is the conjecture true or false? If it is false, provide a
counterexample.

24. Given that $\triangle QSR \cong \triangle TUV$, find the area of each triangle.
(30)

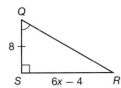 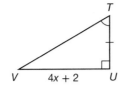

25. Write the converse of the following statement.
(17)

If a triangle is a right triangle, then it has two acute angles.

*26. **Algebra** Find the radius of a circle with an arc length of 40 centimeters
(35) and an arc measure of 60°.

*27. Is $\triangle EFG$ an obtuse, acute, or a right triangle?
(33)

28. **Verify** For a circle with a radius of 7 inches, Vanessa calculated the area to be about
(23) 43.98 square inches. Is Vanessa's calculation accurate? If not, what is the actual
area of the circle to the nearest hundredth?

29. Give two Pythagorean triples that are related to the triple (5, 12, 13).
(29)

30. *ABCD* is a parallelogram. Find the measure of each angle.
(34)

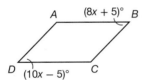

Right Triangle Congruence Theorems

Warm Up

1. **Vocabulary** The _____ Triangle Congruence Postulate states
   (28)
   that if two sides and the included angle of one triangle are congruent to
   two corresponding sides and the included angle of another triangle, then
   the triangles are congruent.

2. In △*UVW*, what is the measure of ∠*W*?
   (30)

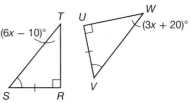

3. **Multiple Choice** If two triangles are congruent, then they have
   (25)
   congruent _____.

   A sides only   **B** right angles
   C sides and angles   **D** angles only

New Concepts

There are four ways to prove triangle congruence: by the SSS Postulate, SAS
Postulate, ASA Postulate, or by the AAS Theorem. If a triangle is a right
triangle however, there are several other ways to prove congruency.

Hint

It is assumed in all right triangle congruence theorems that the measure of the right angle—90°—is already known, so it only takes two other congruent parts to prove congruency.

Theorem 36-1: Leg-Angle (LA) Right Triangle Congruence Theorem

If a leg and an acute angle of one right triangle are congruent to a leg and an acute angle of another right triangle, then the triangles are congruent.

In the diagram, △*ABC* ≅ △*DEF*.

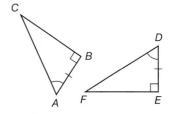

The Leg-Angle Right Triangle Congruence Theorem follows from the
ASA Postulate and the AAS Theorem. Notice that in the diagram,
marking the right angle shows that the triangles are also congruent by
the ASA Postulate.

Example 1   Using the Leg-Angle Triangle Congruence Theorem

a. Use the LA Congruence Theorem to prove
that △*GHJ* and △*KLM* are congruent.

SOLUTION
△*GHJ* and △*KLM* are both right triangles, so
the LA Right Triangle Congruence Theorem
can be used. The legs $\overline{GH}$ and $\overline{KL}$ are
congruent as given. Acute angles ∠*G* and ∠*K* are also congruent.
Therefore, by the LA Congruence Theorem, △*GHJ* ≅ △*KLM*.

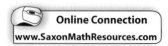

Online Connection
www.SaxonMathResources.com

Theorem 36-2: Hypotenuse-Angle (HA) Right Triangle Congruence Theorem

If the hypotenuse and an acute angle of one right triangle are congruent to the hypotenuse and an acute angle of another right triangle, then the triangles are congruent.

In the diagram, $\triangle GHI \cong \triangle DEF.$

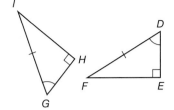

Example 2 **Using and Proving the Hypotenuse-Angle Triangle Congruence Theorem**

a. Use the HA Congruence Theorem to prove that $\triangle MNO \cong \triangle PQR.$

SOLUTION
The diagram tells us that both triangles are right triangles. In addition, they have congruent hypotenuses and acute $\angle M$ is congruent to acute $\angle P$. Therefore, by the HA Triangle Congruence Theorem, $\triangle MNO \cong \triangle PQR.$

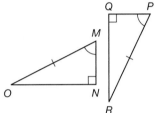

b. Use a paragraph proof to prove the HA Triangle Congruence Theorem.

SOLUTION
Sketch a diagram like the one shown. By the Right Angle Congruence Theorem, $\angle D \cong \angle A$. It is given that $\overline{BC} \cong \overline{EF}$ and $\angle E \cong \angle B$. From the diagram, two angles and a non-included side of $\triangle ABC$ are congruent to two angles and a non-included side of $\triangle DEF$. Therefore, by the AAS Triangle Congruence Theorem, $\triangle ABC \cong \triangle DEF.$

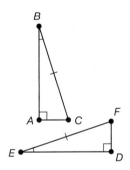

Hint

Unless the question specifies, choose whichever proof method seems easiest. This paragraph proof is compact, but a two-column proof might be easier to follow.

Theorem 36-3: Leg-Leg (LL) Right Triangle Congruence Theorem

If the two legs of one right triangle are congruent to the two legs of another right triangle, then the triangles are congruent.

In the diagram, $\triangle RST \cong \triangle UVW.$

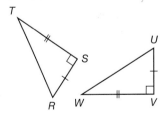

Example 3 Using the Leg-Leg Triangle Congruence Theorem

Use the LL Congruence Theorem to prove that $\triangle JKL \cong \triangle RST$.

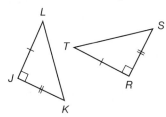

SOLUTION

Statements	Reasons
1. $\triangle JKL$ and $\triangle RST$ are right triangles	1. Given
2. $\overline{JL} \cong \overline{RT}$ and $\overline{JK} \cong \overline{RS}$	2. Given
3. $\triangle JKL \cong \triangle RST$	3. LL Triangle Congruence Theorem

Theorem 36-4: Hypotenuse-Leg (HL) Right Triangle Congruence Theorem

If the hypotenuse and a leg of one right triangle are congruent to the hypotenuse and a leg of another right triangle, then the triangles are congruent.

In the diagram, $\triangle HIJ \cong \triangle MNO$.

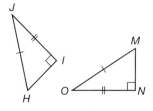

Math Reasoning

Analyze In order to prove right triangle congruence by the SAS, ASA, AAS, or SSS theorems, the right angle is considered part of the proof. How is the HL Theorem different?

Example 4 Using the Hypotenuse-Leg Congruence Theorem

a. In $\triangle UVW$ and $\triangle YZX$, $\angle U$ and $\angle Y$ are right angles. Use the HL Congruence Theorem to prove that $\triangle UVW \cong \triangle YZX$.

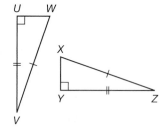

SOLUTION

Statements	Reasons
1. $\triangle UVW$ and $\triangle YZX$ are right triangles	1. Given
2. $\overline{VW} \cong \overline{ZX}$ and $\overline{UV} \cong \overline{YZ}$	2. Given
3. $\triangle UVW \cong \triangle YZX$	3. HL Triangle Congruence Theorem

b. Prove the Hypotenuse-Leg Triangle Congruence Theorem.

Given: $\triangle ABC$ and $\triangle DEF$ are right triangles. $\overline{AC} \cong \overline{FD}$ and $\overline{BC} \cong \overline{EF}$.

Prove: $\triangle ABC \cong \triangle DEF$

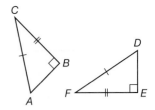

SOLUTION

Statements	Reasons
1. $\triangle ABC$ and $\triangle DEF$ are right triangles	1. Given
2. $\overline{AC} \cong \overline{DF}$ and $\overline{BC} \cong \overline{EF}$	2. Given
3. $AC = DF$ and $BC = EF$	3. Definition of congruent line segments
4. $AC^2 = AB^2 + BC^2$ $DF^2 = DE^2 + EF^2$	4. Pythagorean Theorem
5. $AB = \sqrt{AC^2 - BC^2}$ $DE = \sqrt{DF^2 - EF^2}$	5. Solve for AB and DE
6. $DE = \sqrt{AC^2 - BC^2}$	6. Substitution Property of Equality
7. $DE = AB$	7. Transitive Property
8. $\triangle ABC \cong \triangle DEF$	8. SSS Triangle Congruence Postulate

Example 5 **Application: Engineering**

Rachel must design a plastic cover to fit exactly over the metal plate shown below. The cover will contain a right angle. Rachel knows that she only needs to pick two other dimensions to make sure that the cover is congruent to the plate. List all the pairs of dimensions Rachel could use to ensure the cover is exactly the same size and shape as the metal plate. For each pair of dimensions, write which right triangle congruence theorem applies.

SOLUTION

QR and m$\angle P$	(LA)	PQ and m$\angle R$	(LA)
PQ and m$\angle P$	(LA)	QR and m$\angle R$	(LA)
PR and m$\angle P$	(HA)	PR and m$\angle R$	(HA)
PQ and QR	(LL)	PR and PQ	(HL)
PR and QR	(HL)		

Use the diagram to answer problems a through d.

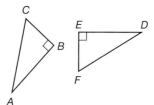

a. Suppose $\overline{AB} \cong \overline{DE}$ and $\angle A \cong \angle D$.
(Ex 1) Use the LA Triangle Congruence
Theorem to prove that $\triangle ABC \cong \triangle DEF$.

b. Suppose $\overline{AC} \cong \overline{DF}$ and $\angle A \cong \angle D$. Use the
(Ex 2) HA Triangle Congruence Theorem to prove
that $\triangle ABC \cong \triangle DEF$.

c. Suppose $\overline{AB} \cong \overline{DE}$ and $\overline{BC} \cong \overline{EF}$. Use the LL Triangle Congruence
(Ex 3) Theorem to prove that $\triangle ABC \cong \triangle DEF$.

d. Suppose $\overline{AC} \cong \overline{DF}$ and $\overline{BC} \cong \overline{EF}$. Use the HL Congruence Theorem
(Ex 4) to prove that $\triangle ABC \cong \triangle DEF$.

e. (**Engineering**) Refer to Example 5. Suppose Rachel provides
(Ex 5) $PR = 14.2$ centimeters and $QR = 8.9$ centimeters as dimensions for the
plastic cover. In this case, which theorem proves that the cover will fit
the metal plate?

Practice **Distributed and Integrated**

* 1. Using the diagram, prove that $\triangle OPQ \cong \triangle TRS$. Name any theorems
(36) used.

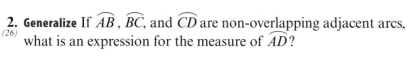

2. **Generalize** If $\overparen{AB}$, $\overparen{BC}$, and $\overparen{CD}$ are non-overlapping adjacent arcs,
(26) what is an expression for the measure of $\overparen{AD}$?

3. (**Arts and Crafts**) The diagram shows a square
(19, 36) piece of paper that is folded to make a kite.
The kite is composed of four triangles.
Prove that the triangles labeled 1 and 2
are congruent.

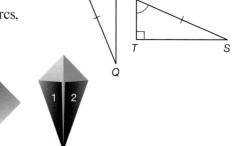

* 4. **Error Analysis** Greta is calculating the area of a sector of a circle with a diameter
(35) of 5. The sector covers 24°. What error has she made?

$$\left(\frac{24°}{360°}\right)\pi(5)^2 = \left(\frac{1}{15}\right)(\pi)(25)$$

$$\frac{5}{3}\pi \approx 5.24$$

5. In a right triangle, the length of one leg is 1.2 yards and the length of the
(29) hypotenuse is 1.3 yards. What is the length of the third side?

* 6. Use the Leg-Leg Congruence Theorem to prove that $\triangle JKL \cong \triangle MNO$.
(36)

7. Calculate the circumference of the circle to the
(26) nearest centimeter.

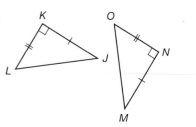

* 8. a. Write the formula for the area of a trapezoid.
(22) **b.** Transform the formula to solve for height.

9. Write a paragraph proof showing that if point X is equidistant from points A and
(31) B in the triangle shown, then it lies along the perpendicular bisector of $\overline{AB}$.
Hint: Draw a line through X *that is perpendicular to* $\overline{AB}$ *at Y.*

10. Calculate the area of the circle shown to the nearest hundredth.
(23)

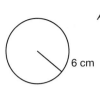

*11. (**Catering**) The catering service is dividing a small pizza,
(35) with a diameter of 8 inches, into four slices. What is the
area of each slice, to the nearest tenth?

12. **Algebra** If $m\overset{\frown}{AB} = (5x + 11)°$ and $m\overset{\frown}{CD} = (7x - 9)°$, what is the value of x
(26) if $\overset{\frown}{AB}$ and $\overset{\frown}{CD}$ are congruent?

13. Using information given in the diagram, determine the area of
(30) each triangle.

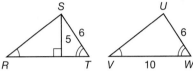

14. **Multiple Choice** Which statement about an equiangular triangle is not true?
(18)

    A Each pair of interior angles   **B** The sum of the interior angle
       is complementary.   measures is 180°.

    C Each exterior angle measures 120°.   **D** Each interior angle measures 60°.

15. Find the value of h in parallelogram $STUV$.
(34)

*16. In the triangle formed by the points $(2, 2)$, $(4, 0)$, and the origin, what is the
(32) equation of the line containing the median passing through the origin?

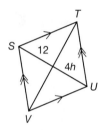

17. In $\triangle ABC$, $AB = 15$, $BC = 20$, and $AC = 25$. In $\triangle DEF$, $DE = 25$, $EF = 15$, and
(25) $DF = 20$. Write the congruency statement for the triangles.

*18. What is the arc measure of one-sixth of a circle? What is the area of one-sixth
(35) of a circle that has a radius of 12, in terms of π?

19. In the diagram given, prove that $\triangle ABE \cong \triangle CDE$.
(30)

20. **Write** If in $\triangle JKL$, $JK = 9$, $KL = 11$, and $JL = 12$, and
(25) in $\triangle PQR$, $PQ = 9$, $QR = 11$, and $PR = 11$, explain why
$\triangle JKL$ and $\triangle PQR$ are not congruent.

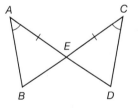

*21. Complete the following proof showing that if a point in the interior of an angle is
(27) equidistant from the sides of the angle, then it is on the bisector of the angle.
    Given: $\overline{VX} \perp \overline{YX}$, $\overline{VZ} \perp \overline{YZ}$, $VX = VZ$
    Prove: V is on the bisector of $\angle XYZ$.

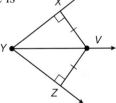

Statements	Reasons
1. $\overline{VX} \perp \overline{YX}$, $\overline{VZ} \perp \overline{YZ}$, $VX = VZ$	1. Given
2.	2. Definition of perpendicular
3.	3. Reflexive Property
4. $\triangle YXV \cong \triangle YZV$	4.
5. $\angle XYV \cong \angle ZYV$	5.
6. $\overrightarrow{YV}$ bisects $\angle XYZ$	6. Definition of angle bisector

22. Find the value of x in the parallelogram.
(34)

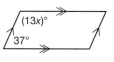

23. (Gardening) A gardener tried to plant a garden in the shape of a right triangle.
(33) The longest side was 14 feet long and the two shorter sides were 5 feet and 13 feet long, respectively. Did the gardener succeed in making the garden a right triangle? If not, what kind of triangle describes the shape of the garden?

24. Write Use the SAS Congruence Postulate to show that a diagonal of a square
(28) divides the square into two congruent triangles.

25. For what values of x are these two triangles congruent?
(28)

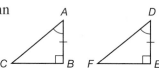

26. (Biology) Consider the following statements.
(20)

An organism is an animal.
An organism has leaves.

a. Make a truth table for the disjunction of these statements.
b. Explain why the disjunction is false.

*27. Use the Leg-Angle Congruence Theorem to prove that $\triangle JKL \cong \triangle MNO$.
(36)

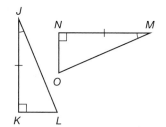

28. Verify Naomi measures the sides of a triangle as 5, 8, and 9 units,
(33) respectively, and declares the triangle to be acute. Is she correct? How do you know?

*29. Analyze Copy and complete this two-column
(27, 36) proof of the LA Congruence Theorem.

Statements	Reasons
1. $\triangle ABC$ and $\triangle DEF$ are right triangles	1. Given
2.	2. Given
3. $\angle A \cong \angle D$	3.
4. $\angle B$ and $\angle E$ are right angles	4.
5.	5.
6. $\triangle ABC \cong \triangle DEF$	6.

30. Write Explain why only two altitudes of a triangle need to be drawn to find the
(32) orthocenter of the triangle.

LESSON 37

Writing Equations of Parallel and Perpendicular Lines

Warm Up

1. (16) **Vocabulary** The _____ of a line is calculated by finding the rise/run.

2. (SB 19) Convert this equation from the point-slope form to slope-intercept form.
$$y - 4 = \frac{1}{2}(x + 4)$$

3. (16) **Multiple Choice** Which equation is a linear equation?

A $y = 12x + 4$ **B** $a^2 + b^2 = c^2$
C $y = x^2 - 4$ **D** $a = \pi r^2$

New Concepts   The coordinate plane provides a connection between algebra and geometry. Postulates 17 and 18 establish a simple way to find lines that are parallel or perpendicular on the coordinate plane.

Math Reasoning

Justify What is the slope of a vertical line? Why do both Postulates 17 and 18 talk separately about vertical lines?

Postulate 17: Parallel Lines Postulate
If two lines are parallel, then they have the same slope. All vertical lines are parallel to each other. 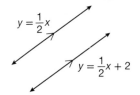

Perpendicular lines can also be found by looking at the slope.

Postulate 18: Perpendicular Lines Postulate
If two nonvertical lines are perpendicular, then the product of their slopes is -1. Vertical and horizontal lines are perpendicular to each other. 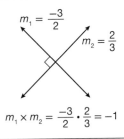

When the product of two numbers is -1, they are **opposite reciprocals**. The opposite reciprocal of a number is the reciprocal of that number with the sign reversed.

For example, the opposite reciprocal of $\frac{1}{2}$ is -2. The opposite reciprocal of -6 is $\frac{1}{6}$. Whenever two lines have slopes that are opposite reciprocals of each other, they are perpendicular lines.

Online Connection
www.SaxonMathResources.com

Lesson 37   **237**

Example 1 Finding the Slopes of Parallel and Perpendicular Lines

a. Find the slope of line a.

SOLUTION
Use the slope formula.
$$m = \frac{y_1 - y_2}{x_1 - x_2}$$

Choose any two points on the line, for example $(1, 0)$ and $(0, -3)$. Substitute the coordinates into the slope formula.
$$m = \frac{0 - (-3)}{1 - 0}$$
$$m = 3$$

b. Find the slope of a line parallel to line a.

SOLUTION
By the Parallel Lines Postulate, parallel lines have the same slope. The slope of a line that is parallel to line a is 3.

c. Find the slope of a line perpendicular to line a.

SOLUTION
By the Perpendicular Lines Postulate, the slopes of perpendicular lines are opposite reciprocals. The reciprocal of 3 is $\frac{1}{3}$. Changing the sign gives the opposite reciprocal, $-\frac{1}{3}$.

Example 2 Identifying Parallel and Perpendicular Lines

a. Are the lines $y = 2x + 4$ and $y = -3 + 2x$ parallel, perpendicular, or neither?

SOLUTION
By looking at the equations we can see that the slope of both lines is 2. Lines with the same slope are parallel, so these two lines are parallel to each other.

b. Are the lines $y = \frac{2}{3}x - 1$ and $y = \frac{3}{2}x$ parallel, perpendicular, or neither?

SOLUTION
The slope of the first line is $\frac{2}{3}$. The slope of the second line is $\frac{3}{2}$. These slopes are reciprocals of each other. They are not, however, opposite reciprocals, since both are positive. These lines are neither perpendicular nor parallel.

The point-slope formula for a line: $y - y_1 = m(x - x_1)$. Sometimes it is helpful to find a line passing through a given point that is parallel or perpendicular to another line. The point-slope formula can be used to solve problems like this, once you have discovered the slope of the parallel or perpendicular line.

Example 3 | Graphing a Line Parallel to a Given Line

Math Reasoning

Formulate How could you use the slope-intercept form of a linear equation to solve this problem, instead of the point-slope form?

a. Find a line that is parallel to $y = x + 2$ and passes through point $(3, 8)$.

SOLUTION

The slope of the given line is 1. Substitute the slope and the given point into the point-slope formula.

$$y - y_1 = m(x - x_1) \qquad \text{Point-slope formula}$$
$$y - 8 = 1(x - 3) \qquad \text{Substitute.}$$
$$y = x + 5 \qquad \text{Solve.}$$

b. Graph the parallel lines from part **a.**

SOLUTION

Both lines are now in slope-intercept form. The diagram shows both lines.

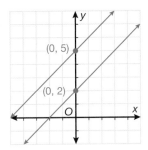

Example 4 | Graphing a Line Perpendicular to a Given Line

a. Find a line that is perpendicular to $y = \frac{2}{3}x$ and passes through the point $(2, 4)$.

SOLUTION

The slope of the given line is $\frac{2}{3}$. A perpendicular line will have a slope that is the opposite reciprocal, or $-\frac{3}{2}$. Substitute this slope and the given point into the point-slope formula.

$$y - y_1 = m(x - x_1) \qquad \text{Point-slope formula}$$
$$y - 4 = -\frac{3}{2}(x - 2) \qquad \text{Substitute.}$$
$$y = -\frac{3}{2}x + 7 \qquad \text{Solve.}$$

b. Graph the perpendicular lines from part **a.**

SOLUTION

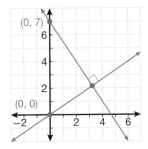

Example 5 Application: Swimming

In a race, one swimmer is swimming at a rate of 21 meters per second. Another swimmer gets a 5-meter head start, and also swims at 21 meters per second. What is the equation that will model the distance, y, that each swimmer has gone after x seconds? Will the first swimmer ever catch up to the second?

SOLUTION

Understand There are two swimmers, both swimming at a speed of 21 meters per second. One of them starts 5 meters ahead of the other. The problem asks us to write two linear equations representing this situation, and determine whether the first swimmer can ever catch up to the swimmer with the head start.

Plan The speed of each swimmer is a rate of change, or a slope. The head start of the second swimmer changes the graph's starting point, or its y-intercept. To determine whether or not the first swimmer can ever catch the second, we need to determine if the lines ever intersect.

Solve Since we know that the slope of each line is 2, and the second swimmer has a y-intercept of 5, the equations for each line are given below.

1st swimmer: $y = 2x$
2nd swimmer: $y = 2x + 5$

These lines both have a slope of 2, so they will never intersect. This tells us that the first swimmer will never catch up to the second swimmer.

Check Graph the lines $y = 2x$ and $y = 2x + 5$. They appear parallel.

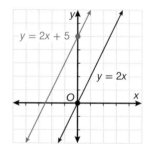

Math Reasoning

Write Explain in words why the first swimmer in this example will never be able to catch the second swimmer.

Lesson Practice

a. Find the slopes of lines that are parallel and perpendicular to line v.
(Ex 1)

b. Are the lines $y = 3x - 2$ and $3y + x = 6$ parallel, perpendicular, or neither?
(Ex 2)

c. Are the lines $4y = 2x + 3$ and $y + 2x = 9$ parallel, perpendicular, or neither?
(Ex 2)

d. Find and graph a line that is parallel to $y = -2x + 7$ and passes through the origin.
(Ex 3)

e. Find and graph a line that is perpendicular to $y = -\dfrac{4}{3}x + 3$ and passes through the point (2, 3).
(Ex 4)

f. (**Sports**) Two rowers are canoeing parallel to one another in a stream.
(Ex 5) The first rower is traveling at a rate of 2 meters per second. The second
rower is traveling at a rate of 2.1 meters per second. The first rower is
4 meters ahead of the second rower. Write an equation for each rower
representing their distance travelled. Will the second rower catch up to
the first? Why or why not?

Practice Distributed and Integrated

1. Analyze Find the unknown length in the triangle shown. Do the side
(29) lengths form a Pythagorean triple?

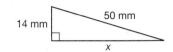

2. What is the area of a 30° sector of a circle with a radius of 9 miles?
(35) Express your answer in terms of π.

3. Kyle says that if Marcy goes to the party, then he will go to the party.
(21) Marcy said she goes to every party. Assuming both statements are true,
will Kyle go to the party?

4. Find the value of a and b in the rhombus $WXYZ$ at right.
(34)

* 5. Write the equation of the line through $(3, 7)$ that is perpendicular to
(37) $y = 3x - 4$.

6. Generalize Can the sum of the central angles for any convex polygon be
(Inv 3) determined? Explain.

7. (**Basketball**) The rim of the basket in a basketball court encloses a circle with
(23) an area of about 0.16 square meters. What is the diameter of the basket, to the
nearest hundredth of a meter?

8. Find the area of the parallelogram at right.
(22)

9. Find the orthocenter of a triangle with vertices at $(-3, 0)$, $(3, 0)$,
(32) and $(-1, 4)$.

*10. Algebra The equations of two parallel lines are $y = \dfrac{3}{k}x + 14$ and $y = \dfrac{h}{2}x + 3$. Find
(37) the values of k and h.

11. Consider the following conjecture.
(14)
If a, b, *and* c *are lines in a plane, then they divide the plane into seven regions.*

    a. What is the hypothesis of the conjecture? What is its conclusion?
    b. Find a counterexample to the conjecture.

*12. Which congruence theorem applies to these triangles?
(36)

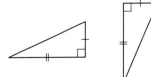

*13. Complete this two-column proof of the LL Right Triangle
(27, 36) Congruence Theorem: If two legs of a right triangle are
congruent to the two legs of another right triangle, then
the triangles are congruent.
Given: $\overline{AB} \cong \overline{DE}$, $\overline{BC} \cong \overline{EF}$, $\triangle ABC$ and $\triangle DEF$ are right triangles.
Prove: $\triangle ABC \cong \triangle DEF$

Statements	Reasons
1. $\overline{AB} \cong \overline{DE}$, $\overline{BC} \cong \overline{EF}$	1. Given
2.	2. Definition of right triangle
3. $\angle B \cong \angle E$	3.
4. $\triangle ABC \cong \triangle DEF$	4.

Complete the following statements:

14. Angle 1 is congruent to $\angle 2$ by the _____.
(Inv 1)

15. Since $\angle 1$ is congruent to $\angle 2$, and $\angle 2$ is congruent to
(Inv 1) $\angle 3$ by the Alternate Interior Angles Theorem, by
the _____, $\angle 1$ is congruent to $\angle 3$.

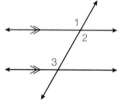

*16. Error Analysis A student says the equation of a line perpendicular to
(37) $y = \frac{12}{13}x + \frac{47}{13}$ is $y = \frac{13}{12}x + \frac{47}{13}$. Explain the student's error.

*17. (**Optics**) In this optical diagram, $\triangle ABC$ and $\triangle DBC$ are congruent.
(36) Explain why.

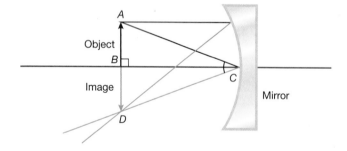

18. In $\triangle HIJ$, $HE = 12$ and $GI = 4.6$. Find GF and GE.
(32)

19. Multiple Choice Which of the following is a Pythagorean triple?
(29)
A (6, 9, 10) **B** (7, 24, 25)
C (3, 6, 9) **D** (1, 2, 3)

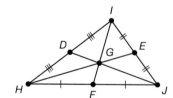

20. The radius of $\odot J$ measures 8 inches. Points K and L lie on $\odot J$. If the
(35) length of $\overline{KL}$ is 1.2 inches, what is the measure of $\overparen{KL}$ to the nearest tenth?

21. (**Design**) A wind power energy company's logo is made up of three
(19) parallelograms as shown. Copy the logo and indicate with tick
marks all pairs or sets of parallel sides.

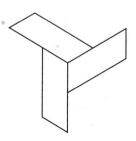

22. Parallelogram *EFGH* has adjacent side lengths of 13 and 17 units. One of its
$\quad$ diagonals is 20 units long. Could the parallelogram be a rectangle?
$\scriptstyle(33)$

23. Generalize If point *B* lies on minor arc $\overset{\frown}{AC}$, and $\overset{\frown}{AB}$ and $\overset{\frown}{BC}$ are non-overlapping
$\scriptstyle(26)$ $\quad$ adjacent arcs, what is an expression for the measure of the major arc associated
$\quad$ with $\overset{\frown}{AC}$ in terms of $\overset{\frown}{AB}$ and $\overset{\frown}{BC}$?

24. In the diagram, *PQRS* is a parallelogram. Find the value of *x* and *y*.
$\scriptstyle(34)$

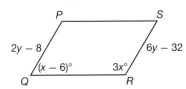

25. Design Shen is drawing a Canadian one-dollar coin, which is a regular 11-sided
$\scriptstyle(Inv\ 3)$ $\quad$ polygon. Approximately what size should he make each angle?

*26. Highways The interstate system consists of divided highways with parallel lanes
$\scriptstyle(37)$ $\quad$ of traffic. Suppose the path of the eastbound lanes is modeled by $y = \frac{4}{3}x$, and
$\quad$ the highway is intersected by a bridge that is perpendicular to the highway. Write
$\quad$ equations to model both the path of the westbound lanes and the path of the
$\quad$ bridge.

27. Coordinate Geometry Parallelogram *JKLM* on the coordinate plane has vertices at
$\scriptstyle(34)$ $\quad$ $J(-1, 2)$, $K(5, 2)$ and $L(2, 6)$. Where is the fourth vertex of *JKLM* located?

*28. Which congruence theorem applies to these triangles?
$\scriptstyle(36)$

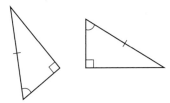

29. Formulate In any isosceles right triangle, what is the ratio of the length of the
$\scriptstyle(33)$ $\quad$ hypotenuse to the length of a leg? Would a higher ratio indicate that the triangle is
$\quad$ acute or obtuse?

30. Find the area of the figure at right, given the length of each diagonal.
$\scriptstyle(22)$

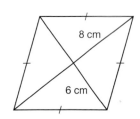

Perpendicular and Angle Bisectors of Triangles

1. **Vocabulary** A segment whose endpoints are a vertex of the triangle and
(32)   the midpoint of the side opposite the vertex is called the _____.
     (*median, altitude, base*)

2. The proof below shows that $\triangle ABC \cong \triangle ADC$.
(30)   Match each step of the proof with the correct justification.
     A justification may be used more than once.

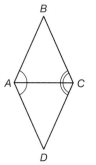

   a. $\angle BAC \cong \angle DAC$ **A** ASA Postulate
   b. $\angle BCA \cong \angle DCA$ **B** Reflexive Property
   c. $\overline{AC} \cong \overline{AC}$ **C** Given
   d. $\triangle ABC \cong \triangle ADC$

3. The measure of $\angle SRU$ is 78°. The angle bisector is $\overrightarrow{RT}$.
(3)   What is $m\angle SRT$?

New Concepts

An angle bisector divides an angle into two congruent angles. When all three angles of a triangle are bisected, the point of concurrency is called the **incenter of the triangle**. The incenter of the triangle is equidistant from all of the sides of the triangle.

Math Language

The **point of concurrency** is the point where things intersect. In this case, the point of concurrency of all three angle bisectors is the incenter of the triangle.

Example 1   Finding an Incenter in the Coordinate Plane

Use a compass and a straightedge to find the incenter of a triangle whose vertices are at $(-2, 1)$, $(1, 2)$, and $(2, -2)$ in a coordinate plane.

SOLUTION

To find the incenter, bisect each of the triangle's angles using the methods described in Construction Lab 3. Once all three angle bisectors have been drawn, the central point where they intersect is the incenter.

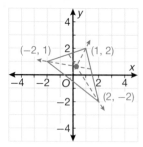

Hint

Refer to Construction Lab 3 to see how to construct an angle bisector.

In addition to finding the incenter of a triangle, angle bisectors can also be used to find the lengths of segments in the triangle. When an angle bisector intersects the side of a triangle, it makes a proportional relationship, given by Theorem 38-1.

Online Connection
www.SaxonMathResources.com

Theorem 38-1: Triangle Angle Bisector Theorem

If a line bisects an angle of a triangle, then it divides the opposite side proportionally to the other two sides of the triangle.

In the diagram, $\dfrac{PM}{PO} = \dfrac{NM}{NO}$.

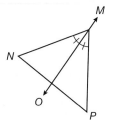

Example 2 Using the Triangle Angle Bisector Theorem

Using the diagram at the right, find BC if $AD = 15$, $DC = 8$, and $AB = 20$.

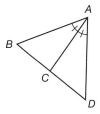

SOLUTION

$$\frac{AD}{DC} = \frac{AB}{BC}$$

$$\frac{15}{8} = \frac{20}{x}$$

$$15x = 160$$

$$x = 10.\overline{6}$$

A perpendicular bisector divides a line segment into two congruent segments and is perpendicular to the segment. If perpendicular bisectors are drawn for every side of a triangle, the point of concurrency is the **circumcenter of the triangle**. The circumcenter of a triangle is equidistant from every vertex in the triangle.

In the diagram below, point P is the circumcenter of the triangle, so $PX = PY = PZ$.

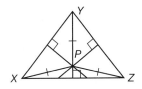

The circumcenter is not always inside a triangle. A right triangle's circumcenter lies on the hypotenuse, and an obtuse triangle's circumcenter is outside the triangle.

The circumcenter lies at the center of the circle that contains the three vertices of the triangle. Any circle that contains all the vertices of a polygon is called a **circumscribed circle**. Any polygon with each vertex on a circle is an **inscribed polygon**.

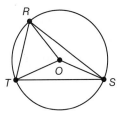

Example **3**   **Finding a Circumcenter in the Coordinate Plane**

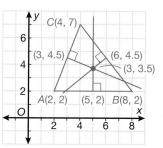

Find the circumcenter of a triangle with vertices at $A(2, 2)$, $B(8, 2)$, and $C(4, 7)$.

SOLUTION The perpendicular bisector of each segment needs to be found. The midpoint of $\overline{AB}$ is (5, 2) and the midpoint of $\overline{AC}$ is (3, 4.5).

For each of these midpoints, a line needs to be found that is perpendicular to the segment on which it lies. Find the slope between the segment's two endpoints, and then use linear equations to find the line.

$\overline{AB}$ lies on the line $y = 2$, and $\overline{AC}$ lies on $y = \frac{5}{2}x - 3$.

Perpendicular lines can be found through the midpoints using the method learned in Lesson 37.

The line perpendicular to $\overline{AB}$ through (5, 2) is $x = 5$.

The line perpendicular to $\overline{AC}$ through (3, 4.5) is $y = -\frac{2}{5}x + 5.7$.

To find the circumcenter, solve the system of equations.

$x = 5$

$y = -\frac{2}{5}x + 5.7$

$y = 3.7$

Finally, substitute this value of y into one of the equations above to find x. The coordinates of the circumcenter are $(5, 3.1\overline{6})$.

Example **4**   **Application: City Planning**

A gas company has three gas stations located at points $R(2, 8)$, $S(6, 2)$, and $T(2, 2)$, as shown. The storage facility is equidistant from the three gas stations. Find the location of the storage facility.

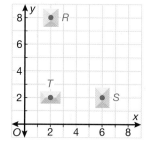

SOLUTION
Since the storage facility is equidistant from the gas stations, it is at the circumcenter of $\triangle RTS$.

By looking at the graph, you can see that the equation for $\overline{TS}$ is $y = 2$, and the equation for $\overline{RT}$ is $x = 2$.

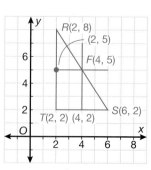

The midpoint of $\overline{TS}$ is (4, 2) and the midpoint of $\overline{RT}$ is (2, 5). The horizontal line $y = 5$ is perpendicular to $\overline{RT}$. The vertical line $x = 4$ is perpendicular to $\overline{TS}$. These two lines intersect on the hypotenuse at (4, 5), so the storage facility is located at (4, 5).

a. Use a compass and a straightedge to find the incenter of a triangle
(Ex 1) whose vertices are at $(-3, 1)$, $(3, -2)$, and $(2, 4)$ in a coordinate plane.

b. Using the diagram at the right, find the length
(Ex 2) of $\overline{TU}$ if $UV = 4$, $TW = 10$, and $WV = 6$.

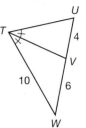

c. Find the circumcenter of a triangle with vertices at
(Ex 3) $(3, -1)$, $(-1, -4)$, and $(-3, -1)$ in a coordinate plane.

d. A restaurant owner wants to place his new restaurant
(Ex 4) equidistant from three nearby grocery stores that will
supply him. They are located at $A(0, 0)$, $B(4, 0)$ and
$C(0, 6)$. Where should he place his restaurant?

Practice Distributed and Integrated

1. Classify the following triangles as acute, obtuse, or right, based on the three side
(33) lengths given.
   a. 5, 9, 12
   b. 6, 17, 17.5
   c. 9, 12, 15

* 2. (**Fire Stations**) Three towns want to build a fire station that can serve them all.
(38) Explain how the towns might go about finding the optimal location for the fire
station.

3. Analyze Find the unknown length in the triangle. Do the side lengths form a
(29) Pythagorean triple?

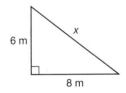

4. Algebra Test the conjecture that the sum of the first n squares is given by the
(7) expression $\frac{1}{6}n(n + 1)(2n + 1)$.

5. Letty is solving the equation $3x - 4 = 21$. Which property of equality should she
(24) use first?

6. What is a common side of $\angle QRS$ and $\angle QRT$?
(6)

* 7. Predict If line x is parallel to line y and line y is perpendicular to line z, describe the
(37) relationship between lines x and z.

8. (**Driving**) A traffic circle is a round intersection. The first one built in the United
(23) States is the Columbus Circle in New York City, and has a circular monument in
the center covering approximately 148,000 square feet. What is the radius of the
monument, to the nearest hundredth?

* 9. Determine the sum of the measures of the angles in the polygon
(Inv 3) shown at right.

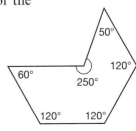

10. Generalize If $\triangle LMN \cong \triangle DEF$, what can you determine about the heights of the
(25) triangles as measured from side $\overline{LM}$ and from side $\overline{DE}$, respectively?

11. Verify Hamid reasoned, "All safety glasses protect your eyes. Sunglasses protect
(21) your eyes. Therefore, all sunglasses are safety glasses." What is wrong with
Hamid's conclusion?

*12. Find the line parallel to $y = -\frac{1}{2}x - 2$ that passes through the origin.
(37)

*13. In the diagram at right, P is the incenter of $\triangle QRS$. Find PT.
(38)

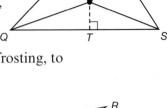

14. (**Baking**) Louis is baking a cake for a party that celebrates the birthdays of
(35) three different people. He wants to use three different kinds of frosting,
making three equal sections on the top of the round cake, each with
a person's name on it. If the cake has a radius of 5 inches, what is the
surface area of each wedge of the cake that Louis needs to cover with frosting, to
the nearest hundredth?

*15. Error Analysis Mira states that she can use the Leg-Angle Congruence
(36) Theorem to prove these triangles congruent. Is she correct about
the triangles being congruent? Is her justification correct? Explain.

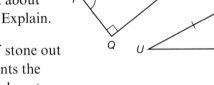

*16. (**Construction**) A contractor wants to cut a circular piece of stone out
(38) of a scrap piece that is shaped like a right triangle. She wants the
circle to be as large as possible. How should she determine how to
cut the piece of stone in order to get the largest possible circle out of it?
Explain your reasoning and support your answer with a drawing.

17. Coordinate Geometry Find the centroid of a triangle with vertices at $(1, 4)$, $(5, 6)$,
(32) and $(5, 0)$.

18. What is the area of the trapezoid in the diagram?
(22)

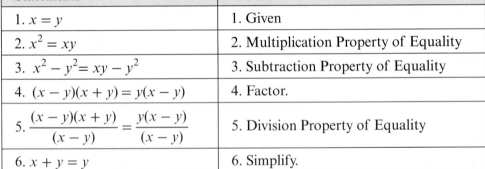

2 in.

3 in.

1.5 in.

19. Analyze Helen wrote the following proof:
(24)

Statements	Reasons
1. $x = y$	1. Given
2. $x^2 = xy$	2. Multiplication Property of Equality
3. $x^2 - y^2 = xy - y^2$	3. Subtraction Property of Equality
4. $(x - y)(x + y) = y(x - y)$	4. Factor.
5. $\dfrac{(x - y)(x + y)}{(x - y)} = \dfrac{y(x - y)}{(x - y)}$	5. Division Property of Equality
6. $x + y = y$	6. Simplify.
7. $x + y - y = y - y$	7. Subtraction Property of Equality
8. $x = 0$	8. Simplify.

Is there something wrong with Helen's proof? If so, which step is wrong and why?

20. Multiple Choice Which statement is *not* true of a rhombus?
(34)
    A Consecutive angles are supplementary.   **B** The diagonals bisect each other.
    C Opposite angles are complementary.   **D** All sides are congruent.

21. In the diagram, $\overline{AD} \parallel \overline{BC}$ and $\overline{DC} \parallel \overline{AB}$. Prove that $\triangle ACB \cong \triangle CAD$
(31) using a flowchart or paragraph proof.

22. (**Home Maintenance**) A window washer needs the top of his ladder
(29) to rest against the bottom of a window that is 25 feet above the
ground. However, the shrubs against the house require him to place
the ladder 10 feet away from the house. Assuming that the ground
and house meet at right angles, what length of ladder will he need,
to the nearest foot?

*23. Using the diagram find *FG* if *G* is the incenter of $\triangle HIJ$.
(38)

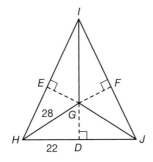

24. In the city of Logica, downtown buses are blue and suburban buses are green.
(20) State the conjunction of the two statements, "A bus is a downtown bus," and
"A bus is green." Is the conjunction true or false?

25. Find the unknown length of the side in the triangle.
(29)

26. In $\odot C$, $\overset{\frown}{AB}$ measures 44° and the radius *AC* is
(35) 5 centimeters. What is the area of sector *ACB*, to
the nearest hundredth?

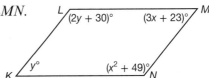

*27. Find the line parallel to $y = \frac{3}{2}x + \frac{1}{2}$ that passes through the point (3, 4).
(37)

*28. In $\triangle JKL$ and $\triangle PQR$, $\angle J$ and $\angle P$ are right angles, $\overline{KL} \cong \overline{QR}$, and
(36) $\angle Q \cong \angle K$. Use the Hypotenuse-Angle Congruence Theorem to prove that
$\triangle JKL \cong \triangle PQR$.

29. Prove that if m$\angle ABC$ = m$\angle ACB$, then $AB = AC$. *Hint: Prove* $\triangle ABC \cong \triangle ACB$.
(30)

30. Find the value of *x* and *y* in parallelogram *KLMN*.
(34)

Circle Through Three Noncollinear Points

Construction Lab 6 (Use with Lesson 38)

In Lesson 38, you learned about perpendicular bisectors of a triangle. This lab shows you how to construct a circle through three noncollinear points using perpendicular bisectors.

1. Begin with three noncollinear points A, B, and C. Draw $\overline{AB}$ and $\overline{BC}$.

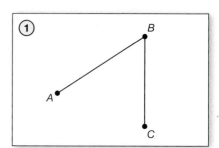

Hint

To construct the perpendicular bisector, use the method you learned in Construction Lab **3**.

2. Construct the perpendicular bisector of $\overline{BC}$ and label the midpoint L.

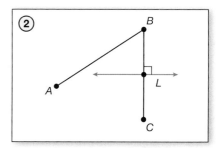

3. Construct the perpendicular bisector of $\overline{AB}$ and label the midpoint M.

4. The point of intersection of the two perpendicular bisectors is the center of the circle. Label it P.

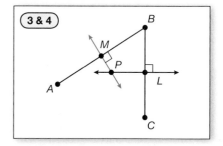

5. Center the compass on P and draw the circle with radius PA. As you draw the circle with this radius, the points A, B, and C will lie on the circle.

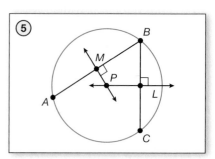

Lab Practice

Practice constructing circles through these sets of given points:

a. (2, 5), (3, 4), and (3, 2)   **b.** (0, 2), (2, 0), and (−1, 1)

Inequalities in a Triangle

1. **Vocabulary** The angle formed by one side of a polygon and the
   extension of an adjacent side is called the _____ angle. (*exterior,*
   corresponding, interior)

   (15)

2. Determine the value of x in the diagram.
   (13)

3. **Multiple Choice** Classify the triangle in the diagram.
   (13)

   A isosceles   **B** obtuse

   C acute   **D** right

New Concepts   The lengths of each side of a triangle are related to the measures of each
angle in the triangle according to Theorems 39-1 and 39-2.

Theorem 39-1
If one side of a triangle is longer than another side, then the angle opposite the first side is larger than the angle opposite the second side.

Math Reasoning

Predict Using Theorems 39-1 and 39-2, what can you say about the angle measures of an isosceles triangle?... of an equilateral triangle?

Theorem 39-2
If one angle of a triangle is larger than another angle, then the side opposite the first angle is longer than the side opposite the second angle.

In other words, a triangle's largest side is always opposite its largest angle,
and it smallest side is always opposite its smallest angle.

Example 1   **Ordering Triangle Side Lengths and Angle Measures**

(**a.**) Order the side lengths in $\triangle ABC$ from least
   to greatest.

SOLUTION
The Triangle Angle Sum Theorem shows that
the missing angle is 80°. Therefore, the side with
the greatest length is $\overline{AB}$ because it is opposite the largest angle. The
shortest side is $\overline{BC}$, as it is opposite the smallest angle. The final order
of sides, from least to greatest length, is $\overline{BC}, \overline{AC}, \overline{AB}$.

(**b.**) Order the measures of the angles in $\triangle XYZ$
   from least to greatest.

SOLUTION
The shortest side of the triangle is $\overline{XZ}$,
therefore the measure of the opposite angle, $\angle Y$, is the least of the three
angles. The longest side is $\overline{YZ}$, so it is opposite the angle with the greatest
measure, $\angle X$. Therefore the order of angles is $\angle Y, \angle Z, \angle X$.

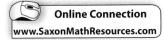

Online Connection
www.SaxonMathResources.com

Recall from Lesson 18 that the measure of the exterior angle of a triangle is equal to the sum of the two remote interior angles. This result leads to the Exterior Angle Inequality Theorem.

Theorem 39-3: Exterior Angle Inequality Theorem
The measure of an exterior angle is greater than the measure of either remote interior angle.

Example 2 **Proving The External Angle Inequality Theorem**

In the given triangle, the exterior angle is labeled as x. Prove that x is greater than the measures of $\angle B$ or $\angle C$.

SOLUTION
By the Exterior Angle Theorem, we know that $m\angle B + m\angle C = x$. We also know, from the definition of an angle, that both $\angle B$ and $\angle C$ have a measure greater than 0°. Rearranging the Exterior Angle Theorem, $m\angle B = x - m\angle C$. Since $m\angle C$ is greater than 0, $m\angle B$ must be less than x.

It is not true that any three line segments can make a triangle. Only line segments of certain lengths can form the three sides needed for a triangle. The requirements are given in the Triangle Inequality Theorem.

Theorem 39-4: Triangle Inequality Theorem
The sum of the lengths of any two sides of a triangle must be greater than the length of the third side.

For example, a triangle could not have side lengths of 3, 5, and 9 because the sum of 3 and 5 is less than 9.

Example 3 **Applying the Triangle Inequality Theorem**

a. Decide if each set of side lengths could form a valid triangle: (3, 4, 5), (5, 11, 6), and (1, 9, 5).

SOLUTION
For the first set, no combination of two sides can be found that will sum to less than the length of the other side, so since $3 + 4 = 7$, and $7 > 5$, these side lengths could represent a triangle.

For the second set, the sum of the two short sides, 5 and 6, equals exactly the length of the third side, 11. Since the sum is not greater than the third side, but only equal to it, this set cannot represent a triangle.

For the third set, the sum of the two short sides, 1 and 5, sum to less than the third side, 9, so this set also cannot represent a triangle.

b. Find the range of values for x in the given triangle.

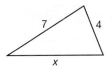

SOLUTION

Using the Triangle Inequality Theorem, the following three statements are true:

$$x + 4 > 7 \qquad x + 7 > 4 \qquad 7 + 4 > x$$
$$x > 3 \qquad x > -3 \qquad 11 > x$$

The second inequality is invalid, since a side length cannot be negative. Combining the two valid statements, we obtain the solution $3 < x < 11$.

Hint

See Skills Bank 15 for more information about using inequalities.

Example 4   Application: Planning a Trip

Simone took a flight from Atlanta to London (a distance of 4281 miles), then flew from London to New York City (a distance of 3470 miles), and then took a flight back to Atlanta. Assuming that all three trips are straight lines, determine the range of distances (from least to greatest) she could have traveled altogether.

SOLUTION

Let x represent the distance Simone traveled back to Atlanta from New York City. Using the Triangle Inequality Theorem, the following three statements are true:

$$x + 3470 > 4281 \qquad x + 4281 > 3470 \qquad 3470 + 4281 > x$$
$$x > 811 \qquad x > -811 \qquad 7751 > x$$

Combining the two valid statements, $811 < x < 7751$. Therefore, the shortest distance Simone could have traveled is $3470 + 4281 + 811 = 8562$ miles. The farthest she could have traveled is $3470 + 4281 + 7751 = 15{,}502$ miles.

Lesson Practice

a. Order the side lengths in $\triangle DEF$ from least to greatest.
(Ex 1)

b. Order the measures of the angles in $\triangle PQR$ from least to greatest.
(Ex 1)

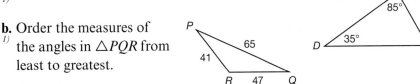

c. Show that in the triangle, the measure of the exterior angle at vertex Z is greater than the angle measure at vertex X or at vertex Y.
(Ex 2)

d. Find the range of values for x in the given triangle.
(Ex 3)

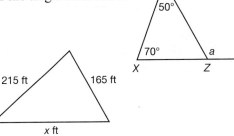

1. In $\triangle GHJ$ and $\triangle KLM$, it is given that $\angle G$ and $\angle K$ are right angles, $\overline{GH} \cong \overline{KL}$,
(36) and $\angle H \cong \angle L$. Prove that $\triangle GHJ \cong \triangle KLM$.

* 2. Write the side lengths of $\triangle ABC$ in order from least to greatest.
(39)

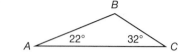

3. Algebra An isosceles right triangle has a hypotenuse of 10 inches.
(29) What is the length of each leg, to the nearest tenth of an inch?

4. (**Security**) To protect a circular area of a park, cameras are placed on a lamppost
(26) in the center of the area. If each camera can observe an arc measure of 75°, how
many cameras are needed to observe the entire area around the lamppost?

5. Model Draw a right triangle and sketch the altitude of each vertex. Where is the
(32) orthocenter of the triangle located? Will the orthocenter be located in the same
place on every right triangle?

* 6. Multiple Choice Which group of line segment lengths can be used to form
(39) a triangle?
A 4, 4, 9   B 3, 7, 12
C 1, 5, 7   D 3, 4, 6

* 7. Analyze Explain whether or not it is possible to construct a triangle with side
(39) lengths of 34 inches, 14 inches, and 51 inches.

8. If rectangle $STUV$ has an area of 27 square units, and $ST = 3(TU)$, find the length
(34) of each side of $STUV$.

9. Error Analysis On $\overline{AD}$, B is the midpoint of $\overline{AC}$ and C is the midpoint of $\overline{BD}$.
(31) The paragraph proof below was written to prove that $AB = CD$. Identify
the mistake in the proof.

*Since B is the midpoint of $\overline{AC}$, AB = BC by the definition of midpoints. Since
C is the midpoint of $\overline{BD}$, BC = CD by the definition of midpoints.
By the Reflexive Property, AB = CD.*

*10. (**Nautical Maps**) Three lighthouses surround a harbor, as shown in the
(38) drawing. The lighthouses are equidistant from a buoy in the harbor.
Copy the diagram and draw the location of the buoy based on
the locations of the lighthouses.

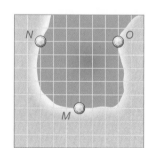

11. (**Engineering**) A bridge uses several girder sections that are right triangles.
(36) Each section is made so that its legs are 10 feet and 12.5 feet long,
respectively. Explain why the sections are congruent to each other.

12. (**Decorating**) Kirsten is buying carpet for a rectangular room that is 11 feet by
(22) 13 feet. How many square feet of carpet does she need?

13. Find the centroid of a triangle with vertices at $(0, -2)$, $(0, -6)$, and $(2, -4)$.
(32)

14. What is the included angle of $\overrightarrow{TS}$ and $\overrightarrow{TR}$?
(28)

15. Algebra Using the diagram at right, find *MN* in terms of *x* if *NO* = 2,
(38) *OP* = 4*x*, and *MP* = *x* + 7.

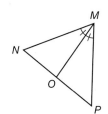

*16. Multiple Choice Which expression represents two parallel lines?
(37)
A $y = 2x + 3$, $y = -2x + 3$ **B** $y = -2x + 4$, $y = -6 - 2x$
C $y = \frac{1}{2}x + 3$, $y = -2x - 3$ **D** $y = 3x + 7$, $\frac{1}{3}y = -x - 7$

17. (Urban Development) A development company plans to build three square
(Inv 2) single-story buildings that completely surround a triangular park.
Buildings *A* and *B* will be situated at right angles to each other and
have an area of 1296 square feet and 2500 square feet, respectively.
What will be the area of building *C*? Explain.

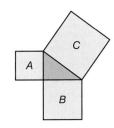

*18. Using the diagram, *P* is the incenter of $\triangle QRS$. Find *PU*.
(38)

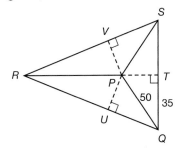

19. Write Explain why the size of each central angle in a regular polygon decreases as
(Inv 3) the number of sides increases.

20. Justify Provide a justification for each step of the proof. The first one is done for you.
(24)

Statements	Reasons
1. $16 = 4(3x - 8)$	1. Given
2. $\dfrac{16}{4} = \dfrac{4(3x - 8)}{4}$	2.
3. $4 = 3x - 8$	3.
4. $4 + 8 = 3x - 8 + 8$	4.
5. $12 = 3x$	5.
6. $\dfrac{12}{3} = \dfrac{3x}{3}$	6.
7. $4 = x$	7.
8. $x = 4$	8.

*21. Find the line parallel to $y = 2x + 3$ that passes through the point $(-1, -1)$.
(37)

22. Use an inequality to indicate the minimum length of the longest side, *x*, of $\triangle ABC$
(33) if the two shorter sides of the triangle measure 8 and 7 and $\triangle ABC$ is obtuse.

23. Calculate the area of sector *QRS* in the diagram to the nearest hundredth
₍₃₅₎ of a centimeter.

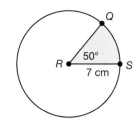

24. Davindra, who is 15 years old, is one year older than twice the age of
₍₂₄₎ her younger sister, Preetha. Find the age of Preetha and justify each step.

25. In the diagram, $\overline{AB} \parallel \overline{DE}$. Write a paragraph to prove that
₍₃₁₎ m∠*CAB* + m∠*ACB* = m∠*DEF*.

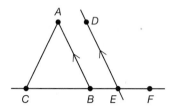

26. A circle with a radius of 6 meters has an arc that measures 35°. If this arc and its
₍₃₅₎ associated sector are completely removed from the circle, what is the length of the
major arc that remains, to the nearest tenth of a meter?

27. Classify the following triangles as acute, obtuse, or right, based on the three side
₍₃₃₎ lengths given.
  a. 4, 4, $\sqrt{34}$
  b. 6, 5, 8
  c. 10, 24, 26

*28. (**Construction**) An important part of many structures is the structural support truss,
₍₃₉₎ a series of interlocking metal rods that form triangles, which gives the structure
its rigidity. The diagram below shows a section of a support truss for a geodesic
dome. What is the shortest length of metal used in this section of a geodesic dome
structural support truss?

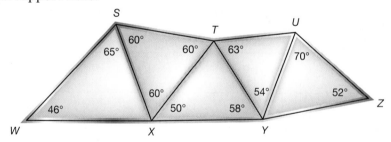

*29. **Predict** If line *a* is perpendicular to line *b* and line *c* is perpendicular to
₍₃₇₎ line *b*, describe the relationship between lines *a* and *c*.

30. Find the value of *x* in the figure shown. Give your answer in simplified
₍₂₉₎ radical form.

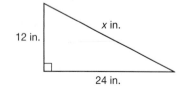

Finding Perimeters and Areas of Composite Figures

Warm Up

1. **Vocabulary** The size of the region bounded by a closed geometric figure
(8) is that figure's _____.

2. Determine the perimeter of a right triangle with a 25-mm base
(13) and a 60-mm height.

3. Which is the correct formula for the area of a trapezoid?
(22)

A $A = \frac{1}{2}bh$ **B** $A = \frac{1}{2}(b_1 + b_2)h$

C $A = bh$ **D** $A = (b_1 + b_2)h$

New Concepts

A **composite figure** is a plane figure made up of simple shapes or a three-dimensional figure made up of simple three-dimensional figures. The perimeter of a plane composite figure is the sum of the lengths of its sides.

> **Math Reasoning**
>
> **Generalize** Explain why the perimeter of a composite figure cannot be found by adding the perimeters of the shapes that compose it.

Example 1 **Finding Perimeters of Composite Figures**

Find the perimeter of the composite figure.

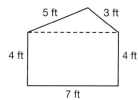

SOLUTION

Start at a vertex and move around the figure, adding the side lengths in order.

$P = 3 + 5 + 4 + 7 + 4$

    $= 23$ ft

The perimeter is 23 feet.

Postulate 19: Area Congruence Postulate
If two polygons are congruent, then they have the same area.

Postulate 20: Area Addition Postulate
The area of a region is equal to the sum of the areas of its nonoverlapping parts.   In the diagram, $A = A_1 + A_2 + A_3$.

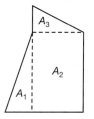

Hint

Sometimes a figure will need to be divided into parts in order to find the area. Look for parts of the figure that appear to be rectangles or triangles and draw dotted lines to indicate how the figure should be divided.

The Area Congruence Postulate and the Area Addition Postulate make it possible to find the area of complex composite figures by breaking them down into simpler shapes and finding the area of each shape.

Example 2 Finding Areas of Composite Figures

Find the area of this composite figure.

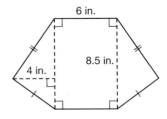

SOLUTION

Left triangle:

$$A_1 = \frac{1}{2}bh$$

$$A_1 = \frac{1}{2}(8.5)(4) = 17 \text{ in}^2$$

Therefore, the area of the left triangle is 17 in².

Rectangle:

$$A_2 = bh$$

$$A_2 = (8.5)(6) = 51 \text{ in}^2$$

Therefore, the area of the rectangle is 51 in².

By the SSS Postulate, the two triangles in the figure are congruent. Therefore, by Postulate 19, the right triangle has the same area as the left triangle, 17 square inches.

By the Area Addition Postulate, the area of the composite figure is

$$A = 17 + 51 + 17 = 85 \text{ square inches.}$$

Example 3 Finding Areas of Composite Figures by Subtracting

Find the area of the shaded region.

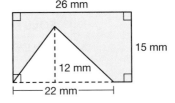

SOLUTION

Outer rectangle:

$$A_1 = bh$$

$$A_1 = (26)(15) = 390 \text{ mm}^2$$

Therefore, the area of the outer rectangle is 390 mm².

"Missing" triangle:

$$A_2 = \frac{1}{2}bh$$

$$A_2 = \frac{1}{2}(22)(12) = 132 \text{ mm}^2$$

Therefore, the area of the triangle is 132 mm².

The Area Addition Postulate implies that the area of the rectangle is the area of the region we want to find plus the area of the triangle:
$A_1 = A + A_2.$
Therefore,

$$390 = A + 132$$
$$A = 258 \text{ mm}^2$$

The area of the shaded region is 258 mm².

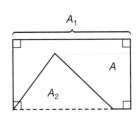

Example 4 Application: Architecture

The diagram shows the plan for a reflecting pool that will form part of a new downtown plaza.

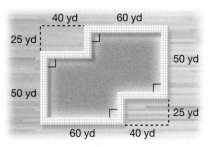

a. The edge of the pool needs to be tiled. What is the perimeter of the pool that will need tiling?

SOLUTION
$P = 25 + 40 + 50 + 60 + 25 + 40 + 50 + 60$
$= 350$ yd
Therefore, the perimeter of the pool is 350 yards.

b. What area of concrete will be needed for the bottom of the pool?

SOLUTION

Outer rectangle:
$A_1 = bh$
$= (40 + 60)(25 + 50)$
$A_1 = (100)(75) = 7500$ yd^2

Therefore, the area of the outer rectangle is 7500 yd^2.

Each "missing" rectangle:
$A_2 = bh$

$A_2 = (40)(25) = 1000$ yd^2

Therefore, the area of each "missing" rectangle is 1000 yd^2.

Let A be the area of the pool bottom. By the Area Addition Postulate,
$$A_1 = A + A_2 + A_2$$
$$7500 = A + 1000 + 1000$$
$$7500 - 2000 = A$$
$$A = 5500 \text{ yd}^2$$

5500 square yards of concrete will be needed.

Math Reasoning

Formulate A semicircle has radius *r*. Give a formula to find the area and the perimeter of the semicircle.

Lesson Practice

a. A composite figure made up of a rectangle and a triangle has a 3-inch horizontal side, two 4.5-inch vertical sides, and two 4-inch sloping sides. What is its perimeter?
(Ex 1)

Use this composite figure to answer problems b and c.

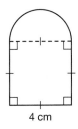

4 cm

b. Determine the area of the figure. Express your answer in terms of π.
(Ex 2)

c. A rectangle with dimensions 3-by-1.5 centimeters is removed from the bottom left corner of the figure. Determine the area of the new figure. Express your answer in terms of π.
(Ex 3)

d. A new office building has one side in the shape of a right triangle on top of a rectangle. The rectangle is 420 feet tall and 120 feet wide. The triangle's base is 120 feet long and its vertical leg is 160 feet long. What is the perimeter of this side of the building?
(Ex 4)

e. How much glass is needed to cover the side of the building in part **d**?
(Ex 4)

1. Multi-Step Find the slope of the line that passes through the points
(37) $X(2, -3)$ and $Y(4, 1)$. Then, find a perpendicular line that passes through point $Z(-2, 2)$.

2. Error Analysis Isabelle has determined that the legs in a right triangle are
(29) 21 centimeters and 75 centimeters long, and the hypotenuse is 72 centimeters long. What error has she made?

3. If a triangle has side lengths of $2x$, $3x$, and $4x$, is it an acute, obtuse, or a right
(33) triangle?

4. Find the line perpendicular to $y = \frac{1}{4}x + 7$ that passes through the origin.
(37)

* 5. These figures are the first two stages of an abstract design. In each
(40) figure, all side are congruent. Determine the perimeter and area of each figure.

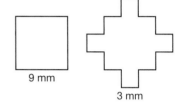

9 mm

3 mm

6. Write For two statements p and q, what can be determined about
(21) the truth value of p when "if p, then q" is a true statement and q is false? Why?

7. What congruence theorem is used to prove that
(36) $\triangle HJK \cong \triangle LNM$? Explain.

8. The U.S. dime has a radius of 8.96 millimeters. Estimate the area of
(23) one side of a dime to the nearest whole square millimeter.

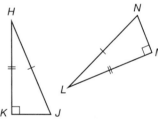

* 9. Use a compass and a straightedge to find the incenter of a triangle
(38) whose vertices are at $(0, 0)$, $(5, -3)$, and $(-1, -4)$ in a coordinate plane.

10. Antoine solved for the length of a hypotenuse in a right triangle and found that
(29) $c^2 = 490$. What is the value of c, in simplified radical form?

11. In the diagram, $QW = 14$ and $TR = 9$. Find QV and TW.
(32)

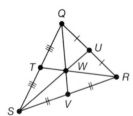

12. (**Track and Field**) Tara is running along a circular track and Petra is standing at the
(26) center of the circle. Petra observes Tara running an arc measure of 56° clockwise, then a measure of 34° counterclockwise, and then a measure of 67° clockwise. What is the measure of the arc from her starting point to where Tara stands now?

*13. If a triangle were constructed with the smallest third side possible, which angle
(39) would measure the greatest, if the two given sides measure 14 yards and
31 yards?

14. In $\odot C$, $\overarc{AB}$ measures 65°, and the area of sector ACB is 9 square units. What is
(35) the radius of $\odot C$, to the nearest hundredth unit?

15. Error Analysis Using the diagram, Lily decided that $\triangle ABE \cong \triangle CDE$. She wrote a
(31) paragraph proof, which is given below. What mistake did Lily make?

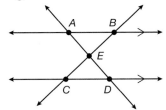

*The diagram shows that $\overline{AB} \parallel \overline{CD}$. Therefore, because alternate interior
angles are congruent, $\angle BAE \cong \angle EDC$ and $\angle ABE \cong \angle ECD$. Since
$\angle AEB$ and $\angle CED$ are vertical angles, $\angle AEB \cong \angle CED$. Since the
corresponding angles of the triangles are all congruent, $\triangle ABE \cong \triangle DCE$.*

*16. A composite figure is made up of a rectangle that is 4 inches wide by 10 inches
(40) long and two semicircles that fit exactly on two adjacent sides of the rectangle.
Determine the figure's perimeter and area. Express your answers in terms of π.

*17. Write the angles of $\triangle SEP$ in order from least to greatest measure.
(39)

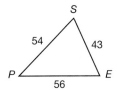

18. In triangles UVW and XYZ, $\angle V$ and $\angle Y$ are right angles, $\overline{UW} \cong \overline{XZ}$,
(36) and $\angle U \cong \angle X$. Use the Hypotenuse-Angle Congruence Theorem to
prove that $\triangle UVW \cong \triangle XYZ$.

19. ⟮ **Security** ⟯ A camera covers an arc measure of 45°. What fraction of
(26) a full rotation does the camera cover? How many cameras would be
needed to be able to look in all directions?

20. In the rectangle $ABCD$, $BD = 2x - 3$. What is the value of x? What is
(34) the length of $\overline{AC}$?

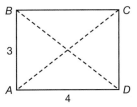

21. ⟮ **Forestry** ⟯ A tree's circular trunk has a circumference of 4 feet 9 inches.
(23) To the nearest inch, how wide is its trunk?

22. Multiple Choice Find the pair of perpendicular lines.
(37)

   A $2x - 3y = 4$, $y = -\dfrac{2}{3}x - 3$

   B $y = \dfrac{1}{2}x + 7$, $y = 2x - 5$

   C $-3y + 2x + 7 = 0$, $-3x + 2y - 4 = 0$

   D $2x + y = 8$, $x - 2y = 1$

23. Write Explain why the size of each exterior angle in a regular polygon decreases as the number of sides increases.
(Inv 3)

*24. Justify Can a triangle have sides with the lengths of 8 feet, 8 feet, and 16 feet?
(39)

Use the diagram to complete the problems.

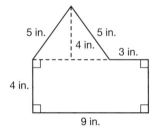

*25. Determine the perimeter of the figure.
(40)

*26. Determine the area of the figure.
(40)

*27. Algebra Find the value of x that makes P the incenter of the triangle shown.
(38)

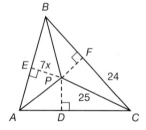

28. (**Driving**) A car drives 20 miles east and then 45 miles south. To the nearest hundredth of a mile, how far is the car from its starting point?
(29)

Use the given diagram. Round to the nearest hundredth.

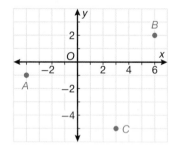

29. Find the length of $\overline{AB}$.
(9)

30. Find the length of $\overline{BC}$.
(9)

Inequalities in Two Triangles

In Lesson 39, you learned to verify whether three segments of given lengths could be used to form a triangle. This involved an inequality in one triangle. In this investigation, you will explore inequalities in two triangles.

1. Consider a door hinge. As shown in the diagram, the doorframe, door, and the distance between them make a triangle. The doorframe and the door itself are always the same length, so two sides of this triangle are fixed. What happens to the third side as the door is opened?

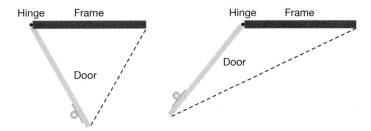

Math Language

An **included angle** is an angle that is formed by two adjacent sides of a polygon. A hinge, like the one on a door, is an included angle of the doorframe and the door itself.

2. Model Use a protractor to construct two triangles. Your triangles should have two pairs of congruent sides, and a third side that changes length depending on how much you open the "hinge," which is the included angle of the fixed sides. The hinge angles on your two triangles should have different measures.

3. Write Consider the door hinge example from step 1. If two triangles have two pairs of congruent sides, but their included angles are not congruent, what conclusion can you make about the third side?

Theorem 40-1: Hinge Theorem
If two sides of one triangle are congruent to two sides of another triangle and the included angle of the first triangle is greater than the included angle of the second triangle, then the third side of the first triangle is longer than the third side of the second triangle.

The Hinge Theorem is useful when comparing two triangles with two pairs of congruent sides. Is the converse of the Hinge Theorem also true? Follow the steps below.

4. Draw an angle with side segments that are 4 and 5 inches long. Connect their endpoints to make a triangle. Label the vertex across from the 4-inch side A, the vertex across from the 5-inch side B, and the last vertex C. Measure $\overline{AB}$ and label its length on your diagram.

5. Draw a second angle, with a different measure from the first, that also has side segments that are 4 and 5 inches long. Connect their endpoints to make a triangle. Label the vertex across from the 4-inch side D, the vertex across from the 5-inch side E, and the last vertex F. Measure $\overline{DE}$ and label its length.

6. Measure the angles at vertices *C* and *F* using a protractor. What do you notice about these angles in comparison to each other and to the lengths of the sides opposite them?

7. The two triangles you drew illustrate the converse of the Hinge Theorem. State the converse of the Hinge Theorem, and determine whether it is true for your triangles.

Math Reasoning

Analyze The Hinge Theorem is a more complex conditional statement than you have seen, taking the form "If (*p* and *q*), then *r*." Write out the statements *p*, *q*, and *r*.

Theorem 40-2: Converse of the Hinge Theorem
If two sides of one triangle are congruent to two sides of another triangle and the third side of the first triangle is longer than the third side of the second triangle, then the measure of the angle opposite the third side of the first triangle is greater than the measure of the angle opposite the third side of the second triangle.

The Hinge Theorem and the Converse of the Hinge Theorem can be used to compare two triangles.

8. **Multi-Step** Concentric circles are circles that have the same center but different radii measures. The circles illustrated here are concentric. The measure of ∠*BAC* is 93°, and the measure of ∠*DAE* is 60°. Explain why *BC* must be greater than *DE*.

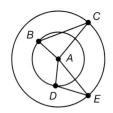

9. (**Ergonomics**) Depending on the task a person is performing, the angle that a person's torso makes with their legs changes. The diagrams below show a student in three different sitting positions: relaxed, writing, and typing. In which position is the angle measure at the hip the greatest? ... the least? Explain how you know.

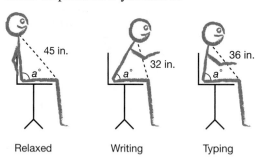

| Relaxed | Writing | Typing |

10. Write an inequality that gives the possible values of *x* in the diagram.

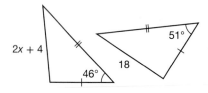

Investigation Practice

a. Use an inequality to compare the lengths of *TV* and *XY*.

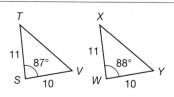

b. Use an inequality to compare the measures of ∠*G* and ∠*L*.

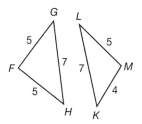

c. Write Describe how the hood of a car illustrates the Hinge Theorem.

d. Multiple Choice Choose the most correct answer for the given diagram.

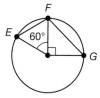

    A $EF = FG$ **B** $EF < FG$

    C $EF > FG$ **D** not enough information

e. Algebra Find the range of values for *x*.

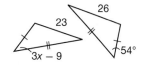

f. (**Door Hinges**) To prevent a door from opening too far and hitting a wall, a doorstop can be placed on the hinge of the door. A door in a small closet swings through a straight-line distance of 40 inches, while another door in a washroom swings through a straight-line distance of 48 inches. Which doorstop needs to be set to open to a larger angle? Explain.

g. (**Playground Equipment**) Kelvin and his friend Theo are swinging on a swing set at the local park. Both swings are the same length. Kelvin swings through an angle of 47° and Theo swings through an angle of 44°. Which of the two friends is swinging through the greatest distance? Explain.

1. **Vocabulary** A statement showing that two ratios are equal is called
(SB)   a(n) _____. (*proportion, inequality, direct variation*)

2. Is this statement true or false?
(25)   *If all three sides of one triangle are congruent to all three sides of another triangle, the two triangles are congruent.*

3. Given $\triangle ABC \cong \triangle DEF$, tell which sides are congruent.
(25)

4. If in $\triangle XYZ$ and $\triangle CAM$, $\angle X \cong \angle C$, $\overline{XY} \cong \overline{CA}$, and $\angle Y \cong \angle A$, which
(30)   postulate or theorem can prove the triangles congruent?

New Concepts A **ratio** is a comparison of two values by division. The ratio of two quantities, a and b, can be written in three ways: a to b, $a{:}b$, or $\frac{a}{b}$ (where $b \neq 0$). A statement that two ratios are equal is called a **proportion**.

Example 1 — Writing Ratios and Proportions

Consider $\triangle MNO$ and $\triangle PQR$.

Math Reasoning

Write Explain the difference between a ratio and a proportion.

a. Write a ratio comparing the lengths of segments $\overline{MN}$ to $\overline{NO}$ to $\overline{OM}$.

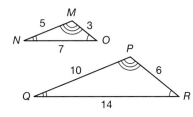

SOLUTION
The quantity that is mentioned first in a ratio is written first.

$MN = 5$, $NO = 7$, $OM = 3$. Therefore $MN{:}NO{:}OM = 5{:}7{:}3$.

b. Write a ratio comparing MN to PQ in three ways.

SOLUTION
$MN = 5$ and $PQ = 10$. The ratio of MN to PQ can be written as 5 to 10, 5:10, or $\frac{5}{10}$. Ratios can be reduced just like fractions. Reducing this ratio results in $\frac{1}{2}$.

c. Write a proportion to show that $MN{:}PQ = NO{:}QR$

SOLUTION
$MN{:}PQ = NO{:}QR$ can be written as $\frac{5}{10} = \frac{7}{14}$. Notice that the proportion is true since both sides of the proportion reduce to $\frac{1}{2}$.

In the proportion $\frac{a}{b} = \frac{c}{d}$, a and d are the **extremes**, and b and c are the **means**. One way to solve a proportion to find a missing value is to use cross products. The **cross product** is the product of the means and the product of the extremes. In other words, if $\frac{a}{b} = \frac{c}{d}$, then $ad = bc$.

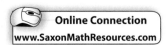

Online Connection
www.SaxonMathResources.com

Example 2 Solving Proportions with Cross Products

Solve the proportion $\frac{3}{15} = \frac{x}{50}$ to find the value of x.

SOLUTION

Find the cross products to solve the proportion.

$$\frac{3}{15} = \frac{x}{50}$$

$$3 \times 50 = 15x$$

$$x = 10$$

Reading Math

The symbol $\sim$ shows that two figures are similar, and should be read "is similar to."

Two figures that have the same shape, but not necessarily the same size, are **similar**. In the diagram, $\triangle ABC$ is similar to $\triangle DEF$. All congruent figures are also similar figures, but the converse is not always true.

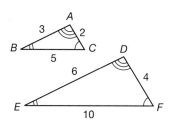

In **similar polygons**, the corresponding angles are congruent and the corresponding sides are proportional. In the diagram, $\triangle ABC$ and $\triangle DEF$ have congruent angles, and each pair of their corresponding sides has the same proportional relationship. A **similarity ratio** is the ratio of two corresponding linear measurements in a pair of similar figures. The following similarity ratios can be written for $\triangle ABC$ and $\triangle DEF$.

$$\frac{DE}{AB} = \frac{6}{3} = 2 \qquad \frac{EF}{BC} = \frac{10}{5} = 2 \qquad \frac{FD}{CA} = \frac{4}{2} = 2$$

Like congruence, similarity is a transitive relation. The Transitive Property of Similarity states that if $a \sim b$, and $b \sim c$, then $a \sim c$.

Example 3 Using Proportion to Find Missing Lengths

Find the unknown side lengths in the two similar triangles.

SOLUTION

The triangles are similar so corresponding sides are proportional: $\frac{VT}{YW} = \frac{TU}{WX} = \frac{UV}{XY}$. Therefore, $\frac{5}{a} = \frac{7}{b} = \frac{4}{16}$.

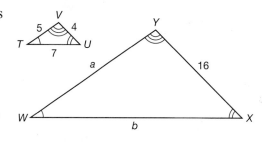

Solve each proportion using a known ratio and a ratio with an unknown.

$$\frac{5}{a} = \frac{4}{16} \qquad\qquad \frac{7}{b} = \frac{4}{16}$$

$$4a = 5 \times 16 \qquad\qquad 4b = 7 \times 16$$

$$a = 20 \qquad\qquad\quad b = 28$$

Therefore, $WX = 28$ and $YW = 20$.

Hint

Using the Symmetric Property, the cross product $ad = bc$ may also be written as $bc = ad$.

A **similarity statement** is a statement indicating that two polygons are similar by listing their vertices in order of correspondence. Much like writing a congruence statement, corresponding angles have to be named in the same order.

Example 4 Finding Missing Measures of Similar Polygons

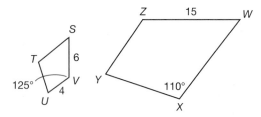

In the diagram, $STUV \sim WXYZ$.

Hint

When setting up proportions, all your ratio's numerators should come from the same figure, and the denominators should all come from the same figure as well. It does not matter which figure's sides you use for the numerator or the denominator.

a. What are the measures of $\angle T$ amd $\angle Z$?

SOLUTION

The quadrilaterals are named in order of corresponding vertices, and corresponding angles of similar figures are congruent. Therefore, $\angle S \cong \angle W$, $\angle T \cong \angle X$, $\angle U \cong \angle Y$, and $\angle V \cong \angle Z$.

$m\angle T = m\angle X = 110°$, and $m\angle Z = m\angle V = 125°$.

b. What is the length of $\overline{YZ}$?

SOLUTION

In the quadrilaterals, $\overline{SV}$ and $\overline{WZ}$ are corresponding sides and $\overline{UV}$ and $\overline{YZ}$ are corresponding sides. Therefore, $\frac{SV}{WZ} = \frac{UV}{YZ}$. Substitute the values:

$$\frac{6}{15} = \frac{4}{YZ}$$
$$6 \times YZ = 15 \times 4$$
$$YZ = 10$$

The length of $\overline{YZ}$ is 10 units.

Example 5 Application: Optics

Math Reasoning

Generalize Are all right triangles similar? Explain.

Siobhan is using a mirror and similar triangles to determine the height of a small tree. She places the mirror at a distance where she can see the top of the tree in the mirror. According to the measures in Siobhan's triangles, what is the height of the tree to the nearest inch?

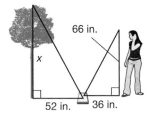

SOLUTION

The two triangles are similar, so corresponding sides are proportional.

$$\frac{x}{66} = \frac{52}{36}$$
$$36x = 66 \times 52$$
$$x \approx 95$$

The tree is about 95 inches tall.

Use the two similar triangles to answer *a* through a through *c*.

a. Write a ratio comparing the lengths of segments $\overline{HA}$ to $\overline{AM}$ to $\overline{MH}$.
(Ex 1)

b. Write a ratio comparing AM to LR in three ways, in simplest form.
(Ex 1)

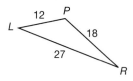

c. Write a proportion to show that $HM{:}PR = AM{:}LR$.
(Ex 1)

d. What is the value of x in the proportion $\frac{8}{7} = \frac{x}{21}$?
(Ex 2)

e. If $\triangle RST \sim \triangle UVW$, find the missing length in $\triangle UVW$.
(Ex 3)

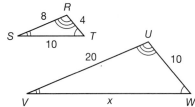

f. If the polygons $MNOP$ and $QRST$ are similar, what are the measures of $\angle O$ and $\angle R$?
(Ex 4)

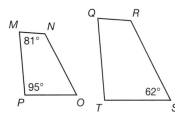

g. If the polygons $ABCD$ and $EFGH$ are similar, what are the values of x, y, and z?
(Ex 4)

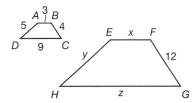

h. Cree uses a 21-foot ladder and a 12-foot ladder while painting the exterior of a house. Each ladder forms the same angle with the ground. If the longer ladder reaches 18 feet up the wall, how high does the other ladder reach, to the nearest foot?
(Ex 5)

1. **Error Analysis** Alexander made a convex pentagon and a convex quadrilateral. He
(Inv 3) measured and then added the measures of all the exterior angles, finding a sum of
900°. Is this possible? Explain.

* 2. If the quadrilaterals at right are similar, what is the measure
(41) of ∠E? What is the length of $\overline{XY}$?

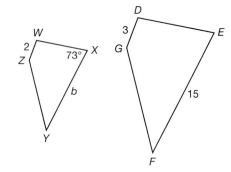

3. (Reunions) Three friends who live far away from each other
(38) want to meet for a reunion. Explain how the friends might
go about finding a location for the reunion that is equidistant
from each friend.

4. Compare the lengths of $\overline{AB}$ and $\overline{DE}$.
(Inv 4)

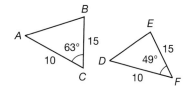

5. **Predict** If line *l* is perpendicular to line *m* and line *n* is parallel to line *m*, describe
(37) the relationship between lines *l* and *n*.

* 6. What is the value of *x* in the proportion $\frac{9}{x+2} = \frac{27}{9}$?
(41)

7. If $x = 5$ and $y = x$, which property of equality can be used to conclude
(24) that $y = 5$?

8. Name this polygon. Is it equiangular? Is it equilateral? Is it regular
(15) or irregular?

9. **Generalize** Given that the sum of all the angle measures in a triangle is
(27) 180°, prove that the interior angles in convex quadrilateral *ABCD* sum to 360°.

10. **Verify** Use the formula for the sum of the interior angles of a convex polygon to
(Inv 3) verify the Triangle Sum Theorem.

*11. **Multi-Step** Determine the area of this figure. Express your answer in
(40) terms of *π*.

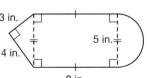

12. (Wheels) A car's wheels are 18 inches across. To the nearest
(23) hundredth of an inch, how far will the car move with four complete rotations
of its wheels?

13. Give a Pythagorean triple that is proportional to the triple (7, 24, 25).
(29)

14. (Engineering) Determine the area of this metal plate.
(40)

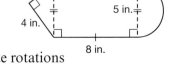

15. Triangle *JKL* is congruent to △*MNO*.
(25) What triangle is △*OMN* congruent to?

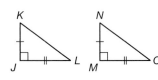

270 *Saxon* Geometry

*16. Complete the flowchart to prove the Pythagorean Inequality Theorem.
(31) **Given:** $(BC)^2 > (AB)^2 + (AC)^2$, $\triangle DEF$ is a right triangle, $\overline{AB} \cong \overline{DE}$, $\overline{AC} \cong \overline{DF}$

Prove: $\triangle ABC$ is an obtuse triangle.

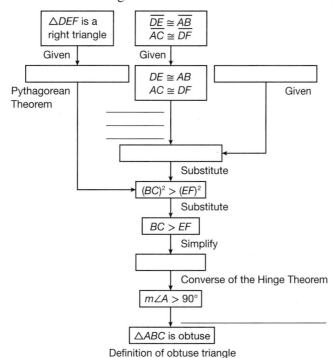

17. In a triangle with vertex D, $\overline{DM}$ is a median, and C is the centroid of the triangle.
(32) What is the length of $\overline{DC}$ if $\overline{DM}$ measures 70.2 centimeters? What theorem or postulate justifies your answer?

18. What is an expression for the area of the trapezoid?
(22)

19. In $\triangle MNO$, m$\angle MNO = 37°$, m$\angle NOM = 53°$, $NO = 50$ feet,
(30) $MO = 40$ feet, and $MN = 30$ feet. In $\triangle RST$, m$\angle RST = 53°$, m$\angle TRS = 37°$, and $RS = 50$ feet. Determine all other lengths.

20. **Geography** Two right triangles have been marked on a map of Seattle,
(36) WA. The distance from Pike Street to Broadway along E. Madison Street is equal to the distance from Marion Street to Boren Avenue along Broadway. Also, Marion Street and E. Madison Street run parallel. Prove that the triangles are congruent.

*21. **Photography** A rectangular 8-by-10-inch photograph has to be reduced
(41) to $\frac{1}{4}$ its original dimensions to be placed in a magazine. What will be the dimensions of the reduced photograph?

Use the figure to find the unknown measures. Quadrilateral *PQRS* is a parallelogram.

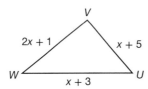

22. Solve for *x*, then find the measure of each interior angle of the parallelogram.
₍₃₄₎

23. Solve for *y*, then find the perimeter if $PQ = 2PS$.
₍₃₄₎

24. Fill in the blanks with the correct words.
_(Inv 4)

If two sides of one _____ are congruent to two sides of another triangle and the included _____ of the first triangle is greater than the included angle of the second triangle, then the third side of the first triangle is longer than the third _____ of the second triangle.

25. Write Can a triangle have a side length that is the sum of the other two side
₍₃₉₎ lengths? Explain.

26. If in $\triangle ABC$, $AB = 10$, $m\angle ABC = 5°$, and $m\angle CAB = 60°$, and in $\triangle DEF$,
₍₃₀₎ $m\angle EFD = 5°$, $m\angle DEF = 60°$, and $FE = 10$, write a congruency statement for these triangles.

27. Algebra If $x = 1$, which angle would have the least measure in the triangle?
₍₃₉₎

28. For two circles, the first circle has a radius of 0.5 units, and the second radius
₍₂₃₎ is the circumference of the first. Using 3.14 for π, what is the area of the second circle?

29. Multiple Choice Which two formulas are used to calculate the area of this
₍₄₀₎ composite figure?

A $A_1 = 4s$, $A_2 = \frac{1}{2}bh$ **B** $A_1 = \frac{1}{2}\ell w$, $A_2 = bh$

C $A_1 = s^2$, $A_2 = bh$ **D** $A_1 = \ell w$, $A_2 = \frac{1}{2}bh$

30. Write Explain why you can use the formula, $A = \frac{1}{2}d_1 d_2$, for both the area of
₍₂₂₎ rhombuses and the area of squares.

LESSON
42

Finding Distance from a Point to a Line

Warm Up

1. **Vocabulary** Two lines that meet at 90° are _____.
   (5)
2. State the formula for the Pythagorean Theorem: _____.
   (8)
3. The formula to find the distance between two points is:
   (9)

   _____.

4. What is the slope of the line $y = 3x + 1$?
   (16)
5. The slope of the line perpendicular to $y = -2x + 1$ is __.
   (37)

New Concepts

Given a line $\overleftrightarrow{AB}$ and a point P, what is the shortest distance between P and $\overleftrightarrow{AB}$? Notice that $\triangle ABP$ is a right triangle, and $\overline{AP}$ is the hypotenuse. The hypotenuse is always the longest side of a right triangle, so $\overline{AP}$ must be longer than $\overline{PB}$.

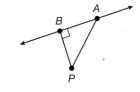

Example 1 Choosing the Closest Point

Which point on the line $y = 0$ is closest to point D—$L(3.6, 0)$, $M(4, 0)$, or $N(4.25, 0)$?

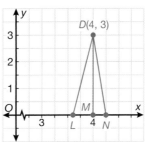

Math Reasoning

Verify Consider the point $(0, 3)$. The perpendicular line from this point to the x-axis is 3 units long. Is there another segment from this point to the x-axis that is shorter? Try using the distance formula to find the distance from $(0, 3)$ to $(1, 0)$ or $(-1, 0)$. Is the distance greater or less?

SOLUTION

Use the distance formula:

$$d = \sqrt{(x_2 - x_1)^2 + (y_2 - y_1)^2}$$

$$DL = \sqrt{(3.6 - 4)^2 + (0 - 3)^2} \approx 3.03$$

$$DM = \sqrt{(4 - 4)^2 + (0 - 3)^2} = 3$$

$$DN = \sqrt{(4.25 - 4)^2 + (0 - 3)^2} \approx 3.01$$

M is the closest point to D.

Theorem 42-1
Through a line and a point not on the line, there exists exactly one perpendicular line to the given line.

Theorem 42-2
The perpendicular segment from a point to a line is the shortest segment from the point to the line.

Theorem 42-1 indicates that there is only one such segment. The length of a perpendicular segment from a point to a line is referred to as the **distance from a point to a line**.

Example 2 **Finding Distance to a Line**

a. Find the distance from $P(6, 4)$ to the line $y = 1$.

SOLUTION
The perpendicular distance is the distance between the y-coordinates of the point and the line.

The distance between point P and the line is 3 units.

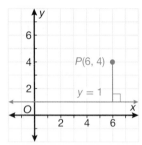

b. Find the distance from $P(6, 4)$ to the line $x = 1$.

SOLUTION
Point P is 5 units to the right of the line $x = 1$. The perpendicular distance is the difference between the x-coordinates. The distance is 5 units.

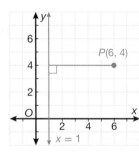

c. Find the distance from $P(6, 4)$ to the line $y = x$.

SOLUTION
Find the line perpendicular to the line $y = x$ that includes point $P(6, 4)$.

The slope of the perpendicular line is the negative reciprocal of the slope of the line $y = x$, which is -1.

Start at point $P(6, 4)$. Use the slope to find other points on the perpendicular line. Draw the line.

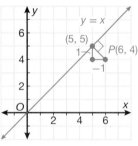

Notice that the lines intersect at $(5, 5)$. Find the distance between $(6, 4)$ and $(5, 5)$ using the distance formula.

$$d = \sqrt{(x_2 - x_1)^2 + (y_2 - y_1)^2} \qquad \text{Distance formula}$$
$$d = \sqrt{(6 - 5)^2 + (4 - 5)^2} \qquad \text{Substitute}$$
$$d = \sqrt{2} \qquad \text{Simplify}$$
$$d \approx 1.414$$

So, the distance from the point to the line is about 1.41 units.

Math Reasoning

Write Describe what happens when you use the distance formula to find the distance between a point and a vertical line.

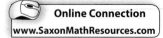
Online Connection
www.SaxonMathResources.com

Example 3 Finding the Closest Point on a Line to a Point

Given the equation $y = 2x + 1$ and the point $S(3, 2)$, find the point on the line that is closest to S. Find the shortest distance from S to the line.

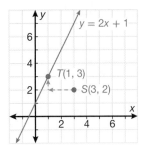

SOLUTION

Draw the line and the point on a coordinate grid.

Next, find the slope of the line. In this case, the slope is 2.

Find the slope of a line perpendicular to the given line. The negative reciprocal of 2 is $-\frac{1}{2}$.

Use the slope to find more points on the perpendicular line. Draw the line. The lines intersect at $(1, 3)$.

Use the distance formula to find the distance between $(3, 2)$ and $(1, 3)$.

$$d = \sqrt{(x_2 - x_1)^2 + (y_2 - y_1)^2}$$ Distance formula

$$d = \sqrt{(1 - 3)^2 + (2 - 3)^2}$$ Substitute.

$$d = \sqrt{(-2)^2 + (1)^2}$$ Simplify.

$$d = \sqrt{5}$$ Simplify.

$$d \approx 2.24$$

So, the distance from the point to the line is about 2.24 units.

> **Hint**
>
> In this example, it is easy to see that the lines intersect at (1, 3). Sometimes, however, it may be necessary to solve the system of equations given by the original line and the perpendicular line to find the exact intersection.

Theorem 42-3
The perpendicular segment from a point to a plane is the shortest segment from the point to the plane.

Because parallel lines are always the same distance from one another, the distance from any point on a line to a line that is parallel is the same, regardless of which point you pick.

Theorem 42-4
If two lines are parallel, then all points on one line are equidistant from the other line.

Example 4 **Proving All Points on Parallel Lines are Equidistant**

Prove that if two lines are parallel, then all the points on one line are equidistant from the other line.

Given: $m \parallel n$
Prove: $\overline{AC} \cong \overline{DB}$

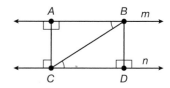

SOLUTION

Draw a diagram like the one above, with two parallel lines cut by a transversal. Draw two perpendicular segments from the endpoints of the transversal to the opposite parallel line, forming $\triangle ABC$ and $\triangle DCB$. To prove Theorem 42-4, we need to show that points A and C are the same distance apart as points B and D, or that $\overline{AC} \cong \overline{DB}$.

Statements	Reasons
1. $m \parallel n$	1. Given
2. $\angle CAB \cong \angle BDC$	2. All right angles are congruent
3. $\angle ABC \cong \angle DCB$	3. Alternate Interior Angles
4. $\overline{BC} \cong \overline{CB}$	4. Reflexive Property
5. $\triangle ABC \cong \triangle DCB$	5. AAS Congruence Theorem
6. $\overline{AC} \cong \overline{DB}$	6. CPCTC

Therefore, all points on one line are equidistant from the other line.

Example 5 **Application: Delivery Routes**

A pizza restaurant only delivers to customers who are less than 3 miles away. The restaurant, located at the origin, receives a call from a customer who lives at the closest point to the restaurant on 5th Street, which can be represented by the line $y = -2x + 3$. If each unit on a coordinate plane represents 1 mile, does this customer live close enough for delivery?

SOLUTION

Draw the restaurant at the origin of a coordinate plane. Graph the line representing 5th Street, as shown in the diagram. To find the shortest distance the customer could be from the restaurant, find a line perpendicular to $y = -2x + 3$ through the origin.

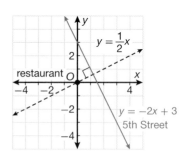

The opposite reciprocal of 5th Street's slope is $\frac{1}{2}$. Use the point-slope formula to find an equation for the line.

$$y - y_1 = m(x - x_1) \qquad \text{Point-slope form}$$
$$y - 0 = \frac{1}{2}(x - 0) \qquad \text{Substitute.}$$
$$y = \frac{1}{2}x \qquad \text{Simplify.}$$

Math Reasoning

Analyze If the restaurant is within 3 miles of a point on 5th Street that is not the closest point, then the closest point will also be within 3 miles. Can you think of such a point on 5th Street?

To find the intersection of these lines, graph them or solve them as a system of equations, as shown below.

$$y = \frac{1}{2}x \qquad\qquad\qquad\qquad y = -2x + 3$$

$$\frac{1}{2}x = -2x + 3$$
$$x = -4x + 6$$
$$5x = 6$$
$$x = 1.2$$

Substituting this value into either equation will reveal that the y-coordinate for this point is 0.6. Use the distance formula to find the distance between (1.2, 0.6) and the origin.

$d = \sqrt{(x_2 - x_1)^2 + (y_2 - y_1)^2}$	Distance equation
$d = \sqrt{(1.2 - 0)^2 + (0.6 - 0)^2}$	Substitute.
$d = \sqrt{1.2^2 + 0.6^2}$	Simplify.
$d \approx 1.34 \text{ mi}$	Simplify.

The distance between these points is less than 3 miles, so the customer is within delivery distance.

Lesson Practice

a. Find the distance between the line $y = 6$ and $(-3, -5)$.
(Ex 2)

b. Find the distance between the line $x = 9$ and $(-3, -5)$.
(Ex 2)

c. Find the distance between the line $y = 2x$ and $(6, 2)$.
(Ex 2)

d. Find the closest point on the line $y = 3x - 1$ to $(5, 4)$.
(Ex 3)

e. Mitchell lives on Woodland Avenue at the closest point to his school.
(Ex 5) The equation $y = 3x + 3$ can be used to represent Woodland Avenue. His school lies at the origin. The school bus will pick him up only if he lives farther than 2 miles from the school. If each unit on a coordinate plane represents 1 mile, find the distance from his house to the school to determine if he will be allowed to ride the school bus.

Practice   Distributed and Integrated

1. **Carpentry** Denzel needs to cut a replacement support for the roof of
(36) his house. The existing supports are 18 inches long and fit against the wall 10 inches below the roof. Explain how Denzel can make sure the triangles in this diagram are congruent.

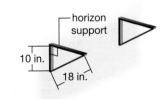

2. The outer circumference of Stonehenge, a ring of standing
(23) stones in England, is about 328 feet. To the nearest tenth of a foot, what is the length of the radius?

3. What is the measure of $\overset{\frown}{AB}$ in this circle?
(26)

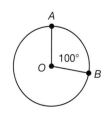

4. The side lengths of a triangle are 5, 12, and c, respectively. If $c = 14$, is the
(33) triangle obtuse, acute, or right? …if $c = 13$? …if $c = 12$?

* 5. **Multi-Step** Which side is longer, $\overline{AB}$ or $\overline{DE}$? Explain.
(Inv 4)

* 6. Complete this two-column proof.
(27) **Given:** ∠1 and ∠2 are straight angles
Prove: ∠1 ≅ ∠2

* 7. **Multiple Choice** The sides of a triangle are 6, 9, and 12 inches long. If the shortest
(41) side of a similar triangle is 19.2 inches, what is the longest side of the triangle?
  A 25.2 in.   **B** 25.6 in.
  C 28.8 in.   **D** 38.4 in.

* 8. In △ABC and △DEF, ∠B and ∠E are right angles, $\overline{BC} ≅ \overline{EF}$, and ∠C ≅ ∠F. Use
(36) the Leg-Angle Congruence Theorem to prove that △ABC ≅ △DEF.

* 9. Find the distance between (4, 5) and the line $y = -\dfrac{3}{2}x - 2$.
(42)

xy^2 **10.** **Algebra** Line j passes through points (4, 6) and (3, 2). Line k passes through points
(37) $(x, -1)$ and $(-3, 3)$. What value of x makes these lines parallel?

11. Determine the area of this figure.
(40)

12. ⟨ **Framing** ⟩ Justine has built a frame for her painting that measures
(29) 27 by 36 inches. In order for the corners of the frame to be at right
angles, what length must the diagonal be?

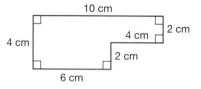

13. Give two Pythagorean triples that are proportional to the triple 5, 12, 13.
(29)

*14. **Analyze** In the diagram shown, prove that m∠BAC = m∠ABC.
(27)

*15. **Analyze** Find the distance between the line $y = 4x - 1$ and the
(42) point (4, 15).

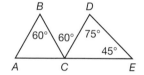

*16. Find the distance from $(-2.15, 3.28)$ to the line $x = 11.3$.
(42)

17. ⟨ **Design** ⟩ Dean wants to make a symmetric bowtie. As in the diagram,
(30) he knows that $AB = CD$, and m∠BAE = m∠DCE. Prove, using a
two-column proof, that △ABE ≅ △CDE.

18. What is the included side of △RST that is between ∠TRS and ∠RST?
(28)

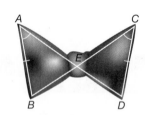

19. Use the Converse of the Corresponding Angles Postulate to prove that
(12) lines q and r in this figure are parallel.

20. **Write** Three side lengths in a triangle are given by the expressions
(39) $2x$, $4x$, and $3x$. Which would be opposite to the angle with the greatest
measure? Explain.

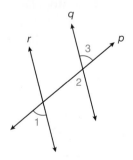

21. **Multi-Step** Find the slope of the line that passes through the points (3, 4)
(37) and $(-1, 3)$. Then find a parallel line that passes through point (4, 2).

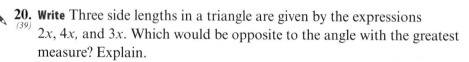

22. The steps of the algebraic proof below are correct. However, the justifications for each step are out of order. Determine the correct order for the justifications.

Statements	Reasons
1. $9x = 3(2x + 3)$	1. Simplify
2. $9x = 6x + 9$	2. Division Property of Equality
3. $9x - 6x = 6x + 9 - 6x$	3. Given
4. $3x = 9$	4. Simplify
5. $\dfrac{3x}{3} = \dfrac{9}{3}$	5. Subtraction Property of Equality
6. $x = 3$	6. Distributive Property

*23. Find the distance from $(-1.2, 1.7)$ to the line $x = 3$.

24. State the converse of the following statement, and write the two statements as a biconditional.

If two numbers are positive, then their product is positive.

25. A triangle has an area of 4, a base of $2x + 4$, and a height of $x - 1$. Solve for x and provide a justification of each step.

*26. **Algebra** If rectangles *KLMN* and *QRST* are similar, and $LM = 4$, $RS = 9$, and *QR* is 5 greater than *KL*, write and solve a proportion to find the measures of $\overline{KL}$ and $\overline{QR}$.

27. In this diagram, find the value of x and the lengths of $\overline{PC}$, $\overline{CM}$, and $\overline{PM}$ if C is the centroid of the triangle.

*28. **Generalize** Explain why each statement is true or false.

  a. *Two squares are always similar.*

  b. *A rectangle and a square are always similar.*

29. **Justify** A triangle has side lengths that are 5, 8, and 11. Classify the triangle by its angles. Cite a theorem or postulate to justify your answer.

30. (**Games and Puzzles**) Determine the area of each tangram piece.

  a.

4 cm

  b.

2 cm

  c.

2 cm

4 cm

  d.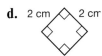

2 cm   2 cm

  e.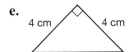

4 cm   4 cm

Perpendicular through a Point Not on a Line

Construction Lab 7 *(Use with Lesson 42)*

Lesson 42 focuses on finding the distance from a given point to a line. This distance is defined by the perpendicular segment from the point to the line. This lab shows you how to construct a line that is perpendicular to a given line through a point not on that line.

① Begin with line $\overleftrightarrow{AB}$ and a point P not on the line.

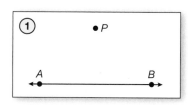

② Use a compass to draw an arc centered at P that intersects $\overleftrightarrow{AB}$ at two points. Label the points of intersection L and M.

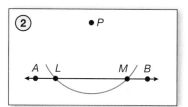

③ Now construct the perpendicular bisector of $\overline{LM}$ by drawing an arc centered at L that extends above and below $\overleftrightarrow{AB}$, and repeating this with an arc centered at M.

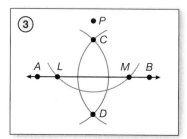

Hint

If you use the same compass setting, one of the points where the two arcs intersect will be P.

④ Label the points where the two arcs intersect C and D.

Notice that $\overleftrightarrow{CD}$ passes through P and is perpendicular to $\overleftrightarrow{AB}$.

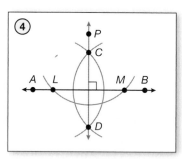

Lab Practice

For each given line and point, graph the line, then construct a line that is perpendicular to it, and which passes through the given point.

a. $y = 2x - 2$; $(0, 3)$ **b.** $y = \frac{1}{3}x + 1$; $(5, 6)$

Chords, Secants, and Tangents

1. **Vocabulary** The distance around a circle is the circle's _____.
(23)

2. A line segment with one endpoint at the center of the circle and the other
(23) endpoint on the circle is called a:

   A chord   **B** secant

   C radius   **D** minor arc

3. The central angle of a circle divides the circle into two arcs. These arcs
(26) are called the:

   A major arc and minor arc   **B** semicircle and minor arc

   C radius and chord   **D** large arc and small arc

4. What is the radius of a circle with an area of 125 cm^2, to the nearest
(23) hundredth of a centimeter?

New Concepts

The diameter and radius of a circle are two special
segments that can be used to find properties of a circle.
There are three more special segments common to every
circle. They are chords, secants, and tangents.

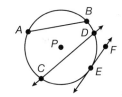

A **chord** is a line segment whose endpoints lie on a circle.
In the diagram, $\overline{AB}$ and $\overline{CD}$ are chords of $\odot P$.

A **secant of a circle** is a line that intersects a circle at two points. In the
diagram, $\overleftrightarrow{CD}$ is a secant of $\odot P$.

A **tangent of a circle** is a line in the same plane as the circle that intersects the
circle at exactly one point, called a **point of tangency**. E is a point of tangency
on $\odot P$, and $\overleftrightarrow{EF}$ is a tangent line.

Example **1** **Identifying Lines and Segments
that Intersect Circles**

Use the figure at right to answer parts **a**, **b**, and **c**.

a. Identify three radii and a diameter.

SOLUTION Three radii are $\overline{PL}$, $\overline{PQ}$ and $\overline{PN}$. $\overline{PM}$
and $\overline{PR}$ are also radii. A diameter is $\overline{NQ}$.

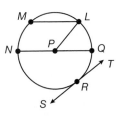

Math Reasoning

Analyze Is a diameter
also a chord? If so, why?

b. Identify two chords.

SOLUTION One chord is $\overline{ML}$. Another chord is the diameter, $\overline{NQ}$.
Since any two endpoints on the circle can make a chord, you could also
answer $\overline{MQ}$, $\overline{QR}$, $\overline{QL}$, $\overline{MR}$, $\overline{MN}$, $\overline{NR}$, $\overline{LN}$ and $\overline{LR}$ are also chords.

c. Name a tangent to the circle and identify the point of tangency.

SOLUTION The tangent is $\overleftrightarrow{ST}$. The point of tangency is R.

These special segments can be used to find unknown lengths in circles with the help of the theorems presented in this lesson.

Theorem 43-1
If a diameter is perpendicular to a chord, then it bisects the chord and its arcs.

Theorem 43-2
If a diameter bisects a chord other than another diameter, then it is perpendicular to the chord.

Example 2 **Finding Unknowns Using Chord-Diameter Relationships**

(a.) Find the length of x.

SOLUTION
Since the diameter is perpendicular to the chord, the chord is bisected by the diameter by Theorem 43-1.

So, $x = 3$.

(b.) Find m$\angle ORP$.

SOLUTION
Since the diameter bisects the chord, the chord must be perpendicular to the diameter by Theorem 43-2.

Thus, m$\angle ORP = 90°$

(c.) The circle shown has a diameter of 14 inches. Chord $\overline{AB}$ is 8 inches long. How far is $\overline{AB}$ from the center of the circle, to the nearest hundredth of an inch?

SOLUTION
Construct $\overline{AP}$ and $\overline{CP}$ so C is the midpoint of $\overline{AB}$. Since C is the midpoint of $\overline{AB}$, $\overline{AC}$ is 4 inches long. $\overline{AP}$ is a radius of the circle, so it is 7 inches long.

Since $\triangle PAC$ is a right triangle, you can apply the Pythagorean Theorem to find the length of $\overline{CP}$.

$$7^2 = 4^2 + x^2$$
$$49 = 16 + x^2$$
$$33 = x^2$$
$$x \approx 5.74$$

The chord is about 5.74 inches from the center of the circle.

Hint

Finding the lengths of chords or other measurements in the circle will often require drawing an auxiliary line that makes a right triangle with the chord and the diameter of the circle.

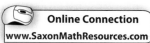
Online Connection
www.SaxonMathResources.com

In fact, any segment that is a perpendicular bisector of a chord is also a diameter of the circle. This leads to Theorem 43-3.

Theorem 43-3
The perpendicular bisector of a chord contains the center of the circle.

Every diameter passes through the center of the circle, so another way of stating Theorem 43-3 is that the perpendicular bisector of a chord is also a diameter or a line containing the diameter.

Example 3 **Proving the Perpendicular Bisector of a Chord Contains the Center**

Given: $\overline{EF}$ is the perpendicular bisector of $\overline{AB}$.
Prove: $\overline{EF}$ passes through the center of $\odot P$.

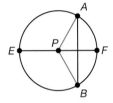

Math Reasoning

Write Write the proof of Theorem 43-3 as a paragraph proof.

SOLUTION

Statements	Reasons
1. $\overline{EF}$ is the perpendicular bisector of $\overline{AB}$.	1. Given
2. $\overline{PA}$ and $\overline{PB}$ are congruent.	2. Definition of a radius
3. Point P lies on the perpendicular bisector of $\overline{AB}$.	3. If a point is equidistant from the endpoints of a segment, then the point lies on the perpendicular bisector of the segment (Theorem 6-6).
4. $\overline{EF}$ passes through the center of $\odot P$.	4. $\overline{EF}$ is the perpendicular bisector of $\overline{AB}$.

Therefore, the center of the circle lies on the perpendicular bisector of a chord.

One final property of chords is that all chords that lie the same distance from the center of the circle must be the same length, as stated in Theorem 43-4.

Theorem 43-4
In a circle or congruent circles: • Chords equidistant from the center are congruent. • Congruent chords are equidistant from the center of the circle.

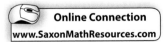

Online Connection
www.SaxonMathResources.com

Math Reasoning

Write Combine the two statements in Theorem 43-4 to make a biconditional statement.

Example 4 **Applying Properties of Congruent Chords**

Find CD, if $AP = 5$ units and $PE = 3$ units.

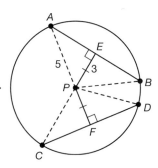

SOLUTION

$\overline{AB}$ and $\overline{CD}$ are equidistant from the center of the circle, so they are congruent by Theorem 43-4. Use the Pythagorean Theorem to find the length of $\overline{AE}$.

$a^2 + b^2 = c^2$
$3^2 + b^2 = 5^2$
$\quad b = 4$

By Theorem 43-1, E is the midpoint of $\overline{AB}$, so $AB = 8$. Since $\overline{AB} \cong \overline{CD}$, $CD = 8$ units.

Example 5 **Application: Plumbing**

Two identical circular pipes have diameters of 12 inches. Water is flowing 1 inch below the center of both pipes. What can be concluded about the width of the water surface in both pipes? Calculate the width.

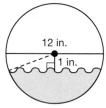

SOLUTION

Taking a cross section of the pipe, the width of the water surface is a chord of the circle. Since it is given that the water surface is one inch below the center in both pipes, we can conclude that the width of the water surface is equal in both pipes.

The radius that has been drawn into the diagram forms a right triangle with the surface of the water and the distance between the water and the center of the pipe. The radius of the pipe is 6 inches. Let x represent half the width of the water surface. Use the Pythagorean Theorem to solve for x.

$a^2 + b^2 = c^2$
$1^2 + x^2 = 6^2$
$\quad x^2 = 35$
$\quad\quad x \approx 5.92$

The width of the water surface is two times x, or approximately 11.84 inches.

Lesson Practice

a. Identify each line or segment that intersects the circle.
(Ex 1)

b. Determine the value of a.
(Ex 2)

c. Determine the value of x.
(Ex 2)

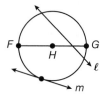

d. Prove the first part of Theorem 43-1: If a diameter is
(Ex 3) perpendicular to a chord, then it bisects the chord.
Given: $\overline{CD} \perp \overline{EF}$
Prove: $\overline{CD}$ bisects $\overline{EF}$

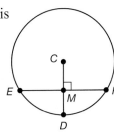

e. If a circle has a diameter of 9 inches and a chord
(Ex 4) that is 5 inches long, what is the distance from the
chord to the center of the circle, to the nearest
hundredth of an inch?

f. A pipe with an 16-inch diameter has water flowing through it. If the
(Ex 5) water makes a chord across the pipe that is 15 inches long, how close to
the center of the pipe is the water?

Practice Distributed and Integrated

1. Which of the following is not a parallelogram?
(34) **A** rhombus **B** rectangle
C square **D** trapezoid

* 2. (**Farming**) A farmer has 264 feet of fencing to enclose a triangular field. If the
(41) farmer wants the sides of the triangle to be in the ratio 3:4:5, what is the length of
each side of the field?

3. **Algebra** A convex polygon has exterior angles $(4x + 10)°$, $(5x + 2)°$, and $(x + 8)°$.
(Inv 3) What is the value of x?

* 4. Find the point on the line $y = -0.25x + 9$ that is closest to $(1, -4)$.
(42)

5. In the diagram, $m\widehat{MN} = 50°$ and $m\widehat{MP} = 80°$. What is the measure of $\widehat{NP}$?
(26)

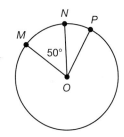

* 6. **Formulate** If $\angle 1$ and $\angle 2$ are adjacent central angles of a circle, show that
(35) the sum of their associated sector areas is equal to the sector area of
their angle sum.

* 7. Identify each line or segment that intersects the circle at right.
(43) What type of intersecting line is each one?

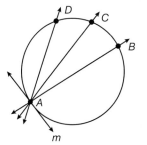

8. **Multi-Step** Graph the line and find the slope of the line that
(37) passes through the points $D(4, 2)$ and $E(-3, 6)$. Then find
the equation of a perpendicular line that passes through
the point $F(-1, 2)$.

9. **Error Analysis** A composite figure is made up of a semicircle and
(40) a rectangle. The width of the rectangle is 3 centimeters, and
the diameter of the semicircle is 5 centimeters. Derek calculated
the perimeter of the figure like this:

$P = 3 + 3 + 5 + \pi d$
$ = (11 + 5\pi) \text{ cm}$

Find and correct Derek's error.

10. (**Carpentry**) A 25-meter-long diagonal piece of wood runs across a rectangular
(29) platform that has a length-to-width ratio of 4:3. What are the dimensions of the
platform?

11. When are all the medians and altitudes of a triangle congruent?
(32)

12. **Multiple Choice** Which congruence theorem applies to these triangles?
(36) **A** Leg-Leg Theorem   **B** Leg-Angle Theorem
C Hypotenuse-Leg Theorem   **D** Hypotenuse-Angle Theorem

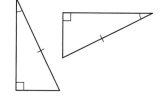

13. **Algebra** Each central angle of a regular polygon is 1°. How many sides
(Inv 3) does the polygon have?

14. Solve the equation $3(2x - 1) = 15$ with a justification for each step.
(24)

*15. A circle has a diameter of 22 centimeters. A chord in the circle is
(43) 11 centimeters long. What is the *exact* distance from the chord to the center of
the circle?

16. **Given:** $\overline{BE}$ is the perpendicular bisector of $\overline{AC}$ at point D.
(38) **Prove:** $\triangle ABD \cong \triangle CBD$

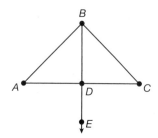

*17. **Algebra** Given the equation $y = 2x + 1$, and $R(4, 4)$, find the point on the line that
(42) is closest to R.

18. **Multi-Step** Find the orthocenter of $\triangle MNP$ with vertices $M(2, 2)$, $N(2, 8)$, and
(32) $P(-6, 4)$.

19. **Formulate** An arc measure of 180° forms a semicircle. Use this fact to derive a
(26) formula for the perimeter of a semicircle.

20. (**Disc Golf**) In disc golf, a player tries to throw a flying disc into a metal basket
(39) target. Four of the disc golf targets are shown in the diagram. Which of the
targets are closest together?

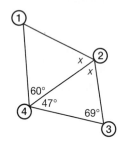

21. Identify the hypothesis and conclusion of the following statement.
(10)

If two planes intersect, then they intersect on exactly one line.

*22. Write Explain the relationship between two chords that are equidistant from the
(43) center of a circle.

23. The area of a circle is 40.6 square meters. If an arc's length is 2.8 meters, what is
(35) the arc's measure?

24. (**Driving**) Esther normally takes Exeter Avenue to work, but it is closed for
(29) repairs. For the detour, she must travel south on Central Street for 2.2 miles
and then east on Court Street for 3.4 miles. How much longer is the detour
than her normal route, to the nearest tenth of a mile?

*25. Find the distance between the two parallel lines $y = 3x + 1$ and $y = 3x - 18$.
(42)

*26. Model Can the following be modeled with straws? If not, explain.
(41)

   a. Two similar equilateral triangles.

   b. Two equilateral triangles that are not similar.

27. Write a paragraph proof.
(31) **Given:** Angle 1 is congruent to ∠2.
Prove: Angle 2 and ∠3 are supplementary.

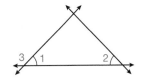

28. (**Skiing**) Developers are designing a new ski hill in the Rocky Mountains.
(29, 33) They use the standard difficulty rating of ski runs as green (easy), blue
(medium), and black (difficult). To determine what color each of the
runs should be, developers use a slope formula. Green runs have a
slope of $\frac{1}{3}$. Blue runs have a slope of $\frac{1}{2}$. Black runs have a slope of 1.
If all the runs start at 150 yards high, what is the length of each run, to
the nearest tenth of a yard?

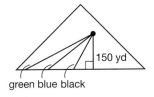

*29. Find the distance from the center of a circle with a diameter of 2.6 inches to a
(43) chord that is 1.2 inches in length, to the nearest hundredth of an inch.

30. Justify Which is the longer side, *x* or *y*? Explain.
(Inv 4)

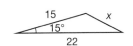

Applying Similarity

Warm Up

1. **Vocabulary** In the proportion $2:5 = 7:x$, the values 2 and x are called
   the _____ of the proportion.
   (41)

2. The angles in a triangle have the ratio 1:2:3. Find the measures of
   (41) the angles.

3. Given the proportion $5:3:2 = 7:2x:4y$, find the values of x and y.
   (41)

4. In the diagram, $\triangle ABC \sim \triangle DEF$. Find the values of x and y to the
   (41) nearest hundredth.

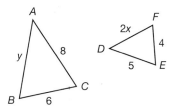

5. **Multiple Choice** The ratio 2 to 4.1 is the same as:
   (41)
   A 4 to 8.1   **B** 3 to 6.1
   C 6 to 12.3   **D** 5 to 10

New Concepts

Recall from Lesson 41 that proportions can be used to find unknown
measures in similar polygons. Any two regular polygons with the same
number of sides are similar. Therefore, all regular polygons with the same
number of sides are similar to each other.

Example **1**   **Using Similarity to Find Unknown Measures**

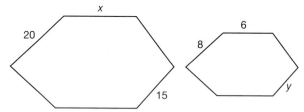

The hexagons in the diagram are similar. Find the values of x and y.

SOLUTION
By looking at the corresponding segments with known lengths, a similarity
ratio can be written: 20:8. Now, use a proportion to solve for x and y.

$$\frac{20}{8} = \frac{x}{6} \qquad\qquad \frac{20}{8} = \frac{15}{y}$$
$$6 \times 20 = 8x \qquad\qquad 20y = 15 \times 8$$
$$120 = 8x \qquad\qquad 20y = 120$$
$$x = 15 \qquad\qquad y = 6$$

Math Reasoning

Verify Are all isosceles
triangles similar? Are all
equiangular triangles
similar? Why or why not?

Example 2 Applying Similarity to Solve for Unknowns

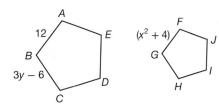

a. Pentagons *ABCDE* and *FGHIJ* are regular pentagons, and are similar to each other. The similarity ratio of *ABCDE* to *FGHIJ* is 3:2. Find the values of *x* and *y*.

SOLUTION
Set up a proportion using the similarity ratio and the ratio of *AB* to *FG*, then find the cross product and solve for *x*.

$$\frac{3}{2} = \frac{12}{x^2 + 4}$$ Set up a proportion.

$3(x^2 + 4) = 12 \times 2$ Cross multiply.

$3x^2 + 12 = 24$ Distribute and simplify.

$x = \pm 2$ Solve.

Since *ABCDE* is a regular pentagon, its sides are congruent. Therefore, *BC* = *AB*.

$3y - 6 = 12$ Substitute.

$3y = 18$ Add 6 to each side.

$y = 6$ Divide each side by 3.

b. What is the ratio of the perimeter of *ABCDE* to *FGHIJ*?

SOLUTION
Since both pentagons are regular, their sides are all congruent. *ABCDE* has five sides, each measuring 12 units, so its perimeter is 60.
Substituting $x = \pm 2$ to find the length of a side of *FGHIJ* shows that each side measures 8 units, so the perimeter of *FGHIJ* is 40.

Therefore, the ratio of the perimeters is 60:40, which reduces to 3:2.

As you can see from Example 2, the perimeters of two similar figures share the same similarity ratio as their sides.

Hint

Often, the negative answer to a root is disregarded in geometry, because a figure cannot have a negative measurement. In this case though, using the value $x = -2$ still results in a positive side length, so both answers are valid.

Theorem 44-1
If two polygons are similar, then the ratio of their perimeters is equal to the ratio of their corresponding sides.

Given $\triangle PQR \sim \triangle STU$, prove that the ratio of their perimeters is 1:2 if the ratio of their corresponding sides is 1:2.

SOLUTION

Use a 2-column proof.

Statements	Reasons
1. $\triangle PQR \sim \triangle STU$	1. Given
2. $\dfrac{PQ}{ST} = \dfrac{QR}{TU} = \dfrac{RP}{US} = \dfrac{1}{2}$	2. Given
3. $2PQ = ST$	3. Cross multiply
4. $2QR = TU$	4. Cross multiply
5. $2RP = US$	5. Cross multiply
6. perimeter of $\triangle STU =$ $ST + TU + US$	6. Definition of perimeter
7. perimeter of $\triangle STU =$ $2PQ + 2QR + 2RP$	7. Substitution Property of Equality
8. perimeter of $\triangle STU =$ $2(PQ + QR + RP)$	8. Simplify
9. perimeter of $\triangle PQR =$ $PQ + QR + RP$	9. Definition of perimeter
10. perimeter of $\triangle STU =$ $2(\text{perimeter of } \triangle PQR)$	10. Substitution Property of Equality

Therefore, the ratio of the perimeter of $\triangle PQR$ to the perimeter of $\triangle STU$ is 1:2.

Example `4` **Applying Similarity to Solve a Perimeter Problem**

Figures $HIJK$ and $LMNO$ are similar polygons. Their corresponding sides have a ratio of 2:5. If the perimeter of figure $HIJK$ is 27 inches, what is the perimeter of figure $LMNO$?

SOLUTION

Because $HIJK$ and $LMNO$ are similar polygons, the ratio of their perimeters is equal to the ratio of their corresponding sides.

Therefore, the ratio of $HIJK$'s perimeter to $LMNO$'s perimeter is 2:5.

Set up a proportion using this ratio to solve for the perimeter of $LMNO$.

$$\frac{27}{x} = \frac{2}{5}$$
$$2(x) = 5(27)$$
$$x = 67.5$$

The figure $LMNO$ has a perimeter of 67.5 inches.

Hint

A 2-column proof is good for proofs that use a lot of geometric symbols or algebra. A paragraph proof of this theorem would be difficult to read.

Example 5   Application: Map Scales

Foxx plans to jog 5000 meters a day in training for a race. The park where Foxx jogs is in the shape of a regular pentagon. The side length of the park is 5 centimeters long on a map with the scale $\frac{1 \text{ cm}}{50 \text{ m}}$. How many times does Foxx need to jog along the perimeter of the park to complete his daily training?

SOLUTION

First, find the perimeter of the park on the map.

Perimeter of a regular pentagon = 5s

$$P = 5(5 \text{ cm})$$
$$P = 25 \text{ cm}$$

The park on the map and the actual park are similar polygons. Therefore, the ratio of their perimeters is the same as the ratio of their corresponding sides.

So, the ratio of the park's perimeter on the map to the perimeter of the actual park is $\frac{1 \text{ cm}}{50 \text{ m}}$.

$$\frac{\text{perimeter on the map}}{\text{perimeter of the actual park}} = \frac{1 \text{ cm}}{50 \text{ m}}$$

$$\frac{25 \text{ cm}}{\text{perimeter of the actual park}} = \frac{1 \text{ cm}}{50 \text{ m}}$$

perimeter of the actual park $\times$ 1 cm = 50 m $\times$ 25 cm

$$\text{perimeter of the actual park} = \frac{50 \text{ m} \times 25 \text{ cm}}{1 \text{ cm}}$$
$$= 1250 \text{ m}$$

Since 1250 is $\frac{1}{4}$ of 5000, Foxx needs to jog along the perimeter of the park 4 times to complete his daily training.

Math Reasoning

Formulate Would Theorem 44-1 also apply to the circumference of a circle? Why or why not?

Lesson Practice

a. $\triangle JKL$ and $\triangle FGH$ are similar triangles. Find the values of x and y.
(Ex 1)

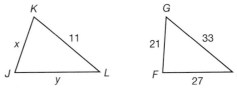

b. In $\triangle ABC$, $AB = x^2 - 7$, $BC = y + 4$, and $CA = 5$. In $\triangle DEF$, $DE = 6$, $EF = 12$, and $FD = 15$. $\triangle ABC \sim \triangle DEF$. Find the values of x and y. Then find the ratio of the perimeters of the two triangles.
(Ex 2)

c. Figures $ABCD$ and $EFGH$ are similar. The ratio of their corresponding sides is 3:5. If the perimeter of $EFGH$ is 45 inches, what is the perimeter of figure $ABCD$?
(Ex 4)

d. Pentagons $ABCDE$ and $FGHIJ$ are similar figures. The perimeter of $ABCDE$ is 32 centimeters. The similarity ratio of $ABCDE$ to $FGHIJ$ is 2:9. What is the perimeter of $FGHIJ$?
(Ex 4)

e. Jana and her brother Jacob are desiging their own tree house with two
separate doors, one that is proportional to Jana's height and one that
(Ex 5) is proportional to Jacob's height. Jacob is 3 feet tall and Jana is 4 feet
tall, so Jana decides that her door should be 5 feet tall by 2 feet wide.
How tall should Jacob's door be, and what will its perimeter be?

Practice   Distributed and Integrated

1. If the triangles shown are similar so that 3:4:5 = *x*:*y*:15, what are the values of
(41) *x* and *y*?

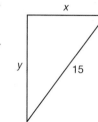

2. (**Design**) Billy is making a kite *STUV*. He knows that $\overline{ST}$ and $\overline{TU}$ are
(28) the same length. Which angles does he need to know are
equal to ensure that the two halves of the kite, △*STV* and △*UTV*,
are congruent?

3. (**Restaurants**) A restaurant owner plans to add a triangular patio to her restaurant,
(38) as shown. She wants to position a fountain on the patio that is the same distance
from each edge. Where should she position the fountain? Explain your reasoning.
Support your answer with a diagram.

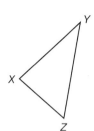

4. Write Explain how the Pythagorean Theorem can be used to ensure that
(29) two pieces of metal in a frame are at right angles to each other.

5. Verify Name two properties of parallelograms that can be used to justify the
(34) congruency marks on this diagram.

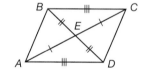

* 6. Given △*FGH* ~ △*MNP*, with *FG* = 8 and *MN* = 16, what is the reduced
(44) ratio of their corresponding sides?

7. a. Determine the slope and *y*-intercept of this graph of a line.
(16) **b.** Write the equation of the line.

* 8. Figure *RSTU* is similar to figure *KLMN*. The ratio of their corresponding
(44) sides is 1:2. If the perimeter of *RSTU* is 10 inches, what is the perimeter
of *KLMN*?

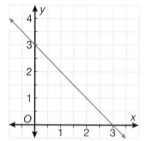

9. Verify Wesley has two triangles, △*ABC* and △*XYZ*. He knows that
(30) m∠*ABC* = 67°, m∠*BAC* = 23°, *AB* = 5 units, *BC* = 13 units, m∠*XYZ* =
67°, m∠*YXZ* = 23°, and *ZY* = 13 units. He concludes that *XY* = 5 units.
Verify that Wesley is correct.

*10. Multi-Step Determine the perimeter of this figure. Express
(40) your answer in terms of *π*.

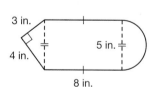

*11. The radius of a circle is 15 centimeters. If a chord measuring
(43) 24 centimeters is perpendicular to the radius and cuts it into
two parts, find the length of each part of the radius.

12. Write When finding a perpendicular or parallel line to a given line, why is the
(37) y-intercept not used?

*13. Elijah wants to change the size of the tires on his bike. The old tires have a radius
(44) of 12 inches. The new tires have a radius of 15 inches. What is the ratio of the
radius of the old tires to that of the new tires?

14. Justify Explain how you know a triangle can have side lengths measuring 11,
(39) 15 and 21.

*15. Given the equation $y = \frac{2}{3}x - 2$ and $G(1, 3)$, find the point on the line that is
(42) closest to G.

Refer to this diagram to answer the next two questions.

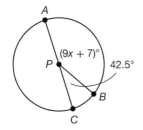

16. a. If $\overline{AC}$ is a diameter, what is the value of x?
(35) **b.** If $PA = 10$ centimeters, what is the length of $\overset{\frown}{AB}$ to the nearest tenth of a
centimeter? Of major arc $\overset{\frown}{ACB}$?

17. If the radius of the circle is 25 inches, what is the area of sector BPC to the
(35) nearest tenth of a square inch?

*18. Find the distance from $(-1, 2)$ to the line $y = \frac{3}{2}x - 3$.
(42)

19. Error Analysis Lionel calculated the sum of the interior angles of a convex polygon
(Inv 3) to be $1710°$. Explain how you know he made an error.

20. Is a triangle with side lengths that measure 3.6, 4.8, and 6.2 a right, obtuse, or
(33) acute triangle?

*21. Using the diagram below, determine the value of x.
(43)

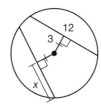

22. Emergency Alert A town wants to place a weather emergency siren where the three
(38) schools can all hear the siren to alert them in case of an emergency. Explain how
the town might go about finding the optimal location for the siren.

23. **Multiple Choice** If the triangles are similar, which of these
(41) is the length of $\overline{BC}$?

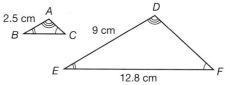

   A 1.3 centimeters   **B** 1.8 centimeters

   C 3.2 centimeters   **D** 3.6 centimeters

24. If the sum of the measures of the interior angles in a regular
(Inv 3) polygon is 1260°, how many sides does the polygon have?

25. **Analyze** Jason is walking around an empty field, from the southwest corner to the
(29) northeast corner. The field is 140 yards long and 90 yards wide. How much farther
will Jason have to walk if he walks around the edge of the field than if he cuts
through the middle of the field, to the nearest yard? _____

*26. $\triangle CDE$ is an equilateral triangle. Its perimeter is 45 inches. If $\triangle CDE \sim \triangle VWX$,
(44) and the ratio of their corresponding sides is 3:1, what are the lengths of the sides
of $\triangle VWX$?

27. A composite figure is made up of a right triangle and three squares. The triangle
(40) has a base length of 8 centimeters and a height of 15 centimeters. Each square fits
exactly along one side of the triangle.
   a. Determine the side length of the largest square.
   b. Determine the perimeter of the figure.
   c. Determine the area of the figure.

28. **Algebra** Line p passes through points (4, 7) and (4, 3). Line q passes through points
(37) (–2, 1) and (5, 1). Are the lines parallel, perpendicular, or neither?

29. **Carpentry** Deklynn and Bonita are making two congruent triangles out of wood.
(25) Deklynn's triangle, $\triangle FGH$ has $m\angle G = 70°$ and $m\angle H = 80°$. If Bonita's triangle is
$\triangle PQR$ and $\triangle FGH \cong \triangle PQR$, how large should she make $m\angle P$?

*30. Identify each line or segment that intersects the circle. What type of
(43) intersecting line is each one?

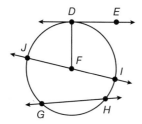

Introduction to Coordinate Proofs

Warm Up

1. **Vocabulary** The formula $d = \sqrt{(x_2 - x_1)^2 + (y_2 - y_1)^2}$ gives the
   (9)
   _____ between two points.

2. Find the length of the line segment joining points $A(1, 3)$ and $B(4, 2)$, to
   (9)
   the nearest tenth.

3. What is the midpoint of a line segment with endpoints $(3, -1)$ and
   (11)
   $(5, 7)$?

4. Find the slope of the line joining points $(-1, 4)$ and $(2, -3)$.
   (16)

5. Find the length of the line segment joining the origin and $(3, 4)$.
   (9)

New Concepts

A **coordinate proof** is a style of proof that uses coordinate geometry and algebra. In a coordinate proof, a diagram is used that is placed on the coordinate plane. Figures can be placed anywhere on the plane, but it is usually easiest to place one side on an axis or to place one vertex at the origin.

Example 1 Positioning a Figure on the Coordinate Plane

Math Reasoning

Error Analysis In Example 1, if you were to place points *B* and *C* at (3, 0) and (0, 4) instead of the coordinates given in the solution, would the solution still be valid? Explain.

Triangle *ABC* has a base of 4 units and a height of 3 units. Angle *A* is a right angle. Position $\triangle ABC$ on the coordinate plane.

SOLUTION

There are various ways to position the triangle on the coordinate plane. A simple way is to use the origin (0, 0) as the vertex for *A*.

Place one of the legs of the triangle on the *x*-axis, and place the other leg on the *y*-axis.

On the *x*-axis, label *B*(4, 0). On the *y*-axis, label *C*(0, 3). Draw the triangle.

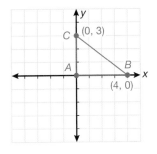

When a figure is placed in a convenient position on the coordinate plane, the equations and values used in a proof will be easier to work with. Below are examples of convenient placement for common figures.

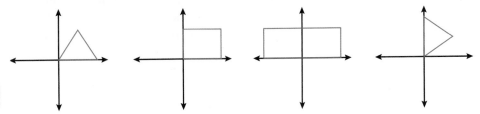

Example 2 Writing a Proof Using Coordinate Geometry

Use a coordinate proof to show that $\triangle HIJ$ is an isosceles triangle.

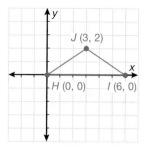

J (3, 2)

H (0, 0) I (6, 0)

SOLUTION

If $\triangle HIJ$ is isosceles then, by definition, two of its sides must have equal length. Calculate each of the side lengths to verify that $\triangle HIJ$ is an isosceles triangle.

$$JI = \sqrt{(x_j - x_i)^2 + (y_j - y_i)^2} \qquad HJ = \sqrt{(x_j - x_h)^2 + (y_j - y_h)^2}$$
$$= \sqrt{(3 - 6)^2 + (2 - 0)^2} \qquad\quad = \sqrt{(3 - 0)^2 + (2 - 0)^2}$$
$$= \sqrt{(-3)^2 + 2^2} \qquad\qquad\quad = \sqrt{3^2 + 2^2}$$
$$= \sqrt{9 + 4} \qquad\qquad\qquad\quad = \sqrt{9 + 4}$$
$$= \sqrt{13} \qquad\qquad\qquad\qquad = \sqrt{13}$$

$$HI = \sqrt{(x_i - x_h)^2 + (y_i - y_h)^2}$$
$$= \sqrt{(6 - 0)^2 + (0 - 0)^2}$$
$$= \sqrt{6^2 + 0^2}$$
$$= \sqrt{6^2}$$
$$= 6$$

Since $\overline{JI}$ and $\overline{HJ}$ are the same length, $\triangle HIJ$ is an isosceles triangle.

Math Reasoning

Analysis Is it necessary to calculate the side length of all three segments of this triangle? Why or why not?

Sometimes a figure's dimensions might be unknown. When placing a figure with unknown dimensions on the coordinate plane, pick a convenient position and label the vertices of the figure using information that is given in the problem.

Example 3 Assigning Variable Coordinates to Vertices

a. A square has a side length, a. Place the square on the coordinate plane and label each vertex with an ordered pair.

SOLUTION

Place one vertex at the origin. Label the vertex at the origin (0,0). Because the vertex on the x-axis is a units away from the origin, its coordinates should be labeled $(a, 0)$. The vertex on the y-axis is a units up from the origin, so its coordinates are $(0, a)$. Finally, the fourth vertex is both a units to the right of the origin and a units up from the origin, at (a, a).

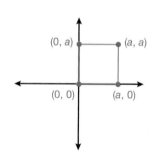

$(0, a)$ ——— (a, a)

$(0, 0)$ $(a, 0)$

b. Given the parallelogram $OPQR$, with one side length labeled c, assign possible coordinates to the vertices.

SOLUTION

Place vertex O at $(0, 0)$ and $\overline{OR}$ along the positive x-axis. Label vertices P, Q, and R. Assign $OR = c$, so the coordinates for R are $(c, 0)$.

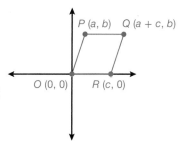

Give P the coordinates (a, b). Because $OPQR$ is a parallelogram, $PQ = OR$. Therefore, the x-coordinate of Q is the x-coordinate of P plus c units, or $a + c$. The coordinates of Q are $(a + c, b)$.

c. Assign coordinates to the vertices of isosceles $\triangle STU$ with a height of 4 from the vertex.

SOLUTION

Place vertex T on the y-axis so that its coordinates are $(0, 4)$. If points S and U are placed such that they are equally distant from the y-axis, then they will form two right triangles with congruent hypotenuses. This ensures that the figure is an isosceles triangle. The coordinates of S and U are $(-x, 0)$ and $(x, 0)$, respectively.

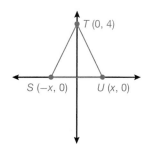

When you assign variable coordinates to a figure used in a proof, remember that the values you choose must apply to all cases. When the dimensions of a figure are not given, variables must be used to ensure the proof is valid for a figure of any size.

Example 4 **Writing a Coordinate Proof**

Prove that the diagonals of a square are perpendicular to one another.

SOLUTION

Assign square $EFGH$ a side length of b. Place E at $(0, 0)$, F at $(0, b)$, G at (b, b), and H at $(b, 0)$. Draw the diagonals $\overline{FH}$ and $\overline{GE}$.

Calculate the slope of diagonals $\overline{FH}$ and $\overline{GE}$.

$$m_{FH} = \frac{0 - b}{b - 0} \qquad m_{GE} = \frac{b - 0}{b - 0}$$
$$= \frac{-b}{b} \qquad\qquad = \frac{b}{b}$$
$$= -1 \qquad\qquad = 1$$

Because the product of the two slopes is -1, $\overline{FH} \perp \overline{GE}$.

Notice that the slopes of the diagonals do not depend on the value of b. Therefore, for all squares, it is true that the diagonals of the square are perpendicular to each other.

Hint

By placing a vertex of the parallelogram on the origin, one fewer variable can be used to diagram the parallelogram.

Hint

Recall that two nonvertical lines are perpendicular if and only if the product of their slopes is -1, and that the formula for slope m of a line is $m = \frac{y_2 - y_1}{x_2 - x_2}$.

Example 5 Application: Constructing a Swimming Pool

A contractor has been hired to build a
swimming pool with a smaller wading pool
beside it. The contractor draws a diagram of
what he plans to build and overlays a
coordinate grid on it, as shown. Show that the
wading pool has a surface area that is one-
eighth the size of the larger pool's surface area.

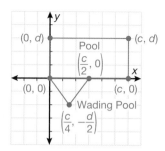

SOLUTION

The area of a rectangle is ℓw. The pool in the diagram has a length of
c and a width of d, so its total area is cd. The wading pool is a triangle.
The area of a triangle is $\frac{1}{2}bh$. The height of the wading pool is $\frac{d}{2}$ and the
length of its base is $\frac{c}{2}$. Substitute these values into the formula for area of
a triangle.

$$A = \frac{1}{2}bh$$
$$A = \frac{1}{2}\left(\frac{c}{2}\right)\left(\frac{d}{2}\right)$$
$$A = \frac{cd}{8}$$

Therefore, the surface area of the wading pool is one-eighth the surface
area of the swimming pool.

Lesson Practice

a. Place a right triangle with leg lengths of 2 and 6 units on the
(Ex 1) coordinate plane so that its legs are on the *x*-axis and *y*-axis. Label the
vertices with their respective coordinates.

b. Prove that $\triangle JKL$ is an isosceles triangle.
(Ex 2)

c. Place a right triangle with leg lengths of *a*
(Ex 3) and *b* units on the coordinate plane. Label
the vertices with their coordinates.

d. Prove that figure *TUVW* is a parallelogram.
(Ex 4)

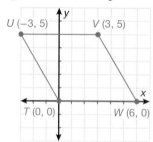

e. (**Traffic Signs**) A yield sign is an equilateral triangle. Draw a yield sign
(Ex 5) on the coordinate plane, using the origin as a vertex. If the length of
one side is *m* units, find the coordinates of the other vertices in terms
of *m*.

1. **Coordinate Geometry** An isosceles triangle has vertices $K(0, 0)$, $L(-2, 2)$, and
(45) $M(x, y)$. Find the coordinates of one possible position of M.

2. Similar pentagons $EFGHI$ and $QRSTU$ are regular and have a similarity ratio
(44) of 5:3. Find the value of x if $FG = 30$ and $TU = x - 11$.

3. If $\overline{CD}$ is a diameter of circle P in the diagram, and it is perpendicular
(43) to $\overline{AB}$, what is the value of x?

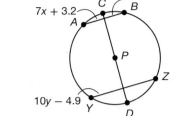

4. In the circle at right if $YZ = 30.2$, what is the value of y?
(43)

5. Find the distance from the line $y = \frac{5}{4}x - 4$ to $(-1, 5)$.
(42)

6. $KLMN$ is a parallelogram. Find the coordinates of $L(x, y)$.
(45)

7. What is the arc length of the minor arc bounded by a 60° central angle
(35) in a circle with a radius of 30 centimeters? Use 3.14 for π.

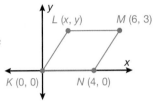

8. Angela draws diagonals on a quadrilateral and notices that they
(34) bisect each other. Without knowing any other information about the
shape, can you classify it as a parallelogram? As a rectangle? Explain.

9. **Coordinate Geometry** Parallelogram $JKLM$ has vertices $J(0, 0)$, $K(2, 2)$, $L(x, y)$, and
(45) $M(5, 0)$. Find (x, y).

10. If $\triangle ABC \sim \triangle XYZ$, what are the measures of $\angle A$ and $\angle Y$?
(41)

11. **Analyze** The side lengths of rectangle A are 16.4 inches and 10.8
(41) inches. The side lengths of rectangle B are 10.25 inches and 6.75
inches. Determine whether the two rectangles are similar. If so, find
the similarity ratio. If not, explain why.

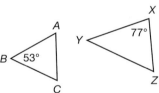

12. (**Architecture**) How much veneer siding is required to cover the front of this house?
(40) Do not include the 3-by-7-foot door or the three 6-by-3-foot windows.

13. (**Packaging**) The cardboard pattern for a gift box is shown at
(40) right. The two trapezoids are congruent. The two end
rectangles are also congruent.
  a. Calculate the area of card stock used for the box.
  b. If the box is cut from a square sheet with side lengths
     of 21 centimeters each, how much card stock is
     wasted?

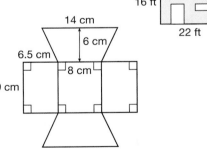

14. Lauren and her grandmother have tables that are similar rectangles. The ratio of
(44) their corresponding sides is 2:3. Lauren's table is 5 feet wide and 6 feet long, and
her grandmother's is the larger table. What is the perimeter of her grandmother's
table?

15. (Construction) The measurements in the diagram were taken from the attic space in a new home. What is the range of possible lengths of the unknown side? Explain.
(39)

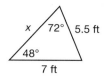

16. Given △LMN ~ △UVW, with LM = 10, and UV = 4, what is the ratio of their corresponding sides?
(44)

17. What is the sum of the measures of the interior angles of a heptagon? If it is a regular heptagon, what is the measure of each interior angle to the nearest hundredth degree?
(Inv 3)

18. Find the line parallel to $2y + 4 = 6x - 5$ that passes through the origin. Write it in slope-intercept form.
(37)

19. Justify If $y = 2x + 3$ and $y = 2x + k$ is parallel to it, what are all the possible values of k? Why does changing the value of k not make the lines intersect?
(37)

20. (Design) The math faculty of Pythagoras College uses this logo. All four triangles are right triangles, the whole logo is a square, and the inner quadrilateral is also a square. Prove that the right triangles are all congruent.
(36)

21. Analyze In △ABC, $AB = (x + 3)$, m∠ABC = $(16x + 8)°$, and $BC = (4x - 8)$. In △DEF, $DE = (2x - 1)$, m∠DEF = $(18x)°$, and $EF = (3x - 6)$. Is it possible for △ABC ≅ △DEF? Explain.
(28)

22. A triangle has side lengths that are 100, 240, and 265 units long. Classify the triangle by side lengths and by angles.
(33)

23. Draw an equilateral triangle using a protractor and straightedge. Locate and label its orthocenter and centroid. What do you notice about the two points?
(32)

24. What is the sum of the interior angle measures of a convex, irregular octagon?
(Inv 3)

25. Error Analysis Mick has two triangles, △MNO and △DEF. He knows that m∠MNO = m∠DEF, m∠NOM = m∠EFD, and NO = DF. Mick concludes that △MNO ≅ △DEF. Is he correct? Explain.
(30)

26. For a circle with center O and diameters $\overline{AB}$ and $\overline{CD}$, prove △AOC ≅ △BOD.
(28)

27. In the parallelogram shown, what is the measure of ∠IFG? m∠FGH? m∠HGE?
(34)

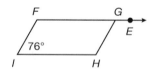

28. Write a coordinate proof showing that the midpoint of a hypotenuse of a right triangle is equidistant from the three vertices.
(45)

29. Algebra If m$\widehat{AB}$ = $(2x + 10)°$ and m$\widehat{BC}$ = $(4x + 5)°$ are non-overlapping adjacent arcs, m$\widehat{XY}$ = $(x + 19)°$ and m$\widehat{YZ}$ = $(2x + 11)°$ are non-overlapping adjacent arcs and $\widehat{AC}$ and $\widehat{XZ}$ are congruent, what is the measure of each arc?
(26)

30. Multiple Choice What is the conclusion of this conjecture?
(10)

If a polygon is regular, then it is equilateral and equiangular.

A A polygon is equilateral. **B** A polygon is regular.
C A polygon is equilateral and equiangular. **D** A polygon is equiangular.

Triangle Similarity: AA, SSS, SAS

Warm Up

1. **Vocabulary** The ratio of two corresponding linear measurements in a pair
(41) of similar figures is the _____.

2. If two angles and a nonincluded side of one triangle are congruent to two
(30) angles and the corresponding nonincluded side of another triangle, then
the two triangles are congruent by _____.

3. **Multiple Choice** If two angles and the corresponding included sides of two
(30) triangles are congruent, then which triangle congruency postulate or
theorem applies?

   A SSS Postulate   **B** SAS Postulate

   C AAS Theorem   **D** ASA Postulate

New Concepts

Two triangles are similar if all their corresponding angles are congruent.
Since the sum of any triangle's angles is 180°, only two angles are required to
prove that two triangles are similar.

Reading Math

The symbol ~ is used to show that two polygons are similar. For example, $\triangle XYZ \sim \triangle KLM$ means, "triangle XYZ is similar to triangle KLM."

Postulate 21: Angle-Angle (AA) Triangle Similarity Postulate
If two angles of one triangle are congruent to two angles of another triangle, then the triangles are similar.

Example 1 Using the AA Similarity Postulate

Show that the two triangles are similar
if $\overline{AB} \parallel \overline{DE}$. Then, find DE.

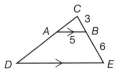

SOLUTION

Math Reasoning

Write Are all pairs of congruent triangles also similar triangles? Explain.

Statements	Reasons
1. $\overline{AB} \parallel \overline{DE}$	1. Given
2. m∠ABC = m∠DEC	2. Corresponding Angles Postulate
3. m∠BAC = m∠EDC	3. Corresponding Angles Postulate
4. $\triangle ABC \sim \triangle DEC$	4. AA Similarity Postulate

Since the two triangles are similar, the ratios of the lengths of
corresponding sides are equal.

$$\frac{AB}{DE} = \frac{BC}{EC}$$
$$\frac{5}{DE} = \frac{3}{9}$$
$$3(DE) = 45$$
$$DE = 15$$

Online Connection
www.SaxonMathResources.com

Exploration **Understanding AA Similarity**

In this exploration, you will construct similar triangles and observe the properties of each.

1. Draw two different line segments on a sheet of paper. Make sure the segments each have a different length.

2. At each end of the first line segment, measure acute angles and draw the rays out to create a triangle.

3. On your second line segment, measure the same two angles with a protractor and draw a second triangle.

4. Measure the unknown angle of each of the triangles you have drawn. What is the relationship between these two angles? What is the relationship between these two triangles?

5. Measure the side lengths of the larger triangle and one side of the small triangle. Now, use what you know about the triangles to predict the side lengths of the smaller triangle without using a ruler. Then, measure the lengths. Were your predictions correct?

It is not always necessary to know a triangle's angle measures to determine similarity. Another way to determine similarity is to verify that the lengths of all the corresponding sides of both triangles are related in the same ratio.

Theorem 46-1: SSS Similarity Theorem
If the lengths of the sides of a triangle are proportional to the lengths of the sides of another triangle, then the triangles are similar.

Example **2** **Using the SSS Similarity Theorem**

Given the two triangles with lengths as shown, show that they are similar triangles.

 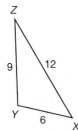

Hint

This example demonstrates a simple way to find out if all the side pairs of two triangles are in the same proportion. If the ratio of each pair of sides reduces to the same fraction, they are proportional.

SOLUTION

Statements	**Reasons**
1. $\dfrac{UW}{XY} = \dfrac{2}{6} = \dfrac{1}{3}$	1. Similarity ratio for $\overline{UW} : \overline{XY}$.
2. $\dfrac{WV}{YZ} = \dfrac{3}{9} = \dfrac{1}{3}$	2. Similarity ratio for $\overline{WV} : \overline{YZ}$.
3. $\dfrac{VU}{ZX} = \dfrac{4}{12} = \dfrac{1}{3}$	3. Similarity ratio for $\overline{VU} : \overline{ZX}$.
4. $\triangle UWV \sim \triangle XYZ$	4. SSS Similarity Theorem

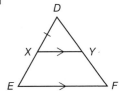

Example 3 **Proving the SSS Similarity Theorem**

Prove Theorem 46-1: If the lengths of the sides of a triangle are proportional to the lengths of the sides of another triangle, then the triangles are similar.

Given: $\dfrac{AB}{DE} = \dfrac{BC}{EF} = \dfrac{CA}{FD}$, $\overline{DX} \cong \overline{AB}$, $\overline{XY} \parallel \overline{EF}$

Prove: $\triangle ABC \sim \triangle DEF$

SOLUTION

Statements	Reasons
1. $\overline{DX} \cong \overline{AB}$, $\overline{XY} \parallel \overline{EF}$	1. Given; Parallel Postulate
2. $\angle DXY \cong \angle DEF$	2. Corresponding angles are congruent
3. $\angle D \cong \angle D$	3. Reflexive Property of Congruence
4. $\triangle DXY \sim \triangle DEF$	4. AA Similarity Postulate
5. $\dfrac{AB}{DE} = \dfrac{BC}{EF} = \dfrac{CA}{FD}$	5. Given
6. $\dfrac{DX}{DE} = \dfrac{XY}{EF} = \dfrac{YD}{FD}$	6. Similarity ratio from $\triangle DXY \sim \triangle DEF$
7. $DX = AB$	7. Definition of congruent segments
8. $\dfrac{AB}{DE} = \dfrac{XY}{EF} = \dfrac{YD}{FD}$	8. Substitute AB for DX in step 6
9. $\dfrac{BC}{EF} = \dfrac{XY}{EF}$, $\dfrac{CA}{FD} = \dfrac{YD}{FD}$	9. Substitute
10. $BC = XY$, $CA = YD$	10. Simplify
11. $\overline{BC} \cong \overline{XY}$, $\overline{CA} \cong \overline{YD}$	11. Definition of Congruent Segments
12. $\triangle ABC \cong \triangle DXY$	12. SSS Triangle Congruence Postulate
13. $\triangle ABC \sim \triangle DEF$	13. Transitive Property of Similarity

One final way to prove triangle similarity is the SAS Similarity Theorem. You will notice that it is similar to one of the congruence postulates you have learned about.

<table>
<tr><th colspan="1">Theorem 46-2: SAS Similarity Theorem</th></tr>
<tr><td>If two sides of one triangle are proportional to two sides of another triangle and the included angles are congruent, then the triangles are similar.</td></tr>
</table>

Caution

To apply SAS similarity, one angle pair has to be congruent, but the side pairs only have to be proportional. Do not confuse this with SAS congruence, where the two pairs of sides must be congruent.

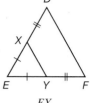

Example 4 Proving Similarity

a. Prove that $\triangle EXY \sim \triangle EDF$.

SOLUTION

By the Reflexive Property, $\angle XEY \cong \angle DEF$. It is given in the diagram, $\overline{EX} \cong \overline{EY}$ and $\overline{XD} \cong \overline{YF}$. The ratio of EX to ED can be given by $\frac{EX}{EX + XD}$. By substituting the congruent segments, it can be rewritten as $\frac{EY}{EY + YF}$, which is also the ratio of EY to EF. So the triangles have two proportional sides and one congruent angle. By the SAS Similarity Theorem, they are similar triangles.

b. If $EX = 6$, $ED = 11$, and $XY = 7$, find DF.

SOLUTION

Use the similarity ratio given by $EX : ED$ and a proportion.

$$\frac{EX}{ED} = \frac{XY}{DF}$$ Definition of similar triangles

$$\frac{6}{11} = \frac{7}{DF}$$ Substitute

$$11 \cdot 7 = 6 \cdot DF$$ Cross product

$$DF = \frac{5}{6}$$ Solve

Example 5 Application: Land Surveying

A surveyor needs to find the distance across a lake. The surveyor makes some measurements as shown. Find the distance across the lake.

SOLUTION

First, determine if the triangles are similar. The angles at C are congruent since they are vertical angles. AB and DE are parallel lines, so $\angle ABC$ and $\angle EDC$ are congruent. Therefore, by the AA Similarity Postulate, $\triangle ABC \sim \triangle EDC$. Now, find the missing value using a proportion.

$$\frac{x}{100} = \frac{20}{25}$$

$$25x = 2000$$

$$x = 80$$

The distance across the lake is 80 meters.

> **Math Reasoning**
>
> **Write** Why might someone use this method to find the distance across the lake?

Lesson Practice

a. Given the two triangles shown, prove they are similar using the AA Similarity Postulate.
(Ex 1)

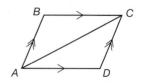

b. Given the two triangles shown, use SSS similarity to prove that they are similar.

(Ex 2)

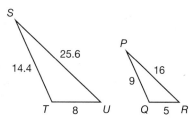

c. Given the two triangles shown, use SAS similarity to prove that they are similar. Find the value of x.

(Ex 4)

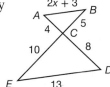

d. Laura wants to find out how tall a tree is. She notices that the tree makes a shadow on the ground. The top of the shadow of the treehouse is 25 feet away from the base of the tree. Laura is 5 feet 8 inches tall and she casts a shadow that is 6 feet 2 inches long. How tall is the tree, to the nearest foot?

(Ex 5)

Practice Distributed and Integrated

1. Find the distance from $(-1, 7)$ to the line $x = 3$.
(42)

* 2. **(Flying)** Two aircraft depart from an airport at $(10, 12)$. The first aircraft travels to an airport at $M(-220, 80)$, and the second aircraft travels to an airport at $N(100, -400)$. If each unit on the grid represents one mile, what is the distance to the nearest mile between the two aircraft after they both land?
(9)

3. **Analyze** The side lengths of rectangle A are 18.3 inches and 24.6 inches. The side lengths of rectangle B are 24.4 inches and 32.8 inches. Determine whether the two rectangles are similar. If so, write the similarity ratio.
(41)

4. **Multiple Choice** Which of the following lines is parallel to $y = 7$?
(37)

    A $y = 2x$ **B** $y = -\dfrac{1}{7}x$

    C $y + 7 = 0$ **D** $y = \dfrac{7}{x}$

5. What is the central angle measure of a regular octagon?
(Inv 3)

* 6. Write a two-column proof to prove the SAS Similarity Theorem: If two sides of one triangle are proportional to two sides of another triangle and the included angles are congruent, then the triangles are similar.
(27)

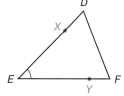
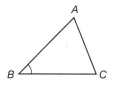

Given: $\angle B \cong \angle E, \dfrac{AB}{DE} = \dfrac{BC}{EF}$

Prove: $\triangle ABC \sim \triangle DEF$

Hint: Assume that $AB < DE$ and choose point X on $\overline{DE}$ so that $\overline{EX} \cong \overline{BA}$. Then choose point Y on $\overline{EF}$ so that $\angle EXY \cong \angle EDF$. Show that $\triangle XEY \sim \triangle DEF$ and that $\triangle ABC \cong \triangle XEY$.

7. Error Analysis A student drew the figure at right and called it a parallelogram. Explain where the student erred.
(34)

* 8. **(Landscaping)** A gardener wants all of the triangular gardens in a yard to be similar shapes. The first garden has sides that are 3, 4, and 6 feet long. If the second garden has to have sides of 12 and 16 feet corresponding to the 3- and 4-foot sides of the other garden, what is the length of the third side?
(46)

9. Given $\triangle JKL \sim \triangle EFG$, with $JL = 25$ and $EG = 15$, what is the ratio of their corresponding sides?
(44)

10. **(Architecture)** The Dome of St. Peters in Vatican City is 119 meters tall. If you are standing 120 meters away from the zenith of the dome, how far are you from standing directly underneath it?
(29, 33)

11. Algebra Suppose the conjunction, "$x^2 < 9$ and $x^2 > 4$" is true. Write a disjunction of two statements, involving x but not x^2, which must be true.
(20)

12. Algebra Find the value of x that makes N the incenter of this triangle.
(38)

*13. A square on the coordinate plane has side lengths that are each 6 units long. A vertex is at the origin. Find the coordinates of each vertex.
(45)

14. Write Explain the difference between a tangent line and a secant line with respect to circles.
(43)

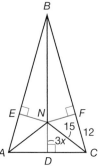

15. The legs in a right triangle are 9 inches and 12 inches long. What is the length of the hypotenuse?
(29)

16. Use the LL Congruence Theorem to prove that $\triangle HJK \cong \triangle LMN$.
(36)

*17. Algebra Given that $\triangle ABC \sim \triangle DEF$, $AB = 12$, $DE = 16$, and $EF = 20$, what is the length of BC?
(46)

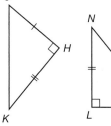

*18. An isosceles triangle on the coordinate plane has side lengths of 8, 5, and 5 units, respectively. The long side is along the x-axis. Find the coordinates of each vertex.
(45)

19. Justify In the two triangles shown, which is greater, x or y? Explain.
(Inv 4)

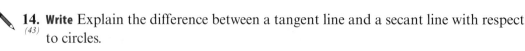

*20. **(Construction)** A house has triangular roof structures, or gables, which are similar to the gables of the attached garage. If the base of the house gable is 25 feet and the base of the garage gable is 40 feet, what is the similarity ratio between the two lengths?
(46)

*21. Explain why the triangles are similar, then find the value of x.
(46)

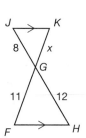

22. **(Wordplay)** Formulate a conjecture that describes the following pattern.
(7)
pop, noon, level, redder, racecar.

23. Using the diagram at right, P is the incenter of $\triangle QRS$. Find RV.
(38)

24. How many sides does a convex polygon have, if each interior
(Inv 3) angle is equal to 179°?

*25. Generalize Line m is given by $ax + by = j$. Line n is given
(37) by $-bx + ay = k$. Is n perpendicular to m?

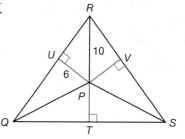

Find the lengths of the minor arcs given. Round to the nearest hundredth of a centimeter.

26. $\overarc{EF}$
(35)

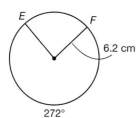

27. $\overarc{GH}$
(35)

28. Algebra Find the length of $\overline{CY}$.
(32)

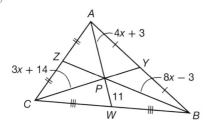

*29. In triangles RST and UVW, $\angle T$ and $\angle W$ are right angles, $\overline{RT} \cong \overline{UW}$, and
(36) $\overline{ST} \cong \overline{VW}$. Prove that $\triangle RST \cong \triangle UVW$.

*30. Write Suppose you know $\triangle ABC \sim \triangle DEF$, and you are given the lengths of all
(44) three sides of one triangle and the length of one side of the other. Explain how
you would find the lengths of all six sides and the ratio of the corresponding sides.

Circles and Inscribed Angles

1. **Vocabulary** The set of all points between the sides of an angle is the
(3)
_____ of the angle. (*interior, exterior, measure*)

2. **Multiple Choice** A central angle separates a circle into two arcs called
(26)
   A the greater arc and the lesser arc.
   B the major arc and the minor arc.
   C the larger arc and the smaller arc.
   D the semi arc and the full arc.

3. An arc with endpoints that lie on the diameter of a circle is called a
(26)
_____ .

New Concepts   Lessons 23 and 26 introduce circles. This lesson also
addresses circles and introduces inscribed angles of
circles. Recall that a central angle is an angle with the
center of a circle as its vertex. Another kind of angle
found in circles is the inscribed angle. An **inscribed angle**
is an angle whose vertex is on a circle and whose sides
contain chords of the circle. In the diagram, ∠*ABC* is an
inscribed angle.

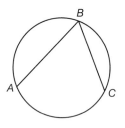

The arc formed by an inscribed angle is the **intercepted arc** of that angle.
In the diagram, $\overset{\frown}{AC}$ is the intercepted arc of ∠*ABC*.

Math Reasoning

Justify A triangle is
inscribed in a circle. Use
Theorem 47-1 to explain
why the sum of the
measures of the angles in
the triangle is 180°.

Theorem 47-1
The measure of an inscribed angle is equal to half the measure of its intercepted arc. $m\angle PRQ = \left(\dfrac{1}{2}\right) m\overset{\frown}{PQ}$ $\overset{\frown}{PQ}$ is an intercepted arc 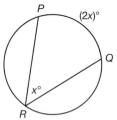

Theorem 47-2
If an inscribed angle intercepts a semicircle, then it is a right angle.   ∠*DEF* intersects the semicircle, so m∠*DEF* = 90°. 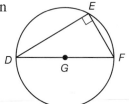

Example 1 | **Proving and Applying Inscribed Angle Theorems**

Use ⊙M to answer each question.

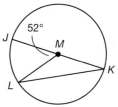

a. Name the inscribed angle.

SOLUTION
The inscribed angle is ∠JKL.

b. Name the arc intercepted by ∠JKL.

SOLUTION
∠JKL intercepts the minor arc $\overset{\frown}{JL}$.

Math Language

Recall that minor arcs are labeled with 2 points, and major arcs are labeled with 3 points.

c. If m∠JML = 52°, find m∠JKL.

SOLUTION
∠JML is a central angle, so m∠JML = m$\overset{\frown}{JL}$. By Theorem 47-1, the measure of inscribed angle ∠JKL is half the measure of $\overset{\frown}{JL}$.

$$m\angle JKL = \frac{1}{2}(52°)$$
$$m\angle JKL = 26°$$

d. Prove Theorem 47-2.
   Given: $\overline{AB}$ is a diameter of ⊙C
   Prove: m∠ADB = 90°

SOLUTION

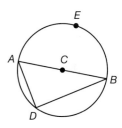

Statements	Reasons
1. $\overline{AB}$ is a diameter	1. Given
2. ∠ACB = 180°	2. Protractor Postulate
3. m$\overset{\frown}{AEB}$ = 180°	3. Definition of the measure of an arc
4. m∠ADB = 90°	4. Theorem 47-1

Example 2 | **Finding Angle Measures in Inscribed Triangles**

Find the measure of ∠1, ∠2, and ∠3.

SOLUTION
The arc intercepted by ∠3 measures 76°.

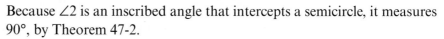

$$m\angle 3 = \frac{1}{2}(76°) \qquad \text{Theorem 47-1}$$
$$m\angle 3 = 38° \qquad \text{Simplify.}$$

Because ∠2 is an inscribed angle that intercepts a semicircle, it measures 90°, by Theorem 47-2.

You can use the Triangle Angle Sum Theorem (Theorem 18-1) to find m∠1.

$$m\angle 1 + m\angle 2 + m\angle 3 = 180° \qquad \text{Triangle Angle Sum Theorem}$$
$$m\angle 1 + 90° + 38° = 180° \qquad \text{Substitute.}$$
$$m\angle 1 = 52° \qquad \text{Solve.}$$

More than one inscribed angle can intercept the same arc. Since both of these inscribed angles measure one-half what the arc does, they have the same measure, and are congruent.

Theorem 47-3

If two inscribed angles intercept the same arc, then they are congruent.

$\angle 1 \cong \angle 2$

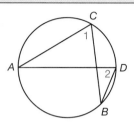

Example 3 Finding Measures of Arcs and Inscribed Angles

a. Find the measures of $\angle FGH$ and of $\overset{\frown}{GJ}$.

SOLUTION

$\angle FGH$ is an inscribed angle with intercepted arc $\overset{\frown}{FH}$. Use Theorem 47-1.

$m\angle FGH = \left(\dfrac{1}{2}\right) m\overset{\frown}{FH}$	Theorem 47-1
$m\angle FGH = \left(\dfrac{1}{2}\right) 36°$	Substitute.
$m\angle FGH = 18°$	Solve.

$\overset{\frown}{GJ}$ is the intercepted arc of $\angle GHJ$. Use Theorem 47-1.

$m\angle GHJ = \left(\dfrac{1}{2}\right) m\overset{\frown}{GJ}$	Theorem 47-1
$48° = \left(\dfrac{1}{2}\right) m\overset{\frown}{GJ}$	Substitute.
$m\overset{\frown}{GJ} = 96°$	Solve.

b. Find the measure of $\angle XYZ$.

SOLUTION

By Theorem 47-3, we know that $\angle XYZ \cong \angle XAZ$.

$\angle XYZ \cong \angle XAZ$	Theorem 47-3
$2c + 9 = 3c$	Substitute.
$c = 9$	Solve.

Substituting $c = 9$ into the expression for $\angle XYZ$ yields $m\angle XYZ = 27°$.

Math Reasoning

Analyze In Example 3b, could you find the value of c if you were given the measure of $\overset{\frown}{XZ}$? Explain.

Theorem 47-4

If a quadrilateral is inscribed in a circle, then it has supplementary opposite angles.

$\angle M + \angle O = 180°$

$\angle P + \angle N = 180°$

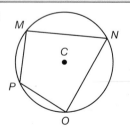

Example 4 **Finding Angle Measures in Inscribed Quadrilaterals**

Find the measure of $\angle U$.

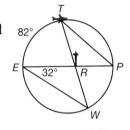

SOLUTION

By Theorem 47-4, $\angle S$ is supplementary to $\angle U$.

$m\angle S + m\angle U = 180°$	Theorem 47-4
$4z + 3z + 5 = 180°$	Substitute.
$z = 25$	Solve for z.

Next, find the measure of $\angle U$.

$m\angle U = 3z + 5$	Given
$m\angle U = 3(25) + 5$	Substitute.
$m\angle U = 80$	Simplify.

The measure of $\angle U$ is 80°.

Example 5 **Application: Air Traffic Control**

A circular radar screen in an air traffic control tower shows aircraft flight paths. The control tower is labeled R. One aircraft must fly from point T to the control tower, and then to its destination at point P. Find $m\angle TRP$.

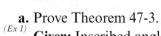

SOLUTION

$\angle WEP \cong \angle WTP$	Theorem 47-3
$m\angle WEP = m\angle WTP$	Definition of Congruence
$m\angle WEP = 32°$	Given
$m\angle WTP = 32°$	Transitive Property of Equality
$m\angle TPE = \frac{1}{2}(82) = 41°$	Theorem 47-1
$m\angle WTP + m\angle TPE + m\angle TRP = 180°$	Triangle Angle Sum Theorem
$32° + 41° + m\angle TRP = 180°$	Substitution Property of Equality
$m\angle TRP = 107°$	Solve

The measure of $\angle TRP$ is 107°.

Lesson Practice

a. Prove Theorem 47-3.
(Ex 1) **Given:** Inscribed angles $\angle ADB$ and $\angle ACB$
Prove: $\angle ADB \cong \angle ACB$

b. Find the value of y in the triangle inscribed
(Ex 2) in $\odot A$.

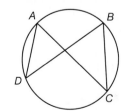

c. Find the value of x.
(Ex 3)

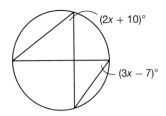

d. Find the measure of $\angle A$.
(Ex 4)

e. [**Air Traffic Control**] A radar screen in an
(Ex 5) air traffic control tower shows flight
paths. The control tower is labeled L.
Points M, L, and P mark the flight path
of a commercial jet. Find the measure
of $\angle MLP$.

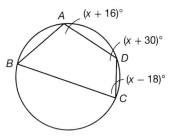

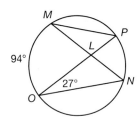

Practice   Distributed and Integrated

* 1. Write a two-column proof, proving that if a quadrilateral is inscribed in a
(27) circle, then it has supplementary opposite angles.
   Given: $ABCD$ is inscribed in a circle.
   Prove: $\angle A$ is supplementary to $\angle C$.

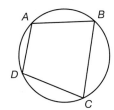

2. [**Electrical Wire**] An electrician cuts a 45-inch cable cord into three pieces,
(41) with a ratio of 2:3:4. What are the lengths of the three pieces of cable?

* 3. Formulate Quadrilateral $ABCD$ is inscribed in a circle. Write two equations that
(47) show the relationships of the angles of the quadrilateral.

4. Justify Ariel is fencing off a triangular area with some caution tape. One side of the
(39) triangle must be 14 feet and a second side must be 22 feet. If the roll of tape is
40 feet long, can he use it to fence off the whole triangular area? Explain.

5. [**Bicycling**] Norma started her cycling trip at $(0, 7)$. Eduardo started his trip at
(45) $(0, 0)$. They crossed paths at $(4, 2)$. Draw their routes on a coordinate plane.

6. Analyze A composite figure is made up of a square and an equilateral triangle. If
(40) the figure has five sides, what is the relationship between the perimeter P and the
square side length s? Explain.

* 7. **Algebra** Find the value(s) of x that make $\angle FGH \sim \angle JKL$.
(46)

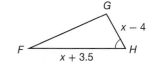

8. **Hardware** If one socket wrench has a $\frac{5}{8}$-inch diameter and the other
(44) has a $\frac{3}{4}$-inch diameter, what is the ratio of the diameter of the bigger wrench to that of the smaller wrench?

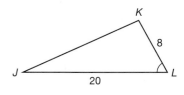

9. If in $\triangle ABC$, $AB = 7$, m$\angle ABC = 55°$, and m$\angle BCA = 60°$,
(30) and in $\triangle DEF$, m$\angle DEF = 60°$, m$\angle DFE = 55°$, and $FD = 10$, is $\triangle ABC \cong \triangle DEF$?

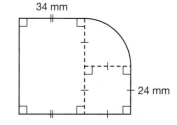

10. a. **Multi-Step** Determine the figure's perimeter. Express your answer in
(40) terms of π.

b. **Multi-Step** Determine the figure's area. Express your answer in terms of π.

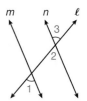

*11. **Analyze** If a triangle is inscribed in a circle such that one edge of the
(47) triangle goes through the center of the circle, what statement can be made about the measure of one of the angles in the triangle?

12. Give an example of a Pythagorean triple that is not a multiple of 3, 4, 5.
(29)

13. There are two theorems or postulates you could use to prove that lines m and n in
(12) this figure are parallel. Which are they? What other facts are you using?

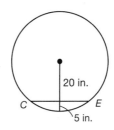

14. A pipe with a diameter of 2.2 centimeters is being used for a drainage system. If a
(43) cross section of the pipe reveals the water level creates a chord 2 centimeters long, how close to the center of the pipe is the water level to the nearest hundredth of a centimeter?

15. What is the measure of each interior angle in a regular heptagon? Round your
(Inv 3) answer to the nearest degree.

16. Find the length of the chord CE.
(43)

17. Consider the following statement.
(17)

If the month is April, then it is spring in the Northern Hemisphere.

a. Write the negation of this statement's hypothesis.

b. Write the negation of the statement's conclusion.

c. Write the contrapositive of the statement.

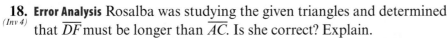

18. **Error Analysis** Rosalba was studying the given triangles and determined
(Inv 4) that $\overline{DF}$ must be longer than $\overline{AC}$. Is she correct? Explain.

19. What is the relationship between the lines $y = \frac{-1}{2}x + b$ and
(37) $2x - y = -\sqrt{b}$?

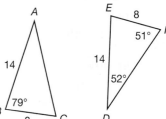
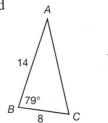

xy² ***20. Algebra** Find the measure of ∠ZWY.
(47)

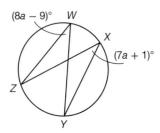

21. What is the value of y in the proportion $\dfrac{2}{y-3} = \dfrac{4}{y}$?
(41)

22. Generalize A square centered at the origin of the coordinate plane
(45) has a side length, k. Find expressions for the coordinates of each vertex.

23. Find the unknown length of the side in the triangle shown.
(29)

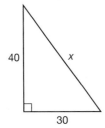

*24. Use a ruler and a protractor to find the approximate circumcenter of a triangle
(38) with vertices at $(-5, -2)$, $(2, 3)$, and $(-2, 4)$.

25. If in $\triangle ABC$ and $\triangle XYZ$, m∠ACB = m∠XYZ, $AC = XY$, and $CB = XZ$, is it true
(28) that $\triangle ACB \cong \triangle XYZ$? How do you know?

*26. (Baking) Tomas is making a layered cake out of 2 individual
(46) triangular cakes. The cakes need to be similar triangles so
Tomas can stack the smaller one on top of the larger cake.
Prove that the two pieces of cake shown are similar.
What is the measure of EF?

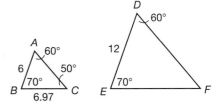

27. The ratio of the angle measures in a triangle is 3:5:10. What are
(41) the measures of the angles?

28. Which congruence theorem applies to these triangles?
(36)

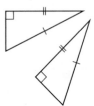

*29. Multiple Choice The measure of an inscribed angle is _____ the measure
(47) of the intercepted arc.
    A half   **B** one third
    C one fourth   **D** twice

*30. (Carpentry) Gwen is building a truss for her roof. She needs to
(46) find the length of segment DE. Use similar triangles to help
Gwen find the measurement she needs.

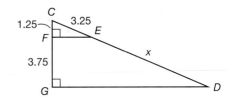

Indirect Proofs

Warm Up

1. **Vocabulary** A style of proof that uses boxes and arrows to show the structure of the proof is called a _____.
(31)

2. Is this statement true or false?
(14)
   A counter example is an example that can help prove a statement true.

3. A proof can be written in a variety of ways. Which is not a method for writing a proof?
(31)
   A algebraic proof   **B** flowchart proof
   C paragraph proof   **D** flow proof

New Concepts

Direct reasoning is the process of reasoning that begins with a true hypothesis and builds a logical argument to show that a conclusion is true. In some cases it is not possible to prove a statement directly, so an **indirect proof** must be used. An indirect proof is a proof in which the statement to be proved is assumed to be false and a contradiction is shown. This is also called **proof by contradiction.**

Follow these three steps to write an indirect proof.

1. Assume the conclusion is false.

2. Show that the assumption you made is contradicted by a theorem, a postulate, a definition, or the given information.

3. State that the assumption must be false, so the conclusion must be true.

> **Math Reasoning**
>
> **Write** What is the primary difference between direct reasoning and the use of an indirect proof?

Example 1 Writing an Indirect Proof

Prove Theorem 4-1: If two lines intersect, then they intersect at exactly one point.

SOLUTION
Since this is an indirect proof, we start by assuming that the statement is not true. In other words, it must be possible for two lines to intersect at more than one point.

Assume that the lines *m* and *n* intersect at both points, *A* and *B*. Now we must show that this contradicts another theorem or postulate. It contradicts Postulate 5, which states that through any 2 points, there exists exactly one line. Since it is not possible for two lines to pass through both point *A* and point *B*, the assumption we have made is contradicted, and Theorem 4-1 must be correct.

In an indirect proof, it is often helpful to draw a diagram, just as you would for a 2-column, paragraph, or flowchart proof. A diagram is helpful in determining what assumptions should be made to prove the statement, and in finding the postulate, theorem, or definition that contradicts the assumption.

Example 2 | **Writing an Indirect Proof**

Use the diagram to prove Theorem 47-2:
If an inscribed angle intercepts a semicircle,
then it is a right angle.
Given: $\overline{AC}$ is a diameter of $\odot M$.
Prove: m$\angle ABC \neq 90°$

SOLUTION

Assume that m$\angle ABC \neq 90°$. By Theorem 47-1, this implies that $\overset{\frown}{ADC} \neq$ 180°, since the arc is twice the measure of the inscribed angle. It is given that $\overline{AC}$ is a diameter. Since $\overline{AC}$ goes through the center of the circle, $\angle AMC$ is the central angle that intercepts $\overset{\frown}{ADC}$. Since $\angle AMC$ is a straight angle, its measure is 180°. An arc's measure is equal to the measure of its central angle, so the measure of $\overset{\frown}{ADC}$ must be 180°. This contradicts our assumption.

Example 3 | **Writing an Indirect Proof**

Use the diagram to prove Theorem 42-1: The perpendicular segment from a point to a line is the shortest segment from the point to the line.

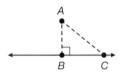

Given: $\overline{AB} \perp \overleftrightarrow{BC}$
Prove: $\overline{AB}$ is the shortest segment from A to $\overleftrightarrow{BC}$.

SOLUTION

First, assume that there is another segment from A to $\overleftrightarrow{BC}$ that is shorter than $\overline{AB}$. In the diagram, this segment is shown as $\overline{AC}$. Our assumption is that $AC < AB$.

$\triangle ABC$ is a right triangle, so by the Pythagorean Theorem, it must be true that $AB^2 + BC^2 = AC^2$. Since $AB > AC$, and both AB and AC are greater than 0, by squaring both sides of the inequality we know that $AB^2 > AC^2$. Using the Subtraction Property of Equality, subtract AB^2 from both sides. Then $BC^2 = AC^2 - AB^2$. Since $AB^2 > AC^2$, $AC^2 - AB^2 < 0$. Substituting shows that $BC^2 < 0$. However, the length of BC must be greater than 0, so this contradicts the definition of a line segment. Therefore, AC is not less than AB, and the theorem is true.

a. State the assumption you would make to start an indirect proof to
(Ex 1) show that m∠X = m∠Y.

b. State the assumption you would make to start an indirect proof to
(Ex 1) show that $\overleftrightarrow{AB} \perp \overleftrightarrow{CB}$.

c. An isosceles triangle has at least two congruent sides. To prove this
(Ex 2) statement indirectly, assume an isosceles triangle does not have at
least two congruent sides. What case needs to be explored to find a
contradiction?

d. Use an indirect proof to prove that a triangle can have at most one
(Ex 2) right angle.

e. Use an indirect proof to show
(Ex 3) that ∠4 ≅ ∠6, if m ∥ n.

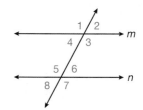

1. **Verify** Choose two points on the line $y = 3x - 4$ and use
(16) them to verify that the value $m = 3$ from the equation
is actually the slope of the line.

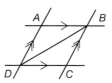

2. In the diagram shown, prove that $AD = BC$.
(30)

3. Find the distance from $(-6, 4)$ to the line $y = x$.
(42)

4. (**Floor Plans**) In Thuy's room, the distance from the door D to the closet C is
(39) 4 feet, and the distance from the door D to the window W is 5 feet. The distance
from the closet C to the window W is 6 feet. On a floor plan, she draws △CDW.
Order the angles from the least to the greatest measure.

* 5. **Justify** If the measures of the angles in a quadrilateral are 132°, 90°, 48°, and 90°,
(47) can the quadrilateral be inscribed in a circle? Explain.

6. **Error Analysis** Arturo wanted to find the orthocenter of the
(32) triangle shown. Explain the error Arturo made.

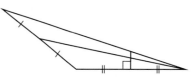

7. Given △MNP ∼ △HJK, with $MN = 3$ and $HJ = 66$,
(41) what is the ratio of the corresponding sides?

8. (**Landscaping**) A lawn for an ornamental garden is designed as shown.
(40) Determine the area of grass needed for the lawn, to the nearest
square meter. The semicircular regions have radius lengths of
2.5 meters each.

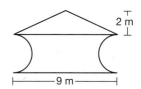

9. Solve the proportion for g. $\frac{6}{16} = \frac{g}{12}$
(41)

*10. Triangle *STU* has vertices *S*(0, 0) and *T*(*h*, *k*). Find the coordinates of vertex *U*
(45) such that the triangle is isosceles.

11. (**Gardening**) Oksana needs to fertilize her gardens for the
(35) winter, so she needs to find out how many square feet
both gardens occupy. If the two gardens are sectors
of a circular area, what is the area to be fertilized,
to the nearest tenth?

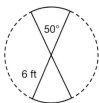

12. If in △*ABC* and △*DEF*, *AB* = *DE* and m∠*ABC* = m∠*EDF*, what other piece of
(28) information is needed to show that △*ABC* ≅ △*EDF* by the SAS postulate?

*13. Use an indirect proof to prove the Converse of the Hinge
(48) Theorem. Refer to △*PQR* and △*XYZ* in the diagram.
Given: $\overline{PQ} \cong \overline{XY}$, $\overline{PR} \cong \overline{XZ}$, *QR* > *YZ*
Prove: m∠*P* > m∠*X*

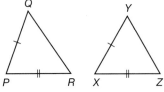

14. Find the length of a circle's radius given a 12-centimeter chord that is
(43) 5 centimeters away from the center of the circle.

15. Verify that △*ABC* ~ △*MNP*.
(46)

16. Multiple Choice The diagonals of parallelogram *RSTU*
(34) intersect at *P*. Which of the following is true?
  A m∠*RSP* = m∠*UTP*   **B** *RP* + *SP* = *RT*
  C m∠*UPT* = m∠*RPU*   **D** 2*TP* = *RT*

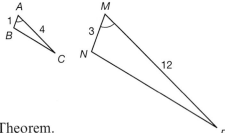

17. Write a flowchart proof of the Right Angle Congruence Theorem.
(31) **Given:** ∠1 and ∠2 are right angles.
Prove: ∠1 ≅ ∠2

*18. Write an indirect proof showing that if two lines intersect, then there exists exactly
(48) one plane that contains them.

19. (**Outdoor Cooking**) Two pairs of tongs are being designed for a barbecue kit, each
(Inv 4) being of the same length. One is to be used to pick up larger items while the
other is to be used to pick up smaller items. Which should be set with a smaller
maximum spread angle? Explain.

*20. (**Skating**) Brianna skates in straight lines across a circular rink until she
(47) reaches a wall. She starts at *P*, turns 75° at *Q*, and turns 100° at *R*. How many
degrees must Brianna turn at *S* to return to her starting point?

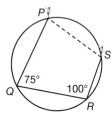

21. A diameter intersects an 8-unit long chord at right angles. What is the length
(43) of each of the two chord segments cut by the diameter?

22. Multi-Step Determine the perimeter and area of this figure.
(40)

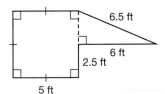

*23. Algebra In an *n*-sided convex polygon, the sum of the interior angles
(Inv 3) is 2*x* + 80. In a 2*n*-sided convex polygon, the sum of the interior
angles is 5*x* + 20. Determine the value of *n* and *x*.

*24. Use an indirect proof to prove the following.
(48) **Given:** $\triangle HIJ \cong \triangle KJL$
 Prove: $\angle KJL \cong \angle MIN$

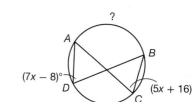

25. Triangles ABC and DEF are right triangles with right
(36) angles B and E, respectively. If $\angle C \cong \angle F$,
 $AC = DF = 25$, and $EF = 7$, what is AB?

xy² **26.** **Algebra** Find the measure of $\overset{\frown}{AB}$.
(47)

27. **Write** Explain how the method for finding arc length is
(35) similar to the formula for the circumference of a circle.

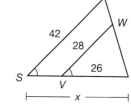

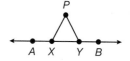

28. Explain why the triangles STU and VWU are similar, then
(46) find the missing length.

29. **Model** Fold a standard 8-by-11-inch sheet of paper
(36) along one diagonal. Carefully cut or tear along
 the diagonal to make two triangles. Explain why
 the hypothesis of the Leg-Leg Congruence Theorem
 is true for these two resulting right triangles.

*30. Write an indirect proof showing that through a line and a point
(48) not on the line, there exists exactly one perpendicular line to the
 given line.
 Given: P not on $\overleftrightarrow{AB}$
 Prove: There exists exactly one perpendicular line through P to $\overleftrightarrow{AB}$.

Introduction to Solids

1. **Vocabulary** A polygon in which all sides are congruent
   (15) is a(n) _____.

2. What is the name of a polygon with 7 sides?
   (15)

3. What is the name of a polygon with 10 sides?
   (15)

4. **Multiple Choice** An equiangular triangle has three angles that are _____.
   (13)   **A** congruent   **B** obtuse
          C vertical   **D** corresponding

New Concepts

The figures discussed in previous lessons are two-dimensional figures. This lesson introduces three-dimensional figures called **solids**. Solids can have flat or curved surfaces.

> **Math Reasoning**
>
> **Write** What are some common objects that are polyhedrons? Spheres?

The surfaces of some solids are polygons. Any closed three-dimensional figure formed by four or more polygons that intersect only at their edges is called a **polyhedron**. Some other solids have circular bases or curved sides. A **cone** is a three-dimensional figure with a circular base and a curved lateral surface that comes to a point. A **cylinder** is a three-dimensional figure with two parallel circular bases and a curved lateral surface that connects the bases. A **sphere** is the set of points in space that are a fixed distance from a given point, called the center of the sphere.

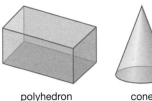
polyhedron

cone

cylinder

sphere

Example 1 **Classifying Solids**

Classify each of the three-dimensional solids shown.

a.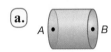

SOLUTION
The figure has two parallel circular bases, and a curved lateral surface. Therefore, the solid is a cylinder.

b.

SOLUTION
The figure is made up of five polygons that meet at their edges. Therefore, the figure is a polyhedron.

c.

SOLUTION
The figure has a circular base and a curved lateral surface that comes to a point. Therefore, the figure is a cone.

 Online Connection
www.SaxonMathResources.com

Hint

One example of a common polyhedron is a cereal box. It is composed of six rectangles joined at the edges. The seams of the box are the edges of the polyhedron, and each rectangle is a face of the polyhedron. The corners of the box are the vertices.

Math Language

An edge of a prism or pyramid that is not an edge of a base is a **lateral edge**.

Each flat surface of a polyhedron is called a **face of the polyhedron**. The segment that is the intersection of two faces of a solid is the **edge of a three-dimensional figure**. The **vertex of a three-dimensional figure** is the point of intersection of three or more faces of the figure.

A **prism** is a polyhedron formed by two parallel congruent polygonal bases connected by lateral faces that are parallelograms. The **base of a prism** is one of the two congruent parallel faces of the prism. A face of a prism that is not a base is called a **lateral face**.

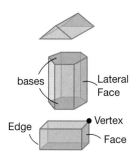

A **pyramid** is a polyhedron formed by a polygonal base and triangular lateral faces that meet at a common vertex. The faces of a pyramid all share a common vertex. The base is the side of the pyramid that does not share a single vertex with all of the other sides.

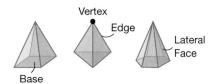

Prisms and pyramids are named by the shape of their bases. For example, a prism with a triangle for a base is called a triangular prism. A pyramid with a hexagon for a base would be called a hexagonal pyramid. A **cube** is the special name for a prism with six square faces.

Example **2** **Classifying Polyhedra**

Classify each polyhedron.

a.

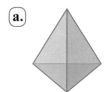

SOLUTION
The polyhedron has one base and the triangular faces meet at a common vertex. Therefore, the polyhedron is a pyramid. Since the base is a triangle, the polyhedron is a triangular pyramid.

b.

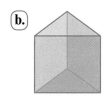

SOLUTION
The polyhedron has two parallel bases and the lateral faces are parallelograms. Therefore, the polyhedron is a prism. Since the bases are triangles, the polyhedron is a triangular prism.

A polyhedron is regular if all of its faces are congruent, regular polygons. A pyramid is regular if its base is a regular polygon and its lateral faces are congruent isosceles triangles. A prism is regular if its base is regular and its faces are rectangles. A cube is both a regular polyhedron and a regular rectangular prism. A triangular prism with equilateral bases is a regular prism but is not a regular polyhedron, since its faces are not congruent to its bases.

A **diagonal of a polyhedron** is a segment whose endpoints are the vertices of two different faces of a polyhedron.

Example 3 | Describing Characteristics of Solids

Classify the polyhedron in the diagram, assuming all the angles of each pentagon are congruent. Is it a regular polyhedron? How many edges, vertices, and faces does it have? Name one diagonal segment of the polyhedron.

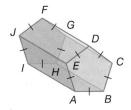

SOLUTION
The figure has two parallel pentagonal bases. Therefore, the polyhedron is a pentagonal prism. The sides of the bases are all congruent, and it is given that the angles are congruent, so it is a regular prism. Since the lateral faces are not congruent to the pentagonal bases, it is not a regular polyhedron. It has 7 faces, 15 edges, and 10 vertices. One diagonal is the segment $\overline{BF}$.

> **Math Reasoning**
>
> **Generalize** A pentagonal pyramid does not have a diagonal. Is this true of all pyramids? Explain.

A unique relationship exists among the number of faces, vertices, and edges of any polyhedron.

Euler's Formula
For any polyhedron with V vertices, E edges and F faces, $$V - E + F = 2.$$

Example 4 | Using Euler's Formula

How many faces does a polyhedron with 12 vertices and 18 edges have?

SOLUTION
Substitute $V = 12$ and $E = 18$ and solve for F.
$$V - E + F = 2$$
$$12 - 18 + F = 2$$
$$F = 8$$
The polyhedron has 8 faces.

Example 5 | Application: Diamond Cutting

> **Math Reasoning**
>
> **Analyze** When a vertex of the pyramid in this example is cut off, why does a triangular face form at the vertex?

Diamonds are cut to change them from a rough stone into a gemstone. The figure below shows two steps in cutting a particular diamond.

If each of the other vertices is cut in the next steps, what is the number of faces, vertices, and edges of the diamond in Step 4? Verify your answer.

SOLUTION

At the start the diamond has 4 faces, 4 vertices, and 6 edges. After cutting in Step 1, the diamond has 5 faces, 6 vertices, and 9 edges. After Step 2, the diamond has 6 faces, 8 vertices, and 12 edges. Since this pattern continues, after Step 3, the diamond will have 7 faces, 10 vertices, and 15 edges. After Step 4, the diamond will have 8 faces, 12 vertices, and 18 edges. Euler's Formula can verify the relationship among the faces, vertices, and edges: $12 - 18 + 8 = 2$.

Lesson Practice

a. Classify the solid. Name its vertices, edges, and bases.
(Ex 1)

b. Classify the solid. How many vertices, edges, and bases does it have?
(Ex 1, 3)

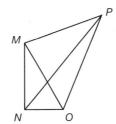

c. Classify the polyhedron. Determine whether it is a regular polyhedron.
(Ex 2, 3)

d. Classify the polyhedron. Determine whether it is a regular polyhedron.
(Ex 2, 3)

e. How many edges does a polyhedron with 14 vertices and 9 faces have?
(Ex 4)

f. **Gemstones** A gemstone is cut in the shape of a cube. Each vertex of the cube is then cut so that there is a triangular facet at each vertex. What is the number of faces, vertices and edges when the first four vertices of the cube are removed? Verify the results with Euler's Formula.
(Ex 5)

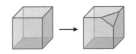

Practice Distributed and Integrated

1. **Verify** Using the information shown, verify that $\triangle DEF \sim \triangle MNP$.
(46)

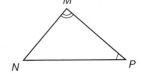

* 2. State the assumption you would make to start an indirect proof if asked to prove the following statement.
(48)

If a polygon is a hexagon, then the sum of the interior angles is 720°.

* 3. **Multiple Choice** Which statement contradicts the fact that $\triangle ABC$ is an equilateral
(48) triangle?

A All angle measures of $\triangle ABC$ are equal.

B The altitude of $\triangle ABC$ is not a median.

C All sides lengths of $\triangle ABC$ are equal.

D All angles of $\triangle ABC$ are acute angles.

4. **Algebra** Using the diagram at the right, find the length of $\overline{NO}$ in
(38) terms of x if $MP = x + 3$, $OP = x$, and $MN = 28$.

* 5. **Generalize** Describe the resulting figures when any prism is cut parallel
(49) to its base.

6. Using information from the diagram, determine the area of
(30) both triangles.

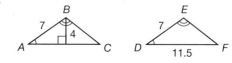

* 7. **Multi-Step** A corner garden has vertices $Q(0, 0)$, $R(0, 2d)$,
(45) and $S(2c, 0)$. A brick walkway runs from point Q to the
midpoint of RS. What is the length of the walkway?

8. Determine the midpoint M of the line segment $\overline{JK}$ with endpoints
(11) $J(7, 5)$ and $K(1, 2)$.

9. What is the included side of $\angle LMO$ and $\angle OMN$?
(28)

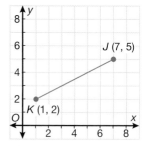

*10. ⬡ **Construction** A construction company wants to pre-cut all
(49) the pieces for the barn in the diagram. Each face, including the
floor, is cut from a special fiberboard. Metal strips will be used
along each edge, and a connector pipe placed at each vertex.
How many fiberboards, metal strips, and connector pipes are
needed to construct the barn?

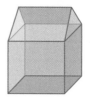

11. In $\odot P$, determine the value of x.
(43)

*12. If you are trying to prove indirectly that the altitude of an equilateral triangle
(48) is also a median, what assumption should you make to start the proof?

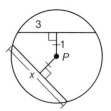

13. ⬡ **Carpentry** Susanna wants to increase the perimeter of her garden shed by a
(44) factor of 1.5. If the shed is 10 feet wide by 12 feet long already, what will be the
perimeter of the new garden shed?

14. ⬡ **Zoos** The habitats of the animals at a local zoo is shown in the
(41) diagram. The zookeeper wants to keep the mammals grouped
nearby each other. How far are the bears from the giraffes, to the
nearest meter?

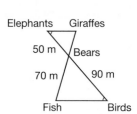

15. The noncongruent angle in an isosceles triangle measures 76°. What are the
(41) measures of the three angles in a triangle that is similar to this triangle?

16. (**Security Cameras**) A surveillance camera company needs to set up cameras that pan
(Inv 4) across three areas as indicated in the given diagram. Which camera pivots through
the largest angle? … the smallest angle? Explain.

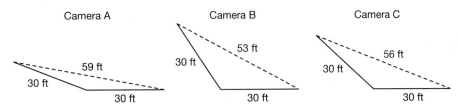

Camera A   Camera B   Camera C

*17. Analyze △*FGH* and △*KLM* are similar scalene triangles. How many of their six
(44) sides do you need numerical values for in order to find all the other side lengths
and the perimeters of both triangles? Explain.

18. a. Determine this figure's perimeter.
(40) **b.** Determine its area.

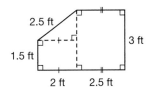

19. Find the closest point on the line
(42) $y = -x - 4$ to $(4, 4)$.

*20. Verify Is it possible to have a solid with 7 faces, 12 edges, and 10 vertices? Explain.
(49)

21. Error Analysis Below are two students' solutions for finding the length of
(29, 33) the hypotenuse. Determine which is incorrect and explain where the error
was made.

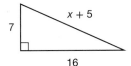

Leonardo's Solution

$16^2 + 7^2 = (x + 5)^2$
$16^2 + 7^2 = x^2 + 5^2$
$256 + 49 = x^2 + 25$
$280 = x^2$
$x \approx 16.7$
$x + 5 \approx 16.7$
$x \approx 11.7$

Florence's Solution

Let $x + 5 = c$
$16^2 + 7^2 = c^2$
$256 + 49 = c^2$
$305 = c^2$
$c \approx 17.5$
$x + 5 \approx 17.5$
$x \approx 12.5$

*22. Probability Explain why a cube is the only prism used in fair probability
(49) experiments.

23. Identify each line or segment that intersects the circle at right.
(43)

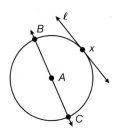

24. Write What conclusion can you make about three inscribed angles
(47) that intercept the same arc?

25. What are the coordinates of the midpoint of $(2, -2)$ and $(-7, 3)$?
(11)

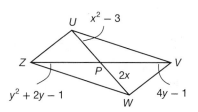

xy² **26. Algebra** $UVWZ$ is a parallelogram. Find each length.
(34)

    a. UW

    b. VP

    c. WP

    d. ZV

Use the diagram to answer the next two questions.

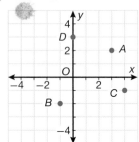

27. Multiple Choice Which distance is closest to 5 units?
(9)

    A AB

    B CD

    C BC

    D AD

28. Multiple Choice What is the length to the nearest hundredth of $\overline{AB}$?
(9)

    A 11.49

    B 8.12

    C 5.66

    D 5.83

29. Find the measure of $\angle ABC$.
(47)

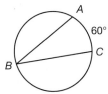

30. **Interior Design** The kitchen sink, countertop, and refrigerator are the vertices of
(39) a triangle in a kitchen. If the distance from the refrigerator to the countertop is
9 feet and from the countertop to the sink is 8 feet, what range of values would
the distance from the sink to the fridge have?

Geometric Mean

Warm Up

1. **Vocabulary** In a proportion, the extremes are the two values at the edges
(41) of the proportion, and the _____ are the two values that are in the
center of the proportion.

2. In a right triangle, the two sides of the triangle that include the right
(Inv 2) angle are the _____.

3. In the proportion 5:8 = 10:16, the means are ____ and ____.
(41)

4. For a proportion such as 3:6 = 5:10, the product of the means will equal
(41) the product of the _____.

New Concepts

When an altitude is drawn from the vertex of a right triangle's 90° angle to
its hypotenuse, it splits the triangle into two right triangles that exhibit a
useful relationship.

Caution

Theorem 50-1 is true only
if the altitude of the right
triangle has an endpoint
on its hypotenuse. An
altitude to either one of
the triangle's legs will
not exhibit this
relationship.

Theorem 50-1
If the altitude is drawn to the hypotenuse of a right triangle, then the two triangles formed are similar to each other and to the original triangle.

In △JKL, for example, △JMK is similar to △LMK,
and both △JMK and △LMK are similar to △JKL.

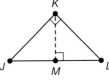

Example 1 **Proving Theorem 50-1**

Given: $\overline{DC}$ is an altitude of △ABC.
Prove: △ABC ~ △CBD, △ABC ~ △ACD, and
△ACD ~ △CBD

SOLUTION
In △ABC, $\overline{CD} \perp \overline{AB}$ by the definition of an altitude.
All right angles are congruent, so ∠BCA ≅ ∠CDA,
∠BDC ≅ ∠CDA, and ∠BCA ≅ ∠BDC.

By the Reflexive Property, ∠B ≅ ∠B. This is sufficient to show that
△ABC ~ △CBD, by the AA Similarity Postulate.

Again, by the Reflexive Property ∠A ≅ ∠A, so
△ABC ~ △ACD, by the AA Similarity Postulate.

By the Transitive Property of Similarity,
△ACD ~ △CBD since both triangles are similar to △ABC.

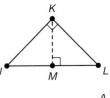

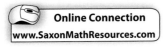

Example 2 Identifying Similar Right Triangles

Find PS and PQ.

SOLUTION

Since $\overline{QS}$ is a segment that is perpendicular to one side of the triangle with one endpoint on a vertex of the triangle, it is an altitude of $\triangle PQR$. By Theorem 50-1, $\triangle PQR \sim \triangle PSQ \sim \triangle QSR$. Set up a proportion to solve for the missing sides.

$$\frac{SQ}{SR} = \frac{PQ}{QR} = \frac{PS}{QS}$$

$$\frac{4}{3} = \frac{PQ}{5} = \frac{PS}{4}$$

$$PQ = 6.\overline{6}$$

$$PS = 5.\overline{3}$$

Sometimes, the means of a proportion are equal to one another. This is a special kind of proportion that can be used to find the geometric mean of two numbers. The **geometric mean** for positive numbers a and b, is the positive number x such that $\frac{a}{x} = \frac{x}{b}$.

Math Reasoning

Write Take the cross product of the definition of the geometric mean and solve for x. What is another way to state the geometric mean of a and b, according to the formula you have found?

Example 3 Finding Geometric Mean

(**a.**) Find the geometric mean of 3 and 12.

SOLUTION

Using the definition of geometric mean, you can obtain the following algebraic expression, where x represents the geometric mean.

$$\frac{3}{x} = \frac{x}{12}$$

$$x(x) = 12(3)$$

$$x^2 = 36$$

$$x = \sqrt{36}$$

$$x = 6$$

(**b.**) Find the geometric mean of 2 and 9 to the nearest tenth.

SOLUTION

Using the definition of geometric mean, you can obtain the following algebraic expression, where x represents the geometric mean.

$$\frac{2}{x} = \frac{x}{9}$$

$$x(x) = 2(9)$$

$$x^2 = 18$$

$$x \approx 4.2$$

Math Reasoning

Formulate Write the answer to part b of Example 3 in simplified radical form.

Two corollaries to Theorem 50-1 use geometric means to relate the segments formed by the altitude of a right triangle to its hypotenuse.

Corollary 50-1-1
If the altitude is drawn to the hypotenuse of a right triangle, then the length of the altitude is the geometric mean between the segments of the hypotenuse.

Corollary 50-1-2
If the altitude is drawn to the hypotenuse of a right triangle, then the length of a leg is the geometric mean between the hypotenuse and the segment of the hypotenuse that is closer to that leg.

Example **4** **Using Geometric Mean with Right Triangles**

a. Given the triangle STU, find the missing value, y.

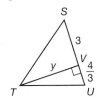

SOLUTION
Since TV is an altitude, by Corollary 50-1-1, y is the geometric mean of the segments of the hypotenuse, which are 3 and $\frac{4}{3}$. Using the definition of geometric mean, you can obtain the following algebraic expression.

$$\frac{3}{y} = \frac{y}{\frac{4}{3}}$$
$$y^2 = 3\left(\frac{4}{3}\right)$$
$$y^2 = 4$$
$$y = 2$$

b. Given the triangle, find the missing values a and b.

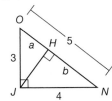

SOLUTION
Since JH is an altitude, there are two relationships that can be derived from Corollary 50-1-2.

$$\frac{a}{3} = \frac{3}{5} \qquad\qquad \frac{b}{4} = \frac{4}{5}$$
$$5(a) = 3(3) \qquad\qquad 5(b) = 4(4)$$
$$5a = 9 \quad \text{and} \quad 5b = 16$$
$$a = \frac{9}{5} \qquad\qquad b = \frac{16}{5}$$
$$a = 1.8 \qquad\qquad b = 3.2$$

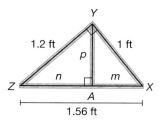

Example 5 **Real World Application**

Jayden is building a truss for a shed, shown in the diagram. Jayden needs to find the lengths of the truss brace $\overline{AY}$ and the lengths of $\overline{XA}$ and $\overline{ZA}$.

SOLUTION

Since $\overline{AY}$ is an altitude to the triangle, then

$$\frac{n}{1.2} = \frac{1.2}{1.56} \qquad\qquad \frac{m}{1} = \frac{1}{1.56}$$

$$1.56n = 1.44 \qquad \text{and} \qquad 1.56m = 1$$

$$n = \frac{1.44}{1.56} \qquad\qquad m = \frac{1}{1.56}$$

$$n \approx 0.92 \qquad\qquad m \approx 0.64$$

These are the lengths of $\overline{XA}$ and $\overline{ZA}$. To find the length of the truss brace $\overline{AY}$, apply Corollary 50-1-1.

$$\frac{m}{p} = \frac{p}{n}$$

$$\frac{0.64}{p} \approx \frac{p}{0.92}$$

$$p^2 \approx 0.5888$$

$$p \approx 0.77$$

So, Jayden needs a brace that is 0.77 feet long, which will divide the truss into two pieces that are 0.64 feet long and 0.92 feet long, respectively.

Hint

If you find it difficult to remember Corollaries 50-1-1 and 50-1-2, use Theorem 50-1 to find the lengths of the segments. The corollaries are useful, but not essential.

Lesson Practice

a. Name the similar triangles.
(Ex 1)

b. Find the values of x and y.
(Ex 2)

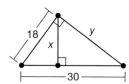

c. Find the geometric mean between 4 and 11 to the nearest
(Ex 3) tenth.

d. Find the geometric mean between 2 and 16 in simplified radical
(Ex 3) form.

e. Find the value of x.
(Ex 4)

f. Find the values of a and b to the nearest tenth.
(Ex 4)

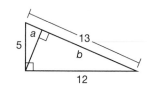

g. To support an old roof, a brace must be installed
(Ex 5) at the altitude. Find the length of the brace
to the nearest tenth of a foot.

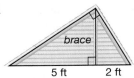
brace
5 ft   2 ft

Practice Distributed and Integrated

* 1. What is the name of a figure that has 5 vertices and 5 faces?
(49)

2. (Interior Design) A wallpaper design is made up of repeating circles that contain
(47) inscribed angles as shown. If m∠ADC = 17°, and m$\overarc{CE}$ = 42°, find m∠ABE.

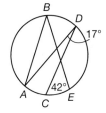

3. Multi-Step Parallelogram *EFGH* has three vertices at *E*(−4, −2), *F*(0, 6), and
(34) *G*(3, 7). Find the coordinates of *H*.

4. Algebra If a chord that is perpendicular to a 6-inch long radius cuts the radius
(43) into two equal lengths, what is the length of the chord?

5. Given that △*ABC* is an isosceles triangle, find (*x*, *y*).
(45)

* 6. Write Explain how the mean of two numbers is different
(50) from their geometric mean.

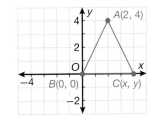

* 7. On the diagram, m∠1 ≠ m∠2. Prove $\overline{XA}$ is not an altitude of △*XYZ*.
(48)

* 8. Algebra Find the geometric mean of *x* and *y*, where *x* and *y* are positive numbers.
(50)

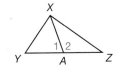

9. Using the diagram given, complete the
(30) proof that △*ABD* ≅ △*BCD*.

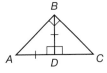

Statements	Reasons
1. m∠ADB = 90°, m∠BDC = 90°, m∠ABC = 90°, AD = DB	1. Given
2.	2. Acute angles in a right triangle are complementary.
3. m∠ABD + m∠DBC = 90°	3.
4. m∠DAB + m∠ABD = m∠ABD + m∠DBC	4. Substitution Property of Equality
5.	5. Subtraction Property of Equality and simplify.
6.	6. Definition of congruent angles
7.	7. All right angles are congruent.
8. $\overline{AD} \cong \overline{DB}$	8.
9. △ABD ≅ △BCD	9.

10. Name the angles that are congruent. If these triangles are similar, explain why.
₍₄₆₎

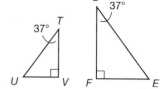

xy² 11. Algebra The sum of the interior angles of a convex polygon is 5400°. How many sides does the polygon have?
_(Inv 3)

*12. Analyze $\triangle XYZ$ and $\triangle MNO$ are similar equilateral triangles. How many of their six sides do you need numerical values for in order to find all the other side lengths and the perimeters of both triangles? Explain.
₍₄₄₎

xy² 13. Algebra Find the measure of $\angle G$ in the figure shown.
₍₄₇₎

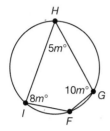

14. Given the line equation $y = -\frac{2}{5}x + 4$ and $H(3, -3)$, find the point on the line that is closest to H.
₍₄₂₎

15. **(Lifeguarding)** A beach inlet has three lifeguards on duty at all times. They are spaced around the inlet so that they are all equidistant from the diving board. Copy this diagram to find the approximate location of the diving board based on the location of the lifeguards.
₍₃₈₎

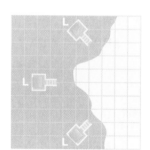

16. Justify Explain why a triangle cannot have side lengths measuring 51, 13, and 31.
₍₃₉₎

*17. **(Carpentry)** A builder is making a triangular eave as shown. What is the length of the brace in the center?
₍₅₀₎

3.21 in.   3.21 in.

*18. Multiple Choice Which of the following *does not* describe a polyhedron?
₍₄₉₎
   A 6 vertices, 9 edges, 5 faces   **B** 6 vertices, 10 edges, 6 faces
   C 8 vertices, 10 edges, 6 faces   **D** 8 vertices, 12 edges, 6 faces

xy² *19. Algebra Suppose you can use the Hypotenuse-Angle Congruence Theorem to prove that these two triangles are congruent. What could the value of z be?
₍₃₆₎

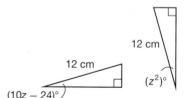

20. **(Landscaping)** City planners have decided to build a new park that is in the shape of a triangle. Along one side of the park will be a 12-foot fence. It will meet at right angles with another fence that is 5 feet long and will run along a second side. What length will be the third side of the park?
₍₂₉₎

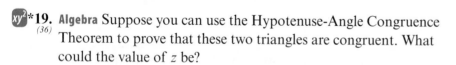

21. A transversal crossing two lines, makes a right angle with one of the lines. When does it make a right angle with the other?
_(Inv 1)

22. **Justify** Give a possible value for the angle at vertex *C* if you
(Inv 4) know that *AB* < *LK*. Explain your answer.

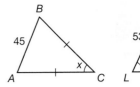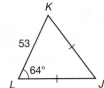

23. A square on the coordinate plane has side lengths of 10 units.
(45) The midpoint of the bottom side is at the origin. Find the
coordinates of the vertices.

24. **Multiple Choice** Which of the following are perpendicular lines?
(37)   i) $x + y = 1$ ii) $-2x - y = 3$ iii) $2y - x = 5$
    A i and ii   **B** i and iii
    C ii and iii   **D** None is perpendicular to either of the other lines.

25. ⬭Baseball⬭ The distance between consecutive bases in a baseball
(29) diamond is 90 feet. How far does the catcher have to throw the ball to get it from
home plate to second base?

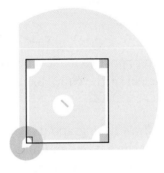

26. **Analyze** Find the measure of an arc so that the associated sector has one-tenth the
(35) area of the circle.

*27. **Error Analysis** Nikki counted 48 edges, 24 vertices, and 22 faces on a polyhedron.
(49) Is it possible for Nikki to be correct? Explain.

*28. Find the value of *x*.
(50)

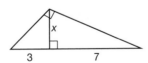

29. **Algebra** If in △*ABC* and △*DEF*, *AB* = *DE*, m∠*ABC* = m∠*EDF*, *BC* = *DF*,
(28) *AC* = (20*x* − 10), and *EF* = (15*x* + 15), what is the value of *x*?

30. Which statement, if true, would contradict the fact that △*JKL* ~ △*PQR*?
(48)   **A** *JK* = *rPQ* for some factor *r*
    B *JK* ≠ *PR*
    C ∠*J* ≠ ∠*R*
    D None of the angles in △*JKL* have the same measure as any
        of the angles in △*PQR*.

Nets

Recall that a polyhedron is a 3-dimensional solid with polygonal faces. A **net** is a diagram of the faces of a three-dimensional figure, arranged so that the diagram can be folded to form the three-dimensional figure.

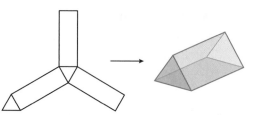

The diagram shows two possible nets of a triangular prism. As shown, a given polyhedron can have more than one net.

Math Reasoning

Model Draw a third net for the triangular prism depicted in the diagram.

1. **Model** Use graphing paper to make a net of a cube. How many squares will comprise the net of the cube? Sketch two possible ways to draw a net for a cube.

Choose one net and draw it on graph paper. Make each side of the cube's faces 5 units long. Cut the net out and fold it along each edge. Fold the net up into the shape of a cube and use tape to secure the edges of the cube.

2. **Analyze** Explain why the nets shown cannot make a cube when folded up.

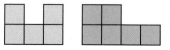

3. **Model** Draw two more possible nets for a cube.

Next, make a regular tetrahedron. A tetrahedron has four faces that are congruent equilateral triangles. Draw an equilateral triangle on grid paper. Make the length of each side 5 units and use a protractor to ensure that each of the triangle's angles measures 60°. To complete the net, draw three more congruent triangles, each sharing one side with the original triangle.

4. Cut out the tetrahedron and fold it up. What kind of solid is it? Classify it based on its faces and its base.

5. Is a cube a regular polyhedron?

6. Are either of the solids made in problem **1** or **4** prisms?

7. Is the tetrahedron a regular polyhedron?

The solid shown in the diagram is a regular octahedron. Think about unfolding one of the pyramids that comprise this shape to create a net. Working in groups with other students, draw two possible nets for the octahedron. Then, pick another solid from the following list, draw a net, and construct it: hexagonal pyramid, pentagonal prism, pentagonal pyramid, rectangular prism, and rectangular pyramid.

8. Are there more than two ways to draw the net of the octahedron?

9. Write Describe one method that can be used to make the net of any regular pyramid.

10. Write Describe one method that can be used to make the net of any regular prism.

11. Is the last solid that was made a regular polyhedron? Why or why not?

Math Reasoning

Generalize Which is the only pyramid that is a regular polyhedron? Which is the only prism that is a regular polyhedron?

Investigation Practice

a. Generalize How many different nets are there for a regular triangular pyramid? Sketch some nets and look for a pattern.

b. Using congruent squares, draw 5 different nets of a cube.

c. Draw a diagram using the connected faces of a cube that cannot be folded into a cube.

d. Draw another net for an octahedron (a regular solid with 8 equilateral-triangle faces).

e. Draw a net for a pentagonal prism.

Properties of Isosceles and Equilateral Triangles

Warm Up

1. **Vocabulary** A triangle with two congruent sides is a(n) _____.
   (13)

2. In $\triangle RST$, $\angle R \cong \angle S$ and $m\angle T = 80°$. Determine $m\angle R$.
   (18)

3. **Multiple Choice** Which statement is always true?
   (13)

   A If a triangle is isosceles, then it is equilateral.

   B If a triangle is equilateral, then it is isosceles.

   C Both **A** and **B**.

   D Neither **A** nor **B**.

New Concepts

In an isosceles triangle, the sides and the angles of the triangle are classified by their position in relation to the triangle's congruent sides.

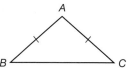

A **leg of an isosceles triangle** is one of the two congruent sides of the triangle. In the diagram, $\overline{AB}$ and $\overline{AC}$ are the legs. The **vertex angle of an isosceles triangle** is the angle formed by the legs of the triangle. The vertex angle is $\angle A$. The **base of an isosceles triangle** is the side opposite the vertex angle. The base of $\triangle ABC$ is $\overline{BC}$. A **base angle of an isosceles triangle** is one of the two angles that have the base of the triangle as a side. In $\triangle ABC$, $\angle B$ and $\angle C$ are base angles.

Math Reasoning

Analyze Does Theorem 51-1 also apply to equilateral triangles? How do you know?

Theorem 51-1: Isosceles Triangle Theorem
If a triangle is isosceles, then its base angles are congruent. $\triangle LMN$ is isosceles. Therefore, $\angle M \cong \angle N$. 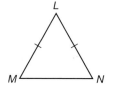

Corollary 51-1-1
If a triangle is equilateral, then it is equiangular.

Example 1 **Proving the Isosceles Triangle Theorem**

Prove the Isosceles Triangle Theorem.

Given: $\triangle ABC$ is an isosceles triangle with $\overline{AB} \cong \overline{AC}$. D is the midpoint of $\overline{BC}$.

Prove: $\angle B \cong \angle C$

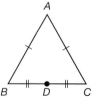

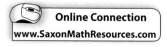

SOLUTION

Statements	Reasons
1. $\triangle ABC$ is isosceles	1. Given
2. $\overline{AB} \cong \overline{AC}$	2. Definition of isosceles triangle
3. $BD = CD$	3. Definition of midpoint
4. $\overline{BD} \cong \overline{CD}$	4. Definition of congruent segments
5. $\overline{AD} \cong \overline{AD}$	5. Reflexive Property
6. $\triangle ABD \cong \triangle ACD$	6. SSS Triangle Congruence Postulate
7. $\angle B \cong \angle C$	7. CPCTC

The Converse of the Isosceles Triangle Theorem is also true, as is the Converse of Corollary 51-1-1.

Theorem 51-2: Converse of the Isosceles Triangle Theorem

If two angles of a triangle are congruent, then the sides opposite those angles are also congruent.

Corollary 51-2-1

If a triangle is equiangular, then it is equilateral.

Example 2 **Using the Isosceles Triangle Theorem and Its Converse**

(**a.**) Triangle DEF is isosceles, and its vertex angle is at E. If m$\angle D = 36°$, determine m$\angle E$ and m$\angle F$.

SOLUTION

The base angles of $\triangle DEF$ are $\angle D$ and $\angle F$, so by the Isosceles Triangle Theorem, $\angle D \cong \angle F$. By the definition of congruent angles, m$\angle F =$ m$\angle D$, so they each measure 36°. Therefore,

m$\angle D +$ m$\angle E +$ m$\angle F = 180°$	Triangle Angle Sum Theorem
$36° +$ m$\angle E + 36° = 180°$	Substitute
m$\angle E = 108°$	Solve

(**b.**) The perimeter of $\triangle GHJ$ is 12 inches, and $\angle G \cong \angle H$. If $GH = 5$ inches, find GJ.

SOLUTION

By the Converse of the Isosceles Triangle Theorem, $\overline{GJ} \cong \overline{HJ}$. Since the perimeter is 8 inches and $GH = 5$ inches,

$P = GH + HJ + GJ$	Formula for perimeter
$12 = 5 + HJ + GJ$	Substitute.
$12 = 5 + GJ + GJ$	Definition of congruent segments
$12 = 5 + 2GJ$	Simplify.
$GJ = 3.5$ in.	Solve.

Hint

If you are not sure which side of the triangle is the base and which sides are the legs, sketch a triangle and label the angles and sides with as much information as possible.

Example 3 Using Relationships in Equilateral Triangles

A triangle is equiangular and has a perimeter of 22.5 centimeters. Determine the length of each side.

SOLUTION

By Corollary 51-2-1, the triangle is equilateral. Let the length of each side be s. The perimeter is the sum of the three sides.

$P = s + s + s$	Formula for perimeter
$22.5 = 3s$	Substitute and simplify.
$s = 7.5$ cm	Solve.

Math Reasoning

Connect How are Theorems 51-3 and 51-4 related to each other as conditional statements?

Theorem 51-3
If a line bisects the vertex angle of an isosceles triangle, then it is the perpendicular bisector of the base.

Theorem 51-4
If a line is the perpendicular bisector of the base of an isosceles triangle, then it bisects the vertex angle.

The diagram illustrates both of these theorems. The altitude $\overline{TU}$ bisects the vertex angle and is a perpendicular bisector of the base of the triangle.

Example 4 Proving Theorems 51-3 and 51-4

a. Prove Theorem 51-3.

Given: $\triangle ABC$ is isosceles, $\overline{AD}$ bisects $\angle A$
Prove: $\overline{AD}$ is the perpendicular bisector of $\overline{BC}$

SOLUTION

Statements	Reasons
1. $\triangle ABC$ is isosceles, $\overline{AD}$ bisects $\angle A$	1. Given
2. $\angle BAD \cong \angle CAD$	2. Definition of angle bisector
3. $\overline{AB} \cong \overline{AC}$	3. Definition of isosceles triangle
4. $\overline{AD} \cong \overline{AD}$	4. Reflexive Property
5. $\triangle ABD \cong \triangle ACD$	5. SAS Triangle Congruence Postulate
6. $\overline{BD} \cong \overline{CD}$	6. CPCTC
7. $BD = CD$	7. Definition of congruent segments
8. $\angle ADB \cong \angle ADC$	8. CPCTC
9. $\overline{AD}$ and $\overline{BC}$ form adjacent angles	9. Definition of adjacent angles
10. $\overline{AD} \perp \overline{BC}$	10. If lines form congruent adjacent angles, they are perpendicular
11. $\overline{AD}$ is $\perp$ bisector of $\overline{BC}$	11. Definition of perpendicular bisector

b. Write a paragraph proof of Theorem 51-4.

Given: △*ABC* is isosceles, $\overline{AD}$ is the perpendicular bisector of $\overline{BC}$
Prove: $\overline{AD}$ bisects ∠*A*

SOLUTION
Since △*ABC* is isosceles, $\overline{AB} \cong \overline{AC}$. By the Reflexive Property, $\overline{AD} \cong \overline{AD}$. Both △*ABD* and △*ACD* are right triangles, since $\overline{AD}$ is the perpendicular bisector of $\overline{BC}$ and forms two right angles at *D*. Therefore, △*ABD* ≅ △*ACD* by the Hypotenuse-Leg Right Triangle Congruence Theorem. By CPCTC, ∠*BAD* ≅ ∠*CAD*. Therefore, by the definition of an angle bisector, $\overline{AD}$ bisects ∠*BAC*.

Example 5 **Application: Infrastructure**

This figure shows the north and east view of a telephone pole that is secured by four cables of equal length.

North View · East View

a. Explain why the base angles, ∠*PAQ* and ∠*PRQ*, are congruent.

SOLUTION
In △*APR*, the cable lengths *AP* and *RP* are equal, so $\overline{AP} \cong \overline{RP}$ by the definition of congruent segments. Therefore, △*APR* is isosceles by definition. Applying the Isosceles Triangle Theorem, the base angles of △*APR* are congruent, so ∠*PAQ* ≅ ∠*PRQ*.

b. Prove that these angles are also congruent to the base angles ∠*B* and ∠*D*.

SOLUTION
By the Reflexive Property of Congruence, $\overline{PQ} \cong \overline{PQ}$. It is given in the problem that $\overline{BP} \cong \overline{AP}$, so by the Hypotenuse-Leg Right Triangle Congruence Theorem, △*BPQ* ≅ △*APQ*. By CPCTC, ∠*B* ≅ ∠*A*. Since △*BPD* is isosceles, ∠*D* ≅ ∠*B* by the Isosceles Triangle Theorem. It is given that ∠*A* ≅ ∠*R*, so by the Transitive Property of Congruence, ∠*R* ≅ ∠*D*.

Lesson Practice

a. For the isosceles triangle shown, determine the
(Ex 2) missing angle measures.

b. The perimeter of △*XYZ* is 15.2 centimeters, and
(Ex 2) ∠*X* ≅ ∠*Z*. If *XY* = 6.3 centimeters, determine *XZ*.

c. If the vertex angle of an isosceles triangle measures 20°,
(Ex 2) what are the measures of each of its base angles?

d. A triangle is equiangular and its perimeter is 7 feet. Determine the
(Ex 3) length of each side.

e. (Ex 5) **Engineering** This diagram shows the side-view profile of a bridge. Determine the angle that each half of the bridge makes with the horizontal.

150 ft   158°   150 ft

Practice   Distributed and Integrated

1. In *FGHJ*, m∠*H* < m∠*FJG*, *GH* = 4*x* − 1, and *FG* = *x* + 8.
(Inv 4) Find the range of values for *x*.

2. From the statement, "If Fabian's socks are clean, then they are in the
(21) dresser," what can you conclude about what will happen when Fabian's socks are clean?

3. a. Determine this figure's perimeter.
(40) **b.** Determine its area.

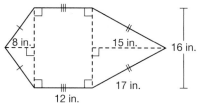

* 4. Write Explain how you would find the geometric mean
(50) of 4 and 7.

5. What is the measure of each exterior angle in a regular hexagon?
(Inv 3)

6. Using the diagram, find the length of $\overset{\frown}{AB}$ to the nearest hundredth
(35) of a centimeter.

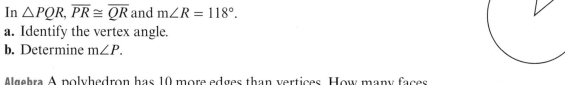

* 7. In △*PQR*, $\overline{PR} \cong \overline{QR}$ and m∠*R* = 118°.
(51) **a.** Identify the vertex angle.
b. Determine m∠*P*.

8. Algebra A polyhedron has 10 more edges than vertices. How many faces
(49) does the polyhedron have?

9. Find the unknown length of the side in the triangle shown.
(29)

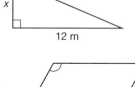

*10. Write Explain why each angle of an equilateral triangle measures 60°.
(51)

11. **Art** In order to design part of her tile pattern correctly, Teresa
(34) needs to make the consecutive angles shown supplementary. How
can she ensure the angles will be supplementary?

12. Algebra Two lines are perpendicular. One line has an equation $y = \frac{12}{m}x - 2$ and the
(37) other line has an equation of $4y = \frac{n}{9}x + 5$. Find one value for *m* and *n*.

*13. **Construction** A builder needs to position a support brace as shown.
(50) What is the length of the support brace *x*, to the nearest tenth?

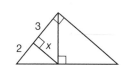

14. Justify Explain why $\triangle ADE \sim \triangle ABC$, then find BC.
(46)

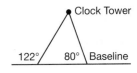

15. Convex pentagon $PQRST$ has vertices $P(0, 0)$, $Q(-2, 2)$, $R(-1, 4)$, $S(x, y)$,
(45) and $T(2, 2)$. $QR = ST$ and $RS = 2$. Find (x, y).

16. (**Surveying**) Kristi, a map surveyor, is using an east-west baseline to
(18) locate various landmarks. She measures the clockwise angle from the baseline to the line passing through her surveying instrument and the landmark. Kristi takes two sightings from a clock tower, as this diagram shows.

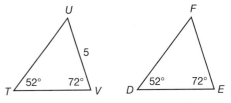

  a. What measure is the angle between the two sightings?
  b. What theorem did you use in part **a**?

17. In the diagram shown at right, what information is needed
(30) to conclude that $\triangle TUV \cong \triangle DFE$?

18. Generalize Triangle GHI has vertices $G(0, 0)$ and $H(a, b)$. Find
(45) coordinates for a point I so that $\triangle GHI$ is a right triangle.

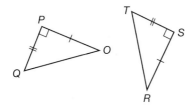

19. Use the LL Congruence Theorem to show that $\triangle OPQ \cong \triangle RST$.
(36)

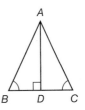

20. What is the equation of a line perpendicular to $y = -4x + 12$, passing through
(37) the origin?

21. (**Cell Phone Towers**) Three cell phone towers are the vertices of
(39) a triangle at right. The measure of $\angle M$ is 10° less than the measure of $\angle K$. The measure of $\angle L$ is one degree greater than the measure of $\angle M$. Which two towers are closest together?

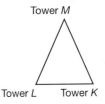

*22. Analyze Give a two-column proof of the Converse of the
(51) Isosceles Triangle Theorem.
  Given: $\angle B \cong \angle C$
  Prove: $\triangle ABC$ is isosceles.
  Hint: Drop an altitude from point A to the base of $\triangle ABC$.

*23. Find the value of x and y on the triangle shown.
(50)

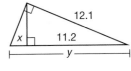

*24. Multi-Step Determine whether each set of numbers can be the side
(33, 39) lengths of a triangle. If they can be a triangle, determine whether it is an acute, obtuse, or right triangle.
  a. $\sqrt{99}$, $3\sqrt{13}$, 21
  b. $\sqrt{7}$, $\sqrt{8}$, 3
  c. 8, 7, 14

25. Determine the perimeter and area of this figure. Give exact answers.
(40)

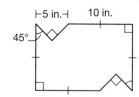

26. If $\triangle KLM \cong \triangle DEF$, what side of $\triangle DEF$ corresponds to side $\overline{KM}$?
(25)

27. Using the diagram, find PY in terms of x if $PV = 3$ and $YU = x$. Point P is the
(38) incenter of the triangle.

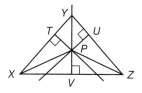

28. Multiple Choice The intercepted arc of an inscribed angle is a semicircle
(47) if and only if the measure of the angle is

   A 20°   **B** 90°

   C 100°   **D** 180°

*29. Find the value of x in this figure.
(51)

30. Prove Theorem 39-2: If one angle of a triangle is larger than another
(48) angle, then the side opposite the first angle is longer than the side
opposite the second angle.

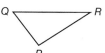

   Given: $m\angle P > m\angle R$

   Prove: $QR > QP$

   *Hint: There are two cases you must consider. One where $QR < QP$ and one
   where $QR = QP$.*

Properties of Rectangles, Rhombuses, and Squares

Warm Up

1. Find the area of the regular quadrilateral.
   (15) Classify this quadrilateral.

2. **Vocabulary** A quadrilateral with two pairs of
   (19) parallel sides is a _____.

3. Find the perimeter of this composite
   (40) figure. Then, name each quadrilateral
   in the figure.

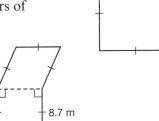

12.5 cm

8.7 m

New Concepts   The diagonals of parallelograms have special properties. Recall that a
rhombus is a parallelogram with four congruent sides, a rectangle is a
parallelogram with four right angles, and a square shares the properties
of both a rectangle and a rhombus. One property of the diagonals of a
parallelogram has already been introduced: they bisect each other. Three
more are introduced in this lesson.

Math Reasoning

Analyze Are the
diagonals of a square
congruent? How do you
know?

Properties of a Rectangle: Congruent Diagonals
The diagonals of a rectangle are congruent. $\overline{PR} \cong \overline{QS}$

P   Q

S   R

If a quadrilateral is a parallelogram, it is a rectangle if and only if the above
property is true.

Example 1   **Using Diagonals of a Rectangle**

A rectangular barn door has diagonal braces.
If AE is 6 feet, what is the length of $\overline{BD}$?

SOLUTION

$\overline{AC} \cong \overline{BD}$ 	 Diagonals of a rectangle are congruent

$AE = EC$ 	 Diagonals of a parallelogram bisect each other

$EC = 6$ 	 Substitute.

$AC = 12$ 	 Segment Addition Postulate

$BD = 12$ 	 Definition of segment congruence

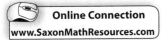

Online Connection
www.SaxonMathResources.com

Exploration **Using Contruction Techniques to Draw a Rhombus**

In this exploration, you will use simple construction techniques to construct a quadrilateral, then classify it. You may wish to review Construction Lab 1 before this exploration.

1. Draw *JK*. Set your compass to *JK*. Place the compass point at *J* and draw an arc above $\overline{JK}$. Choose and label a point *L* on the arc. What is the relationship between *JK* and *JL*?

2. Place the compass point at *L* and draw an arc to the right of *L*.

3. Place the compass point at *K* and draw an arc that intersects the arc you drew in step 2. Label the point of intersection *M*. How are *JK*, *KM*, *ML*, and *LJ* related?

4. How do you know that the quadrilateral you have drawn is a rhombus?

5. Draw the diagonals $\overline{JM}$ and $\overline{LK}$ and label their point of intersection *P*. Measure ∠*LPM*. What can you determine about the diagonals?

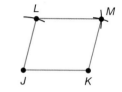

6. By measuring angles, determine the relationship between the diagonals and the angles of the rhombus.

Properties of a Rhombus: Perpendicular Diagonals
The diagonals of a rhombus are perpendicular. $\overline{HJ} \perp \overline{IK}$ 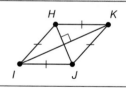

If a quadrilateral is a parallelogram, it is a rhombus if and only if the above property is true. Since a square is both a rhombus and a rectangle, its diagonals are both perpendicular and congruent.

Properties of a Rhombus: Diagonals as Angle Bisectors
Each diagonal of a rhombus bisects opposite angles. Because opposite angles of a rhombus are equal, when they are bisected by a diagonal, four congruent angles result. 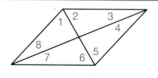 ∠1 ≅ ∠2 ≅ ∠5 ≅ ∠6, and ∠3 ≅ ∠4 ≅ ∠7 ≅ ∠8

If a quadrilateral is a parallelogram, it is a rhombus if and only if the above property is true.

Example 2 Using Properties of Diagonals of a Rhombus

BCDF is a rhombus. Find the measure of each angle.

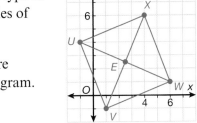

a. m∠*EBC*

SOLUTION

Since m∠*BEC* is 90°, then we know that
m∠*EBC* + m∠*ECB* = 90°

$$(3x + 12)° + (x + 10)° = 90°$$ Substitute.
$$4x + 22 = 90°$$ Simplify.
$$x = 17$$ Solve.

Now substitute the value of *x* to find the measure of ∠*EBC*.

m∠*EBC* = 3*x* + 12
m∠*EBC* = 3(17) + 12 Substitute for *x*.
m∠*EBC* = 63° Simplify.

b. m∠*ECD*

SOLUTION

Since the diagonals of a rhombus bisect the angles, m∠*ECD* = m∠*ECB*.
m∠*ECD* = *x* + 10
m∠*ECD* = 17 + 10
m∠*ECD* = 27°

Example 3 Using Properties of Parallelograms

Hint

It may be possible to classify a parallelogram using more than one of the properties in this lesson.

UVWX is a parallelogram. Decide what type of parallelogram it is by using the properties of rectangles and rhombuses.

a. Determine whether the diagonals are congruent and classify the parallelogram.

SOLUTION

$$UW = \sqrt{(-1 - 6)^2 + (4 - 1)^2} = \sqrt{58}$$
$$VX = \sqrt{(1 - 4)^2 + (-1 - 6)^2} = \sqrt{58}$$

Since *UW* = *VX*, then the diagonals are congruent. By the Congruent Diagonals Property of a Rectangle, the shape must be a rectangle.

b. Determine whether the diagonals are perpendicular and classify the parallelogram.

SOLUTION

$$\text{slope of } \overline{UW} = \frac{4 - 1}{-1 - 6} = -\frac{3}{7} \qquad \text{slope of } \overline{VX} = \frac{-1 - 6}{1 - 4} = \frac{7}{3}$$

Since $-\frac{3}{7} \times \frac{7}{3} = -1$, $\overline{UW}$ is perpendicular to $\overline{VX}$.

This implies that the parallelogram is a rhombus. Since the shape is both a rectangle and a rhombus, it is also a square.

Example 4 Application: Architecture

A rectangular building is designed with steel support braces placed diagonally in the interior. Determine the length of the steel brace that will be used for diagonal $\overline{BD}$.

SOLUTION

$$a^2 + b^2 = c^2 \qquad \text{Pythagorean Theorem}$$
$$50^2 + 120^2 = c^2 \qquad \text{Substitute}$$
$$c = 130 \text{ ft} \qquad \text{Solve}$$
$$EF = 130 \text{ ft} \qquad \text{Substitute}$$
$$\overline{EF} \cong \overline{BD} \qquad \text{Diagonals of a rectangle}$$
$$\qquad\qquad\qquad \text{are congruent}$$
$$BD = 130 \text{ ft} \qquad \text{Substitute}$$

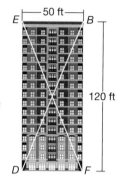

Lesson Practice

a. In rectangle *MNOP*, *MO* = 5.4 inches. What is the length of $\overline{NP}$?
(Ex 1)

WXYZ **is a rhombus. Using the diagram, answer the questions that follow.**

b. Find m∠*OXY*.
(Ex 2)

c. Find m∠*OYZ*.
(Ex 2)

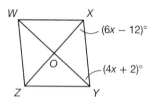

d. Quadrilateral *RSTU* has a center point, *V*. If $\overline{RT} \cong \overline{SU}$, and $\overline{RT} \perp \overline{SU}$,
(Ex 3) classify the quadrilateral.

e. (**Architecture**) A building is made with
(Ex 4) a rhombus-shaped courtyard. If the longer diagonal walkway is 50 feet and the shorter one is 40 feet, what is the perimeter of the courtyard to the nearest foot?

Practice Distributed and Integrated

1. Draw a net for this polyhedron.
(Inv 5)

2. Multi-Step Find the orthocenter of △*JKL* with
(32) vertices *J*(3, 6), *K*(3, −9), and *L*(−5, −5).

3. Write Compare a regular octagonal prism and a cylinder. Consider what
(49) happens as the number of sides of the base of a prism increases.

4. (**Carpentry**) A cabinetmaker needs to position a support
(50) brace as shown. What is the length *x* of the support brace?

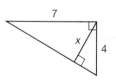

* **5.** (**Art**) Joni wants to use a square piece of paper for an art project.
(52) Explain how she could easily determine if a piece of paper is square.

6. The circumference of a circle is 113 feet. If a sector has an area of 12 ft², what is
₍₃₅₎ the measure of the arc to the nearest degree?

7. **Error Analysis** Henrietta is trying to find the slope of the line that passes through the
₍₃₇₎ point (8, 0) and is perpendicular to the line $y = 2x + 1$. Her answer is $y = \frac{1}{2}x + 4$.
Is she correct? Why or why not?

8. Are these two triangles similar? Give a reason to support
₍₄₆₎ your answer.

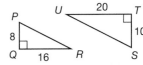

9. **Generalize** Explain how a parallelogram could be similar to a rectangle.
₍₄₁₎

*10. **Algebra** A rhombus has two angles that each measure
₍₅₂₎ $(9x + 1)°$ and two angles that each measure $(20x + 5)°$. Find the
measure of each of the four angles in the rhombus.

*11. Find the geometric mean of $\sqrt{11}$ and $\sqrt{5}$.
₍₅₀₎

*12. (**Chemistry**) This diagram shows the atoms and bonds in a water molecule. The
₍₅₁₎ bond angle at the oxygen (O) atom is always 104.5°. The distances from the
oxygen atom to each hydrogen (H) atom are equal. What are the measures of
the other two angles, to the nearest tenth?

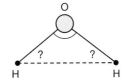

13. **Multiple Choice** Which statement has a false converse?
₍₁₀₎
A If two angles are supplementary, then they sum to 180°.
B If a triangle has two congruent angles, then it is isosceles.
C If a number is an even prime, then the number is 2.
D If two angles are complementary, then they are both acute.

14. **Analyze** For a triangle with vertices $L(0, 0)$, $M(6, 0)$, and $N(3, y)$ to be equilateral,
₍₄₅₎ what must be the value of y?

*15. **Multi-Step** The vertices of a square *KLMN* are $K(0, 1)$, $L(4, -2)$, $M(1, -6)$ and
₍₅₂₎ $N(-3, -3)$. Show that the diagonals are congruent perpendicular bisectors of
each other.

16. If a circle has a circumference of $6x$, and an arc in that circle has a length of x,
₍₃₅₎ what is the angle measure of the associated sector?

17. **Algebra** In the circle shown, find the measure of $\angle I$.
₍₄₇₎

18. In isosceles triangle *PQR*, $PQ = 28$, $QR = 28$, and $m\angle Q = 86°$.
₍₅₁₎
a. Identify the vertex angle.
b. Determine $m\angle R$.

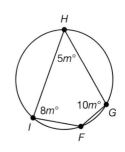

19. What is the relationship between angles intercepted by the same
₍₄₇₎ arc?

*20. Write a paragraph proof showing that if one side of a triangle
₍₃₁₎ is longer than another side, then the angle opposite the first
side is larger than the angle opposite the second side.
Given: $RS > RQ$
Prove: $m\angle RQS > m\angle S$

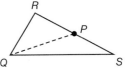

21. In isosceles triangle ABC, $\overline{AB} \cong \overline{BC}$ and each base angle measures 40°.
(51) Determine $m\angle B$.

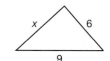

22. **Algebra** Write an inequality to show the largest value for x that
(33) makes the triangle obtuse if the longest side is 9 units.

23. In a right triangle, one leg is 36 feet long, and the hypotenuse is
(29) 39 feet long. What is the length of the third side?

*24. ⬭ Fence Construction ⬭ A homeowner wants to build a fence gate with reinforcing
(52) diagonal braces. How can he make sure the gate is rectangular without measuring
the angles?

*25. **Predict** This figure is formed from three congruent rectangles.
(40)   **a.** Determine the perimeter of the figure.
  b. Consider the figure formed by extending the pattern to four congruent
  rectangles. Determine the perimeter of the new figure.
  c. Determine a formula for the perimeter of a figure in the same pattern
  formed from n congruent rectangles.

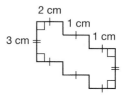

26. To measure the distance EF across the lake, a surveyor at S locates
(46) points E, F, G, and H as shown. What is the length of $\overline{EF}$?

27. **Write** Describe how to circumscribe a circle around an obtuse triangle.
(38)

28. $ABCD$ is a parallelogram. If $\overline{AB}$ is 3 times longer than $\overline{AD}$ and the
(34) perimeter of $ABCD$ is 12, how long is each side?

29. Suppose you are to prove $\overline{AB} \cong \overline{CD}$. State the assumption you would
(48) make to start an indirect proof.

*30. ⬭ Hiking ⬭ Yvette was planning on going around a thick, circular clump of trees,
(43) but she found a shortcut through it. If the diameter of the circular area is
100 meters, how much shorter is her direct path, to the nearest meter, than
traveling around the trees?

45°-45°-90° Right Triangles

Warm Up

1. **Vocabulary** The name given to a triangle with one 90° angle
⁽¹³⁾ is _____.

2. Find the area of a right triangle with legs that are 11 inches and
⁽¹⁸⁾ 17 inches long.

3. A right triangle has legs that are 8 units and 2 units long. What is the ratio
⁽²⁹⁾ of the triangle's hypotenuse to its shortest leg?

New Concepts

Some right triangles are used so frequently that it is helpful to remember some of their particular properties. These triangles are called **special right triangles**. The two most common special right triangles are the 30°-60°-90° triangle and the 45°-45°-90° triangle.

Since the 45°-45°-90° triangle has two angles with equal measures, it is also an isosceles right triangle.

Properties of a 45°-45°-90° Triangle: Side Lengths
In a 45°-45°-90° right triangle, both legs are congruent and the length of the hypotenuse is the length of a leg multiplied by $\sqrt{2}$.

> **Hint**
>
> The name of each special right triangle gives the measure of its angles. 45°-45°-90° triangles are often used because they are one half of a square.

Example 1 Finding the Side Lengths in a 45°-45°-90° Triangle

a. Use the properties of a 45°-45°-90° right triangle to find the length of the hypotenuse of the triangle.

SOLUTION
The length of the hypotenuse is equal to the length of a leg times $\sqrt{2}$. Since the leg is 2 inches long, the hypotenuse has a length of $2\sqrt{2}$ inches.

b. Use the properties of a 45°-45°-90° right triangle to find the length of a leg of the triangle.

SOLUTION
The length of the hypotenuse is equal to the length of the leg times $\sqrt{2}$. To find the length of a leg when given the hypotenuse, divide by $\sqrt{2}$ instead.

The length of a leg of the triangle is $\frac{3}{\sqrt{2}} = \frac{3\sqrt{2}}{2}$ feet.

> **Caution**
>
> When the denominator of a fraction has a square root in it, it must be rationalized. In this example, multiplying both the top and bottom of the fraction by $\sqrt{2}$ eliminates the root in the denominator.

Find the perimeter of the triangle.

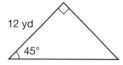

SOLUTION

The length of the hypotenuse is equal to the length of the leg times $\sqrt{2}$. Therefore, the hypotenuse is $12\sqrt{2}$ yards long.

The perimeter can be found by adding the lengths of the three sides together.

$P = 12 + 12 + 12\sqrt{2}$
$P = 24 + 12\sqrt{2}$
$P \approx 40.97$

Therefore, the perimeter is approximately 41 yards.

Though it is often faster to use the properties of 45°-45°-90° triangles to find unknown lengths, the Pythagorean Theorem can still be used to determine lengths in special right triangles.

Example 3 **Applying the Pythagorean Theorem with 45°-45°-90° Right Triangles**

Find the length of the missing sides to the nearest foot, using the Pythagorean Theorem.

SOLUTION

Since the legs are congruent, let x represent the length of the legs of the triangle. Therefore,

$a^2 + b^2 = c^2$
$x^2 + x^2 = 125^2$
$\qquad 2x^2 = 15625$
$\qquad x^2 = \dfrac{15625}{2}$
$\qquad x^2 = 7812.5$
$\qquad x \approx 88.4$

Therefore, the missing side length is approximately 88 feet.

Math Reasoning

Verify Use the properties of 45°-45°-90° triangles to find x. Is the result the same?

Online Connection
www.SaxonMathResources.com

Example 4 Application: Park Construction

A square park is to be fenced around the perimeter with a snow fence for an upcoming outdoor concert. There is a diagonal path that is 430 feet long through the park. How much snow fence is required? Use the 4-Step Problem-Solving Process.

SOLUTION

Understand The fence is to be placed around the perimeter, so the perimeter must be found. The diagonal of the square is given. A diagram would be a helpful visual aid to understand this problem.

Plan First, draw a diagram. Identify the lengths that need to be found and use the properties of 45°-45°-90° triangles to solve for them. Add the length of each side together to find the perimeter.

Solve This involves finding the length of the legs of the 45°-45°-90° triangle created by two adjacent sides of the park and the diagonal path.

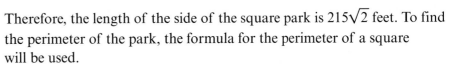

$$430 = \sqrt{2}x$$

$$\frac{430}{\sqrt{2}} = \frac{\sqrt{2}x}{\sqrt{2}}$$

$$\frac{430\sqrt{2}}{2} = x$$

Therefore, the length of the side of the square park is $215\sqrt{2}$ feet. To find the perimeter of the park, the formula for the perimeter of a square will be used.

$$P = 4l$$
$$P = 4(215\sqrt{2})$$
$$P = 860\sqrt{2}$$
$$P \approx 1216.22$$

Therefore, the perimeter of the park, and thus the amount of fencing needed, is approximately 1216 feet.

Check Here, since the diagonal is longer than a side, each side must be less than 430 feet. So the perimeter must be less than 4 × 430, or 1720 feet. The answer of 1216 feet seems to make sense because it is less than 1720 feet.

Hint

A good way to check your work is to consider the possible bounds (minimum and maximum) the answer could have. Does your answer fall between the bounds of the minimum and maximum? In this case, any answer over 1720 feet or under 0 feet would clearly be incorrect.

Lesson Practice

a. Find the length of this triangle's hypotenuse.
(Ex 1)

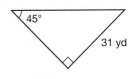

b. Find the length of this triangle's missing side.
(Ex 1)

c. Find the perimeter of the triangle, to the nearest tenth of an inch.
(Ex 2)

d. Find the length of the missing sides to the nearest mile.
(Ex 3)

e. A square building has a diagonal length of 150 feet. What would be the square footage of one floor of the building?
(Ex 4)

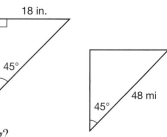

Practice   Distributed and Integrated

1. (**Construction**) The measurements in the diagram are from the attic space in a new home. If the angle measure of 72° is an inaccurate label on the diagram, what is the range of degrees that the mislabeled angle could be? It is given that the side opposite the mislabeled angle is the longest side of the triangle, and the side opposite the 48° is the shortest side.
(39)

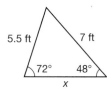

2. A figure has 12 congruent edges and 8 vertices. Classify the figure.
(49)

* 3. Find the exact value of the length of leg a in the triangle.
(53)

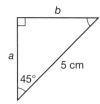

4. (**Center of Gravity**) Are the orthocenter and the centroid both centers of gravity? If not, which one is?
(32)

* 5. Is the following statement always, sometimes, or never true?
(52)

A parallelogram is a rectangle.

6. Find the value of x and y in the figure at right to the nearest tenth.
(50)

7. A composite figure is formed by a rectangle with a square removed at one corner. The rectangle measures $6\frac{1}{2}$ inches by 5 inches.
(40)

The removed square has side lengths of $3\frac{1}{2}$ inches.
a. Determine the perimeter of the figure.
b. Determine the area of the figure.

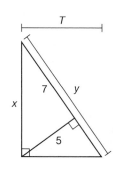

* 8. **Write** Is every right isosceles triangle a 45°-45°-90° triangle? Explain.
(53)

9. Identify each line or segment that intersects the circle shown.
(43)

*10. **Algebra** The value of the side length in a 45°-45°-90° triangle is x inches. What would be the algebraic expression for the length of the triangle's hypotenuse?
(53)

11. Write a similarity statement to explain why the two triangles shown are similar.
(46)

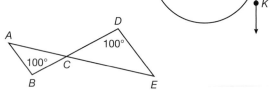

12. Analyze What information must be known about two similar triangles in order to
(44) find the ratio of their perimeters?

*13. Analyze Write a flowchart proof showing that if a triangle is equilateral, then
(51) it is equiangular.

14. What is the total area of the shaded sectors if the diameter of the circle is
(35) 8 meters? Write your answer in terms of π.

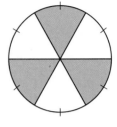

15. Find the distance from $(-5, 2)$ to the line $x = 13$.
(42)

16. Algebra If in $\triangle ABC$, $AB = (12x + 11)$, $m\angle ABC = 12°$, $m\angle BCA = 21°$, and
(30) $AC = 7$ units, and in $\triangle DEF$, $DE = (5x + 18)$, $m\angle DEF = 12°$, $m\angle EFD = 21°$,
and $DF = 7$ units, what is the value of x?

17. Justify Give a possible value for the length of $\overline{XY}$.
(Inv 4) Explain your answer.

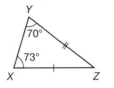 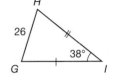

18. What is the value of b in the proportion $\frac{2}{5} = \frac{7}{b}$?
(41)

19. (**Art**) An artist is making a round ceramic plate with a pattern of lines on it.
(47) If $m\angle KLM = 20°$, and $m\widehat{MP} = 30°$, find $m\angle KNP$.

20. Suppose a chord of a circle is 10 inches long, and the radius of the circle is
(43) also 10 inches. What is the measure from the chord to the center of the circle?

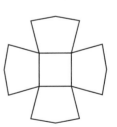

*21. Multiple Choice What would be the perimeter of a 45°-45°-90° triangle with a
(53) hypotenuse of 73 feet, to the nearest foot?
  A 52 feet   **B** 176 feet
  C 104 feet   **D** 198 feet

22. (**Design**) A gift box has the net shown in this figure.
(14, Inv 5) Use the net to provide a counterexample to the
following statement.

*If the net of a three-dimensional figure has all lateral
faces congruent, then the figure is closed.*

*23. *WXYZ* is the rectangle shown. Find *XY*.
(52)

24. Multi-Step Find the centroid of $\triangle DEF$ with vertices $D(-3, 2)$,
(32) $E(9, 8)$, and $F(3, -4)$.

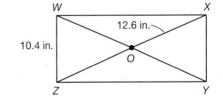

25. It is given that $\triangle ACE \sim \triangle BDF$. If *AC* is 3 units, *CE* is 12 units,
(41) and *BD* is 11 units, what is the length of $\overline{DF}$?

26. If two sides of one triangle are proportional to two sides of another triangle,
(46) and if their corresponding included angles are congruent to each other, then the
triangles are similar by _____.

*27. Find the exact value of the length of the hypotenuse in the triangle.
(53)

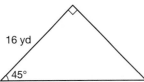

16 yd

45°

28. Use the Hypotenuse-Angle Congruence Theorem to prove that $\triangle ABC \cong \triangle DEF$.
(36)

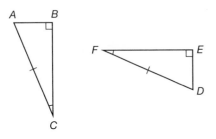

*29. Write Explain why the following statement is true.
(52)

If a quadrilateral is a square, then it is a rhombus.

*30. (**Surveying**) Renée takes the bearings of landmarks at *A*, *B*, and *C* from the same
(51) position, *D*.

Landmark	Bearing from *D*
A: Church Steeple	322°
B: Clock Tower	2°
C: Water Tower	42°

Suppose Renée chose point *D* to be equidistant from the church steeple and the
water tower. What must be true about the clock tower's distance from each of the
other two landmarks? Why?

Representing Solids

Warm Up

1. **Vocabulary** A prism with six square faces is called a _____.
(49)

2. Name each of the pictured solids. If the solid
(49) is a prism or pyramid, classify it.

3. According to Euler's Formula, if a polyhedron
(49) has 7 faces and 10 vertices, how many edges does
it have?

New Concepts

In a **perspective drawing**, nonvertical parallel lines appear to meet at a point
called a vanishing point. If you look straight down a highway, it appears
that the edges of the highway eventually come together at a vanishing point,
like point *A* in the diagram. In a perspective drawing, the **horizon** is the
horizontal line that contains the vanishing point(s). A drawing with just one
vanishing point is called **one-point perspective**.

Math Language

The **vanishing point**
is the point in a
perspective drawing
on the horizon where
parallel lines appear
to meet.

Example 1 Drawing in One-Point Perspective

Draw a rectangular prism in one-point perspective.
Use a pencil with an eraser.

SOLUTION

Step 1 Draw a square and a horizontal line above it
representing the horizon. Mark a vanishing
point on the horizon.

Step 2 Draw a dashed line from the vanishing point to
each of the four corners of the square.

Step 3 Using the dashed lines drawn in Step 2, draw
the sides of a smaller square.

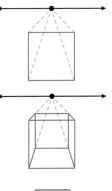

Step 4 Connect the two squares and erase the reference
lines and the horizon that are located behind the prism.
This prism is drawn from a one-point perspective.

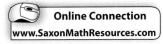
Online Connection
www.SaxonMathResources.com

A drawing with two vanishing points is said to have two-point perspective. Look at the following example to see how a drawing can be made from a two-point perspective.

Example 2 **Drawing in Two-Point Perspective**

Draw a rectangular prism in two-point perspective in which the vanishing points are above the prism.

SOLUTION

Step 1 Draw a horizontal line that represents the horizon. Place two vanishing points on the horizon. Draw a vertical line segment below the horizontal line and between the two vanishing points, representing the front edge of the prism.

Math Reasoning

Model Could you also make a two-point perspective drawing by placing the vanishing points below the original line segment?

Step 2 Draw dashed lines from each vanishing point to the top and bottom of the vertical line as shown.

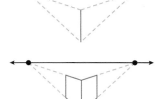

Step 3 Draw vertical segments between the dashed lines from Step 2 as shown and draw segments to connect them to the first segment.

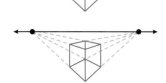

Step 4 Draw dashed perspective lines from the segments drawn in Step 3 to each of the vanishing points as shown.

Step 5 Draw a dashed vertical line between the two intersections of the perspective lines just drawn. Sketch the segments that make the top of the prism.

Step 6 Erase the horizon line and the dashed perspective lines. Keep the dashed lines inside the prism that represent the edges that are hidden.

This prism is drawn from a two-point perspective.

An **isometric drawing** is a way of drawing a three-dimensional figure using isometric dot paper, which has equally spaced dots in a repeating triangular pattern. The drawings can be made by using three axes that intersect to form 120° angles, as shown in the diagram.

Example 3 **Creating Isometric Drawings**

Create an isometric drawing of a rectangular prism.

SOLUTION
Draw the three axes on the isometric dot paper as shown above. Use this vertex as the bottom corner of the prism. Draw the box so that the edges of the prism run parallel to the three axes. Shading the top, front, and side of the prism will add the perception of depth.

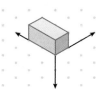

Example 4 **Application: Drafting**

An architecture firm is planning to construct a rectangular building on a corner lot. The client would like a drawing that shows the building as though someone is looking at it from one edge. Should the drawing be from a one-point or two-point perspective? Make a sketch of what the drawing should look like.

SOLUTION

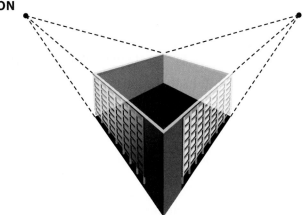

Since the front of the drawing will be an edge of the building, a two-point perspective drawing is appropriate. The diagram shows a completed view of the building.

Hint

In a two-point perspective drawing, it appears that one edge of the solid is the front of the diagram. In a one-point perspective drawing, it appears that a face of the solid is the front.

Lesson Practice

a. Draw a rectangular prism in one-point perspective in which the vanishing point is to the left of the square.
(Ex 1)

b. Draw a cube in two-point perspective with the vanishing points and horizon below the vertical line.
(Ex 2)

c. Make an isometric drawing of a triangular prism.
(Ex 3)

d. (**Drafting**) Morgan wants to make a wooden bookshelf with two shelves. The bookshelf will be 1 meter wide, 1 meter deep, and 1.5 meters tall. To decide how much wood to buy, Morgan will draw his plans for the bookshelf. Should the drawing be from a one-point or two-point perspective? Sketch what Morgan's drawing should look like.
(Ex 4)

* 1. Draw a triangular prism in one-point perspective so that the vanishing point is
(54) below the prism.

2. **Write** Explain why the following statement is true.
(52) *If a quadrilateral is a square, then it is a rectangle.*

3. **Algebra** Find the length of $\overline{ZP}$ in the diagram.
(32)

4. What is the shortest distance from (5, 3) to the line
(42) $y = -2x + 8$?

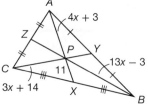

* 5. **(Architecture)** An architect is creating different perspective drawings for a new
(54) building. The building is a rectangular prism and the client would like a drawing
that focuses on the front façade of the building. Should the architect create the
drawing using a one-point or two-point perspective? Sketch a sample drawing of
the building.

6. A figure has a hexagonal base and triangular lateral faces. Classify the figure.
(49)

7. **Multi-Step** Graph the line and find the slope of the line that passes through the
(37) points $L(4, 1)$ and $M(3, -1)$. Then find a perpendicular line that passes through
point $N(-2, -2)$.

8. Find the value of x and y in the triangle shown to the
(50) nearest tenth.

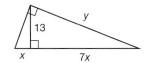

9. What is the sum of the exterior angles of a convex
(Inv 3) 134-sided polygon?

10. Is the following statement always, sometimes, or never true?
(52) *A parallelogram is a rectangle.*

*11. Trace the figure at right on your paper. Then locate the
(54) vanishing point and the horizon line.

12. **Algebra** In $\triangle ABC$, m$\angle ABC = 90°$, $AB = (3x - 7)$, and
(30) m$\angle BCA = 60°$, and in $\triangle DEF$, m$\angle DEF = 90°$, $DE = (5x - 17)$, and m$\angle EFD = 60°$. What value of x will
make $\triangle ABC \cong \triangle DEF$?

13. The point where three or more lines intersect is the _____.
(32)

*14. Use the Hypotenuse-Angle Congruence Theorem to prove
(36) that $\triangle RST \cong \triangle UVW$.

*15. Find the exact length of the hypotenuse of a 45°-45°-90° right
(53) triangle with a leg that is 57 feet long.

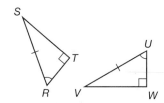

16. Formulate Four congruent circles are cut out of a square as shown. Write an
(40) expression for the area of the shaded region in terms of the radius of each
circle, *r*.

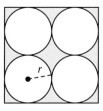

17. (Aviation) Four jet aircraft are flying in a triangular formation. Jets *A*, *B*, and *C*
(51) form a line perpendicular to the flight heading, while jet *B* is midway between
the other two. Jet *D* flies directly in front of jet *B*. If m∠*ADB* = 37°, what does
the vertex angle of the triangular formation measure? Which theorem did you
use?

*18. Use an indirect proof to prove that if two altitudes, $\overline{BX}$ and $\overline{CY}$ of △*ABC* are
(48) congruent, then the triangle must be isosceles.
Given: $\overline{BX} \cong \overline{CY}$, $\overline{BX}$ and $\overline{CY}$ are altitudes.
Prove: △*ABC* isosceles.

*19. Find the area, to the nearest hundredth, of a 45°-45°-90° right triangle with a
(53) hypotenuse of 17 centimeters.

20. Algebra If a chord perpendicular to a radius cuts the radius in two pieces that are
(43) 7 and 2 inches long, respectively, what are the two possible lengths of the chord to
the nearest tenth?

*21. Justify How does a two-point perspective differ when the vanishing points are
(54) located close together compared with when they are located further apart? Justify
your reasoning with drawings.

22. Find the geometric mean of $\sqrt{2}$ and 5.
(50)

23. Multiple Choice If the diagonals of parallelogram *JKLM* intersect at *P*, which of
(34) the following is true?
  A *JP* = *LP*   **B** *JP* = *KP*
  C *JL* = *KM*   **D** *JM* = *KM*

*24. (Construction) The support of a shelf forms a 45°-45°-90° right triangle,
(53) with the shelf and the wall as the legs. Exactly how long
is this support?

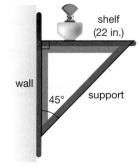

*25. Analyze △*NPQ* and △*STV* are similar isosceles triangles. How many
(44) of their six sides do you need numerical values for in order find all the
other side lengths and the perimeters of both triangles? Explain.

26. Using the diagram, find the length of $\overline{MP}$ if $OP = 5$,
(38) $NO = 8$, and $MN = 18$.

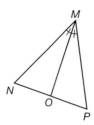

27. (**Cycling**) Katya and Sareema start from the same location and bicycle in opposite
(Inv 4) directions for 2 miles each. Katya turns to her right 90° and continues for another
mile. Sareema turns 45° to her left and continues for another mile. At this point,
who is closer to the starting point?

28. **Error Analysis** Darius drew this net of a number cube. Explain his error.
(Inv 5)

29. **Analyze** Square *RSTU* has vertices at $R(0, 4)$ and $S(0, 0)$. What are the possible
(45) coordinates of *T* and *U*?

30. (**Design**) A white triangle with vertices at $(0, 0)$, $(4, 0)$, and $(0, 4)$ is used to create
(11) a logo. A blue triangle is added to the design so that its vertices are the midpoints
of the sides of the white triangle. The blue triangle divides the white triangle into
three smaller white triangles. Smaller blue triangles are placed in each small white
triangle so that their vertices are the midpoints of the sides of the small white
triangles.

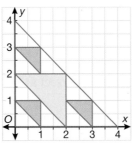

    a. Find the coordinates of the vertices of the large blue triangle.

    b. Find the coordinates of the vertices of each small blue triangle.

    c. Which of the triangles are congruent, if any? Justify your answer.

Triangle Midsegment Theorem

Warm Up

1. **Vocabulary** Two triangles with congruent corresponding angles and
   (41) corresponding sides that are proportional in length are _____.

Determine if the triangles in each pair are similar. If they are, state the theorem or postulate that proves it.
(46)

2.

3.

4.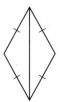

New Concepts

A **midsegment of a triangle** is a segment that joins the midpoints of two sides of the triangle. Every triangle has three midsegments.

Math Language

The **midpoint** of a segment is the point that divides a segment into two congruent segments.

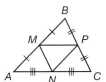

The midsegment is always half the length of the side that does not have a midsegment endpoint on it.

Theorem 55-1: Triangle Midsegment Theorem

The segment joining the midpoints of two sides of a triangle is parallel to, and half the length of, the third side.

$\overline{RQ} \parallel \overline{PM}$ and $RQ = \frac{1}{2}PM$

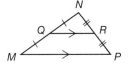

Example 1 **Using the Triangle Midsegment Theorem**

In the diagram, $\overline{DE}$ is a midsegment of $\triangle ABC$. Find the values of x and y.

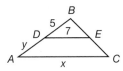

SOLUTION
From the Triangle Midsegment Theorem,

$DE = \frac{1}{2}AC$, so $AC = 2DE$.

$AC = 2(7)$

Therefore, $x = 14$.

From the definition of a midsegment, $AD = DB$. So, $y = 5$.

Online Connection
www.SaxonMathResources.com

Example 2 **Proving the Triangle Midsegment Theorem**

Given: D is the midpoint of $\overline{AB}$ and E is the
midpoint of $\overline{AC}$.

Prove: $\overline{DE} \parallel \overline{BC}$ and $DE = \frac{1}{2}BC$

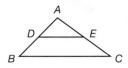

SOLUTION

Statements	Reasons
1. D is the midpoint of $\overline{AB}$; E is the midpoint of $\overline{AC}$	1. Given
2. $AD = DB$; $AE = EC$	2. Definition of midpoint
3. $AD + DB = AB$; $AE + EC = AC$	3. Segment Addition Postulate
4. $AD + AD = AB$; $AE + AE = AC$	4. Substitute
5. $AD = \frac{1}{2}AB$; $AE = \frac{1}{2}AC$	5. Solve
6. $\angle A \cong \angle A$	6. Reflexive Property of Congruence
7. $\triangle ABC \sim \triangle ADE$	7. SAS Triangle Similarity Theorem
8. $\angle AED \cong \angle ACB$	8. Definition of similar polygons
9. $\overline{DE} \parallel \overline{BC}$	9. If corresponding angles are congruent, lines cut by a transversal are parallel
10. $DE = \frac{1}{2}BC$	10. Definition of similar polygons and step 5

Math Reasoning

Analyze Is Theorem 55-2 a conditional statement that is related to 55-1? If so, what is their relationship?

Theorem 55-2
If a line is parallel to one side of a triangle and it contains the midpoint of another side, then it passes through the midpoint of the third side. Since $\overline{UV} \parallel \overline{RT}$ and $\overline{RU} \cong \overline{US}$, $\overline{SV} \cong \overline{VT}$. 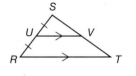

The measure of $\overline{QT}$ can be determined using Theorem 55-2. Since $\overline{RU} = \overline{US}$
in triangle QRS, then U is the midpoint of $\overline{RS}$. By Theorem 55-2, since
$\overline{TU} \parallel \overline{QS}$ and U is the midpoint of $\overline{RS}$, then T is the midpoint of $\overline{QR}$. Since
T is the midpoint of $\overline{QR}$, then $QT = TR$. The measure of $\overline{QT}$ is 13 units.

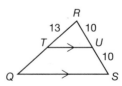

Example 3 Identifying Midpoints of Sides of a Triangle

Triangle MNP has vertices $M(-2, 4)$, $N(6, 2)$, and $P(2, -1)$. $\overline{QR}$ is a midsegment of $\triangle MNP$. Find the coordinates of Q and R.

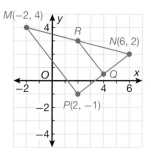

SOLUTION

R and Q are the midpoints of $\overline{MN}$ and $\overline{NP}$. Use the Midpoint Formula to find the coordinates of Q and R.

$$Q\left(\frac{6+2}{2}, \frac{2+(-1)}{2}\right) = Q\left(4, \frac{1}{2}\right)$$

$$R\left(\frac{-2+6}{2}, \frac{4+2}{2}\right) = R(2, 3)$$

The midsegment of a triangle creates two triangles that are similar by AA-Similarity. In the diagram, since $\overline{DE} \parallel \overline{AC}$, then $\angle BAC \cong \angle BDE$ and $\angle BED \cong \angle BCA$. This shows that $\triangle ABC \sim \triangle DBE$.

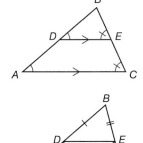

A **midsegment triangle** is the triangle formed by the three midsegments of a triangle. Triangle DEF is a midsegment triangle. Midsegment triangles are similar to the original triangle and to the triangles formed by each midsegment. In the figure, $\triangle ABC \sim \triangle EFD \sim \triangle ADF \sim \triangle DBE \sim \triangle FEC$.

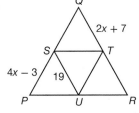

Example 4 Applying Similarity to Midsegment Triangles

Triangle STU is the midsegment triangle of $\triangle PQR$.

a. Show that $\triangle STU \sim \triangle PQR$.

SOLUTION

Since $\triangle STU$ is the midsegment triangle of $\triangle PQR$, by the definition of midsegment:

$ST = \frac{1}{2}PR$; $SU = \frac{1}{2}QR$; $TU = \frac{1}{2}QP$

Therefore, $\triangle STU \sim \triangle PQR$ by SSS similarity.

b. Find PQ.

SOLUTION

QR is twice SU, and T is the midpoint of $\overline{QR}$, so $QT = SU$.

$2x + 7 = 19$

$x = 6$

Since S is the midpoint of $\overline{PQ}$, $PQ = 2PS$.

$PQ = 2(4x - 3) = 2[4(6) - 3]$

$= 42$

The length of $\overline{PQ}$ is 42 units.

Example 5 Application: Maps

A student determined that Toledo Street is the midsegment of the triangle formed by Columbus Avenue, Park Avenue, and William Street. The distance along Park Avenue between Columbus Avenue and William Street is 160 meters, and the distance along Toledo Street in the same span is 80 meters. William Street is 240 meters long. Find the distance from the corner of Columbus and William to the corner of William and Toledo (*x* in the diagram) to the nearest meter.

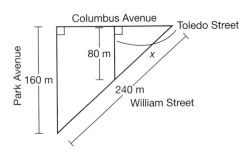

SOLUTION

Since Toledo Street is the midsegment of the triangle, it creates two similar triangles. Use the lengths of Toledo Street and Park Avenue to find the similarity ratio of 160:80. Name the shorter segment *x* and write a proportion.

$$\frac{160}{80} = \frac{240}{x}$$
$$160x = (80)(240)$$
$$x = 120 \text{ m}$$

Lesson Practice

a. $\overline{TU}$ is a midsegment of $\triangle QRS$. Find the
(Ex 1) values of *x* and *y*.

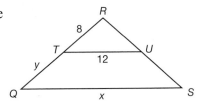

b. Prove Theorem 55-2.
(Ex 2) **Given:** $\overline{DE} \parallel \overline{BC}$ and *D* is the midpoint of $\overline{AB}$.
Prove: *E* is the midpoint of $\overline{AC}$.

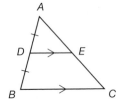

Hint

This proof is similar to the proof of Theorem 55-1, given in Example 2. Start by showing that the two triangles in the diagram are similar.

c. Triangle *FGH* has vertices at *F*(−2, 4),
(Ex 3) *G*(6, 2), and *H*(2,−1). $\overline{DE}$ is a midsegment of
△*FGH*. Find the coordinates of *D* and *E*.

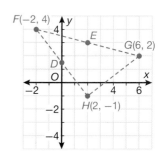

d. Find the perimeter of the midsegment
(Ex 4) triangle, △*XYZ*.

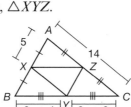

e. In the diagram, 4th Street is a midsegment
(Ex 5) of the triangle formed by Baker, Lowry,
and 5th Streets. Jeremiah leaves his house
at the corner of 5th and Baker and walks
down 5th Street. He then turns left and
walks up Lowry Street until he reaches the
corner at Lowery and 4th Street. How far
has Jeremiah walked?

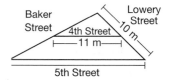

Practice Distributed and Integrated

* 1. Draw a triangular prism in one-point perspective so that the vanishing point
(54) is below the prism.

2. **Multiple Choice** Which of the following statements is *not* true of a
(52) rhombus?
   A All sides are congruent.   **B** The diagonals are perpendicular.
   C The diagonals are congruent.   **D** The diagonals bisect the angles.

3. (**Model Trains**) Gary is building a track for his model trains. He wants to
(Inv 1) ensure that the two sides of the tracks run parallel to each other, so he
places crossbeams at regular intervals along them. What could Gary
check to ensure the tracks are parallel?

4. **Error Analysis** After reading the Triangle Inequality Theorem, which states that the
(39) sum of the lengths of any two sides of a triangle is greater than the length of the
third side, Ken reasons that the theorem would mean the same if it were written,
"The sum of the lengths of any two sides of a triangle cannot be less than the
length of the third side." Is Ken correct? Explain.

* 5. Trace the figure shown on your paper. Then draw the vanishing points and the
(54) horizon line of the figure.

6. **Analyze** In which regular polygon is each exterior angle equal to each interior
(Inv 3) angle?

7. How many edges does a figure with 8 vertices and 6 faces have?
(49)

8. (Aviation) A pilot determines that after flying 200 kilometers on one leg of a
(39) trip and 425 kilometers on a second leg of a trip, that she has enough fuel to fly
another 715 kilometers. Does she have enough fuel to get back to her starting
point? Explain.

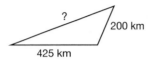

* 9. **Analyze** Give a paragraph proof showing that if a triangle is equiangular, then it is
(51) equilateral.

10. (Golf) In his first shot of a golf tournament, illustrated here, Mitch has hit the ball
(53) too far to the left from where it should be. How far does he need to hit the ball to
get it straight to the green from where it is now? Give your answer to the nearest
tenth of a yard.

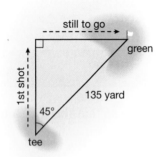

*11. Draw a cube in two-point perspective.
(54)

*12. Segment DE is a midsegment of $\triangle ABC$.
(55) Refer to the diagram to determine the
coordinates of D and E.

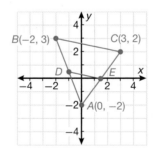

13. In $\triangle QRS$, $m\angle Q = 55°$, and $m\angle R = 86°$.
(46) In $\triangle TUV$, $m\angle T = 55°$, and $m\angle V = 39°$.
Is $\triangle QRS \sim \triangle TUV$? Explain.

14. The ratio of the angle measures in a quadrilateral is 1:4:5:6. Find the measure
(41) of each angle.

*15. Segment UT is a midsegment of $\triangle XYZ$. Find the values of x and y.
(55)

16. **Multi-Step** Find the orthocenter of $\triangle GHI$ with vertices $G(-8, -9)$, $H(-2, -1)$,
(32) and $I(-2, -9)$.

17. Find the geometric mean of 1 and 5.
(50)

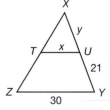

18. **Algebra** Solve the equation $2(x + 3) = \frac{5x - 1}{3}$. Provide a justification for
(24) each step.

19. Analyze A triangle has vertices $L(0, 0)$, $M(8, 0)$, and $N(4, y)$. What value of
₍₄₅₎ y makes $\triangle LMN$ equilateral?

*20. Write Explain why the median and the altitude from the vertex angle of an
₍₅₁₎ isosceles triangle are identical. Refer to any theorems you need to justify
your explanation.

*21. Below is the beginning of a paragraph proof of the Triangle Inequality
₍₃₁₎ Theorem. Write the rest of the proof. *Hint: Use the Isosceles Triangle*
Theorem.
Given: $\triangle ABC$
Prove: $AB + BC > AC$, $AB + AC > BC$, $AC + BC > AB$
One side of $\triangle ABC$ is as long or longer than each of the other
sides. Let this side be $\overline{AB}$. Then $AB + BC > AC$ and $AB + AC > BC$.
Therefore what remains to prove is $AC + BC > AB$. Locate D on $\overrightarrow{AC}$ such
that $BC = DC$.

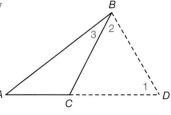

22. If a chord is bisected 2 inches from the center of a circle with a diameter of
₍₄₃₎ 6 inches, what is the length of half of the chord to the nearest tenth of an
inch?

23. Algebra Find the measure of $\overset{\frown}{ST}$ in the circle at right.
₍₄₇₎

*24. The midpoints of the sides of $\triangle ABC$ are as follows: midpoint of
₍₅₅₎ AC: $D(4, 1)$; midpoint of AB: $E(1, 3)$; and midpoint of BC: $F(1, 0)$.
Draw the midsegment triangle. Find the coordinates of A, B, and C,
and then draw $\triangle ABC$.

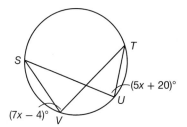

25. (Travel) Marco travels from Yuma to Alamo and then from Alamo to
₍₅₃₎ Gadsden. Each unit on the triangle represents 30 miles. How far does he
need to travel to get from Gadsden to Yuma, if he were to travel a straight
path between them, rounded to the nearest mile?

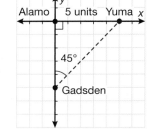

26. (Landscaping) A square field is to be hydro-seeded. The field has a
₍₅₃₎ diagonal length of 225 yards. How many square feet need to be
hydro-seeded?

27. Determine if these two triangles are similar. If so, state the similarity
₍₄₆₎ and the reason.

28. Find the geometric mean of 11 and 1.5.
₍₅₀₎

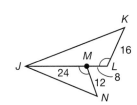

29. Write Sketch this situation or explain why it is impossible.
_(Inv 1) *Two parallel lines are intersected by a transversal so that the same-side interior*
angles are complementary.

30. Algebra Points Q, R, and S are midpoints of $\triangle FGH$. If $FH = x + 12$
₍₅₅₎ and $QR = 2x - 3$, what is the length of $\overline{QR}$?

30°-60°-90° Right Triangles

Warm Up

1. Vocabulary A(n) _____ triangle has three sides that are congruent.
₍₅₁₎

2. If a right triangle has a hypotenuse of 13 centimeters and a base of 5 centimeters, what is its height?
₍₂₉₎

3. If an isosceles triangle has a vertex angle of 20°, what is the measure of a base angle?
₍₅₁₎

    A 60°   **B** 80°

    C 75°   **D** 20°

New Concepts

The 30°-60°-90° triangle is another special triangle. Like the 45°-45°-90° triangle, properties of the 30°-60°-90° triangle can be used to find missing measures of a triangle if the length of one side is known.

In the diagram, two 30°-60°-90° triangles are shown next to each other, with the shorter legs aligned.

Placing the two triangles together so that they share a common leg makes an equilateral triangle. Since all the equilateral triangle's sides are congruent, this shows that the hypotenuse of the triangle is twice the length of the shortest leg.

> ### Math Reasoning
>
> **Connect** A 45°-45°-90° triangle can be used to prove attributes in a square. What quadrilaterals can be formed using 30°-60°-90° triangles?

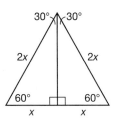

Properties of 30°-60°-90° Triangles
In a 30°-60°-90° triangle, the length of the hypotenuse is twice the length of the short leg, and the length of the longer leg is the length of the shorter leg times $\sqrt{3}$.

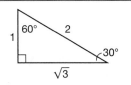

Algebraically, these relationships can be written as follows.

$$PR = a \qquad\qquad PQ = 2a \qquad\qquad QR = a\sqrt{3}$$

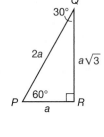

Example **1** **Finding Side Lengths in a 30°-60°-90° Triangle**

Find the values of x and y. Give your answer in simplified radical form.

SOLUTION

The shortest leg must be opposite the smallest angle, so the leg with a measure of 2 is the short leg. The hypotenuse is twice the short leg, so $y = 4$. The long leg is $\sqrt{3}$ times the short leg, so $x = 2\sqrt{3}$.

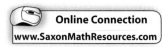

Online Connection
www.SaxonMathResources.com

Example 2 **Finding the Perimeter of a 30°-60°-90° Triangle with Unknown Measures**

Find the perimeter of the triangle. Give your answer in simplified radical form.

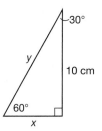

SOLUTION
First, find the length of x and y. The short leg is x. Since the length of the short leg times the square root of 3 equals the long leg, an equation can be written to solve for x.

$$x\sqrt{3} = 10$$
$$x = \frac{10}{\sqrt{3}}$$
$$x = \frac{10\sqrt{3}}{3}$$

Since the length of the short leg is now known, the length of the hypotenuse can be found by multiplying the short leg by 2.

$$y = 2x$$
$$y = 2\left(\frac{10\sqrt{3}}{3}\right)$$
$$y = \frac{20\sqrt{3}}{3}$$

Now that the lengths of all the sides are known, calculate the perimeter.

$$P = s_1 + s_2 + s_3 \qquad \text{Formula for perimeter}$$
$$P = 10 + \frac{10\sqrt{3}}{3} + \frac{20\sqrt{3}}{3} \qquad \text{Substitute.}$$
$$P = 10 + 10\sqrt{3} \qquad \text{Simplify.}$$

> **Caution**
>
> As in Lesson 53, be sure to rationalize any fraction with a square root in the denominator by multiplying both the numerator and denominator by that root.

Instead of memorizing the algebraic expressions for each side of a 30°-60°-90° triangle, it may be helpful to just remember the triangle in the diagram, with side lengths 1, $\sqrt{3}$, and 2.

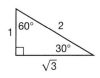

Example 3 Applying the Pythagorean Theorem with 30°-60°-90° Right Triangles

Each tile in a pattern is an equilateral triangle. Find the area of the tile. Use the Pythagorean Theorem and give your answer in simplified radical form.

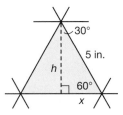

SOLUTION

$$a^2 + b^2 = c^2 \qquad \text{Pythagorean Theorem}$$
$$x^2 + \left(x\sqrt{3}\right)^2 = 25 \qquad \text{Substitute.}$$
$$4x^2 = 25 \qquad \text{Simplify.}$$
$$\frac{4x^2}{4} = \frac{25}{4} \qquad \text{Divide both sides by 4.}$$
$$x^2 = \frac{25}{4} \qquad \text{Simplify.}$$
$$x = \frac{5}{2} \qquad \text{Simplify.}$$

Since h is the longer leg of the right triangle, its length is equal to the length of the shorter leg times $\sqrt{3}$. So, $h = \frac{5}{2}\sqrt{3}$, or $\frac{5\sqrt{3}}{2}$.

Find the area of each tile.

$$A = \frac{1}{2}bh \qquad \text{Triangle Area Formula}$$
$$A = \left(\frac{1}{2}\right)\left(\frac{5}{2}\right)\left(\frac{5\sqrt{3}}{2}\right) \qquad \text{Substitute.}$$
$$A = \frac{25\sqrt{3}}{8} \qquad \text{Simplify.}$$

Double the answer because only half the area of the tile has been found.

$$A_T = 2\left(\frac{25\sqrt{3}}{8}\right)$$
$$A_T = \frac{25\sqrt{3}}{4} \text{ in}^2$$

Notice that the sides match the 30°-60°-90° ratios.

Math Reasoning

Analyze If the side length of each tile were to double, what would happen to the area?

Example 4 Application: Engineering

A museum exhibit contains a model of a pyramid. It has equilateral triangles for faces, and each side of a face is 3 feet long. For a restoration project, the exhibit designers need to know the area of one face of the pyramid. Find this area.

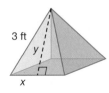

SOLUTION

First, find the dimensions of the 30°-60°-90° triangle. Set up an equation to solve for x.

$2x = 3$ — Hypotenuse of a 30°-60°-90° is twice the shorter leg

$x = \dfrac{3}{2}$ — Simplify.

Use the value for x to find y, the longer leg of the triangle.

$y = x\sqrt{3}$ — Longer leg of a 30°-60°-90° is $\sqrt{3}$ times the shorter leg

$y = \dfrac{3\sqrt{3}}{2}$ — Substitute.

Next, find the area of the triangle.

$A = \dfrac{1}{2}bh$ — Triangle Area Formula

$A = \dfrac{1}{2}\left(\dfrac{3}{2}\right)\left(\dfrac{3\sqrt{3}}{2}\right)$ — Substitute.

$A = \dfrac{9\sqrt{3}}{8}$ — Simplify.

$A_T = 2\left(\dfrac{9\sqrt{3}}{8}\right)$ — Multiply by 2 to find the area of one face.

$A_T = \dfrac{9\sqrt{3}}{4}\ \text{ft}^2$

Math Reasoning

Model This gives the area for one face of the pyramid. How can the area of all the visible faces of the pyramid be found?

Lesson Practice

a. Find the length of each side of the triangle.
(Ex 1)

b. Find the perimeter of the triangle in simplified
(Ex 2) radical form.

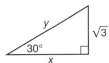

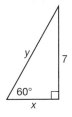

c. A school's banner is an equilateral triangle shown here.
(Ex 3) Use the Pythagorean Theorem to find the area of the equilateral triangle.

14 in.

d. Find the area of one triangular face of a pyramid. The
(Ex 4) faces of the pyramid are equilateral triangles with sides that are 12 centimeter each.

Practice Distributed and Integrated

  1. **Algebra** Find m∠F in the figure at right.
(47)

  2. **Write** Explain why the following statement is true.
(52)

If a quadrilateral is a square then it is a rhombus.

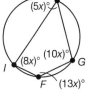

* 3. Find the hypotenuse of a 30°-60°-90° triangle if the shorter leg is
(56) 4 centimeters long.

* 4. Trace the figure at right on your paper. Then locate and
(54) draw the vanishing points and the horizon line.

5. **Multiple Choice** What is the area of a net for a cube with edges that are
(Inv 5) 3 centimeters long?

   A 72 cm^2 **B** 54 cm^2

   C 36 cm^2 **D** 27 cm^2

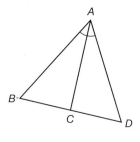

6. Find the length of $\overline{BC}$ if $AD = 9$, $DC = 6$, and $AB = 12$.
(38)

* 7. The midpoints of the sides of $\triangle PQR$ are as follows: midpoint of
(55) $\overline{PQ}$: $S(3, 1)$; midpoint of $\overline{QR}$: $T(1, -1)$; and midpoint of $\overline{RP}$: $U(-1, 1)$.
Draw the midsegment triangle. Find the coordinates of P, Q, and R,
then draw $\triangle PQR$.

8. **Algebra** Find the measure of $\angle B$ in the diagram.
(47)

* 9. **Multiple Choice** Which set of sides does not make a
(56) 30°-60°-90° triangle?

   A $4, 8, 4\sqrt{3}$ **B** $\sqrt{3}, 3, 2\sqrt{3}$

   C $3, 6, 3\sqrt{3}$ **D** $2\sqrt{3}, 4\sqrt{3}, \sqrt{3}$

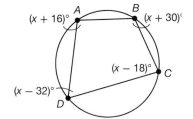

10. What is the sum of the interior angles of a convex
(Inv 3) dodecagon?

11. **Justify** Explain how you know the two triangles shown are similar, and then
(46) write the similarity statement.

12. Find the geometric mean of 1.55 and 5.22.
(50)

13. Find the exact value of the length of the side of a
(53) 45°-45°-90° right triangle with a hypotenuse of 15 miles.

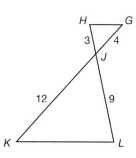

14. In $\triangle DEF$, $\angle D \cong \angle F$.
(51)
   a. Write a congruence statement involving the sides of $\triangle DEF$.
   b. The perimeter of $\triangle DEF$ is 35 millimeters, and $DF = 14$ millimeters.
      Determine DE.

15. **Error Analysis** Triangle PQR has vertex P at the origin. Vertices Q and R are
(45) equidistant from P, and R has coordinates $(2, -6)$. Nasim says that Q must
have coordinates $(6, 2)$. Is Nasim correct? Explain.

16. (**Survey Science**) The diagram shows a new bridge that is being built
(53) across a section of coastline. Given the surveyors' measurements of the
existing road and the bridge span, by how much distance will the bridge
reduce the trip, to the nearest tenth of a mile?

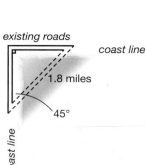

17. **Algebra** Each base angle of an isosceles triangle measures $2(67 - 3y)°$.
(51)
   a. Classify the base angles.
   b. Write and solve an inequality for y.

*18. Draw a triangular prism in one-point perspective so that the vanishing point is
(54) above the prism.

*19. Segment *VW* is a midsegment of △*XYZ*. Find the coordinates of
(55) *V* and *W*.

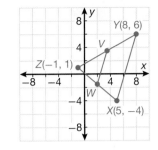

*20. (**Playgrounds**) A slide forms the hypotenuse of a 30°-60°-90° triangle
(56) with the ground. If the slide is 15 feet long, how long are the other
two sides of the triangle?

21. **Justify** Write formulas for finding arc length and area of a sector of given
(35) central angle *m*. Explain why the coefficient is $\frac{m}{360°}$.

*22. **Multi-Step** A triangle has coordinates of (1, 1), (4, 1), and (3, 4). Determine whether
(56) or not this is a 30°-60°-90° triangle.

23. **Analyze** If the angle bisector of one angle of a triangle is also the perpendicular
(38) bisector of the opposite side, what can you conclude about the triangle?

24. (**Plant Growth**) A scientist conducted an
(7) experiment to see which of two plants grew
faster. She recorded her results as shown.
   a. What conjecture could be reasonably
      drawn from the data?
   b. What alternate conjecture could you draw
      that is consistent with the data but incompatible with the first conjecture?

Day	1	5	10	15	20	25
Plant *A* Height	1 cm	2 cm	4 cm	6 cm	7 cm	8 cm
Plant *B* Height	1 cm	2 cm	3 cm	4 cm	5 cm	6 cm

*25. Segment *DE* is a midsegment of △*ABC*. Find the values of *x* and the lengths
(55) of $\overline{DE}$ and $\overline{BC}$.

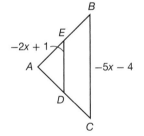

26. **Generalize** The general form of a nonvertical line in slope-intercept form is
(37) $y = mx + b$. What is the general form of a line perpendicular to $y = mx + b$,
in slope-intercept form?

27. Determine whether the set of measures (14, 15, 16) can be the side lengths
(33, 39) of a triangle. If these measures could be a triangle, determine whether it is
an acute, obtuse, or a right triangle.

28. (**Photography**) A photographer is developing a picture of a building which is
(44) 108 feet tall.
   a. What is the relationship between the building and the image of the building in
      the photograph?
   b. If the image of the building in the photograph is 12 inches tall, what is the
      similarity ratio between the building and its image?

*29. **Algebra** The hypotenuse of a 30°-60°-90° triangle has a length of 2*r*. What are the
(56) lengths of the other two legs?

30. (**Design**) The math faculty of Pythagoras College uses this square
(18) logo. One acute angle measure in each right triangle is 37°.
   a. Determine the measure of ∠1.
   b. Determine the measure of ∠2.

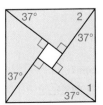

Finding Perimeter and Area with Coordinates

1. **Vocabulary** The _____ is the distance a number is from zero.
   (2)

2. Find the distance between points A and B
   (2) on the number line.

   A ————————————————————— B
   -4 -2   0   2   4   6

3. Find the distance between the points $(-2, 5)$ and $(4, 8)$ to the nearest
   (9) hundredth.

New Concepts

When a figure is on a coordinate plane and no measurements are given, the distance formula can be used to determine the measurements necessary to find the area or perimeter of the figure.

Hint

The distance formula for finding the distance between points (x_1, y_1) and (x_2, y_2) is: $d =$
$\sqrt{(x_2 - x_1)^2 + (y_2 - y_1)^2}$

Example 1 **Finding Perimeter with Coordinates**

Find the perimeter of the rectangle with coordinates $A(2, 4)$, $B(2, -2)$, $C(3, -2)$ and $D(3, 4)$.

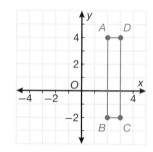

SOLUTION

Plot the given points on a coordinate plane and draw the rectangle.

To find the perimeter, find the length of each side. The distance formula can be used, but since all the segments that compose the rectangle are vertical or horizontal, the length of each segment can be calculated as if it were on a 1-dimensional number line.

$AB = |y_1 - y_2|$
$AB = |4 - (-2)|$
$AB = 6$

Since the figure is a rectangle, $\overline{DC}$ also has a length of 6. The shorter sides, $\overline{AD}$ and $\overline{BC}$, have lengths of 1.

$P = 6 + 6 + 1 + 1$
$P = 14$

Example 2 **Finding Perimeter with the Distance Formula**

Find the perimeter of rectangle $EFGH$ with coordinates $E(1, 3)$, $F(2, 0)$, $G(-4, -2)$, and $H(-5, 1)$. Give your answer in simplified radical form.

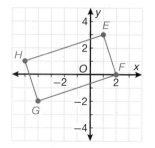

SOLUTION

Plot the points on a coordinate plane.

Then use the distance formula to calculate the length of each side.

$$d = \sqrt{(x_2 - x_1)^2 + (y_2 - y_1)^2}$$
$$d_{EF} = \sqrt{(2 - 1)^2 + (0 - 3)^2}$$
$$d_{EF} = \sqrt{10}$$

Again, because the figure is a rectangle, $\overline{EF}$ and $\overline{HG}$ are congruent. Next, use points F and G to find the length of the rectangle's longer side.

$$d_{FG} = \sqrt{(-4 - 2)^2 + (-2 - 0)^2}$$
$$d_{FG} = 2\sqrt{10}$$

Finally, find the perimeter by summing the four sides.

$$P = 2\sqrt{10} + 2\sqrt{10} + \sqrt{10} + \sqrt{10}$$
$$P = 6\sqrt{10}$$

Area can be found on a coordinate plane in the same way. It is important to know what kind of polygon a figure is before attempting to find its area. Though a figure may look like a rectangle, you cannot assume it is unless you are given that information in the problem or you can prove that the adjacent sides of the figure are perpendicular to each other.

Math Reasoning

Formulate How can you verify that $\angle BAC$ is a right angle?

Example 3 | **Calculating Area with Coordinates**

Find the area of right triangle ABC with right angle $\angle BAC$.

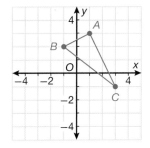

SOLUTION

Since we know the triangle is a right triangle, finding the length of the two legs is sufficient to find the area. Since $\angle BAC$ is the right angle, the legs are $\overline{AB}$ and $\overline{AC}$. Use the distance formula to find the length of each segment.

$$d = \sqrt{(x_2 - x_1)^2 + (y_2 - y_1)^2} \qquad d_{AC} = \sqrt{(3 - 1)^2 + (-1 - 3)^2}$$
$$d_{AB} = \sqrt{(-1 - 1)^2 + (2 - 3)^2} \qquad d_{AC} = 2\sqrt{5}$$
$$d_{AB} = \sqrt{5}$$

Now the formula for area of a triangle can be applied using AC and AB.

$$A = \frac{1}{2}bh$$
$$A = \frac{1}{2}(\sqrt{5})(2\sqrt{5})$$
$$A = 5$$

Sometimes it may be necessary to find the area of an irregular polygon. A coordinate plane makes this possible, because each square on the grid can be counted and added together to find the area of a figure. When a square is not entirely inside a polygon, it may be necessary to estimate and obtain an approximate area.

Example 4 Estimating Area with Coordinates

a. Estimate the area of the polygon.

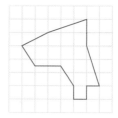

SOLUTION
First, count all the squares that lie completely inside the polygon. There are 10 squares that are completely covered, as shown in the diagram.

Next, estimate the area of the remaining space. One way to do this is to look for triangles, like the right angle shown in the diagram. The triangle's legs measure 1 and 2, so the area of this triangle is 1 square unit.

Two identical right triangles together make a rectangle. Each of these right triangles has one leg that is three squares long and one leg that is one square high. Together, they make a rectangle with an area of 3 square units.

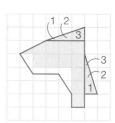

Estimate the remaining area. By looking at the remaining rectangle, which appears to have a height of approximately 0.5 and length of 3, plus the two remaining triangles, it appears that the remaining area is about 3 square units.

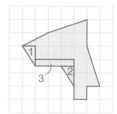

Add all these items together. The total area of the polygon is approximately 17 square units.

b. Estimate the area of the figure.

SOLUTION
First, count complete square units. There are 14. Then, estimate the area of the remaining area. The curved area covers around one square unit on either side. Therefore, the total area is approximately 16 square units.

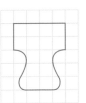

Example 5 | Application: Farming

A farmer wants to estimate how much seed she needs to buy for her land. She cannot farm in the river or on the riverbank, which is shaded in the diagram. For every acre, she needs 2 bags of seed. Estimate how many bags of seed she will need. Each square unit on the grid represents one fourth of an acre.

$\square = \frac{1}{4}$ acre

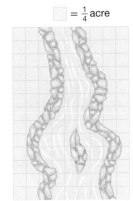

SOLUTION
The entire plot of land is 9 by 13, or 117 units. Every four units equals one acre. Including the river, the farmer has $\frac{117}{4} = 29.25$ acres.

There are 49 full units that are unusable because of the river and riverbank, as shown in the diagram. Next, identify triangles and their approximate measurements.

$\square = \frac{1}{4}$ acre

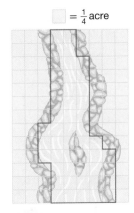

The remaining parts of the river and riverbank add to approximately 14 square units. Your answers may vary, but anything between 13 and 15 units is a reasonable estimation.

Therefore, the total amount of land that is not available for farming is approximately 63 square units.

Every four units equals one acre, so there are $\frac{63}{4} = 15.75$ acres that are unusable.

In total, she can use $29.25 - 15.75 = 13.5$ acres.

The farmer will need $13.5(2) = 27$ bags of seed.

> **Caution**
>
> Acres measure area, not length. Therefore, there is no such thing as a "square acre."

a. Find the perimeter of rectangle *LMNO* with coordinates *L*(−5, 3),
(Ex 1) *M*(−5, −1), *N*(4, −1), and *O*(4, 3).

b. Find the perimeter of triangle *PQR* with coordinates *P*(4, −1), *Q*(9, 3),
(Ex 2) and *R*(6, −4). Round your answer to the nearest hundredth.

c. Calculate the area of △*XYZ*.
(Ex 3)

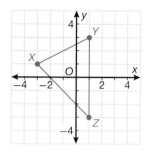

d. Estimate the area of the figure.
(Ex 4)

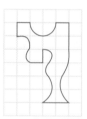

e. The school committee wants to put gravel
(Ex 5) on its new running path. Estimate the
amount of gravel they must buy if every
square meter of path requires 40 kilograms
of gravel.

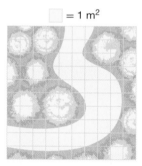

☐ = 1 m²

Practice **Distributed and Integrated**

1. **Generalize** In a 30°-60°-90° right triangle, what is the ratio of the length of the leg
(56) opposite the 30° angle to the length of the hypotenuse? What is the ratio of the
length of the leg opposite the 60° angle to the length of the hypotenuse? Express
your answers both exactly and to the nearest thousandth.

2. What would be necessary information to add to the Side-Side-Side Similarity
(46) Theorem to transform it into the Side-Side-Side Congruence Theorem?

3. (**Blueprints**) An architect is drawing up plans for a house. The house should
(44) have a storage closet with a triangular floor area. If the actual closet's area is to
have dimensions of 10 feet, 6 feet, and 8 feet, and the architect's drawing has a
longest side of $6\frac{2}{3}$ inches, what are the side lengths of the other two sides of his
drawing?

4. Draw a cube in one-point perspective with a vanishing point below
(54) the cube.

*5. Find the area of $\triangle ABC$ with vertices $A(2, 2)$, $B(2, 5)$ and $C(-5, 1)$.
(57)

 6. Algebra Find the distance from the center of a circle with a 10-centimeter radius
(43) to a chord that is 13 centimeters long to the nearest tenth.

7. Show that lines m and n in this figure are parallel.
(12)

*8. Find the perimeter of quadrilateral $WXYZ$ if
(57) $W(2, 4)$, $X(-2, 3)$, $Y(-1, 7)$, and $Z(4, 9)$. Give
your answer in simplified radical form.

9. Multiple Choice What is the approximate area of a sector of a
(35) circle with a radius of 6 units and an arc measure of 60°?

   A 3.14   **B** 113.10
   C 18.85   **D** 6.00

10. Justify Find x in the parallelogram at right, and justify each step.
(34)

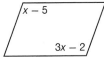

11. (Framing) A photographer is building a frame for a square picture.
(53) In order to have the picture frame stand upright, a collapsible stand
must be attached to the back that folds exactly across the diagonal of
the back. If the frame is 8 inches on a side, how long must the stand be?
Give your answer in simplified radical form and rounded to the nearest tenth.

 12. Algebra Find the measure of $\angle G$ in the circle at right.
(47)

13. Analyze Prove the Converse of the Hypotenuse-Angle
(36) Congruence Theorem.

14. Find the distance from $(-1, 14)$ to the line $y = 5$.
(42)

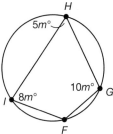

15. Verify Segment $\overline{JK}$ is a midsegment of $\triangle FGH$. Find the coordinates of
(55) J and K. Use the distance formula to verify that $\overline{JK}$ is half the length
of $\overline{GH}$.

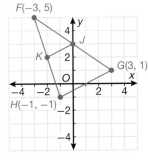

16. (Automotive Engineering) An engineer needs to make a model of
(44) a car's triangular window that has side lengths of 24, 21, and
18 inches long, respectively. If the designer's model has the shortest
corresponding side of the window measuring 4 inches, what are the
lengths of the other two sides of the window?

*17. What is the perimeter of a 30°-60°-90° triangle with a shortest side
(56) length of 3.2 inches?

18. What is the negation of, "If q, then p?"
(17)

19. Verify In $\triangle ABC$, show that $\overline{AC} \parallel \overline{DE}$ if D and E are the midpoints of
(55) their respective sides by finding the slopes of the segments.

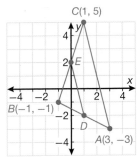

*20. **Multi-Step** Find the area of a polygon with vertices $J(2, 4)$, $K(-3, 2)$, $L(0, -4)$
(57) and $M(4, 1)$.

21. Determine the value of x in the circle at right.
(43)

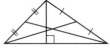

22. **Multiple Choice** Which statement contradicts the fact that
(48) $\triangle HIJ \sim \triangle DEF$?

A $HI = rDE$ for some factor r

B $\angle H = \angle D$

C $\angle H \neq \angle F$

D None of the angles in $\triangle HIJ$ have the same measure as
any of the angles in $\triangle DEF$.

23. **Error Analysis** Carina marked the triangle at right in order to find the centroid.
(32) Explain the error she made.

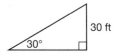

*24. (**Industrial Painting**) A painter needs to paint a triangular ramp.
(56) The ramp is inclined at 30° to the ground and is 30 feet high.
If it takes the painter 1 minute to paint 4 square feet on average,
how long will it take him to paint the side of the ramp, to the
nearest minute?

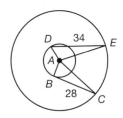

*25. **Justify** Which is the larger angle in the figure shown, $\angle DAE$ or $\angle BAC$?
(Inv 4) Explain your answer.

*26. (**Urban Planning**) A city block is on a grid. If the corner of a rectangular
(57) building is at $(9, 0)$ and another corner is 7 units to the left and 6 units up
from that corner, what are the other coordinates of the building and
what is its perimeter?

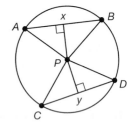

27. **Analyze** What is the fewest number of regular triangular tiles necessary to form
(49) a polyhedron, and what is the name of the polyhedron formed?

*28. Write a paragraph proof showing that in a circle or congruent circles, chords
(31) equidistant from the center are congruent.

Given: $\overline{AB}$ and $\overline{CD}$ are equidistant from P. (i.e. $PX = PY$).

Prove: $\overline{AB} \cong \overline{CD}$

29. Prove that lines d and e in this figure are parallel if $m\angle 1 = 63°$ and
(12) $m\angle 2 = 117°$.

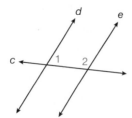

*30. **Write** The hypotenuse of a right triangle is 32 meters and one leg is 16 meters.
(56) A second right triangle has a 60° angle. How can you use this information to find
whether these triangles are similar? Can it be determined whether these triangles
are congruent?

LESSON
58

Tangents and Circles, Part 1

Warm Up

1. **Vocabulary** A segment whose endpoints lie on a circle is a _____ of
$^{(43)}$ the circle.

2. $\overline{CA}$ and $\overline{CB}$ are radii of $\odot C$. Classify $\triangle ABC$ by its sides.
$^{(13)}$

3. Which line is a tangent to $\odot C$? Which line is a secant
$^{(43)}$ to $\odot C$?

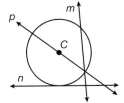

New Concepts A tangent line lies in the same plane as a circle and intersects the circle at exactly one point. A radius of a circle drawn to a point of tangency meets the tangent line at a fixed angle.

Math Language

A **point of tangency** is a point where a tangent line intersects a circle.

Theorem 58-1

If a line is tangent to a circle, then the line is perpendicular to a radius drawn to the point of tangency.

$$\overleftrightarrow{EF} \perp \overleftrightarrow{DG}$$

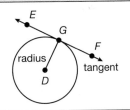

Example **1**   **Tangent Lines and Angle Measures**

Line n is tangent to $\odot C$ at point P, and line m passes through C. Lines n and m intersect at point Q.

a. Sketch $\odot C$ and lines n and m. Mark C, P, and Q on your sketch.

SOLUTION

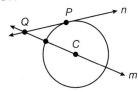

b. If m$\angle CQP = 36°$, determine m$\angle PCQ$.

SOLUTION
By Theorem 58-1, $\overline{CP}$ is perpendicular to n, so $\triangle CPQ$ is a right triangle. Therefore, $\angle CQP$ and $\angle PCQ$ are complementary.

$$\text{m}\angle CQP + \text{m}\angle PCQ = 90°$$
$$36° + \text{m}\angle PCQ = 90°$$
$$\text{m}\angle PCQ = 54°$$

Online Connection
www.SaxonMathResources.com

Theorem 58-2

If a line in the plane of a circle is perpendicular to a radius at its endpoint on the circle, then the line is tangent to the circle.

$\overleftrightarrow{ST}$ is tangent to $\odot V$.

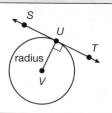

Math Reasoning

Write State the inverse of Theorem 58-1. Is it true or false?

Theorem 58-2 is the converse of Theorem 58-1. Together, they can be used to show that this biconditional statement is true: *A line in the plane of a circle is tangent to the circle if and only if it is perpendicular to a radius drawn to the point of tangency.*

Example 2 **Identifying Tangent Lines**

If m∠BEA = 45°, show that $\overleftrightarrow{BD}$ is tangent to $\odot C$.

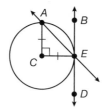

SOLUTION
To show that $\overleftrightarrow{BD}$ is tangent to $\odot C$, it has to be shown that ∠BEC is a right angle. From the diagram, △AEC is an isosceles right triangle, so ∠CAE ≅ ∠CEA. The acute angles of a right triangle are complimentary, so both ∠CAE and ∠CEA are 45° angles. By the Angle Addition Postulate, ∠CEA + ∠BEA = ∠BEC. Substituting shows that ∠BEC = 45° + 45°, so ∠BEC is a right angle. Therefore, by Theorem 58-2, $\overleftrightarrow{BD}$ is tangent to $\odot C$.

Example 3 **Proving Theorem 58-2**

Prove Theorem 58-2.
Given: $\overleftrightarrow{DE} \perp \overline{FD}$ and D is on $\odot$.
Prove: $\overleftrightarrow{DE}$ is tangent to $\odot$.

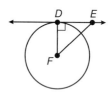

SOLUTION
We will write an indirect proof. Assume that $\overleftrightarrow{DE}$ is not tangent $\odot F$. Then $\overline{DE}$ intersects $\odot F$ at two points, D and point P. Then FD = FP. But if △DFP is isosceles, and the base angles of an isosceles triangle are congruent, then ∠DPF is a right angle. That means there are two right angles from F to $\overline{DE}$, which is a contradiction of the theorem which states that through a line and a point not on a line, there is only one perpendicular line, so the assumption was incorrect and $\overline{DE}$ is tangent to $\odot F$

If two tangents to the same circle intersect, the tangent segments exhibit a special property, stated in Theorem 58-3.

Theorem 58-3

If two tangent segments are drawn to a circle from the same exterior point, then they are congruent.

$\overline{MQ} \cong \overline{NQ}$

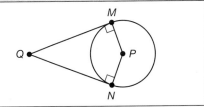

Example 4 Applying Relationships of Tangents from an Exterior Point

In this figure, $\overline{JK}$ and $\overline{JM}$ are tangent to $\odot L$.

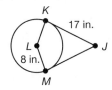

Determine the perimeter of quadrilateral *JKLM*. What type of quadrilateral is *JKLM*?

SOLUTION
Since $\overline{ML}$ and $\overline{LK}$ are radii of the same circle, $KL = 8$ in. By Theorem 58-3, tangents to the same circle are congruent, so $JM = 17$ in. To get the perimeter, add the lengths of the four sides.
$P = JK + KL + LM + JM$
$P = 17 + 8 + 8 + 17 = 50$ in.

Since *JKLM* has exactly two pairs of congruent consecutive sides, it is a kite.

Example 5 Application: Glass Cutting

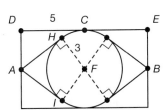

An ornamental window has several glass panes oriented to look like an eye. The radius of the eye's iris is 3 feet, and *DC* is 5 feet. What are *AI* and *AH*?

SOLUTION
The right angles on the diagram indicate that the four segments that form the corners of the eye are tangent to the circle. Draw in the segment $\overline{FA}$. This forms right triangles, *FAI* and *FAH*. The hypotenuse of the triangles are 5 feet long, and their shorter legs are each 3 feet long. By using the Pythagorean Theorem, *AI* and *AH* are each 4 feet long.

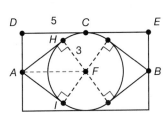

a. Line *a* is tangent to ⊙*R* at *D*, and line *b* passes through *R*. Lines *a*
(Ex 1) and *b* intersect at *E*. If m∠*RED* = 42°, determine m∠*DRE*.

b. Let $\overline{CA}$ be a radius of ⊙*C*. Let line *m* be tangent to ⊙*C* at *A*. Let *B* be
(Ex 2) an exterior point of ⊙*C*, with m∠*BAC* < 90°. Is $\overleftrightarrow{AB}$ a tangent to ⊙*C*?
Why or why not?

c. Give a paragraph proof of Theorem 58-3.
(Ex 2) *Hint: Draw $\overline{PA}$, $\overline{PB}$ and $\overline{PC}$.*
Given: $\overline{AB}$ and $\overline{AC}$ are tangent to ⊙*P*
Prove: $\overline{AB} \cong \overline{AC}$

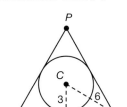

d. Circle *C* has a 5-inch radius. $\overline{XZ}$ and $\overline{YZ}$ are
(Ex 4) tangents to ⊙*C* and *Z* is exterior to ⊙*C*.
If ∠*XCY* is a right angle, what is the area of quadrilateral *CXZY*?

e. A decorative window is shaped like a triangle
(Ex 5) with an inscribed circle. If the triangle is an
equilateral triangle, the circle has a radius
of 3 feet, and *CQ* is 6 feet, what is the
perimeter of the triangle in simplified
radical form?

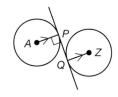

Practice **Distributed and Integrated**

* 1. In this figure, *A* and *Z* are the centers of the circles. $\overline{AP}$ is parallel to $\overline{ZQ}$ and
(58) perpendicular to $\overline{PQ}$. Is $\overline{PQ}$ tangent to ⊙*A* and/or ⊙*Z*? Explain.

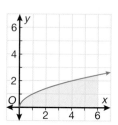

* 2. Find *ER* if *FG* = 36.
(58)

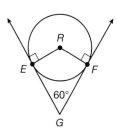

* 3. Estimate Estimate the area under the curve $y = \sqrt{x}$ for $0 \leq x \leq 6$.
(57)

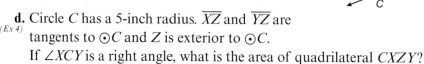

* 4. Multi-Step Find the perimeter of a 30°-60°-90° triangle in which the shorter
(56) leg is 72 feet. Give your answer in simplified radical form.

5. Three fire stations are located at the vertices of a triangle as shown.
(39) Which two stations are located farthest apart?

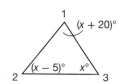

6. Write In $\triangle UVW$, $UW = 3.5$ inches and V is a right angle. In $\triangle XYZ$, $m\angle Z = 90°$, $m\angle X = 33°$ and $m\angle Y = 57°$. The Hypotenuse-Angle Congruence Theorem can be applied if two additional measures are known. What could they be? Explain.
(36)

7. (Dividing Cost) Hubert and three friends bought a rectangular pizza that cost $10 for each of them to share, but the slices were not cut evenly. If the shaded region of this figure represents the portion of the pizza Hubert ate, what amount of money should he pay for his portion of the pizza?
(40)

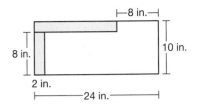

8. Generalize Two segments of different lengths are attached perpendicularly at the midpoint of each segment. If you draw in the segments that connect all 4 of the endpoints of the 2 segments, what kind of a quadrilateral appears? How would the shape differ if both segments were of the same length?
(52)

9. (Indirect Measure) Angelique wants to measure the diameter of a large circular pool, but she cannot reach the center. Angelique has two large pieces of rope. Explain how she can use her knowledge of chords to find the diameter.
(43)

10. Algebra The lengths of three line segments are given by the expressions $3x + 2$, x^2 and $2x$. Can these segments be used to create a triangle if $x = 4$? ... if $x = 6$?
(39)

11. Algebra Points J, K, and L are midpoints of $\triangle FGH$ as shown. If $GH = 5x + 2$ and $KL = 3x - 2$, what is the length of $\overline{GH}$?
(55)

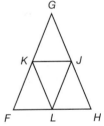

12. (Framing) Derek made a frame for a picture that is 15 inches long and 8 inches wide. He wanted the frame to be a square, but now he thinks his earlier measurements were wrong. In order to determine if the frame he made is a square, Derek measured the diagonals to be 17.26 and 16.78 inches each. With these dimensions, is the frame a square? If not, by how many inches are Derek's measurements off?
(9)

*13. Line p is tangent to $\odot N$ at X, and line q passes the center of $\odot N$. Lines p and q intersect at Y.
(58)
  a. Sketch $\odot N$, lines p and q, and points X and Y.
  b. If $m\angle XNY = 13°$, determine $m\angle NYX$.

*14. $\triangle XYZ$ has vertices $Y(-5, 2)$ and $Z(0, 2)$. $\angle ZYX$ is a right angle, and $m\angle YZX$ is 60°. Find the coordinates of X if X is in Quadrant II.
(56)

15. Multi-Step In equiangular $\triangle LMN$, $LN = 17$ inches. Determine the perimeter of $\triangle LMN$. Give your answer in feet and inches.
(51)

16. Draw a cube in two-point perspective so that the horizon line and vanishing points are below the vertical line.
(54)

17. Multiple Choice Which of the following is perpendicular to $2x + y = 6$?
(37)
  A $2x + 4y + 7 = 0$ **B** $7x = 15y + 6$
  C $2(2x + 6) - 8y = 0$ **D** $2x - y = 4$

*18. Determine the perimeter of this quadrilateral if *U* is the center of
(58) the circle.

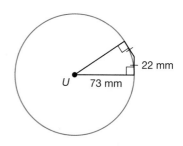

19. Jose started at his home, *H*(0, 0) and walked to a park located at
(57) *P*(4, 2). He then went to Jacintha's house located at *F*(8, 0). Later,
he walked straight home. If each unit on the coordinate plane
represents 50 feet, what is the distance that Jose walked, to the
nearest foot?

20. Error Analysis Noyemi said the following statement is sometimes true.
(52) Is she correct? Explain.

If a quadrilateral is both a rectangle and a rhombus, then it is a square.

21. Find the measure of ∠*CDB* in ⊙*M* given that m∠*AMB* = 74°.
(47)

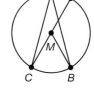

22. Multiple Choice An isosceles triangle has a base and height that are
(51) 2 centimeters each. The length of each leg is:
   A 2 centimeters   **B** √3 centimeters
   C √5 centimeters   **D** 3 centimeters

23. ⟮**Floor Plans**⟯ A floor plan for a square shed was left unfinished.
(53) However, Lisa can see that the diagonal of the shed is 72 feet.
How many 6-foot long pieces of lumber does she need to form the
perimeter of the shed?

24. A quadrilateral inscribed in a circle has interior angles of 55°, 90°, 25*x*°, and 18*x*°,
(47) in clockwise order. Find *x*.

*25. ⟮**Traffic Signs**⟯ An equilateral yield sign has a height of 1 foot. What is the area
(56) of the sign in simplified radical form?

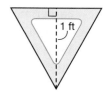

26. What is the included angle of sides $\overline{XY}$ and $\overline{ZX}$ in a triangle?
(28)

*27. Estimate the area of the irregular shape in the diagram.
(57)

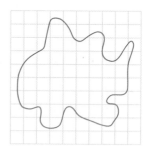

✎ **28. Write** In parallelogram *ABCD*, $\overline{AC}$ is drawn to create
(25) △*ABC* and △*CDA*. Are these triangles congruent?
Explain.

29. Predict If two congruent septagonal prisms are joined together at their
(49) bases, with their bases aligned, what is the resulting shape?

30. Justify Given the information in this diagram, could *RT* be greater than
(Inv 4) *TS*? If so, find the range of values when this is true. If not, explain
your answer.

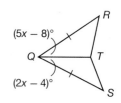

Tangents to a Circle

Construction Lab 8 (Use with Lesson 58)

Lesson 58 shows you how to identify lines tangent to a circle. This lab demonstrates how to construct lines tangent to a circle through a point on the circle.

1. To construct a tangent line through a given point on the circle, begin with ⊙A and a point on the circle, B.

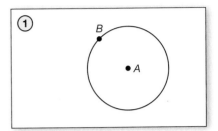

2. Draw $\overrightarrow{AB}$.

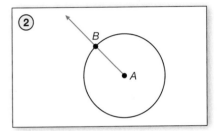

3. Using the method from Construction Lab 2, construct the line perpendicular to $\overrightarrow{AB}$ through B.

The perpendicular line is tangent to ⊙A at B.

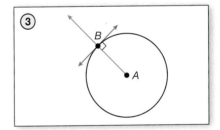

This lab demonstrates how to construct lines tangent to a circle through a point not on the circle.

1. To construct two tangent lines through a point external to a circle, draw a circle and label the center C. Choose a point exterior to the circle, and label it P. Draw $\overleftrightarrow{CP}$.

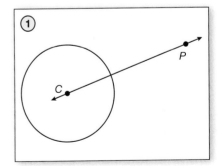

Hint

To construct the midpoint, use the method you learned in Construction Lab **3**.

(2.) Construct the midpoint of $\overline{CP}$. Label the midpoint M.

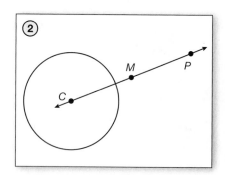

(3.) Draw a circle with a radius, CM centered on M. Notice that P is also on this circle.

(4.) Label the points of intersection of $\odot C$ and $\odot M$ as points X and Y.

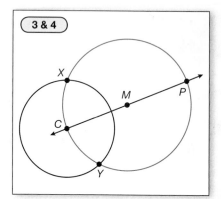

(5.) Draw $\overleftrightarrow{XP}$. This line is tangent to $\odot C$ at point X. Draw $\overleftrightarrow{YP}$. Notice that $\overleftrightarrow{YP}$ is tangent to $\odot C$ at point Y.

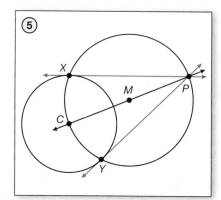

Lab Practice

Use a compass to draw $\odot D$ and $\odot E$. Draw point A on $\odot D$. Draw points F and G outside, but near to, $\odot D$ and $\odot E$. Perform each construction indicated below.

a. a line tangent to $\odot D$ at point A

b. a line tangent to $\odot D$ from point F

c. a line tangent to $\odot E$ from point F

d. a line tangent to $\odot E$ from point G

LESSON 59

Finding Surface Areas and Volumes of Prisms

Warm Up

1. **Vocabulary** A polyhedron formed by two parallel congruent polygonal
(49)   bases connected by lateral faces that are parallelograms is called a _____.

2. Find the area of the triangle.
(13)

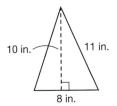

10 in.   11 in.

8 in.

3. Find the area of a 12-by-7-foot rectangle.
(22)

New Concepts

The **surface area** of a solid is the total area of all its faces and curved surfaces. The surface area of the pentagonal prism shown in the diagram, for example, is the sum of the area of the two pentagons and the five rectangles that compose the prism.

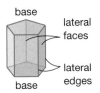

base
lateral faces
lateral edges
base

Lateral area is the sum of the areas of the lateral faces of a prism or pyramid, or the surface area, excluding the base(s), of the lateral surface of a cylinder or cone. In the diagram, the lateral area of the pentagonal prism is the sum of the area of all the rectangular faces.

> **Math Reasoning**
>
> **Formulate** The bases of a cylinder are circles. What would the formula be to find the lateral area of a cylinder?

Lateral Area of a Prism
The lateral area L of a prism can be found using the following formula, where P is the perimeter of the base and h is the prism's height. $$L = Ph$$

Example 1 Finding Lateral Area of a Prism

Find the lateral area of the regular hexagonal prism.

SOLUTION
First find the perimeter of the base, then multiply it by the height. The base is a regular hexagon with 6 side lengths of 12 feet each, so the perimeter is 72 feet. Next, multiply the perimeter by the height, or 72×18.

The lateral area of the prism is 1296 square feet.

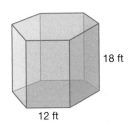

18 ft

12 ft

> **Online Connection**
> www.SaxonMathResources.com

Surface area of a prism can be calculated by finding the sum of the areas of each face. Sometimes it is easiest to first determine the lateral surface area and then add the area of the two bases to the lateral surface area.

Surface Area of a Prism
The surface area of a prism is the sum of the lateral area and the area of the two bases, where B is the area of a base. $$S = L + 2B$$

Example 2 **Finding Surface Area of a Prism**

Find the surface area of the regular pentagonal prism.

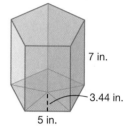

7 in.

3.44 in.

5 in.

SOLUTION
The base is a regular pentagon, so its perimeter is given by multiplying the length of each side by the number of sides. Therefore, the perimeter is $5 \times 5 = 25$ inches. Substitute the perimeter's value and the height 7 into the lateral area formula.

$L = ph$
$L = (25)(7)$
$L = 175 \text{ in}^2$

Now find the area of the base, which is a pentagon. The pentagon can be divided into five congruent triangles as shown. For one triangle:

$A = \dfrac{1}{2}bh$
$A = \dfrac{1}{2}(5)(3.44)$
$A \approx 8.6 \text{ in}^2$

So the area of the base is about $(5)(8.6) = 43$ square inches.

To find the total surface area of the prism, use the formula for surface area.

$S = L + 2B$
$S = 175 + (2)(43) = 261 \text{ in}^2$
So the surface area of the prism is 261 square inches.

Surface area can be considered as the number of square units it takes to exactly cover the outside of a solid. **Volume** is the number of unit cubes of a given size that will exactly fill the interior of a solid.

Volume of a Prism
The volume of a prism can be found using the formula below, where B is the area of the base and h is the height of the prism. $$V = Bh$$

A **right prism** is a prism whose lateral faces are all rectangles and whose lateral edges are perpendicular to both bases. In a right prism the height is the length of one edge that separates the bases.

Example 3 Finding the Volume of a Right Prism

Find the volume of the right prism.

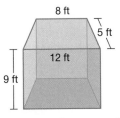

SOLUTION
The base of a prism is a trapezoid, so calculate the area of the base first, then the volume of the prism:

$$A = \frac{1}{2}(b_1 + b_2)h$$

$$A = \frac{1}{2}(8 + 12)5$$

$$A = 50 \text{ ft}^2$$

$$V = Bh$$

$$V = (50)(9) = 450 \text{ ft}^3$$

The volume of the prism is 450 cubic feet.

An **oblique prism** is a prism that has at least one nonrectangular face. An oblique prism is like a prism that has been tilted to one side. The surface area and the volume of an oblique prism are found using the same formulas that are used with a right prism. Instead of using the height of a lateral edge of the prism, an altitude of the prism must be used. An **altitude of a prism** is a segment that is perpendicular to, and has its endpoints on, the planes of the bases.

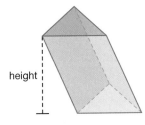

Oblique triangular prism

Example 4 Finding the Volume of an Oblique Prism

Find the volume of the oblique prism shown.

SOLUTION
The volume of an oblique prism is the area of the base times the height. Do not use the slanted height of the lateral face. Instead, use the altitude, which is 8 inches.

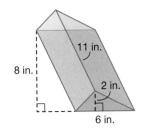

$$V = Bh$$

$$V = \frac{1}{2}(6)(2) \times 8$$

$$V = 48 \text{ in}^3$$

The volume of the oblique triangular prism is 48 cubic inches.

Example 5 Application: Packaging

Rick has a gift that he needs to wrap. He has
15 square feet of wrapping paper. Does Rick
have enough wrapping paper to cover the
gift shown in the diagram?

SOLUTION
First find the lateral surface area by multiplying the perimeter of
the 2-by-3-foot base by the height.

$L = ph$

$L = (2 + 2 + 3 + 3)(1)$

$L = 10 \text{ ft}^2$

Now add the area of the two bases.

$S = L + 2B$

$S = (10) + 2(2)(3)$

$S = 22 \text{ ft}^2$

Since the surface area of the gift is 22 square feet, Rick will need to buy
more wrapping paper to wrap the entire gift.

Lesson Practice

Use the figure at right to answer problems a and b.

a. Find the lateral area of the prism.
(Ex 1)

b. Find the surface area of the prism.
(Ex 2)

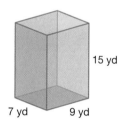

c. Find the volume of the right prism.
(Ex 3)

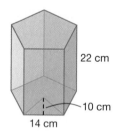

d. Find the volume of the oblique prism.
(Ex 4)

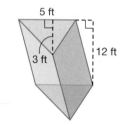

e. Jennifer is filling her swimming pool, shown in the diagram. How many cubic meters of water will it take to fill the pool?

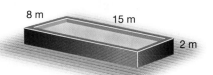

(Ex 5)

8 m 15 m 2 m

Practice Distributed and Integrated

1. **Estimate** Estimate the area under the curve of $y = x^2$ for $0 \le x \le 3$.
(57)

2. Solve the proportion $\dfrac{9}{p} = \dfrac{36}{28}$
(41)

3. Find the values of x and y in the triangle shown to the nearest tenth.
(50)

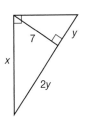

7 y x 2y

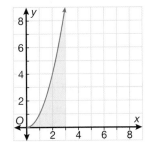

4. Draw a cube in two-point perspective so that the horizon line and vanishing points are above the vertical line.
(54)

* 5. **Write** Armine needs to find the area of a square but does not have the measurements of its sides; however, he does know the diagonal length of the square. Explain how Armine can use this to find the area of the square.
(53)

6. **(Design)** A t-shirt designer is painting this pattern on a shirt. If each square on the grid represents 1 inch on the shirt, approximately what is the area he must cover with paint?
(57)

* 7. Complete this indirect proof showing that if a line is tangent to a circle, then the line is perpendicular to a radius drawn to the point of tangency.
(48)

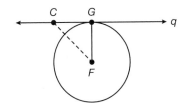

C G q F

Given: Line q is tangent to $\odot F$ at point G.
Prove: $q \perp \overline{FG}$

Assume that q is not perpendicular to $\overline{FG}$. Then it is possible to draw $\overline{FC}$ such that $\overline{FC} \perp q$. If this is true, then $\triangle FCG$ is a right triangle. $FC < FG$ because _____. Since q is a tangent line, it can only intersect $\odot F$ at ___, and C must be in the exterior of $\odot F$, so $FC > FG$. Thus the assumption is false, and _____.

8. **Error Analysis** To find the orthocenter of a triangle, Marina drew the segments shown. Explain the error she made.
(32)

9. **Multiple Choice** Which of the following statements is *not* true of a square?
(52)
  A All sides are congruent.
  B The diagonals are congruent.
  C The diagonals are perpendicular.
  D The interior triangles created by the diagonals make two obtuse and two acute triangles.

10. In this triangle, $\overline{ZD}$ is tangent to $\odot A$ at Z. Determine m$\angle DAZ$.
(58)

11. State the conjunction of these statements. Is the conjunction true or false?
(20)

The triangle is acute. The triangle has exactly two acute angles.

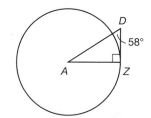

12. Find the value of x and of y in the triangle shown. Give your answers in
(56) simplified radical form.

13. **Generalize** Line m is tangent to $\odot P$ at point A. Points B and C are two other
(58) distinct points on m.

   a. Classify $\angle PBC$ as acute, right, or obtuse when the order of A, B, and C along line m is B, A, C.

   b. Classify $\angle PBC$ when the order of A, B, and C along line m is A, B, C.

   c. Can $\angle PBC$ ever be a right angle? Why or why not?

14. Find the lateral area of the prism shown.
(59)

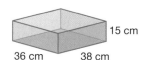

15. (**Construction**) Kai is trying to build a right triangle out of plastic to use as a brace
(33) for furniture. After constructing the brace, he measures its sides as 5.25 inches, 6.33 inches, and 8.25 inches. Does Kai need to cut more plastic off the hypotenuse, or does he need to find a longer piece to use for the hypotenuse? How do you know?

16. Find the volume of the right prism shown.
(59)

17. **Model** Draw and label two similar polygons whose corresponding
(44) sides have a ratio of 1:4.

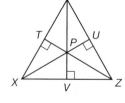

18. Using the diagram, find PU in terms of x if $PX = 8$ and $UY = x$.
(38) Point P is the circumcenter of the triangle.

19. (**Accessibility**) A ramp to the library makes a 30°-60°-90° triangle
(56) with the longer leg on the ground. If the ramp is 3 feet high, how long is the ramp's surface?

20. Find the surface area of the prism shown.
(59)

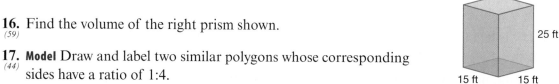

21. **Algebra** Find the area and perimeter of the triangle formed by the
(57) lines $y = -x$, $y = 2x$, and $y = 2$.

22. **Multi-Step** Isosceles $\triangle STU$ has base angles measuring 60° and a base
(51) length of 1 foot, 7 inches long. Determine the perimeter of $\triangle STU$, in feet and inches.

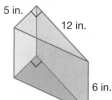

23. <u>Water Treatment</u> A pipe with an 8-foot radius has water flowing through it, but
⁽⁴³⁾ for safety reasons the water level must be kept at least 2.5 feet from the top of
the pipe. If the water makes a chord that is 12.5 feet long, how close to the
top of the pipe is the water flowing? Is the water at a safe level?

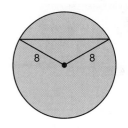

24. If a quadrilateral is inscribed in a circle, then its opposite angles
⁽⁴⁷⁾ are _____.

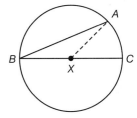

*25. Write a two-column proof showing the first case of this theorem:
⁽²⁷⁾ The measure of an inscribed angle is equal to half of the measure of
its intercepted arc.
Hint: Draw $\overline{XA}$.
Given: $\angle ABX$ is inscribed in $\odot X$.
Prove: $m\angle ABC = \frac{1}{2}m\widehat{AC}$

26. Explain why these two triangles are similar and give
⁽⁴⁶⁾ a similarity statement.

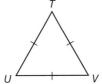

27. Find the height of a right rectangular prism whose surface area is 352 square
⁽⁵⁹⁾ inches. The length of the base of the prism is 8 inches and the width of the base is
4 inches.

28. <u>Lawnmowing</u> Emma is mowing her neighbors' lawns. Emma
⁽⁴⁰⁾ charges 1.5 cents per square foot of grass. How much should she
charge to mow the lawn pictured here?
Hint: Emma does not mow the driveway.

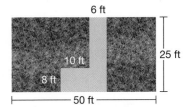

29. **Justify** In two triangles, SBT and WEX, there are right angles at
⁽¹⁸⁾ B and E.
a. If $\angle S \cong \angle W$, what can be said about $\angle T$ and $\angle X$?
Justify your answer.
b. If $\angle T \cong \angle X$, what can be said about $\angle S$ and $\angle W$?

30. Identify the tangent line(s) in this figure.
⁽⁵⁸⁾

Proportionality Theorems

1. **Vocabulary** Lines that lie in the same plane but never intersect are
   (5)   called _____ lines.

2. Rewrite this statement to make it true.
   (37)   *Parallel and perpendicular lines have equal slopes.*

3. **Multiple Choice** Which of the following is *not* true about the angles formed
   (34)   when a transversal intersects parallel lines?
   A Alternate interior angles are supplementary.
   B Alternate exterior angles are congruent.
   C Same-side interior angles are supplementary.
   D Corresponding angles are congruent.

New Concepts

Previous lessons have discussed some of the proportional relationships that exist within triangles when they are divided by a midsegment. A similar relationship exists for any line that intersects two sides of a triangle and is parallel to one side.

Theorem 60-1: Triangle Proportionality Theorem

If a line parallel to one side of a triangle intersects the other two sides, it divides those sides proportionally.

$$\frac{XA}{AY} = \frac{XB}{BZ}$$

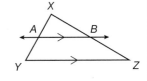

Example 1   Using Triangle Proportionality to Find Unknowns

(a.) Find the length of $\overline{AE}$.

SOLUTION

$\dfrac{AE}{EC} = \dfrac{AD}{DB}$ Triangle Proportionality Theorem

$\dfrac{AE}{5} = \dfrac{2}{3}$ Substitute.

$AE = \dfrac{10}{3}$ Simplify.

(b.) Find the value of x.

SOLUTION

Write a proportion relating the segments based on the Triangle Proportionality Theorem.

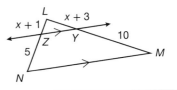

$$\frac{x+1}{5} = \frac{x+3}{10}$$ Triangle Proportionality Theorem

$$10(x+1) = 5(x+3)$$ Cross multiply.

$$x = 1$$ Solve.

The Converse of the Triangle Proportionality Theorem is true, and can be used to check whether a line that intersects 2 sides of a triangle is parallel to the triangle's base.

Theorem 60-2: Converse of the Triangle Proportionality Theorem

If a line divides two sides of a triangle proportionally, then it is parallel to the third side. In $\triangle XYZ$, if $\dfrac{XA}{AY} = \dfrac{XB}{BZ}$, then $\overleftrightarrow{AB} \parallel \overline{YZ}$.

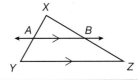

Example 2   **Proving Lines Parallel**

Is $\overline{ST}$ parallel to $\overline{PR}$?

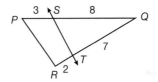

SOLUTION

If $\overline{ST}$ divides $\overline{PQ}$ and $\overline{RQ}$ proportionally, then $\overline{ST} \parallel \overline{PR}$ by Theorem 60-2. Set up a proportion.

$$\frac{PS}{SQ} = \frac{RT}{TQ}$$ Triangle Proportionality Theorem

$$\frac{3}{8} \stackrel{?}{=} \frac{2}{7}$$ Substitute.

$$21 \stackrel{?}{=} 16$$ Cross multiply.

The statement is false, so $\overline{ST}$ is not parallel to $\overline{PR}$.

The Triangle Proportionality Theorem is closely related to Theorem 60-3, which uses the same proportional relationship to relate the segments of transversals that are intersected by parallel lines.

Hint

Theorem 60-3 is very similar to the Triangle Proportionality Theorem. When $\overleftrightarrow{AE}$ and $\overleftrightarrow{BF}$ meet, they will make a triangle with any segment of the parallel lines shown in the figure.

Theorem 60-3

If parallel lines intersect transversals, then they divide the transversals proportionally.

$$\frac{AC}{CE} = \frac{BD}{DF}$$

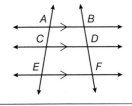

If parallel lines divide a transversal into congruent segments, then the segments are in a 1:1 ratio. By Theorem 60-3, any other transversal cut by the same parallel lines will be divided into segments that also have a 1:1 ratio, so they will also be congruent.

Theorem 60-4

If parallel lines cut congruent segments on one transversal, then they cut congruent segments on all transversals.

In the diagram, if $UV = VW$, then $XY = YZ$.

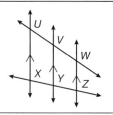

Math Reasoning

Connect Parallel lines form many congruent and supplementary angles. From Example 3, name a pair of corresponding angles, and a pair of same-side interior angles.

Example 3 Proving Theorem 60-4

Use a paragraph proof to prove Theorem 60-4.
Given: $\overline{AB} \cong \overline{BC}, \overleftrightarrow{AD} \parallel \overleftrightarrow{BE}, \overleftrightarrow{BE} \parallel \overleftrightarrow{CF}$
Prove: $\overline{DE} \cong \overline{EF}$

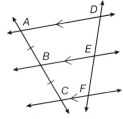

SOLUTION

By Theorem 60-3, we know that $\frac{AB}{BC} = \frac{DE}{EF}$. Since $\overline{AB} \cong \overline{BC}$, the ratio of AB to BC by substitution is the same as $AB{:}AB$, which is equal to 1. Substitute into the proportion given and obtain $1 = \frac{DE}{EF}$. Taking the cross product yields $DE = EF$. By the definition of congruent segments, $\overline{DE} \cong \overline{EF}$.

Example 4 Finding Segment Lengths with Intersecting Transversals

(a.) Find the length of segment $\overline{AB}$.

SOLUTION

The parallel lines cut $\overline{DF}$ into congruent segments. Therefore, $\overline{AC}$ is also cut into congruent segments.

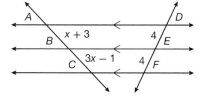

$3x - 1 = x + 3$
$2x = 4$
$x = 2$

Substitute this value into the equation that gives AB.
$AB = x + 3$
$AB = 2 + 3$
$AB = 5$

(b.) Determine whether $\overleftrightarrow{UV}$, $\overleftrightarrow{WX}$, and $\overleftrightarrow{YX}$ are parallel when $x = 3$.

SOLUTION

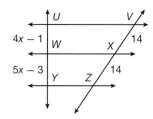

Because $VX = XZ$, $\overline{UW}$ must equal $\overline{WY}$.
$UW = 4x - 1$
$ = 4(3) - 1$
$ = 11$
$WY = 5x - 3$
$ = 5(3) - 3$
$ = 12$

$UX \neq WY$. Therefore the lines $\overline{UV}$, $\overline{WX}$, and $\overline{YZ}$ are not parallel.

Math Reasoning

How is Example 4b related to Theorem 60-4?

Example 5 **Application: Art**

Perspective is a method artists use to make an object appear as if it is receding into the distance. If the fence posts are parallel, then what is the length of $\overline{AB}$ if $EH = 22$, $BC = 4$, $CD = 6$, and $FH = 18$?

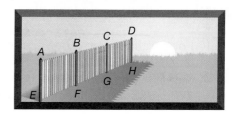

SOLUTION

Use Theorem 60-3 to write a proportion relating the segments given.

$$\frac{EH}{FH} = \frac{AD}{BD}$$

$$\frac{22}{18} = \frac{AD}{(4+6)}$$

$$AD = 12.\overline{2}$$

Use the Segment Addition Postulate:

$$AB = AD - BC - CD$$
$$AB = 12.\overline{2} - 4 - 6$$
$$AB = 2.\overline{2}$$

Lesson Practice

a. Find the length of $\overline{EB}$.
(Ex 1)

b. Find the length of $\overline{PQ}$.
(Ex 1)

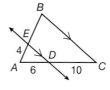

c. Use a paragraph proof to prove the Triangle Proportionality Theorem.
(Ex 3)
Given: $\overline{DE} \parallel \overline{BC}$
Prove: $\dfrac{AD}{DB} = \dfrac{AE}{EC}$

d. Find the length of $\overline{AC}$.
(Ex 4)

e. Determine whether $\overline{KL}$, $\overline{MN}$, and $\overline{OP}$ are parallel.
(Ex 2)

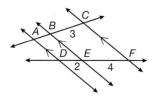

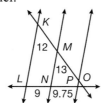

f. (**Art**) A road is drawn with perspective. Find the length of $\overline{AE}$ if $AC = 10$, $BF = 20$, and $BD = 8$.
(Ex 5)

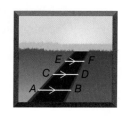

1. For parallelogram *KLMN*, prove the triangles congruent by SAS Congruency.
(28)

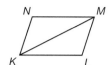

2. **Algebra** Using the diagram, find *WV* in terms of *x* if
(38) *TW* = 28*x*, *UV* = 15, and *TU* = 25*x*.

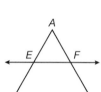

3. A school is located at grid point (−1, 2). If Jordan's
(42) home is on a street along the line $y = \frac{3}{2}x - 3$, what is
the shortest distance Jordan could possibly live from
the school?

* 4. Fill in the blanks of the paragraph proof of the Converse of the
(60) Triangle Proportionality Theorem: If a line divides two sides of a
triangle proportionally, then it is parallel to the third side.

Given: $\frac{EB}{AE} = \frac{FC}{AF}$, so $\frac{AE}{AB} = \frac{AF}{AC}$.
Prove: $\overleftrightarrow{EF} \parallel \overline{BC}$

It is given that $\frac{EB}{AE} = \frac{FC}{AF}$. By the _____ Property of _____, ∠*A* ≅ ∠*A*, so △*AEF* ~
△*ABC* by _____. By definition of similar triangles, ∠*AEF* ≅ ∠*ABC*. Finally, by the
_____, $\overleftrightarrow{EF} \parallel \overline{BC}$.

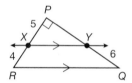

* 5. **Multi-Step** In △*PQR*, ∠*P* is a right angle. Find the perimeter of △*PQR* to the
(60) nearest tenth.

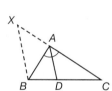

* 6. A right rectangular prism has a length, width, and height that are all equal.
(59) The volume of the prism is 343 cubic meters. Find the surface area.

* 7. Write a paragraph proof of the Triangle Angle Bisector Theorem: If a line
(31) bisects an angle of a triangle, then it divides the opposite side proportionally
to the other two sides of the triangle. *Hint: Draw $\overline{BX}$ parallel to $\overline{AD}$ and
extend $\overline{AC}$ to X.*
Given: In △*ABC*, $\overline{AD}$ bisects ∠*BAC*.
Prove: $\frac{BD}{DC} = \frac{AB}{AC}$

8. **Analyze** Find the measure of each angle in the inscribed quadrilateral.
(47)

9. (**Sleep Patterns**) Unless Cedrick has at least eight hours of sleep per night, he
(21) is tired all day. This morning he woke early and did not get eight hours of
sleep. What can we conclude about Cedrick today?

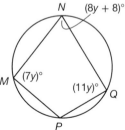

10. **Multi-Step** Find the orthocenter of △*RTV* with vertices
(32) *R*(−2, 6), *T*(1, 3), and *V*(7, 3).

11. **Error Analysis** This circle shown has center *A*, and $\overline{AB}$ is a radius.
(58) Mariah has calculated that ∠*A* in △*ABD* is 65°. Why is her
conclusion invalid?

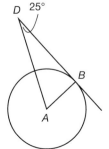

12. Find the geometric mean of 33 and 27.
(50)

13. **Multiple Choice** Which set of side lengths could not form a triangle?
(39)
   A 4, 5, 10   **B** 1, 2, 2
   C 11, 15, 20   **D** 55, 41, 37

*14. Find the surface area of the prism at right.
(59)

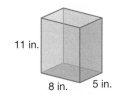

15. (Floor Refinishing) Four congruent floors of a square office building need
(53) to be refinished. The diagonal of one of the floors is 175 feet. If one can
of refinisher covers 8000 square feet, how many cans are needed to finish the
floors?

*16. **Write** Are $\overline{JK}$ and $\overline{LM}$ parallel? Explain how you know.
(60)

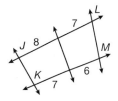

17. **Verify** $\overline{DE}$ is a midsegment of $\triangle FGH$. Find the coordinates of D and E.
(55) Use the distance formula to verify that $\overline{DE}$ is a midsegment of $\triangle FGH$.

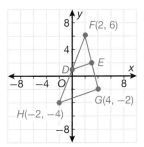

*18. **Multiple Choice** Which dimensions make $\overline{PQ} \parallel \overline{ST}$?
(60)
   A $SR = 14$, $TR = 19$ **B** $SR = 10$, $TR = 13.3$
   C $SR = 17.5$, $TR = 25$ **D** $SR = 13$, $TR = 20$

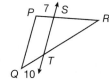

*19. Find the volume of the oblique prism shown.
(59)

20. Line ℓ is tangent to $\odot C$ at A, and line m passes
(58) through C. Lines ℓ and m intersect at B.
   a. Classify $\triangle ACB$ by its angles.
   b. If m$\angle ACB = 53°$, determine m$\angle CBA$.

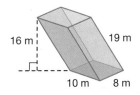

21. **Generalize** Is every solid a polyhedron? Explain.
(49)

22. There are just five regular polyhedra, sometimes called the Platonic solids after
(Inv 5) the Greek philosopher Plato. Which of these polyhedra does *not* have square or
triangular faces, and what shape are its faces?

Tetrahedron   Cube   Octahedron   Dodecahedron   Icosahedron
4 faces   6 faces   8 faces   12 faces   20 faces

*23. **Justify** Use an indirect proof to prove that an altitude of an equilateral triangle
(48) is also a median.

    Given: $\triangle ABC$ is equilateral with altitude $\overline{AX}$.
    Prove: $\overline{AX}$ is a median of $\overline{BC}$.

24. **(Physics)** In an experiment, wooden blocks with different masses are
(16) attached to a force meter and dragged along a rough surface. This
graph shows a line that models the results. What is the equation of
this line? What are the correct units for the slope?

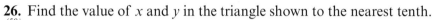

25. An octahedron is a regular polyhedron. If the bases of two congruent
(49) square pyramids are glued together, the result is an octahedron.
Describe the polygons that comprise the faces of an octahedron,
and state the number of vertices, edges, and faces.

26. Find the value of x and y in the triangle shown to the nearest tenth.
(50)

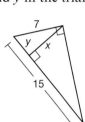

27. **(Geology)** Large meteor craters have been found that are nearly perfectly circular.
(35) A geologist is walking around one such crater. After walking one mile, she finds
that she has covered 85 degrees of arc. What is the radius of the crater, to the
nearest foot?

28. **Justify** Find the value of x in the figure, and justify your answer.
(46)

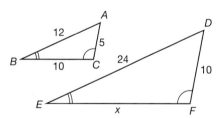

29. **Formulate** Write a formula that uses only the shorter leg to find the area of a
(56) $30°$-$60°$-$90°$ triangle. Let the shorter leg be equal to x.

30. **(Traffic)** A shop owner built his own stop sign according to the diagram. Find
(57) the perimeter of the stop sign to the nearest unit. Is this a regular octagon?

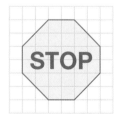

Geometric Probability

Math Reasoning

Predict If it is 40% likely to rain on any of the next five days, how many days would you predict it to be rainy?

Recall that the probability of an event happening is equal to the number of outcomes in which that event happens divided by the total number of possible outcomes.

$$P\ (\text{event}) = \frac{\text{Number of desired outcomes}}{\text{Number of possible outcomes}}$$

Experimental probability is a process where trials are conducted to test the probability of an event and the results are recorded. Experimental probability is determined by dividing the number of trials where the desired event occurred by the total number of trials that were conducted.

Geometric probability is a form of probability that is defined as a ratio of geometric measures such as lengths, areas, or volumes.

For example, a spinner that is divided into two congruent sections shaded red and blue has a 50% chance of landing on red. This is known because the red section makes up 50% of the spinner's area. If the red section comprised just 25% of the spinner's area, the chance that it would land on red would be 25%.

In groups, assemble a spinner and conduct a probability experiment. With a crayon or colored pencil, shade sectors 1-5 red. Shade sectors 6-7 blue. Shade sectors 8-10 yellow. Cut a strip of paper or cardstock into the shape of an arrow, and fasten the arrow to the spinner base with a brad.

1. Use a protractor to measure the central angle of each of the three colored sectors of your spinner. What are the measures of the red, blue, and yellow central angles, respectively?

2. What percent of the time would you expect the spinner to land on red? Why?

Math Reasoning

Connect Why does the proportional area of the spinner sector give the theoretical probability?

The probability of landing on red can be determined because it covers half the spinner. To determine the probability of landing on blue or yellow, write a proportion. For example, blue measures 72° out of 360° in the entire circle.

$$\frac{\text{Degrees in sector}}{\text{Total degrees in circle}} = \frac{72}{360} = \frac{1}{5}$$

As a percentage, this is 20%, so you would expect the spinner to land on blue about 20% of the time.

Now conduct a probability experiment with the spinner. Draw a table like the one shown. Include at least ten rows for results, with each row recording 10 spins in addition to the previous tallies brought down from the row above.

	Results			Experimental Probability		
Spins	Blue	Yellow	Red	*P*(Blue)	*P*(Yellow)	*P*(Red)
10						
20						

To fill out the first row of the table, spin the spinner 10 times and record a tally mark in the corresponding column for each time it lands on a color. Then, find the experimental probability of landing on each color by dividing the number of times the spinner landed on the color by the total number of spins. Fill out the last three columns.

Spin the spinner 10 more times and update the tallies for the next row of the table. Continue finding the experimental probability for each row until 100 spins have been tallied.

3. What are the theoretical probabilities of getting blue, yellow, and red?

4. How close are the experimental probabilities in the first row of the table to the theoretical probabilities? How close are they in the fifth row of the table?

5. What trend occurs in the difference between the experimental probabilities and the theoretical probabilities in the table as the number of spins increases?

Suppose a plane figure *P* can be separated into nonoverlapping regions *Q*, *R*, and *S*. The theoretical probability for one of the smaller regions, say *Q*, is the ratio of the area of *Q* to the area of *P*.

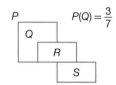

$$P(Q) = \frac{A_Q}{A_P}$$

As long as an event that could happen in any region of *P* is random, the experimental probability in a set of repeated trials should approach the theoretical probability.

Using graph paper, draw the shapes as shown in the art at right. Color each shape a different color.

Cut each of the parallelograms and triangles created by the gridlines. Place all of the triangles in a bag. With all of the remaining parallelograms, cut each along a diagonal to make congruent triangles, and place these triangles in the bag with the others. Draw a table like the one shown. Then randomly pull triangles from the bag and record the results in the table.

Math Reasoning

Formulate If a pin is equally likely to be stuck anywhere in a square, what is the probability that it is stuck on the right side of the square? In the middle third?

Shape	Number of Triangles	Theoretical Probability
Shape 1	16	$\frac{16}{70}$
Shape 2	12	

Complete the first column by listing the 8 shapes that made up the original figure. Make sure to note the color of each shape. In the second column, record the number of triangles that comprised each shape. In the third column, calculate the theoretical probability of drawing a part of each shape from the bag.

Begin taking shapes from the bag. Make a table to tally the results, as in the spinner experiment. Do as many trials as time allows. When finished, find the experimental probability of drawing a triangle from each of the colored shapes.

6. Compare the experimental probability to the theoretical probability. Are the trends the same as in the spinner experiment?

7. What is the theoretical probability that a triangle will be drawn that is part of Shape 1 or Shape 2?

8. Based on observations from both activities, what can be concluded about the experimental and theoretical probabilities as more and more trials are conducted?

Investigation Practice

a. Design and make a spinner so that $P(\text{red}) = \frac{1}{2}$, $P(\text{blue}) = \frac{1}{4}$, and $P(\text{green}) = \frac{1}{4}$.

b. Conduct a probability experiment with the spinner from part **a**. Comment on the results.

c. Use square grid paper to design a simple set of shapes that comprise a rectangle. Use only the grid lines as edges. Make a table like the one used in the second experiment of this investigation for the set of shapes.

d. Write a problem about the theoretical probabilities for the experiment conducted in part **c**. Exchange the probability problem with another student. Solve each other's problems.

e. Compare your experiment in part **c** with the experiment in the second part of the investigation. In each activity, was the probability proportional to the number of triangles or squares in the shape, the area, or both? What key ideas about probability did you observe in both activities?

Determining if a Quadrilateral is a Parallelogram

1. **Vocabulary** A parallelogram that has perpendicular diagonals is either a
(34) _____ or a _____.

2. If a quadrilateral is a parallelogram, then its opposite sides are congruent.
(34) True or false?

3. **Multiple Choice** Which is not a quadrilateral?
(19)
A rhombus   **B** trapezoid
C cube   **D** kite

New Concepts

If a quadrilateral has certain characteristics, it can be identified as a parallelogram. This lesson introduces four methods of identifying parallelograms.

Math Reasoning

Analyze In Lesson 34, it was proven that if a quadrilateral is a parallelogram, then its opposite sides are congruent. What is the relationship between that property and the first method of identifying parallelograms?

Identifying Parallelograms
If both pairs of opposite sides of a quadrilateral are congruent, then the quadrilateral is a parallelogram. 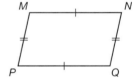   *MNQP* is a parallelogram.

Example 1 **Proving a Quadrilateral is a Parallelogram Using Opposite Sides**

In quadrilateral *WXYZ*, $\overline{WX} \parallel \overline{ZY}$ and $\angle Z \cong \angle X$. Is *WXYZ* a parallelogram?

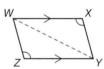

SOLUTION
The diagonal $\overline{WY}$ has been added to create $\triangle WXY$ and $\triangle WZY$. Since $\overline{WX} \parallel \overline{ZY}$, the alternate interior angles $\angle XWY$ and $\angle ZYW$ are congruent. Segment *WY* is congruent to itself by the Reflexive Property of Congruence. Therefore, $\triangle WXY \cong \triangle YZW$ by the AAS Triangle Congruence Theorem. By CPCTC, $\overline{WX} \cong \overline{ZY}$ and $\overline{WZ} \cong \overline{XY}$. Since both pairs of opposite sides of *WXYZ* are congruent, it is a parallelogram.

Identifying Parallelograms
If both pairs of opposite angles of a quadrilateral are congruent, then it is a parallelogram. 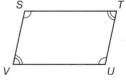   *STUV* is a parallelogram.

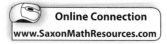

Example 2 Proving a Quadrilateral is a Parallelogram Using Opposite Angles

In quadrilateral $PQRS$, $\overline{PQ} \cong \overline{SR}$. Is $PQRS$ a parallelogram?

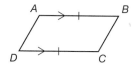

SOLUTION

Since $\overline{PQ} \cong \overline{SR}$, $PQ = SR$. Substitute the given values and solve for x.

$PQ = SR$	Given
$3x + 2 = 5x - 38$	Substitute.
$x = 20$	Solve.

Now that x is known, substitute it into the expression for the measure of each angle. We find that $\angle P = 120°$, $\angle R = 120°$, $\angle S = 60°$, and $\angle Q = 60°$. Since both pairs of opposite angles in $PQRS$ are congruent, $PQRS$ is a parallelogram.

Identifying Parallelograms

If one pair of opposite sides of a quadrilateral is both parallel and congruent, then the quadrilateral is a parallelogram.

$ABCD$ is a parallelogram.

Example 3 Proving a Quadrilateral is a Parallelogram Using One Pair of Sides

In quadrilateral $JKLM$, $\angle J$ and $\angle M$ are supplementary and $\overline{JK} \cong \overline{ML}$. Is $JKLM$ a parallelogram?

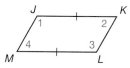

SOLUTION

Since $\angle J$ and $\angle M$ are supplementary, then by the Converse of the Same-Side Interior Angles Theorem, we know that $\overline{JK} \parallel \overline{ML}$. Since the opposite sides $\overline{JK}$ and $\overline{ML}$ are both parallel and congruent, $JKLM$ is a parallelogram.

Identifying Parallelograms

If the diagonals of a quadrilateral bisect each other, then the quadrilateral is a parallelogram.

QRST is a parallelogram.

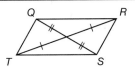

Math Reasoning

Analyze What types of parallelograms have diagonals that are perpendicular bisectors?

Example 4 **Proving a Quadrilateral is a Parallelogram Using Diagonals**

In quadrilateral *RSTU*, $\overline{RU} \cong \overline{ST}$. Is *RSTU* a parallelogram?

SOLUTION
Since $\overline{RU} \cong \overline{ST}$, $RU = ST$. Substitute the given values and solve for x.

$RU = ST$	Given
$9 = 2x + 3$	Substitute.
$x = 3$	Solve.

Now that x is known, it can be used to find the lengths of each diagonal segment in the quadrilateral. We find that $RV = 6$, $VT = 6$, $UV = 7$, and $VS = 7$. The segments of each diagonal are equal, so V is the midpoint of each one. Therefore, the diagonals bisect each other, which proves that *RSTU* is a parallelogram.

Example 5 **Application: Gardening**

A gardener wants to know how much fencing to buy for the perimeter of her garden, shown below. The garden has two paths that bisect each other to form an "X." How much fencing does the gardener need?

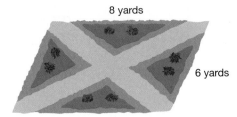

8 yards

6 yards

SOLUTION
The diagonals bisect each other, so the quadrilateral is a parallelogram and opposite sides are equal.

Calculate the perimeter.

$P = 2(8) + 2(6)$
$P = 28$

The gardener needs 28 yards of fencing.

a. In quadrilateral *ABCD*, $\overline{AD} \cong \overline{BC}$ and $\overline{AB} \cong \overline{DC}$. Prove that the diagonals of *ABCD* bisect each other.
(Ex 1)

b. In quadrilateral *EFGH*, $\angle E \cong \angle G$ and $\angle F \cong \angle H$. Prove that the opposite sides are congruent.
(Ex 2)

c. In quadrilateral *WXYZ*, $\triangle WXY \cong \triangle YZW$. Prove that *WXYZ* is a parallelogram by showing that $\overline{WX} \parallel \overline{ZY}$ and $\overline{WX} \cong \overline{ZY}$.
(Ex 3)

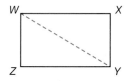

d. In the diagram, $\triangle AED \cong \triangle CEB$. Prove that quadrilateral *ABCD* is a parallelogram.
(Ex 4)

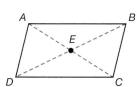

e. A school has a railing on the front staircase. If $\angle 1 \cong \angle 2$ and $\angle 3 \cong \angle 4$, prove that the top railing and the bottom railing are parallel.
(Ex 5)

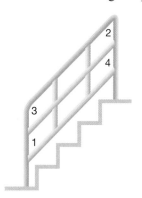

1. Find the value of *x* and *y* in the triangle shown.
(50)

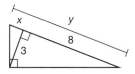

2. **Crafts** Two artisans are sharing a square table measuring 1 meter on each side to display their crafts at a crafts fair. They need to divide its area equally between them before they can arrange their crafts on the table. If they mark the dividing line with tape, how much more tape will they use if they mark it diagonally than if they mark it horizontally?
(53)

3. If the shorter leg of a 30°-60°-90° triangle is 17, what are the lengths of the other leg and hypotenuse?
(56)

4. Find the value of *x* and *y* in the triangle shown. Give your answers in simplified radical form.
(56)

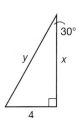

* 5. **Model** Design a spinner with 4 sectors so that one sector is 3 times as probable as the others.
(Inv 6)

* 6. Determine whether lines $\overline{EF}$ and $\overline{XY}$ are parallel.
(60)

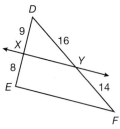

* 7. (**Urban Planning**) A city block forms a parallelogram.
(61) If 3 sides are 86 meters, 86 meters, and 156 meters long, respectively, what is the only length possible for the fourth side?

8. If $\triangle ABC$ and $\triangle DEF$ are similar by Angle-Angle Similarity,
(46) then the lengths of their respective sides are _____.

9. Segment QR is a midsegment of $\triangle MNP$. Find the coordinates
(55) of Q and R.

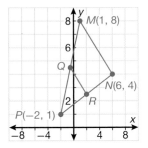

10. (**Radio Communications**) A set of radio communicators are rated to work
(53) up to 1.35 miles apart. Smith and Claude start at the same point, and each takes a communicator. If Smith walks 1 mile east and Claude walks 1 mile north, will their communicators be able to reach each other? Explain.

11. Estimate Estimate the area of the irregular
(57) shape at right.

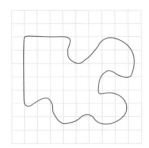

12. Multiple Choice What is the midpoint of the line segment with endpoints $P(-7, 3)$
(11) and $Q(0, 8)$?

   A $(-3.5, 5.5)$ **B** $(-3.5, 4.5)$
   C $(-2.5, 5.5)$ **D** $(-1.5, 1.5)$

13. Find the surface area of the prism at right.
(59)

14. (**Painting**) Aiden wants to paint the walls of his living room which
(59) measure 16 feet wide, 22 feet long, and 8 feet high. How many square feet must Aiden paint?

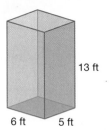

15. $\overline{ZX}$ and $\overline{ZY}$ are tangents to $\odot C$ at X and Y, and $\triangle XYZ$ is equiangular.
(51) The radius of $\odot C$ is 5 centimeters. What is the exact perimeter of quadrilateral $CXZY$?

16. Figure $ABCD$ is similar to figure $EFGH$. The ratio of their corresponding
(44) sides is 4:5. If the perimeter of $EFGH$ is 30 inches, what is the perimeter of $ABCD$?

17. Multi-Step Write an inequality to show the values for x
(33) that make the triangle shown acute. Write an inequality to show the values for x that make the triangle obtuse. Round to the nearest tenth.

18. The area of a circle is 40.6 square meters. If a sector has an arc length of
(35) 2.8 meters, what is the approximate arc measure of the central angle?

*19. (Games and Puzzles) Determine the geometric probabilities for randomly
(Inv 6) drawing each piece of this tangram.

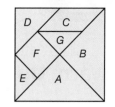

*20. **Multiple Choice** The coordinates of three vertices of a parallelogram are
(61) $(-2, 1)$, $(3, -1)$, and $(-1, -4)$. What are the coordinates of the fourth vertex?
  A $(2, 5)$ **B** $(4, 5)$
  C $(-4, -8)$ **D** $(-6, -2)$

*21. Find the length of $\overline{XA}$ in the triangle shown.
(60)

22. Line ℓ is tangent to $\odot C$ at A. Line m passes through C
(58) and intersects line ℓ at B.
  a. Classify $\triangle ACB$ by its angles.
  b. If m$\angle ACB = 53°$, determine m$\angle CBA$.

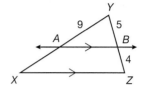

23. **Justify** Write a paragraph proof of the following.
(31) **Given:** $\angle 4 \cong \angle 3$
  Prove: m$\angle 1 =$ m$\angle 2$

24. Is the following statement always, sometimes, or never true?
(52) *A rhombus is a rectangle.*

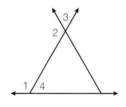

*25. **Algebra** If $x = 4$, then is $ABCD$ a parallelogram?
(61) Explain why it is, or draw a diagram to prove it is not.

*26. **Write** If you connect the midpoints of two sides of a triangle,
(60) will the line you make always be parallel to the third side?
Explain.

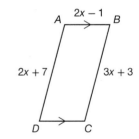

27. **Algebra** Rectangle $QRST$ has vertices $Q(0, 0)$, $R(0, 2w)$, $S(2\ell, 2w)$,
(45) and $T(2\ell, 0)$. Find the coordinates of the midpoint of $\overline{QR}$.

28. Name the shortest line segment in the diagram.
(39)

*29. If two equilateral triangles are joined along one side,
(61) will the result always be a parallelogram? If so, what type
of parallelogram would be formed?

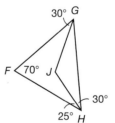

30. **Analyze** What can the intersection of a sphere and
(49) a plane be?

Finding Surface Areas and Volumes of Cylinders

Warm Up

(59) **1.** **Vocabulary** The total area of all faces and curved surfaces of a three-dimensional figure is its _____.

(59) **2.** Find the surface area of the prism.

(59) **3.** Find the volume of the prism.

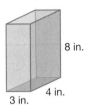

8 in.

4 in.

3 in.

New Concepts

A cylinder is a three-dimensional figure with two parallel circular bases and a curved lateral surface that connects the bases.

The **base of a cylinder** is one of the two circular surfaces of the cylinder. The **altitude of a cylinder** is the segment that is perpendicular to, and has its endpoints on the planes of the bases. The length of the altitude is the height of the cylinder. The **radius of a cylinder** is the distance from the center of the cylinder's base to any point on the edge of the base.

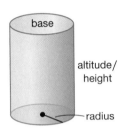

base

altitude/ height

radius

Exploration **Analyzing the Net of a Cylinder**

In this exploration, you will create and analyze the net of a cylinder.

1. What plane figures comprise the net of a cylinder?

2. Draw circle *P* with a radius of 2.5 cm. This will be one base of the cylinder. How can you draw the other base of the cylinder?

3. What is the total area of both bases to the nearest hundredth square centimeter?

4. What would be the length and width of the rectangular piece of the net?

5. Calculate the appropriate length for the rectangular piece to the nearest millimeter. The height of the finished cylinder should be 8 cm. Use a ruler and a protractor to draw the lateral surface of the cylinder. What is the area of the lateral surface?

6. Cut out the bases and the lateral surface for your cylinder. Use a small piece of tape to attach the bases to the lateral surface as shown in the figure below. What is the total area of the net?

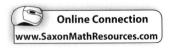

7. Use small pieces of tape to construct the cylinder from your net. How is the surface area of the cylinder related to the total area of the net?

The **lateral area of a cylinder** is the area of the curved surface of a cylinder. The diagram shows the net of a cylinder. When the cylinder is unfolded, the lateral area is actually a rectangle that has a length equal to the circumference of the cylinder's base.

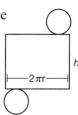

Lateral Area of a Cylinder

Use the following formula for the lateral area of a cylinder where r is the radius and h is the height of the cylinder.

$$L = 2\pi rh$$

Example 1 Finding the Lateral Area of a Cylinder

Find the lateral area of the cylinder in terms of π.

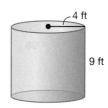

SOLUTION
Use the formula for lateral area.

$L = 2\pi rh$	Lateral Area
$L = 2\pi(4)(9)$	Substitute
$L = 72\pi \text{ ft}^2$	Simplify

To find the total surface area of a cylinder, find the lateral area and add it to the area of the two circular bases.

Surface Area of a Cylinder

Use the following formula to find the total surface area of a cylinder where B is the area of a base and L is the lateral area.

$$S = 2B + L$$

If the formula for the area of each circular base and the formula for lateral area are substituted into the formula for surface area, it becomes:

$$S = 2\pi r^2 + 2\pi rh$$

Example 2 Finding the Surface Area of a Cylinder

Find the total surface area of the cylinder in terms of π.

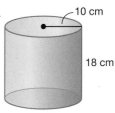

SOLUTION
Use the formula for surface area.

$S = 2\pi r^2 + 2\pi rh$	Surface Area
$S = 2\pi(10)^2 + 2\pi(10)(18)$	Substitute
$S = 560\pi \text{ cm}^2$	Simplify

Taking the volume of a cylinder can be described as taking the base and dropping it through the height.

Volume of a Cylinder
The volume of a cylinder can be found by multiplying the area of the base by the height. Since the base is a circle, use the formula: $$V = \pi r^2 h$$

The cylinders in the examples above are **right cylinders**. A right cylinder's bases are aligned directly above one another. If the bases of a cylinder are not aligned directly on top of each other, it is an **oblique cylinder**. The height of an oblique cylinder can be found by dropping an altitude from one base to the plane that contains the second base.

Caution

The length along the edge of an oblique cylinder is not the height of it. The height is a vertical line that is perpendicular to the cylinder's base.

Example 3 Finding the Volume of a Cylinder

Find the volume of the right cylinder in terms of π.

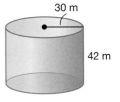

SOLUTION
Use the formula for volume of a cylinder:

$V = \pi r^2 h$ Volume of a cylinder
$V = \pi(30)^2(42)$ Substitute
$V = 37{,}800\pi \text{ m}^3$ Simplify

Math Reasoning

Formulate If the level of water in the tank decreases by one inch every hour, how much water is being used per hour, in cubic feet? Write an expression that solves this problem, but do not evaluate the expression.

Example 4 Application: Water Towers

The city of Lewiston has a cylindrical water tower that is 45 feet tall. The radius of the tower's base is 55 feet. How many cubic feet of water can the tower hold? Use 3.14 to approximate π.

SOLUTION
Find the volume of the cylindrical water tank.

$V = \pi r^2 h$
$V = \pi(55)^2(45)$
$V = 136{,}125\,\pi$
$V \approx 427{,}432.5 \text{ ft}^3$

The volume of the storage tank is approximately 427,432.5 cubic feet.

Lesson Practice

a. Find the lateral area of the cylinder in terms of π.
(Ex 1)

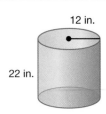

b. Find the total surface area of the cylinder to the
(Ex 2) nearest square centimeter.

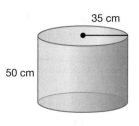

35 cm

50 cm

c. Find the volume of the right cylinder to the
(Ex 3) nearest cubic foot.

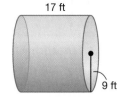

17 ft

9 ft

d. A farmer uses a cylindrical silo to store grain.
(Ex 4) The silo has a radius of 75 feet and is 150 feet
tall. What is the storage capacity of the silo to
the nearest cubic foot?

Practice   Distributed and Integrated

1. Using the figure at right, complete this ratio: $\frac{AC}{CE} = \frac{}{DF}$
(60)

2. Draw a cube in one-point perspective so that the
(54) vanishing point is to the right of the cube.

3. **Algebra** In $\triangle ABC$, $AB = 3x + 2$ and $BC = 5x - 7$. In
(25) $\triangle FED$, $DE = 4x + 13$. If $\triangle ABC$ is congruent to $\triangle FED$, then
what is AB?

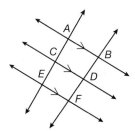

* 4. Find the volume of the oblique cylinder to the nearest tenth of a cubic yard.
(62)

5. What is the radius of a circle in which a 9.2-inch long chord is 4.3 inches
(43) from the center of a circle?

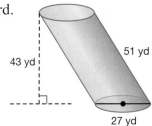

51 yd

43 yd

27 yd

6. Which solid does a basketball most resemble?
(49)

7. Calculate the theoretical probabilities for each sector
(Inv 6) of this spinner.

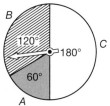

8. What is another name for an equiangular
(52) quadrilateral?

9. **Error Analysis** Alicia wanted to find the centroid in the
(32) diagram shown. Explain the error she made.

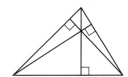

10. **Multiple Choice** In triangles *UVW* and *XYZ*, *U* and *Z* are right angles, $\overline{UV} \cong \overline{YZ}$,
(36) and $\overline{VW} \cong \overline{XY}$. Which theorem proves these triangles are congruent?
   A HA Congruence Theorem   **B** LA Congruence Theorem
   C LL Congruence Theorem   **D** None of these

11. **Furniture Making** The braces for a shelf are 30°-60°-90° triangles, as shown.
(56) What is the perimeter of each brace?

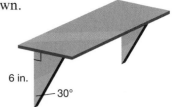

12. In isosceles △ABC, ∠A is the vertex angle and m∠A = 3m∠B.
(51)   **a.** Write a congruence statement involving the angles of △ABC.
       Name a postulate or theorem to justify your statement.
       b. Determine the measure of each angle.

6 in.

30°

13. **Formation** Lucy and her friends are walking to school as shown in the
(52) diagram. Lucy notices that she has to turn her head 90° to look from one
friend to the next. Lucy decides that her friends are arranged in a square.
Is this a valid conclusion? Why or why not?

*14. Find the volume of the cylinder to the nearest hundred cubic inches.
(62)

15. **Algebra** Find the distance from the center of a circle to a 12-inch cord if the
(43) circle has a 20-inch diameter.

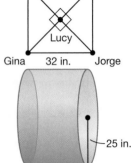

25 in.

*16. **Labeling** A certain soup can measures 4 inches tall and 3 inches in diameter.
(62) If the soup company needs to fill an order for 1092 cans, what is the total area
of labels the company will need to print?

17. Find the value of x and y in the triangle shown.
(50)

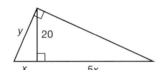

*18. **Analyze** In quadrilateral ABCD, m∠A + m∠B = 180°.
(61) Can it be proven that ABCD is a parallelogram?

*19. **Justify** If $x = 12$, then is ABCD at right a parallelogram? Explain.
(61)

*20. **Model** Explain how the Triangle Proportionality Theorem can be
(60) thought of as a special application of Theorem 60-3 (parallel lines
divide transversals proportionally). Is there a situation in which
one of the theorems could *not* be used to make a triangle?

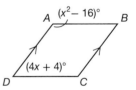

21. **Engineering** A circular rod is clamped between two plates
(58) as shown. The plates have to be 20 inches wide. What is the
least area of metal needed for the plates?

22. Find the geometric mean of 17 and 13.
(50)

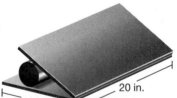

20 in.

11 in.

23. **Justify** Explain how to find x in the diagram shown. What is x?
(Inv 1)

24. A spinner has three colored sectors, with these central angles:
(Inv 6)   purple: 45°; yellow: 105°; orange: 210°
For each color, calculate the theoretical probability that the spinner
lands on it.

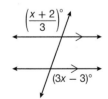

25. **Algebra** Find the measure of each angle in quadrilateral *QRST*.
(47)

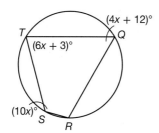

26. (**Packaging**) Mrs. Jenkins has a jar that holds 360 cubic inches of spicy roasted
(59) almonds. She wants to transfer the almonds to gift boxes that measure 2 inches
tall, 3 inches wide, and 5 inches long. How many gift boxes can she fill?

27. **Multi-Step** Given the line $2y - x = 0$ and $F(-6, 2)$, find the point on the line that is
(42) closest to *F*.

*28. Find the surface area of the cylinder shown to the nearest hundredth.
(62)

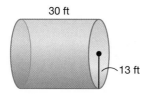

30 ft

13 ft

*29. Is this statement sometimes, always, or never true?
(61)

A trapezoid is a parallelogram.

*30. **Write** Explain how the statement, "If a triangle is equilateral, then it is
(51) equiangular," follows from the Isosceles Triangle Theorem.

Introduction to Vectors

Warm Up

1. **Vocabulary** A _____ is a part of a line consisting of two endpoints and all the points between them.
   (2)

2. A right triangle has legs with lengths of 12 centimeters and 5 centimeters. Use the Pythagorean Theorem to find the length of the hypotenuse.
   (29)

3. **Multiple Choice** Malia walked from her house 30 meters north and 8 meters east to the library. She then walked 8 meters south and 16 meters east to the park. How far is she from home?
   (9)
   A 44.9 meters   **B** 32.6 meters
   C 48.7 meters   **D** 21.3 meters

4. What is the difference between a line and a line segment?
   (2)

New Concepts

A **vector** is a quantity that has both magnitude and direction. The **direction of a vector** is the orientation of the vector, which is determined by the angle the vector makes with a horizontal line.

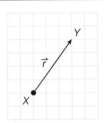

In contrast to vectors, a quantity that consists only of magnitude and has no direction is called a **scalar**.

> **Reading Math**
>
> A vector can also be named by its initial point and terminal point. For example, the vector in the diagram could also be called $\overrightarrow{XY}$.

Vectors are named by an italicized, lowercase letter with the vector symbol. For example, the vector above is named $\vec{r}$. The **initial point of a vector** is the starting point of a vector. The **terminal point of a vector** is the endpoint of a vector. In the diagram, X is the initial point and Y is the terminal point of $\vec{r}$. The arrow at Y indicates the direction of the vector.

Example 1   Identifying Vectors and Scalars

Name each vector shown. Identify the terminal points of each vector, if applicable.

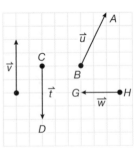

SOLUTION
Each vector should be named and then the terminal points given, in that order. The vectors, therefore, are $\vec{v}$, $\vec{u}$ with terminal point A, $\vec{t}$, with terminal point D, and $\vec{w}$, with terminal point G.

The **magnitude of a vector** is the length of a vector. Since magnitude is a length, absolute value bars are used to represent the magnitude of a vector. The magnitude of $\vec{v}$, for example, would be written $|\vec{v}|$.

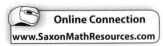

Online Connection
www.SaxonMathResources.com

The location of a vector on the coordinate plane is not fixed. It can be placed anywhere, so for simplicity the initial point of a vector is usually placed on the origin of the coordinate plane. To find the magnitude of a vector, place the initial point of the vector on the origin and use the distance formula.

Example 2 Finding the Magnitude of a Vector

Find $|\vec{v}|$.

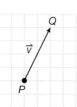

SOLUTION

The initial point of this vector is P. If P is placed on the origin, Q will be the point located two units to the right and four units up from P, so the coordinates of Q are $(2, 4)$.

Use the distance formula to find the distance between $P(0, 0)$ and $Q(2, 4)$.

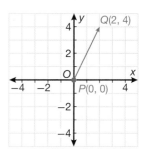

$$d = \sqrt{(x_2 - x_1)^2 + (y_2 - y_1)^2}$$
$$d = \sqrt{(2 - 0)^2 + (4 - 0)^2}$$
$$d = \sqrt{2^2 + 4^2}$$
$$d = 2\sqrt{5}$$

So $|\vec{v}|$ is $2\sqrt{5}$.

Reading Math

The brackets $\langle\ \rangle$ used in component form show that the pair indicates a vector, instead of coordinates on a grid.

The **component form of a vector** lists its horizontal and vertical change from the initial point to the terminal point.

For example, $\vec{x}$ written in component form would be $\langle 2, 5 \rangle$. The horizontal change is listed first, followed by the vertical change.

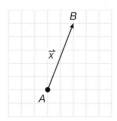

Two vectors with opposite components are called opposite vectors. **Opposite vectors** are vectors that have the same magnitude but opposite directions. The opposite vector of $\langle 2, 5 \rangle$ is $\langle -2, -5 \rangle$.

Any two vectors can be added together by summing their components. The vector that represents the sum or difference of two given vectors is a **resultant vector**.

Example 3 Adding Vectors

a. Add vectors $\vec{r}$ and $\vec{t}$.

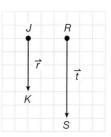

SOLUTION

First, write each vector in component form.

The component form of $\vec{r}$ is $\langle 0, -4 \rangle$, because there is no horizontal distance between J and K, but there is a negative vertical change of 4 units.

The component form of $\vec{t}$ is $\langle 0, -6 \rangle$.

Add the components: $\langle 0 + 0, -4 + -6 \rangle = \langle 0, -10 \rangle$

The resultant vector from adding these two vectors is $\langle 0, -10 \rangle$.

b. Add vectors $\vec{u}$ and $\vec{v}$.

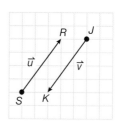

SOLUTION

First, write each vector in component form.

The component form of $\vec{u}$ is $\langle 3, 4 \rangle$.

The component form of $\vec{v}$ is $\langle -3, -4 \rangle$

Since $\vec{u}$ and $\vec{v}$ are opposite vectors, their components sum to 0. The resultant vector is $\langle 0, 0 \rangle$.

Equal vectors are vectors that have the same magnitude and direction. An easy way to add equal vectors is to multiply the vector by a constant. This is known as **scalar multiplication of a vector.** For example, to add $\langle 1, 2 \rangle$ and $\langle 1, 2 \rangle$, simply multiply $\langle 1, 2 \rangle$ by the scalar 2. The resultant vector is $\langle 2, 4 \rangle$, which has a magnitude that is twice that of $\langle 1, 2 \rangle$.

Example 4 **Adding Equal Vectors**

Add the equal vectors $\vec{a}$, $\vec{b}$, and $\vec{c}$.

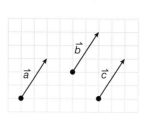

SOLUTION

In component form, all three of these vectors are $\langle 2, 3 \rangle$. Since there are three equal vectors, multiply the component form of the vectors by the scalar 3. The resultant vector is $\langle 3 \times 2, 3 \times 3 \rangle = \langle 6, 9 \rangle$.

Example 5 **Application: Currents**

A rower on a lake is rowing a boat at a rate of 5 miles per hour. A current is moving at 2 miles per hour in the opposite direction as the boat. How fast is the rower traveling over the ground below?

SOLUTION

Use the four-step problem-solving plan.

Understand Sketch the vectors for the rower and the current. The direction of the rower's vector does not matter, as long as the current's vector is pointing in the opposite direction. The magnitude of the rower's vector is 5, and the magnitude of the current's vector is 2.

Plan As in Examples 3 and 4, the vectors need to be added. First, find the component form of the vectors. Then, add them together.

Solve The component form of the rower's vector is $\langle 5, 0 \rangle$, and the current's vector is $\langle -2, 0 \rangle$. Add the vectors.

$\langle -2 + 5, 0 + 0 \rangle = \langle 3, 0 \rangle$

So the rower is traveling at 3 miles per hour.

Check Does it make sense that the current would be slowing the boat's progress? It does, because the current is flowing in the opposite direction.

a. Name the vectors and identify the initial point of each one.
(Ex 1)

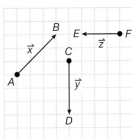

b. Find the magnitude of the vector $\langle 5, 3 \rangle$ in simplified radical form.
(Ex 2)

c. Add vectors $\vec{a}$ and $\vec{b}$.
(Ex 3)

d. Add vectors $\vec{b}$ and $\vec{c}$.
(Ex 3)

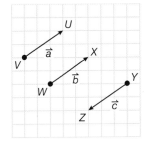

e. Add the four vectors.
(Ex 4)

f. A canoe is traveling down a river. In still water, the canoe would be traveling at 2 miles per hour. The river is flowing 1.5 miles per hour in the same direction as the canoe. How fast is the canoe actually traveling?
(Ex 5)

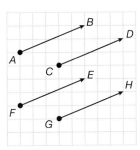

1. Quadrilateral *KLMN* is a rhombus with *X* at its center. If m∠*XKL* = 57°, what is
(52) m∠*NKL*?

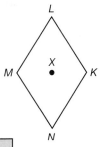

* 2. Maurice has a cylindrical jar with an approximate capacity of 75.36 cubic inches.
(62) He knows the jar is 6 inches tall. What is the jar's radius?

* 3. **Algebra** The opposite vector of $\langle 4, 6 \rangle$ is $\langle 2x + 2, 2x \rangle$. What is the value of *x*?
(63)

4. This grid is cut up into 16 squares, which are placed in a hat.
(Inv 6) What is the probability of drawing a blue square? a white
square?

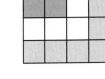

* 5. **Multiple Choice** A swimmer is swimming in a river. He is swimming
(63) in the same direction as a current that is flowing 2 miles per hour.
How fast must he swim if he wants to travel 4 miles per hour?

A 6 mi/hr   **B** 4 mi/hr

C 2 mi/hr   **D** 1 mi/hr

6. Is the following statement sometimes, always, or never true?
⁽⁶¹⁾
   A regular quadrilateral is a parallelogram.

7. Multi-Step In the triangles shown, what range of values for *x* would
⁽ᴵⁿᵛ ⁴⁾ cause ∠*P* to be larger than ∠*S*? ∠*P* to be smaller than ∠*S*?

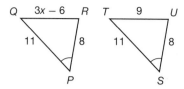

8. Analyze Use an indirect proof to prove that if △*ABC* is isosceles
⁽⁴⁸⁾ where *AB* = *AC* with a point *D* on *BC* such that *D* is not the
   midpoint of *BC*, then △*ABD* ≠ △*ADC*.
   Given: △*ABC* isosceles and *AD* ≠ *DB*.
   Prove: △*ABD* ≠ △*ADC*.

9. (**Farming**) A farmer wants to plant sorghum on a trapezoidal area of his field,
⁽⁵⁵⁾ shown by the triangle midsegment below. What is the area of the field he wishes
   to plant?

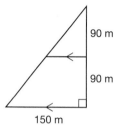

10. Multiple Choice Which of the following is not a parallelogram?
⁽³⁴⁾
   A rhombus   **B** trapezoid
   C rectangle   **D** square

xy² **11. Algebra** Determine the value of *x* in the triangle shown.
⁽³³⁾ Write your answer in simplified radical form.

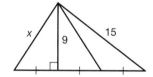

12. Analyze When can two distinct lines which are each tangent to
⁽⁵⁸⁾ the same circle be parallel? Explain.

13. Error Analysis In the figure at right, Rafael found that *BC* = 6.
⁽⁶⁰⁾ Where did he make an error?

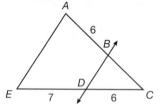

***14.** What is the sum of the vectors ⟨3, 2⟩, ⟨3.7, −8.2⟩, ⟨3, 2⟩,
⁽⁶³⁾ and ⟨−3.7, 8.2⟩?

***15.** Find the lateral area of the cylinder below, to the nearest whole square yard.
⁽⁶²⁾

16. Figure *JKLM* is similar to figure *TUWV*. The ratio of their corresponding sides is
⁽⁴⁴⁾ 6:1. If the perimeter of *TUWV* is 12 inches, what is the perimeter of *JKLM*?

xy² 17. **Algebra** In this figure, what value of z would allow you to apply the Same-Side
(12) Interior Angles Theorem to conclude that lines j and k are parallel?

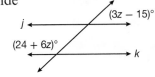

*18. (**Flight**) A small airplane is flying at 120 mph. If the wind is blowing at
(63) 30 mph from directly behind the airplane, at what speed is the airplane
flying over the ground?

19. In this figure, $\overline{AB}$ and $\overline{AD}$ are tangent to circle C. Determine the perimeter
(58) of quadrilateral $ABCD$. What type of quadrilateral is $ABCD$?

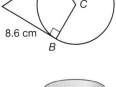

20. **Formulate** Write a formula that uses only the shorter leg to find the area of
(56) a 30°-60°-90° triangle. Let the shorter leg be equal to x.

*21. Find the volume of the cylinder shown to the nearest hundredth.
(62)

22. **Generalize** How many lateral faces can a prism have? As the number of sides
(54) of the base increase, what solid does the figure look more and more like?

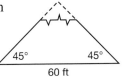

xy² 23. **Algebra** In a certain regular polygon, each exterior angle is half the size of each
(Inv 3) interior angle. How many sides does this polygon have?

*24. (**Archaeology**) An archaeologist is studying the ruins of a small pyramid as shown
(53) in the diagram. How can she find the original height of the pyramid? What was
the original height of the pyramid?

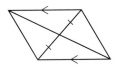

25. Triangle UVW is equilateral with sides measuring 2 inches, and X is the
(51) midpoint of $\overline{VW}$.
   a. Determine the exact height of $\triangle UVW$.
   b. Determine the exact area of $\triangle UVW$.

26. Can it be concluded that the quadrilateral at right is a parallelogram? Explain.
(61)

27. (**Driveways**) Carolina is going to pave her new driveway with concrete. The space
(59) that has been dug out for the driveway is 0.5 feet deep, 20 feet long and 12 feet
wide. What volume of concrete will it take to pave Carolina's driveway?

28. **Estimate** Estimate the area of the heart design at right.
(57)

29. For a triangle with vertices $L(0, 0)$, $M(6, 0)$, and $N(3, y)$ to
(45) be an equilateral triangle, what must be the value of y?

30. (**Jogging**) About how many seconds does it take Henrietta
(35) to jog one-third of the way around a circular track with a
500-meter radius, if she jogs at a speed of 2 meters per second?

Angles Interior to Circles

Warm Up

1. **Vocabulary** An angle whose vertex is on a circle and whose sides contain
   (47) chords of the circle is called a(n) _____.

2. Find the value of *x*.
   (47)

3. Find the value of *x*.
   (47)

New Concepts

A segment or arc is said to **subtend** an angle if the endpoints
of the segments or arc lie on the sides of the angle. In the
diagram, $\angle EDF$ is subtended by $\overgroup{EF}$ or $\overline{EF}$.

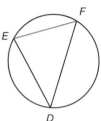

Inscribed angles are one type of subtended angle. Another
type of subtended angle is one formed by a tangent to the
circle and a chord of the circle.

Theorem 64-1

The measure of an angle formed by a tangent
and a chord is equal to half the measure of
the arc that subtends it.

$$m\angle ABC = \frac{1}{2}m\overgroup{BEC}$$

$$m\angle CBD = \frac{1}{2}m\overgroup{BC}$$

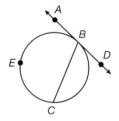

Hint

To review the
relationship between an
inscribed angle and the
arc that subtends it, refer
to Lesson 47. To review
tangents of circles, refer
to Lesson 58.

Example 1   Finding Angle Measures with Tangents and Chords

Find the indicated measure, given that $\overline{BC}$ and $\overline{SR}$ are tangents.

a. $m\angle ABC$

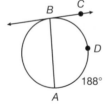

b. $m\overgroup{PR}$

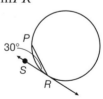

SOLUTION

In the first example, $\angle ABC$ is subtended by $\overgroup{ADB}$, so its measure will be
half the measure of $\overgroup{ADB}$. Since $\overgroup{ADB}$ measures 188°, $\angle ABC$ measures 94°.

In the second example, $\overgroup{PR}$ subtends $\angle PRS$, so $\angle PRS$ is half the measure
of $\overgroup{PR}$. Since the measure of $\angle PRS$ is 30°, $\overgroup{PR}$ measures twice that, or 60°.

Online Connection
www.SaxonMathResources.com

Theorem 64-2

The measure of an angle formed by two chords intersecting in a circle is equal to half the sum of the intersected arcs.

$$m\angle 1 = \frac{1}{2}(m\widehat{AD} + m\widehat{BC})$$

$$m\angle 2 = \frac{1}{2}(m\widehat{AB} + m\widehat{DC})$$

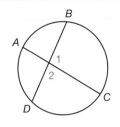

Math Reasoning

Write As one of the intersecting chords of a circle gets smaller and smaller, how do Theorem 64-2 and 64-1 become similar?

Example 2 Proving Theorem 64-2

Given: $\overline{AD}$ and $\overline{BC}$ intersect at E.

Prove: $m\angle 1 = \frac{1}{2}(m\widehat{AB} + m\widehat{CD})$

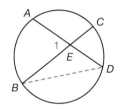

SOLUTION

Statements	Reasons
1. $\overline{AD}$ and $\overline{BC}$ intersect at E	1. Given
2. Draw $\overline{BD}$	2. Two points determine a line
3. $m\angle 1 = m\angle EDB + m\angle EBD$	3. Exterior Angle Theorem
4. $m\angle EDB = \frac{1}{2}m\widehat{AB},$ $m\angle EBD = \frac{1}{2}m\widehat{CD}$	4. Inscribed Angle Theorem
5. $m\angle 1 = \frac{1}{2}m\widehat{AB} + \frac{1}{2}m\widehat{CD}$	5. Substitution Property of Equality
6. $m\angle 1 = \frac{1}{2}(m\widehat{AB} + m\widehat{CD})$	6. Distributive Property

Example 3 Finding Angle Measures of the Intersection of Two Chords

Find x.

SOLUTION

Theorem 64-2 says that the value of x will be equal to half the sum of the two arcs that subtend it. Apply the formula from 64-2.

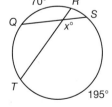

$$x = \frac{1}{2}(m\widehat{QR} + m\widehat{ST})$$

$$x = \frac{1}{2}(70° + 195°)$$

$$x = 132.5°$$

Example 4 Application: Tiling

Albert is laying tile in his kitchen in a circular pattern as shown. He knows the $\mathrm{m}\overset{\frown}{AB} = 50°$ and $\mathrm{m}\overset{\frown}{CD} = 86°$. He wants to know the measure of angle 1 so he can cut the tile accordingly.

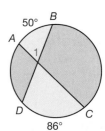

SOLUTION

$$\mathrm{m}\angle 1 = \frac{1}{2}(\mathrm{m}\overset{\frown}{AB} + \mathrm{m}\overset{\frown}{CD})$$

$$= \frac{1}{2}(50° + 86°)$$

$$= 68°$$

So, $\mathrm{m}\angle 1 = 68°$.

Lesson Practice

a. Find the measure of angle x in the figure. Line m is tangent to the circle.
(Ex 1)

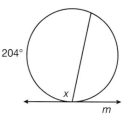

b. Find the measure of $\overset{\frown}{MNO}$ in the figure. Line n is tangent to the circle.
(Ex 1)

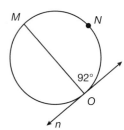

c. Prove Theorem 64-1.
(Ex 2)
Given: Tangent $\overleftrightarrow{BC}$ and secant $\overrightarrow{BA}$.
Prove: $\mathrm{m}\angle ABC = \frac{1}{2}\mathrm{m}\overset{\frown}{AB}$

Hint: There are two cases you must prove: one where $\overline{AB}$ is a diameter and one where $\overline{AB}$ is not a diameter.

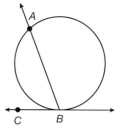

d. Find the measure of angle x.
(Ex 3)

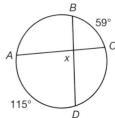

e. An artist is drawing a design for a company logo that has a capital "R" inside a large circle as shown. She first draws a baseline at the top of the R. The R is supposed to be at a 60° angle in relation to the baseline. What is the measure of the arc m, which extends leftward from the top of the R?
(Ex 4)

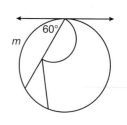

1. **Predict** In this set of pentominoes, there are 12 non-congruent shapes
(Inv 6) made from 5 unit squares each.

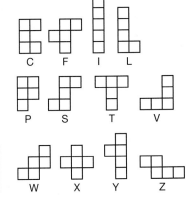

   a. A pentomino *solution* is any way of fitting the 12 pentominoes
together exactly in a rectangle. What must the area of the
solution rectangle be?

   b. What are the geometric probabilities for each pentomino that a
square chosen at random from a solution will be in that
pentomino?

   c. In this *partial solution*, some of the pentominoes are fit
together in a smaller rectangle. What is the geometric
probability in this rectangle of the V-pentomino? the
T-pentomino?

2. **a.** **Generalize** What is the fewest number of sides a polygon can have?
(49) **b.** What is the fewest number of faces a polyhedron can have?

3. (**Household**) A child spilled a glass of grape juice on a square white pillow, as
(57) shown. Approximately what percentage of the pillow is not stained?

4. **Multiple Choice** Which angles would be supplementary in parallelogram *KLMN*?
(34)   **A** $\angle K$ and $\angle M$ **B** $\angle L$ and $\angle N$

      C $\angle L$ and $\angle M$ **D** Any nonconsecutive angle pair

5. **Justify** Name the angles that are congruent in the triangles shown.
(46) If these triangles are similar, state why.

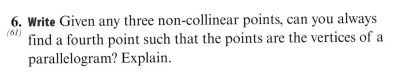

6. **Write** Given any three non-collinear points, can you always
(61) find a fourth point such that the points are the vertices of a
parallelogram? Explain.

7. **Multi-Step** Find *EF* in the compound figure. Round to the
(60) nearest tenth.

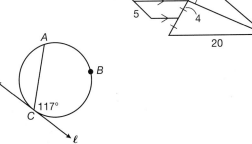

8. Find the geometric mean of $4\sqrt{3}$ and $5\sqrt{2}$.
(50)

* 9. Find the measure of $\overset{\frown}{ABC}$ at right. Line ℓ is
(64) tangent to the circle.

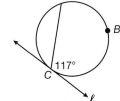

*10. Add the vectors $\langle 4, 8 \rangle$ and $\langle -4, -8 \rangle$. What
(63) kinds of vectors are these?

*11. (**Industrial Mechanics**) In order to allow a series of metal cylinders with 2-inch
(43) diameters to lay flat without rolling, a machine slices off the bottom of
each. If the machine cuts the cylinders so that the flat side is 1 inch across,
how far down from the center of the cylinder should the cut be made?

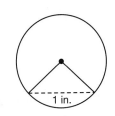

12. Draw two triangles that are similar by Side-Angle-Side Theorem.
(46)

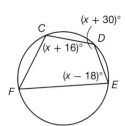

13. **Algebra** Find the measure of $\angle C$ and $\angle F$ in the figure.
(47)

*14. (**Design**) A carpenter wants to make a coffee table in the shape of an
(56) equilateral triangle. She doesn't have a way to measure 60°, but she can
measure a right angle and the length. If she wants the table to have a
perimeter of 9 feet, how should she design the table?

*15. Find the measure of angle x in the circle below.
(64)

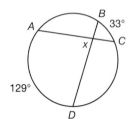

16. **Error Analysis** Jessie used the properties of a 45°-45°-90° right
(53) triangle to find the length of a hypotenuse, while her friend
Matthew used the Pythagorean Theorem to find the same
length. Jessie found the answer to be $34\sqrt{2}$ miles, and Matthew
found the answer to be about 48 miles. Who is correct?

17. Find the surface area of the cylinder shown, to the nearest tenth.
(62)

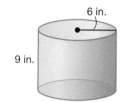

18. **Algebra** Two lines are perpendicular. One line has an equation of
(37) $2y = px + 7$ and the other line has an equation of $y = \frac{q}{4}x - 4$.
Find a possible set of values for p and q.

19. **Multiple Choice** Which of these nets could form a regular polyhedron?
(Inv 5)

A  **B**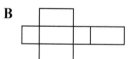

C 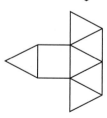   **D**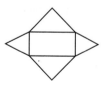

20. The ratio of the perimeters of two similar figures is 3:1. If the larger figure has
(44) a 6-unit side length, what is the length of the corresponding side on the second
figure?

*21. (**Stained Glass**) An artist wants to use stained glass to cover the area in the
(40) rectangle and semicircle at right. If the glass he wants to use comes in
8-by-10-inch panes, what is the minimum number of panes he will need
to buy to complete his project?

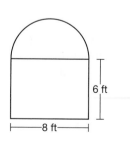

22. Find the surface area, to the nearest hundredth, of a cylinder that is
(62) 27 centimeters tall and has a 12-centimeter diameter.

23. Segment *DE* is a midsegment of △*ABC*. Find the values of *x* and *y*.
(55)

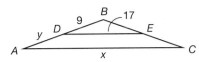

*24. Find the magnitude of vector $\langle 1, -2 \rangle$. Round your answer to
(63) the nearest tenth.

25. **Multi-Step** Find the centroid of △*ABC* with vertices $A(-2, 4)$, $B(1, -6)$, and
(32) $C(4, -4)$.

*26. Find the measure of angle *x* in the figure. Line ℓ is tangent to the circle.
(64)

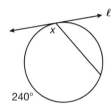

27. **Generalize** If a prism is cut parallel to its base to form two prisms, is the sum of the
(59) two prisms' volumes less than, greater than, or equal to that of the original prism?
Is the sum of the two prisms' lateral areas less than, greater than, or equal to that
of the original prism? Is the sum of the two prisms' total surface areas less than,
greater than, or equal to that of the original prism?

*28. Find the measure of angle *x* in the circle at right.
(64)

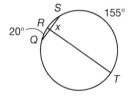

29. **Biology** The estimated number of species on Earth
(16) has been declining in recent decades. One of many
models for this is the equation $y = 5 - 0.025x$,
where *x* is the number of years after 1980, and
y is the number of species, in millions.
  a. How many species were there in 1980?
  b. Predict the number of species in the year 2100.
  c. How many species are expected to become extinct in a 10-year period,
  according to this model?

*30. Write $\overrightarrow{f}$ in component form.
(63)

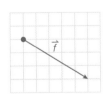

Distinguishing Types of Parallelograms

1. **Vocabulary** A(n) _____ is any four-sided polygon.
   (19)

2. Is the following statement always true, sometimes true, or never true?
   (52)
   A rhombus has two obtuse angles.

3. **Multiple Choice** Which of the following does not prove that a quadrilateral is
   (61) a parallelogram?
   A Both pairs of opposite angles are congruent.
   B Both pairs of adjacent sides are congruent.
   C One pair of opposite sides is both parallel and congruent.
   D The diagonals bisect each other.

New Concepts

Lesson 61 presented several methods for determining if a quadrilateral is a parallelogram. The properties presented in this lesson make it possible to determine if a parallelogram is a rectangle, square, or rhombus.

Hint

Remember that a square has the properties of both a rectangle and a rhombus. If you can use two of the properties in this chapter to show that a parallelogram is both a rectangle and a rhombus, then the parallelogram must be a square.

Properties of Parallelograms
If an angle in a parallelogram is a right angle then the parallelogram is a rectangle.   Since $\angle B$ is a right angle, $ABCD$ is a rectangle. 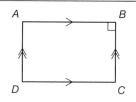

Properties of Parallelograms
If consecutive sides of a parallelogram are congruent, then the parallelogram is a rhombus.   Since $\overline{WZ} \cong \overline{ZY}$, $WXYZ$ is a rhombus. 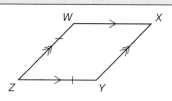

Example 1   Proving Parallelograms Are Rhombuses

Is this parallelogram a rhombus if $x = 11$?

SOLUTION

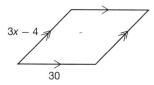

To be a rhombus, two consecutive sides must be congruent. Substitute for x in the expression for the length of the side.

$$3x - 4 = 30$$
$$3(11) - 4 = 29$$

Since this side is not congruent to the side that measures 30 units, the quadrilateral is not a rhombus.

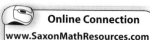
Online Connection
www.SaxonMathResources.com

Properties of Parallelograms

If the diagonals of a parallelogram are congruent then it is a rectangle.

Since $\overline{AC} \cong \overline{BD}$, $ABCD$ is a rectangle.

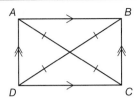

Example 2 Proving Parallelograms are Rectangles

Is parallelogram $HIJK$ a rectangle?

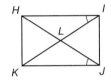

SOLUTION

Since $\angle HLI$ and $\angle KLJ$ are vertical angles, they are congruent. Opposite sides in a parallelogram are congruent, so $\overline{HI} \cong \overline{KJ}$. By Angle-Angle-Side Triangle Congruence, $\triangle HLI \cong \triangle KLJ$. By CPCTC and the definition of congruent segments, $LI = LJ$ and $LH = LK$. By the Addition Property of Equality $LI + LK = LJ + LK$, and by substitution, $LI + LK = LJ + LH$. Therefore, the two diagonals are congruent and the parallelogram is a rectangle.

Properties of Parallelograms

If the diagonals of a parallelogram are perpendicular then it is a rhombus.

Since $\overline{WY}$ is a perpendicular to $\overline{ZX}$, $WXYZ$ is a rhombus.

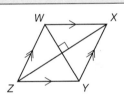

Example 3 Proving Parallelograms are Rhombuses

Is parallelogram $KLMN$ a rhombus?

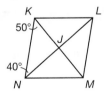

SOLUTION

Use the Triangle Angle Sum Theorem in $\triangle KJN$ to determine the angle measure of $\angle KJN$.

$$50° + 40° + m\angle KJN = 180°$$
$$m\angle KJN = 90°$$

Since they form a right angle, $\overline{KM}$ and $\overline{NL}$ are perpendicular, which means $KLMN$ is a rhombus.

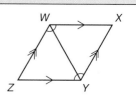

Properties of Parallelograms

If a diagonal in a parallelogram bisects opposite angles, then it is a rhombus.

Since $\angle XWY \cong \angle ZWY$ and $\angle XYW \cong \angle ZYW$, $WXYZ$ is a rhombus.

Math Reasoning

Analyze If the diagonal of a quadrilateral bisects only one pair of opposite angles, is the quadrilateral always a rhombus? Draw a counter example.

Example 4 **Proving Parallelograms are Rhombuses**

Is parallelogram $PQRS$ a rhombus?

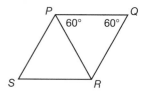

SOLUTION

From the diagram, $\triangle PQR$ is an equilateral triangle, with m$\angle PRQ = 60°$. Since $PQRS$ is a parallelogram, the Alternate Interior Angles Theorem can be used to show that $\angle QPR \cong \angle PRS$ and $\angle PRQ \cong \angle RPS$. Therefore, $\overline{PR}$ bisects both $\angle P$ and $\angle R$, and $PQRS$ is a rhombus.

Example 5 **Application: Signs**

A sign maker is commissioned to make a rectangular sign. The sign needs to be a perfect rectangle. Given the measurements shown in the diagram, is the sign a rectangle? How do you know?

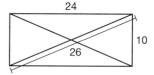

SOLUTION

The length of one diagonal is given. The length of the other one can be determined using the Pythagorean Theorem.

$$a^2 + b^2 = c^2 \qquad \text{Pythagorean Theorem}$$
$$10^2 + 24^2 = c^2 \qquad \text{Subsitute.}$$
$$c = 26 \qquad \text{Solve.}$$

Since the lengths of the two diagonals are the same, they are congruent and the sign is a perfect rectangle.

Lesson Practice

a. Is this parallelogram a rectangle?
(Ex 1)

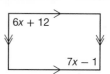

b. Is this parallelogram a rhombus?
(Ex 1)

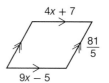

c. Is this parallelogram a rectangle?
(Ex 2)

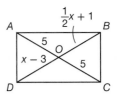

d. Is this parallelogram a rhombus?
(Ex 3)

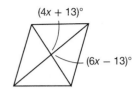

e. Is this parallelogram a rhombus?
(Ex 4)

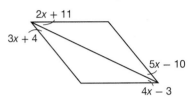

f. A sign in the shape of a parallelogram has diagonals that create an equilateral triangle as shown. Is the sign a perfect rectangle? Explain how you know.
(Ex 5)

Practice Distributed and Integrated

xy^2 *** 1. Algebra** Find the value of x that would make this parallelogram a rhombus.
(65)

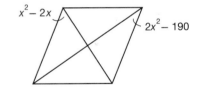

2. A round pie with a 10-inch diameter is 3 inches thick, and it has been cut into eight pieces. If three pieces have already been taken, what is the volume of the pie remaining to the nearest cubic inch?
(35, 62)

* 3. Error Analysis Find and correct any errors in the flowchart proof.
(31) **Given:** $\angle 1$ and $\angle 2$ form a linear pair; $\angle 1 \cong \angle 3$
Prove: $\angle 2$ and $\angle 3$ are supplementary.

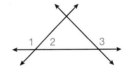

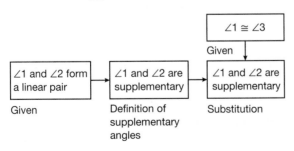

xy^2 **4. Algebra** This set of lines forms a triangle. Find its area and perimeter.
(57) $y = \frac{1}{2}x + 1$, $y = x - 1$, $x = 0$

5. Find the volume of the right prism shown.
(59)

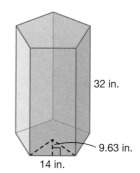

32 in.

9.63 in.

14 in.

6. (**Sailing**) The speed at which a sailboat moves over the water depends
(63) on both the current of the water and the wind. Suppose the wind
is blowing at 12 miles per hour and the current is flowing at 3 miles
per hour in the opposite direction. At what speed is the boat
traveling relative to the shore?

* 7. Find the perimeter of the rhombus shown.
(52)

* 8. Multi-Step The vertices of quadrilateral *LMNP* are
(65) *L*(2, 2), *M*(−4, 1), *N*(−3, −5), and *P*(3, −4). Classify
this quadrilateral.

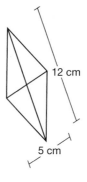

12 cm

5 cm

9. (**Structures**) A crossbeam on a barn gate goes
(29) diagonally across a door 3 feet by 50 inches.
How long is the diagonal line in feet? Round to the
nearest hundredth.

*10. Find the measure of ∠*AXD* in the circle at right.
(64)

11. Multiple Choice A circle with its center at (2, 4) passes
(57) through point (−1, −1). What is the area of the circle in
simplified radical form?

   A 34π **B** $\pi\sqrt{34}$

   C $2\pi\sqrt{34}$ **D** 68π

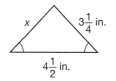

12. Find the missing side length and determine the
(51) perimeter of this triangle.

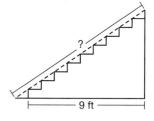

x $3\frac{1}{4}$ in.

$4\frac{1}{2}$ in.

13. (**Construction**) A construction worker is restoring an old staircase, and
(46) needs to find its length. She measures along the floor, from the bottom
step to the wall, and finds that it is 9 feet long. She already knows that
each step measures 1 foot deep and 8 inches high, and that each forms
a right angle with the step above it. She conjectures that the triangular
space formed by each step is similar to the triangle formed by the
staircase and the wall. If she is correct, how long is the staircase,
to the nearest hundredth?

14. List the angles of △*STU* in order from least to greatest measure.
(39)

15. Determine whether {9, 40, 41} can be the set of measures of
(33) the sides of a triangle. If it is a triangle, determine whether it is
an acute, obtuse, or a right triangle.

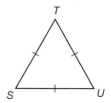

16. **Multiple Choice** In △PQR, m∠Q = 35° and m∠R = 113°. The exterior angle at P measures:
(18)

  A 32°   **B** 148°

  C 55°   **D** 78°

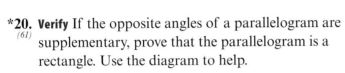

17. (**Painting**) A farmer's cylindrical grain silo needs to be painted. If the silo is 75 feet tall and has a diameter of 25 feet, approximately how many square feet need to be painted?
(62)
Hint: The bottom of the silo does not need to be painted.

18. What is the resultant vector of $\vec{e}$ and $\vec{f}$ in component form?
(63)

19. **Multi-Step** The center of rhombus *DEFG* is *P*. If *DP* = 12 inches, and *EP* = 16 inches, find *DE*.
(52)

*20. **Verify** If the opposite angles of a parallelogram are supplementary, prove that the parallelogram is a rectangle. Use the diagram to help.
(61)

21. (**Shadows**) The flagpole in front of a school casts a 16-foot shadow. At the same time, the student who is 5.6 feet tall casts a 3-foot shadow. How tall is the flagpole?
(41)

*22. Find the value of *x* in the figure.
(64)

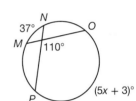

23. **Analyze** What is the radius of a cylinder with a 3-millimeter height that has the same volume as a 6-millimeter cube? Round your answer to the nearest hundredth.
(59)

24. **Justify** Determine whether lines $\overleftrightarrow{GH}$ and $\overleftrightarrow{KL}$ are parallel and explain how you know.
(60)

25. Parallelogram *EFGH* has vertices *E*(0, 0), *F*(4, 4), *G*(x, y), and *H*(10, 0). Find a possible location of (x, y).
(45)

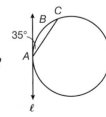

*26. (**Design**) A sweater uses rhombuses to make its design, as shown. Find ∠1.
(65)

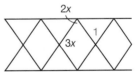

27. **Write** What is unique about the circumcenter of an isosceles right triangle?
(53)

*28. Find the measure of $\overset{\frown}{AC}$ in the figure. Line ℓ is tangent to the circle.
(64)

29. **Algebra** If a chord perpendicular to a radius cuts it in two pieces that measure 6 inches and 4 inches, what are the two possible lengths of the chord?
(43)

*30. **Write** In the diagram, *GJ* is a radius of ⊙*G*. Is $\overleftrightarrow{HJ}$ tangent to ⊙*G*? Explain how you know.
(58)

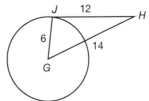

Finding Perimeters and Areas of Regular Polygons

Warm Up

1. **Vocabulary** The height of a triangle is its _____.
(32) (*median, altitude, perpendicular bisector*)

2. Find the area of this triangle.
(8)

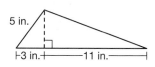

3. **Multiple Choice** Which of the following
(15) polygons is a regular polygon?

A   B

C   D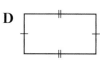

4. Find the area and perimeter of this composite figure.
(40)

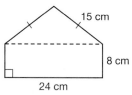

New Concepts

Each regular polygon has a point within it that is equidistant from all vertices. This point is called the **center of a regular polygon**.

The **central angle of a regular polygon** is the angle whose vertex is the center of a regular polygon and whose sides pass through consecutive vertices.

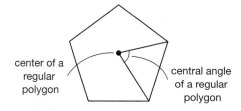

The perpendicular distance from the center of a regular polygon to a side is the **apothem**.

You can use the formula $P = ns$ to find the perimeter of a regular polygon. In the formula, P represents the perimeter, n represents the number of sides, and s represents the side length.

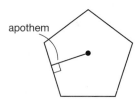

Math Reasoning

Generalize In Example 1, could you use the formula $P = ns$ if the polygons were not regular polygons? Explain.

Example 1 Finding Perimeters of Regular Polygons

a. Find the perimeter of the polygon.

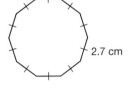

SOLUTION

$P = ns$ Formula for perimeter of a regular polygon

$P = (6)(3)$ Substitute.

$P = 18$ ft

b. Find the perimeter of the polygon.

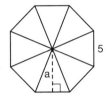

SOLUTION

$P = ns$ Formula for perimeter of a regular polygon

$P = 10 \cdot 2.7$ Substitute.

$P = 27$ cm

Math Reasoning

Formulate Suppose you divide a regular n-sided polygon into n congruent isosceles triangles. If you know the apothem length and the side length of the polygon, how could you determine the two lengths of the congruent side of the isosceles triangles?

You can find the area, A, of a regular polygon using only the apothem and perimeter. Consider an n-sided regular polygon with a side length of s. Divide the polygon into n triangles so the vertices of each triangle are the center of the polygon and the endpoints of a side as shown. By definition, the height of each triangle is the apothem, a. The base of each triangle has a length of s. So, the area of each triangle is $\frac{1}{2}as$. The total area of the polygon is n times the area of one triangle, or $A = \frac{1}{2}nas$. The formula for the perimeter of a regular polygon is $P = ns$. By substitution, the area of a regular polygon is $A = \frac{1}{2}aP$.

Area Formula for Regular Polygons
The area, A, of a regular polygon is half the apothem length a and the perimeter P of the regular polygon. $$A = \frac{1}{2}aP$$

Example 2 Using the Area Formula

Find the area of a regular octagon with an apothem about 18 inches.

SOLUTION

$P = ns$ Formula for perimeter of a regular polygon

$P = (8)(15)$ Substitute.

$P = 120$ Simplify.

$A = \frac{1}{2}aP$ Area formula for regular polygons

$A = \frac{1}{2}(18)(120)$ Substitute.

$A = 1080$ Simplify.

The area is 1080 square inches.

Example 3 Finding the Area of a Regular Hexagon

Use the apothem and perimeter to find the area of this regular hexagon.

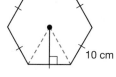
10 cm

SOLUTION

First, find the apothem length of the regular hexagon. Draw an isosceles triangle whose vertices are the center of the hexagon and the endpoints of a side. The triangle contains a central angle whose measure is 60°. The apothem bisects the central angle and the side, forming a 30°-60°-90° triangle. The shorter leg of the triangle is 5 centimeters long. Therefore, the apothem measures $5\sqrt{3}$ centimeters.

Next, find the perimeter of the hexagon.

$P = ns$	Formula for perimeter of a regular polygon
$P = (6)(10)$	Substitute.
$P = 60$	Simplify.

Finally, find the area of the hexagon.

$A = \frac{1}{2}aP$	Area formula for regular polygons
$A = \frac{1}{2}(5\sqrt{3})(60)$	Substitute.
$A = 150\sqrt{3}$	Simplify.

The area is $150\sqrt{3}$ centimeters squared.

Example 4 Finding the Area of an Equilateral Triangle

Find the area of an equilateral triangle with 18-inch sides.

SOLUTION

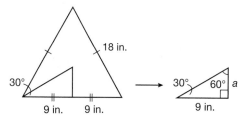

Use your knowledge of 30°-60°-90° triangles to find the apothem, a, and then use your result to find the area.

$$\frac{a}{9} = \frac{1}{\sqrt{3}}$$

$$a = \frac{9}{\sqrt{3}}$$

$$a = 3\sqrt{3}$$

$$A = \frac{1}{2}aP$$

$$A = \frac{1}{2}(3\sqrt{3})(54)$$

$$A = 81\sqrt{3}$$

Hint

Remember that the length of the longer leg in a 30°-60°-90° triangle is the length of the shorter leg times $\sqrt{3}$.

Math Reasoning

Generalize What kinds of triangles have apothems? Explain.

Example 5 Application: Land Survey

A plot of land is in the shape of a regular octagon with 10-mile side lengths and apothem of about 12 miles. The plot needs be divided into eight equal parcels of land. What will the area of land be in each parcel?

SOLUTION

First, find the area of the plot of land. Because the plot is in the shape of an octagon where each side is 10 miles long, its perimeter is 80 miles.

$A = \frac{1}{2}aP$ Area formula for regular polygons

$A = \frac{1}{2}(12)(80)$ Substitute.

$A = 480$ Simplify.

The area of the octagonal plot is about 480 mi².

Next, divide the total area by 8 to find the area in each equal parcel.

$\frac{480}{8} = 60$

Each of the 8 parcels has an area of about 60 mi².

Lesson Practice

a. Find the perimeter of this octagon.
(Ex 1)

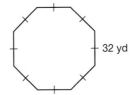

32 yd

b. Use the area formula for regular polygons to find the area of this pentagon.
(Ex 2)

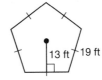

13 ft   19 ft

c. Find the area of this hexagon.
(Ex 3)

24 ft

a

d. Find the area of this equilateral triangle.
(Ex 4)

12 m

a

e. The shape of a playground is a regular hexagon where each side length is 78 feet long. The playground is to be resurfaced with a nonslip rubber material. What is the total area that must be surfaced?
(Ex 5)

1. In the diagram m∠1 ≠ m∠2. Prove $\overline{XA}$ is not an altitude of △XYZ using as
(48) indirect proof.

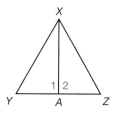

* 2. ⎡ **Rock Formations** ⎤ A rock formation in Ireland is called the Giant's Causeway. It
(66) is made up of about 40,000 columns of rock that formed as lava cooled after a
volcanic eruption. Most of the columns are hexagonal with average side lengths
of 14 inches and an apothem that is $7\sqrt{3}$ inches long. What is the approximate
area covered by the top surface of the 40,000 stone columns?

* 3. Is the statement sometimes, always, or never true?
(65)
    A rhombus is a rectangle.

4. Find the values of x and y. Give your answers in simplified radical form.
(56)

5. Analyze What could you conclude about the bisectors of two consecutive
(61) angles of a rhombus?

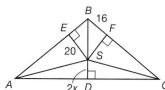

* 6. ⎡ **Archery** ⎤ Assuming that an amateur archer's arrow is equally likely to strike
(Inv 6) anywhere on the target *and* has a 50% chance of missing the target altogether,
what is the probability of the archer hitting a bull's eye on this target?

7. How many edges does a polyhedron with 10 vertices and 10 faces have?
(49)

8. Algebra Find the value of x that makes S the incenter of the triangle shown.
(38)

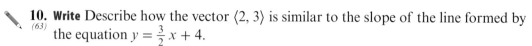

9. ⎡ **Space Exploration** ⎤ A probe is about to use the atmosphere of Mars to aerobrake
(58) (reduce its velocity). The probe's trajectory is an almost straight line that just
grazes the Martian atmosphere. The probe is currently 9600 miles from the center
of Mars, and the radius of Mars is approximately 5460 miles. How far is the probe
from its aerobrake maneuver? Round your answer to the nearest ten miles.

10. Write Describe how the vector ⟨2, 3⟩ is similar to the slope of the line formed by
(63) the equation $y = \frac{3}{2}x + 4$.

11. Find the value of x in the circle at right.
(64)

12. ⎡ **Orbits** ⎤ The orbit of a satellite is decaying, and it will reenter Earth's atmosphere
(Inv 6) at a random time during its next orbit. The satellite's orbit is 26,000 miles long,
17,000 miles of which is over ocean. Find the probability that fragments of the
satellite will land in the ocean.

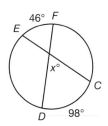

13. Segment *JK* is a midsegment of △*FGH*. Find the values of *x* and the lengths
(55) of $\overline{JK}$ and $\overline{FH}$.

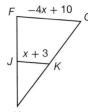

14. If an apple pie is 8 inches in diameter and a pizza is 14 inches in diameter,
(44) what is the ratio of the pie's diameter to the pizza's diameter?

15. Justify Which segment in the figure is greater, *LM* or *PN*?
(Inv 4) Explain how you know. Assume the circles are concentric.

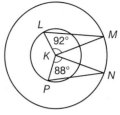

*16. Find the perimeter of a regular heptagon with sides that are
(66) each 77 feet long.

17. Algebra Right triangle *A* has leg lengths of *u* and *v*. Right
(36) triangle *B* has leg lengths of $2v - 1$ and $4 - u$. Given that triangles
A and *B* are congruent, what are the possible values of *u* and *v*?

Use the diagram to answer the next two questions.

*18. If $AB = AD$ and $DB > AC$, what kind of parallelogram is the figure?
(65) List all possibilities.

*19. If m∠*ADC* = 90° and ∠*APB* is an acute angle, what kind of parallelogram
(65) is the figure?

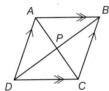

*20. Multiple Choice Which formula is equivalent to the formula $A = \frac{1}{2}aP$?
(66)

A $a = \dfrac{A}{2P}$ **B** $P = \dfrac{2A}{a}$

C $P = \dfrac{A}{2a}$ **D** $a = \dfrac{2A}{P}$

*21. Error Analysis Find and correct any errors in the flowchart proof below.
(31) **Given:** $AB = CD$, $BC = DE$
Prove: *C* is the midpoint of $\overline{AE}$.

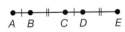

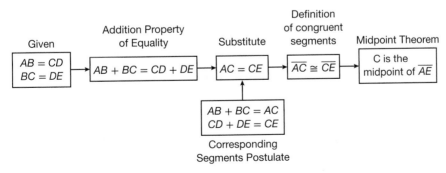

22. Multiple Choice Which of the following is *not always* true about parallelogram
(34) *LMNP*?

A m∠*L* = m∠*N*   **B** m∠*L* = m∠*P*

C m∠*L* + m∠*P* = 180°   **D** m∠*M* = m∠*P*

23. Find x in the figure. Line ℓ is tangent to the circle.
(64)

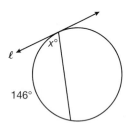

24. **Algebra** In $\triangle STU$, m$\angle TSU = 32°$, $ST = (2y - 4)$, and m$\angle STU = (5x + 12)°$.
(30) In $\triangle DEF$, m$\angle EDF = 32°$, $DE = (x + y)$, and m$\angle DEF = (2x + 24)°$. If
$\triangle STU \cong \triangle DEF$, determine the values of x and y.

*25. **Error Analysis** The center of the circle is A. Josie says that she has determined
(58) that if $\overleftrightarrow{FY}$ and $\overleftrightarrow{XH}$ are tangent lines at Y and X, respectively, then they must
be parallel, because $\angle AYF$ and $\angle AXH$ would both be right angles. Why is her
conjecture false?

26. **Multi-Step** Find the area of the shaded region in the figure.
(34, 35)

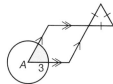

27. (Packaging) A manufacturer distributes a cylindrical tin that is 12 inches across
(62) and 6 inches tall. If the manufacturer fills each tin with mints that are assumed
to be $\frac{1}{8}$ cubic inches each, to the nearest hundred, approximately how many
mints are in the tin?

28. Find the distance from $(-1, 6)$ to the line $y = 2x - 7$.
(42)

29. **Justify** Determine whether $\triangle JKL$ is a right triangle, given that $\overleftrightarrow{DE}$ and $\overleftrightarrow{JK}$ are
(60) parallel. Explain how you know.

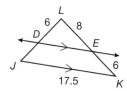

30. What are the perimeter and area of a regular hexagon with apothem 34 feet?
(66) Use exact values.

Regular Polygons

Construction Lab 9 *(Use with Lesson 66)*

In Lesson 66 you learned how to find the perimeter and area of regular polygons. In this lab you will learn how to construct two regular polygons: a hexagon and a pentagon. First, we will construct a regular hexagon.

Analyze

How can you use the fact that each side of the hexagon is as long as the radius of the circle to show that the hexagon is regular?

1. To construct a regular hexagon, begin with a circle and label the center *P*. Set a compass to the radius of the circle.

2. Choose any point *A* on the circle, and with your compass setting from step 1, mark off an arc centered at *A* that intersects the circle. Label this point *B*.

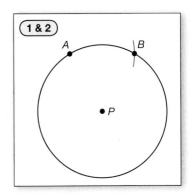

3. Starting from *B*, repeat this process to find and label points *C*, *D*, *E*, and *F*.

4. Draw $\overline{AB}$, $\overline{BC}$, $\overline{CD}$, $\overline{DE}$, $\overline{EF}$, and $\overline{FA}$. Figure *ABCDEF* is a regular hexagon.

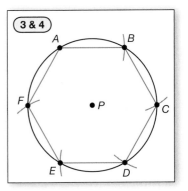

Next, we will construct a regular pentagon.

1. To construct a regular pentagon, begin with a circle *P* with diameter $\overline{AB}$. (Any line segment long enough to intersect the circle twice and which passes through the center is a diameter.)

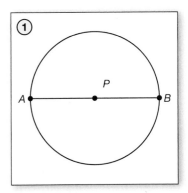

2. Using the method from Construction Lab 3, construct the perpendicular bisector of $\overline{AB}$. Label either point where the bisector intersects the circle as *J*.

3. Construct the midpoint of the radius $\overline{PA}$ and label it *X*.

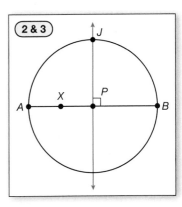

4. Set the compass to the length of $\overline{JX}$, and, starting at J, mark off successive arcs intersecting circle P. Label these points of intersection K, L, M, and N.

5. Draw $\overline{JK}$, $\overline{KL}$, $\overline{LM}$, $\overline{MN}$, and $\overline{JN}$. Figure $JKLMN$ is a regular pentagon.

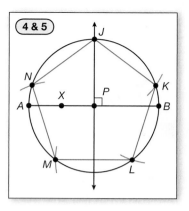

Lab Practice

Construct a regular dodecagon using the regular hexagon you constructed. *Hint: Use a bisector construction method from Construction Lab 3.*

Introduction to Transformations

1. **Vocabulary** If two triangles are the same shape and size, then
(25) they are _____.

2. Line *l* is the perpendicular bisector of $\overline{AA'}$. It intersects $\overline{AA'}$ at *P*. How
(11) are the distances *AP* and *A'P* related?

3. **Multiple Choice** A vector can be described by its
(63) **A** magnitude and direction   **B** length and magnitude
C *x*- and *y*-coordinates   **D** any of these

New Concepts   A change in position, size, or shape of a figure is called a **transformation**. Translations, reflections, and rotations are examples of a special class of transformation called isometries.

The original figure in a transformation is called the **preimage** and the shape that results from the transformation is called the **image**.

An isometry maps a figure to a congruent figure.
An **isometry** is a transformation that does not change the size or shape of a figure. That is, the image of an isometry is congruent to its preimage. This diagram shows an isometry with preimage $\triangle TUV$ and image $\triangle T'U'V'$. 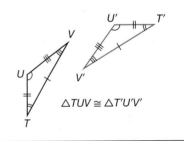

The small ′ marks next to *T*, *U*, and *V* are **primes**: a symbol used to label the image in a transformation.

An isometry is also called a **congruence transformation** or **rigid transformation**.

Hint

When performing a reflection, think of what the image would look like in a mirror, if the mirror were positioned exactly on the line of reflection.

A **translation** or **slide** is a type of transformation that shifts or slides every point of a figure the same distance in the same direction as shown with parallelogram *JKLM*.

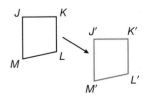

A **reflection** or **flip** is a transformation across a line (the line of reflection) such that the line is the perpendicular bisector of each segment joining each point and its image (If a point lies on the line of reflection, the point and its image will be the same.) In this diagram, the figure has been reflected across $\overleftrightarrow{AD}$. Each point of the preimage is the same distance from $\overleftrightarrow{AD}$ as its matching point on the reflected image.

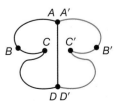

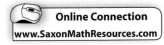
Online Connection
www.SaxonMathResources.com

A **rotation** or **turn** is a transformation about a point (the point or center of rotation) such that each point and its image are the same distance from that point, and angles formed by a point, its image, and the point of rotation (as the vertex) are congruent. In this diagram, *ABCDE* has been rotated clockwise about *E*. Notice that $EA = EA'$, $EB = EB'$, $EC = EC'$, and $ED = ED'$; notice also that $\angle AEA'$, $\angle BEB'$, $\angle CEC'$, and $\angle DED'$ are all congruent. Since E is the point of rotation, *E* and *E'* are the same point.

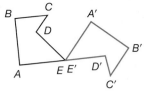

Example 1 | Identifying Transformations

a. Identify the type of transformation illustrated below.

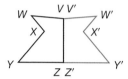

SOLUTION

The figure *VWXYZ* is reflected across $\overleftrightarrow{VZ}$. Reflecting the figure flips the figure across the line of reflection. Notice that each distance from a point of the preimage to its image, other than *V* and *Z*, which are on the line of reflection, is bisected by $\overleftrightarrow{VZ}$.

b. Identify the type of transformation illustrated below.

SOLUTION

Triangle *RST* is rotated about the fixed point *R*. Rotating the figure turns the figure around a fixed point. Notice that the triangle remains the same size and shape as before the rotation.

c. Identify the type of transformation illustrated below.

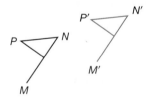

SOLUTION

The figure is translated up and to the right. In a translation the entire figure moves a specific distance in a specific direction.

Hint

Refer to Construction
Lab 3 for a reminder
on constructing
perpendicular bisectors.

Example 2 Performing Transformations

Perform the indicated transformations.

a. Rotate the figure about point L.

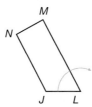

SOLUTION
To rotate the figure about point L, keep L fixed and turn each point
on a circular path around L as indicated.

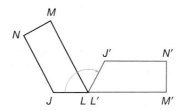

b. Translate the figure as indicated.

SOLUTION
To translate the figure, move each point of the preimage the distance
and direction as indicated.

c. Reflect the figure across $\overleftrightarrow{FG}$.

SOLUTION
To reflect the figure across $\overleftrightarrow{FG}$ move each point across the line of reflection
so that the point and its image are equidistant from the line of reflection.

Example 3   Application: Stained Glass Design

Often stained glass designers use vertical or horizontal symmetry to reduce the time it takes to design a project. Reflect this template across the vertical line $\overleftrightarrow{AB}$ to complete the design.

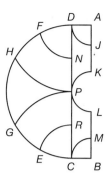

SOLUTION

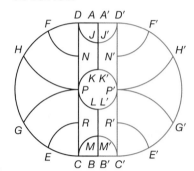

a. Identify the type of transformation which takes $\triangle XYZ$ to $\triangle X'Y'Z'$.
(Ex 1)

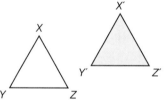

b. Reflect rectangle *DEFG* across $\overleftrightarrow{GF}$. Label the image.
(Ex 2)

c. Rotate $\triangle PQR$ clockwise about point Q, so that
(Ex 2) Q' and P' are collinear with $\overline{QR}$.

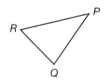

d. This simplified blueprint shows the first two floors of the front of a
(Ex 3) new civic hall. The third floor will be a translation of the second floor so it is directly above the 2nd floor. Complete the plan by performing the translation.

* 1. Find the area of this regular pentagon, to the nearest tenth.
(66)

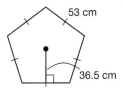

53 cm

36.5 cm

2. (**Meal Preparation**) Mr. Jones is making sandwiches, which he wants to
(53) cut diagonally. If he uses square bread that is six inches on a side,
how long a knife must he use to be able to cut each sandwich in
one cut? Round your answer to the nearest half-inch.

3. Find the value of x in the figure.
(64)

4. Find the distance between the line
(42) $y = \frac{-4}{3}x + 3$ and (4, 6).

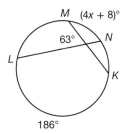

M $(4x + 8)°$

$63°$

N

L

K

$186°$

* 5. Find the perimeter and area of a
(66) regular octagon with 30-foot-long sides
and an apothem that is 36.2 feet long.

* 6. What type of transformation takes square $STUV$ to square $S'T'U'V'$?
(67)

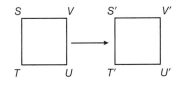

S V S' V'

T U T' U'

7. (**Flight**) A dragonfly is flying at a rate of 26 feet per second.
(63) The wind is blowing at 8 feet per second in the opposite direction.
How fast is the dragonfly traveling over the ground?

8. Determine whether the quadrilateral at right must be a square.
(65)

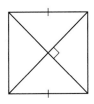

9. **Algebra** Find a counterexample to the following conjecture.
(14)

If the equation $ax^2 - b = 0$ has a rational solution, then b is a perfect square.

10. $\triangle ABC \sim \triangle DEF$. The ratio of their corresponding sides is 5:3.
(44) Given $AB = 3$, what is the length of $\overline{DE}$?

11. Determine m$\angle X$ and m$\angle Z$ in the triangle at right.
(51)

12. (**Painting**) Cecilia is painting the walls of two rooms that are
(59) the same size. The rooms are 14 feet wide, 9 feet long, and
8 feet high. How many square feet does she need to paint?

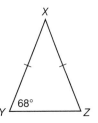

X

$68°$

Y Z

13. If two chords are congruent, and one is 5 inches from
(43) the center of the circle, then what is the distance from the
center to the other chord?

*14. **Multiple Choice** Identify the type of transformation that takes $\triangle EFG$ to
(67) $\triangle E'F'G'$.

  A reflection   **B** translation
  C rotation   **D** rigid transformation

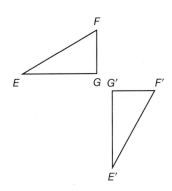

F

E G G' F'

E'

*15. **Analyze** Which of the following scenarios describes a random event?
(Inv 6) Explain why geometric probability cannot be used for the other.

  a: A tennis player hits a ball to the opponent's side of the court.
  b: An astronomer scans a region of the sky looking for meteors.

16. Formulate A trapezoid is formed by the midsegment of a triangle, as in the diagram shown. Find a formula for its area in terms of its height and its longer base, b.

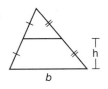

17. (Painting) A cylindrical water tower needs to be painted on all sides, including the base. The tower is 115 feet tall and it has a radius of 40 feet. To the nearest square foot, how many square feet are there to be painted?

18. Multiple Choice Which line or lines appear to be tangent to $\odot C$?

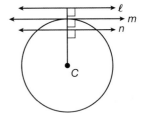

  A ℓ **B** m

  C m and n **D** none of these

19. Find the line perpendicular to $7x + 7y = 49$ that passes through the point $(4, 3)$. Write its equation in slope-intercept form.

20. Multi-Step Classify the polygon with vertices $C(3, -3)$, $D(-5, -3)$ and $E(3, 3)$. Find the perimeter.

21. Error Analysis Charlie found k in rhombus $QRST$ to be roughly 3.464. What error has he made? What is the actual value of k?

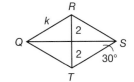

*22. Generalize A certain isometric transformation is performed twice and the resulting figure is in the same location as the original figure. What type or types of transformation could this be? What type or types of transformation could this *not* be?

*23. Reflect quadrilateral $PQRS$ across line n.

24. Use an indirect proof to prove that if no two angles in a triangle are congruent, then no two sides are congruent.

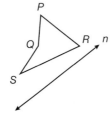

25. Formulate What expression for m$\angle L$ makes $JKLM$ a rhombus?

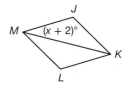

26. (Labels) A soup can is 10 centimeters tall and has a radius of 4 centimeters. What is the area of the label that will be placed on the cans, to the nearest tenth?

27. Multiple Choice Which choice is closest to the length of a 48°-arc of a circle that has a radius of 8.5?

  A 30.26   **B** 3.56

  C 7.12   **D** 60.53

28. Write the resultant vector of $\vec{a}$ and $\vec{b}$ in component form.

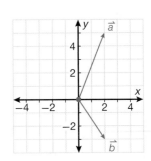

*29. To the nearest tenth, find the area and perimeter of an equilateral triangle with sides that are 32 feet long.

30. Generalize Is the centroid of a triangle always in the interior of the triangle? Can the orthocenter be outside the triangle?

Introduction to Trigonometric Ratios

Warm Up

1. **Vocabulary** A _____ is a comparison of two quantities by division.
   (41)

2. Find *a*.
   (56)

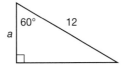

3. **Multiple Choice** Which expression correctly represents
   (53) the perimeter of this triangle?

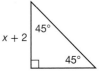

   A $3x + 6$ **B** $2x + 4 + \sqrt{2}(x + 2)$

   C $2x + 4 + \sqrt{2}$ **D** $(2 + \sqrt{2})x + 6$

New Concepts

Trigonometry is the study of the relationship between sides and angles of triangles. There are three basic ratios in trigonometry that can be used to find measures in right triangles.

The three ratios are the sine of an angle, the cosine of an angle, and the tangent of an angle. A **trigonometric ratio** is a ratio of two sides of a right triangle.

Reading Math

When used in equations, the sine, cosine, and tangent ratios are often abbreviated 'sin', 'cos', and 'tan'

Trigonometric Ratios
In a right triangle, the **sine of an angle** is the ratio of the length of the leg opposite the angle to the length of the hypotenuse.
$$\sin A = \frac{\text{Opposite}}{\text{Hypotenuse}}$$
In a right triangle, the **cosine of a triangle** is the ratio of the length of the leg adjacent to the angle to the length of the hypotenuse.
$$\cos A = \frac{\text{Adjacent}}{\text{Hypotenuse}}$$
In a right triangle, the **tangent of an angle** is the ratio of the length of the leg opposite the angle to the length of the leg adjacent to the angle.
$$\tan A = \frac{\text{Opposite}}{\text{Adjacent}}$$

Math Reasoning

Formulate What is the ratio of sin *A* to cos *A*?

For example, the sine of $\angle S$ in $\triangle STU$ is $\frac{a}{c}$. The sine of $\angle T$ is $\frac{b}{c}$.

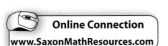
Online Connection
www.SaxonMathResources.com

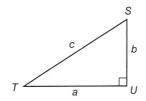

Example 1 Calculating Trigonometric Ratios

a. Give the sine, cosine, and tangent of $\angle G$.

SOLUTION
Find the hypotenuse of the triangle using the
Pythagorean Theorem. The hypotenuse is 5.

$$\sin G = \frac{3}{5}$$

$$\cos G = \frac{4}{5}$$

$$\tan G = \frac{3}{4}$$

A calculator can be used to evaluate the cosine, sine, and tangent of an angle.

Caution

Your calculator needs to
be in degree mode, not
in radian mode. If you
do not get the answers
shown in the solutions
of this example, your
calculator is probably in
radian mode.

Example 2 Calculating Trigonometric Ratios

Use a calculator to evaluate each expression. Round the answer to the
nearest hundredth.

a. $\cos 72°$

SOLUTION
$\cos 72° = 0.31$

b. $\sin 30°$

SOLUTION
$\sin 30° = 0.5$

c. $\tan 70°$

SOLUTION
$\tan 70° = 2.75$

Trigonometric ratios can be used to solve for unknown side lengths in right
triangles. An equation can be divided or multiplied by a trigonometric ratio,
just as it can with any real number.

Example 3 Solving for Side Lengths Using Trigonometry

Use the tangent ratio to find e to the nearest
hundredth.

SOLUTION

$$\tan 72° = \frac{e}{13} \qquad \text{Tangent function}$$

$$13 \tan 72° = e \qquad \text{Multiply both sides by 13.}$$

$$e \approx 40.01 \qquad \text{Simplify.}$$

Example 4 More Solving for Side Lengths

Use the sine ratio to find x to the nearest hundredth.

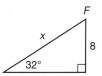

SOLUTION

$$\sin 32° = \frac{8}{x}$$ Sine function

$$x \sin 32° = 8$$ Multiply both sides by x.

$$x = \frac{8}{\sin 32°}$$ Divide both sides by sin 32°.

$$x \approx 15.10$$ Simplify.

Example 5 Application: Art

Artists who make stained glass windows use right triangles in their patterns. If an artist is making a stained glass window for a square window with sides that are 26 inches long, what is the value of x and y in the diagram? Give answers to the nearest hundredth.

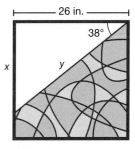

SOLUTION

Use trigonometric ratios to solve for x and y.

$$\tan 38° = \frac{\text{Opposite}}{\text{Adjacent}} \qquad \cos 38° = \frac{\text{Adjacent}}{\text{Hypotenuse}}$$

$$\tan 38° = \frac{x}{26} \qquad \cos 38° = \frac{26}{y}$$

$$x \approx 20.31 \qquad y \approx 33.00$$

The other leg of the right triangle measures approximately 20.31 inches, and the hypotenuse is approximately 33 inches.

Math Reasoning

Formulate How could the sine ratio have been used to find the length of the hypotenuse in Example 5?

Lesson Practice

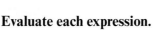

Use the figure to answer problems a and b.

a. What is the sine of $\angle T$?
(Ex 1)

b. What is the tangent of $\angle U$?
(Ex 2)

c. Find x to the nearest hundredth.
(Ex 3)

Evaluate each expression.

d. sin 30°
(Ex 2)

e. cos 90°
(Ex 2)

f. tan 45°
(Ex 2)

g. A playground has a slide that is at a 38° angle with the ground. If the slide is 16 feet long, what is the height?
(Ex 5)

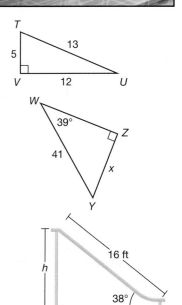

1. (Swimming) A swimmer wants to travel at a rate of 3 miles per hour while
(63) swimming in a river. The river flows at a rate of 2 miles per hour. If she goes with
the current, how fast does she need to swim? How fast does she need to swim if
she goes against the current?

2. **Algebra** Determine the value of x in the figure shown. Write your answer in
(33) simplified radical form.

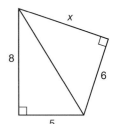

* 3. (Building) It is unsafe to lean a ladder at less than a 70° angle with the ground.
(68) If a ladder is 8 feet tall, at least how far should the ladder be from the wall, to
the nearest hundredth of a foot?

4. Find h if line l is tangent to circle C.
(58)

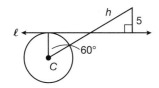

5. **Multiple Choice** $\triangle DEF$ has vertices $D(0, 0)$, $E(3, 3)$, and $F(6, 0)$. What type of
(45) triangle is $\triangle DEF$?

   A obtuse triangle   **B** equilateral triangle
   C scalene triangle   **D** right triangle

6. If $WXYZ$ is a parallelogram, what is the measure of $\angle Y$ if $\angle W$ is two-thirds
(34) the size of $\angle X$?

* 7. Write expressions for the values of sin Q, cos Q, and tan Q in this figure.
(68)

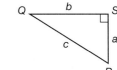

8. The diagonals of parallelogram $ABCD$ meet at O. Find a measure for
(65) $\angle AOD$ that makes $ABCD$ a rhombus.

9. **Predict** Predict the number of spins that would land on sector 4 in 1000 trials
(Inv 6) of the spinner at right.

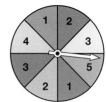

10. **Multi-Step** Find the perimeter and area of a square with a diagonal length
(52) of 8 inches. Express your answer in simplified radical form.

11. Find the surface area of the prism at right.
(59)

*12. **Generalize** Can a rotation of square $ABCD$ and a
(67) reflection of a square $ABCD$ ever have the same image?
*Hint: Label the vertices of a square and notice their
changing positions while rotating and reflecting.*

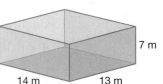

13. Elias is trying to find a coin he dropped which rolled into a
(40, Inv 6) dark, rectangular room. What is the probability he will find
the coin within 1 foot of the wall?

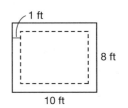

14. In the diagram below, G is the incenter of $\triangle HIJ$. Find EG.
(38)

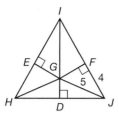

15. **Error Analysis** Marilou found the perimeter of this irregular polygon to be
(57) 19 units. What mistake has she made, and what is the actual perimeter?

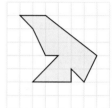

16. Determine whether $BARK$ is a parallelogram. Explain how you know.
(61)

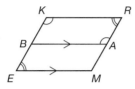

17. **Justify** Is the following a valid application of the Law of Detachment? Explain.
(21)

All birds have feathers. A whale is not a bird. Therefore, a whale does not have feathers.

18. (**Tiling**) A floor tile is in the shape of a regular octagon with sides that are
(66) 6 inches long. What is the perimeter of the tile?

19. (**Manufacturing**) To the nearest centimeter, how tall should a manufacturing
(62) company make its soup cans if the standard diameter is 7.5 centimeters and each
can must hold 575 milliliters? *Hint: Recall that 1 milliliter is equivalent to 1 cubic
centimeter.*

*20. Rectangle $MNOP$ is rotated clockwise about point Q. Explain why
(67) $\triangle MM'Q$ is similar to $\triangle PP'Q$.

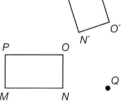

21. Draw a net for a rectangular prism with a length of 9, a width of 3, and a
(Inv 5) height of 12. Label each dimension.

22. **Write** Can a square ever have a numerical value for its perimeter that is the
(66) same as the numerical value for its area? Explain.

23. Prove that $\triangle LMN \cong \triangle QPO$.
(36)

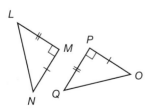

*24. Multiple Choice Which of the following is not a pair of equivalent functions in
(68) $\triangle ABC$ if $\angle C$ is a right angle?

   A $\sin A$ and $\tan B$

   B $\sin A$ and $\cos B$

   C $\tan A$ and $\dfrac{1}{\tan B}$

   D $\sin B$ and $\cos A$

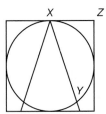

25. (**Architecture**) This drawing depicts the cross section of the air ducts
(64)   and the support beams for a very large building. If $m\widehat{XY} = 100°$,
   what is the measure of the angle for the support beam, $\angle YXZ$?

26. Find the values of x and y in the triangle shown.
(50)

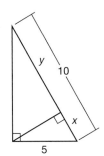

27. Write a paragraph proof of the following.
(31)   **Given:** Ray TR bisects $\angle QTS$, $m\angle QTR = 45°$
   Prove: $\angle QTS$ is a right angle.

*28. Write $\sin M$ as a fraction and then as a decimal rounded to the nearest
(68)   hundredth.

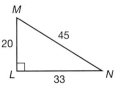

29. Justify Determine whether lines $\overline{WX}$ and $\overline{YZ}$ are parallel. How do you know?
(60)

*30. Multiple Choice Which word *best* describes the transformation
(67)   of $ABCD$ into $A'B'C'D'$?

   A translation   **B** rotation

   C reflection   **D** None of the above

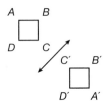

Properties of Trapezoids and Kites

Warm Up

1. **Vocabulary** A quadrilateral with exactly two nonadjacent pairs of congruent adjacent sides is a _____.
   (19)

2. Find the area of this trapezoid.
   (22)

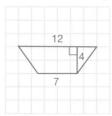

3. If the area of a trapezoid is 24 square inches, its height is 3 inches, and it
   (22) has one 12-inch base, what is the length of its other base?

New Concepts The **bases of a trapezoid** are its two parallel sides. A **base angle of a trapezoid** is one of a pair of consecutive angles whose common side is a base of the trapezoid. Trapezoids have two pairs of base angles. The **legs of a trapezoid** are the two nonparallel sides.

Figure $QRST$ is a trapezoid.

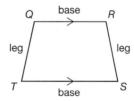

> **Math Reasoning**
>
> **Formulate** The formula for the area of a trapezoid is $A = \frac{1}{2}(b_1 + b_2)h$. Write a formula for the area of a trapezoid in terms of the length of its midsegment, z.

$\overline{QR}$ and $\overline{TS}$ are bases,

$\angle Q$ and $\angle R$ are base angles,

$\angle T$ and $\angle S$ are base angles,

and $\overline{QT}$ and $\overline{RS}$ are legs of the trapezoid.

The **midsegment of a trapezoid** is the segment whose endpoints are the midpoints of the legs of the trapezoid.

Theorem 69–1: Trapezoid Midsegment Theorem
The midsegment of a trapezoid is parallel to both bases and has a length that is equal to half the sum of the bases. Therefore, if $\overline{UV}$ is the midsegment of trapezoid $QRST$, then $\overline{UV} \parallel \overline{QR}$, $\overline{UV} \parallel \overline{TS}$, and $UV = \frac{1}{2}(QR + TS)$. 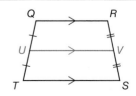

Example 1 **Applying Properties of the Midsegment of a Trapezoid**

The midsegment of trapezoid *ABCD* is $\overline{EF}$. Find the length of $\overline{EF}$.

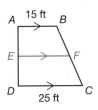

SOLUTION

$EF = \frac{1}{2}(AB + DC)$

$EF = \frac{1}{2}(15 + 25)$

$EF = 20$

The length of $\overline{EF}$ is 20 feet.

An **isosceles trapezoid** is a trapezoid with congruent legs. Like isosceles triangles, isosceles trapezoids have congruent base angles.

Hint

As with isosceles triangles, the converse of this property is also true. That is, if one pair of base angles of a trapezoid is congruent, then the trapezoid is an isosceles trapezoid.

Properties of Isosceles Trapezoids	
Base angles of an isosceles trapezoid are congruent. If trapezoid *HIJK* is isosceles, then $\angle H \cong \angle I$, and $\angle J \cong \angle K$.	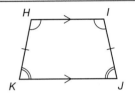

Example 2 **Applying Properties of the Base Angles of an Isosceles Trapezoid**

Find the measures of $\angle N$, $\angle O$, and $\angle P$ in isosceles trapezoid *MNOP*.

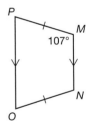

SOLUTION

Because the trapezoid is isosceles, its base angles are congruent.

Therefore, $\angle M \cong \angle N$ and $\angle P \cong \angle O$.

Therefore, m$\angle N = 107°$.

Notice that $\overline{MP}$ is a transversal that intersects two parallel lines. Therefore, $\angle M$ and $\angle P$ are supplementary.

m$\angle P = 180° - 107°$

m$\angle P = 73°$

m$\angle O = 73°$

Properties of Isosceles Trapezoids	
The diagonals of an isosceles trapezoid are congruent. In isosceles trapezoid *STUV*, $\overline{SU} \cong \overline{TV}$.	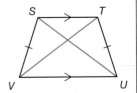

Example 3 **Applying Properties of the Diagonals of an Isosceles Trapezoid**

ABCD is an isosceles trapezoid. Find the length of $\overline{CE}$ if $AC = 22.3$ centimeters and $AE = 8.9$ centimeters.

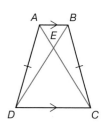

SOLUTION
Because $\overline{AC}$ and $\overline{BD}$ are the diagonals of an isosceles trapezoid, they are congruent.
$CE = AC - AE$
$CE = 22.3 - 8.9$
$CE = 13.4$
The length of $\overline{CE}$ is 13.4 centimeters.

Recall that kites are quadrilaterals with exactly two pairs of congruent adjacent sides.

Properties of Kites
The diagonals of a kite are perpendicular. $\overline{EG} \perp \overline{FH}$ 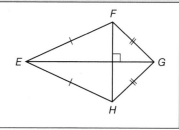

Example 4 **Applying Properties of the Diagonals of a Kite**

Find the lengths of the sides of kite *WXYZ*. Round to the nearest tenth.

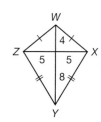

SOLUTION
Because the diagonals of a kite are perpendicular to each other, the Pythagorean Theorem can be used to find the length of each side.
$WX^2 = 4^2 + 5^2$
$WX^2 = 41$
$WX \approx 6.4$

Since *WXYZ*, is a kite, *WX* and *WZ* are congruent. Therefore, *WZ* is also approximately 6.4.
$YZ^2 = 8^2 + 5^2$
$YZ^2 = 89$
$YZ \approx 9.4$
$\overline{YZ}$ and $\overline{YX}$ are also congruent, so *YX* is approximately 9.4.

Example 5 | Application: Woodworking

A carpenter is making an end table with a trapezoid-shaped top. There will be three glass panels on the top of the table, as shown in the diagram. In the trapezoid $BDEG$, $\overline{CF}$ is a midsegment. In the trapezoid $ACFH$, $\overline{BG}$ is a midsegment. What are the lengths of $\overline{CF}$ and $\overline{DE}$?

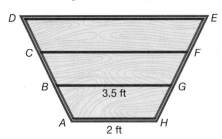

Analyze

Where would the midsegment of trapezoid $ADEH$ be located?

SOLUTION

Since $\overline{BG}$ is a midsegment of $ACFH$, its length is half the sum of CF and AH.

$BG = \dfrac{1}{2}(AH + CF)$	Midsegment of a trapezoid
$3.5 = \dfrac{1}{2}(2 + CF)$	Substitute.
$CF = 5$ feet	Solve.

$\overline{CF}$ is the midsegment of $BDEG$, so:

$CF = \dfrac{1}{2}(DE + BG)$	Midsegment of a trapezoid
$5 = \dfrac{1}{2}(DE + 3.5)$	Substitute.
$DE = 6.5$ feet	Solve.

Lesson Practice

a. In the diagram, $\overline{EF}$ is the midsegment of trapezoid $ABCD$. Find the length of $\overline{CD}$.
(Ex 1)

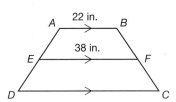

b. Find the measures of $\angle Q$, $\angle S$, and $\angle T$ in trapezoid $QRST$.
(Ex 2)

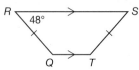

c. In isosceles trapezoid $MNOP$, find the length of $\overline{MQ}$ if $NP = 17.5$ yards and $PQ = 9.6$ yards.
(Ex 3)

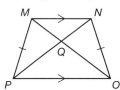

d. Find the lengths of the sides of kite *FGHJ*. Round the lengths to the nearest tenth.

(Ex 4)

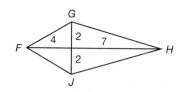

e. The side of a building is shaped like a trapezoid. The base of a row of windows runs along the midsegment of this trapezoid. What is the length of the building's roof?

(Ex 5)

Practice Distributed and Integrated

* 1. Trapezoid *FHJL* has midsegment $\overline{GK}$. Find *FL*.
(69)

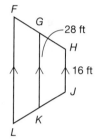

2. Verify Show that quadrilateral *WXYZ*, with vertices *W*(4, −9), *X*(5, −1), *Y*(0, 2), and *Z*(−1, −6) is a parallelogram.
(61)

3. Use the fact that m∠*DEF* < 140° to write an inequality for *x* in the triangle.
(39)

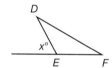

4. (**Swimming**) A salmon is swimming against the flow of the current. Suppose the salmon is swimming at 21 miles per hour (mph) and the current is flowing at 7 mph. How fast is the salmon traveling over the riverbed below?
(63)

5. Write In the circle, the chord $\overline{AB}$ is longer than chord $\overline{EF}$. Which angle is larger, ∠*APB* or ∠*EPF* ? Explain.
(Inv 4)

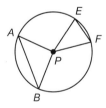

6. Identify the hypothesis and conclusion of the following statement.
(10) *If x² + 16 = 25, then x = 3.*

* 7. What is the sine of a 30° angle if the hypotenuse is 10? What is the cosine of a 60° angle if the hypotenuse is 100?
(68)

8. (**Comparison Shopping**) Elissa is buying brake fluid for her car. There are two containers that are both priced the same, shown below. Is there more fluid in the rectangular container or in the cylindrical can?
(59, 62)

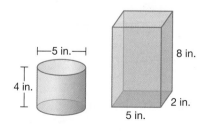

9. **Analyze** How many reflections across the same line must be performed on a
(67) figure to restore it to its original state? How many rotations by 90° (in the same
direction) must be performed to do the same?

10. If $\overline{AB}$ is tangent to circle C at A, what is m∠K?
(58)

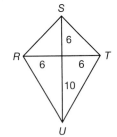

11. **Multiple Choice** What is the approximate length of a 288°-arc in a
(35) circle with a diameter of 12 centimeters?

  A 30.16 cm **B** 15.08 cm
  C 90.48 cm **D** 361.91 cm

12. (Tiling) A floor tile is in the shape of a regular hexagon with
(66) sides that are each 12 inches long. How many tiles would be
needed to cover a 25-by-32-feet floor?

13. A day is chosen at random from a yearly calendar. Are the probabilities for each
(Inv 6) day of the week being chosen the same? Explain.

*14. Find the lengths of the sides of kite *RSTU*. Round to the nearest tenth.
(69)

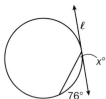

15. **Error Analysis** Jermaine found the perimeter of this irregular
(57) polygon to be 30. Explain what error he made and find the
correct exact perimeter.

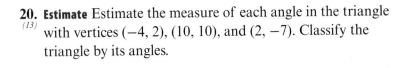

16. How many vertices does a polyhedron with 6 faces and
(49) 12 edges have?

17. Find the value of x in this circle. Line ℓ is tangent to the circle.
(64)

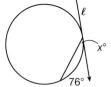

18. **Multi-Step** Classify the polygon with vertices $S(7, 2)$, $T(1, 2)$, $U(3, -3)$ and
(57) $V(5, -3)$. Find the area.

*19. In isosceles trapezoid *TUVW*, find *UW* if $TX = 23$ inches and
(69) $VX = 28.7$ inches.

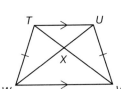

20. **Estimate** Estimate the measure of each angle in the triangle
(13) with vertices $(-4, 2)$, $(10, 10)$, and $(2, -7)$. Classify the
triangle by its angles.

21. Find *AD* in the triangle at right.
(60)

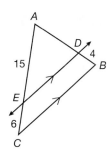

22. **Kitchen Skills** Howard is making frozen orange juice from concentrate. The mix
(62) comes in a cylindrical can that is 6 inches tall with a diameter of 3 inches. If the proper ratio of concentrate to water is 1:8, how much water does Howard need, to the nearest ten cubic inches?

23. Using the diagram at right, find *PZ* in terms of *y* if *TP* = 7 and
(38) *PM* = *y*. Point *P* is the circumcenter of the triangle.

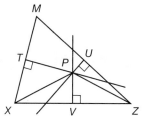

24. **Algebra** In △*JKL*, *JK* = 5 and m∠*JKL* = 60°. In △*DEF*, *DE* = (2*x* − 3*y*) and
(28) m∠*DEF* = (*x* + *y*)°. If *KL* = *EF*, what are the values of *x* and *y* that make △*JKL* ≅ △*DEF* true?

*25. Find the measures of ∠*A*, ∠*B*, and ∠*D* in trapezoid *ABCD*.
(69)

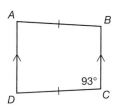

*26. Find sin *b*, cos *b*, and tan *b* in the triangle.
(68)

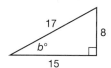

27. An altitude of a triangle is the _____ line segment from a vertex to the
(32) line containing the opposite side of the triangle.

28. **Generalize** Do you need to know the side lengths of a 30°-60°-90° triangle to find
(68) the sine or cosine of its angles? Explain.

29. Find the value of *x* in the figure.
(64)

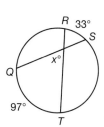

30. **Multiple Choice** Which of the following, when performed on an equilateral
(67) triangle, might result in a nonequilateral triangle?
  A rotation **B** reflection
  C dilation **D** none of these

Finding Surface Areas and Volumes of Pyramids

Warm Up

1. **Vocabulary** The perpendicular distance from the center of a regular
   (66) polygon to one of its sides is called the _____.

2. The formula for the perimeter, P, of a regular polygon is $P = ns$. What do
   (66) n and s stand for?

3. Which of the following is the formula for volume of a rectangular prism?
   (59) **A** $V = lw$ **B** $V = Bh$

   C $V = \frac{1}{2}Bh$ **D** $V = \frac{Bh}{3}$

New Concepts

The **vertex of a pyramid** is the common vertex of the pyramid's lateral faces. The **base of a pyramid** is the face of the pyramid that is opposite the vertex. A **regular pyramid** is a pyramid with a regular polygon as a base and with lateral faces that are congruent isosceles triangles.

The **slant height of a regular pyramid** is the distance from the vertex of a regular pyramid to the midpoint of an edge of the base.

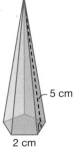

Caution

This formula only applies to regular pyramids. In an irregular pyramid, each lateral face's area must be calculated separately.

Lateral Area Formula for Regular Pyramids

The lateral area, L, of a regular pyramid is given by the formula below, where P is the perimeter of the base and l is the slant height.

$$L = \frac{1}{2}Pl$$

Example 1 Calculating Lateral Area of a Pyramid

What is the lateral area of this regular pentagonal pyramid?

SOLUTION
Since the base is a regular pentagon with side lengths of 2 centimeters, its perimeter is 10 centimeters.

Substitute the perimeter and the slant height into the lateral area formula for regular pyramids.

$L = \frac{1}{2}Pl$ Lateral surface area for regular pyramids

$L = \frac{1}{2}(10)(5)$ Substitute.

$L = 25 \text{ cm}^2$ Simplify.

The lateral area is 25 square centimeters.

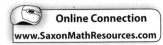
Online Connection
www.SaxonMathResources.com

The sum of the base area and the lateral area is the total surface area of the pyramid.

Surface Area of a Pyramid
The total surface area, S, of a pyramid is given by the formula below, where L is the lateral surface area and B is the area of the base. $$S = L + B$$

Example 2 Calculating Surface Area of a Pyramid

Calculate the total surface area of a regular hexagonal pyramid with a slant height of 12 centimeters and a base side that is 3 centimeters long.

SOLUTION
Since the base is a regular hexagon with 3-centimeter side lengths, its perimeter is 18 centimeters. Use the formula for the area of a regular hexagon that was discussed in Lesson 66, where s is the side length.

$$B = \frac{1}{2}\left(\frac{s\sqrt{3}}{2}\right)P$$

$$B = \frac{1}{2}\left(\frac{3\sqrt{3}}{2}\right)(18)$$

$$B \approx 23.38 \text{ cm}^2$$

Now calculate the total surface area of the pyramid:

$S = L + B$	Surface area of a pyramid
$S = \frac{1}{2}Pl + B$	Substitute.
$S \approx \frac{1}{2}(18)(12) + 23.38$	Substitute.
$S \approx 131.38 \text{ cm}^2$	Simplify.

Therefore, the surface area of the pyramid is approximately 131.38 square centimeters.

The **altitude of a pyramid** is the perpendicular segment from the vertex to the plane containing the base. The length of the altitude is the height of the pyramid. The volume of a pyramid can be found using the height and the area of the base.

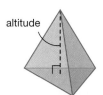

altitude

Volume of a Pyramid
The volume, V, of a pyramid is given by the formula below, where B is the area of the base and h is the height. $$V = \frac{1}{3}Bh$$

Recall that the volume of a prism is given by $V = Bh$. The volume of a pyramid is one third the volume of a prism with equal height and a congruent base.

Example 3 Calculating Volume of a Pyramid

Find the volume of the pyramid. The height is 9 inches and the base is a right triangle with legs that are 5 inches and 8 inches long, respectively.

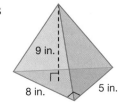

9 in.

8 in.   5 in.

SOLUTION
First find B, the base area.

$B = \frac{1}{2}bh$

$B = \frac{1}{2}(8 \text{ in.})(5 \text{ in.}) = 20 \text{ in}^2$

Then find the volume, V.

$V = \frac{1}{3}Bh$

$V = \frac{1}{3}(20 \text{ in}^2)(9 \text{ in.}) = 60 \text{ in}^3$

The volume of the pyramid is 60 cubic inches.

Example 4 Application: The Louvre Pyramid

The Louvre Pyramid, a regular square pyramid, is the main entrance of the Musée du Louvre in Paris. It has an approximate height of 70 feet and its square base has sides that are 115 feet long. What is the lateral area of the pyramid?

Math Reasoning

Predict If the height of a regular pyramid is decreased, how does the slant height change?

SOLUTION
Since the base has four congruent sides of 115 feet each, the perimeter of the base is 460 feet. To find the slant height, use the Pythagorean Theorem. One leg is the height, the other is the apothem of a square with a side length of 115 feet, which is simply half the length of the side.

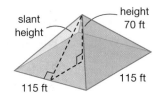

slant height   height 70 ft

115 ft   115 ft

$l = \sqrt{70^2 + \left(\frac{115}{2}\right)^2}$

$l = \sqrt{4900 + \frac{13{,}225}{4}}$

$l \approx 90.58 \text{ ft}$

Now substitute into the formula for lateral surface area.

$L = \frac{1}{2}Pl$

$L \approx \frac{1}{2}(460)(90.58)$

$L \approx 20{,}883.4 \text{ ft}^2$

Thus, the lateral area of the Louvre Pyramid is approximately 20,833 square feet.

a. What is the lateral area of a regular octagonal pyramid with a side
(Ex 1) length of 5 centimeters and a slant height of 7 centimeters?

b. What is the surface area of a regular hexagonal pyramid with a slant
(Ex 2) height of 8 inches and a base side length of 4 inches, to the
nearest hundredth of a square inch?

c. What is the volume of a square pyramid with side lengths of 5 feet
(Ex 3) and a height of 10 feet, to the nearest tenth of a square foot?

d. The Pyramid Arena in Memphis, Tennessee, is the third-largest
(Ex 4) square pyramid in the world. It is approximately 321 feet tall
and the length of one side of the base is about 600 feet. What
is its surface area?

Practice Distributed and Integrated

1. Find m∠P and m∠R in kite *PQRS*.
(69)

* 2. What is the volume of a regular hexagonal pyramid with
(70) a height of 10 feet and base side lengths of 7 feet? Round
your answer to the nearest cubic foot.

3. Multi-Step Determine the perimeter of this figure.
(40)

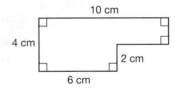

4. Find the measures of the numbered angles that make the figure shown
(65) a rectangle.

5. (**Bermuda Triangle**) The Bermuda Triangle is a region in the
(55) Atlantic Ocean near the southeastern coast of the United States.
The vertices of the triangle are Miami, Florida; San Juan, Puerto
Rico; and Bermuda. In the figure, the dashed segments are
midsegments. Find the perimeter of the midsegment triangle
within the Bermuda Triangle. How does it compare to the
perimeter of the Bermuda Triangle?

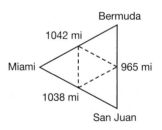

6. (**Sign Art**) A sign maker is welding a sun-shaped design by dividing a regular
(66) hexagon into six equal triangles and arranging them along their base vertices to
form rays of the sun. If the side length of the hexagon is 8 inches, what is the total
area of one triangle?

7. (**Density**) A certain crystal has a density of 34 grams per cubic centimeter.
(59) Find the mass, to the nearest hundredth of a gram, of the sample shown at right if the triangular base is an equilateral triangle.

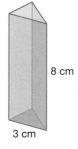
8 cm

3 cm

8. Complete the following statement. In a 45°-45°-90° triangle,
(68) the _____ and _____ are equal. (***sine, cosine, tangent, hypotenuse***)

9. **Write** How are 30°-60°-90° triangles and 45°-45°-90° triangles alike?
(56) How are they different?

10. Find the value of *x*.
(50)

x

5   2

*11. What is the lateral area of a regular octagonal pyramid with base side lengths of
(70) 4 inches and a slant height of 10 inches?

12. **Multiple Choice** The disjunction of *p* and *q* is true when
(20) **A** *p* and *q* are both true, but not when either is false.
B either *p* is true, or *q* is true, or both.
C *p* and *q* are both false.
D either *p* is true, or *q* is true, but not both.

13. **Multiple Choice** Which of the following is *not* parallel to $x = 2$?
(37) **A** $0 = x$ **B** $x = 3y$
C $2x = 3$ **D** $x + 7 = 0$

14. **Analyze** Write a two-column proof of the Converse of the Alternate Exterior
(27) Angles Theorem, "If two lines are cut by a transversal and alternate exterior angles are congruent, then the lines are parallel."
Given: $\angle 1 \cong \angle 2$
Prove: $j \parallel k$

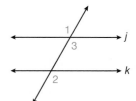

15. **Algebra** Trapezoid *MNOP* has a midsegment $\overline{QR}$. The parallel sides are
(69) $\overline{MN}$ and $\overline{OP}$. If $QR = 3x$, $MN = x + 3$, and $OP = 2x + 6$, find *x* and the length of each segment.

16. (**Games**) A particular dart game is played using darts that can suction to a flat
(Inv 6) surface. A group of students is throwing these darts onto the tiled surface shown. If a dart lands on exactly one tile, what is the probability that it might land on one of the blue tiles?

6 in.	7 in.	7 in.	4 in.	
				2 in.
				7 in.
				5 in.

17. Multi-Step An equilateral triangle has an apothem that is $7\sqrt{3}$ feet long. What is the
(66) area of the triangle?

18. Verify Explain how you know these two triangles are similar, and then
(46) write the similarity statement.

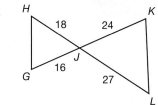

*19. What is the surface area of a regular hexagonal pyramid with a slant
(70) height of 8 centimeters and a base side length of 2 centimeters?
Round your answer to the nearest centimeter.

20. Error Analysis Fernando said that $\triangle ABC$ was translated to form
(67) the image $\triangle A'B'C'$. Raquel said that it had to be rotated, not
translated. Who is correct? Explain.

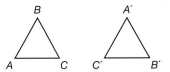

21. Coordinate Geometry $\triangle RST$ is a 30°-60°-90° triangle with
(56) $S(-1, -4)$ and $T(9, -4)$, and m$\angle S = 90°$. What are possible
coordinates of R?

22. Find the measure of $\overset{\frown}{MNO}$ in the figure. Line ℓ is tangent to the circle.
(64)

*23. (**Structures**) When the Great Pyramid of Giza was first built, the side
(70) length of its square base was about 231 meters and the height of the
pyramid was about 147 meters. What was its approximate lateral area?
Round your answer to the nearest hundred square meters.

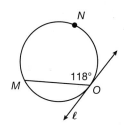

24. Find the measures of $\angle F$, $\angle G$, and $\angle J$ in trapezoid $FGHJ$.
(69)

25. Verify Show that the geometric mean of 12 and 300 is 60 by
(50) writing and solving a proportion.

26. Model Draw a rectangle with a diagonal that is twice as long as its width.
(56) Write an equation to find the length of the rectangle.

27. Use a calculator to find the length of MN. Round to the nearest hundredth.
(68)

28. Algebra The perimeter of parallelogram $ABCD$ is 84. Find the
(34) length of each side if $AB = 3BC$.

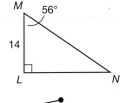

29. Formulate Use the given two-column proof to write a flowchart proof.
(31) **Given:** V is the midpoint of $\overline{SW}$, and W is the midpoint of $\overline{VT}$.
Prove: $\overline{SV} \cong \overline{WT}$

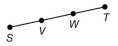

Statements	Reasons
1. V is the midpoint of $\overline{SW}$.	1. Given
2. W is the midpoint of $\overline{VT}$.	2. Given
3. $\overline{SV} \cong \overline{VW}$, $\overline{VW} \cong \overline{WT}$	3. Definition of midpoint
4. $\overline{SV} \cong \overline{WT}$	4. Transitive Property of Congruence

*30. (**Construction**) Steven is making two square pyramids. The first pyramid has a base
(70) length of 10 centimeters and its height is 20 centimeters. The second pyramid has
twice the height of the first but only half the base length. What is the volume of
the second pyramid? Round your answer to the nearest cubic centimeter.

Trigonometric Ratios

The sine, cosine, and tangent ratios for a right triangle are:

$$\sin x = \frac{\text{opposite}}{\text{hypotenuse}} = \frac{BC}{AC}$$

$$\cos x = \frac{\text{adjacent}}{\text{hypotenuse}} = \frac{AB}{AC}$$

$$\tan x = \frac{\text{opposite}}{\text{adjacent}} = \frac{BC}{AB}$$

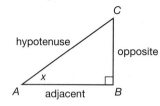

Notice that tan x is the quotient of sin x and cos x:

$$\tan x = \frac{\text{opposite}}{\text{adjacent}}$$

$$\tan x = \left(\frac{\text{opposite}}{\text{hypotenuse}}\right) \bigg/ \left(\frac{\text{adjacent}}{\text{hypotenuse}}\right)$$

$$\tan x = \frac{\sin x}{\cos x}$$

> **Caution**
>
> Remember that if a fraction has a radical in its denominator, you should rationalize it.

In this investigation, you will observe the values of sin x, cos x, and tan x as x varies from 0° to 90° in increments of 15°. Copy this table to record your results.

x	sin x	cos x	tan x
0°			
15°			
30°			
45°			
60°			
75°			
90°			

1. Imagine a right triangle with one angle measuring 0°. The side opposite this angle would be 0 units and the hypotenuse and adjacent side of the triangle would be congruent. Use this information to fill out the first row. Check your answers using a calculator.

2. Draw a diagram of a 30°-60°-90° triangle with the shortest side being 1 unit long. Use your diagram to fill out the 30° and 60° rows of the table.

3. Draw a diagram of a 45°-45°-90° triangle with legs that are 1 unit long. Use your diagram to fill out the 45° row of the table.

4. Use a calculator to fill out the 15° and 75° rows of the table.

5. What do you notice about the sine and cosine values of the 30° and 60° angles? Use this observation to fill out the final row of the table by comparing it to the 0° row. Find the tangent by dividing the sine of x by the cosine of x.

Review your table.

6. What do you notice about the sine of an angle, sin x, and the cosine of its complement, cos $(90° - x)$? Write a conjecture relating the sine and cosine of complementary angles.

7. Describe the ranges of values for cosine and sine, based on your table.

8. What is the value of sin x + cos x for $x = 0°$ and 90°? Is this relationship true for the rest of the table?

9. Find the value of $\sin^2 x + \cos^2 x$ for several values of x. What do you notice about the value of $\sin^2 x + \cos^2 x$?

10. What is the range of the tangent function? Make a conjecture based on your table and test it by calculating the tangent of some other angles with your calculator. Explain why this is the range of the tangent function.

Math Symbols

The square of sin x is written as $\sin^2 x$.

$\sin^2 x = (\sin x)^2$

The same applies to the other trigonometric functions.

Investigation Practice

a. In $\triangle DEF$, $\angle E$ is a right angle, m$\angle D = 45°$, and $DE = 1$. What is m$\angle F$? Use the Converse of the Isosceles Triangle Theorem to relate DE and EF, and then use the Pythagorean Theorem to determine DF and EF. Then give exact values for sin 45°, cos 45°, and tan 45°.

b. In $\triangle GHJ$, $\angle G$ is a right angle, m$\angle H = 60°$, and $GJ = 3$. How are GH and HJ related? Determine GH and HJ. Then, give exact values for sin 60°, cos 60°, and tan 60°.

c. Use your response to **a** to draw a 45°-45°-90° triangle. Include all angle measures and side lengths.

d. Use your response to **b** to draw a 30°-60°-90° triangle. Include all angle measures and side lengths.

Translations

Warm Up

1. **Vocabulary** When a transformation is applied to a figure, the original position is called the _____. The new position is called the _____.
   (67)

2. **Vocabulary** A transformation that shifts or slides every point of a figure the same distance in the same direction is a _____.
   (67)

3. Given a line segment with endpoints $R(4, 7)$ and $S(7, 13)$, find
   (9, 16)
   a. the slope of $\overline{RS}$.
   b. the length of $\overline{RS}$.

New Concepts

A translation shifts every point of a figure the same distance in the same direction. A figure that is transformed by a translation remains congruent to its preimage. Its side lengths, angle measures, and other properties remain the same. Translation changes nothing but the location of a figure.

Math Language

An **isometry** is any transformation that does not change the size and shape of the transformed figure.

Translation
A translation is an isometry, meaning the preimage and its translated image are the same shape and size.

Example 1 **Translations in One Dimension**

A square has vertices $A(1, 1)$, $B(4, 1)$, $C(4, 4)$, and $D(1, 4)$.

(a.) Find the coordinates of the vertices of the image of square $ABCD$ after a translation of 5 units to the right. Show the preimage and image on the same coordinate grid.

SOLUTION

The x-coordinates of the vertices of the image of the square after it is translated 5 units to the right are 5 greater than the x-coordinates of the vertices of the preimage. The y-coordinates are unchanged.

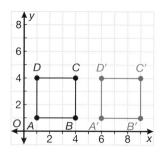

Math Language

Generalize When a point A is transformed, the image is labeled A' (read A prime). How would the translated image of $\triangle ABC$ be labeled?

(b.) Show that the area of $A'B'C'D'$ is equal to the area of $ABCD$.

SOLUTION

In $ABCD$, $AB = 3$ and $BC = 3$, so the area of the square is 9 centimeters squared. In $A'B'C'D'$, $A'B' = 3$ and $B'C' = 3$, so the area of the square is also 9 centimeters squared.

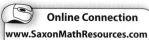

Online Connection
www.SaxonMathResources.com

Mapping notation is used to indicate the way in which a point or several points should be transformed. An example of translation mapping notation is given below.

$$T: (x, y) \rightarrow (x + 1, y - 2)$$

This mapping says that in a transformation, the original pair of coordinates, (x, y), will be changed into $(x + 1, y - 2)$. That is, the x-coordinate will increase by 1 and the y-coordinate will decrease by 2.

Example 2 Translations in Two Dimensions

The vertices of a triangle are $X(-2, 0)$, $Y(-2, -4)$ and $Z(1, -4)$. Find the image of $\triangle XYZ$ after the translation $T: (x, y) \rightarrow (x + 5, y + 4)$. Show the preimage and image on the same coordinate grid.

SOLUTION

Graph the preimage triangle. The translation moves every point 5 units to the right and 4 units up. The transformation mapping is shown below.

$X(-2, 0) \rightarrow X'(3, 4)$
$Y(-2, -4) \rightarrow Y'(3, 0)$
$Z(1, -4) \rightarrow Z'(6, 0)$

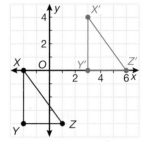

Plot X', Y', and Z' and connect them to form $\triangle X'Y'Z'$.

A translation for a polygon can also be represented using a vector. Placing the initial point of the vector on each point of the preimage will indicate the position of the point in the image.

Example 3 Showing Translations with Vectors

Find the image of $ABCD$ under the translation vector $\langle 2, 3 \rangle$.

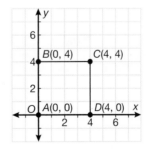

SOLUTION

Place the initial point of the vector on point A and draw the vector, which moves each point 2 units right and 3 units up. Drawing the vector shows that the new image point A' will be at $(2, 3)$.

Repeat this process for points B, C, and D. The image is shown in the diagram. B' is at $(2, 7)$, C' is at $(6, 7)$, and D' is at $(6, 3)$.

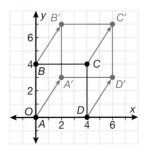

Example 4 | **Application: Computer Animation**

A character in a new animated movie will move from the point $(4, -5)$ first to the point $(16, 3)$, then to point $(13, -2)$. Find the vectors the animators need to apply to the character to make these two translations.

SOLUTION

For the first translation, the change in x-values is 12, and the change in y-values is 8, so the first vector is $\langle 12, 8 \rangle$.

For the second translation, the change in x-values is -3, and the change in y-values is -5, so the second vector is $\langle -3, -5 \rangle$.

Lesson Practice

a. A line segment has endpoints $A(11, 5)$ and $B(4, 9)$. It is translated
(Ex 1)   5 units up. What are the coordinates of A' and B'?

b. A triangle has vertices $K(2, 5)$, $L(1, 11)$ and $M(5, 7)$. Find the image
(Ex 2)   of $\triangle KLM$ after the translation. $T: (x, y) \rightarrow (x - 2, y + 1)$ Show the preimage and image on the same coordinate grid.

c. Find the coordinates of the vertices of the
(Ex 3)   image of parallelogram $ABCD$ after translation by the vector $\langle 3, 1 \rangle$.

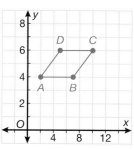

d. In an animated cartoon, the UFO shown
(Ex 4)   will move across the screen to the right. The animator translates two points of the UFO from $(1, 3)$ and $(4, 3)$ to an image at $(9, 4)$ and $(12, 4)$. Give the component form of the vector that describes this translation.

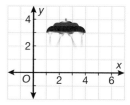

Practice   Distributed and Integrated

1. **Analyze** Using the diagram, explain how you know that the bisectors of
(61)   two consecutive angles of a parallelogram are perpendicular.

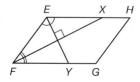

2. A pyramid has a base area of 15 square centimeters and a height of
(70)   10 centimeters. What is the volume of the pyramid?

3. Determine whether these measurements $(1.6, 1.8, 3.4)$ can be the side lengths
(33, 39)   of a triangle. If it is a triangle, determine whether it is an acute, obtuse, or a right triangle.

4. (**Bottled Water**) A company wants to sell cylindrical bottles of water. If each bottle
(62)   is to hold 0.5 liters, and will be 20 centimeters tall, what should the radius of each bottle be, to the nearest hundredth of a centimeter? *Hint: 1 cm³ = 1 mL.*

5. To create a wallpaper pattern, this triangle will be translated as indicated
(71) by vector *u*. Apply the translation to the original triangle and then again to
the image to create two triangle images for the pattern.

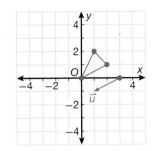

6. Algebra The image of $(5, y)$ after the transformation $v = \langle x, 11 \rangle$ is $(1, -6)$.
(71) Find the values of x and y.

7. The points $P(5, 5)$ and $Q(-5, -5)$ are translated 4 units up and 3 units to
(71) the left. What are the coordinates of the images of the points?

8. Is the parallelogram a rectangle, a rhombus, both, or neither?
(65)

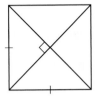

9. A trapezoid has vertices at the points $E(2, 3)$, $F(3, 1)$, $G(2, -2)$,
(69) and $H(-1, 4)$. Use the distance formula to find the length of the
midsegment of the trapezoid.

10. What is the surface area of a regular pyramid with a lateral area of
(70) 10 square feet and a base area of 5 square feet?

11. If trapezoids *RSTU* and *GHIJ* are similar, what is *GH*?
(41)

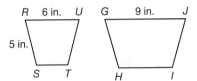

12. Find the equation of the line perpendicular to $\frac{1}{2}y + 3x = 4$ that
(37) passes through the point $(6, 5)$. Write in slope-intercept form.

13. Write Write the inverse of the following statement.
(17)

If a triangle is equilateral, then the sides are congruent.

14. Design The figure at right forms one quadrant of a logo.
(67) The completed logo is also shown. Identify the
transformations performed on the original figure
to form the completed logo.

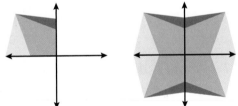

15. Structures In Paris, there is a glass square-based
(70) pyramid in front of the Louvre Museum. The pyramid
is about 70 feet tall and the side lengths of the base are about 115 feet.
What is the approximate volume of the pyramid? Round your answer to
the nearest cubic foot.

16. Multiple Choice $\triangle DEF$ is translated so that the coordinates of E'
(71) are $(0, 3)$. What are the coordinates of F' after the same translation?

  A $(1, -1)$ **B** $(4, -2)$

  C $(-2, -2)$ **D** $(-2, 6)$

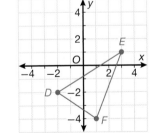

17. Model Draw a 30°-60°-90° triangle with the longer leg measuring 1.
(Inv 7) Include all angle measures and side lengths.

18. Generalize For what angle measures is the tangent ratio less than 1?
(68) greater than 1?

19. Determine the missing side lengths in $\triangle STU$. Give exact answers.
(Inv 7) Approximate any values involving radicals.

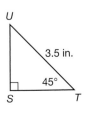

20. **Interior Design** The part of a kitchen containing the sink, countertop, and
(66) refrigerator is in the shape of an equilateral triangle with 12-foot side lengths.
What is the area of this triangle?

21. **Justify** Which of the side lengths of △ABC is greatest? Explain how you know.
(39)

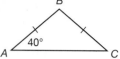

22. **Multiple Choice** Which statement would contradict the claim that
(48) ∠C and ∠D are alternate exterior angles?

A Angles C and D are on the same side of the transversal.

B m∠C < 180°

C m∠C = m∠D

D Angles C and D are on opposite sides of the transversal.

23. Segment $\overline{VY}$ is the midsegment of trapezoid UWXZ. Find the length
(69) of $\overline{WX}$.

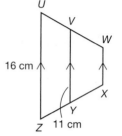

24. **Analyze** A transformation takes △PQR to △P'Q'R'. The vertices of the
(67) triangle are P, Q, and R are in clockwise order and the vertices P', Q',
and R' are also in clockwise order. Which type of transformation could
this *not* be?

A rotation   **B** translation

C reflection   **D** could be any of these

25. What is the resultant vector of c and d, in component form?
(63)

26. In △DEF, F is a right angle, m∠D = 45°, and DE = 3 centimeters.
(Inv 7) **a.** Draw a diagram of △DEF.

b. Determine exact and approximate values for DF.

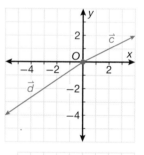

27. **Surveying** A pond is overlaid with a coordinate grid with each square
(57) representing 5 square meters. Estimate the area of the pond.

28. **Multi-Step** Quadrilateral KLMN has vertices at K(3, 3), L(7, 0),
(52) M(3, −3), and N(−1, 0). Classify the quadrilateral.

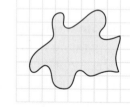

29. Using the given triangles, list the sides x, y, and
(Inv 4) z in order from least to greatest.

30. Use the spinner at right to answer **a.** and **b.**
(Inv 6) **a.** What is the probability of the pointer landing on gray?
Express your answer as a fraction.

b. What is the probability of the pointer landing on white?
Express your answer as a decimal.

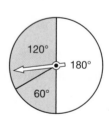

Tangents and Circles, Part 2

Warm Up

1. **Vocabulary** A line that is in the same plane as a circle and intersects the
(43) circle at exactly one point is called a _____.

2. In the diagram, $\overline{RS}$ and $\overline{RT}$ are tangents to the circle.
(58) Find the length of each segment.

3. Draw a circle with a tangent line, a chord, and a secant.
(43) Label each line or segment.

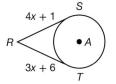

New Concepts

A tangent is a line that intersects a circle at exactly one point. The point
of intersection is called the point of tangency. A **common tangent** is a
tangent to two circles. Common tangents can be internal tangents or
external tangents.

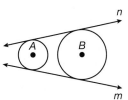

External Common Tangents

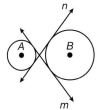

Internal Common Tangents

Recall Theorem 58-3: If two tangent segments are drawn to a circle from the
same exterior point, then they are congruent.

Example 1 | Solving Problems with Common Tangents

Given that $\overleftrightarrow{MR}$ and $\overleftrightarrow{PN}$ are internal common
tangents to $\odot A$ and $\odot B$, find the length
of $\overline{MQ}$.

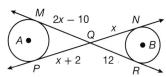

SOLUTION

Since two segments tangent to a circle from the same exterior
point are congruent, $\overline{NQ} \cong \overline{RQ}$ and $\overline{MQ} \cong \overline{PQ}$.

$NQ = RQ$ Definition of congruent segments

   $x = 12$ Substitute

Substitute the value of x into the expression for the length of
$\overline{MQ}$.

$MQ = 2x - 10$

$MQ = 2(12) - 10$

$MQ = 14$

Online Connection
www.SaxonMathResources.com

Tangent circles are coplanar circles that intersect at exactly one point. Tangent circles can be internally tangent or externally tangent. In both cases, the radii of the two circles are collinear.

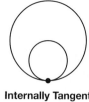

Internally Tangent Circles

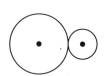

Externally Tangent Circles

Example 2 Solving Problems with Tangent Circles

In the diagram, $\odot Q$ is tangent to $\odot M$ and $\overline{NP}$ is tangent to $\odot Q$. The radius of $\odot Q$ is 5 centimeters and the radius of $\odot M$ is 2 centimeters. Find the area of $\triangle QNP$ to the nearest square centimeter.

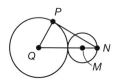

SOLUTION

Since the circles are tangent, they intersect at only one point, and their radii are collinear. Since $\overline{QN}$ is composed of a radius of $\odot Q$ and a diameter of $\odot M$, its length is 9 centimeters.

Since $\overline{NP}$ is tangent to $\odot Q$, m$\angle QPN = 90°$. So $\triangle QNP$ is a right triangle with a 9-centimeter hypotenuse and one 5-centimeter leg.

Use the Pythagorean Theorem to find the length of the other leg.

$QP^2 + PN^2 = QN^2$	Pythagorean Theorem
$5^2 + PN^2 = 9^2$	Substitute.
$PN = 2\sqrt{14}$	Solve.

Now the legs of the triangle can be used to find the area.

$A = \dfrac{1}{2}bh$	Area
$A = \dfrac{1}{2}(5)(2\sqrt{14})$	Substitute.
$A \approx 18.7$	Simplify.

So the area of the triangle is about 19 square centimeters.

Hint

To solve problems involving tangent circles, start by identifying any radius segments and labeling them on the diagram accordingly.

Example 3 Application: Mechanics

A car has a timing belt that consists of two pulleys and a belt, as shown in the diagram. The belt runs around the two pulleys and is tangent to both of them. The dotted segments, $\overline{JI}$ and $\overline{JK}$, have been drawn into the diagram to assist in finding the distance between the two pulleys. Find IH and KL.

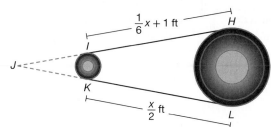

SOLUTION

Since the pulley is tangent to the circle, and the tangent lines meet at point J, $\overline{JH} \cong \overline{JL}$ and $\overline{JI} \cong \overline{JK}$. Therefore, $\overline{IH} \cong \overline{KL}$.

$$IH = KL$$
$$\frac{1}{6}x + 1 = \frac{x}{2}$$
$$x = 3$$

Substituting the value of x back into the expressions in the diagram, IH and KL both equal 1.5 feet.

Lesson Practice

a. In the diagram, $\overline{RT}$ and $\overline{QU}$ are tangents to
(Ex 1) the circles. Find the lengths of $\overline{RS}$, $\overline{ST}$, and $\overline{SU}$.

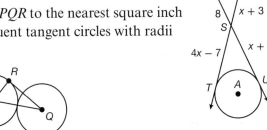

b. Determine the area of $\triangle PQR$ to the nearest square inch
(Ex 2) if $\odot P$ and $\odot Q$ are congruent tangent circles with radii of 6 inches each.

c. **Pulleys** A system of pulleys is set up as shown. Find the value of x.
(Ex 3)

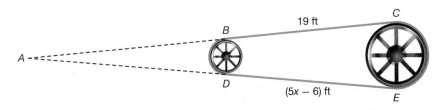

Practice Distributed and Integrated

1. **Painting** An artist will paint a detailed mural on the wall in the
(40) diagram at right, but cannot paint on the 1-by-8-foot window. If it takes him half an hour to paint one square foot, how long will it take him to paint the whole mural?

2. **Multi-Step** Noncollinear segments $\overline{PQ}$ and $\overline{PR}$ are tangent to $\odot C$,
(58) which has a radius of 1 foot, 6 inches. If $CQ = 2PQ$, determine the perimeter of quadrilateral $CQPR$ in feet and inches.

3. **Error Analysis** The quadrilateral shown was transformed, and Heather thinks
(67) the transformation was a translation. Explain why Heather is wrong.

4. Write the negation of the hypothesis and the conclusion in the statement.
(17)

   If a polygon has six sides, then it is a hexagon.

5. Assume the segments that appear to be tangent are tangent. Find *BE*, *CE*, and *DE*.
(72)

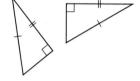

6. **Model** Draw a diagram of a 30°-60°-90° triangle with a hypotenuse that is
(Inv 7) 6 inches long. Label the exact lengths.

7. The translation $T:(x, y) \rightarrow (x - 2, y - 1)$ is applied to a circle with a center
(71) at (5, 4). What are the coordinates of the image of the center?

8. Find the surface area of the cylinder to the nearest whole square meter.
(62)

9. **Multiple Choice** Which congruence theorem applies to these triangles?
(36)   **A** HL Congruence Theorem   **B** LL Congruence Theorem
       C HA Congruence Theorem   **D** LA Congruence Theorem

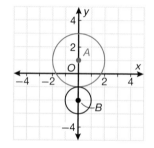

10. **Algebra** Line *a* passes through the points (3, 2) and (5, 2). Line *b* passes
(37) through points (−2, 1) and (−8, 1). Are the lines parallel, perpendicular, or neither?

11. Determine the point of tangency and the equation of the line that is
(72) tangent to both circles. Find the radius of each circle.

12. **Analyze** In a regular hexagonal pyramid, the slant height is 2 inches, and
(70) the height of the pyramid is 1 inch. Determine the lateral area of the pyramid.

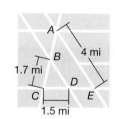

13. (**Riding Trails**) The diagram at right represents horseback riding trails. Point *B* is
(55) the halfway point along $\overline{AC}$. Point *D* is the halfway point along $\overline{CE}$. The paths
along $\overline{BD}$ and $\overline{AE}$ are parallel. If riders travel from *A* to *B* to *D* to *E*, and then
back to *A*, how far do they travel?

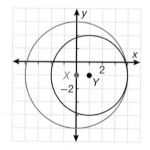

14. Point *K*(4, 11) is translated 5 units to the left. What are the coordinates of the
(71) image?

15. Find the circumcenter of a triangle with vertices at (0, 0), (10, 0), and (5, 4).
(38)

16. Determine the point of tangency and the equation of the tangent line.
(72) Find the radius of each circle.

17. What translation, in mapping notation, would move the point $H(3, -5)$
$^{(71)}$ to $H'(0, -2)$?

18. Find the value of x in the figure. Line ℓ is tangent to the circle.
$^{(64)}$

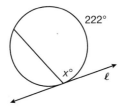

19. **Write** What is the relationship between the apothem length of a square
$^{(66)}$ and the side length of the square?

20. $\overline{EF}$ is the midsegment of trapezoid $ABCD$. Find EF.
$^{(69)}$

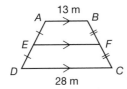

21. Any closed three-dimensional figure formed by four or more
$^{(49)}$ polygons that intersect only at their edges is called a _____.

22. Given the point $(1, 4)$, find the distance to the line $y = \frac{-3}{2}x - 1$.
$^{(42)}$

23. Write a paragraph proof of the following.
$^{(31)}$ **Given:** N is the midpoint of MP; Q is the
      midpoint of RP; $\overline{PQ} \cong \overline{NM}$
Prove: $\overline{PN} \cong \overline{QR}$

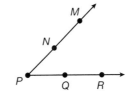

24. (**Design**) Martha wants to make a regular triangular pyramid out of construction
$^{(70)}$ paper. The base is an equilateral triangle. If the equilateral triangle has 10-inch
side lengths, and she wants the slant height to be 5 inches, how much material
does she need? Round your answer to the nearest square inch.

25. Is the parallelogram shown a rectangle, a rhombus, both, or is there not enough
$^{(65)}$ information to know?

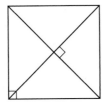

26. The ratio of the length of a rectangle to its width is 7:3. If the width of
$^{(41)}$ the rectangle is 42 centimeters, what is the length?

27. **Algebra** In $\triangle PQR$, $\angle SQR \cong \angle SRQ$, $PQ > PR$, $\text{m}\angle PSR = (4y + 9)°$ and
$\text{m}\angle QSP = (6y - 24)°$. Find the range of values for y.

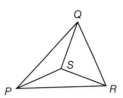

28. **Construction** A square section of a building frame is braced by two diagonal
beams. If the section is 7 feet, 8 inches high, what is the length of each beam
in feet and inches?

29. **Railroad Cars** If the price of gravel is $25 per cubic yard, what is the approximate
value of the gravel filling this railroad car?

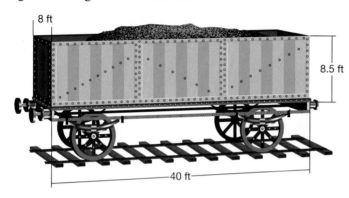

30. **Model** Draw two circles with two common exterior tangents.

Applying Trigonometry:
Angles of Elevation and Depression

Warm Up

1. **Vocabulary** In a right triangle, the ratio of the length of the adjacent leg
(68) to the length of the hypotenuse is the _____.

2. Find the value of x using a trigonometric ratio.
(68) Round your answer to the nearest hundredth.

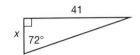

3. **Multiple Choice** What is the ratio of a leg to the
(53) hypotenuse of a 45°-45°-90° triangle?

A $\sqrt{2}$

B 2

C $\dfrac{\sqrt{2}}{2}$

D $\dfrac{2}{\sqrt{2}}$

New Concepts

An **angle of elevation** is the angle formed
by a horizontal line and the line of sight to a
point above. In the diagram, ∠1 is the angle
of elevation. It is the angle to which the bird
watcher raises (or elevates) his line of sight
from the horizontal to see the bird.

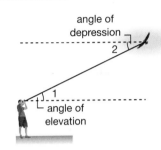

An **angle of depression** is the angle formed by
a horizontal line and a line of sight to a point
below. In the diagram, ∠2 is the angle of depression. It is the angle that the
bird would lower (or depress) its line of sight from the horizontal to see the
bird watcher.

You can use trigonometry to find the angle of elevation and the
angle of depression.

Math Language

The trigonometric ratios
are:

$\sin \theta = \dfrac{\text{opposite}}{\text{hypotenuse}}$

$\cos \theta = \dfrac{\text{adjacent}}{\text{hypotenuse}}$

$\tan \theta = \dfrac{\text{opposite}}{\text{adjacent}}$

Example 1   Angle of Elevation

Use the angle of elevation between
the kite and the child to find the horizontal
distance between the kite and the child.

SOLUTION

The horizontal distance is x. It is adjacent
to the angle of elevation. We also know the
hypotenuse, so we should use the cosine
ratio.

$\cos 36° = \dfrac{\text{adjacent}}{\text{hypotenuse}}$

$\cos 36° = \dfrac{x}{80}$

$x \approx 64.72 \text{ ft}$

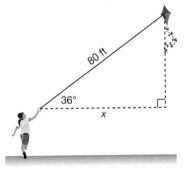

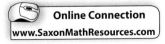
Online Connection
www.SaxonMathResources.com

Hint

Since the angle of elevation and the angle of depression are congruent, you can solve for whichever one seems easiest, given the information in the problem.

The horizontal lines used to measure the angles of elevation and depression are parallel, so the angle of elevation, $\angle 2$, and the angle of depression, $\angle 1$, are congruent, because they are alternate interior angles.

Example 2 **Angle of Depression**

The pilot in a plane cruising at 33,000 feet sees a lake. If the angle of depression from the plane to the lake is 30°, how far is the plane from the lake?

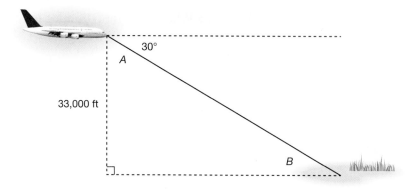

SOLUTION

Since the angle of elevation is congruent to the angle of depression, the angle of elevation is 30°.

$$\sin 30° = \frac{\text{opposite}}{\text{hypotenuse}}$$

$$\sin 30° = \frac{33,000}{x}$$

$$x = 66,000 \text{ feet}$$

The plane is 66,000 feet from the lake.

Example 3 **Angles of Elevation from Multiple Points**

A window washer is working 60 feet above the ground. About how far away is each person pictured at street level?

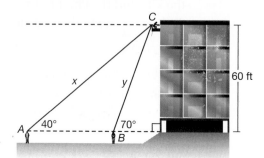

SOLUTION

In both cases, the unknown value is the hypotenuse of the triangle, and the side opposite the angle of elevation is known, so the sine ratio should be used.

Find the length of $\overline{AC}$.

$$\sin A = \frac{\text{opposite}}{\text{hypotenuse}}$$

$$\sin 40° = \frac{60}{x}$$

$$x \approx 93.34$$

Find the length of $\overline{BC}$.

$$\sin B = \frac{\text{opposite}}{\text{hypotenuse}}$$

$$\sin 70° = \frac{60}{y}$$

$$y \approx 63.85$$

Person B is about 64 feet away from the window washer, and person A is about 93 feet away.

Example 4 **Application: Surveying**

A surveyor on the Credit Union building measures an angle of elevation of 15° to the top of the Business Park building. The Credit Union building is 600 feet tall, and the Business Park building is 800 feet tall. What is the distance between the buildings, to the nearest foot?

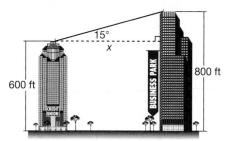

SOLUTION

Draw a triangle showing the difference in height between the buildings and the angle of elevation.

Find the difference in height. $800 - 600 = 200$.

Find the horizontal distance between the buildings using the tangent ratio.

$$\tan 15° = \frac{\text{opposite}}{\text{adjacent}}$$

$$\tan 15° = \frac{200}{x}$$

$$x \approx 746 \text{ feet}$$

In this exploration, you will construct a hypsometer. A hypsometer is used to measure an angle of elevation or an angle of depression.

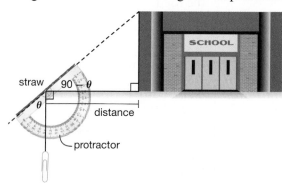

1. To make a hypsometer, you will need a drinking straw, a protractor, string, a paperclip (or small weight), and tape. First, attach the straw to the straight edge of the protractor with tape. Tie one end of the string to the paper clip. Attach the other end of the string to the center point of the protractor.

2. Work with a partner to use the hypsometer to measure the height of your school. While one person looks through the straw at the top of the school, the other person measures the angle of the string on the protractor. As you can see in the diagram, the angle that the string makes with the protractor will be the complement of the angle of elevation.

3. Measure the distance to the school.

4. Use the tangent function to find the height of the school. The adjacent side is the distance between you and the school. The opposite side is the school's height. Remember to add the height of your eyes from the ground to the calculated height of the school to get a more accurate measurement. After you calculate the height of the school, compare your answer with those of other students.

5. Use the same method to find the heights of trees or other buildings.

> **Caution**
>
> When measuring the distance to the school, be sure to measure the distance from the point where you used the hypsometer to the point on the school directly below where you viewed the school's height, and not the point on the school closest to where you used the hypsometer.

Lesson Practice

a. Standing 300 feet away from a building, the angle of elevation to the roof is 67°. Find the height of the building to the nearest foot.
(Ex 1)

b. A family sees their house from a window of a tall building. The angle of depression is 55°. The window where the family is standing is 160 meters above the street. How far away is their house from the window, to the nearest meter?
(Ex 2)

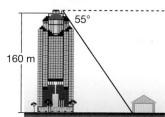

c. Jocelyn sees a hot-air balloon at an angle of elevation of 30°. Anthony sees the same hot-air balloon at an angle of elevation of 50°. The balloon is 140 meters off the ground. They are starting on the same side of the balloon. How far apart are Jocelyn and Anthony, to the nearest meter? Who is farther away from the balloon?

(Ex 3)

d. **Surveying** A surveyor on the top of Blake Hill measures an angle of depression of 16° to the top of Pike Hill. The height of Blake Hill is 400 meters, and the height of Pike Hill is 300 meters. What is the distance between the hilltops, to the nearest meter?

(Ex 4)

Practice   Distributed and Integrated

 *1. **Algebra** A kite is flying at a 45° angle of elevation above the ground. Find the length of string if $x = 20$.

(73)

2. **Multiple Choice** Choose the best answer that compares $m\angle Y$ and $m\angle M$.

(Inv 4)

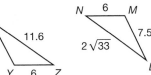

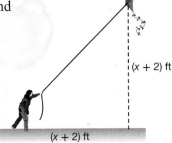

A $m\angle Y = m\angle M$ **B** $m\angle Y > m\angle M$
C $m\angle Y < m\angle M$ **D** Not enough information is given.

* 3. **Model** Draw three circles that share two common tangents.
(72)

4. **Insecticides** A company has developed a new type of insecticide for outdoor patios, which is based on attracting mosquitoes to the figure shown. Which sectors have the greatest probability of attracting mosquitoes?

(Inv 6)

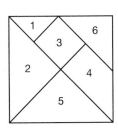

5. **Design** Jameson is making a set of regular pyramids as props for a school play. He makes the bases of all the pyramids with a 1-meter side length and a slant height of 2 meters. Determine a formula for the lateral area of the regular pyramid with n-sided base.

(70)

6. Justify Determine whether △*DEF* is a right triangle, if $\overline{AB} \parallel \overline{DE}$. Explain
(60) how you know.

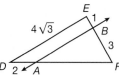

* 7. Assume the segments that appear to be tangent are tangent. Find *x*.
(72)

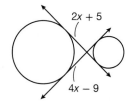

8. Find the length of the radius of a circle given an 8-centimeter chord that
(43) is 4 centimeters from the center of the circle.

9. What arc measure does the needle on a compass create as it goes from north
(26) to northeast?

10. (**Landscaping**) A shop owner wants to build a brick patio, as shown
(40) at right. If each brick covers an area of 18 square inches, how many bricks
will the shop need to cover the entire patio in brick?

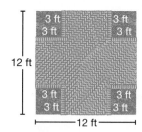

*11. A building's shadow is 136 meters long, and the sun is at an angle of
(73) elevation of 60° above the horizon. Find the height of the building.
Leave your answer in radical form.

12. Find the value of *x* in the figure.
(64)

13. Verify Find the cosines of 8°, 10°, 25°, 30°, 60°, 70° and 90°.
(68) Why is cosine never greater than 1?

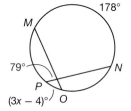

14. In △*ABC*, *A* is a right angle, m∠*B* = 60°, and *BC* = 5 inches.
(Inv 7) **a.** Draw a diagram of △*ABC*.
b. Determine *AB*.
c. Determine exact and approximate values for *AC*.

15. Find the length of $\overset{\frown}{MD}$. Round to the nearest hundredth of an inch.
(35)

16. Verify Quinton was asked to apply the vector translations $u = \langle 3, 8 \rangle$ and
(71) $v = \langle -3, -8 \rangle$ to a series of points. He reasoned that the two vectors
would cause the points to move away from and then back to their
original locations. Is he correct? Explain.

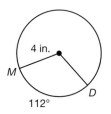

*17. Multiple Choice A pilot in a plane sees a lake at an angle of depression of 31°. The
(73) plane is cruising at 20,000 feet. How far away is the lake from the plane?
    A 17,143 ft   **B** 10,301 ft
    C 23,333 ft   **D** 38,832 ft

18. Find the measures of ∠*B*, ∠*C*, and ∠*D* in trapezoid *ABCD* at right.
(69)

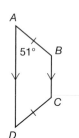

19. (**Flying**) A pilot of a small airplane wants to travel at 225 kilometers per hour over the ground. She knows that she is flying directly into a wind that is blowing at 38 kilometers per hour. What should the speedometer of her airplane read, in order to maintain a ground speed of 225 km/hr?
₍₆₃₎

20. Multi-Step The translation vectors $u_1 = \langle 3, -2 \rangle$ and $u_2 = \langle -1, -3 \rangle$ are applied successively to each of the points $A(-2, 5)$, $B(-1, 3)$, and $C(1, 5)$. Find the coordinates of the image of each point after the translations.
₍₇₁₎

21. List the side lengths of $\triangle JKL$ in order from least to greatest.
₍₃₉₎

22. Model On a coordinate plane, show how to locate the midpoint of the hypotenuse $\overline{AB}$ of a right triangle, using the midpoints of its legs.
₍₁₁₎

*23. (**Surveying**) A surveyor measures a 65° angle of elevation to the top of an 800-foot building. How far is the surveyor from the base of the building, to the nearest foot?
₍₇₃₎

Use the diagram to determine each length.

24. Given that $AP = 123$, determine DP.
₍₃₂₎

25. Given that $BD = 63$, determine BT.
₍₃₂₎

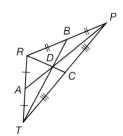

26. Error Analysis Giselle states that since the tile on the right is divided into three sections, there is a one-in-three chance of a fly randomly landing on any section. Is she correct? Explain.
_(Inv 6)

27. Algebra What are the coordinates of the midpoint of $\overline{DE}$ with endpoints $D(5x - 3, 2 - 6y)$ and $E(2x + 3, 2y + 1)$?
₍₁₁₎

28. Justify Is the quadrilateral shown a parallelogram? How do you know?
₍₆₁₎

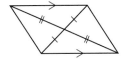

*29. Assume the segments that appear to be tangent are tangent. Find CD, CE, and BC.
₍₇₂₎

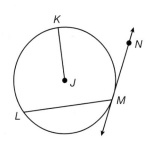

30. Identify each line or segment that intersects the circle.
₍₄₃₎

Warm Up

1. **Vocabulary** Reflections, rotations, and translations are all examples of
(67) transformations that are _____.

2. Reflect rectangle *PQRS* across $\overline{QR}$.
(67)

3. **Multiple Choice** To describe a reflection, you
(67) have to identify

    A the amount of reflection.
    B the center of reflection.
    C the line of reflection.
    D all of the above.

New Concepts

A **reflection** is a transformation that reflects every point in a figure over a given line. After reflection, the image of the figure is congruent to the preimage, but has a different orientation.

Property of Reflection
A reflection is an isometry, meaning the preimage and its reflected image have the same shape and size.

Hint

To review the definition of isometry and the various types of transformations, refer to Lesson 67.

To reflect a point across a horizontal or vertical line, imagine that the line is a mirror, and visualize the reflected location of the point. The figure shows a triangle reflected over the *y*-axis.

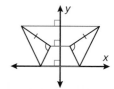

Example 1   **Reflecting Across an Axis**

Reflect △*ABC* across the *y*-axis. Find the coordinates of the vertices of the reflected image and write the transformation in mapping notation.

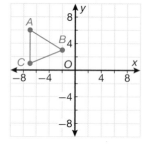

SOLUTION

Imagine each point reflected across a mirror sitting on the *y*-axis. Each point will end up opposite from where it is now in Quadrant II. The *y*-coordinates will not change, but the signs of the *x*-coordinates are reversed. Each point (x, y) will be mapped to $(-x, y)$.

In mapping notation:
$T: (x, y) \rightarrow (-x, y)$.
$T: A(-7, 6) \rightarrow A'(7, 6)$
$T: B(-2, 3) \rightarrow B'(2, 3)$
$T: C(-7, 1) \rightarrow C'(7, 1)$

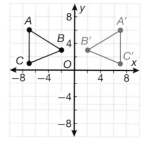

Online Connection
www.SaxonMathResources.com

Example 2 | Reflecting Across a Horizontal Line

Reflect the rectangle *STUV* across the line $y = 4$. Identify the coordinates of the vertices of the reflected image.

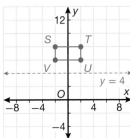

SOLUTION

After a reflection, each point will be the same distance from the mirror as it is now. For example, *S* is 4 units away from the mirror. After reflection, it will still be 4 units away, but in the opposite direction. So it is reflected to (–2, 0), where it is 4 units from $y = 4$. In mapping notation: $T: (x, y) \rightarrow (x, -y + 8)$.

$T: T(2, 8) \rightarrow T'(2, 0)$

$T: U(2, 6) \rightarrow U'(2, 2)$

$T: V(-2, 6) \rightarrow V'(-2, 2)$

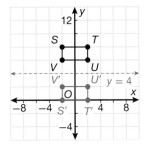

Notice that when a point is reflected across a horizontal line, its *x*-coordinate does not change. When a point is reflected across a vertical line, its *y*-coordinate does not change.

To find the reflection of a point across any line in the coordinate plane, draw a perpendicular line from the point to the line of reflection. The point's reflection will be equidistant from the line of reflection on both sides.

Example 3 | Reflecting Across a Line

Reflect quadrilateral *JKLM* across the line $y = x$. Identify the coordinates of the vertices of the reflected image.

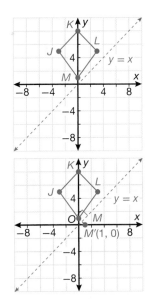

SOLUTION

The perpendicular line to $y = x$ is $y = -x$. In the second diagram, the perpendicular line from *M* to its reflection, *M'* is shown. When a point is reflected over the line $y = x$, it follows the transformation: $T: (x, y) \rightarrow (y, x)$. Apply this to the vertices of the quadrilateral shown.

$T: J(-3, 5) \rightarrow J'(5, -3)$

$T: K(0, 8) \rightarrow K'(8, 0)$

$T: L(3, 5) \rightarrow L'(5, 3)$

$T: M(0, 1) \rightarrow M'(1, 0)$

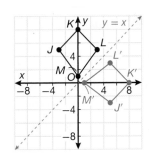

| Example 4 | Application: Visual Arts |

Marina is creating a work of art using part of a photograph and its reflection. In a coordinate grid, the corners of the photograph fragment are located at (−3, 2), (2, 8), and (10, 2). Reflect the fragment across the line $y = 2$.

SOLUTION

Points that lie on $y = 2$ do not move at all, since they are on the line of reflection. The third point (2, 8) is 6 units from the line of reflection. When it is reflected, it will lie 6 units from the line of reflection on its other side, at (2, −4). The transformation is: $T: (x, y) \rightarrow (x, 4 - y)$. Verify that the other 2 points do not move.

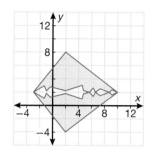

$T: (-3, 2) \rightarrow (-3, 4 - 2) = (-3, 2)$

$T: (10, 2) \rightarrow (10, 4 - 2) = (10, 2)$

Lesson Practice

Rectangle *ABCD* has vertices at *A*(1, 1), *B*(5.5, 1), *C*(5.5, 3.5), and *D*(1, 3.5). Reflect *ABCD* as described in parts a through c.

a. Reflect *ABCD* across the *y*-axis.
(Ex 1)

b. Reflect *ABCD* across the line $y = 2$.
(Ex 2)

c. Reflect *ABCD* across the line $y = x$.
(Ex 3)

d. (**Visual Arts**) This figure shows half of an optical illusion. Complete the figure by reflecting it across the line $x = 4$.
(Ex 4)

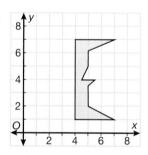

Practice   Distributed and Integrated

1. Find the measure of $\overset{\frown}{JKL}$ in the figure. Line ℓ is tangent to the circle.
(64)

2. Formulate Write a transformation mapping for a reflection *T* across the line $y = -x$.
(74)

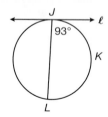

3. Find the distance between points (4, 0) and (5, 6).
(9)

4. Find *DE*. Round to the nearest hundredth.
(68)

5. Multiple Choice Which of the following sets of numbers could be the side lengths of a right triangle?
(33)

A 6, 7, 8

B 5, $3\sqrt{3}$, $2\sqrt{13}$

C 7, 6, $\sqrt{89}$

D 5, 5, 9

6. Algebra In $\triangle PQR$, $\angle P$ is a right angle, $PQ = x$, and $QR = 2x$. In $\triangle STU$, $\angle S$ is a
(36) right angle, $ST = 10 - 3x$, and $TU = y + 2$. For certain values of x and y, you can
use the HL Congruence Theorem to prove that $\triangle PQR \cong \triangle STU$. Determine these
values of x and y.

7. If $\triangle JKL \sim \triangle PQR$, what is the value of a?
(41)

8. $\triangle ABC \sim \triangle WXY$, and the ratio of their
(44) corresponding sides is 4:1. If $BA = 12$,
what is XW?

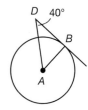

9. (**Average Rainfall**) Sanjay buys a rain barrel to store rainwater for his garden. If
(62) the average rainy day leaves 3 cubic feet of water in his barrel, and the barrel is
4 feet tall with a 1.5-foot radius, how many rainy days, on average, would it take to
completely fill the rain barrel?

10. Error Analysis This circle has center A and radius $\overline{AB}$. Ava has calculated
(58) that $\angle A$ in $\triangle ABD$ is 65°. Is Ava correct? How do you know?

11. (**Computer Generated Images**) A computer-generated triangular prism is made
(71) by constructing a triangle with vertices $(0, 0)$, $(2, 5)$, $(4, 0)$, translating it
using $T: (x, y) \rightarrow (x + 4, y + 6)$ and connecting the vertices of the preimage
to the corresponding vertices of the image. Sketch the triangular prism.

12. Analyze An equilateral triangle is reflected across a line containing one of its sides.
(51, 74) What figure is formed by the image and preimage?

13. What trigonometric ratio can a hiker use to
(73) determine the horizontal distance to the
lake, x.

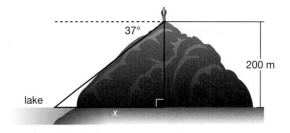

14. Triangle DEF has vertices $D(3, 4)$, $E(7, 2)$, and
(74) $F(6, 6)$. Graph the reflection of $\triangle DEF$ across the
line $x = 2.5$.

15. Find the distance from the line $y = 3x - 1$ to $(5, 4)$.
(42)

16. Determine the missing side lengths in $\triangle PQR$. Give approximate answers to the
(Inv 7) nearest hundredth of a unit.

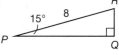

17. Multi-Step A triangle has sides that are 15 feet and 21 feet long, respectively.
(39) What range of values could be the perimeter?

18. Justify The measures of the angles of a triangle are $x°$, $(2x + 5)°$, and $(8x + 10)°$.
(24) Find the value of x. Provide a justification for each step.

19. Reflect $PQRS$ across the line $y = x$.
(74)

 20. Graphing Calculator Use a graphing calculator to verify that the sine function
(68) varies between 0 and 1 by finding the sine of 7°, 14°, 23°, 56°, 78°, and 90°.
Explain why sine cannot be greater than 1.

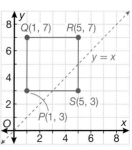

21. Multi-Step Classify the polygon with vertices $C(-2, -1)$, $D(4, 1)$, $E(5, -2)$,
(57) and $F(-1, -4)$. Find its area.

22. **Track Meet** The track at the high school is shown at right. What is the perimeter, to the nearest yard, of the track?
⁽⁴⁰⁾

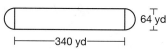

64 yd

340 yd

23. **Algebra** Line g passes through points $(3, -2)$ and $(4, 7)$. Line h passes through
⁽³⁷⁾ points $(1, 5)$ and $(-x, 4)$. What value of x makes these lines perpendicular?

24. Write the equation of the image of the line $y = 3x + 2$ after a reflection across the
⁽⁷⁴⁾ lines in parts **a** and **b**.
   a. Across the line $y = x$.
   b. Across the line $x + y = 0$.

25. **Write** In the figure shown, $\overline{AC}$ is tangent to both circles. $\overline{AD}$ and $\overline{AB}$ are also
⁽⁷²⁾ tangents. Explain why $\overline{AB} \cong \overline{AD}$ even when the circles vary greatly
in size.

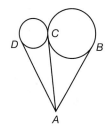

26. If quadrilateral $MNPR$ has three sides of equal length—MN, NP, and
⁽³⁴⁾ PR—must the quadrilateral be a parallelogram?

27. A building is 40 feet tall. If a person stands 40 feet from the bottom of the
⁽⁷³⁾ building and sees the top, what is the angle of elevation from the person to the top of
the building?

28. **Parasailing** A parasail is attached by a 40-foot rope to the boat that is towing it.
⁽⁷³⁾ If the angle of elevation from the boat to the parasail is 34°, about how far above
the water is the parasailer?

29. Determine the point of tangency and the equation of the tangent line.
⁽⁷²⁾ Find the radius of each circle.

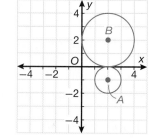

30. **Archaeology** The front of an ancient Greek temple includes an isosceles
^(Inv 7) triangle with base angles that each measure 30°. The legs of the triangle
are each 13 yards long. Determine the height and the base length of the
triangle, to the nearest tenth of a yard.

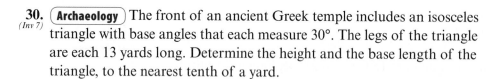

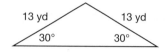

13 yd   13 yd
30°   30°

Writing the Equation of a Circle

1. **Vocabulary** If P is a point on $\odot Q$, then $\overline{PQ}$ is a _____ of the circle.
   (23)

2. If $\odot A$ is centered at the origin of the coordinate plane and has a radius of
   (23) 2, name 4 points that lie on $\odot A$.

3. **Multiple Choice** Which of these numbers are possible values of a, b, and c in
   (8) the equation $a^2 + b^2 = c^2$?

   A 4, 5, 8   **B** 3, 4, 5

   C 2, 4, 6   **D** 5, 8, 13

New Concepts

To analyze a circle on a coordinate plane, it is necessary to derive an equation for the graph of a circle. A simple version of this equation can be found by looking at circles that are centered at the origin.

Math Language

The **unit circle** is a circle with a radius of 1, centered at the origin. The unit circle is frequently used in trigonometry.

Example 1 Analyzing a Circle Centered at the Origin

Suppose $P(x, y)$ is a point on the circle with a 1-unit radius that is centered at the origin. What is QP? How is QP related to x and y?

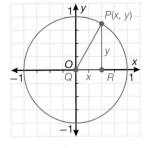

SOLUTION
Since $\overline{QP}$ is a radius of the circle, $QP = 1$. In the diagram, it is shown that $\triangle PQR$ is a right triangle. Therefore, by the Pythagorean Theorem, $x^2 + y^2 = 1$.

The equation given above, $x^2 + y^2 = 1$, gives a relationship between x and y that is true of all points that lie on the circle. Therefore, it is an equation for the circle with a radius of 1 that is centered at the origin. Any circle centered at the origin has an equation given by $x^2 + y^2 = r^2$, where r is the radius.

If the circle is not centered at the origin, this alters the equation. Examine the circle centered at (2, 0) with a radius of 2. For the point (x, y), the distance from the center of the circle to (x, y) is $\sqrt{(x - 2)^2 + (y - 0)^2} = 2$, by the distance formula. So the equation for this circle is $(x - 2)^2 + y^2 = 4$.

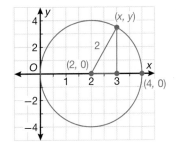

Online Connection
www.SaxonMathResources.com

The Equation of a Circle

The equation of a circle is given by the following formula, where (h, k) is the center of the circle and r is the radius of the circle.

$$(x - h)^2 + (y - k)^2 = r^2$$

Example 2 Writing the Equation for a Circle from a Graph

a. If A is a point on $\odot D$, write the equation of $\odot D$.

SOLUTION
The center of the circle is $D(1, 3)$. Since A is a point on the circle, the radius of the circle is 3 units. Substitute these values into the equation for a circle.

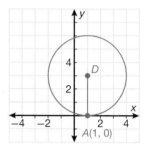

$(x - h)^2 + (y - k)^2 = r^2$ Equation of a circle

$(x - 1)^2 + (y - 3)^2 = 3^2$ Substitute.

b. Circle E is concentric with $\odot D$ and has a radius of 5. Write the equation of $\odot E$.

SOLUTION
Since $\odot E$ has the same center as $\odot D$, only the radius needs to be changed. Substitute in the radius of 5 where 3 was in the equation from part **a.**

$(x - 1)^2 + (y - 3)^2 = 5^2$

Math Language

Two circles are **concentric** if they have the same center.

Example 3 Graphing a Circle Given its Equation

a. The equation of $\odot P$ is $x^2 + y^2 = 25$. Graph $\odot P$.

SOLUTION
Since h and k are both 0 in this equation, the circle is centered at the origin. The radius is $\sqrt{25} = 5$.

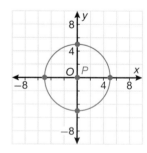

To graph $\odot P$, find some simple points. The endpoints of the horizontal and vertical radii are $(0, 5)$, $(5, 0)$, $(-5, 0)$, and $(0, -5)$. Using these points as guides, draw the circle, as shown in the diagram.

b. The equation of $\odot Q$ is $(x - 2)^2 + (y + 1)^2 = 16$. Graph $\odot Q$.

SOLUTION
Rewrite the equation in the form $(x - h)^2 + (y - k)^2 = r^2$.

$(x - 2)^2 + (y - (-1))^2 = 4^2$

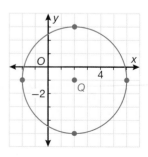

The center of $\odot Q$ is $(2, -1)$ and its radius is 4. Four points on $\odot Q$ are the endpoints of the horizontal and vertical radii: $(2, 3)$, $(6, -1)$, $(2, -5)$, and $(-2, -1)$.

Math Reasoning

Formulate Write the equation for a circle with center (h, k) and a diameter of d.

Example 4 Application: Astronomy

This coordinate grid shows a satellite's orbit around Earth, which is located at the origin. If the satellite's distance from Earth is 23,000 miles, write an equation that describes the satellite's circular orbit.

SOLUTION

The center of the orbit is the origin, and its radius is $OP = 23,000$. Therefore:

$$(x - h)^2 + (y - k)^2 = r^2$$
$$(x - 0)^2 + (y - 0)^2 = 23,000^2$$
$$x^2 + y^2 = 529,000,000$$

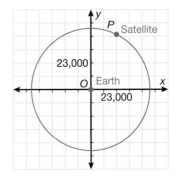

Lesson Practice

a. If $M(x, y)$ is a point on a circle centered at the origin with a radius of 3, what is PM, and what is the equation of the circle?
(Ex 1)

b. Write an equation to relate all the x- and y-coordinates of points that lie on $\odot A$ with a radius of $\sqrt{2}$, which is centered at the origin.
(Ex 1)

c. Write the equation for $\odot B$.
(Ex 2)

d. Circle C is concentric with $\odot B$ and has a radius of 3.5. Write the equation of $\odot C$.
(Ex 2)

e. The equation of $\odot D$ is $x^2 + y^2 = 6.25$. Graph $\odot D$.
(Ex 3)

f. The equation of $\odot E$ is $(x + 1)^2 + (y - 3)^2 = 4$. Graph $\odot E$.
(Ex 3)

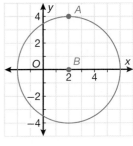

g. (**Sports**) This coordinate grid shows the position of the lines on a basketball court. The center circle crosses the halfcourt line at $A(47, 19)$ and $B(47, 31)$, so $\overline{AB}$ is a diameter. What is the equation of the center circle?
(Ex 3)

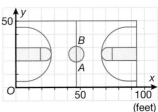

Practice Distributed and Integrated

* 1. Write the equation for $\odot A$.
(75)

2. (**Navigation**) A sailor on a ship sees the top of a distant hill at an 8° angle of elevation. The hill is 450 meters tall. How far is the sailor from the hill, to the nearest meter?
(73)

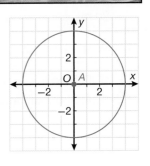

3. Multiple Choice Which is the most correct choice, given the two triangles?
(Inv 4)

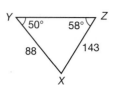 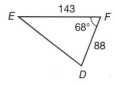

A $YZ > ED$ **B** $m\angle X > m\angle F$
C both **A** and **B** are correct   **D** neither **A** nor **B** are correct

4. Find the geometric mean of 7 and 5.
(50)

5. Error Analysis Minh-li claimed that $m\angle N = 98°$, because base angles of an
(69) isosceles trapezoid are congruent. Determine her error and find the correct
value of $m\angle N$.

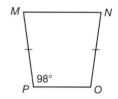

6. (Computer-Generated Images) A computer-generated, square prism is made by
(71) constructing a square with vertices $(0, 0)$, $(0, 2)$, $(2, 2)$, and $(2, 0)$, and then
translating it using $T: (x, y) \rightarrow (x + 3, y + 4)$, and connecting the vertices
of the preimage to the corresponding vertices of the image. Find the volume
of the prism.

* 7. Reflect $\triangle XYZ$ across the line $y = -1$.
(74)

8. (Construction) Yoshi wants to build a square-based pyramid. He wants
(70) to make the slant height of the pyramid twice as long as the height
of the pyramid. If the side lengths of the base are 1 unit each and the
magnitude of the lateral area is 11 units greater than the volume, how
tall should Yoshi make his pyramid?

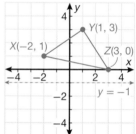

9. Find the length of chord $\overline{LN}$ in the circle at right.
(43)

10. Graphing Calculator Use a calculator to complete a table of
(Inv 7) values for sin x, cos x, and tan x, as x varies from 0° to 90°
in increments of 10°.
  a. Complete this sentence: sin x _____ from __ to __ as
  x increases from 0° to 90°.
  b. Write similar sentences for cos x and tan x.

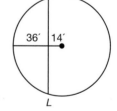

11. If a circle has an area of 144 square meters and a sector of that circle has
(35) an area of 12 square meters, what is the angle measure of the sector?

12. (Landscaping) A hexagonal area in a park has been cleared for a picnic area and
(66) a gazebo to be installed. The park director wants to construct a path from the
gazebo in the center to the edge of the clearing. If the length of the path is
20 meters, how many square meters of park have been cleared?

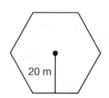

*13. Write the equation for $\odot Q$ with a radius of 3, centered at $Q(2, 4)$.
(75)

14. If a circle has a radius of 7.5 inches, and a chord that is 7 inches long,
(43) what is the distance from the chord to the center of the circle?

15. Find the measure of $\angle D$.
(47)

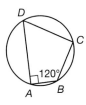

16. Find the magnitude of $\langle -3, 5 \rangle$ to the nearest tenth.
(63)

17. Analyze Determine the number of common tangents
(72) for the circles shown at right, then draw the tangents.

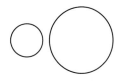

*18. The equation of $\odot B$ is $x^2 + y^2 = 9$. Graph $\odot B$.
(75)

19. Find the slope of line f and then find the slopes of the lines
(37) that are parallel and perpendicular to f.

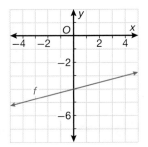

*20. Multiple Choice A reflection across the line
(74) $y = 3$ maps $(4, 0)$ onto:

A $(4, 6)$ **B** $(-2, 0)$

C $(7, 0)$ **D** $(0, 4)$

21. Find the length of $\overline{XY}$. Round to the nearest hundredth.
(68)

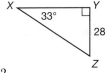

22. (**Aviation**) A plane is flying at an elevation of 10,000 meters and a speed of
(73) 300 kilometers per hour. The pilot sees a lake at an angle of depression of 60°.
In how much time, to the nearest minute, will the plane be directly over the lake?

23. Algebra Rectangle $DEFG$ has vertices $D(x_1, 3)$, $E(0, 0)$, $F(3, 0)$, and $G(x_2, y_2)$.
(45) Find the values of x_1, x_2, and y_2.

24. Is this parallelogram a rectangle, a rhombus, both, or neither?
(65)

*25. The unit circle and $\odot C$ with a radius of r are concentric. Circle D with a radius
(75) of s is centered at $(8, 0)$. If circles C and D touch at one point, write all possible
relations between r and s.

26. Analyze Write a paragraph proof of the following:
(31) **Given:** Ray FD bisects $\angle EFC$, ray FC bisects $\angle DFB$
Prove: $\angle EFD \cong \angle CFB$

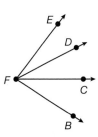

27. Multi-Step The perimeter of $\triangle ABC$ is 54 inches, and its side
(44) lengths are x, $x + 1$, and $x + 2$. If $\triangle ABC \sim \triangle PQR$, and the
ratio of their corresponding sides is 1:2, what are the lengths
of the sides of $\triangle PQR$?

28. Algebra Find the distance from the line $y = \dfrac{5}{3}x + 2$ to $(-5, 5)$.
(42)

*29. Square $ABCD$ with vertices $A(3, 5)$, $B(1, 7)$, $C(-1, 5)$, and $D(1, 3)$ is reflected
(74) across the line $x = 3$. Verify that the reflection image has vertices $A'(3, 5)$,
$B'(5, 7)$, $C'(7, 5)$, and $D'(5, 3)$.

30. Write Can a triangle have sides with lengths of 0.5 centimeters, 0.7 centimeters, and
(39) 0.3 centimeters? Explain why or why not.

Symmetry

Warm Up

1. **Vocabulary** A _____ is a transformation that shifts or slides every
   (71)
   point of a figure or graph the same distance in the same direction.

2. Describe the transformation shown.
   (67)

3. Which equation gives a reflection of $y = x^2$ over
   (74)
   the y-axis?

   A $y = x^2$ **B** $y = 2x$

   C $y = \dfrac{1}{x^2}$ **D** $y^2 = x$

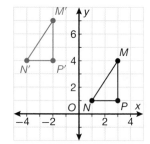

New Concepts

Symmetry is a property illustrated when the image of a transformation of a figure coincides with the preimage. There are several different kinds of symmetry.

Math Language

If two figures coincide, it means they lie directly on top of one another.

A figure has **line symmetry** if the figure can be reflected across a line so that the image coincides with the preimage. The line that divides the figure into two congruent, reflected halves is a **line of symmetry**. The figure at the right has a vertical line of symmetry.

Example 1   Identifying Lines of Symmetry

Identify whether each figure has a line of symmetry. If it does, draw the line of symmetry.

a.

SOLUTION
The heart has one vertical line of symmetry, as shown in the diagram.

b.

SOLUTION
The quadrilateral is not symmetric across any line.

Rotational symmetry is a type of symmetry which describes a figure that can be rotated about a point by an angle less than 360° so that the image coincides with the preimage. The smallest angle through which a figure can be rotated in order to coincide with itself is an **angle of rotational symmetry**. The diagram shows half of a spade that has been rotated 180°. The angle of rotational symmetry is 180°.

The regular hexagon shown has both rotational symmetry and line symmetry. It is symmetrical across a vertical line, a horizontal line, and any line drawn through opposite vertices. It also has a 60° angle of symmetry.

Example 2 Creating Symmetrical Figures Using Transformations

Rotate △*RST* 180° around point *R*. Does the new composite figure have symmetry? What type? Does it matter if you rotate the figure clockwise or counterclockwise?

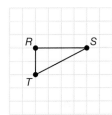

SOLUTION

The new figure has rotational symmetry of 180°. Regardless of whether you rotate the preimage clockwise or counterclockwise, the image will look the same.

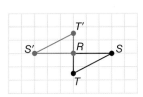

The **order of rotational symmetry** is the number of times a figure with rotational symmetry coincides with itself as it rotates 360°.

A square has an order of 4, as shown in the diagram. The order of rotational symmetry is equal to 360° divided by the angle of rotational symmetry.

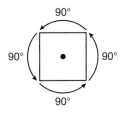

Example 3 Finding Orders of Rotational Symmetry

Tell whether each figure has rotational symmetry. If so, give the angle of rotational symmetry and the order.

a.

b.

c.

SOLUTION

a. yes; 180°; order 2

b. yes; 90°; order 4

c. no rotational symmetry

Example 4 Application: Tiling

a. Tiles often have many lines of symmetry so they can fit easily into patterns. How many lines of symmetry does each regular polygon with up to ten sides have?

SOLUTION

Make a chart to answer this question.

Math Language

Model For any regular polygon with an odd number of sides, what points on the polygon do the lines of symmetry pass through?

Regular Polygon	Number of Lines of Symmetry
Triangle	3
Quadrilateral	4
Pentagon	5
Hexagon	6
Heptagon	7
Octagon	8
Nonagon	9
Decagon	10

b. How many lines of symmetry does a 15-sided tile have?

SOLUTION

The table suggests that a regular polygon has as many lines of symmetry as it has sides, so a 15-sided tile would have 15 lines of symmetry.

Lesson Practice

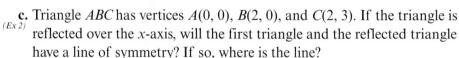

a. Does the star have any lines of symmetry? If so, how many?
(Ex 1)

b. What is the order of rotational symmetry of the star? What is the angle of rotational symmetry?
(Ex 3)

c. Triangle *ABC* has vertices *A*(0, 0), *B*(2, 0), and *C*(2, 3). If the triangle is reflected over the *x*-axis, will the first triangle and the reflected triangle have a line of symmetry? If so, where is the line?
(Ex 2)

d. If there were a regular polygon with 50 sides, how many lines of symmetry would it have?
(Ex 4)

Practice Distributed and Integrated

1. **Fountains** A woman turns her back to a circular fountain with a radius of 3 meters and randomly tosses a penny over her shoulder into the fountain. What is the probability that her penny lands on the triangular piece of art in the fountain?
(Inv 6)

* 2. Complete this chart. Can you make a conjecture about rotational symmetry based
_(76)_ on the information you recorded?

Polygon	Triangle	Quadrilateral	Pentagon	Hexagon	Heptagon	Octagon	Nonagon	Decagon
Order of Rotational Symmetry	3							

3. **Analyze** Genevive reflected $\triangle JKL$ with $J(-2, 3)$, $K(-2, 5)$, and $L(-4, 4)$ across
_(74)_ the line $y = x$. Then she reflected $\triangle J'K'L'$ across the same line. What are the
coordinates of $\triangle J''K''L'''$?

4. **Multiple Choice** Which of the following is *not* perpendicular to $12y = 4x - 6$?
_(37)_ **A** $y = -3x + 4$ **B** $-y + 3x = 12$
C $2y + 6x = 4$ **D** $-9x = 3y - 7$

* 5. Graph the equation $x^2 + y^2 = 36$.
_(75)_

6. In isosceles trapezoid $QRST$, find QS if $TU = 7.8$ feet and $RU = 3.5$ feet.
_(69)_

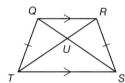

7. **Generalize** On which diagonal can you cut a kite and have two congruent
_(25)_ triangles?

8. **Multi-Step** Complete this indirect proof by writing the indirect proof for
_(48)_ the second case.
    Given: $m\angle N > m\angle M$
    Prove: $LM > LN$
    Indirect Proof: First, assume LM is not greater than LN.
    So, $LM < LN$ or $LM = LN$.
    Case 1: $(LM < LN)$: If $LM < LN$, then $m\angle N < m\angle M$ because the longer side
    would be opposite the larger angle (Theorem 39-1). The inequality $m\angle N < m\angle M$
    contradicts the given information, so LM is not less than LN.
    Case 2: $(LM = LN)$: _____
    The assumption LM is not greater than LN is false, so $LM > LN$.

* 9. **Error Analysis** Akeelah identified this symbol as having 180° rotational symmetry.
_(76)_ Explain her error.

10. (**Ballooning**) A hot air balloon is rising at a rate of 6 feet per second. There is an
_(63)_ updraft of 2 feet per second. If there were no updraft, how fast would the balloon
be rising?

11. What is the lateral area of a square pyramid where the length of one side of the
_(70)_ square is 5 centimeters and the slant height is 10 centimeters?

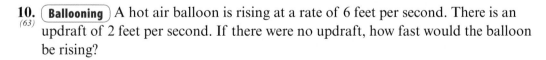

12. Verify Use what you know about areas of composite figures to verify that you
(40) can use the formula $A = 3(as)$ to find the area of the regular polygon in the
figure.

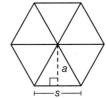

13. If two triangles have two pairs of corresponding congruent angles
(46) and the included sides are proportional, what similarity theorem
or postulate applies?

xy² *14. Algebra Does the function $y = x^2 + 7$ have a line of symmetry? If so,
(76) name the line of symmetry.

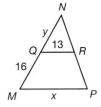

15. Segment QR is a midsegment of $\triangle MNP$. Find the values of x and y.
(55)

16. Multiple Choice In this figure, $\angle 5$ and which other
(12) angle can prove the lines parallel using the
Converse of the Corresponding Angles Theorem?

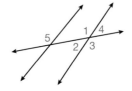

    A 1   **B** 2
    C 3   **D** 4

*17. Verify Check by calculation that the points $P(6, 1)$ and $Q(4, 5)$ lie on $\odot C$ with
(75) center $(1, 1)$ and radius 5.

xy² 18. Algebra Use this triangle to determine the value of z.
(Inv 7)

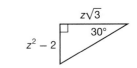

19. Find the area of a regular decagon with 14-centimeter side
(66) lengths and an apothem that is 21.5 centimeters long.

20. (**Hiking**) A hiker looks down from the top of a hill to a lake. The hill is 400 meters
(73) tall, and the angle of depression to the lake is 50°. How far away is the lake, to the
nearest meter?

*21. (**Flags**) Tell whether each flag has rotational symmetry. If yes, determine
(76) the angle of rotation and the order.

22. Rectangle $ABCD$ has vertices $A(0, 4)$, $B(6, 4)$, $C(6, 0)$, and $D(0, 0)$. If E is the
(45) midpoint of DC, find the coordinates of E.

23. The diagonals of parallelogram $DEFG$ intersect at P. Find GE, GF, and
(34) m$\angle GDE$ if $DE = 15$, $EP = 8$, and m$\angle EFG = 130°$.

*24. Multiple Choice Which is the equation of a circle with center $(-3, 4)$ and a radius
$^{(75)}$ of 6?

A $(x + 3)^2 + (y - 4)^2 = 16$ **B** $(x + 3)^2 + (y - 4)^2 = 36$

C $(x - 3)^2 + (y + 4)^2 = 36$ **D** $(x - 3)^2 + (y - 4)^2 = 36$

25. Analyze Explain why the tangent of 90° is undefined. Name one other degree
$^{(Inv 7)}$ measure for which the tangent is undefined.

26. In the proportion $\frac{15}{6} = \frac{8}{y}$, what is the value of y?
$^{(41)}$

27. (**Streets**) On the map, 5th Ave., 6th Ave., and 7th Ave. are parallel.
$^{(60)}$ What is the length of Main St. between 5th and 6th Avenue?

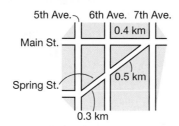

28. Algebra Find the point on the line $y = -x - 4$ that is closest to $(4, 4)$.
$^{(42)}$

29. Write Could opposite vectors also be defined as vectors with the
$^{(63)}$ same direction but opposite magnitude? Why or why not?

30. Reflect square *KMRW* across the line $y = x + 2$.
$^{(74)}$

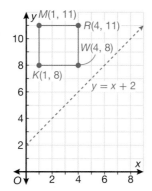

Finding Surface Areas and Volumes of Cones

Warm Up

1. **Vocabulary** The side of a pyramid that is opposite the pyramid's vertex is the _____.
   (70)

2. What is the volume of a pyramid with a height of 5 centimeters and a base area of 12 square centimeters?
   (70)

3. The distance from the vertex of a regular pyramid to the midpoint of an edge of the base is called the:
   (70)
   A altitude   **B** base
   C height   **D** slant height

New Concepts

A cone is a three-dimensional figure with a circular base and a curved lateral surface that connects the base to a point called the vertex. The circular face of a cone is the **base of the cone**. The **vertex of a cone** is the point opposite the base. The **altitude of a cone** is the perpendicular segment from the vertex to the plane containing the base.

If the altitude of a cone has an endpoint on the center of the base, then it is a **right cone**. In an **oblique cone**, the altitude does not intersect the center of the base.

The **slant height of a cone** is the distance from the vertex of the cone to a point on the edge of the base. The **radius of a cone** is the radius of the base of the cone.

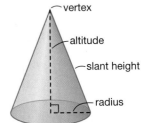

Math Language

As with other solids, the **lateral surface area of a cone** is the total surface area minus the area of the base.

Lateral Surface Area of a Cone
The lateral surface area, L, of a right cone is given by this formula, where r is the radius of the cone, and l is the slant height. $$L = \pi r l$$

Example 1   Calculating the Lateral Area of a Cone

a. Calculate the lateral area of the cone to the nearest hundredth of a square inch.

SOLUTION
The radius of the cone is 3 inches and the slant height of the cone is 7 inches. Use the formula for lateral surface area.

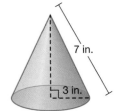

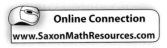
Online Connection
www.SaxonMathResources.com

$L = \pi r l$ Formula for lateral surface area

$L = \pi(3)(7)$ Substitute.

$L \approx 65.97 \text{ in}^2$ Simplify.

The lateral area of the cone is approximately 65.97 square inches.

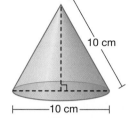

b. Calculate the lateral area of the cone to the nearest hundredth of a centimeter.

SOLUTION

The diameter of the cone is 10 centimeters, so the radius of the cone is 5 centimeters. The slant height of the cone is 10 centimeters. Use the formula for lateral surface area.

$L = \pi r l$ Formula for lateral surface area

$L = \pi(5)(10)$ Substitute.

$L \approx 157.08 \text{ cm}^2$ Simplify.

The lateral area of the cone is approximately 157.08 square centimeters.

Math Reasoning

Formulate How much would the lateral area of this cone increase if the slant height were doubled?

The surface area of a cone is the sum of the area of the base and the lateral area.

Surface Area of a Cone
The total surface area of a right cone is given by this formula, where L is the lateral area, and B is the area of the base. $$S = L + B$$

Example 2 **Calculating the Surface Area of a Cone**

Calculate the surface area of a cone, in terms of π, with a slant height of 5 centimeters and a base radius of 4 centimeters.

SOLUTION

First, use the formula for lateral surface area of a cone.

$L = \pi r l$ Formula for lateral surface area

$L = \pi(4)(5)$ Substitute.

$L = 20\pi$ Simplify.

Now substitute the lateral surface area and the formula for the area of a circle into the formula for total surface area.

$S = L + B$ Formula for surface area

$S = 20\pi + 16\pi$ Substitute.

$S = 36\pi$ Simplify.

The surface area of the cone is 36π square centimeters.

Hint

Recall that the area of a circle is given by the formula, $A = \pi r^2$.

Just as the volume of a pyramid is one third the volume of a prism with the same base and height, the volume of a cone is one third the volume of a cylinder with the same base and height.

Volume of a Cone
The volume, V, of a cone is given by this formula, where B is the area of the base, and h is the length of the cone's altitude. $$V = \frac{1}{3}Bh$$

Example 3 **Calculating the Volume of a Cone**

Math Reasoning

Formulate Write a formula for the volume of a cone in terms of the radius and the slant height.

What is the volume of a right cone with a radius of 3 and a height of 5? Round your answer to the nearest tenth.

SOLUTION
The base of a cone is a circle.

$V = \frac{1}{3}Bh$	Volume of a cone
$V = \frac{1}{3}\pi r^2 h$	Substitute area of a circle.
$V = \frac{1}{3}\pi (3)^2(5)$	Substitute.
$V \approx 47.1$	Simplify.

The volume is approximately 47.1 cubic units.

Example 4 **Application: Cooking**

Funnel cake is made by filling a funnel with batter and letting the batter run out the bottom of the funnel. If the funnel is a right circular cone that is 3 inches tall and has a radius of 1.5 inches, how much batter can it hold, to the nearest hundredth of a cubic inch?

SOLUTION
To find out how much batter the funnel can hold, determine its volume.

$V = \frac{1}{3}Bh$	Volume of a cone
$V = \frac{1}{3}\pi r^2 h$	Substitute area of a circle.
$V = \frac{1}{3}\pi (1.5)^2(3)$	Substitute.
$V \approx 7.07$	Simplify.

The funnel can hold approximately 7.07 cubic inches of batter.

Lesson Practice

a. Calculate the lateral area of a right cone with a radius
(Ex 1) of 2 feet and a slant height of 7 feet.

b. Calculate the total surface area of the cone in the
(Ex 2) diagram.

c. Calculate the volume of a right cone with a height
(Ex 3) of 10 inches and a radius of 6 inches.

d. (**Safety**) A traffic cone is about 70 centimeters high and has a diameter
(Ex 4) of 20 centimeters. What is the volume of the cone?

* 1. Calculate the lateral surface area of a cone with a radius of 3 centimeters and a
(77) slant height of 5 centimeters.

* 2. Multiple Choice Which inequality gives the range of the cosine function?
(Inv 7)
    A $0 < y < 1$ **B** $y > 0$
    C $0 \le y \le 1$ **D** $y \le 1$

3. **(Rockets)** A model rocket's fuel tank is in the shape of a cylinder with a radius of
(62) 1 foot and a height of 3 feet. To the nearest hundredth, how much fuel can the
rocket hold if one cubic foot holds 7.5 gallons of fuel?

4. Find the range of possible lengths of the third side of a triangle if the other two
(39) side lengths are 115 feet and 266 feet.

5. Multiple Choice If _____ are congruent, then lines p and q are parallel.
(12)
    A $\angle 1$ and $\angle 2$ **B** $\angle 2$ and $\angle 4$
    C $\angle 1$ and $\angle 8$ **D** $\angle 1$ and $\angle 7$

6. **(Municipal Planning)** On a grid of a city, the fire station is located at the origin,
(75) $(0, 0)$. One square on the grid represents one square mile. The fire station's new
siren can be heard in an area described on the grid as $x^2 + y^2 = 10$. Would a
house located at $(0, 5)$ be able to hear the new siren? What is the farthest distance,
to the nearest hundredth mile, that a house could be from the fire station and still
hear the siren?

7. Model Draw three internally tangent circles that are tangent to each other at
(72) exactly one point.

8. Algebra $\triangle PQR \sim \triangle KLM$. Given $PQ = x - 2$, $QR = y^2$, $RP = 6$, $KL = 9$, $LM = 12$,
(44) and $MK = 18$, find the values of x and y. Then, find the perimeter of $\triangle PQR$.

9. What is the radius of a circle if the area of a $60°$ arc is 6π?
(35)

10. Compare PS and PQ.
(Inv 4)

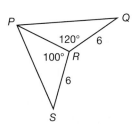

*11. Calculate the surface area of a cone with a radius of 7 inches and a slant height of
(77) 10 inches. Round your answer to the nearest square inch.

12. Triangle PQR has vertices $P(-2, 1)$, $Q(2, 2)$, and $R(0, 5)$. What are the coordinates
(74) of $\triangle P'Q'R'$ after $\triangle PQR$ is reflected across the line $x = 1$?

*13. Multi-Step Does quadrilateral $HJKL$ with the given vertices have line symmetry or
(76) rotational symmetry? $H(3, 3)$, $J(2, -1)$, $K(0, -1)$, and $L(-1, 3)$.

14. Which congruence theorem applies to these triangles?
(36)

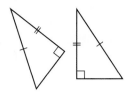

15. (**Surveying**) A surveyor knows the height of a building is 70 meters.
(73) If the angle of elevation to the top of the building from where he stands measures 17°, how far is the surveyor from the building, to the nearest meter?

16. Estimate A triangular loom is made from wooden pieces that form a 45°-45°-90°
(53) triangle. Pegs are placed every half-inch along the hypotenuse and every $\frac{1}{4}$ inch along each leg. If the loom has a 15-inch hypotenuse, estimate how many pegs are on the loom.

17. (**Design**) Blake is making a kite. The frame is a quadrilateral with diagonals that
(69) are perpendicular. Does this guarantee that Blake is making a kite-shaped object? Explain.

*18. (**Cooking**) A waffle cone has a diameter of 7 centimeters and a height of
(77) 14 centimeters. What area of waffle dough is needed to make the cone, to the nearest square centimeter?

19. Find the measure of $\overset{\frown}{JKL}$ in the figure. Line ℓ is tangent to the circle.
(64)

20. Write the formula for $\odot Q$ with a radius of 2, centered at $Q(0, 3)$.
(75)

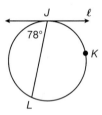

21. The lengths of the sides of a pentagon are 6, 7, 8, 9, and 10 units. A similar
(41) pentagon has side lengths of 21, 24.5, 28, 31.5, and 35 units, respectively. By what factor are the side lengths of the smaller pentagon multiplied to get the side lengths of the larger pentagon?

*22. Write Describe the lines of symmetry and the rotational symmetry of a
(76) parallelogram.

23. Justify The vertices of parallelogram *LMNP* are $L(-4.5, -3.5)$, $M(-2, -1)$,
(65) $N(1.5, -1.5)$, and $P(-1, -4)$. Is it a rhombus or a rectangle? Explain.

24. Algebra Write an inequality to show the possible values for *x*
(33) that make this triangle obtuse.

*25. Name a type of triangle that has exactly one line of symmetry.
(76)

26. Write Explain why the AA Postulate could also be called the AAA Postulate.
(46)

*27. Calculate the volume of a cone with a 1-meter radius and a height of 2 meters.
(77) Round your answer to the nearest tenth of a cubic meter.

28. What is the relationship between inscribed angles that intercept the same arc?
(47)

29. A circle has a diameter of 21 inches. A chord is 5 inches from the center. How
(43) long is the chord to the nearest tenth of an inch?

30. Formulate Write a paragraph proof.
(31) **Given:** ∠*JKL* is a right angle.
Prove: ∠1 and ∠2 are complementary angles.

LESSON 78

Rotations

New Concepts   A rotation is a transformation about a point. The **center of rotation** is the fixed point around which a figure is rotated. Rotations are assumed to be measured as a counterclockwise turn unless otherwise stated.

Rotations
A rotation is an isometry, meaning that the preimage and its rotated image are the same shape and size.

In the diagram, $\triangle ABC$ has been rotated around the center of rotation R. The resulting image, $\triangle A'B'C'$, is congruent to $\triangle ABC$.

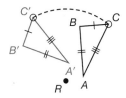

Math Reasoning

Formulate Write a transformation mapping for a rotation around the origin of 360°.

For rotations around the origin with angles of rotation that are multiples of 90°, the following transformation mapping notations apply.

If a point (x, y) is rotated 90° about the origin: $T: (x, y) \rightarrow (-y, x)$.

If a point (x, y) is rotated 180° about the origin: $T: (x, y) \rightarrow (-x, -y)$.

If a point (x, y) is rotated 270° about the origin: $T: (x, y) \rightarrow (y, -x)$.

Example 1 Rotating About the Origin

If $\triangle MNP$ has vertices $M(1, 1)$, $N(4, 3)$, and $P(5, 2)$, graph the triangle and its rotation 180° counterclockwise about the origin.

SOLUTION

Graph $\triangle MNP$ on a coordinate plane. Use the rule for a 180° rotation to find the vertices of the rotated triangle, $\triangle M'N'P'$.

$T: (x, y) \rightarrow (-x, -y)$

$M(1, 1) \rightarrow M'(-1, -1)$

$N(4, 3) \rightarrow N'(-4, -3)$

$P(5, 2) \rightarrow P'(-5, -2)$

Graph $\triangle M'N'P'$.

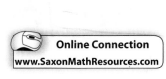

Rotating a figure around a point that is not the origin can be more difficult. For a 180° rotation, there is still a simple transformation mapping, given below.

If a point (x, y) is rotated 180° about the point (a, b):
$T: (x, y) \rightarrow (2a - x, 2b - y)$.

To rotate a figure 90° around a point that is not the origin, consider the diagram of the point E. To rotate E around the point (a, b), notice that the two points have the same y-coordinate. After a 90° turn, the new point, F, will lie directly above the point of rotation and therefore will have the same x-coordinate. It will remain the same distance away from the point of rotation. If, for example, E was originally 2 units to the right of the point of rotation, F will be 2 units above the point of rotation.

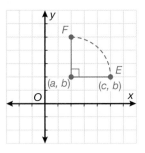

Example 2 Rotating About a Chosen Point

a. Rotate the point $(-3, 4)$ 90° counterclockwise about the point $(-3, 6)$.

SOLUTION
Plot the points as shown in the diagram. After a 90° rotation, the rotated image of $(-3, 4)$ will lie to the right of the center of rotation.

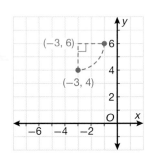

The point $(-3, 4)$, by application of the distance formula, is 2 units away from the center of rotation. The image will also be 2 units away, but will lay 2 units to the right instead of 2 units below the center of rotation. Count 2 units over from the center of rotation or add 2 to the center of rotation's x-coordinate. The rotated point is $(-1, 6)$.

b. Rotate the point $(7, 8)$ 180° around the center of rotation $(-2, 3)$.

SOLUTION
Use the transformation mapping given above: $T: (x, y) \rightarrow (2a - x, 2b - y)$.

$T: (7, 8) \rightarrow (2(-2) - 7, 2(3) - 8)$
$T: (7, 8) \rightarrow (-11, -2)$

Example 3 Application: Amusement Parks

Look at the Ferris wheel at right. Find the angle of rotation that transforms M to move to M'. Explain.

SOLUTION
There are 20 supporting arms, or spokes, on the Ferris wheel. Find the measure of the angle made by two adjacent spokes.

$360° \div 20 = 18°$

The spokes divide the wheel into 20 angles of 18°. M' is 7 places counterclockwise from M.

$18° \times 7 = 126°$

The Ferris wheel must rotate 126° for a seat in position M to move to M'.

Lesson Practice

a. $\triangle ABC$ has vertices $A(-2, -3)$, $B(1, 1)$, and $C(2, -1)$. Graph $\triangle ABC$
(Ex 1) and its image after a 90° rotation. List the coordinates of the vertices of $\triangle A'B'C'$.

b. Triangle DEF has vertices $D(0, -2)$, $E(1, 0)$, and $F(3, -1)$. Find the
(Ex 2) coordinates of the vertices of the image if $\triangle DEF$ is rotated 180° about the point $R(-1, 1)$?

c. **Automotive** Look at the rim of a car tire. To the
(Ex 3) nearest degree, find the angle of rotation required for support spoke A to move counterclockwise to position B. Explain.

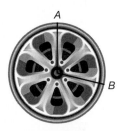

Practice Distributed and Integrated

1. Write an equation for $\odot A$.
(75)

2. **Multiple Choice** Brennan's house has a perimeter of 2500 feet. The perimeter
(44) of Shaniqua's house is 0.8 times as large. Which is the correct ratio of the perimeter of Brennan's house to the perimeter of Shaniqua's house?

A 4 to 5   **B** 1 to 8
C 5 to 4   **D** 25 to 0.8

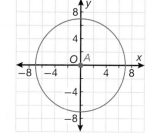

3. Assume the segments that appear to be tangent are tangent. Find x.
(72)

4. **Error Analysis** Robin finds that a hexagon has an apothem length of $\frac{\sqrt{3}}{2}s$,
(66) where s is the side length, but Gustavo finds that it is $\sqrt{3}\,s$. Which student is correct? What error has the other made?

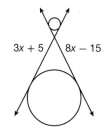

5. **Multiple Choice** Which of these diagrams could not be the net of a
(Inv 5) triangular pyramid?

A

B

C

D

* 6. △*FGH* has vertices *F*(4, 5), *G*(1, 2), and *H*(5, 1). Graph △*FGH* and its image after
(78) a rotation of 270° counterclockwise about the origin. List the coordinates of the
vertices of △*F′G′H′*.

7. Write Describe the lines of symmetry and the rotational symmetry of a non-
(76) isosceles trapezoid.

8. Find the line parallel to $y - 7 = 3(x + 4)$ that passes through the point (2, 3).
(37) Write the equation in slope-intercept form.

9. Coordinate Geometry Use coordinate geometry to show that segments that join the
(45) midpoints of sides of a quadrilateral form a parallelogram.

10. (**Design**) An ecology organization is having a new logo designed. The
(74) logo and its reflection across the line $x = 4$ are shown at right. Complete
the logo.

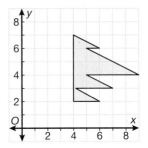

*11. △*ABC* has vertices *A*(−2, 4), *B*(3, 6), and *C*(1, 3). Graph △*ABC* and its
(78) image after a rotation of 90° counterclockwise about the origin. List the
coordinates of the vertices of △*A′B′C′*.

12. Find the value of *x*, *y*, and *z* in the parallelogram.
(34)

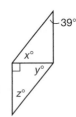

*13. Generalize A special cone has the property that its
(77) radius is the same length as its height. Determine a
formula for the volume of such a cone in terms of
the radius, *r*.

14. (**Refreshments**) A large cylindrical cooler is $2\frac{1}{2}$ feet high and has a diameter of
(62) $1\frac{1}{2}$ feet. It is $\frac{3}{4}$ full of water for athletes to use during their soccer game. Estimate
the volume of water in the cooler, to the nearest gallon. *Hint: 1 gallon = 231 in³*

15. Multi-Step The height of a regular square pyramid is equal to the side length of the
(70) base. The expression $3x - 17.5$ represents the height and $6y - 3$ represents the
side length of the base. If $y = 7$, what is *x*? What is the volume of the pyramid?
What is its surface area?

16. Model For what type of triangle can the orthocenter be in the triangle's exterior:
(32) acute, right, or obtuse? Sketch an example.

17. (**Aviation**) A pilot in a plane notices a lake at an angle of depression of 13°. The
(73) plane is flying at an altitude of 10,000 feet. How far is the pilot from the lake, to
the nearest foot?

18. Compare the measures of ∠*I* and ∠*L*.
(Inv 4)

*19. (**Design**) Rissa is making a cone, including the base, out of paper.
(77) She wants the height of the cone to be 12 centimeters and the
radius of the cone to be 5 centimeters. How much paper will
Rissa need?

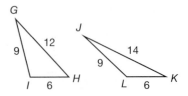

20. Assume the segments that appear to be tangent are tangent. Find *AC*, *CD*,
⁽⁷²⁾ *CE*, and *BC*.

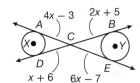

*21. △*RST* has vertices *R*(1, 4), *S*(2, 1), and *T*(4, 5). Graph △*RST* and its image
⁽⁷⁸⁾ after a rotation of 90° about the origin.

22. Find the distance from (5, 3) to the line $y = \frac{4}{3}x$.
⁽⁴²⁾

23. Analyze Maribeth and Claudia are standing on opposite sides of a building.
⁽⁷³⁾ Maribeth is looking at the roof of the building with a 45° angle of elevation
and Claudia is looking at the roof of the building with a 30° angle of elevation.
If the building is 104 feet tall, how far away is Maribeth? How far away is
Claudia? Round both answers to the nearest foot.

24. (Masonry) A mason is laying bricks for a fountain. The water is to
⁽⁴⁰⁾ pour out of a semicircular gap in the structure with a 3-foot
diameter, and there will be a brick trapezoidal patio around the
fountain, as shown in the diagram. If each brick covers 24 square
inches, estimate the number of bricks needed to build the patio
and fountain.

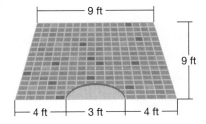

25. Write a valid conclusion from these statements using either the
⁽²¹⁾ Law of Syllogism or the Law of Detachment, then state which law you used.

If it is Sunday or a holiday, then the mail will not be delivered today.
It is Sunday or a holiday.

26. Analyze For what natural-number values of *n* does the equation $y = x^n$ have a line
⁽⁷⁶⁾ of symmetry?

27. Algebra If in △*ABC*, m∠*ABC* = 62°, *AB* = 11, and m∠*BCA* = (22*x* − 11)°, and in
⁽³⁰⁾ △*DEF*, m∠*DEF* = 62°, *DE* = 11, and m∠*EFD* = (25*x* − 44)°, what value of *x* will
make △*ABC* ≅ △*DEF* by AAS Congruence?

28. Find the measure of ∠*FGI*.
⁽⁴⁷⁾

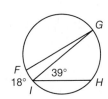

*29. △*JKL* has vertices *J*(−5, −3), *K*(−2, −5), and *L*(−3, 1). Graph △*JKL* and its
⁽⁷⁸⁾ image after a rotation of 180° about the origin.

30. Write Explain why the slant height of a cone is always greater than the height
⁽⁷⁷⁾ of a cone.

Transformations Using Geometry Software

Technology Lab 10 *(Use with Lesson 78)*

So far, you have learned about three different transformations: translations, reflections, and rotations. In this lab, you will use geometry software to construct the transformations you have studied.

(1.) The first section of this lab will demonstrate how to translate a triangle 10 centimeters horizontally. Begin by using the segment tool to construct △*ABC*. Click on each point and choose *Label Points* from the *Display* menu.

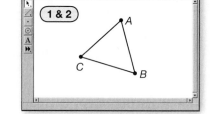

(2.) Select your triangle and choose *Translate* from the *Transform* menu.

(3.) Choose *Rectangular* as the translation vector. Enter "10 cm" in the horizontal distance field and "0 cm" in the vertical distance field.

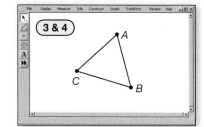

(4.) Measure each side and angle of △*ABC* and its image.

Caution

To measure an angle, select the three points that define the rays of the angle. You must select the vertex of the angle as the second point to ensure you are measuring the correct angle.

(5.) What is the relationship between the original triangle and the translated triangle? Drag a vertex or side of △*ABC*. Does the relationship between the two triangles change?

Next, you will observe the effect of rotating a figure. Begin by constructing △*PQR* using the segment tool.

(1.) Select △*PQR* and choose *Rotate* from the *Transform* menu. Rotate the triangle 90°.

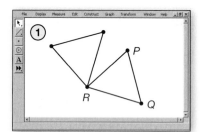

(2.) Measure each side and angle of △*PQR* and its image.

(3.) What is the relationship between the original triangle and the rotated triangle? Drag a vertex or side of △*PQR*. Does the relationship between the two triangles change?

The reflection image of a figure can be described as a translation followed by a rotation. To reflect a figure, choose *Show Grid* from the *Graph* menu.

(1.) Construct △*GHI*. Place each vertex using whole-number coordinates.

(2.) Find the coordinates of each vertex of △*GHI* and fill in the first column of the chart on the next page.

(3.) Select the *x*-axis. Choose *Mark Mirror* from the *Transform* menu.

(4.) Select △*GHI* and choose *Reflect* from the *Transform* menu.

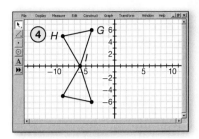

(5.) Fill out the second column of the chart below. What effect does reflection over the *x*-axis have on the coordinates of each vertex?

(6.) Using the line tool, draw the line $y = x$ on your coordinate plane.

(7.) Reflect △*GHI* across the line $y = x$. Fill out the third column of the table. What effect does reflection over the linear parent function have on the coordinates of each vertex?

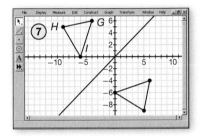

	Original Ordered Pair	**Ordered Pair after Reflection over *x*-axis**	**Ordered Pair after Reflection over $y = x$**
Vertex *G*			
Vertex *H*			
Vertex *I*			

Lab Practice

Construct the triangle with vertices at (3, 2), (7, 5), and (1, 8). Reflect this triangle over the *y*-axis. What are the new coordinates of each vertex? Reflect it over the linear parent function. What are the coordinates of each vertex after this transformation?

Angles Exterior to Circles

Warm Up

1. **Vocabulary** A _____ is a line that intersects a circle at two points.
   (43)

2. What is the circumference of a circle with a 7-inch diameter, in terms
   (23) of π?

3. **Multiple Choice** Which is closest to the length of the minor
   (35) arc shown?

   A 28.7 in. **B** 14.3 in.

   C 75.4 in. **D** 37.7 in.

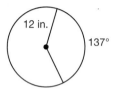

New Concepts

The **exterior of a circle** is the set of all points outside the circle. An exterior angle has its vertex outside of a circle.

Theorem 79-1

The measure of an angle whose vertex is outside of a circle is equal to half the difference of the intercepted arcs.

$$m\angle 1 = \frac{1}{2}(m\widehat{CE} - m\widehat{BD})$$

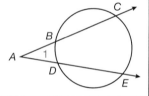

Example 1 Proving Theorem 79-1

Math Language

A **secant** is a line that intersects a circle at two points. A **tangent** intersects a circle at just one point.

There are three different cases to consider when applying Theorem 79-1. The first case is when a secant and a tangent intersect outside the circle.

Given: tangent $\overrightarrow{CD}$ and secant $\overrightarrow{CA}$

Prove: $m\angle ACD = \frac{1}{2}(m\widehat{AD} - m\widehat{BD})$

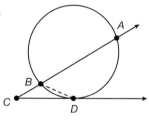

SOLUTION

Draw $\overline{BD}$. By the Exterior Angle Theorem, $m\angle ABD = m\angle ACD + m\angle BDC$, so $m\angle ACD = m\angle ABD - m\angle BDC$. By the Inscribed Angle Theorem, $m\angle ABD = \frac{1}{2}m\widehat{AD}$. The measure of an angle formed by a tangent and a chord is equal to half the measure of the intercepted arc, so $m\angle BDC = \frac{1}{2}m\widehat{BD}$. By substitution, $m\angle ACD = \frac{1}{2}m\widehat{AD} - \frac{1}{2}m\widehat{BD}$. Therefore, by the Distributive Property of Equality, $m\angle ACD = \frac{1}{2}(m\widehat{AD} - m\widehat{BD})$.

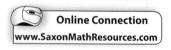

Online Connection
www.SaxonMathResources.com

The proof above demonstrates the first of three ways to apply Theorem 79-1: when the rays of the angle are a secant and a tangent. The second is when the rays of the angle are two tangents, and the third is when they are two secants.

Example 2 Applying Theorem 79-1

a. Find m∠E.

SOLUTION

Use the formula given in Theorem 79-1.

$$m\angle E = \frac{1}{2}(m\widehat{CG} - m\widehat{DF})$$

$$m\angle E = \frac{1}{2}(49° - 13°)$$

$$m\angle E = 18°$$

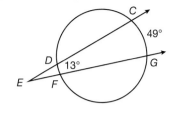

b. Find m∠Q.

SOLUTION

The major arc $\widehat{PSR}$ measures 250° since the sum of its measure and $m\widehat{PR}$ is 360°.

$$m\angle Q = \frac{1}{2}(m\widehat{PSR} - m\widehat{PR})$$

$$m\angle Q = \frac{1}{2}(250° - 110°)$$

$$m\angle Q = 70°$$

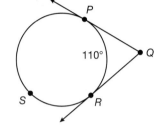

c. Find m∠T.

SOLUTION

First, find the measure of the minor arc $\widehat{SV}$. By the Arc Addition Postulate, $m\widehat{SU} + m\widehat{UV} + m\widehat{SV} = 360°$, so $m\widehat{SV}$ is 138°.

$$m\angle T = \frac{1}{2}(m\widehat{SV} - m\widehat{SU})$$

$$m\angle T = \frac{1}{2}(138° - 68°)$$

$$m\angle T = 35°$$

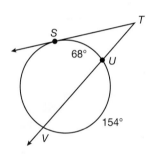

Example 3 Application: Eclipses

A lunar eclipse covers an arc of the earth in the moon's shadow. If observers in the 168° arc experience a full or partial eclipse of the moon, solve for x.

SOLUTION

The major arc of the earth measures 192°. Use Theorem 79-1 to find the value of x.

$$x = \frac{1}{2}(192° - 168°)$$

$$x = 12°$$

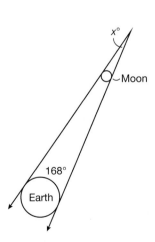

a. Prove the second case of Theorem 79-1.
(Ex 1) **Given:** tangents $\overrightarrow{FE}$ and $\overrightarrow{FG}$

Prove: $m\angle F = \frac{1}{2}(m\widehat{EHG} - m\widehat{EG})$

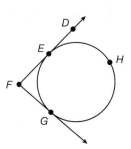

b. Find $m\angle P$.
(Ex 2)

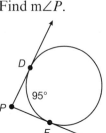

c. Find $m\angle 1$.
(Ex 2)

d. Find $m\angle 2$.
(Ex 2)

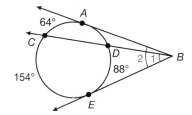

e. If a flashlight illuminates a 154°–arc on a spherical object, how many
(Ex 3) degrees wide is the beam as it leaves the flashlight?

* 1. **Multiple Choice** Which of the following is true of $\angle X$?
(79)
A $m\angle X$ is greater than $m\widehat{VZ} - m\widehat{WY}$.
B $m\angle X$ is less than $m\widehat{WY}$.
C $m\angle X$ is less than $m\widehat{VZ}$.
D $m\angle X$ is greater than $m\widehat{VZ}$.

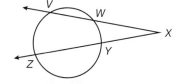

2. **Flying** A pilot is flying his airplane in winds blowing south at 20 kilometers
(63) per hour. What is the vector the pilot needs to fly along in order to have a
velocity of 80 kilometers per hour north?

* 3. Sketch the image of the figure after a rotation of 270° counterclockwise about
(78) the origin.

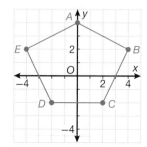

4. Multiple Choice Which is a counterexample to the following conjecture?
(14)

If a triangle is scalene, then it is obtuse.

- **A** triangle with side lengths of 3, 4, and 4.5 units
- **B** triangle with side lengths of 3, 4, and 6 units
- **C** triangle with angle measures of 20°, 40°, and 120°
- **D** triangle with angle measures of 45°, 45°, and 90°

5. Write the equation for the circle at right.
(75)

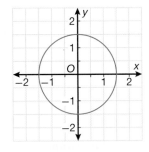

6. Error Analysis Maddie measured the side lengths of a triangle and found them to be 2.4 centimeters, 3 centimeters, and 0.55 centimeters. How does she know that her measurements are incorrect without checking again?
(39)

7. Generalize A property of a special cone is that its radius is the same as its height. Determine a formula for the total surface area of such a cone in terms of the radius, r.
(77)

* 8. Verify What is m∠BAC?
(79)

9. Find the perimeter of a regular hexagon when the endpoints of one side are (2, 4) and (2, 2).
(45)

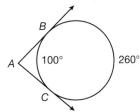

10. Algebra Find the range of values for x using the Hinge Theorem.
(Inv 4)

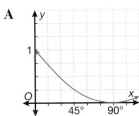

11. Write Describe the lines of symmetry and the rotational symmetry of a kite.
(76)

12. Multiple Choice Which could be the graph of cos x?
(Inv 7)

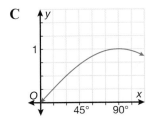

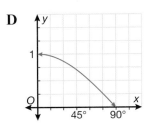

13. **Chess** A chessboard can be a model of a coordinate grid, where each
(71) space is 1 unit from its neighbors. In two moves, a chess piece starts at
point *X* and moves to point *Y*. What vector represents this
movement?

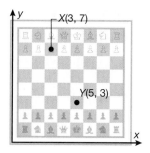

14. What is the volume of a cone with a radius of 8 millimeters
(77) and a slant height of 17 millimeters, to the nearest tenth of a
millimeter?

15. **Verify** What are two similarities that show △*ABC* ~ △*ADE*?
(46)

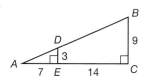

16. Find the distance between the line $y = x$ and the point (2, 8).
(42)

*17. △*DEF* has vertices *D*(3, −3), *E*(0, −1), and *F*(−2, −4). Graph △*DEF* and rotate it
(78) 180° counterclockwise about the origin. List the vertices of its image △*D′E′F′*.

18. **Painting** Kevin is painting a trapezoid in a mural. The height of the trapezoid is
(69) 3 feet and the length of the midsegment is 6 feet. If he can paint 9 square feet with
one ounce of paint, how many ounces of paint does he need?

19. Given that *C*(−1, 2) is on circle *A* and *D*(4, −1) is on circle *B*, determine the point
(72) of tangency of the two circles and the equation of the tangent line. Find the
radius of each circle.

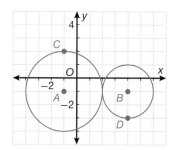

20. **Logos** This logo consists of a circle tangent to the sides of a
(58) rhombus. The diagonal $\overline{AE}$ of the rhombus is 156 millimeters
long, and the diagonal $\overline{CG}$ of the rhombus is 60 millimeters long.
a. Prove that △*AOB* is similar to △*BOC*.
b. What is the perimeter of the whole logo?

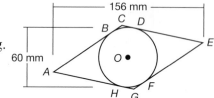

*21. Find m∠*Q*. $\overrightarrow{QR}$ is tangent at *R*.
(79)

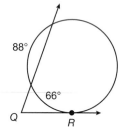

*22. Line segment $\overline{AB}$ has endpoints $A(-4, -2)$ and $B(-1, 5)$. Graph the line segment,
(78) and then rotate it 180° counterclockwise about the origin. List the coordinates of
the endpoints of its image, $\overline{A'B'}$.

23. Use the Leg-Angle Congruence Theorem to show that $\triangle BCD \cong \triangle EFG$.
(36)

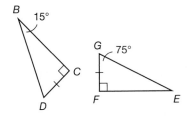

24. Write the equation of a circle with a center at $(5.2, -3.4)$ and a diameter of 7.
(75)

*25. Analyze Write a paragraph proof of the third case of
(31) Theorem 79-1.
Given: secants $\overline{LJ}$ and $\overline{LN}$
Prove: $m\angle JLN = \frac{1}{2}(m\widehat{JN} - m\widehat{KM})$
Hint: Draw $\overline{JM}$.

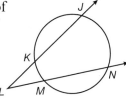

26. Write Explain how to find the other two side lengths of a 30°-60°-90°
(56) right triangle if you only know the length of the shorter leg.

*27. Find $m\angle 1$ in the figure below. Rays that appear to be tangent are tangent.
(79)

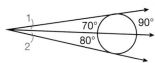

28. Multi-Step What is the lateral surface area of a right square pyramid if the
(70) area of its base is 64 square inches, and its height is 3 inches?

*29. Gemstones A piece of raw peridot is cut into a rectangular prism. Then, a
(49) triangular face is cut at each vertex. Find the number of faces, edges, and vertices
after all eight vertices of the prism have been replaced by triangular faces. Verify
your answers using Euler's Formula.

30. Algebra Find the value of x that makes the triangle a right triangle. Assume the side
(33) labeled $3x - 2$ is the largest side.

Finding Surface Areas and Volumes of Spheres

1. **Vocabulary** A three-dimensional figure is called a _____.
(SB 13)

2. Find the surface area and the volume of the cube.
(59)

3. Find the surface area and the volume of the cylinder
(62)   to the nearest hundredth of an inch.

7 ft

15 in.

4 in.

New Concepts

A sphere is a set of points in space that are a fixed distance from a given point called the **center of the sphere**. A **radius of a sphere** is a segment whose endpoints are the center of the sphere and any point on the sphere. The term radius can also refer to the length of this segment.

> **Math Language**
>
> A line, plane, or other figure that is **tangent to a sphere** intersects the sphere at only one point.

r

> **Math Reasoning**
>
> **Analyze** Can the exact surface area of a sphere be calculated?

Surface Area of a Sphere
The surface area, S, of a sphere is given by the following formula, where r is the radius of the sphere. $$S = 4\pi r^2$$

Example 1 **Calculating Surface Area of a Sphere**

Find the surface area of the sphere in terms of π.

SOLUTION

$S = 4\pi r^2$ Surface area of a sphere

$S = 4\pi(5)^2$ Substitute.

$S = 100\pi \text{ in}^2$ Simplify.

5 in.

Volume of a Sphere
The volume, V, of a sphere is given by the following formula, where r is the radius of the sphere. $$V = \frac{4}{3}\pi r^3$$

Online Connection
www.SaxonMathResources.com

Example 2 **Calculating Volume of a Sphere**

Find the volume of the sphere to the nearest hundredth.

13 ft

SOLUTION

$V = \dfrac{4}{3}\pi r^3$ Volume of a sphere

$V = \dfrac{4}{3}\pi (13)^3$ Substitute.

$V \approx 9202.77 \text{ ft}^3$ Simplify.

The volume of the sphere is approximately 9202.77 cubic feet.

A **hemisphere** is half of a sphere. A **great circle** is a circle on a sphere that divides the sphere into two hemispheres. For example, the great circle that divides the Earth into two congruent hemispheres is called the Equator. The volume of a hemisphere is equal to half the volume of the sphere with the same radius. To find the surface area of a hemisphere, the area of the great circle has to be added to half the surface area of a sphere with the same radius.

great circle

r

Hemisphere

Example 3 **Calculating Surface Area and Volume of a Hemisphere**

Find the surface area and the volume of a hemisphere with a 12-centimeter diameter to the nearest hundredth.

SOLUTION

The radius is 6 centimeters. Find the volume of a sphere with a 6-centimeter radius and divide by 2.

$V = \dfrac{4}{3}\pi r^3$ Volume of a sphere

$V \approx 904.78$ Substitute and simplify.

$904.78 \div 2 = 452.39$

So the volume of the hemisphere is approximately 452.39 cubic centimeters. To find the surface area of the hemisphere, find the surface area of a sphere with a 6-inch radius and divide by 2.

$S = 4\pi r^2$ Surface area of a sphere

$S \approx 452.39$ Substitute and simplify.

$452.39 \div 2 = 226.20$

Since the great circle is also part of the hemisphere's surface area, its area has to be calculated and added as well.

$A = \pi r^2$ Area of a circle

$A = 113.10$ Substitute and simplify.

Now add 226.20 and 113.10. The total surface area of the hemisphere is approximately 339.30 square centimeters.

Caution

The formulas for volume and surface area of a sphere depend on the radius of the sphere. Check to see if the problem gives the radius or the diameter of a sphere.

Example 4 Application: Sports

A regulation men's basketball is about 30 inches in circumference. How many cubic inches of air does it take to inflate a basketball of this size to the nearest hundredth?

SOLUTION

Since the problem mentions inflating the ball, we know it is asking for the volume of the object. Use the circumference to find the radius of the sphere.

$$C = 2\pi r$$
$$30 = 2\pi r$$
$$r \approx 4.77$$

Now the volume of the sphere can be found.

$$V = \frac{4}{3}\pi r^3$$
$$V = \frac{4}{3}\pi(4.77)^3$$
$$V \approx 454.61$$

It takes approximately 454.61 cubic inches of air to inflate the basketball.

> **Hint**
>
> The problem asks for the number of cubic inches of air needed to inflate the basketball. Volume is expressed in cubic units, so the volume of the ball must be found.

Lesson Practice

a. Find the surface area of a sphere with an 8-meter diameter to the
(Ex 1) nearest hundredth of a square meter.

b. Find the volume of the sphere shown to the nearest hundredth of a
(Ex 2) cubic inch.

22 in.

c. Find the surface area and volume of a hemisphere with a 16-foot
(Ex 3) radius to the nearest hundredth of a foot.

d. (**Decorations**) A snowglobe's circumference is 20 centimeters. How much
(Ex 4) water does it take to fill the globe, to the nearest hundredth of a cubic
centimeter?

1. Find the value of x.
(50)

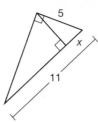

* 2. Find the surface area of a sphere with a radius of 25 inches, to the nearest tenth.
(80)

3. Find the value of x and the measure of $\overparen{CD}$ in the figure. Line ℓ is tangent to
(64) the circle.

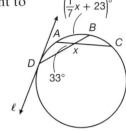

4. Which pair of lines are parallel and which are perpendicular?
(37)
$$y = 3x + 4 \qquad 3y = x - 3$$
$$9y = -3x + 7 \qquad y = \frac{1}{3}x - 6$$

5. Find the measure of an angle formed by two tangents to a circle if
(79) they intercept an arc that measures 238°.

6. Write Explain the relationship between the angle of elevation from
(73) point A to point B and the angle of depression from point B to point A.

7. Justify Can it be concluded that this quadrilateral is a parallelogram?
(61) Explain.

* 8. Find the approximate surface area and volume of a hemisphere with a
(80) diameter that is 23 yards long.

9. (**Vacuuming**) Randy must vacuum his apartment, the floorplan of which is
(40) shown at right. If he vacuums one square foot per second, how long will
it take him to vacuum his whole apartment, in minutes?

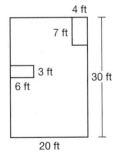

10. Line segment $\overline{MN}$ has endpoints $M(-3, 3)$ and $N(4, -1)$. Graph the line
(78) segment, and then rotate it 270° counterclockwise about the origin. List
the endpoints of its image, $\overline{M'N'}$.

11. (**Manufacturing**) A factory manufactures kites that are each half blue and
(69) half red. Each kite has sides that are 8 inches and 1.5 feet long. The shortest
diagonal of each kite is 10 inches long. What area of the kite is dyed red,
to the nearest square inch?

12. Algebra Find the angle measures of quadrilateral $MNPQ$ in the figure at right.
(47)

13. Algebra Write the inverse of the conditional statement. Is the inverse true?
(17)
$$If\ 3x + 2 = 11,\ then\ x - 6 = -3$$

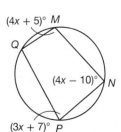

14. Find the magnitude of $\langle 6, -10 \rangle$ to the nearest tenth.
(63)

15. Find m∠*TPR* in the figure at right. $\vec{PR}$ is tangent
(79) to the circle at *R*.

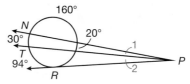

16. Write You are given two points on a circle. Is it possible to find an equation of the
(75) circle? Is it possible if you are given three non-collinear points? Explain.

*17. (**Globes**) A globe has a circumference of 3 feet. Find the approximate volume of
(80) the globe.

18. Write What information would you need to know to conclude that the probability
(Inv 6) of a randomly chosen point on the figure shown here being blue is exactly 50%?

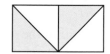

19. Algebra Square *DEFG* has vertices $D(x_1, 3)$, $E(6, 3)$, $F(6, 7)$, and $G(x_2, y_2)$.
(45) Find the values of x_1, x_2, and y_2.

20. Algebra Find the distance from the center of a circle with a 14-inch diameter
(43) to a chord that is 11 inches long.

21. Find m∠1 in the figure at right.
(79)

22. △*BCD* ~ △*QRS*. The ratio of their corresponding side lengths is 1:6.
(44) Given *BD* = 2, what is *QS*?

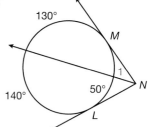

23. (**Props**) For a school play, Anita needs to build a regular, hexagonal
(70) pyramid. If the side lengths of the base are 3 inches and the slant height
is 10 inches, what is the area of the piece of cardboard that Anita needs to use to
make the pyramid, including the base?

24. Multiple Choice Which hypothesis below belongs to the converse of the following
(17) statement?

If two numbers are both odd or both even, then the sum of the numbers is even.
  A *Two numbers are odd and even*   **B** *The sum of two numbers is odd*
  C *The sum of two numbers is even*   **D** *Two numbers are both odd or both even*

25. (**Food Preparation**) What volume of ice cream will fit inside a cone that has a
(77) diameter of 2 inches and a lateral area of 15.7 square inches?

26. Quadrilateral *LMNO* is similar to quadrilateral *BCDE*. If m∠*B* = 65°,
(41) m∠*C* = 125°, m∠*D* = 50°, and m∠*E* = 120°, what is the measure of each angle in
quadrilateral *LMNO*?

*27. Find the approximate volume of the sphere at right.
(80)

28. Find the radius of the circle in which the area of a sector with a 45° interior
(35) angle is 2π.

29. Analyze Make a conjecture for what natural values of *n* does the graph of the
(76) equation $y = x^n$ have rotational symmetry.

30. Quadrilateral *JKLM* has vertices *J*(1, 1), *K*(2, 6), *L*(5, 4), and *M*(4, 0). Graph the
(78) quadrilateral, and then rotate it 180° counterclockwise about the origin. List the
coordinates of the vertices of its image *J'K'L'M'*.

Patterns

Finding patterns is a valuable problem-solving skill. In this investigation, you will study patterns made by transforming a figure. The basic figure we will work with is an isosceles triangle like the one shown here. Copy this triangle by connecting the points (0, 0), (8, 0), and (4, 3) on a coordinate plane, and cut the triangle out so you can trace it onto paper.

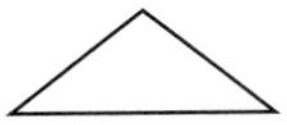

Trace the cutout triangle onto a blank sheet of paper. First, transform the triangle by rotating it. With the triangle oriented as shown in the diagram above, rotate it 90° counterclockwise around either one of its acute angles. Trace the triangle again in its new position. You should now have a design like the one shown.

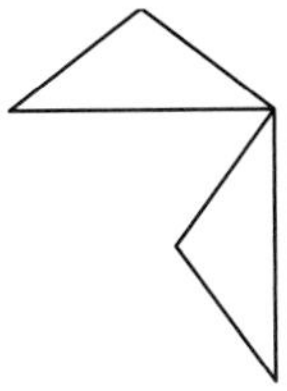

Continue to rotate the triangle 90° and trace it each time. Since the triangle will return to its original orientation after four rotations, the final design will look like the one shown.

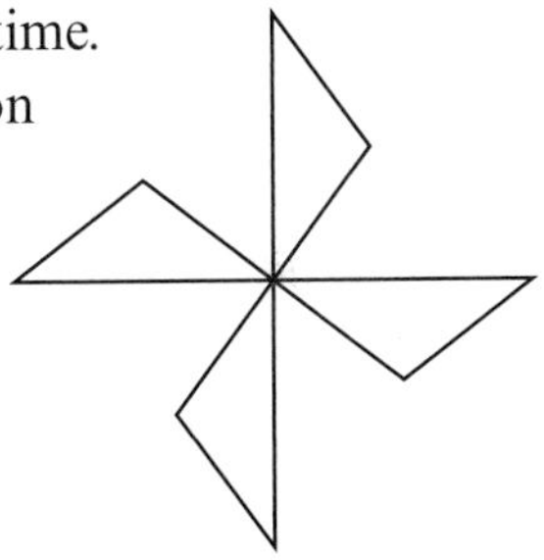

1. What is the order of the rotational symmetry in the final pattern?

2. Does the final pattern have any lines of symmetry? If so, how many?

Trace your cutout triangle onto a blank sheet of paper. This time you will make a pattern by reflecting the triangle. Draw *x*- and *y*-axes and orient the triangle so its base lies on the *x*-axis and one of its acute angles lies on the origin.

3. Reflect the triangle over the *y*-axis and draw the resulting pattern.

4. Reflect the pattern from step 3 over the *x*-axis and draw the resulting pattern.

5. Does the final pattern have rotational symmetry? If so, what is the order of rotational symmetry?

6. Does the final pattern have any lines of symmetry? How many?

Math Reasoning

Model Could you have used different transformations of the triangle to obtain the same image as the one in step 4? Explain how.

Translation symmetry is a type of symmetry describing a figure that can be translated along a vector so that the image coincides with the preimage. A **frieze pattern** is a pattern that has translation symmetry along a line.

Trace your cutout triangle onto a blank sheet of paper. Orient the triangle so its base is parallel to the bottom edge of the paper. Translate the triangle to the right until the vertices of the opposite acute angles lie on the same point, as in the figure shown.

Translate the triangle to the right again. Continue this process until you have 4 triangles in a row. This is a frieze pattern.

7. What other transformation(s) could have been used to create this same pattern?

8. Does the final pattern have rotational symmetry? If so, what is the order of rotational symmetry?

9. Does the final pattern have any lines of symmetry? How many?

Now we will explore some geometric patterns. Draw two points. There is only one segment that can be drawn connecting these two points. What about 3 points? Draw 3 noncollinear points and draw segments connecting them. You find that there are three segments that can be drawn.

10. Draw four non-collinear points. Draw a line segment between each pair of points. How many line segments do you have?

11. Predict Based on the pattern you have seen so far, predict how many line segments you can draw connecting 5 noncollinear points. What about 6 noncollinear points? … 7 noncollinear points?

12. Formulate Write a rule for the number of line segments, L, between n points, in terms of the number of line segments between $n - 1$ points (denoted L_{n-1}).

The numbers in the series you have just discovered are called triangular numbers. Triangular numbers are numbers that are equal to the sum of the first n whole numbers. The first 8 triangular numbers are: 1, 3, 6, 10, 15, 21, 28, and 36.

Can an algebraic expression be written for the nth triangular number? The nth triangular number is given by the formula below.
$$x = 1 + 2 + 3 + 4 + \ldots + n - 1 + n$$

To find an expression for the nth triangular number, take this series and add it to itself. Instead of adding the terms in order though, add the first term to the last term, the second term to the second to last term, and so on.

$$
\begin{aligned}
x &= 1 + 2 + 3 + 4 \ldots + n - 1 + n \\
+\ x &= 1 + 2 + 3 + 4 \ldots + n - 1 + n \\
\hline
2x &= (1 + n) + (2 + n - 1) + (3 + n - 2) + \ldots
\end{aligned}
$$

Now notice that in the sum, the expressions in the parenthesis can be simplified. After being simplified, the sum becomes
$$2x = (n + 1) + (n + 1) + \ldots$$

Each expression in parenthesis is the same. Moreover, we know that the series has n terms. So it can be simplified further, resulting in $2x = n(n + 1)$. To solve for x, which is the nth triangular number, divide by 2.
$$x = \frac{n(n + 1)}{2}$$

13. Using the expression above, what is the 50th triangular number? What is the 100th triangular number?

Investigation Practice

a. Using the right triangle given here, sketch the result of rotating the figure 90° counterclockwise about the point *A*.

b. Continue to rotate the triangle 90° until it coincides with itself, sketching the result of each rotation. What is the order of rotational symmetry in the final figure? Does it have any lines of symmetry?

c. Return to the initial figure and reflect it over the vertical leg of the triangle. Then reflect it over the horizontal leg of the triangle. What kind of polygon is the resulting figure?

d. Does the resulting figure have any lines of symmetry? Does it have rotational symmetry?

e. Square numbers are whole numbers that could be the area of a square. The series begins: 1, 4, 9, 16, 25, …. Write an equation to find the nth square number.

f. What is the 30th square number?

Graphing and Solving Linear Systems

Warm Up

1. Vocabulary _____ are lines that intersect at a 90° angle.
(37)

2. Find the equation of the line.
(16)

3. Find the distance from the point P to the line.
(42)

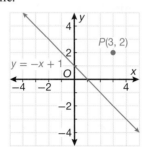

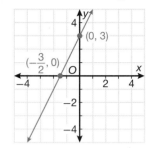

New Concepts

A **system of equations** is a set of two or more equations that have two or more variables. A system of linear equations can be solved algebraically or by graphing the lines.

If the lines are graphed, the solution to the system is the coordinates of the point where the lines intersect. In this example, the solution to the system is point P. To solve a system algebraically, solve both equations for the same variable and use substitution.

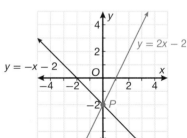

Math Reasoning

Generalize Why can you substitute into either equation to find the value of the second variable? Why are the answers always the same?

> ### Example 1 | Solving Linear Systems Algebraically
>
> Solve this system of equations algebraically.
> $$y = \frac{1}{2}x - 1 \qquad y = -\frac{3}{2}x + 3$$
>
> **SOLUTION**
> Both equations have been solved for y, so x can be found by substituting for y.
>
> $\frac{1}{2}x - 1 = -\frac{3}{2}x + 3$ Substitute
>
> $x = 2$ Solve
>
> Substitute the value for x into one of the original equations to find y.
>
> $y = \frac{1}{2}x - 1$
>
> $y = \frac{1}{2}(2) - 1$
>
> $y = 0$
>
> So the solution to this system is the ordered pair (2, 0).

Example 2 **Solving Linear Systems Graphically**

Estimate the solution to this linear system by graphing the lines.

$$y = -\frac{1}{2}x - 3 \qquad y = \frac{3}{2}x + 1$$

SOLUTION
Graph each line on a coordinate grid.
The lines appear to intersect at the
point $(-2, -2)$.

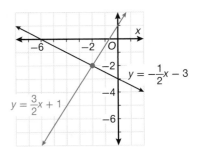

Check the answer by substituting
$x = -2$ and $y = -2$ into both of the
original linear equations.

$$y = -\frac{1}{2}x - 3 \qquad\qquad\qquad y = \frac{3}{2}x - 1$$

$$-2 \stackrel{?}{=} -\frac{1}{2}(-2) - 3 \qquad\qquad -2 \stackrel{?}{=} \frac{3}{2}(-2) + 1$$

$$-2 = -2 \qquad\qquad\qquad\qquad -2 = -2$$

The pair satisfies both equations, so $(-2, -2)$ is the correct solution
to the system.

If two lines are parallel, they do not intersect. A system of equations that
represents two or more parallel lines has no solution.

Example 3 **Analyzing Unsolvable Systems**

Graph this linear system to determine if there is a solution.

$$y = 2x - 1 \qquad\qquad y = \frac{4}{2}x + 2 \qquad\qquad 2y - 2 = 4x - 2$$

SOLUTION
Simplify each equation, then graph the lines on the same coordinate grid.

$$y = 2x - 1 \qquad\qquad y = 2x + 2 \qquad\qquad y = 2x$$

Since all three lines have the same slope, they
are parallel. Graphing them demonstrates that
they have no point of intersection. Since all three
lines are parallel, there is no solution to the
system of linear equations that consists of these
three equations.

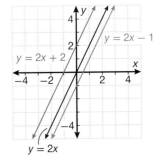

Note that solving a system of equations graphically is not as accurate
as solving the same system algebraically.

Math Reasoning

Estimate In Example 3,
suppose the third line
had slope 2.1. Would this
line intersect each of the
other two lines? If yes, in
which quadrant would
the lines intersect?

Example 4 Application: Economics

An economist is trying to determine the optimum price for a new product. He knows the supply of the product is represented by the function $y = \frac{2}{3}x + 50$ and the demand curve for the product is represented by the function $y = -\frac{1}{3}x + 200$, where y is the price of the product and x is the number of units sold. What is the optimum price of the product? How many units will sell at this price?

Math Reasoning

Verify How can you use the graph to check your algebraic solution?

SOLUTION

Solve the system algebraically by substituting for y.

$$-\frac{1}{3}x + 200 = \frac{2}{3}x + 50$$

$$x = 150$$

Now use this value of x to find y.

$$y = \frac{2}{3}(150) + 50$$

$$y = 150$$

The solution is (150, 150). This means that the optimum price of the product is $150 and 150 units will be sold at this price.

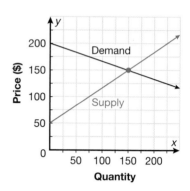

Lesson Practice

a. Solve the system of equations algebraically.
(Ex 1)
$$y = \frac{2}{3}x - 8 \qquad\qquad y = \frac{1}{4}x + 2$$

b. Solve the system of equations algebraically.
(Ex 1)
$$y = -\frac{2}{3}x - 3 \qquad\qquad y = \frac{1}{2}x + 2$$

c. Solve the system of equations by graphing.
(Ex 2)
$$y = 3x + 4 \qquad\qquad y = -x + 8$$

d. Determine if there is a solution for this system. If not, explain why.
(Ex 3)

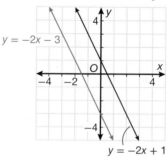

e. The supply curve for a product is represented by $y = 2x + 20$ and the
(Ex 4) demand curve for the product is represented by $y = -\frac{1}{2}x + 80$ where y is the price of the product, and x is the number of units sold. What is the optimum price of the product and how many units will be sold at this price?

1. Justify Two secants intersect in the exterior of the
(79) circle shown. Explain why it is true that for all cases,
m∠*UZY* will be greater than m∠*W*.

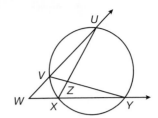

* 2. Solve the system of linear equations.
(81) $y = 5x - 4$
$y = \frac{2}{3}x + 1$

3. Generalize Suppose statement *p* is true, but statement *q* is not true.
(20) **a.** Which is true: the statement "If *p*, then *q*," its converse, both, or neither? Use a
       truth table to answer.
  b. If a biconditional is formed from the statement and its converse, is the
       biconditional true? Explain why or why not.

4. Segment *JK* is a midsegment of △*FGH*. Find the values of *x* and *y*.
(55)

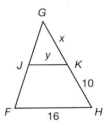

5. Multiple Choice Which graph is the graph of tan *x*?
(Inv 7) *Hint: make a table of values.*

A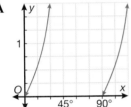

B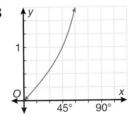

C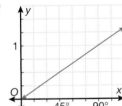

D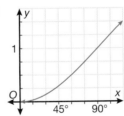

6. Manufacturing A manufacturing company makes cone-shaped drinking cups. The
(77) height of the cone is 15 centimeters. What is the radius if the cone has a volume
    of 45π cubic centimeters?

* 7. a. Sketch this equilateral triangle and its rotation 60° clockwise
(Inv 8)   about point *F*.
  b. Repeat part **a** for clockwise rotations of 120°, 180°, 240°,
       and 300° about the apex of the original figure.
  c. Describe the type and order of symmetry of your pattern
       from part **b**.

8. Generalize Can the area of a trapezoid be calculated from its height, *h*, and the
(69) length of its midsegment, *m*? If so, determine a formula for the area in terms of
    h and *m*.

9. What is the volume of a cone with a base area of 6π square centimeters and a
(77) height of 7 centimeters? Leave your answer in terms of π.

*10. **Multi-Step** A circle with an area of 121π square inches is spun about its diameter
(80) to make a sphere. What is the volume of the sphere to the nearest cubic inch?

11. Determine the missing side lengths in $\triangle JKL$ to the nearest
(56) tenth of a centimeter.

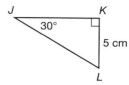

*12. (Pizza) One pizzeria sells a regular pizza for $6 plus $1.50 for each additional
(81) topping. Another pizzeria sells a regular pizza for $8 plus $0.50 for each
additional topping. What number of toppings would make the regular pizzas
from both places the same price? How much would a regular pizza cost with
this number of toppings?

13. **Write** Is every equilateral triangle isosceles? Is every isosceles triangle equilateral?
(51) Explain.

14. (Theater) Elsie needs to make a paper hat shaped like a right cone. The hat must
(77) be 18 inches tall, and have a base radius of 4 inches. How much paper does she
need, to the nearest square inch? *Hint: She does not need paper for the base of the
cone.*

15. Find m$\angle C$ in the figure at right if m$\angle AFB = 100°$.
(79)

16. How many faces are there on a solid with 5 vertices
(49) and 8 edges?

17. Where is the orthocenter of a right triangle located?
(32)

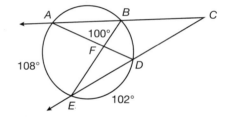

18. **Multiple Choice** What is the hypothesis of the following
(14) conjecture?

*If a pair of corresponding angles formed by a transversal ℓ are congruent,
then lines m and n are parallel.*

A *A pair of angles are corresponding angles*   **B** *A pair of angles are formed by a transversal*
C *Lines m and n are parallel*   **D** *A pair of corresponding angles formed by a
transversal ℓ are congruent*

*19. **Algebra** The solution of a system of equations is $(-1, 1)$. One equation is $y = -x$
(81) and the other equation is $y = \frac{2}{3}x + b$. What is the value of b?

20. Given that $\triangle JKL \sim \triangle PQR$, $JK = 21$, $PQ = 28$, and $QR = 48$, what is KL?
(46)

*21. **Write** Explain how the surface area and the volume of a sphere are different.
(80)

22. **Algebra** Find the range of values of x that would cause m$\angle S$ to be larger than
(Inv 4) m$\angle P$.

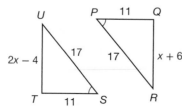

23. Model Draw a net of a pentagonal pyramid.
(Inv 5)

24. Find the distance to the nearest hundredth from the origin to the line
(42) $y = x + 6$.

*25. Error Analysis Hector subtracted two equations, shown below. When he substituted
(81) into the original equations, he obtained two different answers. Where did he make
his mistake? What is the correct answer?

$$y = \frac{5}{2}x + 1 \qquad\qquad y = \frac{5}{2}\left(\frac{4}{7}\right) + 1 \qquad\qquad y = -\left(\frac{4}{7}\right) - 3$$
$$\underline{-(y = -x - 3)} \qquad\quad = \frac{17}{7} \qquad\qquad\qquad = -\frac{25}{7}$$
$$0 = \frac{7}{2}x - 2$$
$$x = \frac{4}{7}$$

26. Coordinate Geometry A parallelogram has vertices $W(-6, -2)$, $X(-4, -2)$,
(45) $Y(x, y)$, and $Z(-4, -3)$. Find all possible values for x and y.

27. In $\triangle ABC$ and $\triangle DEF$, $\angle B$ and $\angle E$ are right angles, $\overline{AC} \cong \overline{DF}$ and $\overline{BC} \cong \overline{EF}$. What
(36) theorem can be used to prove that $\triangle ABC \cong \triangle DEF$?

*28. Find the volume of a sphere with a radius of 7 meters to the nearest square
(80) meter.

29. (Propellers) The propeller of an airplane rotates counterclockwise. Find the angle
(78) of rotation required for propeller blade P to move to position P'. Explain.

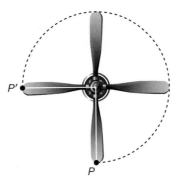

30. Find the range of possible lengths for the third side of a triangle with two sides
(39) that are 22 centimeters and 14 centimeters long, respectively.

More Applications of Trigonometry

1. **Vocabulary** Cosine of an angle is the ratio of the length of the leg that is
(68) _____ to an angle to the length of the _____.

2. A pilot flying a plane sees a waterfall in a river below at a 43° angle of
(73) depression. If she knows that she is flying at 10,000 feet, how far away is
the waterfall to the nearest foot?

3. If a right triangle has an angle of 30°, what is the ratio of the length of the
(56) shorter leg to the length of the longer leg?

New Concepts The sine, cosine, and tangent ratios can be used to find the length of a side
of a right triangle. These trigonometric ratios can also be used to find the
measure of an angle given two side lengths. To do this, the inverse of each
trigonometric function is needed.

Reading Math

The Greek letter θ (theta)
is used to denote the
unknown measure of an
angle.

The **inverse sine** is the measure of an angle where the sine ratio is known.
The **inverse cosine** is the measure of an angle where the cosine ratio is known.
The **inverse tangent** is the measure of an angle where the tangent is known.

The inverse of the sine function is written $\sin^{-1}$.
In the diagram, the measure of the unknown
angle, θ, can be determined using an inverse
trigonometric function as shown below.

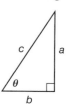

**Graphing
Calculator Tip**

The inverse trigonometric
functions are entered
on your calculator by
pressing **2nd** and the
corresponding function
key.

$$\theta = \sin^{-1}\frac{\text{opposite}}{\text{hypotenuse}} \qquad \theta = \cos^{-1}\frac{\text{adjacent}}{\text{hypotenuse}} \qquad \theta = \tan^{-1}\frac{\text{opposite}}{\text{adjacent}}$$

$$\theta = \sin^{-1}\frac{a}{c} \qquad\qquad \theta = \cos^{-1}\frac{b}{c} \qquad\qquad \theta = \tan^{-1}\frac{a}{b}$$

Example **1** **Using Inverse Sine**

Find θ to the nearest degree.

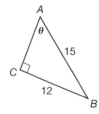

SOLUTION

$$\theta = \sin^{-1}\frac{\text{opposite}}{\text{hypotenuse}}$$

$$\theta = \sin^{-1}\frac{12}{15}$$

$$\theta \approx 53°$$

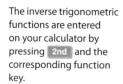

Online Connection
www.SaxonMathResources.com

Example 2 **Using Inverse Cosine**

Find θ to the nearest degree.

SOLUTION

$$\theta = \cos^{-1}\frac{\text{adjacent}}{\text{hypotenuse}}$$

$$\theta = \cos^{-1}\frac{15}{37}$$

$$\theta \approx 66°$$

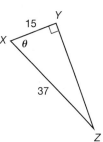

Math Reasoning

Write If you enter $\sin^{-1}(1.2)$ into your calculator, it gives an error. Explain why.

Example 3 **Using Inverse Tangent**

Find θ to the nearest tenth of a degree.

SOLUTION

$$\theta = \tan^{-1}\frac{\text{opposite}}{\text{adjacent}}$$

$$\theta = \tan^{-1}\frac{16}{11}$$

$$\theta \approx 55.5°$$

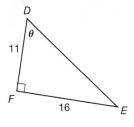

Example 4 **Application: Architecture**

Below is the design for a bridge. Find θ_1 and θ_2 to the nearest tenth of a degree.

SOLUTION

Since the lengths of the two legs of the right triangle are known, use the inverse tangent function.

$$\theta_1 = \tan^{-1}\left(\frac{\text{opposite}}{\text{adjacent}}\right)$$

$$\theta_1 = \tan^{-1}\frac{17}{15}$$

$$\theta_1 \approx 48.6°$$

$$\theta_2 = \tan^{-1}\left(\frac{\text{opposite}}{\text{adjacent}}\right)$$

$$\theta_2 = \tan^{-1}\frac{15}{17}$$

$$\theta_2 \approx 41.4°$$

Lesson Practice

a. Find θ_1 to the nearest degree.
(Ex 1)

b. Find θ_2 to the nearest degree.
(Ex 2)

c. In a right triangle, the side that is opposite
(Ex 3) angle θ measures 72 centimeters and the side adjacent to angle θ is 90 centimeters. Find θ.

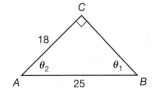

d. A children's piano is made with right triangle supports. The hypotenuse is 2 feet long and the height of the piano is 18 inches. What is θ, to the nearest degree?
(Ex 4)

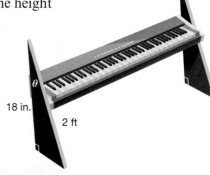

18 in.

2 ft

Practice Distributed and Integrated

1. Find θ_1 and θ_2 to the nearest degree.
(82)

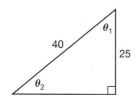

40

25

2. Multi-Step A trapezoid has vertices at the points $P(1, 3)$, $Q(4, 6)$, $R(6, 6)$, and
(69) $S(0, 0)$. Determine which segments are the bases of the trapezoid. Explain.

3. (**Carnival Games**) A carnival game involves dropping a ping-pong
(Inv 6) ball down a chute so that it randomly lands on a platform that is
shown on the right. The player wins a prize if the ball lands in the
square, or an extra game token if the ball lands in the trapezoid.
Both the square and the trapezoid are 5 centimeters tall. What is the
probability of a player winning a prize? …winning a token? …winning nothing?

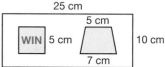

25 cm
5 cm
WIN 5 cm
10 cm
7 cm

xy² * 4. Algebra The solution to the linear system $x = -3$, $y = mx - \frac{7}{2}$ is $(-3, 1)$.
(81) What is the value of m?

5. Quadrilateral $KLMN$ is a rhombus with its center point labeled P.
(52) If $m\angle PKL = 57°$, what is $m\angle PLK$?

6. Write Explain how to use trigonometry to find the length of the radius of
(43, 68) a circle, given a 9-inch chord that is 2.2 inches from the center of the circle.
What is the length of the radius?

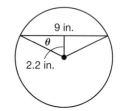

9 in.

θ

2.2 in.

7. Find the approximate surface area of a sphere whose great circle has an
(80) area of 85.2 square centimeters.

8. Find $m\angle SRT$ in the figure at right.
(47)

9. The vertices of a triangle are $J(0, 1)$, $K(4, 2)$ and $L(3, -1)$. The triangle
(71) is translated according to the vector $\langle -4, 2 \rangle$. Find the coordinates of
the vertices of the image.

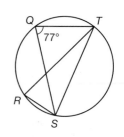

Q

T

77°

R

S

*10. Multiple Choice Which statement is *not* true about the right
(82) triangle shown?

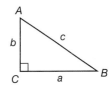

A $\tan^{-1}\frac{b}{a} = \tan^{-1}\frac{a}{b}$ **B** $\tan^{-1}\frac{b}{a} = \sin^{-1}\frac{b}{c}$

C $\cos^{-1}\frac{a}{c} = \sin^{-1}\frac{b}{c}$ **D** $\tan^{-1}\frac{a}{b} = \cos^{-1}\frac{b}{c}$

11. How many lines of symmetry does a regular pentagon have?
(76)

12. Justify Determine whether lines $\overline{NO}$ and $\overline{PQ}$ are parallel. How do
(60) you know?

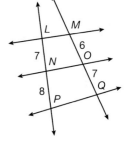

13. Multiple Choice Which quadrilateral has diagonals of equal lengths?
(34)   **A** parallelogram   **B** kite
  C rhombus   **D** rectangle

14. (**Retail Displays**) Cans of soup are stacked in a triangular shape as a
(Inv 8) display in a supermarket. If there are 8 rows of soup cans, how many
cans of soup are in the display?

*15. (**Surveying**) A surveyor needs to calculate the angle of the sun
(82) using the shadow cast by the building. Use the diagram to find
θ to the nearest degree.

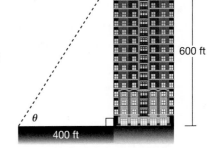

16. Error Analysis Reuben calculates that a reflection of $\triangle TLV$ across
(74) the line $x + y = 0$ maps the three points as follows:
$T(3, 1) \rightarrow T'(-3, -1)$
$L(4, 2) \rightarrow L'(-4, -2)$
$V(-2, 4) \rightarrow V'(2, -4)$
  a. Plot Reuben's reflection. What error has he made?
  b. Plot the correct reflection image $\triangle A'B'C'$ and $\triangle ABC$ on the
     same graph.

17. Justify In $\triangle LMN$, $m\angle L = 140°$ and $m\angle M = 20°$.
(51)   **a.** Determine $m\angle N$, and classify $\triangle LMN$ by its sides and angles.
  b. Which angle is the vertex angle, and why?

*18. (**Landscaping**) A garden is constructed with two paths that
(72) are tangent to each of the circular flower beds. Are there
other possibilities for common tangent paths in the
garden? If so, sketch the possibilities.

*19. Find the solution for the following system of equations.
(81) $y = -2x - 5$, $y = x + 1$

20. Find $m\angle 2$ in the figure at right.
(79)

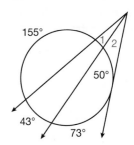

21. Generalize Substitute (x, y) for (x_2, y_2) into the slope formula,
(16) and transform it into a linear equation. Describe this form of
the equation of a line. How can the slope-intercept form be
found from it?

*22. Find the solution for the system of equations. $y = \frac{3}{2}x + 4$, $y = -\frac{1}{3}x - \frac{10}{3}$
(81)

23. (Art) Javier is tracing a circular pattern for an art project. He places the two
(75) endpoints of a diameter at (3, 4) and (9, 12). What is the equation of the circle
Javier traces out?

24. Find the value of x and y in the figure shown.
(50)

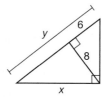

25. Find the approximate surface area and volume of a
(80) hemisphere with a radius of 20 inches.

26. **Algebra** Find the value of x that makes $\triangle RST$ a right triangle.
(33)

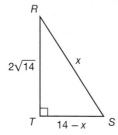

27. **Justify** Fill in the missing justifications in this proof of the
(27) Converse of the Same-Side Interior Angles Theorem.
Given: $\angle 1$ is supplementary to $\angle 2$.
Prove: $p \parallel r$

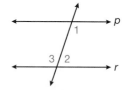

Statements	Reasons
1. $\angle 1$ is supplementary to $\angle 2$	1.
2. $m\angle 1 + m\angle 2 = 180°$	2.
3. $m\angle 2 + m\angle 3 = 180°$	3.
4. $m\angle 1 + m\angle 2 = m\angle 2 + m\angle 3$	4. Transitive Property of Equality
5. $m\angle 1 = m\angle 3$	5.
6. $p \parallel r$	6. Converse of the Corresponding Angles Postulate

*28. **Algebra** The perimeter of a right triangle is 12 inches and the legs are $x + 2$ and
(82) $x + 1$. The hypotenuse is $x + 3$. What are the measures of the two acute angles, to
the nearest degree?

29. Determine the perimeter of this figure.
(40)

30. The perimeter of equilateral $\triangle WXY$ is 24 inches. If
(44) $\triangle WXY \sim \triangle RST$, and the ratio of their corresponding
sides is 4:1, what are the lengths of the sides of $\triangle RST$?

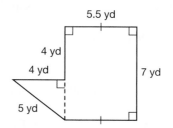

Vector Addition

Warm Up

1. **Vocabulary** A _____ is a quadrilateral with two pairs of
(34) parallel sides.

2. Is the following statement sometimes, always, or never true?
(63) For two different points A and B, $\overrightarrow{AB}$ is identical to $\overrightarrow{BA}$.

3. **Multiple Choice** Which of the following is true?
(82)

    A $\tan^{-1}\left(\frac{2y}{x}\right) = \tan^{-1}\left(\frac{2x}{x}\right)$ **B** $\tan^{-1}\left(\frac{x}{y}\right) = 90 - \tan^{-1}\left(\frac{y}{x}\right)$

    C $\sin^{-1}\left(\frac{x}{y}\right) = \frac{1}{\sin\left(\frac{x}{y}\right)}$ **D** $\sin^{-1}\left(\frac{x}{y}\right) = \cos^{-1}\left(\frac{y}{x}\right)$

New Concepts

You have added vectors that are in the same or opposite directions, but how are vectors added when their directions are not so simple? There are two commonly used methods for adding any two vectors together. The **parallelogram method** is a method of adding two vectors by drawing a parallelogram using the vectors as two of the consecutive sides; the sum is a vector along the diagonal of the parallelogram.

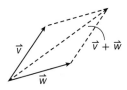

In the diagram, both $\vec{v}$ and $\vec{w}$ are drawn with the same initial point. A vector parallel to $\vec{v}$ and a vector parallel to $\vec{w}$ complete the parallelogram, and the diagonal is the resultant vector.

Hint

Remember that equal vectors have the same direction.

Example 1   Using the Parallelogram Method to Add Vectors

Use the parallelogram method to add the two vectors.
$\vec{a} = \langle 2, 3 \rangle, \vec{b} = \langle 4, -1 \rangle$

SOLUTION

Draw the vectors on a coordinate plane starting at the origin. Complete the parallelogram by drawing identical parallel vectors that start at the terminal points of the original vectors.

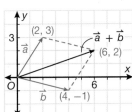

Draw the diagonal of the parallelogram from the origin. This diagonal is the resultant vector. Counting its vertical and horizontal change shows that the component form for the resultant vector is $\langle 6, 2 \rangle$.

$$\vec{a} + \vec{b} = \langle 2, 3 \rangle + \langle 4, -1 \rangle$$
$$\vec{a} + \vec{b} = \langle 2 + 4, 3 + (-1) \rangle$$
$$\vec{a} + \vec{b} = \langle 6, 2 \rangle$$

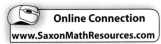

Online Connection
www.SaxonMathResources.com

The **head-to-tail method** is a method of adding two vectors by placing the tail of the second vector on the head of the first vector. The sum is the vector drawn from the tail of the first vector to the head of the second vector. One benefit to using this method is that while the parallelogram method can only be used to add two vectors, the head-to-tail method can be used to add any number of vectors.

Example 2 Adding Vectors Head-to-Tail

Add the vectors $\vec{c} = \langle 5, 4 \rangle$, $\vec{d} = \langle -2, 3 \rangle$, $\vec{e} = \langle -1, -1 \rangle$ using the head-to-tail method.

SOLUTION

First, draw a diagram. The first vector should begin at the origin. The vectors can be placed on the diagram in any order, but each vector must start at the terminal point of the vector before it.

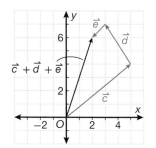

The diagram shows that the resultant vector is $\vec{c} + \vec{d} + \vec{e} = \langle 2, 6 \rangle$.

Adding the vectors by components gives the same answer.
$$\vec{c} + \vec{d} + \vec{e} = \langle 5, 4 \rangle + \langle -2, 3 \rangle + \langle -1, -1 \rangle$$
$$\vec{c} + \vec{d} + \vec{e} = \langle 2, 6 \rangle$$

Remember that the direction of a vector must also be specified. To find a vector's direction, measure the angle it forms with the horizontal. By convention, when a vector's direction is given with an angle, the angle is measured counterclockwise from the positive x-axis.

Example 3 Adding Vectors with Trigonometry

Add the vectors $\vec{a} = \langle 6, 0 \rangle$ and $\vec{b} = \langle 0, 4 \rangle$, and find the magnitude and direction of the resultant vector.

SOLUTION

First, add the vectors using the head-to-tail method. The resultant vector is $\langle 6, 4 \rangle$. Notice that $\vec{a}$ and $\vec{b}$ make a right angle. Use the Pythagorean Theorem to find the magnitude of the resultant vector.

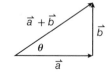

$$\left|\vec{a}\right|^2 + \left|\vec{b}\right|^2 = \left|\vec{a} + \vec{b}\right|^2$$
$$|6|^2 + |4|^2 = \left|\vec{a} + \vec{b}\right|^2$$
$$\left|\vec{a} + \vec{b}\right| \approx 7.21 \text{ units}$$

Now find the angle θ that the resultant vector makes with the horizontal.
$$\theta = \tan^{-1}\frac{\text{opposite}}{\text{adjacent}}$$
$$\theta = \tan^{-1}\frac{4}{6}$$
$$\theta \approx 33.69°$$

Example 4 Application: Aviation

A plane has traveled a horizontal distance that can be represented by the vector $\langle 8000, 0 \rangle$, and a vertical distance represented by the vector $\langle 0, 4000 \rangle$, where the magnitude of both vectors is measured in meters. What is the magnitude of the distance it has traveled to the nearest meter?

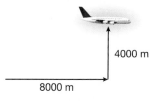

4000 m

8000 m

SOLUTION

Add the two vectors using the head-to-tail method. The magnitude of the resultant vector can be found using the Pythagorean Theorem. Call the resultant vector $\vec{x}$.

$$\left| \langle 8000, 0 \rangle \right|^2 + \left| \langle 0, 4000 \rangle \right|^2 = \left| \vec{x} \right|^2$$
$$8000^2 + 4000^2 = \left| \vec{x} \right|^2$$
$$x \approx 8944$$

The plane has traveled approximately 8944 meters.

Lesson Practice

a. Use the parallelogram method to add vectors $\vec{x} = \langle 3, 4 \rangle$ and
(Ex 1) $\vec{y} = \langle 4, 1 \rangle$.

b. Add the vectors $\vec{a} = \langle 1, 2 \rangle$, $\vec{b} = \langle 4, -5 \rangle$, and $\vec{c} = \langle -1, 7 \rangle$, using the
(Ex 2) head-to-tail method.

c. Add the vectors $\vec{u} = \langle 3, 0 \rangle$ and $\vec{v} = \langle 0, -2 \rangle$. Use trigonometry to find
(Ex 3) the magnitude and direction of the resulting vector.

d. A plane has traveled 2000 meters east while climbing to a height of
(Ex 4) 3000 meters. Write vectors to represent these two translations and find the magnitude of the resultant vector that gives the total distance the plane has traveled.

Practice Distributed and Integrated

1. Given $\tan X = \frac{5}{4}$, which angle in $\triangle RST$ is $\angle X$, θ_1 or θ_2?
(82)

2. What is the perimeter of a regular hexagon with an apothem length of
(66) $28\sqrt{3}$ meters?

3. Analyze How many lines of symmetry does a semicircle have? How many lines of
(76) symmetry does a circle have?

4. In $\triangle PQR$ and $\triangle STU$, $PQ = UT$, $PR = TS$, and $m\angle RPQ = m\angle UTS$.
(28) Which triangles are congruent?

5. Compare $m\angle ADB$ and $m\angle DBC$ in the figure shown.
(Inv 4)

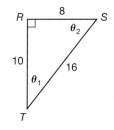

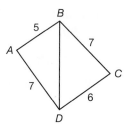

* 6. Find the magnitude of the sum of the vectors $\langle 4, 5 \rangle$
(83)   and $\langle -2, 2 \rangle$. Round your answer to the nearest hundredth.

7. Find the exact length of the hypotenuse of a 45°-45°-90° right triangle with a side
(53)   length of 5 inches.

8. (**Candles**) Kyle bought a candle in the shape of a right cone. The candle has a
(77)   height of 7 inches and a base radius of 1 inch. What is the volume of the candle
       to the nearest hundredth?

* 9. Classify this prism.
(49, Inv 5)

*10. (**Running**) A runner runs diagonally across a field. She runs 80 meters north and
(83)   50 meters east. How far did she run and what angle did her path make with the
        horizontal, to the nearest degree?

11. A reflection across the line $x = 4$ maps $\odot C$ to $\odot C'$. Find the
(74)   coordinates of the center and the radius of $\odot C'$.

12. **Estimate** Estimate the volume of a cylindrical fuel drum with a height
(62)   of 68 centimeters and a radius of 33 centimeters.

*13. **Multiple Choice** The direction of each vector is given in degrees
(83)   counterclockwise from the positive x-axis. Which of the
        following is incorrect?
   A $\langle 3, 4 \rangle$, 53°   **B** $\langle 7, 2 \rangle$, 16°
   C $\langle 4, 7 \rangle$, 30°   **D** $\langle 2, 5 \rangle$, 68°

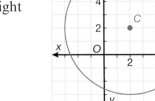

14. **Justify** Cyndie knows that r and s are parallel, so she writes "$\angle 4 \cong \angle 8$".
(Inv 2) What postulate or theorem justifies her statement?

xy^2 **15.** **Algebra** A system of equations has no solution. One equation is
(81) $y = 6x + 4$ and the other equation is $y = \frac{a}{2}x - 1$. Find the
       value of a.

16. (**Crafts**) Students at Franklin School are cutting out and decorating
(40)   cardboard letters for a display. If they remove two squares from a
       rectangular piece of cardboard to form the letter E, what is the area
       of the letter they must decorate?

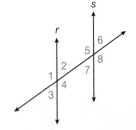

xy^2***17.** **Algebra** Find the resultant vector sum of $\vec{a}$, $\vec{b}$, and $\vec{c}$ if $\vec{a} = \langle 2, 3 \rangle$,
(83) $\vec{b} = 2\vec{a}$, and $\vec{c} = -\vec{a}$.

18. **Coordinate Geometry** A square is placed on the coordinate plane so that two vertices
(45)   are located at $(0, 0)$ and $(0, b)$. Give possible locations for the other two vertices.

19. (**Bird Watching**) Vic sees a bird resting on a telephone pole. He is standing 40 meters
(73)   from the base of the pole and he sees the bird at an angle of elevation of 17°. How
       far is Vic from the bird, to the nearest meter?

20. Write Why can the radius and volume of a sphere never both be whole numbers?
(80)

21. Multiple Choice Which of these figures is not the net of a cube?
(Inv 5)

A   B C D

22. Simplify the expression $\tan^{-1}[\tan(15°)]$.
(82)

23. What is the distance from $(-1, -4)$ to the line $y = 0$?
(42)

24. (**Hiking**) A hill makes an angle of elevation of 34°. If a hiker climbs to an
(73) altitude of 40 meters, what distance has he walked? Round your answer to the nearest meter.

25. Find the measures of the acute angles in the right triangle. Round
(82) your answer to the nearest degree.

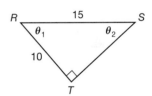

26. Analyze How many solutions will the system of equations $y = 2$ and
(81) $y = 4$ have? Why?

27. Algebra Two lines are perpendicular. One line has an equation of $rx + 3y = 4$
(37) and the other line has an equation of $\frac{2}{s}x + y = 6$. Find a linear equation to describe the relationship between r and s.

28. Multi-Step Use the head-to-tail method to find the resultant vector of the sum of
(83) $\vec{a} = \langle 6, 5 \rangle$, $\vec{b} = \langle 3, -2 \rangle$, and $\vec{c} = \langle 7, -1 \rangle$. Then find its magnitude to the nearest hundredth.

29. Find the volume, surface area, and lateral surface area of a cube with edge lengths
(59) of $4\sqrt{3}$ inches.

30. Error Analysis A student was asked to find $m\angle C$.
(79)

$$m\angle DHE = \frac{1}{2}\left(m\widehat{DE} - m\widehat{FG}\right)$$

$$98° = \frac{1}{2}\left(108° - m\widehat{FG}\right)$$

$$44° = -\frac{1}{2}m\widehat{FG}$$

$$-88° = m\widehat{FG}$$

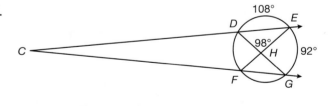

The student didn't know how to finish. Explain the student's error and calculate $m\angle C$.

Dilations

Warm Up

1. **Vocabulary** A change in position, size, or shape of a figure or graph is called a _____.
(67)

2. Triangle *ABC* has vertices *A*(3, 4), *B*(−1, 2), and *C*(0, −5). The triangle
(71) is translated so that *B'* is located at (2, −4). Determine the location of *A'* and *C'*.

3. In the diagram, △*DEF* ∼ △*GHI*.
(41) What is *DF*?

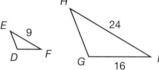

New Concepts

A **dilation** is a transformation that changes the size of a figure but not its shape. The multiplier used on each dimension of a figure to change it into a similar figure is the **scale factor**.

Dilations
A dilation maps a figure to a similar figure.

A dilation that results in an image smaller than its preimage is called a **reduction** or a **contraction**. A dilation that results in an image larger than its preimage is called an **enlargement** or an **expansion**.

Dilations require a center and a scale factor. The **center of dilation** is the intersection of lines that connect each point of the image with the corresponding point of the preimage. In the diagram, △*XYZ* was enlarged by a scale factor of 2, with the center of dilation *C* to create the image △*X'Y'Z'*.

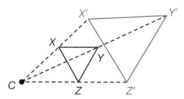

Hint

When a dilation is applied, it also affects the figure's distance from the center of dilation. For example, if a dilation of scale factor 2 is applied to a single point that is 3 units from the origin, the image will be 6 units from the origin.

Example 1 Enlarging by Dilation

Find the image of $\overline{AB}$ after a dilation with a scale factor of 2 and center *C*.

SOLUTION
The scale factor is greater than 1, so the dilation is an enlargement.

Draw lines from the center *C* through the endpoints of the line segment.

Because the scale factor is 2, *CA'* = 2*CA* and *CB'* = 2*CB*. Mark *A'* and *B'*.

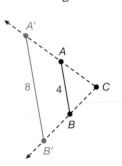

Connect *A'* and *B'* to form $\overline{A'B'}$. Since *AB* = 4, and it was enlarged by a factor of 2, *A'B'* = 8.

Example 2 Contracting by Dilation

Apply a dilation to △*JKL* using a scale factor of $\frac{1}{2}$ and center *C*.

SOLUTION

This dilation is a reduction. Draw lines from the center of dilation *C* to each of the vertices in △*JKL*. Find the distance between *C* and each vertex.

Because the scale factor is $\frac{1}{2}$, $CJ' = \frac{1}{2}CJ$, $CL' = \frac{1}{2}CL$, and $CK' = \frac{1}{2}CK$. Mark and label vertices *J'*, *K'*, and *L'*. Draw △*J'K'L'*.

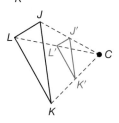

The scale factor can be used to find the coordinates of an image after a dilation. The notation $D_{O,k}$ indicates a dilation that is centered at the origin *O* of the coordinate plane and that has a scale factor of *k*. In mapping notation, $D_{O,k}(x, y) \rightarrow (kx, ky)$.

Example 3 Dilating on the Coordinate Plane

Triangle *DEF* has vertices located at *D*(4, 6), *E*(6, 2), and *F*(2, 4). Graph the image after a dilation centered at the origin and with a scale factor of $\frac{1}{2}$.

SOLUTION

Apply the transformation mapping given above.

$D_{O,k}(x, y) \rightarrow (kx, ky)$

$D_{O,\frac{1}{2}}(4, 6) \rightarrow (2, 3)$

$D_{O,\frac{1}{2}}(6, 2) \rightarrow (3, 1)$

$D_{O,\frac{1}{2}}(2, 4) \rightarrow (1, 2)$

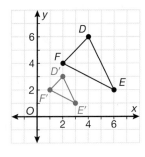

Plot the points and draw △*D'E'F'* on the coordinate plane.

Example 4 Application: Photocopiers

A student wants to scan and enlarge a piece of art that is 6 inches by 8 inches. If the student selects the 150% enlargement function, what will the lengths of the sides of the copy be? How does the perimeter of the original art compare to the perimeter of the copy?

SOLUTION

An enlargement of 150% indicates the scale factor is 1.5. The student should multiply each side of the original piece of art by 1.5. The copy will be 9 inches by 12 inches.

The perimeter of the original art is 6 + 6 + 8 + 8 = 28 inches.

The perimeter of the copy is 9 + 9 + 12 + 12 = 42 inches.

The original art has a perimeter that is $\frac{2}{3}$ the perimeter of the enlarged copy.

> ### Hint
>
> Any change in size greater than 100% of an image is an enlargement. Any change less than 100% is a reduction.
>
> Recall that a percentage can be converted to a decimal by dividing it by 100.

a. Apply a dilation to $\triangle QRS$ using
(Ex 1) a scale factor of 3 and center C.

b. Apply a dilation to rectangle $JKLM$
(Ex 2) using a scale factor of $\frac{1}{2}$ and center C.

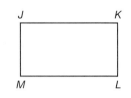

c. Triangle MNP has vertices $M(-2, 1)$,
(Ex 3) $N(-1, -2)$, and $P(-3, -2)$. Apply a dilation
with the center at the origin of the coordinate
plane and a scale factor of 3.

d. (**Architecture**) An architect is drawing plans for a building. The
(Ex 3) drawing for the front of the building is 4 feet long by 2.5 feet high.
If the drawing is a reduction by a scale factor of $\frac{1}{20}$, what will the
actual dimensions of the front of the building be? How do the areas
of the drawing and the building compare?

Practice Distributed and Integrated

* 1. Apply a dilation to $\triangle ABC$ using a scale factor of 2 and center D.
(84)

2. Find the solution for the system of equations. $y = 2x + 6$, $y = 2$
(81)

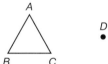

3. (**Cooking**) One cup is equivalent to 14.44 cubic inches. If a one-cup cylindrical
(62) measuring cup has a radius of 1.75 inches, what is its height to the nearest
tenth?

4. Predict If $\odot B$ is enlarged, but continues to pass through point A, will it remain
(72) tangent to $\odot C$? If not, when will it *not* be tangent to $\odot C$?

5. (**Snow Sculpture**) Lee is making a figure out of snow. If he makes his figure
(80) out of three spheres of snow with radii that are 1 foot, 2 feet, and 3 feet,
respectively, what total volume of snow does he use, to the nearest cubic
foot?

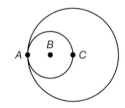

6. Algebra If two vectors have the sum of $\langle 3, a \rangle$ and one vector is $\langle 6, a \rangle$, what is the
(83) value of the other vector?

7. Multiple Choice The paths in a park are shown. Which of these routes
(39) represents the longest walk?

    A A to B to D **B** A to D to B
    C C to B to D **D** C to D to B

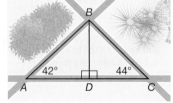

8. Write Explain how to determine whether a parallelogram
(52) is a rectangle.

9. Rectangle $MNPQ$ has vertices $M(6, 6)$, $N(6, 3)$, $P(-3, 3)$, and $Q(-3, 6)$.
(84) Graph its image after a dilation with the center at the origin and a scale
factor of $\frac{1}{3}$.

10. Given $\cos X = \dfrac{3}{5}$, what is the measure of $\angle X$, θ_1 or θ_2?
₍₈₂₎

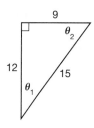

11. **Verify** Show, using an indirect proof, that it is impossible for
₍₇₀₎ the area of the base of a regular pyramid to equal the lateral surface area of the pyramid.

12. Use Euler's Formula to determine how many faces a polyhedron
₍₄₉₎ with 15 edges and 10 vertices has.

13. (**Astronomy**) The Moon's orbit is not exactly circular, but the average distance from
_(Inv 2) its surface to Earth's surface is 384,000 kilometers. The diameter of the Moon is 3476 kilometers. Find the distance to the nearest kilometer from the surface of Earth to the visible edge of the Moon, if the Moon is directly above the observer. *(Note: the figure is not drawn to scale.)*

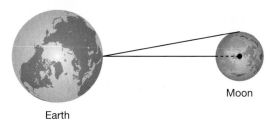

Moon

Earth

14. $\triangle MNP$ has vertices $M(6, 1)$, $N(2, 1)$, and $P(1, 4)$. Graph $\triangle MNP$ and rotate it
₍₇₈₎ 270° counterclockwise about the origin. List the coordinates of the vertices of $\triangle M'N'P'$.

15. Rhombus $STUV$ has vertices $S(3, -1)$, $T(6, 3)$, $U(x, y)$, and $V(7, -4)$. Find
₍₄₅₎ x and y.

*16. **Multi-Step** Find the angle the resultant vector makes with the horizontal if
₍₈₃₎ $\vec{a} = \langle 4, 3 \rangle$ and $\vec{b} = \langle -2, 3 \rangle$. Round your answer to the nearest degree.

17. Use the Venn diagram to write a conditional statement.
₍₁₀₎

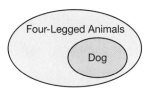

18. (**Radar**) A radar device scans an arc of 80° out to a distance of
₍₃₅₎ 10 miles. What is the minimum number of such devices needed to scan all directions around an area? What would the area of the overlap be? Express your answer in terms of π.

*19. (**Photocopiers**) A rectangular design with sides that are 12 centimeters long and
₍₈₄₎ a width of 8 centimeters is enlarged on a photocopier by 175%. What are the dimensions of the enlarged design? How do the areas of the original design and the copy compare?

xy² **20.** **Algebra** Find the value of a if $\overline{JL}$ is a diameter.
₍₄₇₎

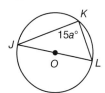

21. **Write** How do you know that one half of a diagonal is congruent to one half of
₍₆₅₎ the other diagonal in a rectangle?

22. What is the equation of a circle centered at $(-1, 5)$, with a diameter of 16?
(75)

23. Find the measures of the acute angles in the right triangle. Round your answers to the nearest degree.
(82)

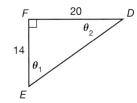

24. If 3.2 is the geometric mean between two values, what is the product of the two values?
(50)

25. For every 13 pennies in a bag of coins, there are 3 dimes and 1 quarter. If there are 325 pennies in the bag, how many dimes and quarters are there?
(41)

26. Use the head-to-tail method to find the resultant vector of the sum of $\vec{d}\langle-2, -1\rangle$, $\vec{e}\langle 5, 4\rangle$, and $\vec{f}\langle-5, -1\rangle$.
(83)

27. a. Identify two pairs of same-side interior angles in this figure.
(12)
b. Given that $\angle 3$ and $\angle 8$ are supplementary, prove that lines a and b are parallel.

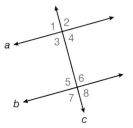

28. Formulate Monostrips are formed by placing black or white square tiles in a row, from left to right. The order of tiles matters: for example, ■■□ and □■■ count as different monostrips.
(Inv 8)
a. How many different 5-tile monostrips can you form?
b. How many of these monostrips have exactly 2 black tiles? *Hint: Count like this:*

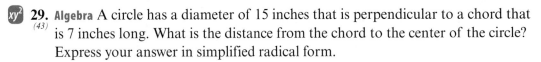

c. How many different n-tile monostrips can you form? How many of these have 2 black tiles?

29. Algebra A circle has a diameter of 15 inches that is perpendicular to a chord that is 7 inches long. What is the distance from the chord to the center of the circle? Express your answer in simplified radical form.
(43)

*30. Draw the image of $\triangle XYZ$ under a dilation with scale factor $\frac{1}{2}$ and center C.
(84)

Cross Sections of Solids

Warm Up

1. **Vocabulary** A(n) _____ is a way of drawing three-dimensional
   figures using paper that has equally spaced dots in a repeating triangular
   pattern.
   (54)

2. Find the surface area and volume of a sphere with a 12-centimeter radius
   in terms of π.
   (80)

3. **Multiple Choice** Which of the following is *not* a solid?
   (49)

   A pyramid   **B** cone
   C circle   **D** cylinder

New Concepts

A **cross section** is the intersection of a three-dimensional
figure and a plane. In the diagram, the first plane intersects
the cylinder to make a circular cross section. The second
plane intersects the cylinder to make a rectangle.

Think of a cross section as the shape that would
be revealed if you cut straight through an object.

Math Reasoning

Analyze Is it possible to
take a cross section of
a cone that is a single
point? Describe how.

Example 1 **Describing and Sketching Cross
Sections of Solids**

Describe and draw the cross section created by each plane.

a.

SOLUTION
The cross section is a circle.

b.

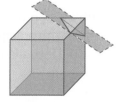

SOLUTION
The cross section is a triangle.

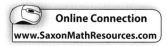
Online Connection
www.SaxonMathResources.com

Example 2 Finding Perimeter of a Cross Section

(a.) If the plane shown is perpendicular to the altitude of the cylinder, what is the perimeter of the cross section?

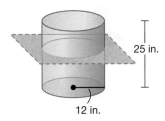

25 in.

12 in.

SOLUTION
The cross section is a circle with the same radius as the cylinder itself. Find the circumference of the circle in terms of π.

$C = 2\pi r$
$C = 2\pi(12)$
$C = 24\pi$ inches

(b.) If the altitude of the cylinder lies on the plane shown, what is the perimeter of the cross section of the cylinder?

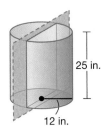

25 in.

12 in.

Hint

If the altitude of the cylinder lies on the cross section, this means that the plane cuts straight through "the middle" of the cylinder. Therefore, the diameter of each circular base of the cylinder also lies on the plane.

SOLUTION
The cross section is a rectangle. The height is 25 inches. Because the cross section travels through the center of the circle, the base of the rectangle is twice the radius.

$P = 2b + 2h$
$P = 2(2)(12) + 2(25)$
$P = 98$ inches

Example 3 Finding Area of a Cross Section

Find the area of this cross section of a square pyramid. The pyramid is 15 inches tall and the base is 6 inches wide. The cross section is perpendicular to the base of the pyramid and passes through the vertex.

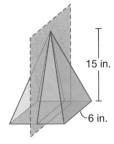

15 in.

6 in.

SOLUTION
Draw the cross section with the appropriate lengths labeled. The cross section is a triangle with a height of 15 inches and a base of 6 inches. Use the formula for area of a triangle.

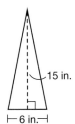

15 in.

6 in.

$A = \frac{1}{2}bh$

$A = \frac{1}{2}(15)(6)$

$A = 45$ square inches

Cavalieri's Principle
If two solids lying between parallel planes have equal heights and all cross sections at equal distances from their bases have equal areas, then the solids have equal volumes.

The two cones in the diagram illustrate Cavalieri's principle. The cones have the same radius and height, so Cavalieri's principle indicates that they will also have the same volume.

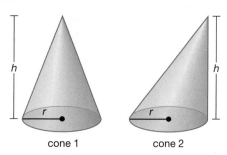

cone 1   cone 2

Imagine cutting the cones into many thin, circular cross sections. Each corresponding circular cross section of the two cones will be congruent.

Math Reasoning

Model A right cylinder has a vertical cross section that travels through the diameter of both circular bases. If an oblique cylinder has the same volume and also has a cross section that travels through the diameter of both circles, will the areas of these two cross sections be equal? Explain.

Example 4   Application: Office Supplies

There is a stack of CDs on a desk. The stack is bumped and makes a 70° angle with the table. What is the volume of the stack? Is it equal to the volume of the stack if it made a 90° angle with the table?

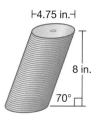

⊢4.75 in.⊣

8 in.

70°

SOLUTION

The stack has the same height before and after it is bumped. Moreover, each cross section represented by a CD has the same area. Therefore, according to Cavalieri's Principle, the cylinder has the same volume before it is bumped to the side as it does afterwards.

$V = \pi r^2 h$

$V = \pi \left(\dfrac{4.75}{2}\right)^2 (8)$

$V \approx 141.76 \text{ in}^3$

Lesson Practice

a. If the plane is parallel to the prism's bases, what is the shape of the cross section?
(Ex 1)

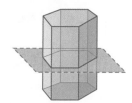

b. Find the perimeter of the cross section of the square pyramid if the
(Ex 2) cross section is parallel to the base. One side of the cross section is
4 inches long.

c. Find the perimeter of the cross section of the square pyramid. The
(Ex 2) cross section is perpendicular to the pyramid's base. The height of the
pyramid is 14 inches and the base of the pyramid is 12 inches on each
side. Round your answer to the nearest hundredth of an inch.

d. Find the area of the cross section. Each edge of the cube is
(Ex 3) 4 centimeters long. Round your answer to the nearest hundredth
square centimeter.

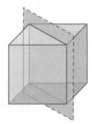

e. Find the exact volume for the oblique cone.
(Ex 4)

13 cm

2.5 cm

1. (**Architecture**) This decorative gate uses right triangles as supports. Find θ to the
(82) nearest degree.

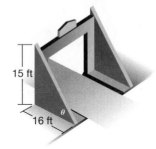

15 ft

16 ft

θ

* 2. Multiple Choice Which of the following solids could not give
(85) a triangular cross section?

    A cube   **B** triangular prism

    C square pyramid   **D** cylinder

3. What vector translates $\triangle ABC$ onto $\triangle A'B'C''$?
(71)

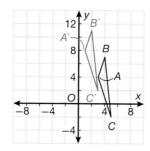

4. Algebra If $\triangle GHJ \sim \triangle LMN$, $GH = x^2 + 2$, $HJ = y - 4$, $JG = 6$, $LM = 36$,
(44) $MN = 2$, and $NL = 12$. Find the values of x and y. Then, find the ratio of
the two triangles' perimeters.

* 5. Apply a dilation to $\triangle QRS$ using a scale factor of $\frac{1}{3}$ and center C.
(84)

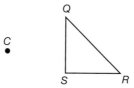

6. Victor has an hourglass made of two 4-inch-tall right circular cones
(77) connected at their vertices. After the sand has run for a time, there is only
0.883 in^3 of sand in the top half, making a cone 1.5 inches tall from the vertex.
What is the radius at the top of the cone of remaining sand?

7. a. Isosceles $\triangle JKL$ has a vertex angle at K. Its perimeter is 26 inches, and
(51) $JK = 7.5$ inches. Determine JL.

    b. The bisector of $\angle K$ intersects $\overline{JL}$ at M. Determine JM.

8. Verify Segment DE is a midsegment of $\triangle ABC$. Find the coordinates
(55) of D and E.

9. Multiple Choice Which of the following sets of numbers could be the
(33) side lengths of an obtuse triangle?

    A $7, 3\sqrt{5}, 2$ **B** $4, 3\sqrt{2}, \sqrt{17}$

    C $11, 19, 17$ **D** $16, 4\sqrt{5}, 8$

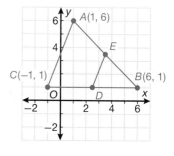

*10. Model Describe the cross section at right and draw it.
(85)

11. **Analyze** Are all right triangles with side lengths that are Pythagorean triples similar
$_{(46)}$ to each other? If so, explain why. If not, are any of them similar?

12. Compare ZY and WZ in the figure at right.
$_{(Inv\ 4)}$

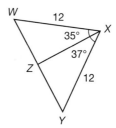

*13. $\triangle ABC$ has vertices $A(-3, 2)$, $B(1, 4)$, and $C(-1, -3)$. Apply a dilation with the
$_{(84)}$ center at the origin of the coordinate plane and sketch the graph. Use a scale
factor of 2.

14. **Generalize** How many edges does a triangular prism have? … a rectangular prism?
$_{(49)}$ … a pentagonal prism? Based on these answers, write a formula for the number of
edges in a prism with an n-sided base.

15. **Multi-Step** Use the head-to-tail method to find the resultant vector of
$_{(83)}$ $\vec{c} = \langle -2, 4\rangle$, $\vec{d} = \langle -6, 1\rangle$, and $\vec{e} = \langle 7, 4\rangle$. Then, find its magnitude
to the nearest hundredth.

16. a. This parallelogram contains two equilateral triangles. Sketch
$_{(Inv\ 8)}$ the figure and translate it right by the distance of one side length.
 b. Repeat part **a** for three more translations. What type of pattern have you
created? What type of symmetry does it have?

17. **Error Analysis** On Todd's last math assignment, he was given a 37° angle formed by
$_{(64)}$ a tangent and a chord of a circle, and he was asked to find the measure of the
arc that subtends it. Recalling a theorem from class, he wrote down the answer as
18.5°, but his teacher marked the question incorrect. What was his mistake?

18. Use the Hypotenuse-Leg Congruence Theorem to show that
$_{(36)}$ $\triangle ABC \cong \triangle DEF$.

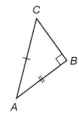
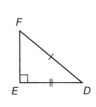

19. Rectangle $GHJK$ has vertices $G(2, 4)$, $H(4.5, 4)$, $J(4.5, -1)$, and $K(2, -1)$.
$_{(74)}$ Reflect $GHJK$ across $y = 2x$ and graph the image.

*20. **Algebra** Find the perimeter of a cross section of the square pyramid shown.
$_{(85)}$ The cross section of the pyramid makes a square and it is taken halfway
between the vertex of the pyramid and the bottom of the pyramid. The
original height of the pyramid is 8 inches and the base side lengths are
4 inches. Round your answer to the nearest whole inch.

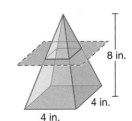

21. **Food Packaging** A cylindrical soup tin has a diameter of 7 centimeters and a
$_{(62)}$ height of 13 centimeters. Ignoring the thickness of the tin's walls, how much
soup is it able to hold?

22. Quadrilateral *WXYZ* is a rectangle. Find *XY* and *WY*.
(52)

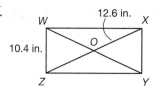

23. Explain why △*PQR* ≅ △*STR*.
(30)

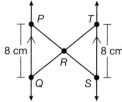

24. (Sailing) A sailboat is traveling directly across a lagoon at a velocity of 15 meters
(83) per second. If the water is flowing perpendicular to the sailboat at a velocity
of 8 meters per second, what is the actual direction of the sailboat to the
nearest degree?

25. Find m∠*A*.
(79)

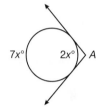

26. **Analyze** What is the order of rotation of a circle? How many lines
(76) of symmetry does a circle have?

*27. (Breakfast) A chef is slicing a bagel into two halves along its side
(85) as shown. Describe the cross section.

28. In the diagram, *ABCD* ~ *EFGH*. Find the measure of
(41) ∠*A* and ∠*C*.

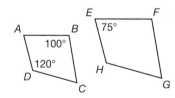

*29. △*M′N′P′* is the image of △*MNP* after a dilation centered at the origin. Identify
(84) the type of dilation and find the scale factor.

30. Triangle *ABC* is rotated 120° in a counterclockwise direction and
(78) then rotated 190° in a clockwise direction about the origin. What
clockwise rotation is equivalent to this set of rotations?

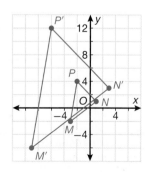

Determining Chord Length

Warm Up

1. **Vocabulary** A chord of a circle is a _____ whose _____ are on the circle.
   (43)

2. Triangle *DEF* is inscribed in ⊙*C*. If $\overline{DF}$ is a diameter of ⊙*C*, what type of angle is ∠*E*?
   (47)

3. **Multiple Choice** If diameter $\overline{AC}$ in a circle is the perpendicular bisector of chord $\overline{BD}$, which is not a diameter, what type of quadrilateral is *ABCD*?
   (61)

   A a rectangle   **B** a kite

   C a rhombus   **D** a square

New Concepts

A chord is a segment whose endpoints lie on a circle. Theorem 86-1 relates the lengths of chord segments when two chords intersect.

Hint

A common mistake is to apply Theorem 86-1 only when chords bisect each other, or only when chords are congruent, but Theorem 86-1 can actually be used for any pair of chords, as long as they intersect.

Theorem 86-1
If two chords intersect in a circle, then the products of the chord segments are equal. In the diagram, $(AE)(EB) = (CE)(ED)$. 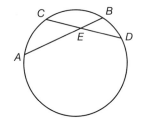

Example 1 **Proving Theorem 86-1**

Given: Chords $\overline{TQ}$ and $\overline{RS}$ intersect at point *P*.

Prove: $(QP)(PT) = (RP)(PS)$

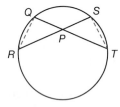

SOLUTION

Since two points determine a line, we can draw $\overline{QR}$ and $\overline{ST}$. Because they intersect the same arc on the circle, ∠*RQT* ≅ ∠*TSR*. By the Vertical Angles Theorem, ∠*QPR* ≅ ∠*SPT*. Therefore, △*QPR* ~ △*SPT* by the AA Similarity Postulate. The corresponding sides of these similar triangles must be proportional, so $\frac{RP}{PT} = \frac{QP}{PS}$. The cross product shows that $(QP)(PT) = (RP)(PS)$.

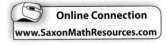

Online Connection

www.SaxonMathResources.com

Finding Chord Lengths

In the circle, chords $\overline{PQ}$ and $\overline{RS}$ intersect at T.
Determine ST.

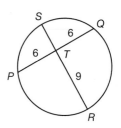

SOLUTION

Use Theorem 86-1 to write an expression
relating the lengths of the chord segments.

$$(PT)(QT) = (RT)(ST)$$
$$(6)(6) = (9)ST$$
$$ST = 4$$

Math Reasoning

Formulate Under what
conditions do two
intersecting chords
form four congruent
segments?

Example 3 **Solving for Unknowns with Intersecting Chords**

In this circle, use the expressions for the segment
lengths to write and solve an equation for x.

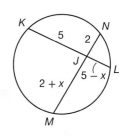

SOLUTION

$$(JK)(JL) = (JM)(JN)$$
$$5(5 - x) = (2 + x)(2)$$
$$25 - 5x = 4 + 2x$$
$$x = 3$$

Example 4 **Application: Aviation**

A "super-heavy" passenger jet has an upper passenger deck that is
located $\frac{3}{4}$ of the way up the cylindrical fuselage. What percentage
of the height of the fuselage is the width of the upper deck?

SOLUTION

Understand Draw a diagram. A cross-section of
the fuselage is circular, as shown.

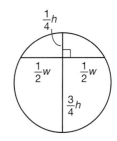

Plan Use Theorem 86-1 to write an equation.

$$\left(\frac{1}{2}w\right)\left(\frac{1}{2}w\right) = \left(\frac{1}{4}h\right)\left(\frac{3}{4}h\right)$$

Solve
$$\left(\frac{1}{2}w\right)\left(\frac{1}{2}w\right) = \left(\frac{1}{4}h\right)\left(\frac{3}{4}h\right)$$
$$\frac{1}{4}w^2 = \frac{3}{16}h^2$$
$$w^2 = \frac{3}{4}h^2$$
$$w = \frac{\sqrt{3}}{2}h$$
$$w \approx 0.87h$$

The width of the upper deck is approximately 0.87, or 87% of the height
of the fuselage.

Check Look at the diagram. Does it look like the upper deck is a little
shorter than, close to the same length as, the height of the fuselage? It
appears to be that way, so the answer seems correct.

In $\odot G$, chords $\overline{AB}$ and $\overline{CD}$ intersect at E. Use this information for parts a and b.

a. Determine DE if $AE = 3$, $BE = 16$, and $CE = 9$.
(Ex 2)

b. Suppose $AE = 7$, $BE = y$, $CE = 4 - y$, and $DE = 2$.
(Ex 3) Write and solve an equation for y.

c. In the diagram, $\overline{LO}$ and $\overline{PM}$ intersect at N.
(Ex 3) Find the value of x.

d. (**Civil Engineering**) A cylindrical natural gas
(Ex 4) pipeline is supported at two points that are 10% of the diameter of the pipeline above its lowest point. If the diameter of the pipeline is 4 feet, 9 inches, how far apart are the supports?

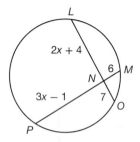

* **1.** In $\odot C$, what is VZ?
(86)

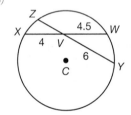

2. (**Map Reading**) When Erin drew a circle around the intersection of two straight
(64) trails on a map, she found the trails intersected an arc of 85° on one side of the intersection and 103° on the opposite side. What is the measure of the angle between the two trails?

3. Generalize A cone has one-third the volume of a cylinder with the same height
(77) and radius. Does such a cone also have less surface area than the circular cylinder? Explain.

4. Find the distance from (5, 3) to the line $x = -2$.
(42)

5. Generalize Describe how to determine the probability of a pointer on a spinner
(Inv 6) *not* landing on a given sector.

6. Apply a dilation to $\overline{ST}$ using a scale factor of 4 and center C.
(84)

7. *DEFG* is a parallelogram. If ∠*D* is one-eighth the size of ∠*G*, what is the measure
(34) of each angle?

8. Find θ_1 and θ_2 to the nearest degree.
(82)

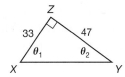

9. Add the vectors ⟨1, 5⟩ and ⟨−1, −5⟩. What kind
(63) of vectors are these?

10. Verify Verify that *A*(4, −1), *B*(1, 0.5), and *C*(−2, 2) are collinear.
(45)

*11. The plane shown intersects a diameter of the cylinder. Find the perimeter
(85) of the cross section. The radius of the cylinder is 5 centimeters and
the height is 18 centimeters.

12. (Gardening) Gordon's outdoor fountain is in the shape of the lower
(80) half of a sphere. If the diameter of the hemisphere is 12 inches,
what volume of water does the fountain hold to the nearest cubic inch?

13. Find the perimeter of the regular polygon shown,
(66) to the nearest foot.

47 ft

*14. Analyze $\overline{AB}$ and $\overline{CD}$ are chords in the same circle, and
(86) they intersect at point *E*. If *AE* = *CE*, What is the
relationship between the two intersecting chords?

15. Multiple Choice What is the best description for ∠*S* in △*STU*?
(18)
   A right angle   **B** remote interior angle
   C complementary angle   **D** exterior angle

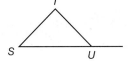

xy^2 *16. Algebra Two chords of a circle, $\overline{FG}$ and $\overline{CD}$, intersect at *E*. *FE* = 2,
(86) *GE* = 3 + 2*x*, *CE* = 6, and *DE* = *x*. Write and solve an equation for *x*.

17. Multi-Step Find the angle that the resultant vector of the vectors ⟨−3, 5⟩ and ⟨6, 4⟩
(83) makes with the horizontal. Round your answer to the nearest degree.

xy^2 18. Algebra Find the measure of each angle in the triangle shown.
(79)

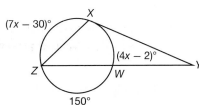

19. Pentagon *ABCDE* has vertices *A*(0, 4), *B*(4, 0), *C*(2, −4), *D*(−2, −4), and *E*(−4, 0).
(84) Apply a dilation with the center at the origin of the coordinate plane. Use a scale
factor of $\frac{1}{4}$.

20. Find the values of *x* and *y*. Express your answers in simplified
(56) radical form.

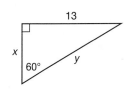

21. **Write** In the diagram, $\angle A \cong \angle B$. Use a paragraph proof to
(31) show that $m\widehat{FE} = m\widehat{HG}$.

22. **Fire Protection** A forest ranger stands in a tower that is
(73) 40 meters tall. If he sees a fire at an angle of depression
of 6°, how far away is the fire from the base of the
tower? Round your answer to the nearest tenth of
a meter.

*23. Find the perimeter of the cross section of the equilateral triangular
(85) prism shown. The sides of the triangle are 8 centimeters long.

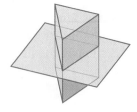

24. **Pet Supplies** Charlie's fish tank is shaped like a regular pentagonal
(70) pyramid. If the area of the base is 200 cm^2 and the height of the fish
tank is 18 centimeters, what volume of water is needed to fill the
fish tank?

25. Use the Law of Detachment to write a conclusion based on the following
(21) statements.

On Sundays, sandwiches are 25% off.
Today is a Sunday.

*26. **Algebra** Two chords of a circle, $\overline{FG}$ and $\overline{HJ}$, intersect at K. $FK = 5$, $GK = 3 - 2y$,
(86) $HK = y - 4$, and $JK = 4$. Write and solve an equation for y.

27. **Railways** A train station's rail tunnel has a semicircular opening large
(40) enough for a train to pass through, as shown. What is the area of the
rail tunnel's opening, to the nearest square meter?

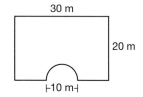

28. **Error Analysis** Jean-Paul drew the net shown for a pentagonal prism.
(Inv 5) What is wrong with his net? How could he fix it?

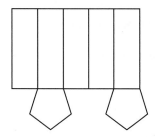

29. Find the length of the hypotenuse of a 45°-45°-90° right triangle with a leg that is
(53) $13\sqrt{2}$ centimeters long.

*30. **Multiple Choice** Which of the following could *not* have a square cross section?
(85) **A** square pyramid   **B** sphere
C cube   **D** triangular prism

Intersecting Chords Using Geometry Software

Technology Lab 11 *(Use with Lesson 86)*

In Lesson 86, you learned how to determine lengths of chords. In this lab, you will use geometry software to explore chords and tangents. When two chords intersect in a circle, the segments formed by their intersection have a special relationship.

1. Use the circle tool to construct ⊙A.

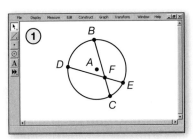

2. Use the segment tool to draw two intersecting chords through ⊙A. Label them $\overline{BC}$ and $\overline{DE}$. Label their point of intersection F.

3. Measure the lengths of $\overline{DF}$, $\overline{EF}$, $\overline{BF}$, and $\overline{CF}$. Record the length of each segment in the chart below.

4. Complete the last two columns of the chart by multiplying the indicated segments' measures and writing their products in the corresponding blanks.

5. Drag F to change the length of your chords and record the new values in the second row of the chart. Repeat this step to fill in the third row of the chart.

6. Using your chart as a guide, what conjecture can you make about the lengths of the segments formed by intersecting chords of a circle?

Math Reasoning

Predict After filling out the first row completely, what do you notice about the final two columns? Do you think this will be true of the next two rows?

DF	EF	BF	CF	DF • EF	BF • CF

In Lesson 72, you learned a relationship between the segments formed by intersecting tangent lines. Geometry software can be used to demonstrate and verify this relationship.

1. Construct ⊙P using the circle tool.

2. Place two points on the circle. Label them Q and R.

3. Use the segment tool to create $\overline{PQ}$ and $\overline{PR}$.

(4.) A tangent line must be perpendicular to the radius of a circle. Select $\overline{PQ}$ and point Q. Choose *Perpendicular Line* from the *Construct* menu. This will create a line perpendicular to $\overline{PQ}$ that intersects the circle at point Q.

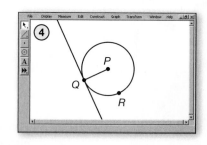

(5.) Follow the directions in step 4 to create another tangent line that passes through point R.

(6.) Place a point at the intersection of the two new tangent lines. Label it S.

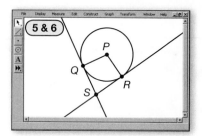

(7.) Select points R and S. Choose *Distance* from the *Measure* menu. Measure $\overline{QS}$ using this same method.

(8.) What is the relationship between $\overline{QS}$ and $\overline{RS}$? Drag point Q around the circle. Does the relationship remain the same?

Lab Practice

Construct a new circle with intersecting chords. Measure all but one of the four segments in the circle. Can you determine the length of the fourth segment from what you know of the other three? If so, how?

Area Ratios of Similar Figures

Warm Up

1. **Vocabulary** A plane figure made up of simple shapes is known as a _____ figure.
   (40)

2. Determine the perimeter and area of the given polygon.
   (22)

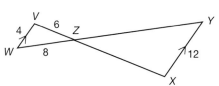

3. **Multiple Choice** The diagram shows two similar figures. What is the length of side $\overline{YZ}$?
   (41)

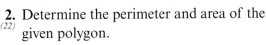

   A 6   **B** 8
   C 18   **D** 24

New Concepts

Recall that polygons are similar if they have the same shape, but differ in size. This difference in size describes their scale factor to each other and can be written as a similarity ratio.

For the squares given, the perimeter of the first square is $4a$ and the second is $4b$. The ratio of their perimeters is $4a{:}4b$, which can be reduced to $a{:}b$. Their areas are a^2 and b^2, so the ratio of their areas is $a^2{:}b^2$. These relationships are true of all similar polygons.

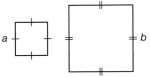

Math Reasoning

Verify Draw two similar triangles, one with a base of b and a height of h and the second with a base of $2b$ and a height of $2h$. Find the area of each triangle and determine their area ratio.

Theorem 87-1
If two similar figures have a scale factor of $a{:}b$, then the ratio of their perimeters is $a{:}b$, and the ratio of their areas is $a^2{:}b^2$.

Example 1 | **Proving Theorem 87-1**

Prove the first part of Theorem 87-1.

Given: $\triangle ABC \sim \triangle DEF$

Prove: $\dfrac{AB + BC + AC}{DE + EF + DF} = \dfrac{AB}{DE}$

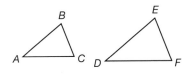

SOLUTION

Let x be the similarity ratio of $AB{:}DE$. In other words, $x = \frac{AB}{DE}$. Then $AB = (DE)(x)$. Furthermore, $BC = (EF)(x)$ and $CA = (FD)(x)$. By the Addition Property of Equality, $AB + BC + CA = (DE)(x) + (EF)(x) + (FD)(x)$. Therefore, $AB + BC + CA = x(DE + EF + FD)$. By the Division Property of Equality, $x = \frac{AB + BC + CA}{DE + EF + FD}$. By substitution, $\frac{AB + BC + AC}{DE + EF + DF} = \frac{AB}{DE}$.

Online Connection
www.SaxonMathResources.com

Example 2   Ratio of Perimeters of Similar Figures

In the given similar figures, the perimeter of the smaller shape is 50 inches. Determine the perimeter of the larger shape.

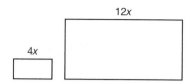

SOLUTION

The similarity ratio of the larger rectangle to the smaller rectangle is $12x{:}4x$, or 3:1 after being reduced.

By Theorem 87-1, the perimeters of the two rectangles will be in the same ratio. Set up a proportion to solve for the larger figure's perimeter.

$$\frac{3}{1} = \frac{P}{50}$$
$$P = 150$$

So the perimeter of the larger rectangle is 150 inches.

Example 3   Ratio of Areas of Similar Figures

The two triangles given have a similarity ratio of 2:5. Determine the ratio of their areas and the area of the smaller triangle.

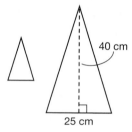

SOLUTION

From Theorem 87-1, the ratio of their areas will be $a^2{:}b^2$. Therefore, the ratio of the smaller triangle's area to the larger triangle's area is 4:25. To find the area of the smaller triangle, first find the area of the larger triangle.

$$A = \frac{1}{2}bh$$
$$A = \frac{1}{2}(25)(40)$$
$$A = 500$$

Now set up a proportion using the ratio of the triangles' areas to find the area of the small triangle.

$$\frac{4}{25} = \frac{A}{500}$$
$$A = 80$$

The area of the smaller triangle is 80 square centimeters.

Example 4 | Application: Landscape Design

A landscape design company has created a plan for a large garden in the shape of an isosceles trapezoid, as illustrated in the diagram. The diagram of the garden is in a 2:355 ratio with the size of the actual garden. Find the perimeter and area of the actual garden.

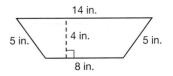

SOLUTION

Add the sides together to find the perimeter of the shape.

$$P_1 = 14 + 5 + 8 + 5 = 32$$

The perimeter of the scale garden is 32 inches. Write a proportion to find the perimeter of the real garden.

$$\frac{2}{35} = \frac{32}{P_2}$$
$$P_2 = 5680$$

The perimeter of the actual garden will be 5680 inches, or approximately 473 feet. Apply the formula for area of a trapezoid.

$$A_1 = \left(\frac{b_1 + b_2}{2}\right)h$$

$$A_1 = \left(\frac{14 + 8}{2}\right)(4)$$

$$A_1 = 44$$

Therefore, the area of the trapezoid in the diagram is 44 square inches. The area ratio will be given by $2^2:355^2$, which is 4:126,025. Applying this ratio, we obtain

$$\frac{4}{126,025} = \frac{44}{A_2}$$
$$A_2 = 1,386,275$$

The area of the garden is 1,386,275 square inches or approximately 9627 square feet.

Caution

To convert the area of the garden into square feet, you cannot simply divide by 12. A square foot measures 12 inches on each side, or 144 square inches, so you must divide the area by 144.

Lesson Practice

a. Prove the second part of Theorem 87-1.
(Ex 1)

Given: $\triangle PQR \sim \triangle WXY$

Prove: $\dfrac{\text{area } \triangle PQR}{\text{area } \triangle WXY} = \dfrac{PR^2}{WY^2}$

b. In the given similar figures, the perimeter of the large hexagon is
(Ex 2) 120 feet. Determine the perimeter of the small hexagon.

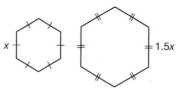

c. The two parallelograms given have a similarity ratio of 2:5.
(Ex 3) Determine the ratio of their areas and the area of the larger
parallelogram.

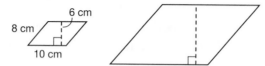

d. The kitchen on a floor plan shows a triangle from the sink to the
(Ex 4) refrigerator to the counter that has an area of 1.5 square feet. If
the floor plan has a scale of 1:10, what will be the actual area of
this triangle when the house is built?

Practice Distributed and Integrated

* 1. (Archeology) A dig crew is asked to start excavating a site that is 4 meters by
(87) 5 meters. After this site is studied and some artifacts are uncovered, the crew is
then asked to expand the site by a factor of 7 in each direction. What will be the
total area of the site after this expansion?

2. Estimate Estimate the surface area of a right circular cone with a slant height of
(77) 13.78 centimeters and a radius of 5.21 centimeters.

3. In the figure at right, $\triangle ABC$ has midsegment $\overline{DE}$. Find the coordinates
(55) of D and E.

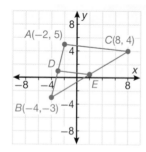

4. Rectangle $JKLM$ has vertices $J(-4, 2)$, $K(-4, -6)$, $L(4, -6)$, and $M(4, 2)$. Apply
(84) a dilation with the center at the origin of the coordinate plane. Use a scale factor
of $\frac{1}{2}$.

5. Write three inequalities that follow from the Triangle Inequality Theorem for a
(39) triangle with side lengths of a, b, and c.

* 6. Two chords of a circle, $\overline{AB}$ and $\overline{CD}$, intersect at E. If AE and $BE = 6$
(86) and $CE = 8$, determine DE.

7. What is the volume of the regular pyramid shown below?
(70)

8 ft

6 ft

8. (**Cost Comparison**) Two cafes are having a sale on coffee. The first one charges $2.50
(81) for a regular, plain coffee and $0.75 for each additional flavoring. The second one
charges $3.25 for a regular, plain coffee and $0.50 for each additional flavoring. If
Brynna buys one regular coffee, how many additional flavorings would make her
drink from each location cost the same? What is that price?

9. Error Analysis Jerrod and Anthony are defining a rectangle. Whose definition is
(52) correct? Explain.

Jerrod: *"A rectangle has a pair of sides that are parallel and it has a right angle."*
Anthony: *"A rectangle is a parallelogram with one right angle."*

*10. Two squares are related by a scale factor of 5. If the smaller square has a
(87) perimeter of 32 inches, what is the perimeter of the larger square?

xy^2 ***11. Algebra** Find the measure of $\angle H$.
(47)

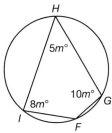

12. (**Baseball**) A baseball is a spherical ball. The hide of a baseball covers the outside
(80) of the ball. If the ball has a radius of 4 centimeters, what area of material is
needed for the hide to the nearest tenth?

13. Write Explain how you could get a triangular cross section from a cube.
(85)

14. If the hypotenuse of a 30°-60°-90° triangle is 4, what are the lengths of the other
(56) two legs?

*15. Multiple Choice Kite $RSTU \sim HIJK$. If $HJ = 2RT$ and $KI = 2SU$, what is
(87) the area of kite $HIJK$, compared to the area of $RSTU$?

A The area is twice as large.
B The area is three times as large.
C The area is four times as large.
D The area is eight times as large.

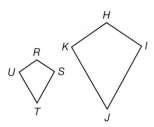

16. **Parks** Oakdale Park takes up one square block that is 300 feet on each side. Two
(53) diagonal sidewalks join the opposite corners. To the nearest foot, how long is each
diagonal sidewalk?

17. Graph the line with equation $x + 2y = 8$.
(16)

*18. In the circle, what is QT?
(86)

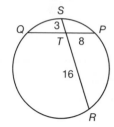

19. If the length of a median of a triangle equals 16.5, what is the distance from the
(32) centroid to the opposite side?

20. **Multi-Step** Use the head-to-tail method to find the resultant vector of $\vec{a} = \langle 2, 2 \rangle$,
(83) $\vec{b} = \langle -3, -2 \rangle$, and $\vec{c} = \langle 4, -3 \rangle$. Then find its magnitude to the nearest
hundredth.

21. **Algebra** Line x passes through points $(1, -3)$ and $(3, k)$. Line y passes through
(37) points $(2, 4)$ and $(4, 7)$. What value of k makes these lines parallel?

22. Find the area of the cross section of the triangular prism to the nearest
(85) hundredth. Each side of the base is 8 centimeters long.

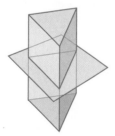

23. **Analyze** Does the midsegment of an isosceles trapezoid divide the trapezoid
(69) into two similar trapezoids? Explain.

*24. A regular hexagon has a perimeter of 36 centimeters and an area of 93.5 square
(87) centimeters. A similar hexagon is constructed with a similarity ratio of 3:7. Find
the perimeter and area of the larger hexagon.

25. **Write** How are a solid's net and its surface area related?
(Inv 5)

26. **Compact Discs** Do the interior circle and exterior circle of a compact disc have
(72) any common tangents? Are they tangent circles?

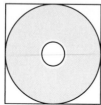

27. **Write** Explain how to find the perimeter of the polygon with vertices at
(57) $(-2, 3)$, $(3, 1)$, $(-2, -1)$, and $(-3, 1)$.

28. Use *sine*, *cosine*, or *tangent* to fill in each blank in this statement:
(Inv 7) The _____ function increases from 0 to 1 as angle measures increase from
$0°$ to $90°$, while the _____ function decreases from 1 to 0.

*29. **Multiple Choice** What is the missing segment length in this figure?
(86)

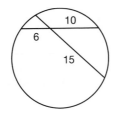

   A $2\sqrt{15}$ **B** 15
   C 4   **D** 25

30. A cube has a surface area of 54 square inches. Find its volume.
(59)

Graphing and Solving Linear Inequalities

Warm Up

1. **Vocabulary** A statement that compares two expressions using $<$, $>$, $\leq$, $\geq$,
$_{(SB)}$ or $\neq$ is a(n) _____.

2. At what point do the lines $y = 2x + 1$ and $y = x - 5$ intersect?
$_{(81)}$

3. Nonparallel lines intersect at _____.
$_{(4)}$
   A zero points   **B** exactly one point
   C exactly two points   **D** infinitely many points

New Concepts

A linear inequality can be rearranged just like a linear equation. The only difference is that multiplying or dividing both sides of the inequality by a negative number changes the direction of the inequality sign. To solve a linear inequality, convert it to slope-intercept form.

Example 1 Solving Linear Inequalities

a. Solve the linear inequality $3x + 2y > 1$ for y.

SOLUTION

Rearrange the linear inequality as if it were a linear equation.

$3x + 2y > 1$	Given
$3x + 2y - 3x > 1 - 3x$	Subtraction Property of Inequality
$2y > 1 - 3x$	Simplify.
$\dfrac{2y}{2} > \dfrac{1 - 3x}{2}$	Division Property of Inequality
$y > -\dfrac{3}{2}x + \dfrac{1}{2}$	Simplify.

b. Solve the linear inequality $2x - 5y < 6$ for y.

SOLUTION

$2x - 5y < 6$	Given
$2x - 5y - 2x < 6 - 2x$	Subtraction Property of Inequality
$-5y < 6 - 2x$	Simplify.
$\dfrac{-5y}{-5} > \dfrac{6 - 2x}{-5}$	Division Property of Inequality
$y > \dfrac{2}{5}x - \dfrac{6}{5}$	Simplify.

Caution

Remember to change the direction of the inequality sign when dividing by a negative number.

The graph of an inequality includes points that are not on the graph of the linear equation. A region of the coordinate grid bounded by the graph of the linear equation is shaded to show the points that satisfy the inequality. For inequalities where the y-values are greater than the y-values on the line, shade the region above the line. For y-values less than the y-values on the line, shade the region below the line.

Online Connection
www.SaxonMathResources.com

When an inequality uses ≥ and ≤, the graph includes the line itself, so a solid line is drawn. For inequalities that use > and <, the graph does not include points on the line, so a dashed line is drawn to show that ordered pairs lying on the line are not part of the solution to the inequality.

Example 2 **Graphing an Inequality**

(a.) Graph the inequality $y < \frac{1}{2}x + 3$.

SOLUTION
Begin by graphing the line $y = \frac{1}{2}x + 3$ using a dashed line because the inequality does not include the line itself. Since y is less than $\frac{1}{2}x + 3$, shade the area below the linear equation.

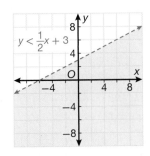

(b.) Graph the inequality $y \geq 2x - 1$.

SOLUTION
Graph the line $y = 2x - 1$. Since y is greater than $2x - 1$, shade the area above the linear equation. Since this inequality is greater than or equal to, the points on the line are included, so a solid line is used.

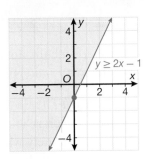

Example 3 **Application: Sports**

Lily and Amit are playing a game. At the end of the game, the sum of their scores will always be at least 21. Let x represent Lily's score and y represent Amit's score. Draw a graph showing the set of possible scores. Is it possible that the game is over when Lily has 13 points and Amit has 8?

SOLUTION
The sum of their scores must be at least 21, so $x + y \geq 21$.

Solve the inequality for y.

$x + y \geq 21$	Given
$x + y - x \geq 21 - x$	Subtraction Property of Inequality
$y \geq -x + 21$	Simplify.

Math Reasoning

Analyze If the scores could only be integers, how would the solution graph change?

Graph the line $y = -x + 21$ using a solid line. Since y is greater than $-x + 21$, shade the area above the line.

The point (13, 8) lies on the line. Since the line is included in the set of possible solutions for this inequality, it is possible for the game to be over when Lily has 13 points and Amit has 8.

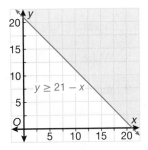

a. Solve the linear inequality $-2x - 4y < -8$ for y.
(Ex 1)

b. Graph $y \geq 2x - 6$.
(Ex 2)

c. Graph $2y - 3 > 5 - x$.
(Ex 3)

d. Graph $x - 4y \leq 7x + 8$.
(Ex 4)

e. (**Electricity**) Trevor has two appliances connected to a power supply
(Ex 5) that provides 15 Joules of power per second (J/s). If x represents the power to the first appliance and y represents the power to the second appliance, graph the region that shows the possible power to each appliance. Use your graph to determine whether Trevor can send 7 J/s to the first appliance and 6 J/s to the second appliance.

Practice Distributed and Integrated

* 1. Determine the area of the smaller parallelogram in the diagram if the
(87) similarity ratio is 3:11.

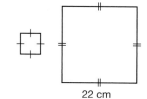

22 cm

2. (**Biology**) The Eastern Auger is a type of snail that has a cone-shaped
(77) shell, which is about 2.25 inches long and has a 0.5-inch radius. Approximately what is the surface area of the Eastern Auger's shell? Assume that the base is flat.

* 3. Solve the linear inequality $6x + 3y > 9$ for y.
(88)

4. **Analyze** A square pyramid has a cross section that passes through the vertex and
(85) intersects two opposite corners of the base. What is the area of the cross section if the height of the pyramid is 70 centimeters and the base is 50 centimeters wide? Round your answer to the nearest hundredth of a square centimeter.

5. Write the converse of the following statement about positive numbers x and y.
(17) Which is true—the statement, its converse, both, or neither?

$If \frac{x}{y} > 1, then\ x > y.$

6. **Justify** Is this expression true for all x? $\sin^{-1} x = \frac{1}{\sin x}$. If so, prove it. If it is not
(82) true, find a counterexample.

7. **Algebra** Use this figure to write and solve an equation for a.
(86)

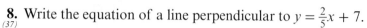

8. Write the equation of a line perpendicular to $y = \frac{2}{5}x + 7$.
(37)

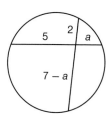

9. **Multiple Choice** Which of the following values represents the area of an
(66) equilateral triangle with 30-centimeter side lengths?

A $450\sqrt{3}$ cm^2

B $225\sqrt{3}$ cm^2

C $\frac{450}{\sqrt{3}}$ cm^2

D $\frac{225}{\sqrt{3}}$ cm^2

10. Find the values of x, y, and z in the parallelogram.
(34)

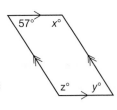

11. (**Economics**) The supply of Brand A is represented by $y = 2x + 15$,
(81) where y represents the price in dollars and x represents the number of units. The demand for Brand A is represented by $y = \frac{1}{3}x + 75$. What are the optimum number of units and the price for Brand A?

*12. Graph $4 + 2y > 8x - 4$.
(88)

*13. (**Scale Models**) The scale model of a barn is shown at right. It has four sides
(87) that are congruent rectangles. It is modeled after a rectangular barn with a length of 24 feet and a height of 12 feet. If the scale factor used is 1:16, what is the total exterior area of the model?

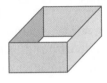

14. $\triangle ABC$ has vertices $A(-4, -3)$, $B(-3, 1)$, and $C(1, -2)$. Graph $\triangle ABC$, and then
(78) rotate it 90° counterclockwise about the origin. List the coordinates of the vertices of $\triangle A'B'C'$.

15. Given $\triangle FGH \sim \triangle XYZ$, find the ratio of the perimeters if the ratio of their
(87) corresponding sides is 1:8.

16. Find the area of the cross section of the cylinder at right, through the cylinder's
(85) diameter. The radius is 5 centimeters and the height is 18 centimeters.

17. Multi-Step Classify the polygon with vertices $R(1, 2)$, $S(-2, 5)$,
(57) and $T(9, 4)$. Find the perimeter. Round your answer to the nearest hundredth.

18. Find the area of a 45°-45°-90° right triangle that has a 14-meter hypotenuse.
(53)

*19. Analyze Jin is measuring the distance between himself and the city monument
(73) shown below from various heights in the 8-story building where he works. He knows that the monument is 50 feet from the base of the building and that each floor is 15 feet higher than the last. As Jin climbs up the stories of the building, will the angle of depression increase at a steady rate? Will his distance from the monument increase at a steady rate? Explain why or why not.

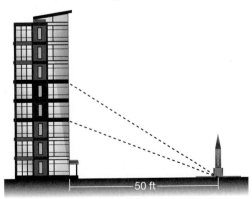

*20. Graph $-3y \geq 2x - 6$.
(88)

21. What is the value of x in the proportion $\frac{6}{x + 2} = \frac{3}{5}$?
(41)

22. **Algebra** A pair of corresponding angles in two congruent triangles measure
(28) $(3x + 80)°$ and $(4x + 70)°$. What is the value of x?

23. (**Architecture**) In this circular stained-glass window, what is the length
(86) of the horizontal crosspiece?

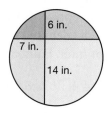

*24. **Error Analysis** Raquel states that since the ratio of the areas of two
(87) similar polygons is 4:9, the ratio of their perimeters must be 4:9.
Is she correct? Explain why or why not.

25. **Analyze** The equation $(x - h)^2 + (y - k)^2 = r^2$ is one way of representing
(75) a circle with radius r and center (h, k). Expand the equation. Then, determine the
radius and center of the circle $x^2 + y^2 + 2x - 6y + 1 + 9 - 25 = 0$.

26. Find the values of x and y to the nearest hundredth in the triangle shown.
(50)

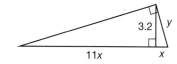

27. If a triangle were constructed with the largest third side possible, which angle
(39) would measure the least, if the two given sides were 8 meters and 5 meters long,
respectively?

*28. (**Landscaping**) Veruka is building a rectangular fence in her backyard. However,
(88) she only has 20 meters of fencing. If x represents the length and y represents the
width of the rectangle in meters, graph the set of possible values for x and y. Use
your graph to determine whether Veruka can build a fence with a length of
10 meters and a width of 2 meters.

*29. **Algebra** Find the measure of each angle in the triangle.
(79)

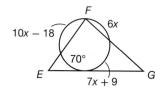

30. Determine the midpoint M of the points $P(-5, 4)$ and
(11) $Q(10, 0)$.

Vector Decomposition

Warm Up

1. **Vocabulary** The vector obtained after adding two or more vectors
   (63) together is called the _____.

2. Add the vectors ⟨5, −3⟩, ⟨−2, 4⟩, and ⟨−1, −6⟩.
   (83)

3. Find all the missing measures of the triangle.
   (73) Give side lengths to the nearest hundredth.

New Concepts

Decomposing a vector means to separate it into two vectors which, when added, result in the original vector. When a vector is decomposed, it is split into a horizontal vector and a vertical vector. In the diagram, $\vec{v}$ is decomposed into vertical $\overrightarrow{v_y}$ and horizontal $\overrightarrow{v_x}$.

Math Reasoning

Formulate If the magnitude of a horizontal vector is 5, what is the component form of that vector?

Notice that the vector makes a right triangle with its vertical and horizontal components. Since it makes a right triangle, trigonometric ratios can be used to find the magnitude of the vertical and horizontal components.

The angle of a vector is always given counterclockwise from the positive x-axis.

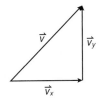

Example 1 | **Decomposing Vectors**

Decompose each vector. Round your answer to the nearest hundredth.

a. ⟨2, 3⟩

SOLUTION

A vector in component form is easily decomposed. The x and y components of the vector are given.

$$\overrightarrow{v_x} = \langle 2, 0 \rangle \qquad \overrightarrow{v_y} = \langle 0, 3 \rangle$$

b. $\vec{v}$ has a 33° angle and a magnitude of 7.

SOLUTION

Draw the vector. Use trigonometry to find the magnitudes of $\overrightarrow{v_x}$ and $\overrightarrow{v_y}$.

$$\cos 33° = \frac{\left| \overrightarrow{v_x} \right|}{7} \qquad \sin 33° = \frac{\left| \overrightarrow{v_y} \right|}{7}$$

$$\left| \overrightarrow{v_x} \right| \approx 5.87 \qquad \left| \overrightarrow{v_y} \right| \approx 3.81$$

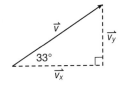

Since $\overrightarrow{v_x}$ is a horizontal vector, its component form is ⟨5.87, 0⟩, and the component form of the vertical vector $\overrightarrow{v_y}$ is ⟨0, 3.81⟩.

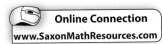
Online Connection
www.SaxonMathResources.com

Vectors can also be added by decomposition. To add two vectors, decompose them and add their vertical and horizontal components. This will give the resultant vector in component form.

Example 2 Adding Vectors by Decomposition

Vector $\vec{v}$ has a magnitude of 4 and makes a 25° angle with the horizontal. Vector $\vec{w}$ has a magnitude of 7 and makes a 61° angle with the horizontal. Add the vectors by decomposition. Find the magnitude and the angle the resultant vector makes with the horizontal to the nearest tenth.

SOLUTION
First, decompose $\vec{v}$.

$$\cos 25° = \frac{|\overrightarrow{v_x}|}{4} \qquad \sin 25° = \frac{|\overrightarrow{v_y}|}{4}$$

$$|\overrightarrow{v_x}| \approx 3.63 \qquad\qquad |\overrightarrow{v_y}| \approx 1.69$$

Then, decompose $\vec{w}$.

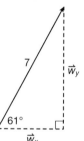

$$\cos 61° = \frac{|\overrightarrow{w_x}|}{7} \qquad \sin 61° = \frac{|\overrightarrow{w_y}|}{7}$$

$$|\overrightarrow{w_x}| \approx 3.39 \qquad\qquad |\overrightarrow{w_y}| \approx 6.12$$

Next, add the magnitudes of the components.

$$|\overrightarrow{v_x}| + |\overrightarrow{w_x}| \approx 3.63 + 3.39 \approx 7.02$$

$$|\overrightarrow{v_y}| + |\overrightarrow{w_y}| \approx 1.69 + 6.12 \approx 7.81$$

Now that we know the magnitudes of the legs of the right triangle, we can use the Pythagorean Theorem to find the magnitude of the resultant vector.

$$a^2 + b^2 = c^2$$
$$(7.02)^2 + (7.81)^2 \approx c^2$$
$$\sqrt{110.28} \approx c$$
$$c \approx 10.5$$

Finally, use one of the inverse trigonometric functions to find the angle measure that the resultant vector makes with the horizontal. Because you know the values of all three sides, you can use any function.

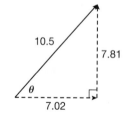

$$\tan \theta = \frac{7.81}{7.02}$$

$$\tan^{-1} \frac{7.81}{7.02} = \theta$$

$$\theta \approx 48°$$

The resultant vector has a magnitude of approximately 10.50 and makes a 48° angle with the horizontal.

Caution

These answers are all approximate, since we are rounding to the hundredths place at each step. For this reason, be sure to use the approximate sign.

Example 3 **Adding More Than Two Vectors by Decomposition**

Vector $\vec{a}$ has a magnitude of 5 and makes a 34° angle counterclockwise.
Vector $\vec{b}$ has a magnitude of 8 and makes a 90° angle counterclockwise.
Vector $\vec{c}$ has a magnitude of 3 and makes a 60° angle counterclockwise.
Add the vectors by decomposition. Find the magnitude and angle of the resultant vector.

SOLUTION
Draw the vectors head to tail. Use the drawing to estimate the magnitude and angle of the resultant vector.

First, decompose each vector.

$$\cos 34° = \frac{|\vec{a_x}|}{5} \qquad\qquad \sin 34° = \frac{|\vec{a_y}|}{5}$$
$$|\vec{a_x}| \approx 4.15 \qquad\qquad |\vec{a_y}| \approx 2.80$$

Because $\vec{b}$ has no horizontal component, $|\vec{b_x}| = 0$
and $|\vec{b_y}| = 8$.

$$\cos 60° = \frac{|\vec{c_x}|}{3} \qquad\qquad \sin 60° = \frac{|\vec{c_y}|}{3}$$
$$|\vec{c_x}| = 1.5 \qquad\qquad |\vec{c_y}| \approx 2.60$$

Add the magnitudes of the horizontal and vertical vectors.

$$|\vec{a_x}| + |\vec{b_x}| + |\vec{c_x}| \approx 4.15 + 0 + 1.5 \approx 5.65$$
$$|\vec{a_y}| + |\vec{b_y}| + |\vec{c_y}| \approx 2.80 + 8 + 2.60 \approx 13.40$$

Next, use the Pythagorean Theorem to find the magnitude of the resultant vector.

$$a^2 + b^2 = c^2$$
$$(5.65)^2 + (13.40)^2 \approx c^2$$
$$\sqrt{211.48} \approx c$$
$$14.54 \approx c$$

Math Reasoning

Verify Check the results for Example 3 against the estimate of the magnitude and angle of the resultant vector you made from the drawing of the summed vectors.

Finally, find the angle θ that the resultant vector makes with the horizontal.

$$\sin \theta = \frac{13.40}{14.54}$$
$$\theta \approx 67°$$

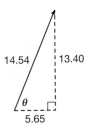

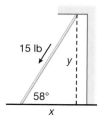

Example 4 **Application: Camping**

You can use vectors to calculate forces acting on an
object. For example, if a tent pole is placed at a 58° angle
and is supporting a load of 15 pounds, what are the
vertical and horizontal loads on the pole? Round your
answers to the nearest hundredth of a pound.

SOLUTION

The magnitude of the vector is 15 pounds. Use the sine
function to find the vertical component.

$$\sin 58° = \frac{y}{15}$$
$$y \approx 12.72$$

The vertical load is about 12.72 lbs.

Use the cosine function to find the horizontal component.

$$\cos 58° = \frac{x}{15}$$
$$x \approx 7.95$$

The horizontal load is about 7.95 lbs.

Hint

Decomposition of
vectors is often used to
solve problems involving
forces or speed. It can be
useful to know what the
vertical and horizontal
forces acting on an
object are.

Lesson Practice

Round to the nearest hundredth or nearest degree.

a. Vector $\vec{d}$ has a magnitude of 100 and makes a 55° angle with the
(Ex 1) horizontal. Decompose the vector into its x- and y-components.

b. Vector $\vec{p}$ makes a 35° angle with the horizontal. Vector $\vec{q}$ makes a
(Ex 2) 40° angle with the horizontal. If $\vec{p}$ has a magnitude of 30 and $\vec{q}$ has
a magnitude of 40, what is the angle and magnitude of the resultant
vector when $\vec{p}$ and $\vec{q}$ are added?

c. Vector $\vec{t}$ has a magnitude of 12 and a direction of 80°. Vector $\vec{u}$ has a
(Ex 3) magnitude of 16 and a direction of 0°. Vector $\vec{v}$ has a magnitude of 8
and a direction of 64°. Find the magnitude and angle of their sum.

d. **(Camping)** The pole for a tent makes an 86° angle with the ground. If
(Ex 4) the load on the pole is 25 pounds, what are the vertical and horizontal
loads?

Practice **Distributed and Integrated**

1. Points A and B are on $\odot C$, with $\angle AOB$ acute. OB is 4 inches
(58) and BD is 3 inches. $\overline{OA}$ is extended to D with $\overline{BD}$ tangent to $\odot C$.
 a. Determine OD.
 b. Determine AD.

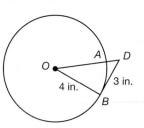

* 2. Find the component form of a vector with a magnitude of 4
(89) and a direction of 35°.

3. **Analyze** Use Theorem 86-1 to prove that if $\overline{AB}$ and $\overline{CD}$ are chords of a circle
(86) and $\overline{CD}$ bisects $\overline{AB}$, then $(CE)(DE) = \frac{1}{4}(AB)^2$.

4. **Multiple Choice** $\triangle ABC$ is translated so that the coordinates of B' are
(71) (1, 5). What are the coordinates of C' after the same translation?

    A (4, 1)   **B** (4, −1)

    C (1, 4)   **D** (−1, 1)

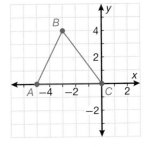

5. **Kayaking** A kayaker is traveling across a river at a velocity of
(83) 5 kilometers per hour. If the current is traveling at a rate of
2 kilometers per hour in a direction perpendicular to the kayaker's
movement, what is the actual speed of the kayaker to the nearest
hundredth?

6. If the coordinates of one endpoint of a segment are (4, 5) and the segment
(9) extends vertically downward 4 units, what would be the coordinates of the
other endpoint?

7. Given $\triangle BCD \sim \triangle JKL$, find KL.
(41)

8. **Analyze** If a regular polygon has an angle of rotational
(76) symmetry of 6°, how many sides does the polygon have?

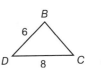
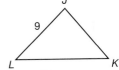

* 9. Sketch the graph of $y \geq -2x - 3$.
(88)

10. **Construction** A builder drills a hole with a 1-inch radius through a
(59) cube of concrete, as shown in the diagram. This cube will be an
outlet for a water tap on the side of a house.

    a. Find an expression involving π that gives the surface area of
    the cube. *Hint: Do not include the hole.*

    b. Find the lateral area of the cylinder to the nearest tenth of a square inch.

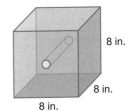

11. **Baking** Warren must cut a carrot cake in half that is in the shape of a wide
(85) cylinder, for a catered event. If the cake is 5 inches tall with a diameter of
12 inches, and he cuts along its diameter, what is the perimeter of the cross
section?

12. **Multiple Choice** The side length of a regular hexagon is 3 times as long as the side
(87) length of a similar hexagon. Which is a true statement about these two hexagons?

    A The smaller hexagon's perimeter is $\frac{1}{9}$ and its area is $\frac{1}{3}$ the size of the larger.

    B The smaller hexagon's perimeter is $\frac{1}{3}$ and its area is $\frac{1}{9}$ the size of the larger.

    C The smaller hexagon's perimeter and its area are each $\frac{1}{3}$ the size of the larger.

    D The smaller hexagon's perimeter and its area are each $\frac{1}{9}$ the size of the larger.

13. Find the value of x in the triangle at right to the nearest tenth.
(50)

14. Find the circumcenter of a triangle with vertices at (6, −2),
(38) (−2, −8), and (−6, −2) in a coordinate plane.

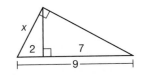

*15. **Algebra** Vector $\vec{v}$ has a magnitude of x and a direction of θ. Find the magnitude and angle of a
(89) vector that is 3 times longer than $\vec{v}$ and that points in the opposite direction.

16. Model On a coordinate plane, draw a parallelogram using the vectors $\langle 3, 5 \rangle$
(83) and $\langle 1, 6 \rangle$ and find the sum of the two vectors.

*17. What is the 15th triangular number? What is the 20th triangular number?
(Inv 8)

*18. (**Transport**) The density of rice is 782 kilograms per cubic meter and the density of
(88) wheat flour is 528 kilograms per cubic meter. A small cargo ship can carry 120,000
kilograms. Write an inequality for the total weight of rice (x), and flour (y), that
the ship can carry.

*19. The hour hand on a clock points to 7. If the vector represented by the hour hand
(89) has a magnitude of 5 inches, what is the magnitude of its horizontal component?

20. Multi-Step Find the perimeter and area of a 45°-45°-90° triangle with a hypotenuse
(53) that is 20 inches long to the nearest hundredth.

21. Find the measure of the angle exterior to a circle made by one tangent and one
(79) secant if the shorter arc measure between the tangent and secant is 54° and the
longer one is 112°.

22. Find the circumference using the given arc length, assuming that $\overline{AB}$ is
(35) a diameter.

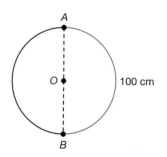

23. Two similar triangles have areas of 76 square centimeters and
(87) 931 square centimeters. Determine the ratio of their perimeters.

24. Justify The endpoints of the diagonals of a quadrilateral are $(-4, -3)$,
(52) $(-5, 8)$, and $(6, 9)$, $(7, -2)$. Determine whether the quadrilateral is a
rectangle. Explain how you know.

25. The center of the dilation at right is the origin of the coordinate plane. The
(84) preimage is rectangle $JKLM$, and the image is rectangle $J'K'L'M'$.
Identify the type of dilation and find the scale factor.

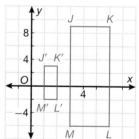

26. (**Construction**) For maximum accessibility, a ramp should have a slope
(82) between $\frac{1}{12}$ and $\frac{1}{16}$. What is the range of the measure of the angle
between the horizontal and the ramp, rounded to the nearest degree?

*27. Error Analysis A student found the components of a vector with an angle θ
(89) and a magnitude of a. The student's answer was $\langle a\sin\theta, a\cos\theta \rangle$. Explain the error.

28. If the hypotenuse of a 30°-60°-90° triangle is 125, what are the lengths of the
(56) two legs?

29. In the given triangles, order the angles x, y, and z from smallest to largest.
(Inv 4)

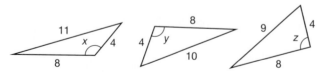

30. Graph $4x + 3 < 2y - 5$.
(88)

Composite Transformations

Warm Up

1. **Vocabulary** A transformation that changes the size of a figure but not its shape is a _____.
   (67)

2. Reflect the object shown across the dashed line.
   (74)

3. What is the image of a figure after reflecting it across the same line twice?
   (74)

New Concepts

More than one transformation can be applied to a figure. A combination of transformations is called a **composite transformation**. If a figure is both translated 3 units to the right and rotated 90°, for example, it has undergone a composite transformation.

Composition of Two Isometries
The composition of two isometries is an isometry.

Recall that translations, rotations, and reflections are all isometries. A composite transformation with two or more of these transformations will also be an isometry.

A **glide reflection** is a composition of a translation and a reflection across a line parallel to the translation vector. Since a glide reflection combines translation and reflection, it is an isometry.

Example 1 **Performing a Glide Reflection**

Reflect △ABC across line m and then translate it along $\vec{v}$.

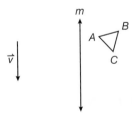

Math Reasoning

Verify In Example 1, what would be the result of performing the translation first, followed by the reflection?

SOLUTION

Draw the reflection of △ABC. The vector $\vec{v}$ indicates a translation downwards. Shift the image △A′B′C′ down to complete the glide reflection. After the composite transformation, the new image is △A″B″C″.

Two Reflections Across Two Parallel Lines

The composition of two reflections across two parallel lines is a translation.

If a figure is reflected twice, the end result is an image that represents a translation from the original preimage. Example 2 demonstrates this.

Example 2   Translating by Composite Reflection

Reflect the line segment across c, and then reflect the image across d. In the diagram, $c \parallel d$.

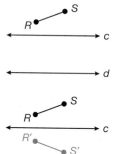

SOLUTION

Draw the first reflection across line c. The result is a segment with a slope that is the reciprocal of the preimage's slope.

Draw the second reflection across line d. The result is the original preimage, translated down.

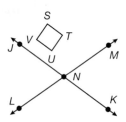

Two Reflections Across Two Intersecting Lines

The composition of two reflections across two intersecting lines is a rotation about the point of intersection. The angle of rotation is twice the angle formed by the intersecting lines.

If a figure is reflected across lines that intersect, it results in a rotation of the figure. Example 3 demonstrates this.

Example 3   Rotating by Composite Reflection

Reflect rectangle $STUV$ across $\overleftrightarrow{JK}$, and then reflect the image across $\overleftrightarrow{LM}$.

Math Reasoning

Analyze If you continue to reflect $STUV$ clockwise across $\overleftrightarrow{JK}$ and $\overleftrightarrow{LM}$, what will the final result be?

SOLUTION

The diagram shows two images. The rectangle $S'T'U'V'$ is the first reflection across $\overleftrightarrow{JK}$. Reflecting this image across $\overleftrightarrow{LM}$ results in $S''T''U''V''$.

Notice that the final image is just a rotation of $STUV$ around N, the intersection of the two lines.

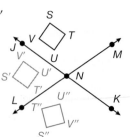

Applying the same transformations to a figure in a different order sometimes results in a different image.

Math Reasoning

Analyze Write one way to perform the composition in this example in a different order and get the same final image as the one in **a**.

Example 4 Analyzing Order of Composition

a. Perform transformations on the rectangle in the following order: reflection across a vertical line, rotation 90° clockwise, rotation 90° clockwise, and reflection across a horizontal line. What is the result?

SOLUTION

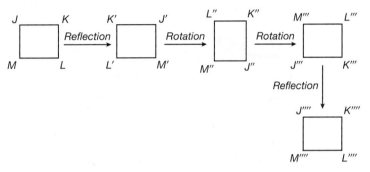

The four transformations returned the rectangle to its original orientation. Such a series of transformations is called an identity transformation.

b. Perform transformations on the original rectangle in part **a** in the following order: reflection across a vertical line, rotation 90° clockwise, reflection across a horizontal line, and rotation 90° clockwise. What is the result?

SOLUTION

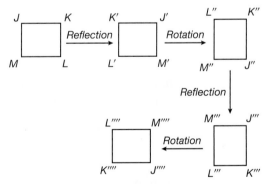

The four transformations resulted in an image that is not oriented the same as the original preimage, so this is not an identity transformation.

c. Compare the final images in parts **a** and **b**. What do the results indicate?

SOLUTION

The final images in parts **a** and **b** are oriented differently. This indicates that the order in which the transformations are performed affects the final result.

Math Reasoning

Analyze Does the order that a composite transformation is performed in matter when every step in the transformation is a translation?

Exploration

In this exploration, you will perform a composite transformation beginning with a given preimage. You will compare your final image to those of your classmates and to a final image provided by your teacher.

1. On grid paper, create △XYZ with X(0, 0), Y(3, 0), and Z(0, 4).

2. Follow the series of translations described by your teacher to generate an image based on composite transformations.

3. Redraw △XYZ on a new grid and follow a second series of translations described by your teacher to generate an image.

4. Compare your results for the two composite transformations with a classmate.

5. Check your results with the results that your teacher posts.

Example 5 **Application: Design**

To create a design, Reina draws a line segment, and reflects it twice as shown. The reflection lines are parallel and are 1 inch apart. Find the distance from A to A′.

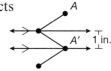

SOLUTION

Since the reflection creates an isometry, the distance from point A′ on the lower reflection line to the upper reflection line is the same as the distance from the upper reflection line to point A. Therefore, the distance from A to A′ is 2 inches.

Lesson Practice

a. Reflect the square over the dashed line and
(Ex 1) translate it by $\vec{v}$.

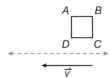

b. Reflect the figure across m, and then reflect
(Ex 2) the image across n.

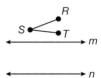

c. Reflect △ABC across $\overleftrightarrow{PQ}$, and then reflect
(Ex 3) the image across $\overleftrightarrow{RS}$.

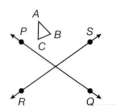

d. Perform transformations on the triangle in the following
(Ex 4) order: Translation to the right, reflection across a
horizontal line, translation to the right, and reflection
across a horizontal line.

e. To create a new logo for a sweatshirt, a designer
(Ex 5) reflects the letter **T** across line *h*. This image is then
reflected across line *j*. What single transformation
could move the letter **T** from its starting point to
its final position?

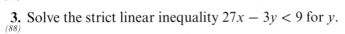

Practice Distributed and Integrated

1. **Multiple Choice** Which corresponding parts of these triangles are shown to be
(36) congruent?
  A hypotenuses, pair of acute angles, pair of right angles
  B pair of legs and pair of right angles
  C pair of legs and pair of acute angles
  D hypotenuses and pair of legs

* 2. Reflect △*DEF* across line *q*, and then translate the image along *u*.
(90)

3. Solve the strict linear inequality $27x - 3y < 9$ for *y*.
(88)

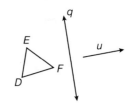

4. **Manufacturing** A toy manufacturer is pouring melted plastic into a right
(77) cone-shaped mold to make a part of a toy. The mold has a slant height of
13 centimeters and a base radius of 5 centimeters. What is the approximate
volume of the conical mold?

5. **Landscaping** A graph showing the top of a circular fountain has its center at
(57) (4, 6). The circle representing the fountain passes through (2, 1). What is
the exact area of the space covered by the fountain?

6. In isosceles trapezoid *MNOP*, find the length of $\overline{MQ}$ if
(69) $NP = 51.7$ centimeters and $PQ = 24.8$ centimeters.

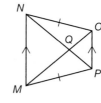

7. A bird is flying 120 feet away from two bird watchers who are standing
(73) together. They estimate that the bird is 60 feet above them. What is the
angle of elevation at which they see the bird?

8. Determine the area of this composite figure.
(40)

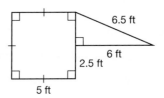

9. Suresh claims, "the expression $4n - 1$ will always result
(7) in a prime number." Find a value of *n* that
disproves his conjecture.

10. **Sailing** The wind moving a sailboat is blowing north at a velocity of 60 feet per
(83) second. The current is moving east at a velocity of 22 feet per second. What is the
magnitude and angle of the sailboat's speed and course?

11. Determine the exact value of AC in this circle.
(86)

12. Generalize Construct a formula for the volume of a sphere
(80) in terms of its diameter, d.

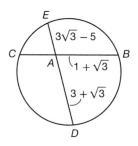

13. Given that $\triangle ABC \sim \triangle DEF$, $AB = 15$, $DE = 20$, and $EF = 16$,
(46) what is the length of BC?

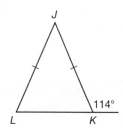

14. Show that the measure of the exterior angle at vertex K is greater than the
(39) measure of either remote interior angles.

15. Algebra The points M, N, and P are midpoints of $\triangle FGH$,
(55) $GH = 4x - 6$, and $MP = 15$. Find x and GH.

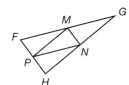

16. Multi-Step Two similar equilateral triangles have a scale
(87) factor of 2:5. Find the perimeter and area of the larger
triangle if the smaller triangle has a side length of 8 inches.

*17. Error Analysis Quan reflected a triangle across line x, followed by a reflection
(90) across line y, as shown. Did Quan perform the reflections correctly? Explain.

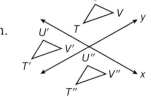

18. Algebra A reflection R across the line $x = a$ maps $(2, 4) \rightarrow (7, 4)$. Write and
(74) solve an equation for a.

19. Apply a dilation to $\triangle ABC$, at right, using a scale factor of 2 and P as the center.
(84)

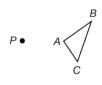

20. Analyze A vector has a magnitude x, where $x > 0$, and an angle θ
(89) such that $90° < \theta < 180°$. Find an expression for the component
form of the vector in terms of θ.

*21. (**Logos**) A skateboarding company is designing a logo for its new line of
(90) skateboards. The company's designers transformed its logo, SK8 as shown.
Which two transformations did the company use?

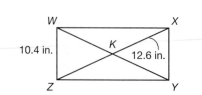

22. Find the geometric mean between 5 and 8 in simplified radical form.
(50)

23. a. State the converse of the following statement.
(20)

  If the time is 9 p.m., then it is evening.

  b. Determine whether the statement is true and whether its converse is true.

24. Quadrilateral $WXYZ$ is a rectangle. Find WK.
(52)

25. Generalize Two circles intersect at exactly two points. How many
(72) common tangents do they share?

26. Find the measure of angle x.
(64)

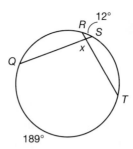

27. Justify A vector has vertical and horizontal components of -3 and 4, but it is not
(89) known which is the x-component and which is the y-component. Name the angle measures this vector could possibly have. Explain your answer.

28. Justify Determine whether lines $\overleftrightarrow{KL}$ and $\overleftrightarrow{MN}$ are parallel. How do you know?
(60)

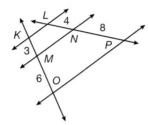

29. ⬭**Fund Raising**⬭ Justin is running a bake sale to raise money for a class trip to the
(88) local science center. He is selling bran muffins for $1.00 and blueberry muffins for $1.50 and needs to have total sales of at least $650. Graph the inequality of the number of bran muffins (x), and blueberry muffins (y), he needs to sell.

*30. Rotate $\triangle XYZ$ 90° counterclockwise about point P, and then
(90) translate the image along v.

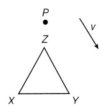

Tessellations

A **tessellation** is a repeating pattern of plane figures that completely covers a plane with no gaps or overlaps. The simplest kind of tessellation is a **regular tessellation**: a repeating pattern of congruent regular polygons.

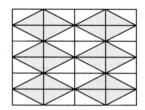

Does every regular polygon tessellate? If not, what kinds of regular tessellations are possible?

1. Draw an equilateral triangle. Use rotated and/or translated copies of the triangle to create a section of a tessellation. Does the triangle tessellate?

2. How many triangles meet at each vertex in the tessellation? What is the measure of each angle of the triangle? What is the total angle measure of all the angles that meet at a vertex?

3. **Write** Based on your observations in **2**, how can you use the interior angle measure of a regular polygon to determine whether it will tessellate?

4. What are the interior angle measures of the regular polygons with 4, 5, 6, and 7 sides?

5. Which of these regular polygons can tessellate?

6. **Model** Make a quick sketch of the tessellating regular polygons you found in **2**.

7. **Predict** Are there any regular polygons with more than 8 sides that can tessellate? How do you know?

8. Write a complete list of possible regular tessellations.

A **semiregular tessellation** is a repeating pattern formed by two or more regular polygons in which the same number of each polygon occurs in the same order at every vertex and it completely covers a plane with no gaps or overlaps.

9. The semiregular tessellation shown is made of congruent regular octagons and congruent squares. How many polygons meet at each vertex? What is the sum of the polygons' interior angles at each vertex?

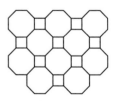

Tessellating with Triangles and Quadrilaterals

10. Create a tessellation by translating this parallelogram.

11. **Analyze** Will every parallelogram tessellate? What properties of parallelograms indicate that they will or will not tessellate?

12. **Predict** Will every quadrilateral tessellate? Why?

Tessellations may feature different kinds of symmetry. The regular polygon tessellations you saw above have several lines of symmetry and also feature rotational symmetry. Look at the three tessellations below.

A 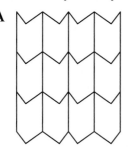 **B** 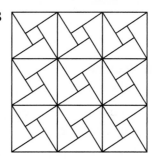 **C**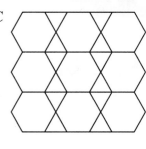

13. Which tessellation above is semiregular? Justify your answer.

14. Which of these tessellations have rotational symmetry around the center of the tessellation? What is the angle of rotation for each of them?

15. Which of these tessellations have lines of symmetry?

Investigation Practice

a. This figure is a trapezium. Will the trapezium tessellate?

b. Rotate the trapezium 180° around its top right vertex, and sketch the resulting figure.

c. Translate the figure from part **b** to create a section of a tessellation.

d. Does this tessellation have rotational symmetry? If so, what is the angle of rotation?

e. Does this tessellation have lines of symmetry?

Introduction to Trigonometric Identities

Warm Up

1. Vocabulary _____ are vectors that have the same slope and
(63) either different or equal directions.

2. Find θ in the figure shown. Round your answer to the
(82) nearest degree.

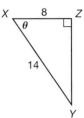

3. Vector *v* makes a 32° counterclockwise angle and has a
(89) magnitude of 5. Vector *w* makes a 17° counterclockwise
angle and has a magnitude of 3. Decompose the
resultant vectors, R_x and R_y. Round your answer
to the nearest hundredth.

New Concepts

It has already been shown in previous lessons that $\tan \theta = \frac{\sin \theta}{\cos \theta}$. This is an
example of a trigonometric identity. Trigonometric identities are expressions
that relate any two trigonometric functions. Several trigonometric identities
will be discussed later in this lesson, but two of the most commonly used
trigonometric identities are two that relate sine and cosine.

Reading Math

The square of a
trigonometric function
can be written two ways,
as shown below.

$\cos^2 x = (\cos x)^2$

Trigonometric Identities
The most commonly used trigonometric identities are below. $$\tan \theta = \frac{\sin \theta}{\cos \theta}$$ $$\sin^2 x + \cos^2 x = 1$$

To prove the identity $\sin^2 x + \cos^2 x = 1$, substitute the trigonometric ratios
for sine and cosine into the expression. The following proof uses the
triangle shown.

$$\sin^2 x + \cos^2 x = 1$$
$$\left(\frac{a}{c}\right)^2 + \left(\frac{b}{c}\right)^2 = 1$$
$$\frac{a^2}{c^2} + \frac{b^2}{c^2} = 1$$
$$\left(\frac{a^2}{c^2} + \frac{b^2}{c^2}\right) \times c^2 = 1 \times c^2$$
$$a^2 + b^2 = c^2$$

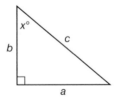

Online Connection
www.SaxonMathResources.com

This demonstrates that the identity given above is identical to the
Pythagorean Theorem, which we already know is true.

Example 1 Relating Sine and Cosine

Find $\sin \theta$ if $\cos \theta = 0.5$.

SOLUTION

$\sin^2\theta + \cos^2\theta = 1$	Trigonometric Identity
$\sin^2\theta + (0.5)^2 = 1$	Substitute.
$\sin^2\theta + 0.25 = 1$	Simplify.
$\sin \theta \approx 0.87$	Solve.

By rearranging the trigonometric identities we already know, several more identities can be created.

Example 2 Building More Identities

(**a.**) Express $\tan \theta$ using only $\sin \theta$.

SOLUTION

Both of the identities are needed to complete this problem. First, rearrange the second identity to solve for $\cos \theta$.

$$\sin^2\theta + \cos^2\theta = 1$$
$$\cos^2\theta = 1 - \sin^2\theta$$
$$\sqrt{\cos^2\theta} = \sqrt{1 - \sin^2\theta}$$
$$\cos \theta = \sqrt{1 - \sin^2\theta}$$

Now, substitute this into the tangent identity.

$$\tan \theta = \frac{\sin \theta}{\cos \theta}$$

$$\tan \theta = \frac{\sin \theta}{\sqrt{1 - \sin^2\theta}}$$

(**b.**) Express $\tan \theta$ using only $\cos \theta$.

Solving the same identity for $\cos \theta$ gives a similar expression.

$$\sin \theta = \sqrt{1 - \cos^2\theta}$$

Substitute this into the tangent identity.

$$\tan \theta = \frac{\sin \theta}{\cos \theta}$$

$$\tan \theta = \frac{\sqrt{1 - \cos^2\theta}}{\cos \theta}$$

Math Reasoning

Verify Confirm that the identity given in part **a** is valid by substituting an angle measure for θ and verifying the identity.

This table outlines the relationships between the trigonometric functions.

Function	In terms of sine	In terms of cosine	In terms of tangent
$\sin \theta =$	$\sin \theta$	$\sqrt{1 - \cos^2\theta}$	$\dfrac{\tan \theta}{\sqrt{1 + \tan^2\theta}}$
$\cos \theta =$	$\sqrt{1 - \sin^2\theta}$	$\cos \theta$	$\dfrac{1}{\sqrt{1 + \tan^2\theta}}$
$\tan \theta =$	$\dfrac{\sin \theta}{\sqrt{1 - \sin^2\theta}}$	$\dfrac{\sqrt{1 - \cos^2\theta}}{\cos \theta}$	$\tan \theta$

Example 3 Application: Estimating Distance

Leopold and Melody are standing on the street near their school, as shown in the diagram. Melody knows she is three times as far from the school as Leopold. What is the approximate ratio of Melody's distance from Leopold to Melody's distance from the school?

Hint

It may be helpful to label the opposite leg, adjacent leg, and hypotenuse in this diagram, from Melody's perspective.

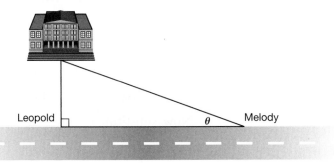

SOLUTION

If Melody is three times as far from the school as Leopold is, this implies that the ratio of the opposite leg to the hypotenuse is $\frac{1}{3}$. So:

$$\sin \theta = \frac{1}{3}$$

The problem asks for the ratio of the adjacent leg to the hypotenuse, which is a cosine function. Use the identity to solve for cosine.

$$\sin^2 x + \cos^2 x = 1$$

$$\left(\frac{1}{3}\right)^2 + \cos^2\theta = 1$$

$$\cos^2\theta = \frac{8}{9}$$

$$\cos \theta = \frac{2\sqrt{2}}{3}$$

So the distance between Leopold and Melody is approximately $\frac{2\sqrt{2}}{3}$ times as far as the distance from Melody to the school.

a. If $\sin^2\theta = 0.67$, what is the value of $\cos\theta$ to the nearest
(Ex 1) hundredth?

b. Express $\cos\theta$ in terms of $\tan\theta$ and show each step.
(Ex 2)

c. As shown in the diagram, Ruby is about twice
(Ex 3) as far from Becky as she is from Ivan. What is
the approximate ratio of Ivan's distance from
Becky to Ruby's distance from Becky?

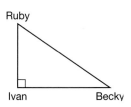

Practice Distributed and Integrated

* 1. Sketch a tessellation using congruent images of this kite.
(Inv 9)

* 2. Write How would you go about finding a value of θ where $\sin\theta = \tan\theta$? What is
(91) that value?

3. Sketch the graph of $f(x) < x + 2$.
(88)

4. Multi-Step A dragonfly is flying parallel to the ground up a hill that has a slope
(89) of 27°. If the dragonfly is flying at a speed of 16 feet per second, and the wind is
blowing up the hill at a speed of 4 feet per second, what are the horizontal and
vertical components of the dragonfly's speed over the ground?

* 5. Draw two lines of reflection that will transform rectangle $STUV$
(90) to rectangle $S'T'U'V'$ as shown.

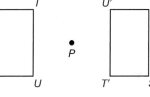

6. Analyze If a regular polygon has an angle of rotational symmetry
(76) of 15°, how many sides does it have?

7. What is the surface area of a sphere with a radius of $\dfrac{4}{\sqrt{\pi}}$
(80) centimeters?

8. Coordinate Geometry A triangle has vertices at $(2, 6)$, $(4, -5)$, and $(-3, 0)$. Classify
(45) the triangle by its sides.

9. A circle's radius is $4b$. Find the area of a 60° sector of this circle in terms
(35) of π and b.

10. If the length of the shorter leg of a 30°-60°-90° triangle is 68, what are the lengths
(56) of the other leg and hypotenuse?

*11. Perform a glide reflection by reflecting $\angle ABC$ across line w
(90) and along $\vec{u}$.

12. Verify A student says that a vector with a magnitude of x and an angle of
(89) 196° will have components that are negative. If $x > 0$, is the student correct?
Explain.

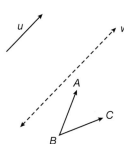

13. Analyze All solids can have a cross section of a point. Explain how this is possible
(85) with a sphere, a cube, and a cylinder.

14. Algebra Use this figure to write and solve an equation for x.
(86)

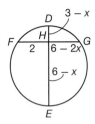

15. What is the ratio of the measure of an arc to the measure of its inscribed
(47) angle?

16. (Tennis) A tennis ball is made of two equally sized patches of material. If the
(80) diameter of the tennis ball is 3 inches, what is the area of each patch to the
nearest hundredth?

17. Compare QR and SR.
(Inv 4)

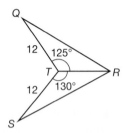

18. (Architecture) Jan is designing a square-based pyramid that will be in front of
(70) a building. If the sides of the square are to be 6 feet long and the height of
the pyramid is to be 4 feet, how much material does Jan need to construct the
pyramid, including the base?

*19. In the diagram, $\triangle ABC$ undergoes a glide reflection. Copy the
(90) diagram and draw in the reflection line and the translation
that occurred.

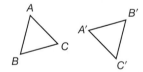

20. Multiple Choice Which statement is not true about all parallelograms?
(61)
A Opposite angles are congruent.
B Consecutive angles are supplementary.
C Diagonals perpendicularly bisect each other.
D Opposite sides are congruent.

21. (Containers) A cylindrical juice container has the dimensions show. To the nearest
(62) cubic inch, how much juice does the container hold?

6 in.

12 in.

22. **Analyze** If the height of a cylinder is 31.4 centimeters, and the area of the
(85) cross section parallel to a circular face has the same area as the cross section
perpendicular to the circular faces through a diameter of each face, what is
the radius of the cylinder?

23. A rectangular table has a length of $3x - 2$, a width of $2x + 1$, and an
(24) area of $2x + 28$. Solve for x and provide a justification for each step.

*24. Write $\tan^2\theta$ in terms of $\cos\theta$.
(91)

25. (Sawmills) To make the best use of the wood from each tree, sawmill operators
(53) must trim each log on four sides to make the largest square beam possible. If a log
has a 20-inch diameter, find the width of the square beam that can be made from
that log.

26. Find the distance from $(-5, 12)$ to the line $x = 6$.
(42)

*27. **Algebra** For what value of θ does $\cos^2\theta = 3\sin^2\theta$?
(91)

28. Roderick wants to know how tall the flagpole in his schoolyard is, but he cannot
(44) measure it directly. He knows that his shadow and the shadow of the flagpole
form similar triangles. The length of his shadow is 2.25 feet and he is 6 feet tall.
If the flagpole's shadow is 20.25 feet, how tall is the flagpole?

29. (Bridges) A cable is suspended diagonally to hold the supports of a bridge. If there
(83) is a 200-pound vertical load and a 500-pound horizontal load, what is the total
tension load of the cable? What is the angle the cable makes with the horizontal?
Round your answers to the nearest tenth.

*30. **Error Analysis** Desi tried to prove that $\tan\theta = \dfrac{\sin\theta}{\sqrt{1 - \sin^2\theta}}$. His work is shown below.
(91) Where did he make his error?

$$\tan\theta = \frac{\sin\theta}{\sqrt{1 - \sin^2\theta}}$$

$$\frac{a}{b} = \frac{a}{c} \cdot \frac{1}{\sqrt{1 - \left(\frac{a}{c}\right)^2}}$$

$$\frac{a}{b} = \frac{a}{c} \cdot \frac{1}{\sqrt{c^2 - a^2}}$$

$$\frac{a}{b} = \frac{a}{c} \cdot \frac{1}{\sqrt{b^2}}$$

$$\frac{a}{b} = \frac{a}{c} \cdot \frac{1}{b}$$

$$\frac{a}{b} = \frac{a}{b} \cdot c$$

Quadrilaterals on the Coordinate Plane

1. **Vocabulary** Two polygons are congruent if their corresponding _____
(15) and _____ are congruent.

2. If the diagonals of a parallelogram are congruent, then it is
(65) a _____.

3. **Multiple Choice** How many pairs of parallel sides does a trapezoid have?
(19)
    A none   **B** at least one

    C two   **D** exactly one

New Concepts If the coordinates of a quadrilateral's vertices are given, it is possible to determine the properties of the quadrilateral. Since the length and slope of each side can be found from the coordinates, it is possible to classify any quadrilateral in the coordinate plane.

Example 1   **Proving Quadrilaterals are Parallelograms**

Hint

To review ways of identifying quadrilaterals and parallelograms, refer to Lessons 61 and 65.

The vertices of quadrilateral $ABCD$ are $A(1, 4)$, $B(5, 7)$, $C(6, 5)$, and $D(2, 2)$.

Is $ABCD$ a parallelogram?

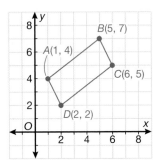

SOLUTION

slope of $\overline{AB} = \dfrac{Y_B - Y_A}{X_B - X_A}$

$= \dfrac{7 - 4}{5 - 1}$

$= \dfrac{3}{4}$

slope of $\overline{BC} = \dfrac{Y_C - Y_B}{X_C - X_B}$

$= \dfrac{5 - 7}{6 - 5}$

$= -2$

slope of $\overline{CD} = \dfrac{Y_D - Y_C}{X_D - X_C}$

$= \dfrac{2 - 5}{2 - 6}$

$= \dfrac{3}{4}$

slope of $\overline{AD} = \dfrac{Y_D - Y_A}{X_D - X_A}$

$= \dfrac{2 - 4}{2 - 1}$

$= -2$

Online Connection
www.SaxonMathResources.com

The slopes of $\overline{AB}$ and $\overline{CD}$ are equal, so $\overline{AB} \parallel \overline{CD}$. Similarly, $\overline{BC} \parallel \overline{AD}$. Therefore, $ABCD$ is a parallelogram.

Example 2 Proving Quadrilaterals are Trapezoids

The vertices of quadrilateral $PQRS$ are $P(1, 2)$, $Q(7, 5)$, $R(5, 6)$, and $S(1, 4)$. Is $PQRS$ a trapezoid?

SOLUTION

Remember that trapezoids have exactly one pair of parallel sides. Determine the slope of each side.

$$\text{slope of } \overline{PQ} = \frac{Y_Q - Y_P}{X_Q - X_P} \qquad \text{slope of } \overline{QR} = \frac{Y_R - Y_Q}{X_R - X_Q}$$

$$= \frac{5 - 2}{7 - 1} \qquad\qquad\qquad = \frac{6 - 5}{5 - 7}$$

$$= \frac{1}{2} \qquad\qquad\qquad\quad = -\frac{1}{2}$$

$$\text{slope of } \overline{RS} = \frac{Y_S - Y_R}{X_S - X_R}$$

$$= \frac{4 - 6}{1 - 5}$$

$$= \frac{1}{2}$$

Since $\overline{PS}$ is vertical, its slope is undefined. Therefore $\overline{PQ}$ and $\overline{RS}$ are parallel, but $\overline{QR}$ and $\overline{PS}$ are not. So $PQRS$ is a trapezoid.

Example 3 Proving Congruence with Coordinates

Are these quadrilaterals congruent?

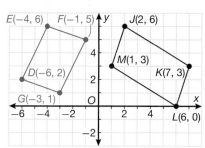

SOLUTION

Use the distance formula to determine if the sides of these quadrilaterals are congruent.

$$DE = \sqrt{(-4 - (-6))^2 + (6 - 2)^2} \qquad JK = \sqrt{(7 - 2)^2 + (3 - 6)^2}$$
$$= \sqrt{20} = 2\sqrt{5} \qquad\qquad = \sqrt{34}$$

$$EF = \sqrt{(-1 - (-4))^2 + (5 - 6)^2} \qquad KL = \sqrt{(6 - 7)^2 + (0 - 3)^2}$$
$$= \sqrt{10} \qquad\qquad\qquad = \sqrt{10}$$

$$FG = \sqrt{(-3 - (-1))^2 + (1 - 5)^2} \qquad LM = \sqrt{(1 - 6)^2 + (3 - 0)^2}$$
$$= \sqrt{20} = 2\sqrt{5} \qquad\qquad = \sqrt{34}$$

$$DG = \sqrt{(-3 - (-6))^2 + (1 - 2)^2} \qquad JM = \sqrt{(2 - 1)^2 + (6 - 3)^2}$$
$$= \sqrt{10} \qquad\qquad\qquad = \sqrt{10}$$

While $EF = KL$ and $DG = JM$, the other sides are not congruent, so $DEFG$ is not congruent to $JKLM$.

Example 4 | **Application: Surveying**

This plan shows the boundary of a plot of land.

a. Determine whether the plot is a parallelogram, a trapezoid, or neither.

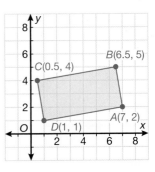

SOLUTION
Check whether $\overline{AB} \parallel \overline{CD}$, and whether $\overline{BC} \parallel \overline{AD}$.

$$\text{slope of } \overline{AB} = \frac{Y_B - Y_A}{X_B - X_A}$$
$$= \frac{5 - 2}{6.5 - 7}$$
$$= \frac{3}{-0.5}$$
$$= -6$$

$$\text{slope of } \overline{CD} = \frac{Y_D - Y_C}{X_D - X_C}$$
$$= \frac{1 - 4}{1 - 0.5}$$
$$= \frac{-3}{0.5}$$
$$= -6$$

$$\text{slope of } \overline{AD} = \frac{Y_D - Y_A}{X_D - X_A}$$
$$= \frac{1 - 2}{1 - 7}$$
$$= \frac{1}{6}$$

$$\text{slope of } \overline{BC} = \frac{Y_C - Y_B}{X_C - X_B}$$
$$= \frac{4 - 5}{0.5 - 6.5}$$
$$= \frac{1}{6}$$

Since $\overline{AB} \parallel \overline{CD}$ and $\overline{BC} \parallel \overline{AD}$, the plot is a parallelogram.

b. What other type of quadrilateral is the plot, if any?

SOLUTION
Notice that the slope of $\overline{AB}$ is the negative reciprocal of the slope of $\overline{BC}$, so $\overline{AB} \perp \overline{BC}$. Since $ABCD$ is a parallelogram, this means that all of its interior angles must be right angles. Therefore, the plot is also a rectangle. However, it is not a square, since $AB = \sqrt{(6.5 - 7)^2 + (5 - 2)^2} = \sqrt{9.25}$ but $AD = \sqrt{(1 - 7)^2 + (1 - 2)^2} = \sqrt{37}$.

Math Reasoning

Analyze What other method will determine if the parallelogram is a rectangle?

Lesson Practice

a. Prove that quadrilateral $CDEF$ is a parallelogram by determining the slope of each side.
(Ex 1)

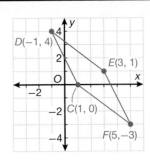

b. Is quadrilateral *PQRS* a trapezoid? Explain *(Ex 2)* how you know.

c. Determine if quadrilateral *JKLM* with *(Ex 3)* vertices *J*(3, 5), *K*(5, 4), *L*(5, 0), and *M*(2, 1) is congruent to quadrilateral *STUV* with vertices *S*(5, 3), *T*(7, 2), *U*(7, −2), and *V*(4, −1).

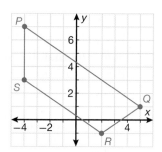

d. **Surveying** Determine if this plot of land *(Ex 4)* is a parallelogram, trapezoid, or neither.

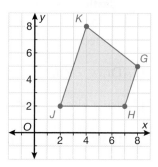

Practice Distributed and Integrated

1. **Multiple Choice** △*DEF* is translated so that the coordinates of *E′* are (−2, 2). *(71)* What are the coordinates of *F′*?

A (0, −1)   **B** (−1, 0)
C (1, 0)   **D** (1, 1)

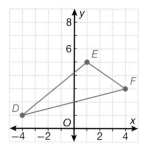

2. Find the horizontal and vertical components of a vector with a magnitude *(89)* of 80 and an angle of 12°.

* **3.** Quadrilateral *ABCD* has vertices *A*(3, 1), *B*(5, 3), *C*(2, 3), and *D*(1, 2). *(92)* Quadrilateral *EFGH* has vertices *E*(−1, 3), *F*(−3, 5), *G*(−3, 2), and *H*(−2, 1). Prove that *ABCD* ≅ *EFGH* by determining corresponding side lengths and slopes.

4. What is the longest side in the figure shown? *(39)*

5. Write the equation for a line that is perpendicular to and a line that is *(37)* parallel to the line *y* = 0.2*x* − 4.

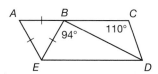

* **6.** **Generalize** Into how many segments do *n* points on a line divide that line? *(Inv 8)* *Hint: Try putting 1, 2, and 3 points on a line and look for a pattern.*

7. **Multi-Step** Find m$\overset{\frown}{BF}$ on the larger circle and m$\overset{\frown}{CE}$ on the smaller circle. *(79)*

* **8.** The vertices of quadrilateral *WXYZ* are *W*(0, 3), *X*(3, 2), *Y*(−2, −3), *(92)* and *Z*(−5, −2). Prove that *WXYZ* is a parallelogram by determining the slope of each side.

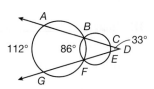

9. **Analyze** Use Theorem 58-2 (if a line in the plane of a circle is perpendicular to a radius at its endpoint on the circle, then the line is tangent to the circle) to show *indirectly* that, if line ℓ intersects $\odot C$ at P but is *not* perpendicular to $\overline{CP}$, then line ℓ intersects $\odot C$ at one other point Q, and $\angle CPQ$ and $\angle CQP$ are congruent acute angles.
(58)

*10. Prove that quadrilateral $PQRS$ is a trapezoid.
(92)

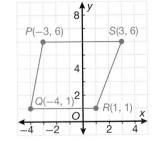

11. (**Boating**) A boat sails along the vector $\langle 3, 4 \rangle$, then sails along the vector $\langle 1, 5 \rangle$. What vector can it use to go directly back to where it started?
(63)

12. Draw the net of a pentagonal pyramid.
(Inv 5)

*13. **Verify** Use a right triangle to prove that $\cos\theta = \dfrac{1}{\sqrt{1+\tan^2\theta}}$.
(91)

14. If $AG = 10$ and $DB = 16$, find GF and BG to the nearest tenth.
(32)

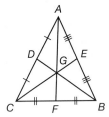

15. (**Dartboards**) A dartboard has a diameter of 20 inches, with a center circle that has a 5 inch diameter.
(Inv 6)
   a. If you assume that throwing a dart is a random event, what is the probability of hitting the center circle?
   b. In reality, is the probability of hitting the center circle different from what you calculated in part **a**?

16. Apply a dilation to $\overline{FG}$ using a scale factor of $\frac{1}{2}$ and center C.
(84)

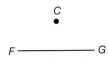

17. (**Aviation**) A plane takes off and flies at an angle of 30° from the ground. After a certain amount of time, the plane has reached an altitude of 3000 meters. Use trigonometry to find the horizontal distance of the plane from its starting point to the nearest hundredth.
(73)

18. **Analyze** A log that is 5 meters long is rolling down a hill. The lateral surface area of the log is 31.4 square meters. How many times will the log revolve if the hill is 62.8π meters long?
(62)

19. Find the circumference of circle O based on the arc given.
(35)

*20. Prove that quadrilateral $JKLM$ with vertices $J(1, 1)$, $K(4, 1)$, $L(5, 4)$, and $M(2, 5)$ is congruent to quadrilateral $NOPQ$ with vertices $N(5, 7)$, $O(8, 7)$, $P(9, 4)$, and $Q(6, 3)$.
(92)

*21. A boat is at coordinates (2, 0). The boat is towing a parasail.
(73) The coordinates for the parasail rider are (4, 6). If the angle
of elevation is x, what is the tangent of x?

22. Reflect the figure across ℓ_1, and then reflect
(90) the image across ℓ_2.

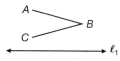

23. Express $\cos\theta \cdot \tan\theta$ in terms of $\sin\theta$.
(91)

24. The chord of a circle is 28.3 inches long, and
(43) is 11.2 inches from the center of the circle.
What is the length of the radius of the circle to the nearest inch?

25. (**Timekeeping**) An hourglass is a device used to measure time. A particular
(70) hourglass is made of two congruent, regular, square-based pyramids attached
at the two vertices of the pyramids. One of the two pyramids is filled with sand.
What volume of sand is needed if the height of each pyramid is 5 inches and
the side lengths of each square is 3 inches?

26. Do the statements below show use of the Law of Syllogism or the Law of
(21) Detachment?

If William wakes up before 5 a.m., then he goes running that day.
Today, William woke up before 5 a.m.
Therefore, he will go running today.

27. **Write** Ray drew equilateral triangle $\triangle EFG$. He then drew two lines, l_1, and l_2, that
(90) made a 60° angle through the center of the triangle. Ray reflected the triangle
across l_1, and then across l_2. Describe the similarities and differences between
the final image and the original triangle.

*28. Continue the tessellation of this pentagon that begins with
(Inv 9) **a.** the given translation; **b.** the given rotation.

*29. (**Playgrounds**) A slide for a preschool's playground should have a slope between
(82) $\frac{1}{4}$ and $\frac{1}{6}$. What is the range of the measure of the angle between the slide and
the ground rounded to the nearest degree?

30. **Algebra** If $\frac{\cos\theta}{\sin\theta} = 12$, what is the value of $\tan\theta$? Then find θ to the nearest degree.
(91)

Distinguishing Types of Quadrilaterals Using Geometry Software

Technology Lab 12 (*Use with Lesson 92*)

In Lesson 92, you learned about quadrilaterals in the coordinate plane. Now you will use geometry software to construct several different types of quadrilaterals using their coordinates, and you will identify them by their angle and side measures.

(1.) Choose *Show Grid* from the *Graph* menu. Then, choose *Snap Points* from the *Graph* menu.

(2.) Plot the following ordered pairs: (−3, 5), (−4, 8), (−7, 7), and (−6, 4).

(3.) Use the segment tool to connect these points and make a quadrilateral.

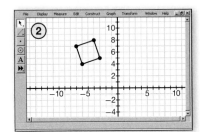

(4.) Measure the quadrilateral's sides and angles.

(5.) What kind of quadrilateral is this? List all the possible names for it.

Lab Practice

Use steps 2–4 to identify each of the quadrilaterals with the vertices given below. Remember that there is sometimes more than one way to classify a quadrilateral. Give all the possible names.

a. (6, 0), (7, 5), (5, 3), and (4, −2).

b. (−5, −6), (−2, −2), (−6, −5), and (−9, −9).

c. (1, −2), (2, 3), (−2, 7), and (−7, 6).

d. (6, −5), (7, 0), (2, −1), and (−3, −10).

e. (−4, −2), (10, 5), (7, 11), and (−7, 4).

Warm Up

1. Vocabulary A _____ is a polyhedron with six congruent faces.
(49)

2. How many edges, vertices, and faces does a cube have?
(49)

3. Draw a three-dimensional sketch of a right cone.
(54)

4. Classify the solid shown here.
(49)

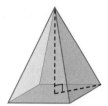

New Concepts

An **orthographic drawing** is a drawing that shows a three-dimensional object in which the line of sight for each view is perpendicular to the plane of the picture. Typically, an orthographic drawing has three views: the front, top, and side. To draw an orthographic view of a solid, imagine you are looking straight at one of its faces and can see nothing else but the flat surfaces facing you.

Example 1 **Creating Orthographic Drawings**

Draw the front, top, and side orthographic views of this solid.

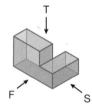

SOLUTION
Draw the front face of the solid.

Front

From the top, two adjacent rectangles can be seen.

Top

Math Reasoning

Model Would the orthographic view of the opposite side of this solid be different from the side view shown here?

A solid line indicates that two parts of the object are not level with each other.

From the side, a rectangle is visible. Just above the left side of the rectangle, a square is visible.

Side

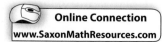

Example 2 Constructing from an Orthographic Drawing

This orthographic drawing shows the front, top, and right-side views of a solid.

Which of these sketches represents the solid shown in the views?

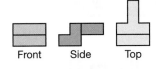

Front   Side   Top

A

B

C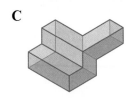

Hint

Imagine visualizing each sketch from the front, the top, and the side. Consider whether each visualization matches the given front, top, and side views.

SOLUTION

Looking at the top view first, notice that the figure in sketch **A** does not have a long double base like the top view does. Figure **C** looks very similar, but the block that comes off of the back is not centered along the back, as in the top view. The answer is **B**.

If you compare the front and side views, you can see that in each case there are at least two figures with the same orthographic views.

Example 3 Application: Consumer Electronics

A speaker stand consists of a base, which is a rectangular prism that is 5 inches along the front edge, 4 inches along the side edge and $\frac{1}{2}$ inch high. Mounted in the center of the base is a cylindrical rod that is 1 inch in diameter and 8 inches high.

Make orthographic drawings of the top, front, and side of the speaker stand.

SOLUTION

Draw the top view of the stand.

Top

Draw the front view of the stand. From the front, the entire height of the cylindrical rod is visible.

Front

Draw the side view of the stand.

Side

Exploration · **Drawing Orthographic Views of Objects**

In this exploration, you create orthographic views of some objects in your classroom.

1. Choose two items in the classroom. Sketch the top, front, and side orthographic views of each object.

2. Swap your drawings with a partner. Walk around the classroom and try to discover what objects their orthographic drawings represent.

3. What kinds of objects are easy to find based on their orthographic views? What kinds of objects are more difficult to identify?

4. Draw orthographic views of another object. Use what you learned in step 3 to choose an object that will be difficult for your partner to identify.

Lesson Practice

a. Draw the front, top, and side orthographic views
(Ex 1) of this solid.

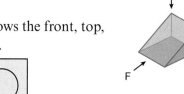

b. This orthographic drawing shows the front, top,
(Ex 2) and right-side views of a solid.

    Front   Side   Top

Identify the correct sketch of the solid.

A   **B**

C   **D**

c. **Architecture** The diagram shows orthographic drawings of a building.
(Ex 3)

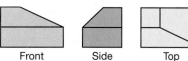

    Front   Side   Top

Which of these sketches represents the building?

A   **B**   **C**

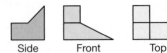

* 1. Model This orthographic drawing shows the front, top, and
(93) side views of a solid. Sketch the solid.

Side   Front   Top

2. Graph the region described by the inequalities
(88) $2x - y > -1$, $x + y < 1$, and $x - y < 1$.

3. Algebra If $\tan^2\theta = 3.2$, what does $\cos\theta$ equal?
(91)

4. (**Compact discs**) Compact discs have the dimensions shown. Each disc
(62) is 1 millimeter thick. Find the volume in cubic centimeters of a stack
of 25 discs. Round to the nearest tenth of a centimeter.

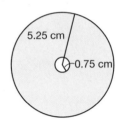

5.25 cm

0.75 cm

5. (**Pottery**) Brandon crafted a clay bowl on a pottery wheel in the shape
(80) of a hemisphere. After the clay hardens, Brandon will paint only the
outside of the bowl. If the radius of the bowl is 6 inches, what surface
area will he paint to the nearest square inch?

6. Multiple Choice Which of these best describes the composite
(90) transformation in the diagram?
  A glide reflection   **B** glide rotation
  C reflection then translation   **D** translation then reflection

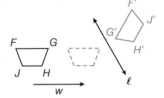

7. Two similar triangles have areas of 76 square centimeters and
(87) 931 square centimeters, respectively. Determine the ratio of
their perimeters.

8. Write Quadrilateral $ABCD$ has vertices $A(-3, 3)$, $B(4, 1)$, $C(-1, -2)$, and
(61) $D(-5, -1)$. Could you make the quadrilateral a parallelogram by moving
any point? Explain why or why not.

*9. Multiple Choice This orthographic drawing shows the front, top, and right-side
(93) views of a solid. Which of these sketches represents the solid?

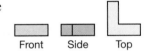

Front   Side   Top

  A   **B**

  C   **D**

10. Multi-Step For the right triangle with vertices $K(-2, -2)$, $L(1, -3)$, and $M(3, 3)$,
(82) find the side lengths to the nearest tenth of a unit and the angle measures to the
nearest degree.

11. Quadrilateral *KLMN* is a rhombus with diagonals that intersect at *S*.
(52) What is m∠*KSL*?

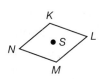

12. Multiple Choice Which of the following are perpendicular lines?
(37)
  i) $x + y = 1$,
  ii) $x - y = 3$, and
  iii) $y - x = 5$

  A i and ii   **B** i and iii
  C ii and iii   **D** i and ii and i and iii

13. There are two triangles, △*AGK* and △*JOV*, with side lengths as follows:
(25) *AG* = 14, *GK* = 15, *KA* = 13, *OJ* = 14, *JV* = 15, and *VO* = 13. Write a
congruency statement for the triangles.

14. Error Analysis Sarita is standing at *P* and wants to travel to *C*. She reasons
(40) that since the area of quadrilateral *ABCD* is greater than the area of
polygon *APQRCD*, she will be taking a shortcut if she travels along
$\overline{PQ}$ and then $\overline{QR}$, rather than by traveling along $\overline{PB}$, then $\overline{BR}$.
Explain why she is wrong.

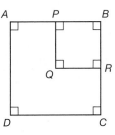

xy² **15. Algebra** Given the equation $y = 3x - 4$ and the point *C*(0, 6), find the point
(42) on the line that is closest to *C*.

16. (**Civil Planning**) The diagram represents train tracks surrounding
(57) a children's zoo. Find the area enclosed by the tracks. The
side lengths of each square represents 1 meter.

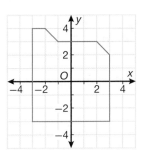

17. What is the surface area of a sphere with a radius of 5.5 meters?
(80)

18. (**Woodworking**) Mark has a cube-shaped block of wood he wants
(85) to cut. He wants to make a cut through the plane of the
midpoints of the three edges adjacent to one of the vertices.
What shape is this cross section?

*19. Prove that *UVWX* is a parallelogram by determining its
(92) side lengths and slopes.

xy² **20. Algebra** The solution to a linear system is (3, 2).
(81) One equation is $y = 3x + b$ and the other
equation is $y = 2$. What is the value of *b*?

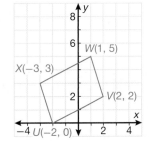

21. If △*QPR* ≅ △*DEF*, and *QP* = 5 inches, *PR* = 7 inches,
(25) and *DF* = 8 inches, what are the lengths of the
other sides?

*22. Model Construct an accurate orthographic drawing of a square pyramid
(93) with slant height and side lengths of 2. Include horizontal and vertical
dimensions.

*23. **(Physics)** Newton's Second Law says the force on an object is the net acceleration,
$_{(63)}$ which is a vector, multiplied by the mass. A block on a ramp is under the force of gravity at about 10 m/s² down and a friction force of 4 m/s² up and 8 m/s² to the left. What is the net acceleration on the block?

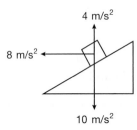

24. **Algebra** Two chords of a circle, $\overline{DE}$ and $\overline{FG}$, intersect at C with segment lengths
$_{(86)}$ $CD = 3$, $CE = x + 2$, $CF = 6 - x$, and $CG = 5$. Write and solve an equation for x.

25. Segment UV is a midsegment of $\triangle RST$. Find RT.
$_{(55)}$

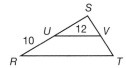

*26. **Verify** The vertices of quadrilateral $MNOP$ are $M(2, 2)$, $N(3, 0)$, $O(6, 4)$, and
$_{(92)}$ $P(5, 6)$. Prove that $MNOP$ is a parallelogram by determining its side lengths.

27. Find an expression for $\cos \theta$ in terms of $\tan \theta$.
$_{(91)}$

*28. **Model** Draw the front, top, and side orthographic views of this solid.
$_{(93)}$

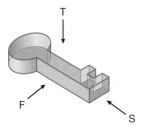

29. Find the horizontal and vertical components of a vector with a magnitude of 44
$_{(89)}$ and an angle of 275°.

30. Find the value of x in the diagram.
$_{(46)}$

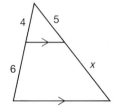

Warm Up

1. **Vocabulary** In a right triangle, the ratio of the opposite side to the adjacent side is the _____ function.
   (68)

2. Is the following equality true or false? $\sin^2\theta = \dfrac{\tan\theta}{\sqrt{1 + \tan\theta}}$
   (91)

3. A pilot looks out of a plane and sees a lake. She calculates an angle of depression of 38°. The plane is at an altitude of 11,000 feet. How far away is the lake, to the nearest foot?
   (73)

New Concepts

Trigonometric ratios are generally used to find unknown measures in right triangles, but they can also be used to find unknown lengths in any triangle using the Law of Sines.

Math Reasoning

Model Draw a scalene triangle. Measure its side lengths and angles. Verify that the ratios given in the Law of Sines are correct.

The Law of Sines

In any triangle, *ABC*:

$$\frac{a}{\sin A} = \frac{b}{\sin B} = \frac{c}{\sin C}$$

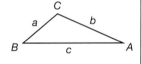

Example 1   **Finding Distance with the Law of Sines**

(a.) Find the length of $\overline{XZ}$. Round your answer to the nearest hundredth.

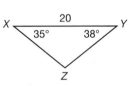

SOLUTION

By the Triangle Angle Sum Theorem, m∠Z = 107°.

Apply the Law of Sines.

$\dfrac{XZ}{\sin Y} = \dfrac{XY}{\sin Z}$	Law of Sines
$\dfrac{XZ}{\sin 38°} = \dfrac{20}{\sin 107°}$	Substitute.
$XZ \approx 12.88$	Solve.

Hint

If you know the measures of two angles of a triangle, you can find the measure of the third angle by using Theorem 18.1, the Triangle Sum Theorem.

(b.) Find the length of $\overline{EF}$. Round your answer to the nearest hundredth.

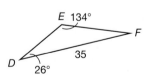

SOLUTION

$\dfrac{EF}{\sin D} = \dfrac{DF}{\sin E}$	Law of Sines
$\dfrac{EF}{\sin 26°} = \dfrac{35}{\sin 134°}$	Substitute.
$EF \approx 21.33$	Solve.

Online Connection
www.SaxonMathResources.com

To find the measure of an angle using the Law of Sines, apply the inverse sine function.

Example 2 **Finding a Missing Angle with the Law of Sines**

Find the measure of ∠L. Round your answer to the nearest degree.

SOLUTION

First, find the measure of ∠N.

$$\frac{LM}{\sin N} = \frac{LN}{\sin M} \qquad \text{Law of Sines}$$

$$\frac{12}{\sin N} = \frac{14}{\sin 75°} \qquad \text{Substitute}$$

$$14(\sin N) = 12(\sin 75°) \qquad \text{Cross product}$$

$$\sin N = \frac{12(\sin 75°)}{14} \qquad \text{Simplify}$$

$$N = \sin^{-1}\left(\frac{12(\sin 75°)}{14}\right) \qquad \text{Inverse Sine}$$

$$N \approx 56°$$

Since the measure of two angles is now known, use the Triangle Angle Sum Theorem to find the third. The measure of ∠L is about 49°.

Hint

In Example 2, you cannot solve directly for m∠L because you do not know the length of $\overline{MN}$. However, you can solve for m∠N, and then use Theorem 18.1 to find m∠L.

Example 3 **Application: Surveying**

Two surveyors are 25 feet apart. How far is each surveyor from the storm drain shown? Find the distances to the nearest foot.

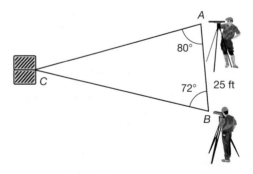

SOLUTION

First, find m∠C using the Triangle Angle Sum Theorem. The measure of ∠C is 28°. Use the Law of Sines to find the lengths of $\overline{AC}$ and $\overline{BC}$.

$$\frac{AC}{\sin B} = \frac{AB}{\sin C} \qquad\qquad \frac{BC}{\sin A} = \frac{AB}{\sin C}$$

$$\frac{AC}{\sin 72°} = \frac{25}{\sin 28°} \qquad\qquad \frac{BC}{\sin 80°} = \frac{25}{\sin 28°}$$

$$AC \approx 51 \text{ ft} \qquad\qquad\qquad BC \approx 52 \text{ ft}$$

So the surveyor at A is approximately 51 feet from the storm drain, and the surveyor at B is approximately 52 feet from the storm drain.

Math Reasoning

Generalize Can you always use the Law of Sines to find measures in a triangle if you know one angle and any two sides?

a. Find the length of $\overline{TU}$. Round your answer
(Ex 1) to the nearest hundredth.

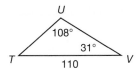

b. Find the measures of $\angle D$ and $\angle E$.
(Ex 2) Round your answers to the nearest degree.

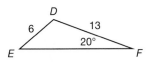

c. Marissa and Hector walk away from each other on straight paths that
(Ex 3) make a 66° angle with each other. When they stop, Hector is 125 feet
away from the starting point, and he is 120 feet away from Marissa.
How far did Marissa walk, to the nearest foot?

Practice Distributed and Integrated

1. What is the similarity ratio between these two
(41) similar trapezoids?

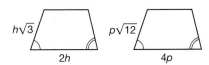

* 2. Error Analysis Humberto sees a bird fly the length of the vector $\langle 4, 10 \rangle$ and then fly
(89) the length of the vector $\langle 14, 14 \rangle$, where 1 unit is 1 kilometer. Humberto said that
since the bird flew for a total of an hour, the bird's velocity for the trip was
30 km/h. What mistake did Humberto make?

3. (**Wheels**) A wheel has a square built in for support. When the wheel is rolling on
(64) a flat road, what will be the measure of the angle between the road and a chord
from one of the sides of the square when the cord meets the road?

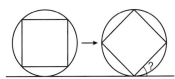

4. (**Softball**) A softball diamond is a square that is 90 feet long on each side. To the
(53) nearest tenth, what is the distance from home plate to second base, at the opposite
corner?

5. The vertices of quadrilateral *ABCD* are *A*(3, 7), *B*(8, 8), *C*(7, 3), and
(92) *D*(2, 2). Prove that *ABCD* is a parallelogram by determining the slope
of each side.

6. Multi-Step What is the volume of a pyramid with an equilateral triangle
(70) for a base and a height equal to the base side length? Express your
answer in terms of *s*, the length of a side of the base.

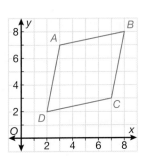

* 7. (Aircraft) An observer sees a helicopter at an angle of elevation of 30°
(94) and an airplane at an angle of elevation of 65°. The pilot of the airplane
sees the helicopter at an angle of depression of 79.87°. The observer uses
a laser to find that the helicopter is 300 meters away. How far is the
helicopter from the airplane, to the nearest meter?

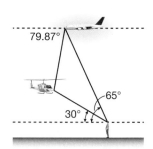

8. **Algebra** Find the range of values for x.
(Inv 4)

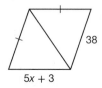

9. Pentagon $QRSTU$ has vertices $Q(1, 1)$, $R(2, 0)$, $S(1, -1)$, $T(-1, -\frac{1}{2})$, and
(84) $U(-1, \frac{1}{2})$. Sketch a dilation of $QRSTU$ with the center at the origin. Use
a scale factor of 6.

*10. Use the Law of Sines to find PR to the nearest whole number.
(94)

11. Find an expression for $\tan^2\theta$ in terms of $\sin\theta$.
(91)

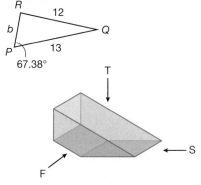

*12. **Model** Draw the front, top, and side orthographic views of this solid.
(93)

13. **Multiple Choice** Which could be the slopes of the sides of a trapezoid,
(92) in clockwise order?

    A $0, 2, 0, 3$ **B** $3, -\frac{1}{2}, \frac{1}{2}, 5$

    C $1, 4, 2, 3$ **D** $1, 5, -1, -\frac{1}{5}$

14. **Algebra** Find the measure of x in the figure.
(47)

15. The volume of a cone is 96π cubic centimeters. If the radius of the cone is 6
(77) centimeters, what is the surface area of the cone to the nearest tenth?

16. A regular pentagon has 45-inch side lengths and an apothem that is 31 inches
(66) long. Find the perimeter and area of the pentagon.

*17. **Algebra** Express the Law of Sines in terms of only the cosine function.
(94)

18. This square tile has rotational symmetry of order 4 about its center but no
(Inv 9) lines of symmetry. Construct a tessellation that has rotational symmetry
about the center of each square and about the vertex of each square.

19. (Painting) Gregory is painting a mural design on his wall. He has already painted a
(75) grid pattern where 1 unit is 1 foot. He wants to paint concentric circles that will be
centered at point $(1, -1)$ on his grid. If the first circle has a radius of 1 foot, what
is the equation of this circle?

*20. Analyze Nadim is given a Side-Side-Angle relationship in
$_{(94)}$ which the angle between sides a and b can either be obtuse
or acute. Can the student find two answers for the length
of side c that might both be correct? Explain.

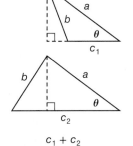

21. Verify Show that the volume and the surface area of a
$_{(80)}$ sphere have the same value only if the radius is 3 units.

22. Determine how many common tangents the circles have
$_{(72)}$ and draw the tangents.

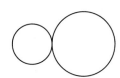

23. If $\triangle ABC \cong \triangle DEF$, $AB = DE$, $BC = EF$, $m\angle ABC = (7x - 36)°$, and
$_{(28)}$ $m\angle DEF = (3x + 16)°$, then what is x?

24. Model Design a spinner with 4 sectors so that the probability of the pointer
$_{(Inv\ 6)}$ landing on one sector is 3 times greater than any of the others.

25. Which congruence theorem applies to these triangles?
$_{(36)}$

26. Multiple Choice Which of the following sets of
$_{(33)}$ numbers does not form a right triangle?

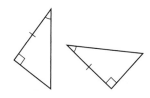

 A 3, 4, 5   **B** $2\sqrt{2}, 2\sqrt{2}, \sqrt{8}$
 C $2\sqrt{2}, 2, 2$ **D** $7, 6\sqrt{2}, 11$

27. (Rock Garden) The illustration depicts a rock garden, with the side
$_{(57)}$ of each square representing 1 foot. If there should be one basalt
rock for every 2 square feet of rock garden, find an estimate for
how many basalt rocks should be used.

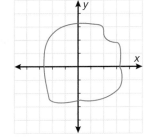

28. Algebra Two chords of a circle, $\overline{RS}$ and $\overline{TU}$, intersect at V.
$_{(86)}$ $RV = 13 - c$, $SV = 5$, $TV = 7$, and $UV = c - 1$.
Write and solve an equation for c.

29. Graph $y > -2x + 3$.
$_{(88)}$

*30. Model This orthographic drawing shows the front, top, and
$_{(93)}$ left-side views of a solid. Sketch the solid.

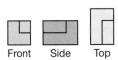

Front   Side   Top

Equations of Circles: Translating and Dilating

Warm Up

1. **Vocabulary** In the equation for the graph of a circle, the point (h, k) is
$^{(75)}$ the _____ of the circle.

2. Find the radius and coordinates of the center of a circle whose equation
$^{(75)}$ is $x^2 + y^2 = 4$.

3. Write the equation of this circle.
$^{(75)}$

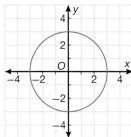

4. **Multiple Choice** If a circle is centered at the point $(1, -1)$ and the length of
$^{(75)}$ its radius is 3 units, what is the equation of the circle?

    A $(x - 1)^2 + (y - 1)^2 = 9$ **B** $(x - 1)^2 + (y + 1)^2 = 9$

    C $(x - 1)^2 + (y + 1)^2 = 3$ **D** $(x + 1)^2 + (y - 1)^2 = 9$

New Concepts

The equation of a circle with radius r and center (h, k) is
$(x - h)^2 + (y - k)^2 = r^2$. To translate a circle on the coordinate plane,
identify the center of the circle, translate the center, and write the
equation for a circle with the same radius at the new center.
Translating a circle does not affect the radius of the circle.

Example 1   Translating a Circle

The equation of a circle is $(x - 3)^2 + (y + 2)^2 = 25$. Translate the circle
4 units to the left and 2 units down. Write the equation of the translated
circle and sketch both circles on the coordinate plane.

Caution

The center of a circle is
found from the equation
by looking at the
numbers in parenthesis
with x and y. Remember
that, because the
equation calls for
subtracting the
center's coordinates,
these numbers are the
opposite sign of the
coordinates of the circle's
center.

SOLUTION
The center of the first circle is $(3, -2)$, and the
length of its radius is 5. Sketch the circle.
Translating the center four units to the left and
2 units down means the new center is at $(-1, -4)$.
The length of the radius is still 5. Therefore, the
equation of the new circle is
$(x + 1)^2 + (y + 4)^2 = 25$.

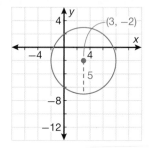

Sketch the new circle. As you can see, every point on the new circle is 4 units left and 2 units down from the corresponding point on the old circle.

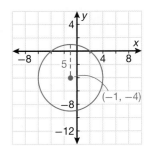

The process for dilating a circle is identical to the process for dilating any other figure on the coordinate plane. It is not practical, however, to apply a dilation to every point on a circle. Instead, apply the dilation to the radius and the center of the circle only.

Example 2 Dilating a Circle Centered at the Origin

The equation of a circle is $x^2 + y^2 = 49$. Apply a dilation centered at the origin with a scale factor of 2. Write the new equation and sketch both circles on the coordinate plane.

SOLUTION
Since this circle is centered at $(0, 0)$, its center will not change.

The radius, however, will become twice as long. Currently, the length of the radius is 7. After the dilation, it will be 14.

Therefore, the new equation is $x^2 + y^2 = 196$.

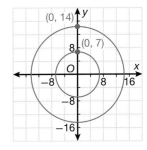

Example 3 Dilating a Circle Not Centered at the Origin

The equation of a circle is $(x - 9)^2 + (y + 6)^2 = 9$. Apply a dilation centered at the origin with a scale factor of $\frac{1}{3}$. Write an equation for the new circle and sketch both circles.

SOLUTION
The length of the radius of the original circle is 3, and it is centered at $(9, -6)$.

First, multipy the radius by the scale factor. The radius of the new circle will be 1 unit long. Next, apply the scale factor to each coordinate in the circle's center.

$$\left(\frac{1}{3} \times 9, \frac{1}{3} \times -6\right) = (3, -2)$$

The new center of the circle is $(3, -2)$. Therefore, the equation of the dilated circle is $(x - 3)^2 + (y + 2)^2 = 1$. Sketch both circles on the coordinate plane.

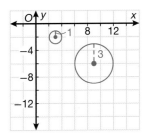

Hint

Recall that a dilation is a transformation that changes the size of a figure but not its shape.

Math Reasoning

Analyze How could you make sure that your suggested adjustment to the siren's volume will reach the corners of the city?

Example 4 Application: Emergency Planning

A coordinate plane is overlaid onto the map of a city to help place a tornado-warning siren. The current location of the siren is shown on the coordinate plane, where each unit represents one mile. At its current volume, it is audible to everyone in a 2 mile radius. Suggest a translation and an approximate adjustment to the siren's volume that would make the siren audible to everyone in the city.

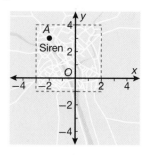

SOLUTION

It makes sense that the siren should be in the center of the city. The city is bounded by the points $(-3, 4)$, $(-3, -1)$, $(2, -1)$, and $(2, 4)$. Use the midpoint formula to find the center of the city.

$$\left(\frac{-3 + 2}{2}, \frac{4 - 1}{2}\right)$$

$(-0.5, 1.5)$

The center of the city is at $(-0.5, 1.5)$. This is a translation of 1.5 units right and 1.5 units down.

The siren at its current strength will not be heard at the north end of town. Amplifying the siren by a factor of 3 would make it extend far beyond the bounds of the city. Increasing its volume by a factor of 2 will approximately cover the city, so this is a good estimate.

Lesson Practice

a. The equation of a circle is $(x + 2)^2 + (y - 1)^2 = 16$. Write the equation
(Ex 1) of this circle after it is shifted 6 units to the right. Draw the original circle and the shifted circle on the coordinate plane.

b. The equation of a circle is $x^2 + y^2 = 25$. Apply a dilation centered at
(Ex 2) the origin with a scale factor of 0.5. Write the equation for and sketch the dilated circle.

c. The equation of a circle is $(x - 2)^2 + (y - 1)^2 = 1$. Find the equation
(Ex 3) of the circle after it is dilated by a scale factor of 4. Draw the original circle and the dilated circle on the coordinate plane.

d. June lives 2 miles north and 2 miles west of a tornado siren that can
(Ex 4) be heard for 3 miles in any direction. The city plans to move the siren 2 miles south but increase its volume by a factor of 1.5. Will the siren still be audible at June's house?

1. (**Furniture**) This orthographic drawing shows front, top, and
(93) left-side views of a child's chair. Sketch the chair.

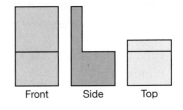

Front   Side   Top

2. Find the measure of the angle exterior to a circle made by two
(79) secants to the circle if the shorter arc is 60° and the longer one
is 150°.

3. (**Football**) A quarterback throws a football at a velocity of 50 miles per hour, at a
(89) 30° angle above the horizontal. What is the horizontal speed of the football, to the
nearest tenth?

4. Algebra Write an inequality to show the least value of x that would
(33) make the triangle acute.

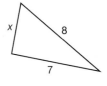

* 5. The equation of a circle is $x^2 + y^2 = 9$. Write the equation of this
(95) circle after it is shifted 8 units to the right. Draw the original circle
and the shifted circle on the coordinate plane.

6. Verify Show that the geometric mean of $\frac{1}{2}$ and $\frac{1}{32}$ is $\frac{1}{8}$.
(50)

7. Reflect the triangle across the horizontal line that goes through point C.
(90) Then reflect the image across line t.

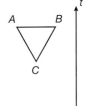

* 8. Multi-Step The equation of a circle is $x^2 + y^2 = 144$. Find the equation of the circle
(95) after it is dilated by a factor of ¼, then is shifted 1 unit to the right and
4 units down.

* 9. Find AC. Round your answer to the nearest hundredth.
(94)

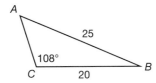

10. (**Skiing**) A skier skis downhill at a velocity of 14 meters per second (m/s). If there
(83) is a wind coming towards him up the hill at a velocity of 2 m/s, what is his actual
magnitude? What would his velocity be if the wind were traveling *with* him down
the mountain at the same velocity?

11. Find the value of x in the diagram.
(46)

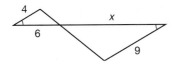

*12. The equation of a circle is $x^2 + y^2 = 36$. Write the equation of this circle after
(95) it is shifted 4 units upward. Draw the original circle and the shifted circle on the
coordinate plane.

13. Justify For what value of y are line m and n parallel? State a
(12) theorem to justify your answer.

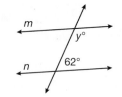

14. Analyze At what value between 0° and 180° is the tangent function
(91) undefined? Explain why it is undefined at this value.

15. Find the length of $\overline{LM}$. Round your answer to the nearest
(94) hundredth.

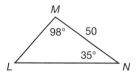

16. Multiple Choice Which of the following are parallel lines?
(37)
i) $y = 2x + 7$ **ii)** $y = 2(x - 4)$ **iii)** $y - x = 7$

A i and ii   **B** i and iii

C ii and iii   **D** i, ii, and iii

17. Error Analysis Noemi says that quadrilateral *CDEF* is a trapezoid.
(92) Explain why she is mistaken.

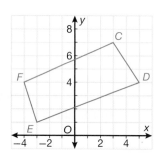

18. Find the exact length of a leg of a 45°-45°-90° right triangle with a
(53) hypotenuse of 31 centimeters.

*19. Formulate Express the Law of Sines using only the tangent function.
(94)

20. If $x = 3$, is this parallelogram a rhombus, a rectangle, both,
(65) or neither?

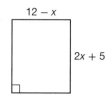

21. If two parallel lines are cut by a transversal, and the alternate
(Inv 1) interior angles formed measure $(3x - 10)°$ and $(-x + 60)°$,
find the value of x.

22. (Carpentry) A carpenter is cutting a cylindrical piece of wood. If he makes a cut
(85) that is parallel to the base, what is the shape of the cross section?

23. (Design) A wallpaper pattern consists of nested squares, as shown
(Inv 8) in the diagram. Does this pattern have rotational symmetry?
If so, what is the order of rotational symmetry?

*24. The equation of a circle is $x^2 + y^2 = 16$. Find the equation of the circle after it
(95) is dilated by a factor of 3. Draw the original circle and the dilated circle on the
coordinate plane.

25. How far apart are the centers of these two circles?
(58)

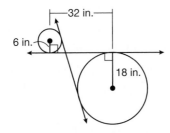

26. Quadrilateral $QRST$ has vertices $Q(5, -6)$, $R(6, -4)$, $S(3, -1)$, and $T(1, -3)$.
(78) Graph the quadrilateral, and then rotate it 90° counterclockwise about the
origin. List the coordinates of the vertices of quadrilateral $Q'R'S'T'$.

27. A chord of a circle is 15 inches long, and 12 inches from the center of the circle.
(43) What is the length of the radius of the circle?

28. Model Draw the front, top, and side orthographic views of this solid.
(93)

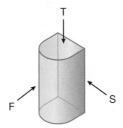

29. Is this statement sometimes, always, or never true?
(61)

    A rhombus is a parallelogram.

30. Write a two-column proof showing that m∠1 = m∠3.
(27)

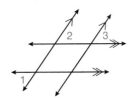

Effects of Changing Dimensions on Perimeter and Area

1. **Vocabulary** A _____ is a repeating pattern of plane figures that
(Inv 9) completely covers a plane with no gaps or overlaps.

2. Find the perimeter of a quadrilateral that has vertices at coordinates
(57) $M(3, 2)$, $N(7, 1)$, $O(2, -1)$, and $P(-4, -2)$. Round your answer to
the nearest hundredth.

3. **Multiple Choice** What dimensions make $\overline{AB} \parallel \overline{DE}$?
(60)
A $AD = 10$, $BE = 11$ **B** $AD = 8$, $BE = 7$
C $AD = 7$, $BE = 8$ **D** $AD = 9$, $BE = 10.5$

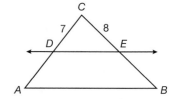

New Concepts

Theorem 87-1 states that two similar figures with a similarity ratio of $a{:}b$ have perimeters in the ratio $a{:}b$ and areas in the ratio $a^2{:}b^2$. This theorem can be used when an entire figure is dilated. Sometimes, however, we may want to find the area or perimeter of a figure when only one dimension is altered, or when the dimensions are changed by different scale factors.

If one dimension of a polygon is changed, the ratio of its original perimeter to its new perimeter can be found by applying the formula for perimeter of a polygon.

Example 1 **Changing Perimeter of a Polygon**

A rectangle is half as tall as it is long. If its height is reduced by half its original height, what is the ratio of the new rectangle's perimeter to the original rectangle's perimeter?

SOLUTION

Let the length of the rectangle be x. Since the rectangle's height is half its length, its height is $0.5x$. Determine its perimeter by adding the sides together.

$$P = x + x + 0.5x + 0.5x$$
$$P = 3x$$

When the height is reduced to one half its original height, it will be half of $0.5x$, or $0.25x$. The length of the rectangle does not change. Find the perimeter of the new rectangle by adding its sides together.

$$P = x + x + 0.25x + 0.25x$$
$$P = 2.5x$$

Therefore, the ratio of the new rectangle's perimeter to the original rectangle's perimeter is 2.5:3, or 5:6.

Math Reasoning

Formulate What would the ratio of the perimeters be, if in addition to halving the length of the rectangle, the length of its height was doubled?

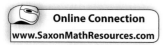

Online Connection
www.SaxonMathResources.com

The same method can be applied to find the ratio of the area of two polygons when one dimension is altered.

Example 2 Changing Area of a Polygon

Find the area of each polygon. Describe how each change affects the area.

a. Triangle *ABC* has a base that is congruent to its height. If the base is dilated by a factor of 2, what is the ratio of the new triangle's area to the original triangle's area?

SOLUTION

The diagram illustrates this problem. Use the formula for area of a triangle to find the area of the original triangle.

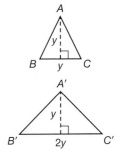

$$A = \frac{1}{2}bh$$

$$A = \frac{1}{2}(y)(y)$$

$$A = \frac{1}{2}y^2$$

Now find the area of the altered triangle.

$$A = \frac{1}{2}bh$$

$$A = \frac{1}{2}(2y)(y)$$

$$A = y^2$$

Compare the two expressions for area. The ratio of the triangles' areas is 2:1.

b. A parallelogram's base is twice as long as its height. If the length of the base is doubled, and the height is halved, what is the ratio of the new parallelogram's area to the original parallelogram's area?

SOLUTION

The diagram illustrates this problem. Use the formula for area of a parallelogram to find the original parallelogram's area.

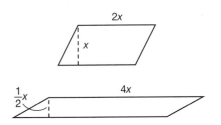

$$A = (2x)(x)$$
$$A = 2x^2$$

Now find the area of the altered parallelogram.

$$A = (4x)(\frac{1}{2}x)$$
$$A = 2x^2$$

The ratio of the areas is 1:1.

Example 3 Altering the Dimensions of a Circle

A circle's radius is increased by a factor of 3. Find the ratio of the circle's new area and circumference to its original area and circumference.

SOLUTION

Call the length of the initial radius x. The new radius will have a length of $3x$. Find the area of each circle.

$A = \pi r^2$ $A = \pi r^2$
$A = \pi x^2$ $A = \pi(3x)^2$
                       $A = 9\pi x^2$

The ratio of the areas is 9:1.

Now, find the circumference of each circle.

$C = 2\pi r$ $C = 2\pi r$
$C = 2\pi x$ $C = 2\pi(3x)$
                       $C = 6\pi x$

The ratio of the circumferences is 3:1. As you can see, circles conform to the ratios given in Theorem 87-1.

Hint

Because all circles are similar to one another, any two noncongruent circles centered at the origin are dilations of each other.

Example 4 Application: Home Improvements

Bev is having a pool installed in her backyard. Her backyard is a rectangle with a length that is twice as long as its width. Bev decides that the pool will also be a rectangle, but it will run only three-fourths the length of the backyard and be half as wide. What is the ratio of the pool's area to the backyard's area?

SOLUTION

Draw a diagram to illustrate this situation. Notice that the length of the pool is three-fourths of $2x$, or $1.5x$. Find the area of the pool and Bev's backyard.

$A = bh$ $A = bh$
$A = (2x)(x)$ $A = (1.5x)(0.5x)$
$A = 2x^2$ $A = 0.75x^2$

So the ratio is 0.75:2. To simplify this, multiply by 4 to eliminate the decimal, which results in the ratio 3:8.

a. One pair of opposite sides of a square is dilated by a factor of 4 while
(Ex 1) the other sides remain the same. What is the ratio of the new figure's
perimeter to that of the original?

b. What is the ratio of the first trapezoid's area to the second
(Ex 2) trapezoid's area?

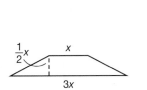 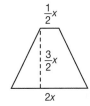

c. The radius of a circle is *x*. If the radius is changed by a factor
(Ex 3) of $\frac{1}{2}$, what is the ratio between the original area and the new area?
What is the ratio between the original circumference and the new
circumference?

d. A block of apartments must be constructed in the shape of a square.
(Ex 4) In the middle of construction, it is discovered that the apartment lot
must be shortened to make way for a road expansion. Due to this fact,
the length of the lot is nine-tenths of what it was before. What is the
ratio of the apartment block's new area to its original planned area?

Practice **Distributed and Integrated**

1. Find the length of $\overline{DE}$. Round your answer to the nearest
(94) hundredth.

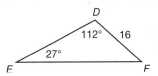

2. Two chords of a circle, $\overline{JK}$ and $\overline{LM}$, intersect at *N*.
(86) $JN = KN$, $LN = 3$, and $MN = 12$.
a. Sketch the circle and the two chords, showing the given information.
b. Determine *JN*.

3. **Algebra** Line *m* passes through points $(5, -3)$ and $(2, -6)$. Line *n* passes through
(37) points $(4, y)$ and $(-2, 3)$. What value of *y* makes these lines perpendicular?

* 4. The equation of a circle is $x^2 + y^2 = 100$. Write the equation of this circle after
(95) it is shifted 3 units upward. Draw the original circle and the shifted circle on
the coordinate plane.

* 5. **Multi-Step** The perimeter of a square is 16 centimeters. If the area is doubled,
(96) what is the new side length?

6. Is the net at right the net of a cube? Explain.
(Inv 5)

* 7. (**Milk**) After fresh milk was stored in a 452-cubic-inch cylindrical jar for
(62) half a day, it separated into a 2-inch band of cream and a 14-inch band
of low-fat milk. What is the radius of the can?

8. **Miniature Golf** A hole on a miniature golf course is in the shape
(40) given in the diagram, with dimensions in feet. What is the exact
amount of artificial turf needed to resurface this hole? What is the
amount rounded to the nearest tenth? The regions that appear to
be semicircles are semicircles.

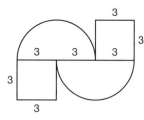

9. Triangle *ABC* has been rotated counterclockwise about the origin. The original
(78) vertices were located at $A(-5, -8)$, $B(-3, 2)$, and $C(-1, -8)$. After the rotation,
the vertices were located at $A'(-8, 5)$, $B'(2, 3)$, and $C'(-8, 1)$. What was the angle
of rotation?

10. **Write** Can you use the Law of Sines to find the other measures of
(94) $\triangle EKG$? Explain.

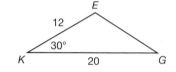

xy² *11. **Algebra** The circumference of a circle is πx. If the radius is
(96) increased by a factor of 2, what is the area of the circle, in
terms of π?

12. Find the value of k in these similar polygons.
(44)

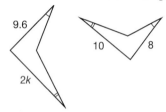

***13.** The equation of a circle is $x^2 + y^2 = 64$. Write the equation of this circle after it
(95) is shifted 1 unit to the right. Draw the original circle and the shifted circle on the
coordinate plane.

14. **Error Analysis** What has Leif done incorrectly in the glide reflection
(90) from *ABCD* to $A''B''C''D''$?

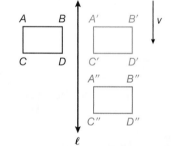

15. **Multiple Choice** Which is the area of a pentagon with side lengths of
(66) 44 feet and a 50.5-foot apothem?
  A 3703.3 ft²   **B** 11,110 ft²
  C 555.5 ft²   **D** 5555 ft²

16. **Shadows** A streetlight is mounted at the top of a 15-foot pole. A 6-foot-tall
(60) man walks away from the pole along a straight path. If it is dark and the
streetlight provides the only light, how long is the man's shadow when
he is 40 feet from the pole, to the nearest tenth of a foot?

17. **Model** This orthographic drawing shows the front, top, and left-side
(93) views of a solid. Sketch the solid. Classify the solid.

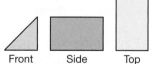

18. **Write** Two cross sections of a cylinder are being compared. One is parallel to
(85) the base and the other is slanted through the base. Which will have the larger area?
What shape will the cross sections have? Explain.

19. In the diagram given, what piece of information is needed to conclude that △ABC ≅ △DEF?
(30)

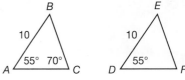

*20. (Converting Units) If a playground covers 10 square yards, how many square feet does it cover? Explain how you can use dilations to answer this question.
(96)

21. **Multi-Step** Find the value of *x* in the figure shown.
(56)

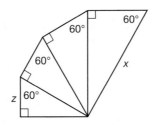

22. (Architecture) A model of a pyramid for an art gallery has glass triangular lateral faces, each with a base of 3 feet and a height of 4 feet. The scale of the model is 2:41. How many square feet of glass will be needed for each triangular glass pane?
(87)

23. Classify the solid. How many vertices, edges, and bases does it have?
(49)

24. Find the surface area of a sphere with a 12-inch radius to the nearest square inch.
(80)

25. Find the circumference of this circle based on the arc given.
(35)

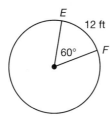

26. **Verify** Show that the area of a right triangle with vertices *A*(0, 0), *B*(0, 5), and *C*(4, 0) under the translation *T*: (*x*, *y*) → (*x* − 2, *y* + 1) remains constant.
(71)

*27. A circle whose equation is $x^2 + y^2 = 25$ is shifted 1 unit down and 2 units to the left. What is the equation for the shifted circle? Graph the shifted circle.
(95)

28. Each exterior angle of a regular polygon is 10°. How many sides does the polygon have?
(Inv 3)

29. **Verify** Use the diagram to show that $\frac{\tan\theta}{\sin\theta}$ is equivalent to $\frac{1}{\cos\theta}$.
(91)

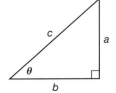

*30. (Architecture) An architect makes a replica of a clock tower. If the replica has dimensions that are ten times smaller than the clock, what is the ratio between the replica's circumference and the actual circumference?
(96)

Concentric Circles

Warm Up

1. **Vocabulary** A transformation that changes the size of a figure but not its
   (84) shape is called a _____.

2. Triangle *ABC* has vertices located at *A*(3, −2), *B*(5, 1), and *C*(1, 3) on a
   (84) coordinate plane. Find the image of △*ABC* after a dilation centered at
   the origin with a scale factor of 2. Draw the image and the preimage on a
   coordinate plane.

3. The circle defined by the equation $x^2 + y^2 = 64$ is shifted 3 units to the
   (95) right and 5 units down. What is the equation of the translated circle?

New Concepts

Concentric circles are coplanar circles with the same center.
Concentric circles do not intersect. They have the same
center, but their radii are different lengths. Concentric
circle are dilations of each other by a factor of the radius
of one of the circles.

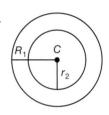

Example 1 **Determining When Two Circles are Concentric**

Determine if the circles in each diagram are concentric. Explain your
reasoning.

a.   b.   c.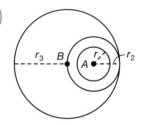

Hint

Concentric circles are
dilations of each other
with a center of dilation
located on the center of
the circle.

SOLUTION

a. The circles are not concentric. The circles are coplanar, but they
   intersect at one point, so they cannot have the same center.

b. Yes, the circles are coplanar and they share the same center, so
   they are concentric.

c. The circles with center *A* are concentric, but ⊙*B* is not concentric
   with either of them.

Given equations of circles in the form $(x - h)^2 + (y - k)^2 = r^2$, concentric
circles have the same *h* and *k* values but different *r* values. This is because *h*
and *k* represent the location of the center of the circle on a coordinate
plane and *r* represents the radius.

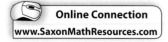
Online Connection
www.SaxonMathResources.com

Example 2 **Equations of Concentric Circles**

Write the equations for the concentric circles centered at $(-2, 3)$. Describe how the equations are alike and how they are different.

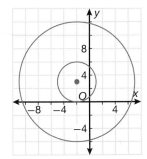

SOLUTION

Both circles have center $(-2, 3)$. So, one side of each equation is $(x + 2)^2 + (y - 3)^2$.
The smaller circle's radius is 3 units long, so the other side of its equation will be $(3)^2$ or 9.
The larger circle's radius is 9 units long, so the other side of the equation will be $(9)^2$ or 81.
Write the equation for each circle.

$(x + 2)^2 + (y - 3)^2 = 9$
$(x + 2)^2 + (y - 3)^2 = 81$

The circles are coplanar and the share the same center. They have different radii. The larger circle is the smaller circle dilated by a scale factor of 3.

The **annulus** is the region between two concentric circles. To find the area of the annulus, subtract the area of the smaller circle from the area of the larger circle. The resulting difference is the area of the annulus.

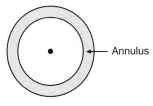

— Annulus

Math Reasoning

Formulate Write an expression to represent the area of the annulus of two concentric circles.

Example 3 **Finding the Area of the Annulus of Concentric Circles**

Find the exact area of the annulus in these concentric circles.

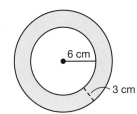

6 cm

3 cm

SOLUTION

Since the problem called for the exact area, leave all calculations in terms of π.
Find the area of the smaller circle with a 6-centimeter radius.

$A = \pi r^2$
$A = \pi(6)^2$
$A = 36\pi$

Find the area of the larger circle which has a 9-centimeter radius.
$A = \pi(9)^2$
$A = 81\pi$

Find the difference of the areas.
$81\pi - 36\pi = 45\pi$

The area of the annulus is 45π square centimeters.

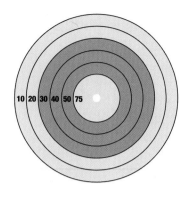

Example 4 | Application: Targets

Eric is using a board painted with concentric circles for target practice. The central circle has a 2-inch radius and each annulus beyond that is 1 inch wide. What is the probability that Eric will hit the darker-shaded portion of the target?

SOLUTION

To find the probability, divide the area of the darker-shaded part by the area of the whole board.

$$P = \frac{\text{Area of the annulus}}{\text{Area of the target}}$$

Since the radius of the central circle is 2 inches and each annulus is 1 inch, the radius of the entire board is 7 inches.

$$A = \pi(7)^2$$
$$A = 49\pi$$

Now find the area of the darker-shaded region. The length of the region's radius, including the bull's eye is 5 inches. The bull's eye, which is not part of the shaded area, has a 2-inch radius.

$$A_{Annulus} = A_{shaded} - A_{bullseye}$$
$$A_{Annulus} = \pi(5)^2 - \pi(2)^2$$
$$A_{Annulus} = 21\pi$$

Therefore, the probability is $\frac{21\pi}{49\pi}$ or $\frac{3}{7}$.

Hint

To review geometric probability, refer to Investigation 6.

Lesson Practice

a. Determine if the circles are concentric. Explain your
(Ex 1) reasoning.

b. Write the equations for these concentric circles. Describe how the
(Ex 2) equations are alike and how they are different.

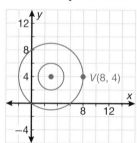

c. Find the area, in terms of π, of the annulus
(Ex 3) between these concentric circles.

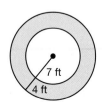

d. Consider the board from Example 4.
(Ex 4) What is the probability that a randomly thrown dart that hits the board will hit outside the bull's eye and outside the darker–shaded region?

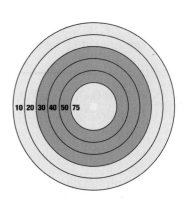

Practice   Distributed and Integrated

1. **Error Analysis** Julius found the sine of $\angle F$ to be $\frac{13}{5}$. Identify and
(Inv 7) correct his mistake.

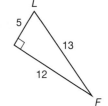

2. Polygon *ABCDE* has vertices $A(0, 1)$, $B(0, 5)$, $C(-2, 2)$, $D(-4, 5)$,
(74) and $E(-4, 1)$. Sketch the reflection of *ABCDE* across the line $y = x$.

* 3. Find the exact area of the annulus between two concentric circles, one
(97) with a 15-inch radius and the other with a 2-inch radius.

4. Write the equation of a circle centered at $(-4, 3)$ with a radius of
(95) 1000.

5. (**Physics**) A camper uses a pole to hold up an awning. The pole is at a 60° angle
(89) and has 30 pounds of force on it. What is the horizontal component of the force in pounds?

6. (**Ballooning**) The passengers on a hot-air balloon see a playground at an angle of
(73) depression of 16°. Later, they see the playground at a 36° angle of depression. At both times they were 150 meters above the ground. Have they moved closer or farther from the playground, and by how many meters? Round your answer to the nearest meter.

* 7. **Probability** This figure is made up of concentric circles on a flat surface,
(97) with dimensions as shown. If a coin is randomly dropped onto the figure, what is the probability it will land on the shaded region?

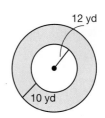

8. $\triangle MNP$ has vertices $M(3, 3)$, $N(5, 1)$, and $P(1, 1)$. Apply a dilation
(84) centered at $(1, 2)$ with a scale factor of 2 to $\triangle MNP$. Find the
coordinates of the vertices of the image.

9. Multiple Choice This trapezoid is made up of congruent equiangular
(51) triangles. What is its perimeter?

   A 104 mm   **B** 117 mm
   C 273 mm   **D** 130 mm

10. Graph the inequality $y \leq \frac{3}{2}x - 3$.
(88)

11. Metalworking A sheet of metal that is 8 feet long and 6 feet wide is to be cut
(62) into cylindrical cans like the one shown. How many lateral surfaces for
the cans can be cut from the metal sheet with as little waste as possible?

12. Write Describe four coplanar points that do not form a quadrilateral.
(45)

*13. Find the perimeter of the new image if each dimension of the
(96) triangle shown is dilated by a factor of 3.

*14. Write How could you change one dimension of a trapezoid to
(96) triple its area?

*15. Find the area, in terms of π, of the annulus between these concentric circles.
(97)

16. Write Simplify the expression $\cos\left[\cos^{-1}\left(\frac{4}{5}\right)\right]$. Explain why your answer is
(82) correct.

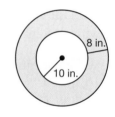

17. What are the possible measures for θ_1 and θ_2?
(94) Round your answers to the nearest degree.

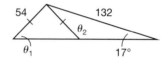

18. Analyze Determine coordinate-based criteria for a quadrilateral to be each of
(92) the following: a rhombus, a rectangle, and a trapezium.

19. Sports This simplified plan shows part of an Olympic complex. The boundary
(67) of the complex will be completed by rotations of the segments $\overline{AB}$ and $\overline{BC}$
through 120° clockwise and 120° counterclockwise about the central fountain F.
Determine the perimeter of the whole complex.

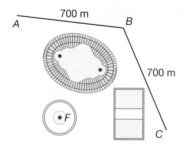

20. Multi-Step Find the area of a rectangle with a diagonal of 13 and a side length
(22) of 5.

21. **Write** Consider this tessellation of a parallelogram.
(Inv 9)
    a. The tessellation has translation symmetries that are based on combinations of two vectors. Explain how these vectors are related to the sides of the original parallelogram.

    b. The tessellation also has rotational symmetry. What angle must the rotation be? Where could its center be?

    c. Explain why the tessellation has no reflection symmetry.

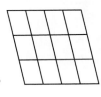

*22. (**Mapping**) A map has a scale of 1:250. If a map represents a community center by
(96) a square with an area of 100 square centimeters, what is the actual area of the community center, in square meters?

23. **Multi-Step** Classify the polygon with vertices $P(2, 1)$, $Q(7, -1)$, $R(2, -3)$, and
(57) $S(-3, -1)$, and find its perimeter. Give your answer in simplified radical form.

*24. **Write** Write the equations for the concentric circles shown with
(97) center $(5, -4)$. How are the equations alike? How are they different?

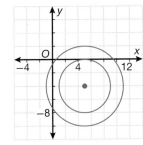

25. The equation of a circle is $x^2 + y^2 = 169$. Write the equation of this
(95) circle after it is shifted 6 units down. Draw the original circle and the shifted circle on the coordinate plane.

26. Quadrilateral *KLMN* is a rhombus with its center
(52) point labeled *S*. What is m∠*NSM*?

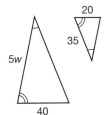

27. Assume the following conditional statement is true.
(17) State its converse. Is the converse true?

    If Noel does not clean her room, she will not be able to find her shoes.

28. These triangles are similar. Find *w*.
(41)

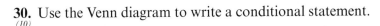

29. **Algebra** If an arc measures 96°, and the measure of its inscribed angle is
(47) $(2x + 3)°$, what is the value of *x*?

30. Use the Venn diagram to write a conditional statement.
(10)

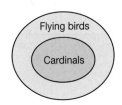

Law of Cosines

1. **Vocabulary** The _____ form of a vector lists the vertical and horizontal change from the initial point to the terminal point.
   (63)

2. Decompose the vector into its vertical and horizontal components. Round your answer to the nearest hundredth.
   (89)

3. Find θ. Round your answer to the nearest degree.
   (82)

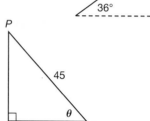

New Concepts

Like the Law of Sines, the Law of Cosines relates the sides and angles of any triangle. The Law of Cosines may sometimes be useful to solve for unknowns when not enough information is given to use the Law of Sines.

Math Reasoning

Formulate Solve each equation in the Law of Cosines for cos A, cos B or cos C.

The Law of Cosines

In any triangle *ABC*:

$$a^2 = b^2 + c^2 - 2bc\,(\cos A)$$
$$b^2 = a^2 + c^2 - 2ac\,(\cos B)$$
$$c^2 = a^2 + b^2 - 2ab\,(\cos C)$$

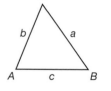

Example 1 **Finding Distance with the Law of Cosines**

Find *c*. Round your answer to the nearest tenth.

SOLUTION
Apply the Law of Cosines.
$$c^2 = a^2 + b^2 - 2ab\,(\cos V)$$
$$c^2 = 80^2 + 60^2 - 2(80)(60)(\cos 22°)$$
$$c^2 \approx 1099.03$$
$$c \approx 33.2$$

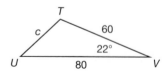

There are two particular cases where the Law of Cosines can be used and the Law of Sines cannot be used. The first case is when two side lengths and the included angle measure are given, and the other is when all three side lengths of a triangle are given.

Online Connection
www.SaxonMathResources.com

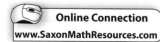

Example 2 Finding a Missing Angle with the Law of Cosines

(a.) Find m∠C. Round your answer to the nearest degree.

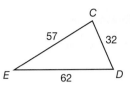

SOLUTION
Use the Law of Cosines. To find the angle measure, use the inverse cosine.

$$c^2 = a^2 + b^2 - 2ab(\cos C)$$
$$62^2 = 32^2 + 57^2 - 2(32)(57)(\cos C)$$
$$-429 = -3648(\cos C)$$
$$\cos C \approx 0.118$$
$$m\angle C \approx 83.22°$$

(b.) Find m∠Y. Round your answer to the nearest degree.

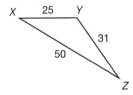

SOLUTION
Use the Law of Cosines.

$$y^2 = x^2 + z^2 - 2xz(\cos Y)$$
$$50^2 = 31^2 + 25^2 - 2(31)(25)(\cos Y)$$
$$914 = -1550(\cos Y)$$
$$\cos Y \approx -0.590$$
$$m\angle Y \approx 126.16°$$

The Law of Cosines and the Law of Sines apply to all triangles, including right triangles. The next example demonstrates the similarities between the Law of Cosines and the Pythagorean Theorem.

Example 3 Using the Law of Cosines with a Right Triangle

(a.) Find y. Round your answer to the nearest tenth.

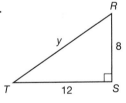

SOLUTION
$$y^2 = r^2 + t^2 - 2rt(\cos S)$$
$$y^2 = 12^2 + 8^2 - 2(12)(8)(\cos 90°)$$
$$y^2 = 12^2 + 8^2 - 0$$
$$y = \sqrt{12^2 + 8^2}$$
$$y \approx 14.4$$

Hint

Remember that trigonometric ratios refer to values in right triangles, but the Law of Sines and the Law of Cosines can be used on any triangles.

Since $\cos 90° = 0$, the third term of the equation is eliminated. Notice that after that term is eliminated, the Law of Cosines is identical to the Pythagorean Theorem.

(b.) Use the Law of Cosines to find m∠M.

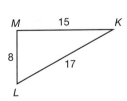

SOLUTION
$$c^2 = a^2 + b^2 - 2ab(\cos M)$$
$$17^2 = 8^2 + 15^2 - 2(8)(15)(\cos M)$$
$$0 = -240°(\cos M)$$
$$\cos M = 0$$
$$m\angle M = 90°$$

The measure of angle M is 90°.

Example 4 **Application: Surveying**

Based on the measurements between the two buildings shown in the diagram, a surveyor wants to approximate the height of the taller building. What is the height of the taller building to the nearest foot?

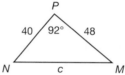

SOLUTION

We are given two sides and the included angle, so we must use the Law of Cosines.

$$a^2 = b^2 + c^2 - 2bc(\cos A)$$
$$a^2 = 580^2 + 320^2 - 2(580)(320)(\cos 99°)$$
$$a \approx \sqrt{496868.47}$$
$$a \approx 705 \text{ ft}$$

The building is approximately 705 feet tall.

Lesson Practice

a. Find c. Round your answer to the nearest whole number.
(Ex 1)

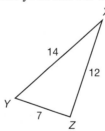

b. Find $m\angle X$. Round your answer to the nearest degree.
(Ex 2)

c. Use the Law of Cosines to find a. Round your answer to the nearest tenth. Then, suggest an alternative way to solve for a.
(Ex 3)

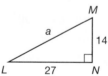

d. Strom and Milan are walking away from each other at a 66° angle. If they stop when Strom is 45 yards from the starting point and Milan is 38 yards from the starting point, how far is Strom from Milan? Round your answer to the nearest hundredth of a yard.
(Ex 4)

xy² *** 1. Algebra** Find θ. Round your answer to the nearest degree.
(98)

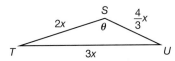

2. Write A circle whose equation is $x^2 + y^2 = 4$ is dilated about its center by a factor
(95) of 3 and shifted 3 units to the left and 2 units down. Franklin performs the
dilation first, then the translation, while Benjamin performs the translation first,
and then the dilation. Will the two students' images be the same? Justify your
reasoning. What is the equation of the new circle?

* 3. (**Track and Field**) A circular track, shown at right, needs to be repaved.
(97) What is the area of the track's surface to the nearest square meter?

4. If two chords are congruent, and one is 3 inches from the center of the
(43) circle, then what is the distance from the center to the other chord?

5. Find the magnitude of vector $\langle 3, -4 \rangle$.
(63)

6. Segment TU is a midsegment of $\triangle QRS$. Find the values of x and y.
(55)

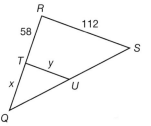

xy² **7. Algebra** If $\frac{\sin \theta}{\tan \theta} = 0.6$, find the value of $\cos \theta$. Then find θ to the
(91) nearest degree.

8. Write a paragraph proof showing that $m\angle 1 = m\angle 2$.
(31)

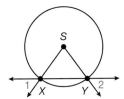

9. (**Architecture**) In a regular square pyramid that has a height of 100 feet
(85) and a base with 200-foot-long sides, there is a second floor located
halfway up the pyramid. What is the area of this floor?

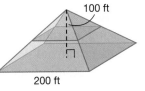

10. Analyze Vector u has a magnitude of 3 and an angle of 30° from the positive
(89) x-axis. How many unique vectors in the same plane with the same initial
point as u have a magnitude of 2 and are perpendicular to u? List them.

*11. Find θ. Round your answer to the nearest degree.
(98)

12. Prove that $\triangle JKL \cong \triangle PQR$.
(36)

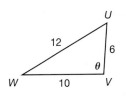

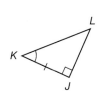

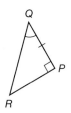

13. Find the perimeter of this square if the length but not the height is
⁽⁹⁶⁾ shrunk by a factor of $\frac{1}{6}$. Write an expression for the perimeter of
the new rectangle based on the original perimeter, P.

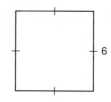

14. If the coordinates of one endpoint of a segment are $(-1, -3)$
⁽⁹⁾ and the segment extends from this point to the left by 7 units,
what is the coordinate of the other endpoint?

15. (Carpentry) A wooden frame is built in the shape of a parallelogram.
⁽⁶⁵⁾ The carpenter knows that $\overline{AB} \cong \overline{CD}$ and $\overline{AC} \cong \overline{BD}$. Using a tape
measure, she determines that $BC = AD$. Is the frame a rectangle,
rhombus, both, or neither?

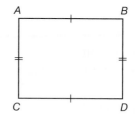

16. What value of x completes the equation $\sin 60° = \cos x$?
⁽⁶⁸⁾

*17. Write the equations for these concentric circles, centered at $(-1, 0)$.
⁽⁹⁷⁾ Explain how the equations are similar and how they are different.

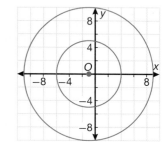

*18. Write Explain why the Law of Cosines is identical to the
⁽⁹⁸⁾ Pythagorean Theorem when it is used in a right triangle.

19. Analyze In the figures below, it is given that two sides and an angle
⁽⁹⁴⁾ are congruent between the two triangles. Explain why it cannot be
concluded that the triangles are congruent.

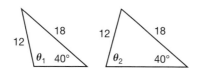

20. In a two-column proof, prove that $KL > HL$, given that $\overline{JK} \parallel \overline{HL}$, $\overline{JK} \cong \overline{HL}$,
^(Inv 4) and $m\angle KML > m\angle HML$.

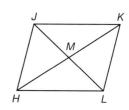

21. Find the exact length of the leg of a isosceles right triangle with a
⁽⁵³⁾ hypotenuse of 75 inches.

*22. (Kites) Shannon is flying a kite. She has released 55 feet of
⁽⁹⁸⁾ string. Adrian sees the kite from 35 feet away. How far is
Shannon from Adrian if the kite makes a 95° angle between
them? Round your answer to the nearest foot.

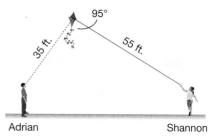

23. Error Analysis Kaelie believes she has found a new trigonometric
⁽⁹¹⁾ identity below based on $\sin^2 \theta + \cos^2 \theta = 1$. Identify any
errors she made and explain why she is wrong.

$$\sin^2\theta + \cos^2\theta = 1$$
$$\sqrt{\sin^2\theta + \cos^2\theta} = \sqrt{1}$$
$$\sin \theta + \cos \theta = \pm 1$$

24. Find the value of x in the figure.
(64)

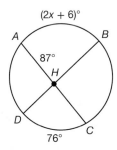

25. **Multiple Choice** The set of probabilities for a spinner experiment are 41% red,
(Inv 6) 34% blue, and 25% green. The measures of the central angles for the sectors
of the spinner are:

    A red: 120°; blue: 120°; green: 120°   **B** red: 148°; blue: 122°; green: 90°

    C red: 120°; blue: 105°; green: 75°   **D** red: 160°; blue: 60°; green: 140°

26. (**Packaging**) Two identical round vases are packed into a box. There is
(72) reinforcement cardboard to keep the vases from smashing into each
other. How many tangents and common tangents are there?

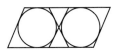

27. **Multi-Step** A square has coordinates $D(1, 2)$, $E(-1, 2)$, $F(-1, 4)$, and $G(1, 4)$. If
(96) a pair of parallel sides of the square is dilated by a factor of 3, what shape does
the image take? What are the perimeter and area of the image?

*28. **Probability** The figure below is made up of concentric circles with dimensions as
(97) shown. If a coin is randomly dropped onto the figure, what is the probability it
will land on one of the shaded regions, assuming it lands in only one region?

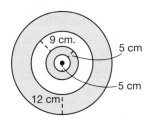

29. **Analyze** Two circles' radii have a ratio of 4:7. How do the measures of the
(87) circumference and area of the larger circle compare to the circumference
and area of the smaller circle?

30. Reflect the triangle across line t, and then reflect the image across line m.
(90)

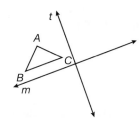

Volume Ratios of Similar Solids

1. **Vocabulary** A polyhedron with a rectangular base and four triangular
(49) faces that meet at a vertex is called a _____.

2. Determine the volume of this solid.
(70)

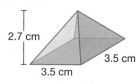

2.7 cm

3.5 cm

3.5 cm

3. **Multiple Choice** What is the height of this solid?
(70)

A 2.7 in.   **B** 3.9 in.

C 4.1 in.   **D** 4.4 in.

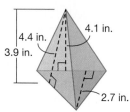

4.1 in.

4.4 in.

3.9 in.

2.7 in.

New Concepts

Theorem 87-1 gives the relationship between similar polygons with
sides that are in an *a:b* ratio. Theorem 99-1 provides a similar rule for
three-dimensional solids.

Theorem 99-1
If two similar solids have a scale factor of *a:b*, then, **1.** the ratio of the perimeters of their corresponding faces is *a:b*. **2.** the ratio of the areas of their corresponding faces is $a^2{:}b^2$. **3.** the ratio of their volumes is $a^3{:}b^3$.

Math Reasoning

Connect What would you
call the transformation
that maps the points of
a solid to the points of a
larger, similar solid?

Notice that the first two parts of this theorem are nearly identical to the
two parts of Theorem 87-1.

Example 1 **Corresponding Perimeters of Similar Solids**

Two similar rectangular pyramids have
a scale factor of 3:2. Determine the
perimeter of the smaller pyramid's base.

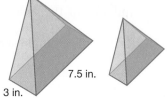

7.5 in.

3 in.

SOLUTION
First, determine the perimeter of the larger
pyramid's base.

$P = 7.5 + 7.5 + 3 + 3$
$P = 21$

By Theorem 99-1, the perimeter of the smaller pyramid's base is in a
3:2 ratio with the perimeter of the larger pyramid's base.

$$\frac{21}{P} = \frac{3}{2}$$

$P = 14$

So the perimeter of the smaller pyramid's base is 14 inches.

The second part of Theorem 99-1 implies that the total surface area of two similar solids with a scale factor of $a{:}b$ will be $a^2{:}b^2$.

Example 2   Finding Surface Areas of Similar Solids

The surface area of the smaller pyramid shown is 54 square centimeters. What is the surface area of the larger pyramid?

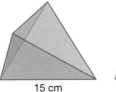

15 cm   6 cm

SOLUTION
Since the pyramids are similar, we can apply Theorem 99-1. From the dimensions given, the ratio of the solids' sides is 15:6, or 5:2. Apply Theorem 99-1.

$$\frac{5^2}{2^2} = \frac{A}{54}$$
$$A = 337.5$$

The surface area of the larger pyramid is 337.5 square centimeters.

Example 3   Finding Volumes of Similar Solids

(**a.**) The two cylinders shown are similar. If the volume of the smaller cylinder is 38 cubic feet, what is the volume of the larger cylinder?

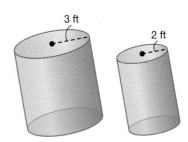

3 ft   2 ft

SOLUTION
Looking at the radii, we can see that the scale factor is 3:2. By Theorem 99-1, the ratio of their volumes will be $\frac{3^3}{2^3}$. Write a proportion.

$$\frac{3^3}{2^3} = \frac{V}{38}$$
$$V = 128.25$$

The volume of the larger cylinder is 128.25 cubic feet.

(**b.**) Prove the third part of Theorem 99-1 for any pair of similar pyramids.

SOLUTION
Suppose similar pyramids K and L have base areas B_K and B_L and heights h_K and h_L. Since the base areas and heights are of corresponding parts, they are in the ratios $a^2{:}b^2$ and $a{:}b$, respectively. Apply the formula for volume of a pyramid.

$$V_K = \frac{1}{3}B_K h_K \qquad \text{Volume of a pyramid}$$

$$V_K = \frac{1}{3}\left(\frac{a^2}{b^2}B_L\right)\left(\frac{a}{b}h_L\right) \qquad \text{Substitute.}$$

$$V_K = \frac{a^2}{b^2}\left(\frac{a}{b}\right)\left(\frac{1}{3}B_L h_L\right) \qquad \text{Simplify.}$$

$$V_K = \frac{a^3}{b^3}V_L \qquad \text{Substitute.}$$

Lesson 99   **643**

Math Reasoning

Connect The radius of a balloon grows at a constant rate of 1 inch per second. How does its volume grow — at a constant rate, an increasing rate, or a decreasing rate?

Example 4 **Application: Space Exploration**

A proposed crew capsule for space exploration is shaped like a square pyramid. A scale model of the capsule has base sides of 26 inches each and a height of 15 inches. Determine the volume of the actual capsule, which will be dilated by a factor of 9, to the nearest cubic foot.

SOLUTION

The scale factor is 9:1. First, find the volume of the square pyramid. The area of the base is 676, and the height is 15.

$$V = \frac{1}{3}Bh$$

$$V = \frac{1}{3}(676)(15)$$

$$V = 3380$$

Now apply Theorem 99-1 by writing a proportion.

$$\frac{9^3}{1^3} = \frac{V}{3380}$$

$$V = 2,464,020$$

This gives the volume in cubic inches. There are 12 inches in a foot, so to find the volume in cubic feet, divide by 12^3. The actual capsule will have a volume of approximately 1426 cubic feet.

Hint

Simply dividing by 12 to convert from inches to feet would give an incorrect answer in this example. Since the answer is in cubed units, you need to divide by the cube of the 12.

Lesson Practice

These two similar right triangular prisms have a scale factor of 3:4.

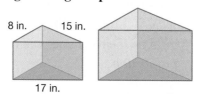

8 in.   15 in.

17 in.

a. Determine the perimeter of the larger prism's base.
(Ex 1)

b. If the surface area of the smaller prism is 182 square inches, find the surface area of the larger prism.
(Ex 2)

c. If the volume of the smaller prism is 323 cubic inches, find the volume of the larger prism.
(Ex 3)

d. (**Space Exploration**) The first booster stage of the Saturn 1B moon rocket consists of a cylindrical tank section and five rocket motors. In a 1:50 scale model of the booster stage, the tank is 59 centimeters high with a diameter of 20 centimeters. Determine the volume of the actual booster stage, to the nearest cubic meter.
(Ex 4)

* 1. Find x. Round your answer to the nearest tenth.
(98)

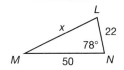

2. Multi-Step Find the perimeter of a 30°-60°-90° triangle where the
(56) longer leg is 630 millimeters. Give your answer in simplified
radical form.

3. (Plumbing) A pipe junction consists of three cylindrical pipe
(93) sections positioned at right angles to one another. Using the
three orthographic views, sketch the pipe junction.

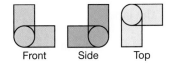

* 4. Two similar right rectangular prisms R and S have a scale
(99) factor of 3:4. The base of R is 3 centimeters by 4.5 centimeters and its
height is 9 centimeters. Determine the area of each face of S.

* 5. Two similar rectangular pyramids P and Q have a scale factor
(99) of 4:5. The base of P is 6 inches long and 4 inches wide. Each
sloping edge of P is 10 inches long. Find the edge lengths of Q and
their sum.

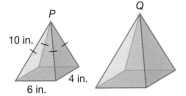

6. Predict A family of lines have equations of the form
(16) $x - ay + a^2 = 1$.
  a. Describe the line corresponding to $a = 0$.
  b. What happens as a increases?

7. Find θ. Round your answer to the nearest degree.
(94)

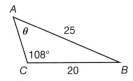

8. The ratio of the lengths of a trapezoid's bases is 2:1. The
(96) trapezoid's height is equal to the length of its long base. If the
height of the trapezoid is doubled, what is the ratio of the new
trapezoid's area to the original trapezoid's area?

* 9. (Mapping) The map shows the distance between three landmarks. Find all three angles.
(98) Round each measure to the nearest tenth.

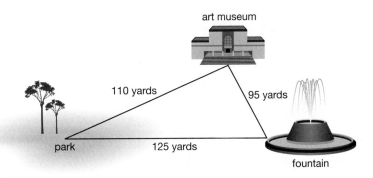

10. The sides of two similar polygons have a similarity ratio of 8:13. What is the ratio
(87) of the perimeter of the smaller polygon to the perimeter of the larger polygon?
What is the ratio of the area of the smaller polygon to that of the larger polygon?

*11. ₍₉₇₎ **Compact Discs** The length of a song on a compact disc is proportional to the area of the disc it takes up. The radius of the central hole, where no music can be stored, is 7.5 millimeters. If a ten-minute song takes up the shaded region in the diagram at right, how many minutes of music can the entire disc hold, to the nearest minute?

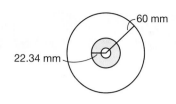

12. ₍₆₂₎ **Canning** A particular can of tuna has a height of 1 inch and a diameter of 3 inches. The label on the can covers the entire lateral side. What is the area of the label?

13. ₍₆₉₎ Find the lengths of the sides of kite *ABCD*. Round to the nearest tenth.

*14. ₍₉₈₎ **Algebra** Find the angles of a triangle if the sides are x, $2x$, and $\frac{3}{2}x$.

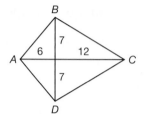

15. ₍₅₀₎ Find the geometric mean of 12.1 and 5.6.

16. ₍₄₀₎ **a.** Determine this figure's perimeter.
  b. Determine its area.

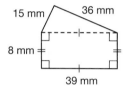

17. ₍₉₁₎ **Formulate** Write an identity for tangent in terms of only cosine.

*18. ₍₉₉₎ **Analyze** Prove that if two regular triangular pyramids have a scale factor of $a{:}b$, then the ratio of the areas of their corresponding faces is $a^2{:}b^2$.

19. ₍₃₉₎ List the side lengths of the triangle from smallest to largest.

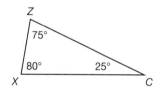

20. ₍₉₇₎ **Baking** A bagel with a diameter of 12 inches has a hole with a radius of 1 inch. What is the area of one face the bagel, to the nearest tenth of a square inch?

21. ₍₃₄₎ **Algebra** Find an expression for the height of parallelogram *PTRW* in terms of x.

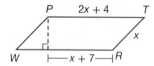

22. ₍₆₁₎ The measures of three interior angles of a parallelogram are 133°, 57°, and 133°. What is the measure of the fourth interior angle?

23. Error Analysis Luka wrote the following as she was trying to prove
(45) that △RTF is equilateral. Find and correct any errors she made.

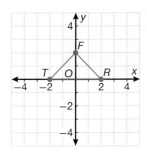

$TF = \sqrt{2^2 + 2^2} = \sqrt{4 + 4} = \sqrt{16} = 4$

$FR = \sqrt{2^2 + 2^2} = \sqrt{4 + 4} = \sqrt{16} = 4$

$TR = 4$

24. Model The price of eggs is at least a dollar less than three times the price
(88) of milk. Write a linear inequality that models this relationship and sketch
the graph of it.

*25. These two similar triangular pyramids have a scale factor 2:5. Determine
(87) their respective surface areas.

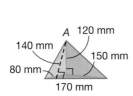
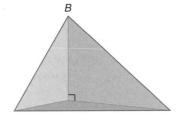

26. Multiple Choice R is an exterior point of $\odot C$, and $\overline{RP}$ and $\overline{RQ}$ are tangent to $\odot C$ at P
(58) and Q. Given that $\angle PCQ$ is a right angle, what type of quadrilateral is $PCQR$?

A kite   **B** parallelogram

C square   **D** none of these

27. The equation of a circle is $x^2 + y^2 = 121$. Write the equation of this circle after it
(95) is translated 6 units to the left. Draw the original circle and the translated circle on
the coordinate plane.

28. Multiple Choice Which is closest to the length of $\overline{RS}$?
(94)

A 24   **B** 14

C 12   **D** 17

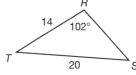

29. Algebra If a regular hexagon with 40-inch side lengths
(66) were divided into six congruent equilateral triangles with
vertices connecting at the center, what would be the length
of their corresponding congruent sides?

30. Prove that this quadrilateral is a trapezoid.
(92)

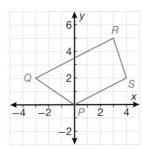

Transformation Matrices

Warm Up

1. **Vocabulary** A combination of transformations is
(90)
called a _____.

2. Add the vectors $\langle 2, 3 \rangle$ and $\langle -1, 0 \rangle$.
(83)

3. **Multiple Choice** A polygon is reflected across the line $x = y$. Which
(74)
transformation mapping shows this transformation?

    A $T: (x, y) \rightarrow (y, -x)$ **B** $T: (x, y) \rightarrow (-x, y)$

    C $T: (x, y) \rightarrow (x, -y)$ **D** $T: (x, y) \rightarrow (y, x)$

New Concepts

A **matrix** is an ordered, rectangular arrangement of numbers. The matrix below is a 3×4 matrix, because it has three rows and four columns.

$$\begin{bmatrix} 1 & 2 & 5 & 0 \\ 3 & 4 & 1 & 2 \\ 0 & -1 & 2 & 3 \end{bmatrix}$$

Corresponding numbers in two or more matrices are those that are in the same row and column in each matrix. To add matrices, simply add the corresponding components together.

Hint

The dimensions of a matrix are always given with the number of rows first and the number of columns second.

Example 1   Adding Matrices

Add the two matrices.

$$\begin{bmatrix} 2 & 3 \\ 0 & -1 \end{bmatrix} + \begin{bmatrix} -1 & 4 \\ 3 & 0 \end{bmatrix}$$

SOLUTION

To solve, add the components of each matrix.

$$\begin{bmatrix} 2 & 3 \\ 0 & -1 \end{bmatrix} + \begin{bmatrix} -1 & 4 \\ 3 & 0 \end{bmatrix} = \begin{bmatrix} 2 + (-1) & 3 + 4 \\ 0 + 3 & (-1) + 0 \end{bmatrix} = \begin{bmatrix} 1 & 7 \\ 3 & -1 \end{bmatrix}$$

A **point matrix** is a matrix that represents the coordinates of the vertices of a polygon or a line segment. The first row of a point matrix contains the x-values, and the second row contains the y-values.

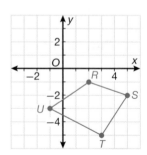

The quadrilateral $RSTU$ has vertices $R(2, -1)$, $S(5, -2)$, $T(3, -5)$, and $U(-1, -3)$.

The point matrix for this quadrilateral is given below.

$$\begin{bmatrix} 2 & 5 & 3 & -1 \\ -1 & -2 & -5 & -3 \end{bmatrix}$$

Example 2 Adding a Matrix to a Point Matrix

Write a point matrix for the line segment $\overline{AB}$. Add the point matrix to the matrix given below.

$$\begin{bmatrix} 1 & 1 \\ 1 & 1 \end{bmatrix}$$

Finally, graph the line represented by the new matrix.

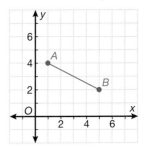

SOLUTION

The point matrix for $\overline{AB}$ is $\begin{bmatrix} 1 & 5 \\ 4 & 2 \end{bmatrix}$

Add the matrices.

$$\begin{bmatrix} 1 & 5 \\ 4 & 2 \end{bmatrix} + \begin{bmatrix} 1 & 1 \\ 1 & 1 \end{bmatrix} = \begin{bmatrix} 1+1 & 5+1 \\ 4+1 & 2+1 \end{bmatrix} = \begin{bmatrix} 2 & 6 \\ 5 & 3 \end{bmatrix}$$

Finally, graph the coordinates given by the new point matrix.

The new line is a translation of the original. The translation is one unit up and one unit right from the original.

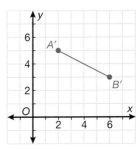

> **Math Reasoning**
>
> **Predict** If you add the point matrix to the matrix $\begin{bmatrix} 2 & 2 \\ 2 & 2 \end{bmatrix}$, what do you predict will happen to the line segment?

You can see from Example 2 that matrices are used to describe transformations on the coordinate plane. Adding a matrix to a point matrix results in a translation of the line segment.

Example 3 Translating with Matrices

The triangle XYZ is translated three units right and two units down. Find the matrix that transforms $\triangle XYZ$ to $\triangle X'Y'Z'$.

SOLUTION

Start by representing each triangle with a point matrix.

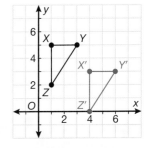

$$\triangle XYZ = \begin{bmatrix} 1 & 3 & 1 \\ 5 & 5 & 2 \end{bmatrix} \qquad \triangle X'Y'Z' = \begin{bmatrix} 4 & 6 & 4 \\ 3 & 3 & 0 \end{bmatrix}$$

Now subtract the matrices by component to find the translation matrix.

$$\begin{bmatrix} 4 & 6 & 4 \\ 3 & 3 & 0 \end{bmatrix} - \begin{bmatrix} 1 & 3 & 1 \\ 5 & 5 & 2 \end{bmatrix} = \begin{bmatrix} 3 & 3 & 3 \\ -2 & -2 & -2 \end{bmatrix}$$

As you can see, all the x-components are translated by 3, and all the y-components are translated by -2. The triangle is translated 3 units right and 2 units down.

| Example | 4 | **Application: Retractable Bridge** |

Two halves of a bridge can be raised to allow boats through.
When the bridge is down, each half extends 40 feet over the water to
meet in the middle. When they are up, they extend 40 feet directly
up into the air. Write a translation vector for the movement of the
end of the left half of the bridge from a fully raised position to a fully
extended position.

SOLUTION

The end of the left half of the bridge is translated 40 units down and 40
units right. Since we are only interested in the translation of one point on
the bridge, the translation matrix will be 2 × 1.

$$\begin{bmatrix} 40 \\ -40 \end{bmatrix}$$

Lesson Practice

a. Add the matrices.
(Ex 1)
$$\begin{bmatrix} 2 & 0 \\ -2 & 1 \end{bmatrix} + \begin{bmatrix} 0 & 4 \\ 3 & -1 \end{bmatrix}$$

b. Find a point matrix for the line segment.
(Ex 2)
Then add the transformation matrix $\begin{bmatrix} -1 & -1 \\ -1 & -1 \end{bmatrix}$
and describe the change.

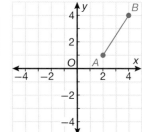

c. A line segment is translated three units left and
(Ex 3) one unit down. Write the transformation in a
matrix.

d. (**Amusement Parks**) A Ferris wheel has seats that
(Ex 4) are always horizontal. A rotation of the wheel
translates one of the cars 6 meters right and 5 meters down.
Write a translation matrix to represent this transformation.

Practice   Distributed and Integrated

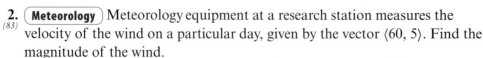

* **1.** In the concentric circles shown, find the area of the annulus in terms
(97) of π.

2. (**Meteorology**) Meteorology equipment at a research station measures the
(83) velocity of the wind on a particular day, given by the vector $\langle 60, 5 \rangle$. Find the
magnitude of the wind.

3. Identify the type of transformation shown at right.
(67)

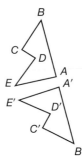

4. Multiple Choice A pair of similar solids A and B have a scale factor of 3:5.
(99) Identify the correct equation for the surface area of B in terms of the surface area of A.

A $S_B = \frac{3}{5}S_A$ **B** $S_B = \frac{25}{9}S_A$

C $S_B = \frac{9}{25}S_A$ **D** $S_B = \frac{125}{27}S_A$

* 5. Find the translation matrix for $\triangle LMN$.
(100)

 6. Algebra If $\cos^2\theta$ equals 0.4, what does $\sin\theta$ equal?
(91)

 7. Algebra What is the arc measure of $\overparen{JK}$ if $v = 3.2$?
(47)

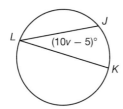

8. Two segments are divided proportionally. The first segment is divided into parts
(60) that are 12, 18, and x units long. The second segment is divided into parts that are
y, 9, and 13 units long. What are the values of x and y?

9. Where is the orthocenter of a right triangle located?
(32)

*10. (**Dancing**) A certain type of dance uses the step pattern of three steps forward and
(100) three steps to the right. If everyone is dancing in the same direction, what 2×1
matrix could be used to define this step?

11. Multiple Choice This orthographic drawing shows the front, top, and
(93) side views of a solid. Which of the sketches below represents the solid?

Front   Side   Top

A   **B**

C   **D**

12. Write an equation for $\odot C$ with a 2.5 radius, centered at the origin.
(75)

*13. These two similar cylinders have a scale factor of 3:2. Determine
(99) the base radius and height of cylinder D.

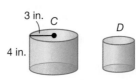

14. Find the horizontal and vertical components of a vector with a
(89) magnitude of 15 and an angle of 37°.

15. Sketch the net of a pentagonal pyramid.
(Inv 5)

*16. **Multi-Step** Find all angle measures in the triangle. Round each answer
(98) to the nearest degree.

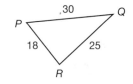

17. A quadrilateral has diagonals that are perpendicular, but are not
(52) the same length. Classify the quadrilateral.

*18. (**Visual Arts**) A sculpture consists of two similar square pyramids with a scale factor
(99) of 3:8. The smaller pyramid has a base that is 30 millimeters long and each of its
triangular faces has a 36-millimeter height. Determine the area of each
face of the larger pyramid.

19. Determine how many common tangents the internally tangent
(72) circles have and draw the tangent(s).

*20. **Write** Which law can be used to find the side lengths of a triangle,
(98) given the three angle measures—the Law of Sines, the Law of
Cosines, both, or neither? Explain.

21. (**Cheese**) This equilateral triangular prism represents a wedge of cheese
(59) that is to be coated with wax to prevent it from drying out. What is the
exact surface area that needs to be coated with wax?

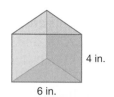

4 in.

22. **Formulate** Determine a formula for the total surface area of a
(70) square pyramid with a side length of s and a slant height of l.

6 in.

*23. **Error Analysis** Donna wrote the matrix $\begin{bmatrix} -3 & 1 \\ -3 & 1 \end{bmatrix}$ for the translation three units
(100) left and one unit up. Explain Donna's error and correct the matrix.

24. How can the equation $y^2 + 1 = x$ be altered to make it symmetrical along
(76) the x-axis?

25. If parallelogram $QRST$ is similar to parallelogram $JKLM$, $JK = 12$, $QR = 45$, and
(44) $LM = 15$, find ST.

*26. **Formulate** A circle is given by the equation $(x + 3)^2 + (y - 2)^2 = 9$.
(95) Another circle is a dilation of the first circle by a factor of 3.
What is the equation of the second circle?

*27. **Algebra** A translation moves a line segment up by $2x$ units and then down by
(100) $x - 1$ units. Write the matrix for this translation.

28. **Multi-Step** Find the surface area of a sphere with a
(80) circumference of 28 inches, to the nearest square inch.

29. What is the value of y in the proportion $\frac{15}{45} = \frac{33}{y}$?
(41)

30. In the given diagram, if P is the center of the circle and $\overline{BC}$ and $\overline{AD}$ are
(27) diameters, prove that $\triangle PAB \cong \triangle PCD$, using a two-column proof.

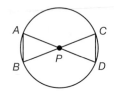

Fractals

Iteration is the repeated application of the same rule. A **fractal** is a figure that is generated by iteration. Each part of the figure is reduced in size compared to the previous part. For example, start with a single line segment and apply the rule: add half-length segments at one end of the original segment, at angles of 135°.

135°

→

In the second iteration, the same rule is applied to the new segments, as shown. The diagrams show the figure through 4 iterations.

135°

Notice that each branch of the fractal is self-similar to the entire fractal. A **self-similar figure** is a figure that can be divided into parts, each of which is similar to the entire figure. All fractals are self-similar.

The rule that generates this similar figure is a combination of transformations. In this case, the line segment is dilated, and then it is rotated and translated to the ends of the original line.

One famous fractal is the **Sierpinski triangle**. Waclaw Sierpinski first described such triangular fractals in 1915. The Sierpinski triangle starts with an equilateral triangle, and is formed by removing an inverted triangle from the interior of the original triangle. This can be done by connecting the original triangle's midsegments, and then removing the central triangle. This process forms four congruent triangles as shown.

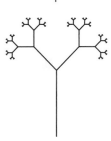

1. Draw a Sierpinski triangle as shown in the diagram. In each iteration, connect the midsegments of the shaded triangles and remove the central triangle that is formed. Use this method to draw the second and third iterations of the Sierpinski triangle.

2. Is the Sierpinski triangle a self-similar figure?

3. Complete this definition: a Sierpinski triangle is a fractal formed from a(n) _____ by removing triangles with vertices at the _____ of the _____ of each remaining _____.

What are the components of a fractal? First, a starting figure is needed. Second, a rule is needed for generating each stage of the fractal, given the previous stage.

 4. Write The diagram shows a rule for generating a fractal from an equilateral triangle. If each new triangle is half the size of the previous one, describe the rule in a complete sentence.

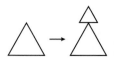

5. Multiple Choice Which of the combinations of transformations would result in a fractal if repeated?

A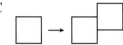

B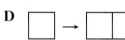

C

D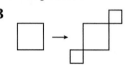

Math Reasoning

Analyze What are some other objects in the natural world that are fractal patterns?

6. The diagram shows a fractal similar to the Sierpinski triangle, where space is removed from the rectangle. Draw the second iteration of this fractal.

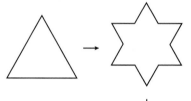

Another well-known fractal is the Koch snowflake. Helge von Koch first described it in 1904. The starting figure is an equilateral triangle. The first iteration and the third iteration are shown.

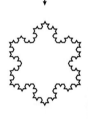

 7. Write Describe the iterating rule used to generate the Koch snowflake.

8. Sketch the second iteration of the Koch snowflake.

9. Complete this definition: the **Koch snowflake** is a fractal formed from a(n) _____ by replacing the middle third of each segment with two _____ that form a _____ angle.

Investigation Practice

a. The diagram shows the first two iterations of a fractal pattern. Draw the third iteration.

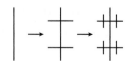

b. Multiple Choice Which of the sets of transformations shown would result in a fractal if repeated?

A

B

C

D

c. Write Describe the series of transformations that form the iterations of the fractal shown.

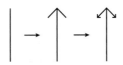

d. The first iteration of a fractal is shown. This fractal is produced by removing figures, much like the Sierpinski triangle. Sketch the second iteration of the fractal.

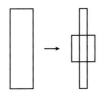

Determining Lengths of Segments Intersecting Circles

1. Vocabulary A line that intersects a circle at two points is called a _____
(43) line. (**tangent, secant, chord**)

2. Find the value of *x* in the diagram.
(41)

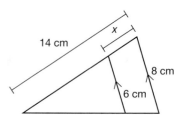

14 cm

x

8 cm

6 cm

3. Multiple Choice What is the value of $\overline{UW}$?
(86)

A 8	**B** 12
C 16	**D** 20

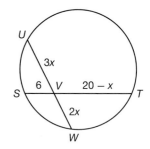

U

3x

S 6 *V* 20 − *x* *T*

2x

W

New Concepts A segment of a secant with at least one endpoint on the circle is a **secant segment**. An **external secant segment** is a secant segment that lies in the exterior of a circle with one endpoint on the circle.

Theorem 101-1
If two secant segments are drawn to a circle from an external point, the product of the length of one secant segment and that of its external segment is equal to the product of the other secant segment length and that of its external segment.   In the diagram, $(AC)(AB) = (AE)(AD)$.

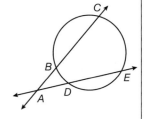

C

B

A *D* *E*

Example 1 Intersection by Two Secants

Determine the value of *x* in the diagram.

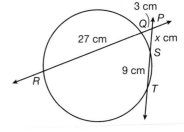

3 cm

Q *P*

27 cm

x cm

S

9 cm

R

T

SOLUTION

From Theorem 101-1:

$$(PT)(PS) = (PR)(PQ)$$
$$(9 + x)x = 30(3)$$
$$9x + x^2 = 90$$
$$x^2 + 9x - 90 = 0$$
$$(x - 6)(x + 15) = 0$$
$$x = 6 \text{ or } x = -15$$

Since a length cannot be negative, *x* = 6 centimeters.

Example 2 Proving Theorem 101-1

Given: Secant segments $\overline{QM}$ and $\overline{PM}$
Prove: $(QM)(NM) = (PM)(RM)$

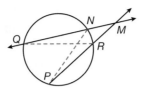

SOLUTION

Draw auxiliary line segments $\overline{QR}$ and $\overline{PN}$. By Theorem 47-3, $\angle MQR \cong \angle MPN$, since both intercept $\overset{\frown}{NR}$. By the Reflexive Property of Congruence, $\angle M \cong \angle M$. Therefore, $\triangle MQR \sim \triangle MPN$ by Angle-Angle Similarity. The corresponding parts of similar triangles are proportional, so $\frac{QM}{PM} = \frac{RM}{NM}$. The cross product is $(QM)(NM) = (PM)(RM)$.

A **tangent segment** is a segment of a tangent with one endpoint on the circle. A special case of Theorem 101-1 is the case where one secant segment is replaced with a tangent segment, as described in Theorem 101-2.

Theorem 101-2

If one secant segment and one tangent segment are drawn to a circle from an external point, the product of the length of the secant segment and that of its external segment is equal to the length of the tangent segment squared.

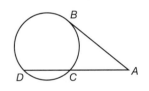

In the diagram, $(AD)(AC) = AB^2$.

Math Reasoning

Formulate If two tangents to a circle were to be constructed from a common exterior point, would the square of the length of one tangent be equal to the square of the length of the second tangent?

Example 3 Intersection by Secant and Tangent

Determine the value of x in the diagram.

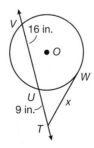

SOLUTION
From Theorem 101-2:

$$TW^2 = (TV)(TU)$$
$$x^2 = (25)(9)$$
$$x^2 = 225$$
$$x = 15$$

The length of the tangent segment is 15 inches.

Example 4 Finding the Distance to the Horizon

In the diagram shown, an observer in a
hot-air balloon is 2.5 miles above the earth's
surface. Assume the earth is a sphere, with a
diameter of approximately 7920 miles. Find
the distance from O to H to determine the
distance between the person in the balloon
and the horizon as shown.

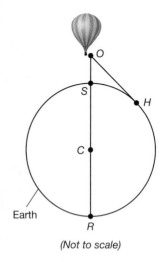

(Not to scale)

SOLUTION

Use Theorem 101-2

$(OH)^2 = (OR)(OS)$

$(OH)^2 = (7922.5)(2.5)$

$(OH)^2 = 19806.25$

$OH = 140.7$

Therefore, the distance from the hot-air balloon to the horizon
is approximately 141 miles.

Hint

The length of *OR* can
be found by adding the
diameter of the Earth to
the height of the balloon.

Lesson Practice

a. Determine the value of x in the diagram.
(Ex 1)

b. Prove Theorem 101-2.
(Ex 2)
  Given: Secant segment $\overline{AC}$, tangent segment $\overline{DC}$
  Prove: $(AC)(BC) = (DC)^2$

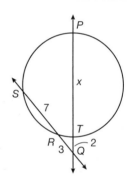

c. Determine the value of x in the diagram.
(Ex 3)

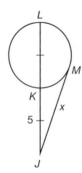

d. If the earth's diameter is 7920 miles, find the distance to the horizon,
(Ex 4)
  to the nearest tenth of a mile, for an unobstructed view from the top
  of a 200-foot apartment building.

* 1. Find the length of $\overline{EG}$ to the nearest tenth of a unit.
(101)

* 2. What shape is made by the graph of the solution to $|x| + |y| \leq 2$?
(88)

3. A boat travels to $\langle 3, 4 \rangle$ using its own power. However, the wind pushes it by $\langle 1, -1 \rangle$. What is the final position of the boat?
(63)

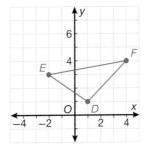

* 4. Multi-Step Find the matrix representation of the triangle shown. Then, write a matrix to translate the triangle 3 units to the right. Finally, write the matrix for the translated triangle.
(100)

5. A quadrilateral is rotated 99° in a clockwise direction, and then rotated 231° in a counterclockwise direction about the origin. What single rotation is equivalent to these rotations?
(78)

6. Algebra Two lines are parallel. One line has an equation of $y = \frac{2}{k}x + 7$ and the other line has an equation of $2y = \frac{h}{6}x + 4$. Find possible values for k and h.
(37)

7. Determine the midpoint M of $\overline{AB}$.
(11)

8. Express $\tan \theta$ in terms of $\sin \theta$.
(91)

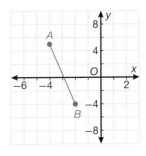

* 9. (**Packaging**) A flour package in the shape of a right triangular prism comes in two sizes. The smaller size has right triangular bases with leg lengths of $\frac{3}{4}$-inch and 1-inch, and is 4.5 inches long. The larger package has a similar shape to the smaller one, but is twice the size. Find the face areas of the larger package.
(99)

*10. Find the exact value of x in the diagram if p is tangent to the circle.
(101)

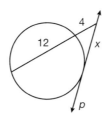

11. Draw a 45°-45°-90° triangle with 1-unit long legs. Include all angle measures and side lengths.
(Inv 7)

*12. Find θ in the triangle shown. Round your answer to the nearest degree.
(98)

13. Find the matrix that represents the translation of a line segment five units left and three units up.
(100)

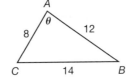

*14. In the concentric circles, find the area of the annulus in terms of π.
(97)

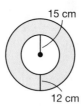

15 cm

12 cm

15. **Analyze** A system of equations with one solution is $y = ax - 1$, $y = bx + 1$, where
(81) $a > b > 0$. Describe where the intersection of these two lines will be on the coordinate plane.

*16. **Multiple Choice** Which is the length of $\overline{AB}$?
(101)

    A $4\sqrt{2}$ centimeters   **B** 32 centimeters

    C 24 centimeters   **D** $24\sqrt{2}$ centimeters

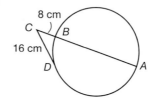

8 cm

16 cm

17. In $\triangle PQR$, $PQ > QR$. Can $\triangle PQR$ be equiangular?
(51) Explain.

*18. **Algebra** If a line is translated $3x$ units right and $2x$ units down,
(100) what is the matrix for the translation?

19. Find the length of $\overline{EF}$. Round your answer to the nearest
(94) hundredth unit.

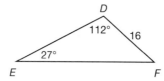

D

112° 16

27°

E   F

20. Four of the interior angles of a convex pentagon measure 100°. What is the
(Inv 3) measure of the fifth angle?

*21. Find the length of $\overline{HJ}$ in the figure.
(101)

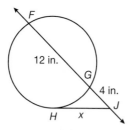

F

12 in.

G

4 in.

J

H x

22. **Multi-Step** Determine the difference in the perimeters of two 45°-45°-90° triangles
(53) if one has a hypotenuse of 20 feet and the second has a hypotenuse of 10 feet.
Round your answer to the nearest foot.

23. These two similar right rectangular prisms have a scale factor of 4:9.
(99) Determine the area of each face of *S*.

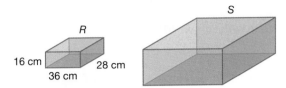

24. Find the distance from $(-6, -2)$ to the line $y = -5$.
(42)

25. Multiple Choice Triangle *LMN* has an area of 13 square units. Given the
(57) coordinates of $L(-1, 2)$ and $M(1, -1)$, which two choices cannot be a possible location for *N*?

A $(7, 3)$ **B** $(5, 6)$
C $(-5, 5)$ **D** $(-6, -2)$

26. The points $H(3, -5)$ and $I(-5, -8)$ are translated 11 units down. Determine
(71) the coordinates of each translation point.

27. Algebra Find the area of the circular cross section of a cone that is parallel to the
(85) cone's base and is two-thirds of the way from the vertex to the base of the cone. The height of the cone is 3 centimeters. The radius of the base is 3 centimeters. Round your answer to the nearest hundredth square centimeter.

*28. Predict Describe the effect on the area of a square if its diagonals were dilated
(96) by $\frac{1}{4}$.

29. Multiple Choice Which point is the midpoint of the line segment
(11) joining *M* and *N*?

A *R*   **B** *S*
C *T*   **D** *U*

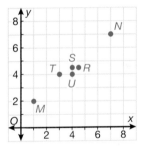

30. (Carpentry) Find the length of metal needed to make the pair of braces
(65) in the shelving unit shown if one diagonal brace is 6 feet long.

Exploring Secant Segments Using Geometry Software

Technology Lab 13 (Use with Lesson 101)

Lesson 101 shows you how to determine lengths of segments that intersect circles. In this lab, you will use geometry software to examine the relationship between segments that are formed by intersecting secants in circles.

(1.) Use the circle tool to construct ⊙A.

(2.) Use the segment tool to construct the secant segment $\overline{BC}$, with point B on ⊙A and point C exterior to the circle. Draw point D where this secant segment intersects ⊙A.

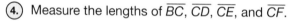

(3.) Construct another secant segment $\overline{EC}$, with E on ⊙A. Draw point F where this secant segment intersects ⊙A.

(4.) Measure the lengths of $\overline{BC}$, $\overline{CD}$, $\overline{CE}$, and $\overline{CF}$.

(5.) Record the lengths of these segments in the chart below.

(6.) Complete the last two columns of the chart by finding the indicated products.

(7.) Drag C to change the lengths of the secant segments and record the new values in the second row of the chart. Repeat this step to fill in the third row of the chart.

(8.) Using your chart as a guide, what conjecture can you make about the lengths of the segments formed by intersecting secants of a circle?

Caution

If you drag C too far, your segments will not intersect the circle. If they no longer intersect the circle, your results will be inaccurate.

BC	CD	CE	CF	BC • CD	CE • CF

Lab Practice

Construct a new circle with secants that intersect outside the circle. Measure the length of the two secant segments and one external secant segment. Can you determine the length of the fourth segment from what you know of the other three?

Dilations in the Coordinate Plane

Warm Up

1. *(67)* **Vocabulary** A transformation that shifts every point of a figure the same distance in the same direction is a _____.

2. *(84)* Draw the image of the triangle shown after a dilation by a scale factor of 2.

3. *(67)* **Multiple Choice** Which of the following transformations does not preserve the size of a figure?

   A rotation   **B** reflection

   C translation   **D** dilation

New Concepts Recall that a dilation is a transformation that changes a figure's size but not its shape. One convenient way to apply a dilation to a figure in the coordinate plane is by using a point matrix.

Remember that the mapping notation for a dilation by a factor of k, with the origin O as the center, is written, $D_{O, k}(x, y) \rightarrow (kx, ky)$.

Example 1 | **Dilating on the Coordinate Plane**

Determine the result of the dilation $D_{0, 3}(x, y)$ on the points $(3, 4)$ and $(-1, 5)$.

SOLUTION

Applying this dilation we have:

$D_{O, 3}(3, 4) \rightarrow (3 \times 3, 3 \times 4)$, which is the point $(9, 12)$.
$D_{O, 3}(-1, 5) \rightarrow (3 \times -1, 3 \times 5)$, which is the point $(-3, 15)$.

A dilation can be applied to a polygon by creating a point matrix of the polygon's vertices and multiplying by the scale factor. To multiply a matrix by a number, simply multiply each of the matrix's elements by the number.

Example 2 | **Dilating by Point Matrix**

Write a point matrix for the triangle shown and use this point matrix to find the coordinates of the image after a dilation with a scale factor of 5.

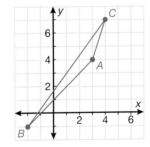

Hint

Recall that a point matrix is a 2-row matrix where each column is the x and then the y coordinate of one point in the figure.

SOLUTION

The point matrix for the triangle is $\begin{bmatrix} 3 & -2 & 4 \\ 4 & -1 & 7 \end{bmatrix}$.

To find the vertices of the image, multiply the point matrix by the scale factor, 5.

$$5 \cdot \begin{bmatrix} 3 & -2 & 4 \\ 4 & -1 & 7 \end{bmatrix} = \begin{bmatrix} 15 & -10 & 20 \\ 20 & -5 & 35 \end{bmatrix}$$

Therefore, the new coordinates for the vertices of the dilated triangle are $(15, 20)$, $(-10, -5)$, and $(20, 35)$.

Example 3 Application: Photographic Enlargement

Basim is enlarging a photograph by a scale factor of 2.5, and then placing a 2-inch-wide mat around the perimeter of the enlarged photograph. If the scale on the graph is in inches, find the area of the mat.

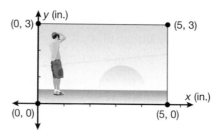

SOLUTION

The point matrix for the photograph is $\begin{bmatrix} 0 & 0 & 5 & 5 \\ 0 & 3 & 3 & 0 \end{bmatrix}$. Multiply by the scale factor:

$2.5 \times \begin{bmatrix} 0 & 0 & 5 & 5 \\ 0 & 3 & 3 & 0 \end{bmatrix} = \begin{bmatrix} 0 & 0 & 12.5 & 12.5 \\ 0 & 7.5 & 7.5 & 0 \end{bmatrix}$. Therefore, the vertices

of the enlarged photo are $(0, 0)$, $(0, 7.5)$, $(12.5, 7.5)$, and $(12.5, 0)$.

The area of the picture can be found by multiplying the length and the width:

$A = l \times w$
$A = (12.5)(7.5)$
$A = 93.75 \text{ in}^2$

A 2-inch mat that surrounds the picture would extend the vertices two inches in all directions, with vertices $(-2, -2)$, $(-2, 9.5)$, $(14.5, 9.5)$, and $(14.5, -2)$. So, the combined area of the picture and the surrounding mat is:

$A = l \times w$
$A = (16.5)(11.5)$
$A = 189.75 \text{ in}^2$

The area of the mat is the difference between the combined area of the picture and the mat and the area of the picture alone:

$189.75 - 93.75 = 96$

Therefore, the area of the mat is 96 square inches.

Hint

Use the distance formula to find the length and width of the enlarged rectangle and the enlarged rectangle with a 2-inch mat.

Lesson Practice

a. Determine the result of the dilation $D_{O, 2.5}(x, y)$ on the points
(Ex 1) $(10, -4)$ and $(0, 8)$.

b. Write a point matrix for the rectangle
(Ex 2) shown, and then use this point matrix to find the new vertices of the shape after a dilation with a scale factor of 2.

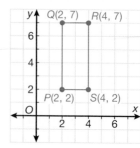

c. Using a point matrix and a scale factor
of 50, determine the actual area of the
planned park in the diagram, if the
scale is in feet.

(Ex 3)

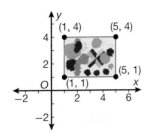

* **1.** **Probability** A target is made up of concentric circles with dimensions
(97) as shown. Assuming it is lying face-up on a flat surface, what is the
probability that a coin dropped onto it will land on the shaded region?

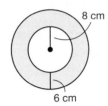

* **2.** Find the image of the point (−4, 5) under a dilation with a scale factor of 7,
(102) centered at the origin.

3. **Algebra** A regular hexagon has an apothem that is $13\sqrt{3}$ feet long. What is
(66) the area of the hexagon?

4. What is the value of *y* in the diagram?
(101)

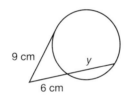

5. **Hiking** A hiker travels along a trail with vertical distance of 6 meters and a
(83) horizontal distance of 50 meters, then climbs up a cliff that rises 75 meters
over a horizontal distance of 12 meters. Write a vector to describe the
displacement of this hike.

* **6.** Find the value of *x*. Round your answer to the nearest hundredth unit.
(98)

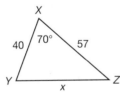

7. **Maps** At right is a partial map of three Texas cities. List the
(39) distances between cities in order from greatest to least.

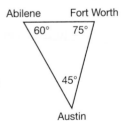

8. **Analyze** A right rectangular prism has a width that is equal to the
(59) length and a height that is twice the length. The volume of the prism
is 432 cubic feet. Find the surface area.

* 9. Find the vertices of the given triangle after the dilation $D_{O,\,0.75}(x, y)$
₍₁₀₂₎ is applied.

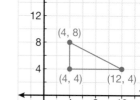

10. Analyze A rectangle is dilated by a factor of 12. If the area of the image
₍₉₆₎ 576 square units, what was the original area?

11. (Water Supply) A large water tank servicing a rural district is in the
₍₆₂₎ shape of a cylinder that is 56 feet tall and has a diameter of 40 feet.
If one cubic foot is equal to about 7.5 gallons, find the volume of the
tank to the nearest thousand gallons.

*12. Find the value of x in the diagram.
₍₁₀₁₎

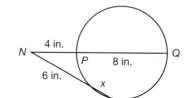

*13. (Space Exploration) A 1:16 scale model of a spacecraft is a
₍₉₉₎ truncated right cone with a base radius of 6.2 inches, a top
radius of 2.1 inches, and a height of 5.0 inches. The height
of the cone before it was truncated was 7.4 inches. Determine
the surface area and volume of the model and the surface area
and volume of the actual spacecraft to the nearest square or
cubic foot, respectively.

*14. Multiple Choice If the triangle shown is dilated with a scale factor centered
₍₁₀₂₎ at the origin so that T' has the coordinates $(9, -6)$, what are the
coordinates of S'?

A $(12, 3)$ **B** $(12, -3)$
C $(-12, 3)$ **D** $(3, 12)$

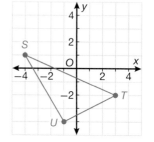

15. Generalize A three-dimensional object is made by adjoining two congruent
₍₇₇₎ cones along their circular faces. If the length from vertex to vertex of the
new shape is h, and the radii of the initial cones is r, determine
an expression for the surface area of this object.

16. Verify In $\triangle ABC$ at right, show that $\overline{BC} \parallel \overline{DE}$, if D and E are the midpoints
₍₅₅₎ of their respective sides. *Hint: Find the slope of each segment.*

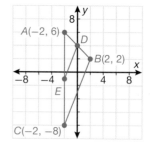

17. If a horizontal vector has a magnitude of x and a vertical vector has a
₍₈₉₎ magnitude of x, what is the angle measure of the resultant vector when
they are summed? What is its magnitude?

18. Find the matrix that represents the translation of a triangle 3 units
₍₁₀₀₎ down and 4 units right.

*19. Multi-Step Find the matrix representation of the line segment. Then,
₍₁₀₀₎ write a matrix to translate the line segment 2 units down. Finally,
write the matrix for the translated line segment.

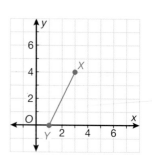

20. Two chords of a circle, $\overline{ST}$ and $\overline{XY}$, intersect at point V. Find VT if
(86) $SV = 5.4$, $XV = 3.9$, and $VY = 1.8$.

21. **Error Analysis** Dantrell wrote a paragraph proof that included
(31) the statement, "By the Vertical Angles Theorem, $\angle 1$ and $\angle 3$ are
supplementary." How can his statement be reworded to make it
a true statement?

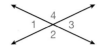

22. Assume the segments that appear to be tangent are tangent. Find
(72) AC, CD, CE, and BC.

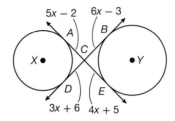

23. Draw two triangles that are similar by Angle-Angle Similarity.
(46)

24. (**Graphics**) Digital images are comprised of thousands of pixels, which are small
(Inv 6) points of light that combine to make the larger picture. A screen that is 8 inches
tall by 13 inches wide displays a photograph of the Earth on a black background.
The radius of the image of the Earth is 3 inches. If a pixel is chosen at random,
what is the probability that it will be a part of the black background?

25. **Multi-Step** For the triangle with vertices $X(0, 0)$, $Y(0, 4)$, and $Z(-2, 0)$, find
(82) the side lengths to the nearest tenth of a unit and the angle measures to the
nearest degree.

xy^2 *26. **Algebra** Show the point matrix multiplication with a scale factor of 3 for
(102) the given image.

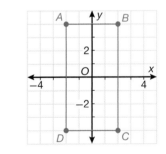

xy^2 27. **Algebra** The angles of a triangle measure $(x^2 - 2x)°$, $(6x + 3)°$, and $12°$.
(18) Write and solve an equation for x. Which solution determines
the angle measures?

28. Find the geometric mean of 35 and 57 to the nearest tenth.
(50)

\ *29. **Write** Explain how the Tangent-Secant Product Theorem is a special case of the
(101) Secant Product Theorem.

30. Which congruence theorem applies to these triangles?
(36)

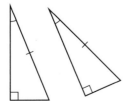

Frustums of Cones and Pyramids

Warm Up

1. **Vocabulary** The number of nonoverlapping unit cubes of a given size
$^{(59)}$ that will exactly fill the interior of a three-dimensional figure is the
_____ of that figure.

2. Find the volume of a cone if the radius is x and the height is $2x$.
$^{(77)}$

3. If the height of a square pyramid is equal to the length of one side
$^{(70)}$ of its base, and its volume is 243 cm^3, find its height.

New Concepts

If the top of a cone or a pyramid is removed, eliminating the figure's vertex, the result is a frustum.

A **frustum of a cone** is a part of a cone with two parallel circular bases.

A **frustum of a pyramid** is a part of a pyramid with two parallel bases.

Math Language

Recall that a **cone** has a circular base and a curved surface that connects the base to a vertex. A **pyramid** has a polygonal base and triangular faces that meet at a common vertex.

Volume of a Frustum

The following formula is used to find the volume of a frustum, regardless of whether it is part of a cone or a pyramid. The variables B_1 and B_2 are the areas of the two bases and h is the height of the frustum.

$$V = \frac{1}{3}h(B_1 + \sqrt{B_1 B_2} + B_2)$$

Example 1 **Finding the Volume of a Frustum of a Pyramid**

Find the volume of the frustum of the pyramid shown.

SOLUTION
Find the area of each base of the frustum.

$B_1 = bh$	$B_2 = bh$
$B_1 = (8)(10)$	$B_2 = (6)(7.5)$
$B_1 = 80$	$B_2 = 45$

Now, apply the formula for volume of a frustum.

$$V = \frac{1}{3}h(B_1 + \sqrt{B_1 B_2} + B_2)$$

$$V = \frac{1}{3}(10)(80 + \sqrt{80 \times 45} + 45)$$

$$V = 616\frac{2}{3}$$

The volume of this frustum is $616\frac{2}{3}$ cubic meters.

Online Connection
www.SaxonMathResources.com

Example 2 Finding the Volume of a Frustum
of a Cone

a. Find the volume of the frustum of the cone to the nearest hundredth
of a cubic inch.

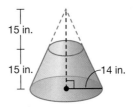

SOLUTION
Notice the two triangles highlighted in the
diagram. Both are right triangles that share
an angle. Therefore, they are similar.
Write a proportion.

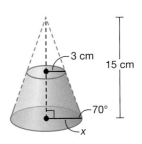

$$\frac{r}{15} = \frac{14}{30}$$

$$r = 7$$

Now find the area of the bases.

$B_1 = \pi r^2$ $B_2 = \pi r^2$
$B_1 = \pi(7)^2$ $B_2 = \pi(14)^2$
$B_1 \approx 153.94$ $B_2 \approx 615.75$

Finally, apply the formula for volume of a frustum.

$$V = \frac{1}{3}h(B_1 + \sqrt{B_1 B_2} + B_2)$$

$$V = \frac{1}{3}(15)(153.94 + \sqrt{153.94 \times 615.75} + 615.75)$$

$$V \approx 5387.84$$

The volume is approximately 5387.84 cubic inches.

b. Find the areas of the frustum's bases.

SOLUTION
Since one angle of the cone's cross section is
given, and the cross section is a right triangle,
the tangent function can be used to find x, the
radius of the lower base.

$$\tan 70° = \frac{15}{x}$$

$$x \approx 5.46$$

Now find the area of the bases.

$B_1 = \pi r^2$ $B_2 = \pi r^2$
$B_1 = \pi(3)^2$ $B_2 = \pi(5.46)^2$
$B_1 \approx 28.27$ $B_2 \approx 93.66$

So the areas of the frustum's bases are approximately
28.27 square centimeters and 93.66 square centimeters.

Hint

Recall that the tangent of
an angle is equal to the
ratio of the length of the
side opposite the angle
to the length of the side
adjacent to the angle.

Example 3 | Application: Farming

A grain silo is shaped like a cone, as shown in the diagram. If the height of the grain in the silo is 40 feet, what volume of grain is in the silo, to the nearest cubic foot?

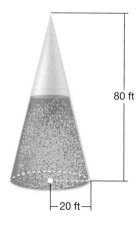

SOLUTION

As in Example 2a, similar triangles will have to be used to find the radius of the top of the frustum made by the grain. The diagram illustrates the similar triangles. Write a proportion.

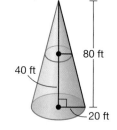

$$\frac{r}{40} = \frac{20}{80}$$
$$r = 10$$

Next, find the area of the frustum's bases.

$$B_1 = \pi r^2 \qquad\qquad B_2 = \pi r^2$$
$$B_1 = \pi(10)^2 \qquad\quad B_2 = \pi(20)^2$$
$$B_1 \approx 314.16 \qquad\quad B_2 \approx 1256.64$$

Finally, apply the formula for volume of a frustum.

$$V = \frac{1}{3}h(B_1 + \sqrt{B_1 B_2} + B_2)$$
$$V = \frac{1}{3}(40)(314.16 + \sqrt{314.16 \times 1256.64} + 1256.64)$$
$$V \approx 29{,}322$$

The volume of grain in the silo is approximately 29,322 cubic feet.

Lesson Practice

a. Find the volume of this frustum of a pyramid. Round your answer to the nearest cubic inch.
(Ex 1)

b. Find the volume of this frustum of a cone to the nearest hundredth.
(Ex 2)

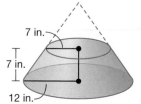

c. The Great Pyramid of Giza is the largest of ancient Egypt's pyramids.
(Ex 3) It is a square pyramid that stands 147 meters tall. The diagram depicts
what the Great Pyramid might have looked like during construction.
Given the dimensions in the diagram, what is the volume of the
pyramid at this point in its construction?

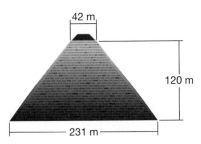

Practice Distributed and Integrated

* 1. Find the point matrix for the given rhombus and apply a scale
(102) factor of 1.5 to find the vertices after dilation.

2. Find the measure of the longer arc between two secants if the
(79) shorter arc is 50° and the exterior angle is 33°.

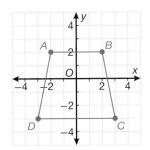

3. **Error Analysis** Farnaz and Samantha each solved for x.
(94) Which student is correct? Explain the error the other one made.

Farnaz:

$$\frac{x}{\sin 38°} = \frac{40}{\sin 55°}$$

$$x = \frac{40(\sin 38°)}{\sin 55°}$$

$$x \approx 30$$

Samantha:

$$\frac{x}{\sin 87°} = \frac{40}{\sin 55°}$$

$$x = \frac{40(\sin 87°)}{\sin 55°}$$

$$x \approx 49$$

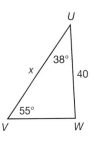

xy^2 * 4. **Algebra** The area of the base of a square pyramid is x^2 and the height is y. If the
(103) frustum is $\frac{5}{8}$ the height of the pyramid, and the smaller base of the frustum has
an area of 16 square units, what is an expression for the volume of the frustum?

5. Use $\triangle XYZ$ to determine the approximate sine, cosine, and tangent
(Inv 7) of 35°.

6. If the acute corners of a right triangle are folded to align with
(61) the right angle, will the result always be a parallelogram?
... a rectangle?

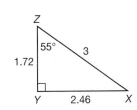

7. Compare *FH* and *AB*.
(Inv 4)

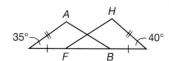

* 8. Write What is the meaning of the subscript *O* in the dilation notation
(102) $D_{o,k}(x, y) \rightarrow (kx, ky)$?

9. Algebra Find the measure of each angle in quadrilateral *JKLM*.
(47)

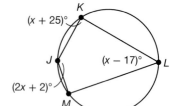

10. (**Robotics**) Leah programs an industrial robot with the following
(21) logical rules:

If there is a green item in the tray, paint it blue.

If there is a blue item in the tray, attach a widget to it.

What can Leah conclude will happen if she puts a green item in the robot's tray?

11. Model Sketch a figure which *cannot* be tessellated.
(Inv 9)

12. In the diagram, which angles are not angles of elevation or depression?
(73)

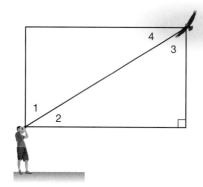

13. Algebra *R* is an exterior point of ⊙*C*, and $\overline{RP}$ and $\overline{RQ}$ are tangent to ⊙*C* at *P*
(58) and *Q*. Given that m∠*CPR* = $(x^2 - x)°$ and m∠*PRQ* = $(9x)°$, prove that *CPRQ*
is a square.

*14. (**Mountaineering**) In the diagram shown, a mountain climber
(101) stands at the top of a mountain that is 2640 feet above sea level.
The diameter of the earth is approximately 7920 miles. Assuming
the earth to be spherical, how far away can the mountain climber
see, to the nearest mile?

*15. Find θ. Round your answer to the nearest degree.
(98)

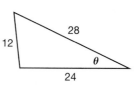

16. A trapezoid has vertices at the points *W*(−3, 2), *X*(1, 0),
(69) *Y*(−1, −2), and *Z*(−3, −1). Use the distance formula
to find the length of the midsegment of the trapezoid.

17. If a circle has a radius of 4.5 inches, and a chord through the circle is 5 inches
(43) long, what is the distance from the chord to the center of the circle?

*18. Find an expression for the volume of the frustum of a cone that is $\frac{1}{6}$ the height
(103) of the cone.

19. A dilation with a scale factor of 3 is applied to the rectangular
(84) prism at right. Find the surface area and the volume of the image
that results.

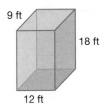

9 ft
18 ft
12 ft

20. Write Explain what it means for a fractal to be self-similar.
(Inv 10)

*21. Find the result of the dilation $D_{0,3.5}(x, y)$ on the points
(102) (4, 6) and (−2, −4).

22. Are the points (1, 2), (2, 9), and (−2, 3) part of the solution set of $y > 2x + 3$?
(88)

23. Find the surface area of a sphere whose great circle has a 52-meter circumference,
(80) to the nearest tenth.

*24. Multi-Step $\overline{BC}$ has endpoints (−6, −3) and (−2, −1). If the line segment is
(100) translated 3 units up and 2 units left, what is the point matrix of the translated
line segment?

*25. (Cooking) A decorative teapot makes a cone shape. If the teapot stands 1 foot tall
(103) with a base diameter of 8 inches, what is the volume of half a pot of tea? Round
your answer to the nearest tenth of a cubic inch.

26. Write If two chords intersect at right angles, what can you say about the measures
(64) of the two opposite arcs? Explain how you know.

27. (Mobiles) Hugh is making a triangular shape for a mobile to hang above his baby's
(32) crib, which should remain level. He measures the medians of the triangle to be
2.4, 3.2, and 3.75 inches. From which point should he suspend the triangle? How
far is this point from each vertex, to the nearest tenth of an inch?

*28. Multiple Choice Which of these could *not* be a formula for finding the volume
(103) of a frustum?

A $V = \frac{1}{3}h(B_1 + \sqrt{B_1 B_2} + B_2)$　　　　**B** $V = \frac{1}{3}(h_1 B_1 - h_2 B_2), h_2 > h_1$

C $V = \frac{1}{3}\pi h(r_1^2 + r_1 r_2 + r_2^2)$　　　　　　**D** none of the above

29. Use a special right triangle to find the value of tan 60°.
(68)

30. (Architecture) An architect designs the circular window shown in the diagram. If
(101) the diameter of the window is 18 inches, and the architect wants exactly half of
the shorter crosspieces to pass in front of the window, what should the total length
of the shorter crosspieces be, to the nearest half-inch?

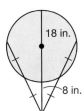

18 in.

8 in.

Relating Arc Lengths and Chords

Warm Up

1. **Vocabulary** An angle with a vertex at the center of a circle is
   (26)
   called a _____.

2. **Vocabulary** An unbroken part of a circle, consisting of two endpoints
   (26)
   and all the points on the circle between them is called a(n) _____.

3. Use the diagram of ⊙*P* to find *x*,
   (26)
   if m$\widehat{AB}$ = (4*x* − 5)°

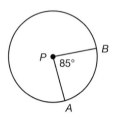

New Concepts

The endpoints of a chord in a circle are also the endpoints of an arc.
Theorem 104-1 relates the subtended arcs of chords that are congruent
to one another.

Math Language

Subtend means to
be opposite of or to
establish the boundaries.
The subtended arc of
a chord is opposite of
the chord and shares is
endpoints.

Theorem 104-1
1. In the same or congruent circles, congruent arcs have congruent chords. In the diagram, if $\widehat{BC}$ ≅ $\widehat{DE}$, then $\overline{BC}$ ≅ $\overline{DE}$. 2. In the same or congruent circles, congruent chords have congruent arcs. In the diagram, if $\overline{BC}$ ≅ $\overline{DE}$, then $\widehat{BC}$ ≅ $\widehat{DE}$.

Example 1 Congruent Arcs and Chords

Hint

Theorem 104-1 has two
closely related parts.
Each part is the converse
of the other.

Use the figure to find the measure of
each arc.

SOLUTION
Since the chords that subtend the arcs are
congruent, the arcs shown must also be
congruent, by Theorem 104-1. Write an
equation to solve for *x*.

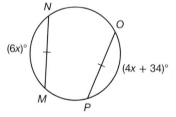

$6x = 4x + 34$
$x = 17$

Substitute into the solution for *x* to find the measure of an arc.
m$\widehat{MN}$ = 6*x*
m$\widehat{MN}$ = 6(17)
m$\widehat{MN}$ = 102°

Since the arcs are congruent, m$\widehat{OP}$ = 102°.

Example **2** **Proving Theorem 104-1**

Use the diagram to prove the
first part of Theorem 104-1.

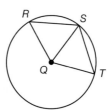

Given: $\odot Q$, $\overset{\frown}{RS} \cong \overset{\frown}{ST}$

Prove: $\overline{RS} \cong \overline{ST}$

SOLUTION

Statements:	Reasons:
1. $\overset{\frown}{RS} \cong \overset{\frown}{ST}$	1. Given
2. $\angle RQS \cong \angle SQT$	2. In the same or congruent circles, congruent arcs have congruent central angles.
3. $\overline{QR} \cong \overline{QS}$, $\overline{QT} \cong \overline{QS}$	3. Radii of the same circle are congruent.
4. $\triangle QRS \cong \triangle QTS$	4. SAS
5. $\overline{RS} \cong \overline{ST}$	5. CPCTC

It follows from the proof that since $\angle RQS \cong \angle SQT$ and $\overline{RS} \cong \overline{ST}$, in the same or congruent circles, congruent central angles have congruent chords. The converse is also true, in the same or congruent circles, congruent chords have congruent central angles.

The relationships formed between congruent arcs, chords, and central angles are summarized in the chart below, and are illustrated in the diagram of $\odot F$ at right.

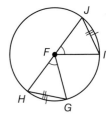

Hint

If two chords are congruent, then the corresponding arcs and central angles are congruent. The same is true if two central angles are congruent, or if two arcs are congruent. Any one implies the other two.

Theorem:	If:	Then:
In the same circle or congruent circles:		
1. Congruent central angles have congruent chords.	$\angle JFI \cong \angle GFH$	$\overline{JI} \cong \overline{GH}$
2. Congruent arcs have congruent central angles.	$\overset{\frown}{JI} \cong \overset{\frown}{GH}$	$\angle JFI \cong \angle GFH$
3. Congruent chords have congruent arcs.	$\overline{JI} \cong \overline{GH}$	$\overset{\frown}{JI} \cong \overset{\frown}{GH}$
4. Congruent arcs have congruent chords.	$\overset{\frown}{JI} \cong \overset{\frown}{GH}$	$\overline{JI} \cong \overline{GH}$

Example 3 Application: Design

To construct a hexagon, Akira marks a 60° central angle on a circular piece of cloth. He draws the chord that subtends the angle and measures it. Then, starting at one endpoint of the arc, he draws another chord of equal length. What kind of hexagon will this process create? What is the length of the circle's radius?

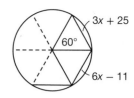

SOLUTION

Since the central angles are congruent, the chords must be congruent. Therefore, the hexagon will be a regular hexagon.

Write an equation to solve for x.

$3x + 25 = 6x - 11$

$\qquad x = 12$

For each arc, the radii that intercept the arc and the chord between them make an isosceles triangle. Since the triangle has one angle that measures 60°, it must also be equilateral. Therefore, the length of the chords is equal to the length of the radii. Find the length of one chord to find the length of the radius.

$r = 3x + 25$

$r = 3(12) + 25$

$r = 61$

Lesson Practice

a. Use the diagram to find the measure of $\overset{\frown}{AB}$ and $\overset{\frown}{DC}$.
(Ex 1)

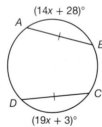

b. Use the diagram to prove the second part of Theorem 104-1.
(Ex 2)

Given: $\overline{BC} \cong \overline{DC}$
Prove: $\overset{\frown}{BC} \cong \overset{\frown}{DC}$

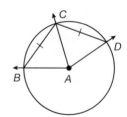

c. (**Millwheel**) A millwheel is divided into sections by chords $\overline{BD}$ and $\overline{EC}$. Find the value of x.
(Ex 3)

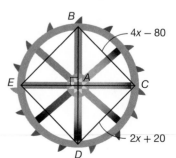

* 1. Use the figure shown and the given, $\overline{AD} \cong \overline{BC}$, to find $m\overset{\frown}{AD}$.
(104)

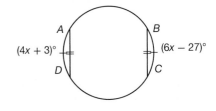

2. **Public Transportation** A bus route runs close to three large
(38) apartment complexes located at $(0, 0)$, $(6, 0)$, and $(0, 2)$.
The transportation agency wants to add a bus stop that is the
same distance from each complex. Where should the stop
be located?

* 3. **Multi-Step** Two similar cylinders, C and D have a scale factor of 9:4. Cylinder C
(99) has a base radius of 3 feet, 6 inches and a height of 5 feet, 9 inches. Determine
the surface area of cylinder D in square meters, to the nearest hundredth. Use
the conversion factor 1 inch = 2.54 centimeters.

4. Reflect rectangle $JKLM$ across the line $x + y = 0$.
(74)

5. Show the point matrix multiplication for a
(102) dilation with a scale factor of 2 if the vertices
of a triangle are $(3, -1)$, $(4, 0)$, and $(0, 9)$.

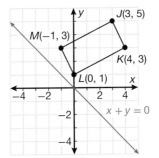

6. Perform a reflection of quadrilateral $ABCD$ across line l_1,
(90) followed by a reflection across line l_2.

7. **Multiple Choice** Which of the following statements is true
(52) of a rectangle but not true of a square?
  A All sides are congruent.
  B The diagonals are perpendicular.
  C The interior triangles created by the diagonals are two obtuse and
    two acute triangles.
  D The diagonals bisect opposite angles.

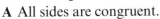

8. Three of the vertices of a parallelogram are located at $(1, 0)$, $(4, 3)$, and $(5, -2)$.
(34) Give all possible answers for the coordinates of the fourth vertex.

* 9. Find the value of x in the diagram.
(101)

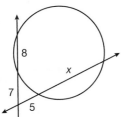

10. **Multiple Choice** Which linear system has no solution?
(81)
  A $y = -x$, $-y = x - 1$ **B** $y = 2x + 4$, $\frac{1}{4}y = x - 3$
  C $4y = x$, $2y = 2x$ **D** $x + y = 6$, $y = x - 4$

*11. Find the value of x if $m\overset{\frown}{RT} = 96°$ and
(104) $m\overset{\frown}{SU} = (x^2 + 4x)°$.

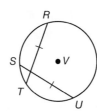

12. A regular hexagon has six congruent sides and six congruent
(45) 120° angles. The figure shows a hexagon with a side length of
L in a coordinate plane with one side on the x-axis and one
vertex on the y-axis. Determine the coordinates of each
vertex in terms of L.

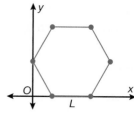

*13. Find the volume of the frustum shown. Round
(103) your answer to the nearest tenth.

14. A circle is dilated by a factor of 4. If the radius of the dilated circle
(96) is 12, what is the exact circumference of the original circle?

*15. Use the figure shown at the right and the given information to
(104) find $m\overset{\frown}{MN}$.

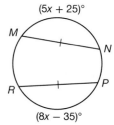

16. **Write** Each side of a square is doubled. Describe how the area of the
(96) resulting square compares with the area of the original square.

17. How many line segments can you draw between 9 noncollinear points?
$(Inv\ 8)$

*18. Find the vertices of the resulting triangle, if the given triangle is dilated
(102) by a factor of 3, centered at the origin.

19. Two segments are divided proportionally. The first segment is divided
(60) into parts that are 36, 42, and x units long, respectively. The second
segment is divided into parts that are 24, y, and 30 units long,
respectively. What are the values of x and y?

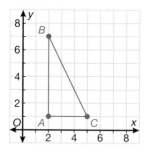

*20. Describe the translation of the graph $y = x^2$, and then write a 2×1
(100) matrix to represent the translation.

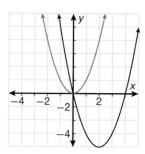

21. **Algebra** If $\frac{1}{\tan^2\theta} = 0.6$, what does $\sin\theta$ equal?
(91)

*22. **Multi-Step** Find the volume of the frustum at right. Round your answer to the
(103) nearest whole number.

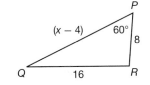

23. What shapes, and how many of each, make up the net of a hexagonal prism?
(Inv 5)

*24. **Write** Circle E is divided as shown in the figure. Describe the parts that are
(104) congruent and explain why they are congruent. Find the measures of $\angle CED$, $\overset{\frown}{AB}$,
$\angle AED$, and $\overset{\frown}{BC}$, given that m$\angle BEA$ is 130°.

25. The vertices of quadrilateral $EFGH$ are $E(-3, -3)$, $F(1, -1)$, $G(3, 5)$, and
(92) $H(-3, 2)$. Prove that $EFGH$ is a trapezoid.

26. **Algebra** Use the Law of Cosines to find the value of x in $\triangle PQR$.
(98)

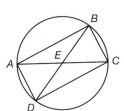

27. **Multi-Step** Find the perimeter of a 30°-60°-90° triangle with a 12-centimeter
(56) hypotenuse. Give your answer in simplified radical form.

28. Find the volume of the frustum shown. Round your answer to the nearest
(103) whole number.

15 in. 10 in. 4 in.

29. Find the distance from (5, 3) to the line $y = 2$.
(42)

30. A hexagonal figure is formed by two congruent isosceles trapezoids joined at their
(40) longer bases. What is the area of the figure if the shorter bases are 10 centimeters
long, the longer bases are 25 centimeters long, and the distance from one shorter
base to the other is 13 centimeters?

Rotations and Reflections in the Coordinate Plane

Warm Up

1. **Vocabulary** A change in position, size, or shape of a figure or graph is
(67) called a(n) _____.

2. **Vocabulary** A transformation that shifts or slides every point of a line or
(67) graph the same distance in the same direction is called a(n) _____.

3. Triangle *ABC* has vertices *A*(3, 2), *B*(−4, 1), and *C*(0, 4). Find the
(78) coordinates of the vertices of the image of △*A′B′C′* after a rotation
of 180° about the origin.

New Concepts

Two matrices can only be multiplied if the number of columns in the first
matrix is equal to the number of rows in the second matrix.

In other words, two matrices must have the dimensions (*a* × *b*) and (*b* × *c*) in
order to be multiplied. The resulting matrix will have the dimensions (*a* × *c*).
Example 1 demonstrates how to multiply two matrices.

Example 1 Multiplying Matrices

Simplify $\begin{bmatrix} 5 & -2 \\ 0 & 2 \\ 3 & 1 \end{bmatrix} \begin{bmatrix} 0 & -3 & 2 \\ 4 & 0 & 4 \end{bmatrix}$.

SOLUTION

The first matrix is 3 × 2, and the second is 2 × 3, so they can be multiplied,
and will result in a 3 × 3 matrix.

Highlight the first row of the first matrix, and the first column of the
second matrix, as shown. Multiply the first pair of each, then the second
pair. Add the products. This sum is the first item in the new matrix.

$$\begin{bmatrix} 5 & -2 \\ 0 & 2 \\ 3 & 1 \end{bmatrix} \begin{bmatrix} 0 & -3 & 2 \\ 4 & 0 & 4 \end{bmatrix} = \begin{bmatrix} (5)(0) + (-2)(4) & ? & ? \\ ? & ? & ? \\ ? & ? & ? \end{bmatrix}$$

To get the second number in the first row, repeat this process with
the first row of the first matrix and the second column of the second
matrix, as shown.

$$\begin{bmatrix} 5 & -2 \\ 0 & 2 \\ 3 & 1 \end{bmatrix} \begin{bmatrix} 0 & -3 & 2 \\ 4 & 0 & 4 \end{bmatrix} = \begin{bmatrix} -8 & (5)(-3) + (-2)(0) & ? \\ ? & ? & ? \\ ? & ? & ? \end{bmatrix}$$

The next several steps are shown. From the first matrix, the rows are used
and from the second matrix, the columns are used.

$$\begin{bmatrix} 5 & -2 \\ 0 & 2 \\ 3 & 1 \end{bmatrix} \begin{bmatrix} 0 & -3 & 2 \\ 4 & 0 & 4 \end{bmatrix} = \begin{bmatrix} -8 & -15 & (5)(2) + (-2)(4) \\ ? & ? & ? \\ ? & ? & ? \end{bmatrix}$$

Hint

Before multiplying two
matrices, write their
dimensions side by side.
For instance:
4 × 3 and 3 × 2. The
two interior values must
be the same in order to
multiply the matrices,
and the resulting
product will have
dimensions equal to the
exterior values. So in this
instance, the matrices
can be multiplied and
will result in a
4 × 2 matrix.

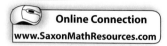
Online Connection
www.SaxonMathResources.com

$$\begin{bmatrix} 5 & -2 \\ 0 & 2 \\ 3 & 1 \end{bmatrix} \begin{bmatrix} 0 & -3 & 2 \\ 4 & 0 & 4 \end{bmatrix} = \begin{bmatrix} -8 & -15 & 2 \\ (0)(0)+(2)(4) & ? & ? \\ ? & ? & ? \end{bmatrix}$$

The resulting matrix is:

$$\begin{bmatrix} -8 & -15 & 2 \\ 8 & 0 & 8 \\ 4 & -9 & 10 \end{bmatrix}$$

Matrix multiplication can be used to find the image of a polygon that is reflected across the *x*- or *y*-axis. To reflect a polygon, create a point matrix from its vertices and multiply it by one of the reflection matrices given below.

Reflection Matrices	
Reflection over the *x*-axis	Reflection over the *y*-axis
$\begin{bmatrix} 1 & 0 \\ 0 & -1 \end{bmatrix}$	$\begin{bmatrix} -1 & 0 \\ 0 & 1 \end{bmatrix}$

Example 2 Reflecting by Matrix Multiplication

The vertices of $\triangle JKL$ are $J(-3, 4)$, $K(2, 4)$, and $L(2, 1)$. Use matrix multiplication to reflect the figure across the *x*-axis. Draw the preimage and the image in a coordinate plane.

SOLUTION
The point matrix for this triangle is $\begin{bmatrix} -3 & 2 & 2 \\ 4 & 4 & 1 \end{bmatrix}$.

Multiply this by the matrix for reflection across the *x*-axis. The reflection matrix must be placed on the left.

$$\begin{bmatrix} 1 & 0 \\ 0 & -1 \end{bmatrix} \begin{bmatrix} -3 & 2 & 2 \\ 4 & 4 & 1 \end{bmatrix} = \begin{bmatrix} (1)(-3)+(0)(4) & ? & ? \\ ? & ? & ? \end{bmatrix}$$

$$\begin{bmatrix} 1 & 0 \\ 0 & -1 \end{bmatrix} \begin{bmatrix} -3 & 2 & 2 \\ 4 & 4 & 1 \end{bmatrix} = \begin{bmatrix} -3 & (1)(2)+(0)(4) & ? \\ ? & ? & ? \end{bmatrix}$$

$$\begin{bmatrix} 1 & 0 \\ 0 & -1 \end{bmatrix} \begin{bmatrix} -3 & 2 & 2 \\ 4 & 4 & 1 \end{bmatrix} = \begin{bmatrix} -3 & 2 & (1)(2)+(0)(1) \\ ? & ? & ? \end{bmatrix}$$

$$= \begin{bmatrix} -3 & 2 & 2 \\ -4 & -4 & -1 \end{bmatrix}$$

This point matrix gives the vertices of the image. A graph of the image and preimage are shown.

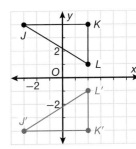

Hint

Notice that matrix multiplication is not commutative. Matrices must be multiplied in a particular order. To reflect a figure, always place the reflection matrix before the point matrix.

Matrix multiplication can also be used to rotate a polygon around the origin. Write a point matrix of the polygon and multiply it by one of the rotation matrices given in the **Rotation Matrices** table.

Rotation Matrices	
90° rotation	180° rotation
$\begin{bmatrix} 0 & -1 \\ 1 & 0 \end{bmatrix}$	$\begin{bmatrix} -1 & 0 \\ 0 & -1 \end{bmatrix}$
270° rotation	360° rotation
$\begin{bmatrix} 0 & 1 \\ -1 & 0 \end{bmatrix}$	$\begin{bmatrix} 1 & 0 \\ 0 & 1 \end{bmatrix}$

Example 3 **Rotating by Matrix Multiplication**

Parallelogram $QRST$ has vertices $Q(-1, 3)$, $R(5, 3)$, $S(4, 1)$, and $T(-2, 1)$. Determine the coordinates of the vertices of the image after a 270° rotation of $QRST$ about the origin.

SOLUTION

Create a point matrix from $QRST$ and multiply it by the rotation matrix for a 270° rotation. The first two steps and the solution are below.

$$\begin{bmatrix} 0 & 1 \\ -1 & 0 \end{bmatrix} \begin{bmatrix} -1 & 5 & 4 & -2 \\ 3 & 3 & 1 & 1 \end{bmatrix} = \begin{bmatrix} (0)(-1) + (1)(3) & ? & ? & ? \\ ? & ? & ? & ? \end{bmatrix}$$

$$\begin{bmatrix} 0 & 1 \\ -1 & 0 \end{bmatrix} \begin{bmatrix} -1 & 5 & 4 & -2 \\ 3 & 3 & 1 & 1 \end{bmatrix} = \begin{bmatrix} 3 & (0)(5) + (1)(3) & ? & ? \\ ? & ? & ? & ? \end{bmatrix}$$

$$= \begin{bmatrix} 3 & 3 & 1 & 1 \\ 1 & -5 & -4 & 2 \end{bmatrix}$$

So, parallelogram $Q'R'S'T'$ has vertices at $Q'(3, 1)$, $R'(3, -5)$, $S'(1, -4)$, and $T'(1, 2)$.

Example 4 **Application: Stained Glass**

When designing the layout for a stained glass window, Cheryl drew the basic shape of a pentagon onto a coordinate grid. In order to complete this design, she must reflect the pentagon $ABCDE$ across the y-axis. Find the reflection's coordinates.

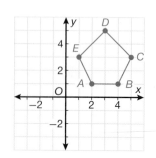

SOLUTION

The reflection matrix across the y-axis is $\begin{bmatrix} -1 & 0 \\ 0 & 1 \end{bmatrix}$.

Create a point matrix from $ABCDE$'s vertices and multiply it by the reflection matrix.

$$\begin{bmatrix} -1 & 0 \\ 0 & 1 \end{bmatrix} \begin{bmatrix} 2 & 4 & 5 & 3 & 1 \\ 1 & 1 & 3 & 5 & 3 \end{bmatrix} = \begin{bmatrix} -2 & -4 & -5 & -3 & -1 \\ 1 & 1 & 3 & 5 & 3 \end{bmatrix}$$

The vertices of the image are located at $A'(-2, 1)$, $B'(-4, 1)$, $C'(-5, 3)$, $D'(-3, 5)$, and $E'(-1, 3)$.

Lesson Practice

a. Simplify $\begin{bmatrix} -2 & 7 \\ 1 & -3 \end{bmatrix} \begin{bmatrix} 2 & 1 \\ 3 & -5 \end{bmatrix}$.
(Ex 1)

b. The vertices of $\triangle DEF$ are $D(2, 1)$, $E(6, 1)$, and $F(4, 5)$. Use matrix
(Ex 2) multiplication to reflect the image across the y-axis. Draw the preimage and the image in a coordinate plane.

c. $\triangle MNO$ has vertices $M(2, -4)$, $N(4, 1)$, and $O(1, 3)$. Determine the
(Ex 3) point matrix of the image for a 90° rotation of $\triangle MNO$ about the origin. Graph $\triangle MNO$ and its image in a coordinate plane.

d. (Art) Isaac is creating a tile mosaic. He is
(Ex 4) going to sketch the design for the mosaic in a coordinate plane to ensure accuracy. He draws a trapezoid as shown in the figure. He wants to rotate the trapezoid 180° about the origin. What is the point matrix of the rotated image?

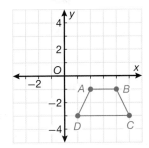

Practice Distributed and Integrated

* 1. Simplify $\begin{bmatrix} 6 & -1 \\ 3 & 2 \end{bmatrix} \begin{bmatrix} 5 & 2 \\ -8 & 4 \end{bmatrix}$.
(105)

2. Write vector $\overrightarrow{HI}$ in component form.
(63)

3. Multi-Step A triangle has vertices $A(3, 2)$, $B(6, 1)$, and $C(2, -4)$. Find all the
(98) lengths and measures in the triangle. Round angle measures to the nearest degree and side lengths to the nearest tenth.

4. Figures $LMNO$ and $PQRS$ are similar. The ratio of their corresponding sides is
(44) 4:7. If the perimeter of $PQRS$ is 87.5 centimeters, what is the perimeter of figure $LMNO$?

5. (Advertising) The figure at right shows the background for a company's
(104) logo. In order for the design to be symmetrical, what must be true about $\overset{\frown}{AB}$, $\overset{\frown}{AC}$, $\overline{AB}$, and $\overline{AC}$? Explain your reasoning.

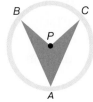

6. The equation for ⊙*A* is $(x + 2)^2 + (y - 3)^2 = 2.25$. Circle *B* has the same center as
(75) ⊙*A* and has a radius of 5. Write the equation of ⊙*B*.

7. Find the volume of this frustum. Round your answer
(103) to the nearest whole number.

3 in.

20 in.

9 in.

8. Multiple Choice This orthographic drawing shows the front, top, and
(93) right-side views of a solid. Which of these sketches represents the solid?

Front   Side   Top

A

B

C

D

9. Solve this system of equations.
(81)
$$y = \frac{3}{2}x - 3, \; x = 2$$

10. Formulate Determine a formula for the volume of a tetrahedron with side length
(70) equal to $2m$.

*11. The vertices of △*JKL* are $J(-6, 2)$, $K(-6, -3)$, and $L(1, -3)$. Use matrix
(105) multiplication to reflect the image across the *x*-axis. Draw the preimage
and the image in a coordinate plane.

12. Two similar pentagonal prisms have a scale factor of 4:9. If the volume of the
(99) smaller prism is 100 cubic meters, find the volume of the larger prism to the
nearest cubic meter.

13. Copy the figure at right. Then, draw the rotation of the figure 180° about
(78) the point shown.

14. Multiple Choice A circle given by $x^2 + y^2 = 49$ is shifted 5 units
(95) to the right and 6 units downward. It is also dilated by a factor of 2.
The equation of the shifted and dilated circle is:
A $(x - 5)^2 + (y + 6)^2 = 196$ **B** $(x - 5)^2 + (y + 6)^2 = 14$
C $(x + 5)^2 + (y - 6)^2 = 196$ **D** $(x + 5)^2 + (y - 6)^2 = 14$

15. Analyze The equation of a circle is $(x - 3)^2 + (y + 4)^2 = 16$. The circle is dilated by
(95) a scale factor of 0.5. What translation of the original figure would have translated
the center of the circle to the same coordinates as the center of the new image
after dilation?

16. **Food Preparation** A deli worker has a large block of cheese in the shape of a cube. She makes one cut through the block parallel to one of the faces. What is the shape of the remaining block?
(49)

17. **Write** For which values of k is a dilation a reduction? For which values of k is a dilation an enlargement?
(102)

*18. Triangle XYZ has vertices $X(-5, -1)$, $Y(-4, 3)$, and $Z(-1, 1)$. Determine the point matrix of the image for a 180° rotation about the origin of $\triangle XYZ$. Graph $\triangle XYZ$ and its image in a coordinate plane.
(105)

19. Find the value of z in the diagram.
(101)

20. In the city of Logica, there are two types of buses. Downtown buses are blue and suburban buses are green. State the disjunction of "A bus is a downtown bus," and "A bus is green." Is the disjunction true or false?
(20)

21. Use the figure shown at right and the fact that $\overline{AB} \cong \overline{CD}$ to find the measure of $\overset{\frown}{AB}$.
(104)

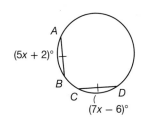

22. What is the sum of the interior angles of a convex decagon?
(Inv 3)

23. **Algebra** In an isosceles triangle, the vertex angle measures $3(x + 5)°$ and the base angles measure $(2x - 47)°$. Determine the angle measures.
(51)

*24. Trapezoid $TUVW$ has vertices at $T(-2, 2)$, $U(-1, 2)$, $V(0, 1)$, and $W(-3, 1)$. Determine the image matrix for a 90° rotation about the origin of trapezoid $TUVW$. Graph trapezoid $TUVW$ and its image in a coordinate plane.
(105)

25. What is the ratio of the areas of two 45°-45°-90° right triangles if one has a hypotenuse of 14 yards and the other a hypotenuse of 7 yards?
(53)

26. **Multiple Choice** Which group of line segment lengths can be used to form a triangle?
(39)
  A 3, 5, 8   **B** 5, 9, 18
  C 4, 7, 9   **D** 6, 8, 16

27. **Coordinate Geometry** One vector is parallel to the y-axis and is perpendicular to a second vector. These two vectors are added and the resultant vector makes a 45° angle with the horizontal and has a magnitude of $5\sqrt{2}$. What is the magnitude of each of the two original vectors?
(83)

28. **Model** Draw four circles, each pair of which share two common tangents.
(72)

29. **Multi-Step** Find the volume of the frustum of the cone shown with a slant height of 25 centimeters and a radius of 10 centimeters. The frustum is $\frac{2}{3}$ the height of the cone. Round your answer to the nearest whole cubic centimeter.
(103)

30. The sides of a rhombus are 5 inches long. One diagonal is 6 inches long. What is the length of the other diagonal?
(65)

Circumscribed and Inscribed Figures

1. *(97)* **Vocabulary** Coplanar circles with the same center are called
_____. (*congruent circles, concentric circles, tangent circles*)

2. *(75)* What is the equation for a circle with a radius of 6 centered at (2, 2)?

3. *(35)* **Multiple Choice** What is the area of sector *P*?

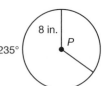

A 131 in²   **B** 16.5 in²

C 69.8 in²   **D** 8.7 in²

New Concepts Recall that an inscribed polygon is a polygon in which
every vertex lies on a circle.

To find the lengths of the sides of a regular polygon
inscribed into a circle with a radius *r*, divide the polygon
into triangles as shown in the diagram. Since two of each
triangle's sides are congruent radii, the triangles are
isosceles. Determine the central angle of the polygon and
use the Law of Cosines to solve for the length of a side.

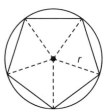

Math Reasoning

Analyze Is this
quadrilateral inscribed?
Explain.

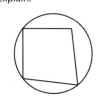

Example 1   **Inscribed Polygons**

Find the perimeter of a regular pentagon that is
inscribed in a circle with a radius of 4. Round your answer
to the nearest hundredth.

SOLUTION

Draw a diagram to help identify what needs to be
determined. The sum of the pentagon's central angles
is 360°, so each central angle is $\frac{360°}{5} = 72°$.

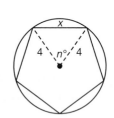

The radii of the circle are 4 units in length, resulting
in the isosceles triangle shown in the diagram. Use the
Law of Cosines to calculate the length of one side of the
pentagon.

$$c^2 = a^2 + b^2 - 2ab(\cos \theta)$$
$$c^2 = 4^2 + 4^2 - 2(4)(4)(\cos 72°)$$
$$c^2 \approx 22.11$$
$$c \approx 4.70$$

Online Connection
www.SaxonMathResources.com

The perimeter of the pentagon is 5 times the length of one
of its sides, so it is 5(4.70) = 23.50 units.

An **inscribed circle** is a circle that is tangent to each side of a polygon.

A **circumscribed polygon** is a polygon whose sides are all tangent to the same circle.

A circle inscribed in a regular polygon will have a radius equal to the length of the circumscribed polygon's apothem.

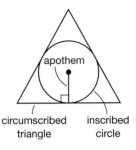

circumscribed triangle   inscribed circle

Example 2   Circumscribed Polygons

a. Find the perimeter of a regular hexagon circumscribing a circle of radius 10. Round your answer to the nearest hundredth.

SOLUTION

First, draw a diagram.

The sum of the hexagon's central angles is 360°, so each central angle is $\frac{360°}{6} = 60°$.

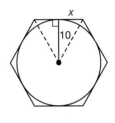

The apothem of the hexagon makes a right triangle with the segment from the center of the circle to a vertex of the hexagon. The right triangle is half the isosceles triangle formed by the rays of the hexagon's central angle, so it is a 30°-60°-90° triangle.

$$\tan 30° = \frac{x}{10}$$

$$10 \tan 30° = x$$

$$x \approx 5.77 \text{ units}$$

Since each side of the hexagon is $2x$, the perimeter is:

$$P = 6s$$
$$P = 6(2x)$$
$$P \approx 6(2)(5.77)$$
$$P \approx 69.24 \text{ units}$$

b. Find the area of a circumscribed square if the inscribed circle has a radius of 15 inches.

SOLUTION

First, draw a diagram.

Since $\theta = 45°$, the triangle shown is an isosceles right triangle. Therefore, $x = 15$ inches.

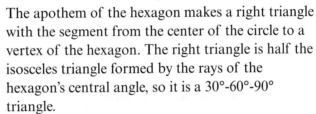

Because half the side of the square is 15 inches, the whole side is 30 inches.

Therefore, the area of the square is:

$$A = bh$$
$$A = (30)(30)$$
$$A = 900 \text{ in}^2$$

Example 3 **Application: Baking**

Tariq is baking bran muffins. He has a pan with square depressions in it as shown. Unfortunately, the paper muffin wrappers he bought are round. If the radius of each muffin wrapper is 4.5 centimeters, will it fit in the square hole with the dimensions shown?

SOLUTION

The radii of the wrappers must be less than or equal to half the length of the square. Sketch the diagonal on the diagram, creating a 45°-45°-90° triangle with a 6-centimeter hypotenuse. Use the Pythagorean Theorem to find the length of each leg.

$$x^2 + x^2 = 6^2$$
$$2x^2 = 36$$
$$x^2 = 18$$
$$x \approx 4.24$$

The muffin wrappers will not fit, as their radii are larger than half the square's length.

Lesson Practice

a. Find the perimeter of a regular octagon that is inscribed in a circle with a radius of 6. Round your answer to the nearest tenth.
(Ex 1)

b. Find the perimeter of an equilateral triangle inscribed in a circle with a radius of 7. Round your answer to the nearest whole number.
(Ex 2)

c. Find the area of a circumscribed square if the inscribed circle has a diameter of 5 centimeters.
(Ex 2)

d. Mariah is baking a casserole. The recipe calls for a square mold, but she only has a round pan. In order to make the casserole square, Mariah has to fashion a square mold out of aluminum foil. If Mariah's round pan has a radius of 8 inches, what must be the length of one side of her square mold? Round to the nearest tenth of an inch.
(Ex 3)

Practice   Distributed and Integrated

* 1. **Multi-Step** Find the circumferences of the inscribed and circumscribed circles if the equilateral triangle has a side length of 12 centimeters.
(106)

2. **Multiple Choice** Which of the following transformation matrices represents a translation?
(100)

A $\begin{bmatrix} 1 & 0 \\ 0 & 1 \end{bmatrix}$ **B** $\begin{bmatrix} 0 & 0 \\ 2 & -2 \end{bmatrix}$ **C** $\begin{bmatrix} -1 & -1 \\ 0 & 0 \end{bmatrix}$ **D** $\begin{bmatrix} 1 & 2 \\ 1 & 2 \end{bmatrix}$

3. Point M has coordinates $(-6, 5)$. What is the vector that translates
(71) $\triangle MNP$ to $\triangle M'N'P'$?

4. What is the geometric mean of 4 and 9?
(50)

 5. Algebra A cylinder has a height that is twice the length of the
(62) diameter of its base. The volume of the cylinder is 100.48 cubic
centimeters. Find the surface area to the nearest square centimeter.

* 6. Triangle ABC has vertices $A(3, 4)$, $B(7, 2)$, and $C(4, -3)$. Determine the
(105) image matrix for a $270°$ counterclockwise rotation of $\triangle ABC$
about the origin. Graph $\triangle ABC$ and its image in a coordinate plane.

7. Multi-Step For the right triangle with vertices $D(-2, 0)$, $E(4, 3)$, and $F(4, 0)$,
(82) find the side lengths to the nearest tenth of a unit and the angle measures
to the nearest degree.

8. Use the figure to find m$\overarc{MN}$.
(104)

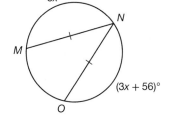

9. Sketch the graph of $y > \frac{1}{2}x - 1$.
(88)

10. Write the contrapositive of the statement "If a quadrilateral has
(17) exactly one pair of parallel sides, then it is a trapezoid." Is the
contrapositive true?

11. Find the volume of the frustum of the square pyramid shown. Round
(103) your answer to the nearest hundredth of a cubic meter.

12. Analyze A parallelogram has coordinates $A(-1, 1)$, $B(-3, -3)$, $C(3, -2)$,
(92) and $D(p, q)$. How many possible coordinates exist for D? Find them.

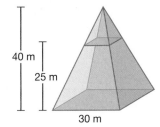

13. A right rectangular prism has a length, width, and height that are
(59) all equal. The volume of the prism is 343 cubic meters. Find the
surface area.

14. How could you make the equation $y = x^2$ symmetrical along the x-axis?
(76)

 15. Algebra Two chords of a circle, $\overline{AB}$ and $\overline{CD}$, intersect at E. This figure's dimensions
(86) are: $AE = 5$, $BE = z$, $CE = 3$, and $DE = 7z - 4$. Write and solve an equation
for z.

*16. (Carpentry) Sawmills use inscribed squares to cut the most economical
(106) beams of wood from logs. Using the illustration at right, find the ratio
between the length of the side of a beam and the radius of the log.

17. (Records) If the record player has a radius of 17 centimeters, what is the
(35) arc length traced by its needle moving half of a rotation?

18. **Justify** Determine if the circles shown are concentric.
₍₉₇₎ Explain your reasoning.

19. Find the area of a regular hexagon with sides that are
₍₆₆₎ 22 inches long.

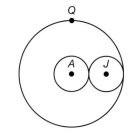

*20. Simplify $\begin{bmatrix} -5 & 0 \\ 2 & -7 \end{bmatrix}\begin{bmatrix} 3 & 1 & -1 \\ -2 & 6 & -3 \end{bmatrix}$.
₍₁₀₅₎

21. Use the figure shown to find m$\overset{\frown}{CD}$ if $\odot P \cong \odot T$.
₍₁₀₄₎

22. In $\triangle JKL$, $JL = 4$, $LK = 2.7$, and
₍₃₉₎ $KJ = 2.9$. List angles J, K, and L from
greatest to least measure.

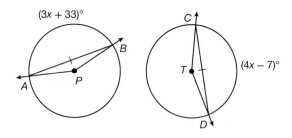

23. **Algebra** Find the area of the polygon bounded by the
₍₅₇₎ equalities $y = x$, $x = 6$, $y = 4$, and $x = 0$.

24. **Error Analysis** Melanie drew an orthographic view of this solid. Determine if
₍₉₃₎ any of these views are incorrect, and if so, draw the correct view.

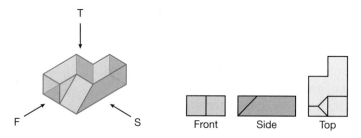

*25. Find the perimeter of a regular decagon inscribed in a circle having a radius of 10.
₍₁₀₆₎ Round your answer to the nearest tenth.

*26. The vertices of $\triangle XYZ$ are $X(3, -1)$, $Y(5, -5)$, and $Z(7, -1)$. Use matrix
₍₁₀₅₎ multiplication to reflect the image across the y-axis. Draw the preimage and
the image in a coordinate plane.

27. Segment JK is a midsegment of $\triangle FGH$. Find the value of x and the
₍₅₅₎ lengths of $\overline{JK}$ and $\overline{FH}$.

28. Find the vertices of the image of $D_{o, \frac{3}{2}}(x, y)$ for a triangle with vertices
₍₁₀₂₎ $(-3, 0)$, $(5, 2)$, and $(2, -4)$.

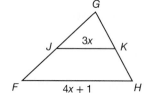

29. **Model** Draw two triangles that are similar by SSS.
₍₄₆₎

*30. **Algebra** A regular polygon is inscribed in a circle. If the circle has a radius of r and
₍₁₀₆₎ the polygon has a side length of r, what type of polygon is it?

Maximizing Area

Warm Up	

1. **Vocabulary** The perpendicular distance from the center of a regular
(66) polygon to one of its sides is called the _____.

2. Which has a greater perimeter—a square with an area of 144 square
(66) inches, or a rectangle with an 8-inch height and an 18-inch base?

3. **Multiple Choice** Which of the following formulas gives the area of the
(22) rhombus?

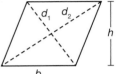

A $A = \frac{1}{2}d_1d_2$ **B** $A = \frac{1}{2}bh$

C $A = \frac{d_1 + d_2}{2}h$ **D** $A = d_1d_2h$

New Concepts Two polygons can have the same perimeter but different areas. What is the greatest amount of area that a given perimeter can enclose? To find out, hold the perimeter of a figure constant while changing its dimensions.

Example 1 | **Differing Areas**

Math Reasoning

Can any given perimeter always correspond with a square? For any given value of k, how can you find the area of a square with perimeter k?

Rectangle $QRST$'s base is b, its height is h, and it has a perimeter of 24 units.

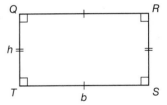

a. Determine the area of the rectangle for $b = 2$, $b = 4$, $b = 6$, and $b = 8$.

SOLUTION

First, solve for h using each perimeter and base. Then, substitute h and b into the formula for the area of a rectangle.

$P = 2b + 2h$	$P = 2b + 2h$	$P = 2b + 2h$	$P = 2b + 2h$
$24 = 2(2) + 2h$	$24 = 2(4) + 2h$	$24 = 2(6) + 2h$	$24 = 2(8) + 2h$
$h = 10$	$h = 8$	$h = 6$	$h = 4$
$A = bh$	$A = bh$	$A = bh$	$A = bh$
$A = 20$	$A = 32$	$A = 36$	$A = 32$

The area of the rectangle is 20 square units for $b = 2$ units, 32 square units for $b = 4$ units, 36 square units for $b = 6$ units, and 32 square units for $b = 8$ units.

b. Based on the results of part **a**, what conjecture can you make about maximizing the area of a rectangle with a fixed perimeter?

SOLUTION

The area of the rectangle increased and then decreased as the length of the base changed. Look at the 3rd rectangle, where the area is greatest. This rectangle is a square. We can make the conjecture that for a given perimeter, a regular polygon has a greater area than an irregular polygon.

Online Connection
www.SaxonMathResources.com

Example 1 suggests that regular polygons have greater areas than irregular polygons. A related question is, what kind of polygon maximizes area for a given perimeter. For example, which has a greater area: an equilateral triangle with a perimeter of 20 units, or a square with the same perimeter?

Example 2 Maximizing Area

The perimeter of each figure below is 30 units. Find and compare their areas. Then make a conjecture about the relationship between the number of sides of a polygon and its area for a fixed perimeter.

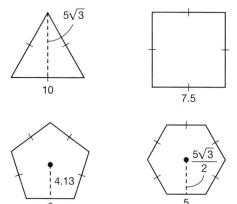

SOLUTION
Find the area of each figure.

Triangle	Square	Pentagon	Hexagon
$A = \frac{1}{2}bh$	$A = bh$	$A = \frac{1}{2}aP$	$A = \frac{1}{2}aP$
$A = \frac{1}{2}(10)(5\sqrt{3})$	$A = (7.5)(7.5)$	$A = \frac{1}{2}(4.13)(30)$	$A = \frac{1}{2}\left(\frac{5\sqrt{3}}{2}\right)(30)$
$A \approx 43.30$	$A = 56.25$	$A = 61.95$	$A \approx 64.95$

All of these polygons have the same perimeter, but the polygons with more sides have increasingly greater area. We can make the conjecture that for a given perimeter, a regular polygon with the greatest number of sides has the greatest area.

Math Reasoning

Write Explain how maximizing the area of a rectangle is related to minimizing its perimeter.

As the number of sides of a polygon increases, it begins to resemble a circle. For a given perimeter, a circle contains the greatest area of any possible shape.

┌─ **Example 3** **Application: Jigsaw Puzzles**

Gillian is putting together a jigsaw puzzle containing 5500 pieces. Each puzzle piece is approximately 1-inch-by-1-inch. Gillian does not know the exact dimensions of the finished puzzle, but knows that it is rectangular. What is the minimum perimeter the finished puzzle could have, to the nearest inch?

SOLUTION

Since a square maximizes rectangular area for a given perimeter, it also minimizes perimeter for a given area. Therefore, the minimum perimeter of the finished puzzle will be the perimeter if it is a square. The puzzle will have an area of 5500 square inches.

$$A = s^2$$
$$5500 = s^2$$
$$s \approx 74.16 \text{ inches}$$

Therefore, the square is about 74-by-74-inches and the minimum perimeter would be 296 inches.

Math Reasoning

Connect Make a conjecture about what sort of rectangular prism will have the greatest volume for a fixed surface area.

Lesson Practice

a. A rectangle has a perimeter of 144 inches. Determine the area of the
(Ex 1) rectangle for $b = 9$ inches, $b = 12$ inches, and $b = 18$ inches.

b. What is the maximum area of a rectangle with a perimeter of
(Ex 1) 144 inches?

c. Explain which of these two triangles has the greater area: A triangle
(Ex 2) with side lengths of 7, 8, and 9, or a triangle with side lengths of 8, 8, and 8?

d. A rectangular patio is to be made from 625 square tiles. The tiles will
(Ex 3) be placed so that the patio will have the least possible perimeter. How many tiles will be placed along the length and the width of the patio?

Practice **Distributed and Integrated**

1. Use this figure to find m$\overset{\frown}{AD}$.
(104)

2. (**Packaging**) A package of rice in the shape of a right triangular prism
(99) comes in two sizes. The smaller size is 4.5 inches tall and has right triangular bases with legs that are 0.75 and 1 inch long, respectively. The larger size is similar to the smaller but with double the dimensions. Find the surface area of the larger package.

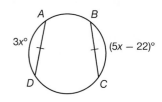

* 3. Triangle A is a right triangle with side lengths of 3, 4, and 5 inches. Triangle B is
(107) an equilateral triangle with side lengths of 4 inches each. Find the perimeters of each of these triangles and predict which will have the greater area. Verify your answer.

4. What is the name of a right solid whose bases are five-sided polygons and whose
(49) lateral faces are rectangles?

5. Find the value of x in the figure.
(64)

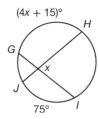

6. Find a counterexample to the following conjecture.
(14)

If the product of two numbers is positive, then both numbers are positive.

7. Triangle *DEF* has vertices $D(-4, -5)$, $E(-1, 5)$, and $F(2, -5)$.
(105) Determine the image matrix for a 360° rotation of $\triangle DEF$ about
the origin. Graph $\triangle DEF$ and its image in a coordinate plane.

* 8. Determine the largest possible area for a hexagon with a perimeter of
(107) 12 centimeters. Round the area to the nearest tenth.

9. Triangles *DEF* and *XYZ* are similar. If $DE = 3(XY)$, what is the ratio of the
(41) perimeters of the triangles?

10. A square has side lengths of 5 centimeters. A second square is drawn inside the
(Inv 6) first by connecting the midpoints of each side. Find the probability that a point
chosen at random within the larger square will fall inside the smaller square.

*11. Describe the diagram in two ways.
(106)

12. What is sin M?
(68)

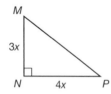

13. Multiple Choice What are the coordinates of the preimage point in the
(102) dilation $D_{O, 4} (x, y) \rightarrow (20, -4)$?
 A $(80, -16)$ **B** $(5, -1)$
 C $(-1, 5)$ **D** $(-16, 80)$

14. What angle has a tangent equal to 1?
(68)

15. Error Analysis Jeremy suggests that the value of x in the figure shown
(101) must be 5 inches. Is he correct? If not, give the correct answer.

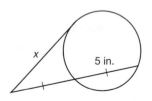

*16. What is the minimum perimeter that will enclose a triangular area of
(107) 100 square feet? Round your answer to the nearest tenth of a foot.

17. Find b in the figure at right.
(86)

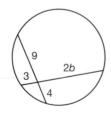

*18. A regular hexagon has a perimeter of 60 inches. What is the area of a circle
₍₁₀₆₎ inscribed in the hexagon?

 19. Algebra $\overleftrightarrow{AB}$ is tangent to $\odot Z$ at B, m$\angle ZAB = (3x + 72)°$, and
₍₅₈₎ m$\angle AZB = (13 - 2x)°$. Determine the value of x.

20. Formulate Write a formula for finding the volume of a frustum of a cone
₍₁₀₃₎ that is half the height of that cone.

21. Translate the triangle along v, and then reflect it across line t.
₍₉₀₎

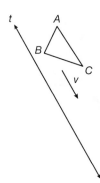

22. Simplify $\begin{bmatrix} 4 & 2 & -7 \\ -1 & -3 & 5 \end{bmatrix} \begin{bmatrix} 9 & 6 \\ 3 & 5 \\ 1 & -2 \end{bmatrix}$.
₍₁₀₅₎

23. Write The congruence theorems for right triangles are derived directly from
₍₃₆₎ the general Triangle Congruence Theorems and Postulates SAS, SSS, AAS and
ASA. Which of these is HA Right Triangle Congruence derived from?
Explain.

 24. Algebra Let $s = \sin x$, $c = \cos x$, and $t = \tan x$. Determine an equation for t
₍₉₁₎ in terms of c and an equation for t in terms of s.

25. A regular pentagon is inscribed in a circle with a radius of 7. Find the
₍₁₀₆₎ perimeter of the pentagon. Round your answer to the nearest tenth.

26. Is the following statement always, sometimes, or never true?
₍₅₂₎
*The diagonals of a square, a rhombus, and a rectangle are
congruent.*

27. (**Targets**) A target sits on a grid. It has a center circle whose equation is
₍₉₅₎ $(x + 3)^2 + (y - 2)^2 = 9$. An outer circle of the target is a dilation of the
center circle by a factor of 3. What is the equation of the outer circle?

*28. (**Corrals**) A farmer wants to build a rectangular corral for his horse. The farmer
₍₁₀₇₎ has 64 feet of fencing. What is the maximum area that the farmer can surround
with his fencing?

29. Generalize A three-dimensional figure is formed by adjoining two congruent cones
₍₇₇₎ along their circular faces. If the length from vertex to vertex of the new shape is h,
and the radius of the initial cones is r, determine a formula for the volume of this
figure.

30. Analyze Which point of concurrency in a triangle is the center of a circle that
_(32, 38) contains all three vertices? How is it found?

Maximizing Area Using Geometry Software

Technology Lab 14 *(Use with Lesson 107)*

In Lesson 107, you learned that squares and equilateral triangles maximize area, given a set perimeter. In this lab, you will compare several regular polygons with the same perimeter.

1. First, to create a regular pentagon, draw two points anywhere. You do not need to label them.

2. Select one of the points and choose *Mark Center* from the *Transform* menu.

3. Select both points and choose *Rotate* from the *Transform* menu.

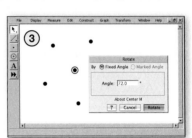

4. Enter "72" into the angle field.

5. Repeat steps 3 and 4 until you have created five points.

6. Select your five points in the order that you drew them, and choose *Pentagon Interior* from the *Construct* menu.

7. Select the interior of your pentagon and choose *Perimeter* from the *Measure* menu. Select it again and select *Area* from the *Measure* menu.

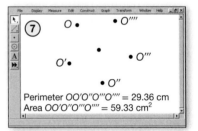

8. Would a square with the same perimeter as your pentagon have greater, lesser, or equal area? Construct a square with the same perimeter as your pentagon, and calculate its area. What did you discover?

> **Math Reasoning**
>
> **Justify** Why does a 72° rotation eventually result in a pentagon? How would you determine the angle needed to construct other polygons?

Would a hexagon with the same perimeter as the pentagon have the same area? Geometry software can aid in answering this question.

1. Create two new points.

2. Follow steps 2 through 5 from above. This time, enter "60" into the angle field. Six points should appear.

3. Select your six points in order and choose *Hexagon Interior* from the *Construct* menu.

4. Following step 7 above, measure the perimeter and area of your hexagon. Drag any point on the hexagon to make it larger or smaller. Adjust its size until it has about the same perimeter as your pentagon.

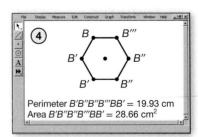

(5.) How does the hexagon's area compare to the pentagon and the square? Would an octagon with the same perimeter as your hexagon have greater, lesser, or equal area?

(6.) What conjecture can you make about how area changes when you hold perimeter constant and increase the number of sides of a polygon? What shape do you think has the greatest area for any given perimeter?

Math Reasoning

Analyze If you had instead held the area of your polygons constant, what shape would have resulted in the largest perimeter for a given area?

Lab Practice

Construct the following polygons. Measure the perimeter and area of each one. Adjust the size of each polygon until it has the same perimeter as the polygons you constructed above. Record the area of each figure. Do the results agree with the conjecture you made earlier?

a. Construct an octagon and record its area.

b. Construct a decagon and record its area.

c. Construct a dodecagon and record its area.

Warm Up

1. **Vocabulary** The plane divided into four regions by the *x*- and
(SB 13) *y*-axis is the _____ plane.

2. **Give the coordinates of each point shown**
(SB 13) **at right.**

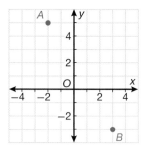

3. **Multiple Choice** Which point is collinear to the
(1) line through points *R* and *S*?

    A (−1, 5)   **B** (5, −1)

    C (2, 4)   **D** (4, 2)

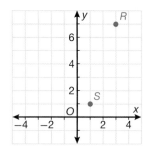

New Concepts

When a coordinate system is extended into three dimensions, a third axis called the *z*-axis is used. This creates a **three-dimensional coordinate system**, which is a space divided into eight regions by an *x*-axis, a *y*-axis, and a *z*-axis. The coordinates of points are given by **ordered triples**. An ordered triple is a set of three numbers used to locate and plot a point (*x*, *y*, *z*) in a three-dimensional coordinate system.

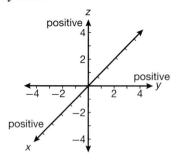

> **Hint**
>
> A three-dimensional coordinate plane is drawn with the *y*-axis as a horizontal line and the *z*-axis as a vertical line. The *x*-axis is isometric, and can be thought of as running into or out of the page.

When drawn on a flat surface, the three-dimensional plane has an isometric perspective.

Example 1 Identifying Ordered Triplets

In the given cube, identify the coordinates of the vertices.

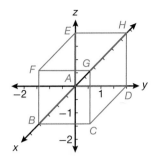

SOLUTION

For each vertex, read the x-coordinate first, then the y-coordinate, and finally the z-coordinate.

Start with the vertices that lie on the axes. Point E, for example, lies at 2 on the z-axes, and 0 on both the x- and y-axes, so it's coordinates are $(0, 0, 2)$. Point A lies at the origin, which is $(0, 0, 0)$. Similarly, we can identify $B(2, 0, 0)$ and $D(0, 2, 0)$.

Consider point F. Since the figure is a cube with side lengths of 2, F lies at $(2, 0, 2)$. Even though F lies on the x-axis, it appears to one side of the axis because of the isometric perspective. The remaining points are $C(2, 2, 0)$, $G(2, 2, 2)$, and $H(0, 2, 2)$.

Example 2 Graphing Points in Coordinate Space

Plot the point $A(3, 1, 5)$ on a three-dimensional coordinate system.

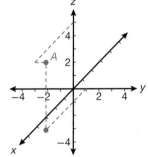

SOLUTION

First, locate the intersection of 1 and 3 on the x- and y-axis. This point is shown on the diagram. Notice that instead of a straight line extending from 1 on the y-axis, the dashed line is parallel to the x-axis. This point is $(3, 1, 0)$.

Next, imagine moving the origin of the graph up 5 units, to $z = 5$. This will translate the point to the location shown in the diagram. The dashed line indicates the same x- and y-coordinates as the first point we located.

In two dimensions, points are collinear if they lie on the same line. This is also true in three dimensions.

Math Reasoning

Connect Explain how the direction vector for points in three dimensions relates to the slope of a line.

To find collinear points in three dimensions, first find a direction vector. A direction vector is found by subtracting the corresponding coordinates of two ordered triples. If the direction vector (d_1, d_2, d_3) is known and any point on the line (x_0, y_0, z_0) is also known, the equation of the line is $(x, y, z) = (x_0, y_0, z_0) + t(d_1, d_2, d_3)$. The coordinates of each point on this line can be found by substituting different values for t, each of which gives a different point.

Example 3 **Finding Collinear Points**

Determine the coordinates of two points that are collinear with the points $R(3, 5, 7)$ and $S(4, 6, 12)$.

SOLUTION

Subtract the coordinates of the given points to find a direction vector for the line.

$$S(4, 6, 12) - R(3, 5, 7) = (4 - 3, 6 - 5, 12 - 7)$$
$$(d_1, d_2, d_3) = (1, 1, 5)$$

Inserting R into the equation for the line yields:

$$(x, y, z) = (3, 5, 7) + t(1, 1, 5).$$

Choose any two values for t to find the coordinates of other points on the line.

If $t = 2$:

$(x, y, z) = (3, 5, 7) + 2(1, 1, 5)$

$(x, y, z) = (3, 5, 7) + (2, 2, 10)$

$(x, y, z) = (5, 7, 17)$

If $t = -3$:

$(x, y, z) = (3, 5, 7) + (-3)(1, 1, 5)$

$(x, y, z) = (3, 5, 7) + (-3, -3, -15)$

$(x, y, z) = (0, 2, -8)$

The coordinates of two collinear points are $(5, 7, 17)$ and $(0, 2, -8)$.

Example 4 **Application: Movement of Electrons in an Electric Field**

In a three-dimensional space, two electrons start from a point located at $(5, -1, 4)$. One electron moves to the point $(9, -4, 6)$ while the other moves to the point $(13, -7, 8)$. Are the two electrons moving in a collinear path? Explain.

SOLUTION

Find the equation of the line for one of the paths, then determine if the other electron's current position is on this line.

First, find a direction vector:

$$(9, -4, 6) - (5, -1, 4) = (9 - 5, -4 + 1, 6 - 4)$$
$$(d_1, d_2, d_3) = (4, -3, 2)$$

Choose the first point as the fixed point.

$$(x, y, z) = (5, -1, 4) + t(4, -3, 2)$$

Substitute the coordinates of the point associated with the second electron to check if it lies on the same line.

$$(13, -7, 8) = (5, -1, 4) + t(4, -3, 2)$$

Write separate statements for x, y, and z using the equation. Solve for t in each statement.

$13 = 5 + 4t$	$-7 = -1 - 3t$	$8 = 4 + 2t$
$13 - 5 = 4t$	$-7 + 1 = -3t$	$8 - 4 = 2t$
$2 = t$	$2 = t$	$2 = t$

Since the value of t is the same for the x, y, and z portions of this linear equation, the second electron is following a path that is collinear to the first electron's path.

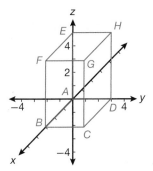

a. Identify the coordinates of the vertices of
(Ex 1) this rectangular prism.

b. Plot the point $J(3, -1, 2)$ on a three-
(Ex 2) dimensional coordinate system.

c. Determine the coordinates of two points
(Ex 3) that are collinear with the line through the
points $A(2, -2, 3)$ and $B(1, -5, 5)$.

d. A charged particle is moving along a line.
(Ex 4) First it passes through the point $(4, 7, 2)$ and
then through $(1, 12, 13)$. Will the particle hit a
target located at $(-11, 32, 60)$? Explain.

Practice Distributed and Integrated

* 1. State the coordinates of the points on the graph.
(108)

2. Two chords of a circle, $\overline{AB}$ and $\overline{CD}$, intersect at P. $AP = 8$, $BP = x$, $CP = 2$,
(86)   and $DP = 2x + 3$. Write and solve an equation for x.

3. Justify If a triangle is represented by the matrix $\begin{bmatrix} -1 & -3 & 3 \\ -3 & 3 & 1 \end{bmatrix}$ and its
(105)

image matrix after a transformation is $\begin{bmatrix} -3 & 3 & 1 \\ 1 & 3 & -3 \end{bmatrix}$, was the triangle

rotated about the origin, or was it reflected? If it was reflected, was it
across the x-axis or the y-axis? If it was rotated about the origin, was it
by 90°, 180°, 270°, or 360°? Explain your reasoning.

4. Farming A water trough for horses is in the shape of a trapezoidal prism. The
(69)   lengths of the parallel sides of the trapezoid faces are 15 inches and 21 inches.
If the water level is such that it just touches the midpoints of each leg of the
trapezoid faces, what is the length of the water level across the face?

5. Find the perimeter of the rectangle that is inscribed in a circle. The radius of the
(106)   circle is 14 centimeters and $\theta = 130°$. Round your answer to the nearest tenth
of a centimeter.

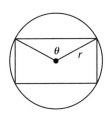

6. Prove that the line containing the points $(3, -1)$ and $(7, 1)$ is perpendicular to
(45)   the line containing the points $(-2, 4)$ and $(0, 0)$.

7. Write Describe how to use the Secant-Tangent Product Theorem to
(101)   solve for the length of the tangent segment in the given diagram.

8. **Write** Given the equations of two circles, how can you tell whether the circles are concentric or not?
(97)

9. **Error Analysis** Oliver drew this graph as the solution to the inequality $x + y > 1$. Find and correct Oliver's mistake.
(88)

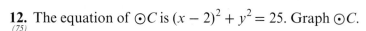

10. Find the geometric mean of $\sqrt{3}$ and $\sqrt{13}$.
(50)

11. **Algebra** Does the point $(5, -2, 6)$ lie on the line defined by the equation $(x, y, z) = (1, 0, -3) + t(2, -1, 5)$?
(108)

12. The equation of $\odot C$ is $(x - 2)^2 + y^2 = 25$. Graph $\odot C$.
(75)

*13. What is the least possible area of the rectangle that contains a circle with a radius of 10 inches?
(107)

14. (Art) Dante is painting two arcs of equal length on a circle. If the length of the chord associated with one arc is 5 centimeters, how long should he make the chord that corresponds to the other arc? Explain how you know.
(104)

15. Two chords of a circle, $\overline{AB}$ and $\overline{CD}$, intersect at P. $AP = x - 1$, $BP = 9$, $CP = 6$, and $DP = x + 1$. Write and solve an equation for x.
(86)

16. **Algebra** Find the value of x in this rhombus.
(52)

*17. Plot the points $K(4, -3, -3)$ and $L(3, -4, -4)$.
(108)

18. (Sports) A soccer ball needs to be filled with air. What is the volume of air needed if the ball is to be a perfect sphere with a radius of 4 inches?
(80)

*19. Determine the largest possible area that can be contained by an octagon with a perimeter of 16 inches. Round your answer to the nearest tenth.
(107)

20. **Multi-Step** Find the perimeter of $\triangle PQR$. Round your answer to the nearest whole number.
(94)

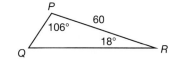

21. **Write** Describe what information is needed to prove that a quadrilateral is a parallelogram.
(61)

22. (Lunch) Alyssa has taken a roughly-semicircular bite out of her sandwich, as shown. If each bite she takes is about the same size as her first bite, how many bites in all will it take for Alyssa to eat her entire sandwich?
(40)

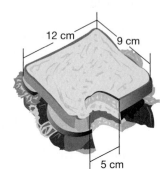

23. A regular pentagon circumscribes a circle with a radius of 8. Find the perimeter of the pentagon to the nearest tenth.
(106)

24. (**Diving**) A scuba diver is in the water, as shown, and he sees a coral at an angle of
(73) depression of 18°. He knows that he is 100 feet from the boat and sees it at a 30°
angle of elevation. If the scuba diver is 12 feet above the coral, how far is the boat
from the coral, to the nearest foot?

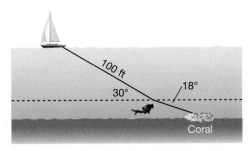

*25. Multiple Choice Which point is collinear with the points $(-1, 4, 4)$ and
(108) $(-10, 1, -8)$?

    A $(5, 6, 10)$ **B** $(2, 2, 8)$

    C $(-7, 2, -4)$ **D** $(10, 8, 20)$

26. The endpoints of $\overline{AB}$ are $A(1, 3)$ and $B(2, 5)$. Graph $\overline{A'B'}$ after $\overline{AB}$ has been
(78) rotated 90° counterclockwise about the origin. Then, graph $\overline{A''B''}$ after
$A'B'$ has been rotated 180° counterclockwise about the point $C(-3, -1)$.

*27. What is the minimum length of wire needed to secure an area of 34 square
(107) centimeters? Round your answer to the nearest tenth.

28. Algebra A system of equations has no solution. One equation is $y = -\frac{1}{3}x - 4$
(81) and the other equation is $x = by + 1$. What is the value of b?

29. Analyze Describe the translation of the graph $y = x^3$, and then write a
(100) 2×1 matrix to represent the translation.

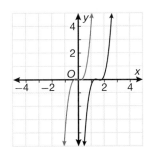

30. Solve for θ. Show your steps.
(24) $\sin \theta + \sin^2 \theta = 1.25 - \cos^2 \theta$

Non-Euclidean Geometry

Warm Up

1. **Vocabulary** Because distance cannot have a negative value, it is calculated
   (9)
   using a(n) _____. (**square root, square, absolute value**)

2. Solve the system of linear equations.
   (81)
   $y = 2x - 2$
   $y = -3x + 3$

3. **Multiple Choice** The distance from a point to a line is 5. What is *not* a
   (42)
   possible coordinate for the point, if the line is $y = \frac{3}{4}x + 2$?

   A $(4, -1)$ **B** $(7, 1)$

   C $(3, -2)$ **D** $(-1, -5)$

New Concepts

All the lessons prior to this one have dealt with **Euclidean geometry**, which
is a system of geometry described by the Greek mathematician Euclid. The
primary difference between Euclidean geometry and other geometries is
that Euclidean geometry assumes the Parallel Postulate holds.

Non-Euclidean geometry is based on figures in a curved surface and is a
system of geometry where the Parallel Postulate does not hold true. One
type of non-Euclidean geometry is **spherical geometry**, a system of
geometry defined on a sphere.

Recall that a circle dividing a sphere into two hemispheres is called a great
circle. On a sphere, the shortest distance between any two points is a minor
arc of a great circle. Therefore, instead of referring to lines in spherical
geometry, we refer to arcs of great circles. An angle on a sphere is the angle
between the planes that two intersecting great circles lie on.

Math Reasoning

Analyze Do two points
on a sphere always
define a single great
circle? Explain.

Example 1   **Naming Figures on a Sphere**

Identify a great circle, a segment, an angle, and a triangle on the sphere.

SOLUTION

There are three great circles: great circle
AB, great circle *CD*, and great circle *AD*.

A segment on the sphere is $\overline{CD}$.

An angle on the sphere is the one formed
by the intersection of great circles *AB*
and *AD*.

A triangle consists of three connected points. Therefore, *ACD* is a triangle.

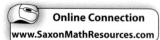
Online Connection
www.SaxonMathResources.com

(Exploration) **Exploring Triangle Angles in Spherical Geometry**

In this exploration, you will investigate spherical geometry using a globe.

1. Choose three cities on a globe. Their positions should form a triangle. Make sure they are far enough apart to easily measure.

2. With a tape measure, measure the distances between the cities as accurately as possible.

3. Use the Law of Cosines to find the angles in the triangle.

4. Use a protractor to measure the angles in the triangle on the sphere. Are they the same as the answers you found with trigonometry?

5. Now add the angle measures of the triangle. Do they add to 180°? If not, is it more or less?

In spherical geometry, the angle measures of a triangle will always sum to more than 180°. Imagine three points on a sphere that are close together. The triangle they form will be relatively flat, and the sum of its angles will be close to 180°. As the points get farther away from each other, the triangle becomes more curved, and the sum of its angle measures gets further from 180°.

Example 2 **Classifying Triangles on a Sphere**

Classify the triangle shown according to its angle measures and side lengths.

SOLUTION
$m\angle A = 90°$
$m\angle B = 90°$
$m\angle C = 90°$

This is an equiangular and equilateral triangle, since all the side lengths and all the angle measures are equal.

Notice that the sum of the angles of the triangle is 270°.

Hint

Recall that in Euclidean geometry you can prove the Triangle Sum Theorem using the Parallel Postulate. Since the Parallel Postulate is no longer a valid reason in spherical geometry, the Triangle Angle Sum Theorem is not either.

The area of a spherical triangle is part of the surface area of the sphere. If the radius of the sphere and the measures of each angle of the triangle are known, then the area of the triangle can be determined.

Area of a Spherical Triangle
The area of a triangle on a sphere is given by the following formula, where L, M, and N are the vertices of the triangle and the sphere has a radius of r. 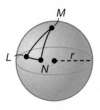 $$A = \frac{\pi r^2}{180°}(m\angle L + m\angle M + m\angle N - 180°)$$

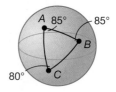

Example 3 **Finding the Area of a Spherical Triangle**

Find the area of the spherical triangle when the radius of the sphere is 3 meters. Round your answer to the nearest hundredth.

SOLUTION

Using the formula for area of a spherical triangle:

$$A = \frac{\pi r^2}{180°}(m\angle A + m\angle B + m\angle C - 180°)$$

$$A = \frac{\pi (3)^2}{180°}(85° + 85° + 80° - 180°)$$

$$A = \frac{7\pi}{2}$$

$$A \approx 11.00$$

The area of the spherical triangle is about 11 square meters.

Lesson Practice

a. Identify a great circle, a segment, and a
(Ex 1) triangle on the sphere.

b. Classify the triangle as equilateral,
(Ex 2) isosceles, or scalene.

c. Find the area of the spherical triangle to the
(Ex 3) nearest tenth. The diameter of the sphere is 24 inches.

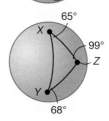

Practice **Distributed and Integrated**

1. (**Wheels**) Some wheels have support spokes equally spaced along the
(104) circumference to keep the wheel from breaking. How can you confirm that the arc length between the first and third spokes is equal to the arc length between the second and fourth spokes?

* 2. Triangle *JKL* is on a sphere with radius 10 centimeters. Its angles
(109) measure 90°, 80°, and 45°. What is the area of △*JKL* to the nearest tenth of a square centimeter?

3. Multiple Choice Sonja always has a salad for lunch on Tuesdays. Which statement
(20) is true?

    A If Sonja has a salad for lunch, then it is a Tuesday.

    B Sonja has a salad for lunch if and only if it is a Tuesday.

    C If it is a Tuesday, then Sonja has a salad for lunch.

    D All of these are true.

4. A dilation with a scale factor of $\frac{1}{4}$ is applied to this rectangular prism.
(84) Find the surface area and the volume of the image that results.

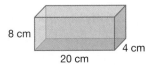
8 cm   4 cm   20 cm

5. The vertices of $\triangle QRS$ are $Q(1, 1)$, $R(5, 3)$, and $S(2, -3)$. Use matrix
(105) multiplication to first rotate the triangle 90° about the origin, and
then reflect the image across the *x*-axis. What is the image matrix?

6. Write Do you need to construct all three perpendicular bisectors of a triangle to
(38) locate its circumcenter? Explain.

7. What point does a reflection across the line $y = x$ map (x, y) onto?
(74)

8. Analyze The frustum of a cone is 10 inches tall. The base radius is 3 inches
(103) and the upper radius is 2 inches. The frustum of a square pyramid is also
10 inches tall. It has a base square with side lengths of 3 inches and an upper
square with side lengths of 2 inches. Which frustum has the greater volume?

* 9. Determine the coordinates of two points that lie on the line
(108) $(x, y, z) = (4, 0, 0) + t(1, 0, 1)$.

10. A regular hexagon can be divided into six equilateral triangles. The height
(66) of each triangle is 6 centimeters. What is the perimeter of the hexagon?
What is the area?

*11. Justify Determine whether $\overline{BC}$ and $\overline{EF}$ are parallel. How do you know?
(60)

12. The sides of two equilateral triangles are in the ratio 1:4. The area of the larger
(87) triangle is 64 square feet. What is the area of the smaller triangle?

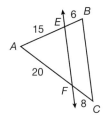

*13. In spherical geometry, the shortest distance between two points is the minor arc
(109) of a _____.

14. Write Marina has a photograph that is 5 inches by 7 inches. She wants to enlarge
(96) the photo so that the length and width are each tripled. Describe how the area of
the photo will change.

15. Use the figure shown at the right to find m$\overset{\frown}{QS}$.
(104)

16. Emma wants to use 100 centimeters of wire to enclose either a triangular
(107) area or a rectangular area. What is the largest possible area that she can
enclose?

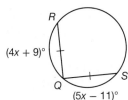
$(4x + 9)°$ $(5x - 11)°$

17. Multiple Choice What is the magnitude of the vector $\langle 5, 7 \rangle$, to the nearest
(63) whole unit?

    A 6 units   **B** 7 units

    C 8 units   **D** 9 units

18. If $\angle AEC$ is a right angle and $\overrightarrow{EB}$ bisects $\angle AEC$, find $m\overset{\frown}{AB}$.
₍₇₉₎

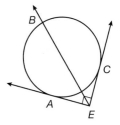

*19. ⟮**Navigation**⟯ A boat sailing on the Atlantic Ocean traveled along a triangular
₍₁₀₉₎ path with vertices at 77°, 56°, and 90°. If Earth has a radius of 3960 miles, what is the area of the triangle that the boat traveled around?

20. **Algebra** If $\sin^2\theta = 0.12$, what does $\cos\theta$ equal?
₍₉₁₎

*21. **Verify** Show that the point $(1, -8, -4)$ lies on the line through the points
₍₁₀₈₎ $(4, -5, -1)$ and $(3, -6, -2)$.

*22. **Error Analysis** Eddie was asked to find the perimeter of the largest square
₍₁₀₆₎ that can be inscribed in a circle with a 7-inch radius. His answer was 56 inches. What is the correct answer and what was Eddie's mistake?

23. ⟮**Closets**⟯ A storage closet is built underneath a staircase. Find the
₍₅₆₎ height of the closet to the nearest tenth.

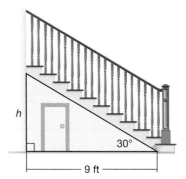

24. ⟮**Art**⟯ Tracy wants to make a circle in her art class using a cone that
₍₈₅₎ has a radius of 4 centimeters and a height of 20 centimeters. If she cuts the cone parallel to the circular face 5 centimeters from the vertex, what is the area of the circle that is formed? Round your answer to the nearest hundredth.

25. **Multiple Choice** Which of these points is collinear with the points
₍₁₀₈₎ $(1, -4, 6)$ and $(2, 5, 8)$?

A $(4, 23, 12)$ **B** $(0, 5, 4)$
C $(3, 22, 10)$ **D** $(-1, -14, 2)$

26. Consider the statements, "A quadrilateral has four right angles," and
₍₂₀₎ "A quadrilateral is not a rectangle." Which is true: their conjunction, their disjunction, both, or neither?

27. ⟮**Ornithology**⟯ An ornithologist, someone that studies birds, put radio tags on
₍₈₉₎ two birds and released them at the same time from the same place to study their migration. After two days, one bird was 27 miles away at 210 degrees southwest and the other was 22 miles away at 120 degrees southeast. How far apart were the birds from each other, to the nearest tenth of a mile?

28. What is the maximum area of an octagon with a perimeter of 24 centimeters, to
₍₁₀₇₎ the nearest tenth?

29. Triangle PQR has been reflected to form $\triangle P'Q'R'$. Is the slope of the
₍₆₇₎ line of reflection positive, negative, zero, or undefined?

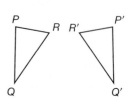

30. What is the diameter of a circle in which a 9.2-centimeter chord
₍₄₃₎ is 4.3 centimeters from the center of a circle?

Scale Drawings and Maps

Warm Up

1. **Vocabulary** A _____ is a transformation where a figure is increased or
(84) decreased in size but retains the same shape. (*dilation, rotation, reflection*)

2. Calculate the perimeter of the rectangle to the
(96) nearest hundredth if it is dilated by a factor of
one-third. Then, find the ratio between the
original perimeter of the rectangle and the
dilated perimeter.

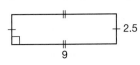

3. **Multiple Choice** Which of the following flags has both rotational and line
(76) symmetry?

A   **B**   **C**   **D**

New Concepts

A **scale** is the ratio between two corresponding measurements. A **scale
drawing** is a drawing that uses a scale to represent an object as being smaller
or larger than the actual object.

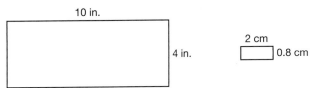

Math Reasoning

Write Name a few
instances where scale
drawings are commonly
used.

In the diagram, the second rectangle is a scale drawing of the first with a
scale of 1 cm : 5 in. In other words, every centimeter on the scale drawing
represents 5 inches of the actual figure.

Example 1 | Making a Scale Drawing

The White House is approximately 168 feet long and 85 feet wide.
Draw a scale representation of the base of the building using a scale
of 1 cm : 30 ft. Give your answer to the nearest tenth.

SOLUTION
Set up a proportion with the scale on one side and the length of the White
House and its scale drawing on the other side, where the length of the scale
drawing is ℓ. Notice that the dimensions given in feet on each side of the
equation cancel, leaving the answer in centimeters.

Length: $\dfrac{\ell}{168 \text{ ft}} = \dfrac{1 \text{ cm}}{30 \text{ ft}}$

$\ell = 5.6 \text{ cm}$

Width: $\dfrac{w}{85 \text{ ft}} = \dfrac{1 \text{ cm}}{30 \text{ ft}}$

$w \approx 2.8 \text{ cm}$

Scale drawings are often used when the object is either too large or too small to be represented on paper.

A **scale model** is a three-dimensional model that uses scale to represent an object smaller or larger than the actual object.

Hint

This question refers to picometers, a very small unit of measure. The rate of conversion between picometers and meters

is 1 pm: 1×10^{-12} m. But, in this example, the unit conversion is unnecessary, because the picometers in the problem cancel out, leaving an answer in centimeters.

Example 2 | Making a Scale Model

The carbon atom, which is roughly spherical, has an atomic radius of 70 picometers. If a student models the carbon atom with a scale of 5 pm : 1 cm, what is the radius of the model?

SOLUTION

$$\frac{1 \text{ cm}}{5 \text{ pm}} = \frac{x}{70 \text{ pm}}$$

$$14 \text{ cm} = x$$

The atom model would have a radius of 14 centimeters.

Example 3 | Application: Maps

Adina lives 145 meters from Darren. On the map, that distance is represented as 2 centimeters. If, on the map, Carmen lives 3.5 centimeters from Darren and 4 centimeters from Adina, what is the actual distance that Carmen lives from Adina and Darren?

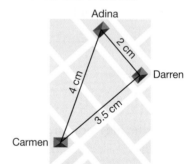

SOLUTION

First, we must determine the scale factor of the map. Since 2 cm : 145 m, reduce the ratio by dividing by 2 to find that 1 cm : 72.5 m.

From Carmen's house to Darren's house on the map, it is 3.5 centimeters.

$$\frac{1 \text{ cm}}{72.5 \text{ m}} = \frac{3.5 \text{ cm}}{x}$$

$$x = 253.75 \text{ m}$$

The actual distance from Carmen's house to Darren's is 253.75 meters.

From Carmen's house to Adina's house on the map, it is 4 centimeters.

$$\frac{1 \text{ cm}}{72.5 \text{ m}} = \frac{4 \text{ cm}}{x}$$

$$x = 290 \text{ m}$$

The actual distance from Carmen's house to Adina's is 290 meters.

a. Make a scale drawing of the base of an apartment building that is
(Ex 1) 30 meters long and 15 meters wide. Use a scale of 1 m : 0.5 cm.

b. The Washington Monument is approximately 555.5 feet tall. If a
(Ex 1) scale drawing of the Washington Monument is 10 inches tall,
what is the scale?

c. The planet Mercury has a diameter of 3031 miles. If a scale model
(Ex 2) is made with a ratio of 200 miles : 1 foot, what is the diameter of
the model? Round your answer to the nearest tenth.

d. On a map of the United States, San Antonio is 3.3 centimeters away
(Ex 4) from Dallas. They are actually 277 miles apart. If Oklahoma City is
2.4 centimeters away from Dallas on the same map, how far apart are
the two cities? What is the scale of the map? Round your answer to the
nearest whole number.

Practice   Distributed and Integrated

1. Error Analysis Based on the information that $\overline{AB} \parallel \overline{CD}$, $\overline{AD} \parallel \overline{BC}$, and that $AC \neq BD$,
(65) Suzanne concludes that the figure on the right must be a rhombus. Yvonne argues
that there is not enough information to know that for sure. Who is correct? Why?

2. Multiple Choice The perimeter of an equiangular triangle with side lengths
(51) *s*, *t*, and *u* is:

A $3s$ **B** $3t$

C $3u$ **D** All of the above

3. What is the total arc length highlighted in blue? Write your answer
(35) in terms of π.

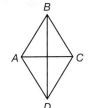

4. Write Explain why the line passing through the points $(-2, 3, 1)$ and
(108) $(5, 1, 1)$ can either be written as $(x, y, z) = (-2, 3, 1) + t(7, -2, 0)$ or
$(x, y, z) = (5, 1, 1) + t(7, -2, 0)$.

5. **Automotive Repair** A mechanic raises a car hood to inspect
(53) the engine. To keep the hood from falling, the mechanic
positions the prop rod from the base to the hood. If the
base of the prop rod is 67 centimeters from where the
hood connects to the car, how long is the rod?

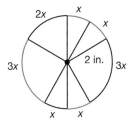

6. Simplify $\begin{bmatrix} 10 & -2 \\ 5 & 1 \end{bmatrix} \begin{bmatrix} -1 & -3 \\ -4 & 0 \end{bmatrix}$.
(105)

* 7. Find the area of a triangle that is on a sphere with a radius of 8 centimeters and
(109) whose angle measures have a sum of 215°. Round your answer to the nearest tenth
of a square centimeter.

8. Algebra Determine the equation of the line that passes through the points
(108) $G(3, 1, -2)$ and $H(6, 9, 3)$.

9. If an arc is 148°, and the measure of its inscribed angle is $(3x - 4)$°, what is the value of x?
(47)

10. **Write** Describe two successive dilations that would cause the preimage to coincide with the image.
(102)

11. **Multi-Step** Find the perimeter of the triangle. Round your answer to the nearest tenth.
(98)

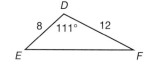

12. **Algebra** Does the function $y = x^3$ have a line of symmetry? Does it have rotational symmetry? If so, what is the order of rotational symmetry?
(76)

13. Find the perimeter of the rectangle that is inscribed in the circle. The radius of the circle is 8 inches and $\theta = 55$°. Round your answer to the nearest tenth.
(106)

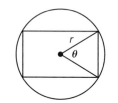

14. **Multiple Choice** For a given perimeter, which of the following shapes has the greatest area?
(107)

   A an equilateral triangle   **B** a square
   C a pentagon   **D** a regular dodecahedron

15. Determine the point of tangency and the equation of the tangent line which is tangent to both circles. Find the radius of each circle.
(72)

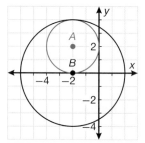

16. **Maps** Philadelphia is 12.1 centimeters away from Washington, D.C. on a map. If their actual distance is 139 miles, what is the scale of the map? Round your answer to the nearest hundredth.
(110)

17. A cylinder has a height that is three times the length of the diameter. The volume of the cylinder is 509 cubic meters. Find the approximate surface area.
(62)

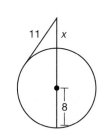

18. **Algebra** Find x to the nearest tenth of a unit.
(101)

19. **Navigation** A ship sets out on a course traveling 6 miles per hour at 60° north of east. If the current is traveling at an equal speed at 30° north of east, what is the actual speed and direction of the ship to the nearest tenth?
(83)

*20. **Astronomy** In a model of the solar system, Mercury is 4 feet from the sun. Its actual distance from the sun is 36.0 million miles. If Jupiter is 483.4 million miles from the sun, how far would it be from the sun in the same model? Round your answer to the nearest foot.
(110)

*21. Find the area of $\triangle LMN$ if the radius of the sphere is 4 centimeters.
(109)

*22. **Model** Draw a scale drawing of your classroom. Include the desks, the chalkboard, and other objects in the classroom. You can either measure the distances or estimate them.
(110)

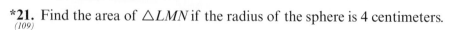

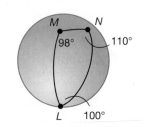

23. (**Fashion**) Helga has 31.5 inches of material that she will use to make two square
(44) designs on a shirt. The similarity ratio for the squares is 2:5. How long will each
side of each square be?

24. Write Two different circles have arcs of equal length. Does this mean that the
(104) chords created by connecting the endpoints of each arc are equal in length?
Explain.

25. Algebra Find the side length of a square if it is inscribed in a circle with a radius, *r*.
(106) What is the side length of a square circumscribing the circle?

26. Multiple Choice This orthographic drawing shows the front, top, and
(93) left-side views of a solid. Which of these sketches represents the solid?

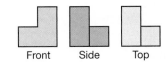

Front   Side   Top

A 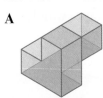   **B**

C   **D**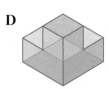

27. Multi-Step Classify the polygon with vertices $W(-3, -1)$, $X(1, 1)$, $Y(0, 3)$,
(57) and $Z(-2, 2)$. Find the area.

*28. Classify the triangle on the sphere as equilateral, isosceles, or scalene.
(109)

29. Generalize Determine a formula for the surface area of a sphere in terms of
(80) its diameter, *d*.

30. If two parallel lines are cut by a transversal, and the alternate interior angles
(Inv 1) formed measure $(-2x + 40)°$ and $(3x - 15)°$, find the measure of each angle.

Golden Ratio

Two numbers display the **golden ratio** when their sum divided by the larger number is equal to the ratio of the two numbers. That is, a and b are in the golden ratio if $\frac{a+b}{a} = \frac{a}{b}$.

The golden ratio is equal to $\frac{1+\sqrt{5}}{2}$ or approximately 1.618. It is commonly represented by the Greek letter phi, ϕ. The golden ratio has the property that its reciprocal is itself minus 1 ($\frac{1}{\phi} = \phi - 1$). The golden ratio has been known since ancient times, and was often used in art and architecture by the Greeks and later by European artists. These artists considered shapes with lengths related by the golden ratio to be pleasing to the eye.

In the Greek Parthenon shown here, nearly every aspect of the front of the building displays the golden ratio, from the distance between the columns to the proportions in the triangle that crowns the structure, to the proportions of the building's height and width.

Hint

To solve problems 1 through 3, remember that if two numbers are related by the golden ratio, $\frac{a+b}{a} = \phi$ and the golden ratio is equal to $\frac{1+\sqrt{5}}{2}$.

1. Suppose two numbers, a and b, are in the golden ratio. If a is the larger number and $a = 1$, what is b, in simplified radical form?

2. Suppose c and d are in the golden ratio. If c is the larger number and $c = \sqrt{5}$, what is d, in simplified radical form?

3. Suppose e and f are in the golden ratio. If f is the smaller number and $f = 2$, what is e, in simplified radical form?

To understand why $\phi = \frac{1+\sqrt{5}}{2}$, consider the proportion given in the first paragraph. We know that for two numbers to be in the golden ratio, they must satisfy the equation. Call this ratio x, so $\frac{a+b}{a} = x$ and $\frac{a}{b} = x$.

4. In the equation $\frac{a}{b} = x$, solve for a.

5. Substitute your expression for a from step 4 into the equation $\frac{a+b}{a} = \frac{a}{b}$ and simplify the resulting expression.

6. Solve the resulting quadratic expression from step 5, using the quadratic formula. Write your answer in simplified radical form.

A **golden rectangle** is a rectangle in which the ratio of the length of the longer side to the length of the shorter side is in the golden ratio.

Let $ABCD$ be a golden rectangle, with $\frac{AB}{BC} = \frac{b}{h} = \phi$, where b is the longer side of the rectangle.

7. Write a proportion relating b and h.

8. Suppose $b = 2\sqrt{5}$. What is the value of h in simplified radical form?

Another common geometric figure that features the golden ratio is a golden spiral. To construct a golden spiral, follow the steps shown in the diagrams here. Start by drawing a square with side lengths of any size. Then draw another identical square next to it.

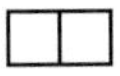

Next, draw a square above these two that has a width equal to the sum of the smaller squares' width.

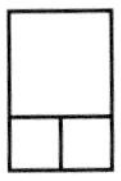

Now draw another square, bordering our existing squares on the right, which is as wide as both the smaller square and the larger square combined.

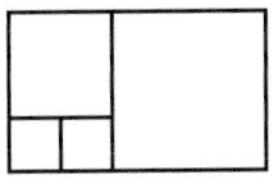

9. Draw the next iteration of this pattern. The next square should continue to spiral outwards, below the squares that have already been drawn.

10. Draw the next iteration. This time, the square should border the left side of your existing squares.

11. Draw one more iteration of the golden spiral.

Math Reasoning

Verify Measure the sides of the large rectangle you created in step 11. Show that the ratio of its sides is approximately equal to the golden ratio.

If a quarter circle is drawn between the opposite vertices of each square in the figure we have just constructed, it will result in the golden spiral, which is shown here. The large rectangle which contains the spiral and each of the smaller rectangles formed by the squares are golden rectangles.

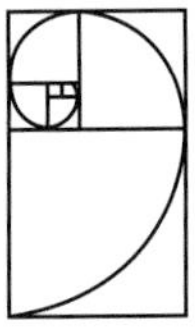

Investigation Practice

The **Fibonacci sequence** is the infinite sequence of numbers beginning 1, 1, 2, 3, 5, … such that each term is the sum of the two previous terms.

a. Determine the first ten Fibonacci numbers. Then, determine the ratios $\frac{1}{1}, \frac{2}{1}, \frac{3}{2}$ … of each consecutive pair of these numbers. Round your answers to four decimal places, where necessary.

b. Based on part **a**, state a hypothesis about the ratios of consecutive Fibonacci numbers.

c. Check your hypothesis by calculating ratios until the value is fixed for the first 5 decimal places.

d. Explain how the squares in the golden rectangle are related to the Fibonacci sequence.

Finding Distance and Midpoint in Three Dimensions

New Concepts

The distance between two points in three-dimensional space can be found using the distance formula for three dimensions. Notice that it is very similar to the distance formula used for coordinate planes, with the addition of a third term, z.

Math Reasoning

Analyze When finding the distance between two points in three-dimensional space, does it matter which point is chosen to be (x_1, y_1, z_1) and which is (x_2, y_2, z_2)? Explain your reasoning.

Distance Formula for Three Dimensions
Given two points in three-dimensional space, $A(x_1, y_1, z_1)$ and $B(x_2, y_2, z_2)$, the distance between them is found by the following equation. $$d = \sqrt{(x_2 - x_1)^2 + (y_2 - y_1)^2 + (z_2 - z_1)^2}$$ 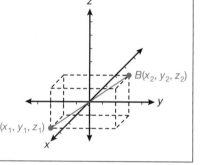

Example 1 Finding Distance in Three Dimensions

a. Find the distance between $M(5, 7, -3)$ and $N(0, -3, 3)$.

SOLUTION

$$d = \sqrt{(x_2 - x_1)^2 + (y_2 - y_1)^2 + (z_2 - z_1)^2} \quad \text{Distance formula}$$

$$MN = \sqrt{(0 - 5)^2 + (-3 - 7)^2 + (3 - (-3))^2} \quad \text{Substitute.}$$

$$MN = \sqrt{25 + 100 + 36} \quad \text{Simplify.}$$

$$MN = \sqrt{161}$$

The distance between M and N is $\sqrt{161}$.

b. Find the distance between $C(0, 5, 2)$ and $D(2, 0, 4)$.

SOLUTION

$$d = \sqrt{(x_2 - x_1)^2 + (y_2 - y_1)^2 + (z_2 - z_1)^2} \quad \text{Distance formula}$$

$$CD = \sqrt{(2 - 0)^2 + (0 - 5)^2 + (4 - 2)^2} \quad \text{Substitute.}$$

$$CD = \sqrt{4 + 25 + 4} \quad \text{Simplify.}$$

$$CD = \sqrt{33}$$

The distance between C and D is $\sqrt{33}$.

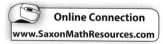

Online Connection
www.SaxonMathResources.com

In Lesson 11, you learned how to find the midpoint of a line segment in two dimensions. You can find the midpoint of a segment in three-dimensional space by adding an additional term to the midpoint formula.

Midpoint Formula in Three Dimensions

Given two points in three dimensional space, $A(x_1, y_1, z_1)$ and $B(x_2, y_2, z_2)$, the midpoint M can be found using the following equation.

$$M = \left(\frac{x_1 + x_2}{2}, \frac{y_1 + y_2}{2}, \frac{z_1 + z_2}{2}\right)$$

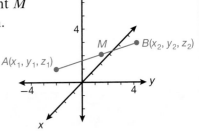

Example 2 Finding Midpoint in Three Dimensions

The endpoints of $\overline{XY}$ are $X(9, 4, -2)$ and $Y(-3, 6, -8)$. Find the midpoint.

SOLUTION

$$M = \left(\frac{x_1 + x_2}{2}, \frac{y_1 + y_2}{2}, \frac{z_1 + z_2}{2}\right) \qquad \text{Midpoint formula}$$

$$M = \left(\frac{9 + (-3)}{2}, \frac{4 + 6}{2}, \frac{(-2) + -8}{2}\right) \qquad \text{Substitute.}$$

$$M = (3, 5, -5) \qquad \text{Simplify.}$$

The midpoint of $\overline{XY}$ is $(3, 5, -5)$.

> **Hint**
>
> Notice that both the midpoint formula and the distance formula for three dimensions are exactly the same as their 2-dimensional counterparts when both points have the same z-coordinate.

Example 3 Application: Mountain Climbing

Two people are climbing a mountain. In terms of each other's position, the first climber is 75 meters north, 25 meters west, and 48 meters higher than the second climber. What is the distance between the climbers? Round your answer to the nearest tenth.

SOLUTION

The first climber's position is given in terms of the second climber's position. Therefore, you can assign coordinates of $(0, 0, 0)$ to the second climber. The coordinates for the first climber are $(25, 75, 48)$.

$$d = \sqrt{(x_2 - x_1)^2 + (y_2 - y_1)^2 + (z_2 - z_1)^2} \qquad \text{Distance formula}$$

$$d = \sqrt{(25 - 0)^2 + (75 - 0)^2 + (48 - 0)^2} \qquad \text{Substitute.}$$

$$d = \sqrt{8554} \qquad \text{Simplify.}$$

$$d \approx 92.5$$

The distance between the two climbers is about 92.5 meters.

a. Find the distance between $R(0, 12, -2)$ and $S(6, 8, -3)$.
(Ex 1)

b. Find the midpoint of $F(22, 14, 9)$ and $G(15, -8, -6)$.
(Ex 2)

c. (**Skydiving**) Two people are skydiving from an airplane. Approximately
(Ex 3) how far apart are they at the instant when the first person is 40 feet west, 25 feet north, and 300 feet below the plane, and the second person is 25 feet west, 10 feet north, and 250 feet below the plane? Round your answer to the nearest tenth.

Practice Distributed and Integrated

1. A circle whose equation is $x^2 + y^2 = 36$ is shifted 5 units up and 2 units to the left.
(95) What is the equation for the shifted circle? Graph the shifted circle.

2. Onto what point does a reflection across the line $y = x$ map the point $(-2x, 3y)$?
(74)

3. Quadrilateral $JKLM$ was rotated counterclockwise about the origin. The original
(78) vertices were located at $J(4, -2)$, $K(6, 1)$, $L(-1, -3)$, and $M(0, 2)$. After the rotation, the vertices were located at $J'(2, 4)$, $K'(-1, 6)$, $L'(3, -1)$, and $M'(-2, 0)$. What was the angle of rotation?

* 4. Find the midpoint of $D(-10, 2, 13)$ and $E(-4, -2, 5)$.
(111)

5. Multi-Step Find expressions for R and r if the height of the equilateral
(106) triangle is h.

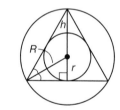

6. A rhombus has diagonals that are 10 and 24 units long,
(65) respectively. Find the perimeter of the rhombus.

7. Error Analysis Donald states that the point $(13, 18, 16)$ must lie on the line defined by
(108) the equation $(x, y, z) = (4, 3, 5) + t(3, 5, 4)$. Is he correct? Explain why or why not.

8. (**Landscaping**) Ingrid is planning to build a plastic greenhouse for her plants. Which
(59) of these frames will require more plastic?

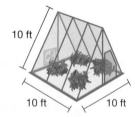

9. Algebra A system of three equations has no solution. One equation is
(81) $y = 2x + 7$, another equation is $2y = mx + 8$, and the third equation is $y = 3nx + p$. What are the values of m, n, and p?

*10. **Space Exploration** Two astronauts are repairing a satellite telescope as shown in
(111) the figure. The first astronaut is 6 feet in front of the satellite, and the second
astronaut is 3 feet behind the satellite. How far apart are the astronauts from each
other? Round your answer to the nearest tenth of a foot.

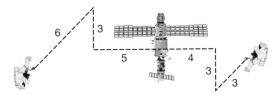

*11. Find the midpoint M and distance d between $V(3, 7, -6)$ and $W(-4, -8, 7)$.
(111)

12. **Algebra** Points J, K, and L are midpoints of $\triangle FGH$, shown at right. If
(55) $FG = 4x + 18$ and $KL = 5x - 12$, what is the length of $\overline{FG}$?

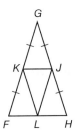

13. Find the matrix that would represent the translation of a
(100) quadrilateral $\frac{1}{2}$ unit up and 4 units left.

14. Draw two triangles that are similar by SAS. Explain how you made
(46) your sketches.

*15. **Chemistry** An iron atom has an atomic radius of 140 picometers (pm). If a model
(110) of the atom has a diameter of 4 centimeters, what is the scale factor?

16. **Justify** Write and solve an equation for the value of x that allows
(12) you to apply the Converse of the Same-Side Interior Angles
Theorem. Explain your answer.

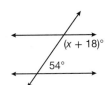

*17. Find two numbers that are in the golden ratio to 15. Round your
(Inv 11) answers to the nearest hundredth.

18. **Milk Cartons** These two milk cartons are similar in shape. If
(99) the smaller carton holds 32 ounces, what volume does the
larger hold?

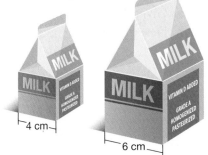

19. **Justify** Rogelio has a set of sticks arranged to form an isosceles
(69) trapezoid. He adjusts them so that the figure has right angles.
Explain why the new figure must be a rectangle.

20. **Formulate** If the sum of the angles of a spherical triangle are
(109) 181°, does that mean that the triangle is large or small, relative
to the size of the sphere? Explain. Use the formula for the area
of spherical triangles to justify your answer.

21. **Algebra** Determine the value of x in the figure at right. Write your answer
(33) in simplified radical form.

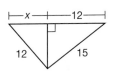

22. **Gift Wrapping** Harris wrapped a present for a friend in the shape of a frustum of
(103) a square pyramid. If the present was 21 centimeters tall and the two square faces
had side lengths of 15 centimeters and 25 centimeters, what was the volume of his
present?

23. Name a line, a segment, and a triangle on the sphere.
(109)

*24. Find the distance between $A(-1, 2, -3)$ and $B(-6, -5, 4)$.
(111)

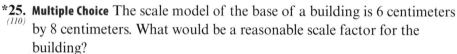

*25. **Multiple Choice** The scale model of the base of a building is 6 centimeters
(110) by 8 centimeters. What would be a reasonable scale factor for the
building?

    A 1 cm : 1 ft   **B** 1 cm : 2 m

    C 1 cm : 100,000 cm   **D** 1 cm : 6 in.

 26. **Algebra** Write an expression for the side length s of a regular n-sided polygon
(66) with area A and apothem length a.

27. What is the lateral area of a regular pentagonal pyramid with a side length
(70) of 6.2 inches and a slant height of 4 inches?

28. **Generalize** Derive a formula to find the minimum perimeter, P, required to enclose
(107) a rectangular area, A.

29. Write three similarity statements for these triangles.
(50)

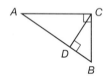

30. **Write** An abstract mosaic design has two types of tile. One tile is octagonal.
(Inv 9)   **a.** Given that the design is a semiregular tessellation, sketch a section of it.
    b. Describe the symmetries of the design.

Finding Areas of Circle Segments

Warm Up

1. **Vocabulary** In a circle, the region bounded by two radii and
   (86) their intercepted arc is a _____ of the circle.

2. Determine the height, *BD*, and the area of $\triangle ABC$.
   (53) Give exact and approximate answers.

3. A circle sector has a central angle of 30°. The
   (35) radius of the circle is 3 centimeters. What is the
   area of the sector?

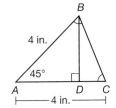

New Concepts   A **segment of a circle** is a region inside a circle that is bounded by a chord and
an arc.

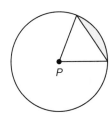

The shaded area in the diagram is a segment of the circle. Since the area of
the segment is the difference between the area of the sector and the area
of the triangle, its area can be found by subtracting the triangle's area from
the sector's area.

Example 1   Finding the Area of a Segment

Determine the area of the shaded segment of $\odot Q$.
Round your answer to the nearest hundredth.

6 in.

Q

SOLUTION

First, find the area of the circle sector and of the
triangle. Since the triangle is an isosceles right
triangle, its height and its base are both 6 inches.

$$A_{sector} = \pi r^2 \left(\frac{m°}{360°}\right) \qquad A_{triangle} = \frac{1}{2}bh$$

$$A_{sector} = \pi(6^2)\left(\frac{90°}{360°}\right) \qquad A_{triangle} = \frac{1}{2}(6)(6)$$

$$A_{sector} = 9\pi \qquad A_{triangle} = 18$$

Finally, subtract the area of the triangle from the area of the sector.

$$A_{segment} = A_{sector} - A_{triangle}$$
$$A_{segment} = 9\pi - 18$$
$$A_{segment} \approx 10.27$$

Math Reasoning

Connect How could
you use the solution
for Example 1 to find
the total area of the
segments formed by a
square inscribed in
the same circle?

Example 2 Using Trigonometry to Find Segment Area

A circle has a radius of 5.2 centimeters. Determine the area of the segment formed by a chord with a central angle of 78°. Give your answer to the nearest hundredth.

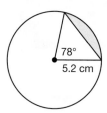

SOLUTION
First find the area of the sector.

$$A_{sector} = \pi r^2 \left(\frac{m°}{360°} \right)$$
$$A_{sector} = \pi (5.2)^2 \left(\frac{78°}{360°} \right)$$
$$A_{sector} \approx 18.41$$

Next, determine the height of the triangle. The altitude of the triangle can be found using the sine function.

$$\sin 78° = \frac{h}{5.2}$$
$$h \approx 5.09$$

Next, substitute the value found for h into the area of a triangle formula.

$$A_{triangle} \approx \frac{1}{2}(5.2)(5.09)$$
$$A_{triangle} \approx 13.23 \text{ cm}^2$$

Therefore, the area of the triangle is approximately 13.23 square centimeters.

Now find the area of the segment.

$$A_{segment} = A_{sector} - A_{triangle}$$
$$A_{segment} \approx 18.41 - 13.23$$
$$A_{segment} \approx 5.18 \text{ cm}^2$$

The area of the segment is approximately 5.18 square centimeters.

Math Reasoning

Predict Is the area of a sector proportional to its central angle? What about the area of a segment?

Example 3 Application: Civil Engineering

A wall placed on either side of a bridge is a segment of a circle that corresponds to a central angle of 45°. The length of the bridge is 40 feet. Find the area of the wall.

SOLUTION
First, determine the radius of the circle. Draw an altitude to the triangle, bisecting the central angle and the bridge. This creates a right angle, so a trigonometric function can be used to solve for the radius.

$$\sin 22.5° = \frac{20}{r}$$
$$r \approx 52.26$$

Next, determine the area of the sector.

$$A_{sector} = \pi r^2 \left(\frac{m°}{360°} \right)$$
$$A_{sector} = \pi (52.26)^2 \left(\frac{45°}{360°} \right)$$
$$A_{sector} = 1072.09$$

The height of the triangle can be found just as the radius was, but using the cosine function. The height is (52.25)(cos 25.5°). Use this to determine the area of the triangle.

$$A_{triangle} = \frac{1}{2}bh$$

$$A_{triangle} = \frac{1}{2}(40)(52.25)(\cos 25.5°)$$

$$A_{triangle} \approx 943.20$$

Using the area of the triangle, find the area of the segment.

$$A_{segment} = A_{sector} - A_{triangle}$$

$$A_{segment} \approx 1072.09 - 943.20$$

$$A_{segment} \approx 128.89 \text{ ft}^2$$

Therefore, the area of the structure is approximately 128.89 square feet.

Lesson Practice

a. In $\odot C$, $\overline{CB}$ and $\overline{CA}$ are perpendicular radii that are each *(Ex 1)* 48 millimeters long. Determine the area of the segment formed by $\overline{AB}$, to the nearest square millimeter.

b. Determine the area of the shaded circle segment, *(Ex 2)* to the nearest square inch.

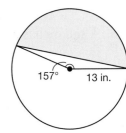

c. (**Civil Engineering**) The base of a dam is in the *(Ex 3)* shape of a segment of a circle. The straight edge of the base is 320 meters long and corresponds to a central angle of 60°. What is the area of the structure, to the nearest square meter?

Practice   Distributed and Integrated

* 1. Determine the area of the shaded circle segment. Round your answer to the *(112)* nearest square centimeter.

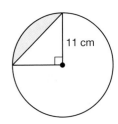

2. (**Design**) Andres is designing a logo for the math club. He wants the symbol *(69)* to include a trapezoid with one base length being 4 centimeters and the midsegment being 10 centimeters. What length should he make the other base?

* 3. Find the exact distance between $P(-1, 7, -5)$ and $Q(12, 11, 10)$. *(111)*

4. Error Analysis A student determines that the maximum area that can be contained by *(107)* a triangle with a perimeter of 30 centimeters is 30 square centimeters, since a right triangle with side lengths 5, 12, and 13 centimeters can be drawn. Is the student correct?

5. Multiple Choice This orthographic drawing shows the front, top,
(93) and right-side views of a solid. Which of these sketches
represents the solid?

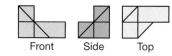

Front   Side   Top

A   B

C   D

6. Model If three of a parallelogram's vertices are located at $(-4, 2)$, $(-1, 3)$,
(34) and $(-1, 6)$, what are all the possible coordinates of the fourth vertex?

7. Use the figure at right to answer the following questions.
(105) **a.** Write the matrix representing the vertices of the polygon.
  b. Find the image matrix for a 90° rotation about the origin.
  c. Graph the image.

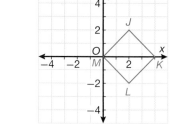

8. A square has an area of 36 square centimeters. If each side of
(96) this square is multiplied by x, the resulting square has an area
of 900 square centimeters. Find the value of x.

9. (Astronomy) A scale model of the sun has a diameter of 8 feet. If the sun's
(110) actual diameter is 865,000 miles, what is the scale factor? Express your
answer in scientific notation with two significant digits.

10. Justify $\overline{AP}$ is parallel to $\overline{BQ}$ and perpendicular to $\overleftrightarrow{PQ}$.
(58) **a.** Circle A's radius is $\overline{AP}$. Is $\overleftrightarrow{PQ}$ tangent to $\odot A$? Why or why not?
  b. A second circle, $\odot D$, has $\overleftrightarrow{PQ}$ as a tangent at point Q. On which line
  does the center of $\odot D$ lie?

*11. Multi-Step Determine the area of the shaded circle segment in this kite,
(112) to the nearest tenth of a square inch.

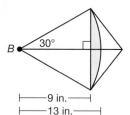

12. In spherical geometry, which postulate does not hold true?
(109)

13. Write Two different circles have chords of equal length. Does this mean that
(104) the arcs that correspond to each of the chords are equal in length? Explain.

14. Multiple Choice Which statement is *not* true?
(82)
A $\tan^{-1}\left(\frac{15}{10}\right) = \tan^{-1}\left(\frac{3}{2}\right)$ **B** $\tan^{-1}\left(\frac{14}{16}\right) = 90° - \tan^{-1}\left(\frac{16}{14}\right)$

C $180° - \tan^{-1}\left(\frac{4}{2}\right) = \tan^{-1}\left(\frac{1}{2}\right)$ **D** $\tan^{-1}\left(\frac{20}{15}\right) = 90° - \tan^{-1}\left(\frac{3}{4}\right)$

15. Describe the diagram in two ways.
(106)

16. **Algebra** Given the point $D(1, -1)$, find the point on the line $y = -\frac{3}{2}x - 6$ that is closest to D.
(42)

*17. A circle has a radius of 5 inches. Determine the area of the segment formed by a chord with a central angle of 36°. Round your answer to the nearest tenth of a square inch.
(112)

*18. A line segment has an endpoint at $(-7, 3, 8)$. If the segment's midpoint is at $(2, -4, 9)$, what are the coordinates of the other endpoint?
(111)

19. (Farming) A farmer wants to fill a conical silo to half its capacity. Would he fill more than half the height or less than a quarter of the height? Explain.
(103)

20. If $\triangle ABC \cong \triangle LMN$, and $AB = 9.3$ centimeters, $BC = 3.2$ centimeters, and $LN = 4.6$ centimeters, what are the lengths of the other sides?
(25)

21. **Verify** Show that the origin lies on the line defined by the equation $(x, y, z) = t(4, 1, 3)$.
(108)

22. Rotate the figure 90° clockwise about point P, and then translate the image along v.
(90)

23. (Packaging) A tin can is designed to have a radius of 2 inches and a height of 3 inches. To the nearest hundredth of a square inch, how much paper is needed to wrap a label around the can?
(62)

24. (Programming) A programmer is making a video game. He wants to program a button with a command that moves a player's character along the vector $\langle -4, 2 \rangle$. If a player's character starts at the point $(8, 3)$ and the button is pressed three times, what should the character's final position be?
(71)

    A $(-8, 3)$ **B** $(-7, -2)$

    C $(-4, 9)$ **D** $(9, 2)$

*25. Determine the area of the shaded part of this figure, to the nearest tenth of a square foot.
(112)

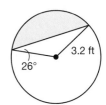

26. **Formulate** Anushka performs a dilation on a figure such that all the points of the figure remain the same. Write the dilation map for her transformation.
(102)

27. Explain whether it is possible or not to construct a triangle with side lengths of $x + 7$ inches, $x - 11$ inches, and $2x + 1$ inches.
(39)

28. **Multiple Choice** Which statement is *not* true of all rhombuses?
(65)

    A Both pairs of opposite angles are equal.

    B The diagonals bisect the opposite angles.

    C All consecutive sides are equal.

    D All consecutive angles are perpendicular.

*29. Find the midpoint of $(12, 3, 3)$ and $(-4, 7, -9)$.
(111)

30. **Algebra** In a regular polygon, the sum of half of the angles is 1260°. How many sides does this polygon have?
(Inv 3)

Symmetry of Solids and Polyhedra

Warm Up

1. ₍₇₆₎ **Vocabulary** The smallest angle through which a figure with rotational symmetry can be rotated to coincide with itself is called the _____.

2. ₍₇₆₎ Find the order of rotational symmetry for the following figures.

a.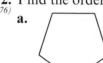

b.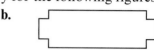

3. ₍₇₆₎ How many of the digits from 1 to 9 exhibit rotational symmetry? Which ones?

New Concepts

Recall that line symmetry refers to a plane figure which can be divided by a line into two congruent, reflected halves.

Plane symmetry describes a three-dimensional solid that can be divided into two congruent, reflected halves by a plane.

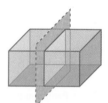

Plane Symmetry

Symmetry about an axis, or rotational symmetry, describes a figure or solid that can be rotated about a line, called the **axis of symmetry**, so that the image coincides with the preimage.

Symmetry about an axis

Math Reasoning

Analyze Look at the solids in Example 1. How many planes of symmetry does each figure have?

Example 1 **Plane Symmetry**

For each figure, sketch one plane through the solid that will divide it into two congruent, reflected halves.

a.

b.

SOLUTION

a. Two possible planes are shown.

b. One possible plane is the plane through the figure's center.

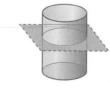

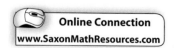
Online Connection
www.SaxonMathResources.com

Like plane figures, solids also have orders of rotational symmetry and angles of rotational symmetry. Every time the cube at right is rotated 90° about the axis shown, it coincides with itself. So, its angle of rotational symmetry is 90°, and it has an order of rotational symmetry of 4.

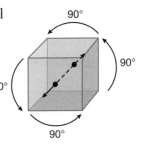

Example 2 Rotational Symmetry in Solids

Determine whether each solid has rotational symmetry over the axis shown. If so, give the order of symmetry and the angle of rotational symmetry.

a.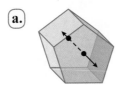

SOLUTION

Yes, the solid has rotational symmetry. The order of symmetry is 5, and the angle of rotational symmetry is 72°.

b.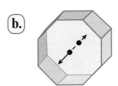

SOLUTION

Yes, the solid has rotational symmetry. The order of symmetry is 8, and the angle of rotational symmetry is 45°.

Example 3 Application: Identifying Symmetries of Solids

Draw two planes through the figure that will divide it into symmetrical halves.

Then, draw an axis through which it has rotational symmetry and give the angle and order of rotational symmetry.

SOLUTION

Possible solutions are shown here, but there are many planes through which this figure is symmetrical, and several axes of symmetry through which the figure displays rotational symmetry.

The degree of rotational symmetry over the axis shown is 180°, and the order of rotational symmetry is 2.

Recall that the order of rotational symmetry is the number of times a figure with rotational symmetry coincides with itself as it rotates 360°.

a. Draw one plane through which the
(Ex 1) figure has plane symmetry.

b. Determine whether the solid has rotational symmetry through the
(Ex 2) given axis. If it does, give the angle of rotational symmetry and the
order of symmetry.

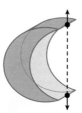

c. Draw two planes through which the figure is symmetrical. Then draw
(Ex 3) an axis through which it has rotational symmetry and give the angle
and order of rotational symmetry.

Practice Distributed and Integrated

* 1. Determine whether the solid has plane symmetry. If it does, show
(113) the symmetry in a drawing.

2. **Algebra** Two rectangles are in the similarity ratio of 4:3. The smaller
(87) rectangle has a length of 9 feet and a width of 3 feet. What is the
perimeter of the larger rectangle?

3. Find the length of the diameter of a sphere whose radius has endpoints
(111) $C(3, 4, -2)$ and $D(6, -3, 8)$, where C is the center of the sphere.

4. **Furniture Restoring** Yuan is restoring the top of his desk, shown
(40) at right. He has cut a circular hole in the top, which has a 2-inch
radius for computer wires to pass through. What area will he
need to paint, to the nearest inch?

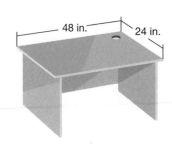

5. Find the perimeter of right $\triangle ABC$ if m$\angle A = 42°$ and the hypotenuse
(68) is 12 centimeters. Round to the nearest hundredth of a centimeter.

6. Multiple Choice Which does *not* represent the geometric probability
(Inv 6) associated with a single region of this tangram?

    A $\frac{1}{2}$ **B** $\frac{1}{4}$

    C $\frac{1}{16}$ **D** $\frac{1}{8}$

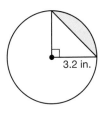

7. The vertices of $\triangle JKL$ are $J(2, 1)$, $K(6, 1)$, and $L(4, 5)$. Use matrix
(105) multiplication to reflect the image across the y-axis and then rotate
the triangle 180° about the origin. What is the image matrix?

* 8. Determine the area of the shaded circle segment, to the
(112) nearest hundredth of a square inch.

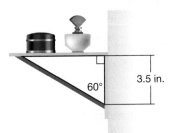

3.2 in.

9. Determine the sums of the first seven Fibonacci numbers
(Inv 11) the first eight Fibonacci numbers and the first nine
Fibonacci numbers.

*10. Formulate A parallelogram has coordinates $W(3, -1)$, $X(a, b)$, $Y(7, 6)$,
(92) and $Z(c, d)$. Write equations for c and d in terms of a and b.

11. (Home Improvement) Michaela is attaching a shelf to a wall and will
(56) use a metal bar for support, as shown. What length of metal does she
need to use?

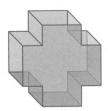

60°   3.5 in.

12. What is the 90th triangular number?
(Inv 8)

*13. Determine whether the solid has rotational symmetry. If it does, give the angle
(113) of rotational symmetry and the order of symmetry.

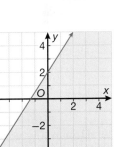

14. (Chemistry) An iron atom has an atomic radius of 140 picometers (pm). If a
(110) model of the atom is a sphere with a volume of 30 cubic inches, what is the
scale factor between the model and an actual iron atom? Round your
answer to the nearest hundredth.

15. Error Analysis Ella attempted to graph the inequality $y < 1.5x + 2$. Identify
(88) any errors she has made.

16. Multi-Step The tangent of one acute angle in a right triangle is 0.75. What
(Inv 7) is the tangent of the other acute angle?

*17. In $\odot C$, $\overline{CA}$ and $\overline{CB}$ are perpendicular radii that are each 28 centimeters
(112) long. Determine the area of the circle segment formed by $\overline{AB}$, to the
nearest square centimeter.

18. Multiple Choice The equation of the line at right is:
(16)
    A $y = 2x - 1$ **B** $y = -2x + 1$
    C $y = 2x + 1$ **D** $y = -2x - 1$

19. Find the midpoint of $J(4.5, 3.25, 8.6)$ and $K(-2.5, 6.75, -2.6)$.
(111)

*20. Determine whether this solid has plane symmetry, symmetry about
(113) an axis, or neither.

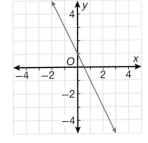

21. The perimeter of △*EFG* is 35.7 inches, and ∠*E* ≅ ∠*G*. If *EF* = 12.2 inches, determine *EG*.
(51)

22. A tangent and a secant are drawn from an external point *P* to a circle. The tangent is 8 meters long and the part of the secant inside the circle is 12 meters long. How long is the secant?
(101)

23. **Verify** Use the relationship $A_{segment} = A_{sector} - A_{\triangle}$ to verify that the area of this segment is given by $\left(\frac{1}{4}\pi - \frac{1}{2}\right)r^2$.
(112)

24. Determine the slope of the line passing through (−4, 0) and (2, 4).
(16)

25. (**Force**) A beam supports a structure that leans against it. The beam carries a 15-pound vertical load and a 25-pound horizontal load. What is the total load that the beam is carrying?
(83)

26. **Generalize** Two cones are similar, but one has twice the volume of the other. If the lengths of their radii are the same, how are the heights of the two cones related?
(77)

27. (**Art**) Bruce wants to build a rectangular picture frame out of clay. He can stretch the clay to a maximum length of 32 centimeters. What is the greatest area the picture frame can enclose?
(107)

*28. Tell whether this solid has plane symmetry, symmetry about the given axis, or neither.
(113)

29. How many vertices does a polyhedron with 22 edges and 10 faces have?
(49)

30. **Analyze** Will the line connecting the points *A*(−2, 4, 6) and *B*(3, 5, 6) ever intersect the line connecting the points *C*(8, −3, −4) and *D*(13, −2, −4)? Explain.
(108)

LESSON 114

Solving and Graphing Systems of Inequalities

Warm Up

1. Vocabulary A statement that compares two expressions using $>$, $<$, $\geq$, $\leq$, or $\neq$ is a(n) _____.
(88)

2. Solve $4x + 5 < -x - 5$.
(88)

3. Write the inequality shown in the graph.
(88)

4. Multiple Choice Which of the following values
(88) satisfies the inequality $3x - 13 < 2x + 2$?

A $\sqrt{10}$ **B** 42

C 15   **D** $\sqrt{253}$

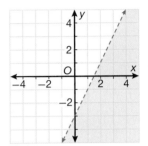

New Concepts   Recall that when an inequality is graphed, the graph must be shaded to show all the points that satisfy the inequality. If, when the inequality is solved for y, it starts with $y >$ or $y \geq$, then the region above the line must be shaded. If it starts with $y <$ or $y \leq$, then the region below the line must be shaded.

A system of two or more inequalities is graphed the same way. In a conjunction, the solution set is the set of points on the coordinate plane that satisfy all of the inequalities in the system. In a disjunction, the solution set is the set of points on the coordinate plane that satisfy any of the inequalities in the system.

Hint

Recall that when an inequality uses $\geq$ or $\leq$, a solid line is used to graph it. When an inequality uses $>$ or $<$, a dashed line is used to graph it.

Example 1   **Systems of Inequalities in Two Dimensions**

Graph the region described by the inequalities $y \leq x + 4$ and $y > 2x - 1$.

SOLUTION

First, graph the lines $y = x + 4$ and $y = 2x - 1$. Since the first inequality uses $\leq$, use a solid line. For the second inequality, use a dashed line. Next, shade the region below $y = x + 4$ and above $y = 2x - 1$. The shaded region is the solution to the system of inequalities.

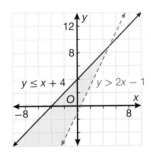

In Example 1, the problem asks for the conjunction of two inequalities. Only points that are in both graphs are part of the solution. The disjunction of two inequalities can also be graphed. In a disjunction, all points that are in either graph are part of the solution.

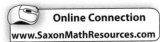

Online Connection
www.SaxonMathResources.com

Example 2 Systems of Inequalities With Disjunction

Graph the region described by the inequalities $y < x - 1$ or $y > -x + 1$.

SOLUTION

First, graph the lines $y = x - 1$ and $y = -x + 1$, then shade the region below $y = x - 1$ and above $y = -x + 1$. Both lines should be dashed lines, since both are strict inequalities. The shaded region is the solution to the system of inequalities.

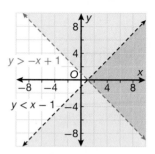

For a system of more than two inequalities, the solution set is the conjunction or disjunction of all three graphs.

Example 3 More Systems of Inequalities in Two Dimensions

Graph the region described by the inequalities $y \geq -2$, $y \leq 3x + 1$, and $y \leq -2x + 6$.

SOLUTION

First graph the lines $y = -2$, $y = 3x + 1$, and $y = -2x + 6$. Since the inequalities are not strict, use solid lines. Shade in the region that is above $y = -2$, below $y = 3x + 1$, and below $y = -2x + 6$. The shaded region is the solution to the system of inequalities.

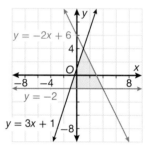

Caution

The solution region for Example 3 is an enclosed area bounded on all sides by the three inequalities. However, not every system of inequalities is bounded on all sides. The region that is described by the system of inequalities $y \leq -2$, or $y \geq 3x + 1$, or $y \geq 2x + 6$, would describe all the points outside of the closed figure.

Example 4 Application: Logic

Leslie's age is equal to or less than twice Bruno's age, and greater than or equal to Bruno's age plus 3. Draw the graph that represents Leslie's possible age.

SOLUTION

Let Bruno's age be x. The problem states that Leslie's age is less than or equal to $2x$ and greater than or equal to $x + 3$. If Leslie's age is y, two equations that represent this situation are $y \leq 2x$ and $y \geq x + 3$. Moreover, both Leslie and Bruno must have a positive age, so the system should also include $y \geq 0$ and $x \geq 0$. Graph all of these lines and find the conjunction of their solutions sets as shown in the graph here.

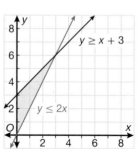

a. Graph the region described by the inequalities $y > x$ and $y < -2x + 5$.
(Ex 1)

b. Graph the region described by the inequalities $y \le x + 1$ or
(Ex 2) $y \ge -2x - 1$.

c. Graph the region described by the inequalities $x > -3$, $y < -2x$,
(Ex 3) and $y > x - 1$.

d. (Security) A house has two security cameras. One camera can observe
(Ex 4) the region above $y = x - 5$ and the other camera can observe the
region above $y = -2x - 4$. Graph the region that can be observed by
both cameras.

Practice   Distributed and Integrated

*1. Determine if the solid has plane symmetry.
(113)

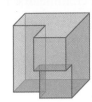

2. The length of the hypotenuse of a right triangle is 17 centimeters.
(9) Name one pair of possible leg lengths.

3. Find θ_1 and θ_2 to the nearest degree.
(82)

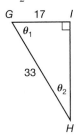

4. Multi-Step Show that $(4, -4, 5)$ is the point of intersection of the line
(108) $(x, y, z) = (-2, 0, 1) + t(3, -2, 2)$ and the line through the points
$(3, 5, 11)$ and $(5, -13, -1)$.

5. Determine the area of the shaded circle segment, to the nearest tenth
(112) of a square centimeter.

* 6. Graph the region described by the inequalities $y \ge x$ and $y \le x - 3$.
(114)

7. (Weather) A particular spherical weather balloon has a radius
(80) of 10 meters at a certain altitude. What is the volume of air in
the balloon?

8. The first iteration of a fractal is shown. Sketch the second
(Inv 10) iteration of the fractal.

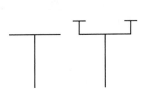

9. A cube with sides that are 6 units long has one vertex at the origin.
(111) Determine the distance between the origin and the vertex that is
diagonally opposite it through the cube.

*10. The drawing shows a part of a solid and its center of rotation.
(113) Draw the complete solid with an order of rotational symmetry of 4.

order: 4

11. Write What is the direction of a zero-magnitude vector?
(89)

*12. (**Lighthouses**) Two lighthouses are placed along a shoreline. The first one can light
(114) the region below the line $y = 2x + 5$ on a map and the other can light the region
above the line $y = -x + 1$ on the same map. Graph the region that at least one
lighthouse can light.

13. Find *LN* in the triangle. Round your answer to the nearest hundredth.
(94)

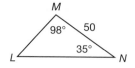

14. Multi-Step The hypotenuse of one 45°-45°-90° right triangle is
(53) 53 feet while a leg of a second, similar triangle is 37.5 feet.
Which triangle has the greater area?

15. (**Anatomy**) In the diagram, *A* is 1.9 centimeters, *B* is 1.17 centimeters,
(Inv 11) *C* is 0.42 centimeters, *E* is 0.68 centimeters, *G* is 1.89 centimeters,
and *H* is 3.06 centimeters. Which ratios of these lengths are
approximately equal to the golden ratio?

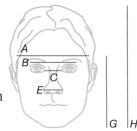

16. Triangle *ABC* has vertices $A(-3, 2)$, $B(-4, -2)$, and $C(2, 1)$. Apply a dilation
(84) centered at $(3, -3)$. Use a scale factor of 3.
Hint: First, translate the preimage, △ABC, so the center is at the origin.
Then, apply the dilation. Finally, translate the figures so the preimage
returns to its original location, and identify the coordinates of A′, B′, and C′.

17. Algebra Find the value of *x* in this rhombus.
(52)

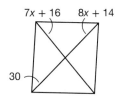

*18. Graph the region described by the inequalities $y > 3x + 2$ or $y > 2x$.
(114)

*19. Formulate What is an expression for the area of a shape composed of a circle with
(97) a radius *g* that has had a circle of radius *r* cut out of it?

20. Error Analysis Patrice wrote this expression for the area of the segment shown:
(112) $\frac{2}{3}\pi(9)^2 - \frac{1}{2}(9)\left(\frac{9\sqrt{3}}{2}\right)$. Correct Patrice's error.

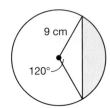

21. Multiple Choice Circle *C* has the equation, $(x - 5)^2 + (y - 2)^2 = 16$. If $\odot C$
(75) is translated right 2 units, what is the equation of its image, $\odot C'$?
    A $(x - 3)^2 + (y - 2)^2 = 16$ **B** $(x - 5)^2 + (y - 4)^2 = 16$
    C $(x - 7)^2 + (y - 2)^2 = 16$ **D** $(x - 5)^2 + (y - 2)^2 = 36$

22. Find the length of $\overline{IF}$ in the triangle at right.
(60)

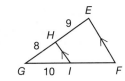

23. If $\cos^2\theta = 0.45$, what is the value of $\sin\theta$
(91) to the nearest hundredth?

24. Determine whether the solid has plane symmetry, symmetry about an axis, or
(113) neither.

25. Generalize If a polygon circumscribes a circle, is the polygon regular?
(106) What if the circle circumscribes the polygon?

26. **(Photography)** A photographer takes a photo of the ridge above him from the
(73) bottom of a gorge. If he knows the gorge is 110 meters deep and his angle of
elevation is 78°, how far should he focus the lens to obtain the clearest photo
of the ridge of the gorge? Round your answer to the nearest meter.

27. **(Buildings)** The front of the building shown is to be surfaced with bricks.
(40) The building has two 3-by-7-foot doors and ten 2.5-by-2.5-foot windows that will
not be covered with brick. If each brick covers 0.5 square feet, how many bricks
will be needed?

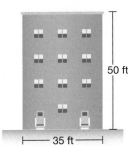

*28. Graph the region described by the inequalities $y \geq 2x$, $y \geq -2x$, and
(114) $y \leq x + 2$.

29. Verify Stacy says that a triangle with side lengths of 8.7, 10.1, and 13.6
(33) is a right triangle. Is she correct?

30. Analyze The coordinates for the image of point $P(x, y)$ under a transformation
(67) are $P'(\frac{4}{5}y - \frac{3}{5}x, \frac{4}{5}x + \frac{3}{5}y)$. Identify the transformation as a rotation, translation,
or reflection, and specify the center of rotation, line of reflection, or translation
vector.

Finding Surface Areas and Volumes of Composite Solids

Warm Up

1. **Vocabulary** A plane figure that is made up of two or more simple shapes is called a _____.
(40)

2. Find the surface area and the volume of this right prism.
(59)

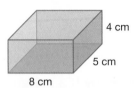

4 cm
5 cm
8 cm

3. Find the surface area and the volume of this right cylinder to the nearest hundredth.
(62)

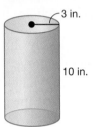

3 in.
10 in.

New Concepts

Recall that the perimeter of a two-dimensional composite figure is the sum of the perimeters of the shapes that make up the figure, minus the lengths of the common sides. Similarly, the surface area of a three-dimensional composite solid is the sum of the surface areas of the simple solids that make up the composite solid, minus the areas of the common faces.

Math Reasoning

Generalize The surface area of a composite solid cannot be found by simply adding the surface areas of the solids that compose it. Explain why this statement is true.

$$S_{\text{composite figure}} = S_{\text{cone}} + S_{\text{cylinder}} - A_{\text{common faces}}$$

Example 1 Finding Surface Areas of Composite Figures

Find the surface area of the solid. Round any decimal answers to the nearest tenth of a square unit.

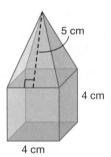

5 cm
4 cm
4 cm

SOLUTION

Determine the surface area of the figure by adding the surface areas of the cube and the pyramid *excluding* the areas of the common faces.

A cube has six congruent faces. In this composite solid, one face is common with the pyramid. Find the surface area of five faces.

$$S_{cube} = 5s^2$$
$$S_{cube} = 5(4)^2$$
$$S_{cube} = 80$$

The pyramid is regular, so the four triangular faces are congruent. In this solid, the base of the pyramid is common with the cube, so it is not included in the surface area. Find the lateral surface area of the pyramid.

$$L_{pyramid} = \frac{1}{2}P\ell$$
$$L_{pyramid} = \frac{1}{2}(4 \times 4)(5)$$
$$L_{pyramid} = 40$$

Reading Math

In the formula

$L_{pyramid} = \frac{1}{2}P\ell$, P is the

perimeter of the pyramid and ℓ is its slant height.

Add the areas to find the surface area of the composite solid.

$$S_{composite} = 80 + 40$$
$$S_{composite} = 120$$

The surface area of the composite solid is 120 square centimeters.

The volume of a composite solid is the sum of the volumes of the individual solids that make up the composite solid.

Example 2 **Finding Volumes of Composite Figures by Adding**

Find the volume of the solid. Assume that the prisms and cylinders are right and that the bottom of the figure is exactly half of a cylinder. Round any decimal answers to the nearest tenth of a square meter.

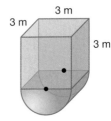

SOLUTION
The volume of this composite solid is the sum of the volumes of the cube and the half-cylinder. First, find the volume of the cube.

$$V_{cube} = Bh$$
$$V_{cube} = (3 \cdot 3)3 = 3^3$$
$$V_{cube} = 27$$

Now, find the volume of the half-cylinder.

$$V_{half-cylinder} = \frac{1}{2}\pi r^2 h$$
$$V_{half-cylinder} = \frac{1}{2}\pi(1.5)^2(3)$$
$$V_{half-cylinder} \approx 10.6$$

Add the volumes to find the volume of the composite solid.

$$V_{composite} \approx 27 + 10.6$$
$$V_{composite} \approx 37.6$$

The volume of the composite solid is approximately 37.6 cubic meters.

Hint

To review the surface area and volume of various solids, refer to these previous lessons.

• Prisms: Lesson 59
• Cylinders: Lesson 62
• Pyramids: Lesson 70
• Cones: Lesson 77
• Spheres: Lesson 80

Example 3 **Finding Volumes of Composite Figures by Subtracting**

Find the volume of the right solid. Round the answers to the nearest tenth of a cubic unit.

SOLUTION

The solid is composed of a large cylinder with a smaller cylinder removed. Find the total volume of the larger cylinder, and subtract the volume of the smaller cylinder.

Hint

Use substcripts, such as $V_{\text{large cylinder}}$ to help keep the quantities organized when the same formula is being used on more than one figure.

$$V_{\text{large cylinder}} = \pi r^2 h \qquad\qquad V_{\text{smaller cylinder}} = \pi r^2 h$$
$$V_{\text{large cylinder}} = \pi(5)^2(12) \qquad V_{\text{smaller cylinder}} = \pi(1)^2(12)$$
$$V_{\text{large cylinder}} = 300\pi \qquad\qquad V_{\text{smaller cylinder}} = 12\pi$$

Subtract the volume of the small cylinder from the volume of the large cylinder to find the volume of the composite solid.

$$V_{\text{composite}} = 300\pi - 12\pi$$
$$V_{\text{composite}} = 288\pi$$
$$V_{\text{composite}} = 904.3$$

The volume of the solid is approximately 904.3 cubic yards.

Example 4 **Application: Volume of a Gas Tank**

Jenna heats her home with propane, which is stored in a tank in her backyard. How many cubic feet of propane does it take to fill the tank to the nearest tenth?

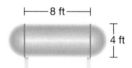

SOLUTION

The tank is composed of a cylinder and two congruent hemispheres. Since the hemispheres on either side of the tank combine to make one full sphere, the volume of the tank is the sum of the volumes of the cylinder and a sphere with a 2-foot radius.

$$V_{\text{tank}} = V_{\text{cylinder}} + V_{\text{sphere}}$$
$$V_{\text{tank}} = \pi r^2 h + \frac{4}{3} r^3 \pi$$
$$V_{\text{tank}} = \pi(2)^2(8) + \frac{4}{3}(2)^3 \pi$$
$$V_{\text{tank}} = 32\pi + \frac{32}{3}\pi$$
$$V_{\text{tank}} \approx 134.0$$

It takes approximately 134 cubic feet of propane to fill the tank.

a. Find the surface area of the solid.
(Ex 1) Assume that the cones shown are right.
Give the answer to the nearest tenth
of a square unit.

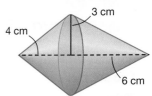

b. Find the volume of the solid. Assume
(Ex 2) that the cone and cylinder are right. Give
the answer to the nearest tenth of a
cubic unit.

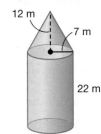

c. Find the volume of the solid. Assume that
(Ex 3) the prisms are right.

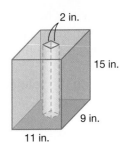

d. A truck has a storage compartment as
(Ex 4) shown. If the width of the truck is 5 feet, how
many cubic feet of cargo can the truck carry
in the compartment?

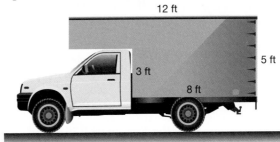

Practice **Distributed and Integrated**

1. Find the surface area of the solid. Assume that the prism and pyramid
(115) are right.

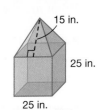

2. Find the matrix that would represent the translation of a triangle 4 units
(100) left and 8 units down.

3. **Analyze** What are two possible shapes for a cross section of a cube?
(85)

4. Determine whether the solid has plane symmetry, symmetry about
(113) an axis, or neither.

5. Find the measure of the smaller arc between two secants if the larger arc
(79) measure is 114° and the exterior angle is 29°.

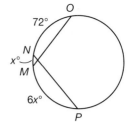

6. In the figure, $\overline{MO} \cong \overline{NP}$. Find m$\widehat{NP}$.
(104)

7. (**Astronomy**) Four moons are in the same circular orbit around a planet.
(64) The arc measure between moons A and B is 55°, and the arc measure
between moons C and D is 33°. If $\overline{AC}$ and $\overline{BD}$ intersect at E on the
planet, what is m∠AEB?

* 8. Analyze Can points A, L, and F on the sphere at right be considered
(109) a triangle? Explain why or why not.

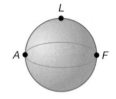

*9. Graph the solution to $y - x > 4$ or $3y + x < 6$ and label the point
(114) $(3, -4)$. Is $(3, -4)$ in the solution set?

10. A circle has a radius of 5.2 meters. Determine the area of the segment
(112) formed by a chord with a central angle of 120°. Round your answer to
the nearest hundredth.

11. Find the volume of the solid to the nearest tenth. Assume that the cylinder
(115) is a right cylinder.

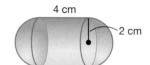

12. In an isosceles triangle, the vertex angle measures $2(x + 7)$° and the base
(51) angles measure $(3x - 37)$°. Determine the angle measures.

13. Multiple Choice The coordinates of three vertices of a parallelogram are (1, 2),
(61) (2, −3), and (−3, −1). Which cannot be the coordinates of the fourth vertex?
    A (0, 6)   **B** (−2, −7)
    C (−2, 8)   **D** (−4, 3)

14. Describe the diagram in two ways.
(106)

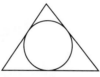

15. Write Write three statements that form a syllogism.
(21)

*16. Is (−1, −4) a solution to the disjunction of $y > 3x - 2$ and $2x - y > 1$? Is (2, 3)?
(114)

17. Find the volume of this solid. The prisms are right prisms.
(115)

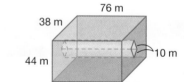

18. $\overline{XY}$ has a midpoint of (4, −3, 6). Point X has coordinates (12, 7, 1).
(111) What are the coordinates of point Y?

19. Multi-Step Draw quadrilateral $JKLM$ with vertices $J(2, 4)$, $K(2, -3)$, $L(-3, -3)$,
(102) and $M(-3, 4)$. In the same coordinate plane, draw the image under the dilation
$D_{O, \frac{1}{2}}(x, y)$. What is the relationship between the areas of the two quadrilaterals?

20. Error Analysis The two triangles are similar. Daphne found the value of
(41) x to be 8. What was her error?

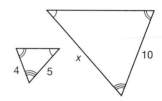

21. Verify Show that quadrilaterals *ABCD* and *EFGH* are congruent,
$_{(92)}$ and identify the type of quadrilateral(s) they are.

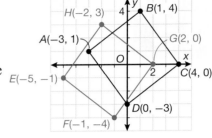

22. (Art) Rhea is making a regular heptagon-based pyramid out
$_{(70)}$ of paper maché. To cover the lateral surface of the pyramid, she
needs 70 square inches of paper maché. If the slant height
of the pyramid is 2 inches, what are the side lengths of the
heptagon?

*23. (Silos) How many cubic feet of grain can the silo hold to the
$_{(115)}$ nearest tenth?

24. Analyze Does the function $y = x^4$ have a line of symmetry?
$_{(76)}$ Does it have rotational symmetry?

25. (Surveying) A surveyor uses trigonometry to find the height of
$_{(98)}$ a building by measuring the distance from the top of one to the
top of the other. Find θ. Round your answer to the nearest degree.

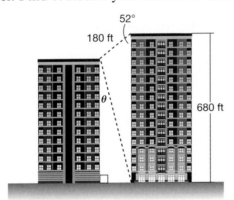

26. Which congruence theorem applies to these triangles?
$_{(36)}$

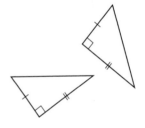

27. A regular polygon has 120 sides. What angle of rotational
$_{(113)}$ symmetry does the polygon have?

28. Algebra The points *D*, *E*, and *F* are midpoints of $\triangle ABC$. If $DF = 2x + 1$
$_{(55)}$ and $BC = 14$, find x and DF.

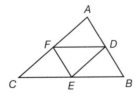

*29. (Baseball) In baseball, the term *foul ball* refers to a ball that is hit and its trajectory
$_{(114)}$ goes outside of two rays, one formed by home base and first base and the other
formed by home base and third base. For a diagram of a baseball diamond
with home base at (3, 2) and first base at (5, 4), write a disjunction of simplified
inequalities whose solution is the area where a foul ball would go.

30. Two congruent parallelograms make the following shape. Describe m∠*JKL*
$_{(34)}$ in terms of *x*.

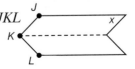

Secant, Cosecant, and Cotangent

1. **Vocabulary** A _____ is a relationship in which every input
(SB 17)   is paired with exactly one output.

2. Find the angle measures of this right triangle
(82)   to the nearest degree.

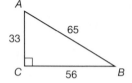

3. Find the value of x in this scalene triangle to
(94)   the nearest tenth.

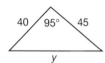

4. Find the value of y in this triangle to the
(98)   nearest tenth.

New Concepts The reciprocals of the sine, cosine, and tangent ratios are the trigonometric ratios called cosecant, secant, and cotangent.

Caution

Since cosecant, secant, and cotangent are reciprocals of trigonometric ratios, they are not defined when the ratio that they are the reciprocal of is equal to 0. For example, cosecant is undefined when $\theta = 0°$, because $\sin 0° = 0$.

More Trigonometric Ratios
In a right triangle, the **cosecant of an angle** is the ratio of the length of the hypotenuse to the length of the side opposite the angle. It is the reciprocal of the sine ratio. $$\csc \theta = \frac{\text{hypotenuse}}{\text{opposite}} = \frac{1}{\sin \theta}$$ In a right triangle, the **secant of an angle** is the ratio of the length of the hypotenuse to the length of the side adjacent to the angle. It is the reciprocal of the cosine ratio. $$\sec \theta = \frac{\text{hypotenuse}}{\text{adjacent}} = \frac{1}{\cos \theta}$$ In a right triangle, the **cotangent of an angle** is the ratio of the length of the side adjacent to the angle to the length of the side opposite the angle. It is the reciprocal of the tangent ratio. $$\cot \theta = \frac{\text{adjacent}}{\text{opposite}} = \frac{1}{\tan \theta}$$

Example 1 Identifying Secant, Cosecant, and Cotangent

a. Find the value of the secant, cosecant, and cotangent of θ in the triangle.

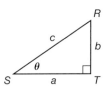

SOLUTION

$$\sec \theta = \frac{1}{\cos \theta} \qquad \csc \theta = \frac{1}{\sin \theta} \qquad \cot \theta = \frac{1}{\tan \theta}$$

$$\cos \theta = \frac{a}{c} \qquad \sin \theta = \frac{b}{c} \qquad \tan \theta = \frac{b}{a}$$

$$\sec \theta = \frac{c}{a} \qquad \csc \theta = \frac{c}{b} \qquad \cot \theta = \frac{a}{b}$$

b. If $a = 65$, $b = 72$, and $c = 97$, find the value of the secant, cosecant, and cotangent of θ.

SOLUTION

$$\sec \theta = \frac{c}{a} \qquad \csc \theta = \frac{c}{b} \qquad \cot \theta = \frac{a}{b}$$

$$\sec \theta = \frac{97}{65} \qquad \csc \theta = \frac{97}{72} \qquad \cot \theta = \frac{65}{72}$$

Hint

Remember that if a is the reciprocal of b, then b is also the reciprocal of a. Therefore, the sine, cosine, and tangent ratios are the reciprocals of the cosecant, secant, and cotangent ratios.

Using trigonometric identities for sine, cosine, and tangent, it is possible to derive two trigonometric identities using secant, tangent, and cosecant.

$$\sin^2\theta + \cos^2\theta = 1 \qquad\qquad \sin^2\theta + \cos^2\theta = 1$$

$$\cos^2\theta = 1 - \sin^2\theta \qquad\qquad \sin^2\theta = 1 - \cos^2\theta$$

$$\frac{\cos^2\theta}{\cos^2\theta} = \frac{1}{\cos^2\theta} - \frac{\sin^2\theta}{\cos^2\theta} \qquad\qquad \frac{\sin^2\theta}{\sin^2\theta} = \frac{1}{\sin^2\theta} - \frac{\cos^2\theta}{\sin^2\theta}$$

$$1 = \sec^2\theta - \tan^2\theta \qquad\qquad 1 = \csc^2\theta - \cot^2\theta$$

Trigonometric Identities
Secant and tangent are related by the identity: $$1 = \sec^2\theta - \tan^2\theta$$ Cosecant and cotangent are related by the identity: $$1 = \csc^2\theta - \cot^2\theta$$

Example 2 Using Trigonometric Identities

In a right triangle, the secant of an angle is 2. What is the tangent of the angle?

SOLUTION

Use the first trigonometric identity given above.

$$1 = \sec^2\theta - \tan^2\theta$$
$$1 = 2^2 - \tan^2\theta$$
$$-3 = -\tan^2\theta$$
$$\tan^2\theta = 3$$
$$\tan \theta = \sqrt{3}$$

The tangent of the angle is $\sqrt{3}$.

By using these two new trigonometric identities and $\sin^2\theta + \cos^2\theta = 1$, you can solve any of the six trigonometric ratios in terms of any of the other five.

Math Reasoning

Predict If $\sin(-\theta) = -\sin\theta$, then what can we predict about $\csc(-\theta)$? Explain your answer.

Example 3 **Application: Directions**

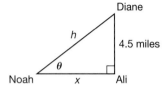

The distance between Diane and Ali's home is 4.5 miles. The cosecant of θ equals 2.66. How far does Noah live from Diane and Ali? Round your answers to the nearest tenth.

SOLUTION

Set up an expression showing the cosecant of the angle.

$$\csc\theta = \frac{\text{hypotenuse}}{\text{opposite}}$$

$$2.66 = \frac{h}{4.5}$$

$$h \approx 12.0 \text{ miles}$$

Then, use the Pythagorean Theorem to solve for x.

$$4.5^2 + x^2 \approx 12.0^2$$

$$x^2 \approx 123.8$$

$$x \approx 11.1 \text{ miles}$$

Therefore, Noah lives 11.1 miles from Ali and 12.0 miles from Diane.

Lesson Practice

a. Find the value of the secant, cosecant, and cotangent of θ in the triangle.
(Ex 1)

b. If the cosecant of an angle is 2, what is the cotangent of that angle?
(Ex 2)

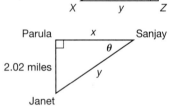

c. Parula and Janet live 2.02 miles apart. Find how far Sanjay lives from Parula and Janet if the $\cot\theta = 1.19$. Round your answers to the nearest hundredth of a mile.
(Ex 3)

Practice **Distributed and Integrated**

1. (**Elasticity**) A certain elastic band will break if stretched so that its total length is more than 22 inches. The plastic band is placed around a spherical balloon with a radius of 1.7 inches. If the balloon is inflated so that its radius doubles, will the elastic band break? Explain.
(96)

2. (**Architecture**) Determine the combined areas of the four circular segments of glass used in this stained-glass window. Round your answer to the nearest square inch.
(112)

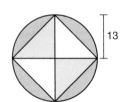

* 3. Multiple Choice Which of the following is not a possible value for sec θ?
(116)
    A 1.46   **B** 8.99
    C −0.89   **D** −19.02

* 4. Algebra The surface area of this composite solid made of two cones
(115) is $(3\sqrt{13} + 15)\pi$. Find the radius, r.

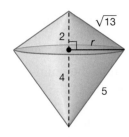

5. (Exercise) Doug is walking on a treadmill. If the conveyor belt
(63) is moving at 3.6 miles per hour backward, how fast and in what
direction must Doug walk in order to stay on the treadmill?

* 6. Write List all the possible reasons why a system of two linear
(81) equations could have no solution and explain why there is no
solution in each case.

7. A region is bound by the vertices (1, 2), (3, 5), and (4, 0). Find the
(114) coordinates of a point inside the region and a point outside the region.

8. Analyze In the diagram, $\angle 1$ and $\angle 2$ are complementary. Write
(31) a paragraph proof to show that $\angle 2$ and $\angle 3$ are complementary.

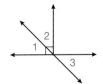

9. Error Analysis Joanna is trying to place a segment with vertices (3, 2) and
(108) (4, 5) in a three-dimensional coordinate system. She decides the new
coordinates are (0, 3, 2) and (0, 4, 5). Explain her error and correct it.

*10. Find the volume of the composite solid. Assume that the pyramid is
(115) right. Give your answer to the nearest tenth of a cubic inch.

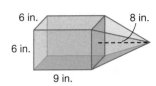

11. Multi-Step Classify the polygon with vertices $L(1, -1)$, $M(-3, -3)$,
(57) and $N(-3, 2)$. Find the perimeter. Give your answer in simplified
radical form.

12. Write How is it that a conjunction of linear inequalities might have no solution,
(114) but a disjunction of the same inequalities would always have a solution?

13. Algebra Determine the perimeter of this quadrilateral.
(58)

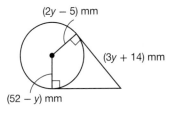

14. (Storage) A rubber stopper for a bottle is in the shape of
(103) a frustum of a cone with a height of 5 centimeters. If the
top circle has a radius of 1 centimeter and the bottom
circle has a radius of 3 centimeters, what is the volume
of the stopper? Round your answer to the nearest tenth
of a cubic centimeter.

15. The diagram shows intersecting chords in a circle. Find the value
(86) of x to the nearest hundredth.

*16. Analyze The sine and cosine functions always give a value between
(116) 1 and −1. Is the same true for secant and cosecant? Explain.

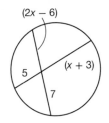

17. Draw the net of a pentagonal pyramid.
(Inv 5)

18. Find the distance between $Y(5, 5, 5)$ and $Z(-5, -5, -5)$.
(111)

*19. **Medicine** Find the surface area of this medicine capsule to the nearest
(115) tenth. Assume that the cylinder is right and each of the ends are
hemispheres.

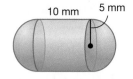

20. The endpoints of line segment $\overline{XY}$ are $X(-3, -2)$ and $Y(-1, 1)$.
(78) Graph $\overline{X'Y'}$ after $\overline{XY}$ has been rotated 180° about the origin.

21. Determine the point of tangency and the equation of the
(72) tangent line in the diagram at right.

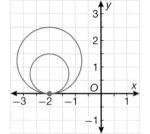

*22. **Algebra** If the hypotenuse of a right triangle is 4 and $\tan \theta = \frac{a}{b}$,
(116) find $\csc \theta$, $\sec \theta$, and $\cot \theta$.

23. Find the value of x to the nearest hundredth if $\overline{AC}$ bisects $\angle A$.
(98)

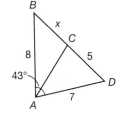

24. **Multi-Step** If $\cos \theta \cdot \tan \theta = 0.8$, find the value of $\cos \theta$. Then, find θ to
(91) the nearest degree.

25. Use inductive reasoning to write the next three steps of this series:
(7) 1, 5, 7, 11, 13, 17, …

26. Given a fixed perimeter, which shape has the greatest area?
(107)

27. Determine whether this wheel has rotational symmetry. If it does, give
(113) the angle of rotational symmetry and the order of symmetry.

*28. **Formulate** Write an expression for $\sin \theta$ in terms of $\sec \theta$.
(116)

29. A circle whose equation is $(x + 2)^2 + (y - 7)^2 = 81$ is shifted
(95) 4 units upward and 8 units to the left. What is the equation for
the shifted circle? Graph the shifted circle.

30. Using the diagram, prove $\triangle BAM \cong \triangle CAM$.
(30)

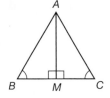

Determining Line of Best Fit

1. **Vocabulary** Two lines with the same slope and different *y*-intercepts
(37) are _____ lines. (***parallel, perpendicular, skew***)

2. What is the midpoint of the line segment with endpoints
(111) (5, 7, −2) and (1, 1, −6)?

3. Determine the slope of the line passing through (4, 8) and (10, 10).
(16)
    A −3   **B** $-\frac{1}{3}$
    C $\frac{1}{3}$ **D** 3

New Concepts

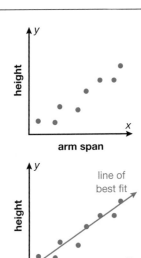

Data is easier to read and understand when it is displayed in graphical form. In one such graph called a scatterplot, points are plotted to show a possible relationship between two sets of data. For example, if the heights and arm spans of people are measured, the data may appear as shown.

The relationship between the two quantities can be better understood by placing a **line of best fit** through the data. The line of best fit is a line that comes closest to all of the points in the data set. The line of best fit approximates the data set and can be used to make predictions about the data.

Caution

A well-chosen line of best fit has a slope that represents the trend of the data and is located within the data points.

Example 1   **Choosing a Line of Best Fit**

Explain why the line in each graph is, or is not, a good choice for the line of best fit.

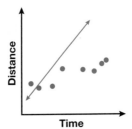
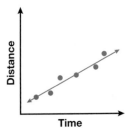

SOLUTION

In the leftmost graph, the line has the correct slope, but it lies too far from the data points. This line is not a good choice for the line of best fit.

In middle graph, the line goes through some of the data points, but the slope is too steep. This line is not a good choice for the line of best fit.

In the rightmost graph, the line goes through the data points and its slope represents the trend in the data. This line is a good choice for the line of best fit.

Online Connection
www.SaxonMathResources.com

The relationship, or correlation, between two quantities in a scatterplot can exhibit a positive correlation, a negative correlation, or no correlation. When both quantities increase together, there is a positive correlation and the line of best fit has a positive slope. When one quantity increases as the other decreases, there is a negative correlation and the line of best fit has a negative slope. When the points are randomly scattered, there is no correlation and a line of best fit cannot be drawn.

If there is a correlation in a scatterplot, the correlation may be described as strong or weak. A strong correlation exists when most of the data points are very close to the line of best fit. As the data points move further away from the line of best fit, the correlation weakens.

Example 2 Identifying Correlation

Describe the correlation between the data points in the scatterplot.

SOLUTION
The line of best fit has a negative slope. Therefore, the data has a negative correlation. The data is not tightly bunched around the line of best fit. Therefore, the correlation can be described as weak. The correlation in this scatterplot would then be described as negative and weak.

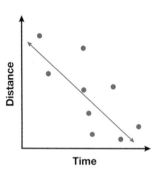

To draw a line of best fit, attempt to find the line that passes through as many data points as possible, with the other points being as close as possible to the line. A good rule to remember is that approximately the same number of points should lie above the line as lie below it.

Example 3 Drawing a Line of Best Fit

Draw the line of best fit in the scatterplot. Then, describe the correlation in the data.

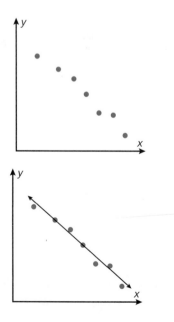

SOLUTION

The line of best fit shows a strong, negative correlation.

A line of best fit can be used to find values that are not a part of the data plotted in a scatterplot. The line approximates the trend in the data. To predict the value of a data point that is not included in the data, find that point on or near the line of best fit.

Example 4 | Application: Tree Growth Patterns

A local power company made a scatterplot to show the growth of the trees in a forested area through which power lines pass. When the trees reach a height of 26 feet, the company trims them so they do not interfere with the power lines. Use the line of best fit in the scatterplot to estimate the time it takes for the trees to reach 26 feet in height.

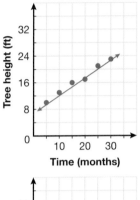

SOLUTION

Extend the line of best fit. Locate 26 feet on the vertical axis and draw a horizontal line until it touches the line of best fit. Then draw a vertical line from this point to the horizontal axis, and read the time. The time is 36 months when the height is 26 feet. Therefore, the trees should take about 36 months, or 3 years, to reach a height of 26 feet.

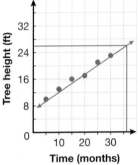

Lesson Practice

a. Explain why the line in each graph is, or is not, a good choice for line of best fit.
(Ex 1)

A

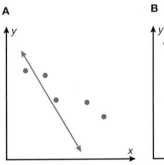

B

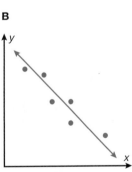

C

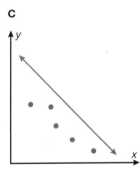

b. Describe the correlation between the data points in the scatterplot.
(Ex 2)

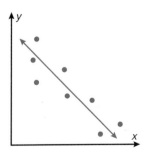

c. Draw the line of best fit in the scatterplot. Then describe the correlation in the data.

(Ex 3)

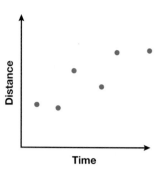

d. Minh is keeping track of the progress as his new home is being built. Every 5 weeks, he records the percent of the home that is complete. He constructs a scatterplot of his findings. If the trend in the data continues, when will his home be 100% complete?

(Ex 4)

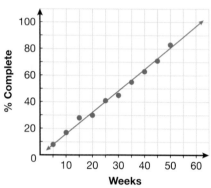

Practice   Distributed and Integrated

1. **Error Analysis** Describe and correct the error in finding the volume of the composite
(115) solid. Give your answer to the nearest tenth of a cubic unit.

$$V = \frac{1}{3}\pi (4^2)(5) + \frac{1}{3}\pi(4^2)(5)$$
$$V = 26\frac{2}{3}\pi + 26\frac{2}{3}\pi$$
$$V = 53\frac{1}{3}\pi$$
$$V = 167.5$$

The volume is approximately 167.5 cubic units.

2. Is the following statement always, sometimes, or never true?
(52)

A square is both a rhombus and a rectangle.

3. (**Barn Construction**) The exterior of the barn shown will be entirely covered with
(115) metal sheeting. To the nearest square foot, how much metal sheeting is needed to cover the building?

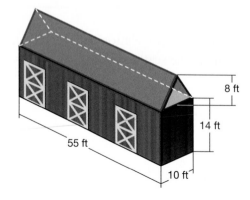

4. Algebra An exterior angle of a triangle measures $(45 - y)°$. The two remote interior angles measure $(2y + 12)°$ and $(31 - y)°$. Determine the value of y.

5. Multi-Step Find all the missing angle measures and lengths in the triangle shown. Round each answer to the nearest whole number.

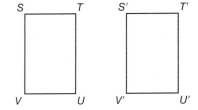

6. A triangle has a base of 1.5 meters and a height of 6 meters. A second, similar triangle has three times the base and height of the original triangle. What is the area of the second triangle?

7. Model Draw two lines of reflection that will transform rectangle $STUV$ to rectangle $S'T'U'V'$.

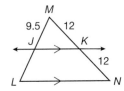

8. What are the vertices of the region that is the solution of $y \le -x + 1$, $y \le 3x + 5$, and $y \ge \frac{1}{3}(x - 1)$? Give a point in the region that is not on the boundary.

9. Find LJ in the triangle at right.

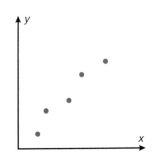

*10. $\triangle JKL$ has vertices $J(1, 3)$, $K(0, 5)$, and $L(-2, -2)$. Apply a dilation centered at $(-3, 4)$. Use a scale factor of $\frac{1}{2}$. Find the coordinates of the vertices of the image. *Hint: Draw axes at $(-3, 4)$ and find the coordinates of J, K, and L as if this was the origin. Apply the dilation, then find the coordinates of each image point with respect to the actual origin.*

*11. Estimate Draw the line of best fit in the scatterplot. Then, describe the correlation.

*12. Express $\tan \theta$ in terms of $\csc \theta$.

13. Algebra As part of their training for a triathlon, Tila and Jamaal must run to and then swim across a circular pond. Find the length of Tila's total route to the nearest meter.

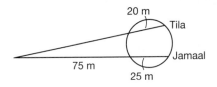

*14. Write Would you expect the data relating the number of days a student is absent
(117) and the student's grades to show a positive or a negative correlation? Explain.

15. Find the radius of this circle to the nearest hundredth of a meter.
(35)

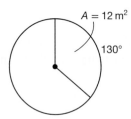

16. (**Architecture**) A building has a base that is twice as long as it is wide, and the area
(44) is 50 square meters. If a contractor wants to make a scale drawing on an $8\frac{1}{2} \times 11$
inch piece of paper, what is the largest scale he could use that would be a whole
number?

*17. According to the scatterplot, approximately what is the average number
(117) of hours of daylight per day one could expect for the month of June?

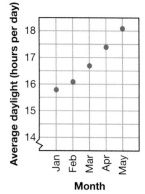

*18. Write Using the equation of a circle and a line, explain why a
(75) circle and a line can intersect in at most two points. *Hint: Solve the
general equation of a line for one variable and substitute into the
general equation of a circle.*

19. How many lines of symmetry does the figure below have? What order is
(76) the rotational symmetry?

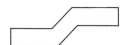

*20. Multiple Choice Which of the following is *not* a possible value for csc θ?
(116)
   A 1.12   **B** 14.24
   C −8.76   **D** −0.12

21. Use the figure shown at right to find WY.
(104)

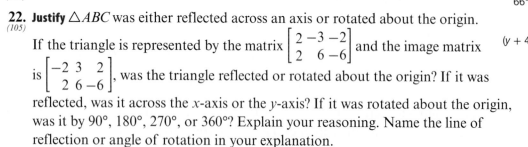

22. Justify $\triangle ABC$ was either reflected across an axis or rotated about the origin.
(105)
   If the triangle is represented by the matrix $\begin{bmatrix} 2 & -3 & -2 \\ 2 & 6 & -6 \end{bmatrix}$ and the image matrix
   is $\begin{bmatrix} -2 & 3 & 2 \\ 2 & 6 & -6 \end{bmatrix}$, was the triangle reflected or rotated about the origin? If it was
reflected, was it across the *x*-axis or the *y*-axis? If it was rotated about the origin,
was it by 90°, 180°, 270°, or 360°? Explain your reasoning. Name the line of
reflection or angle of rotation in your explanation.

23. How many plane symmetries does a regular square pyramid have?
(113) Describe them.

24. Write an indirect proof showing that a right triangle cannot have an obtuse angle.
(48)

25. What is the lateral area of a regular pyramid with a base perimeter of
$_{(70)}$ 10 inches and a slant height of 10 inches?

26. If the volume of the smaller of these two similar solids is 17.2 cubic meter,
$_{(99)}$ what is the volume of the larger?

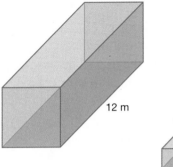

12 m

4 m

*27. Verify Use trigonometric identities to verify $\tan \theta = \dfrac{\sin \theta}{\sqrt{1 - \sin^2\theta}}$.
$_{(91)}$

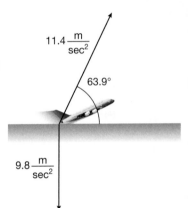

11.4 $\dfrac{m}{sec^2}$

63.9°

9.8 $\dfrac{m}{sec^2}$

28. ⟮**Flight**⟯ The diagram shows a plane taking off. If the acceleration due
$_{(83)}$ to lift is 11.14 m/sec^2 at 63.9° to the horizontal, and gravity pulls
straight down at 9.8 m/sec^2, what is the resultant acceleration vector
felt by the passengers?

29. ⟮**Horticulture**⟯ Sunflower seeds grow in a special pattern: each
$_{(Inv\ 11)}$ spiral of seeds has the same number of seeds as the previous
two spirals combined. If the first two spirals have 3 and 4 seeds,
respectively,
a. how many seeds are in the tenth spiral?
b. how many seeds are there in the first ten spirals altogether?

*30. Multiple Choice Which scatterplot shows a correct line of best fit?
$_{(117)}$

A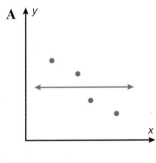

B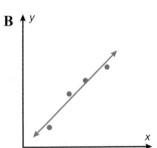

C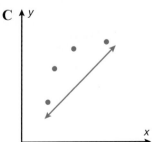

D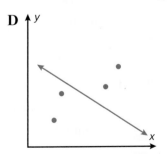

Determining Line of Best Fit Using a Graphing Calculator

Technology Lab 15 *(Use with Lesson 117)*

In this lab, a graphing calculator will be used to examine a set of data and predict values that are missing from the data. The chart shows a test that was applied to a number of different bookshelves. The bookshelves were loaded with the weight given in the first column, and the time until the bookshelf collapsed is recorded in the second column. Use the data in the chart for this lab.

Bookshelf Test	
Test Weights	**Time to Collapse**
200 lbs	1460 seconds
400 lbs	1213 seconds
600 lbs	1108 seconds
800 lbs	930 seconds
1000 lbs	104 seconds
1200 lbs	807 seconds
1400 lbs	642 seconds
1600 lbs	491 seconds

1. Press **STAT** on the calculator. Press **ENTER**. Now enter the coordinates for a scatterplot into the calculator. Enter the test weights into List 1 (L1) and the times into List 2 (L2).

2. Press **2nd** and then press **Y=**. To plot the points entered onto a graph, press **ENTER** twice, then press ⬇ to highlight the scatterplot icon and press **ENTER** again.

Graphing Calculator Tip

Choose as small a window as possible that still shows all the points on the scatterplot. This will help the patterns be seen in the data and verify that the line of best fit is accurate.

3. Adjust the window using the test weights for the interval of *x* values and the time to collapse for the interval of *y* values. By pressing **GRAPH**, the points entered can be seen plotted onto the coordinate plane.

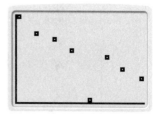

By fitting a line to the scatterplot, the time to collapse can be predicted for the load weights that are not given on the chart.

4. Press **STAT**. Then press ▶ and choose *LinReg(ax + b)* from the menu.
Press **ENTER** twice.

5. The calculator will output values for a line that fits the scatterplot. Write these values down, then press **Y=** and enter the equation in slope-intercept form.

6. Press **GRAPH**. The graph should now display the scatterplot with a line over it.

7. What does this data tell about the time to collapse under different loads? Explain what may have caused the dip in the 1000-pound test.

An outlier is a point in a set of data that is considerably displaced from the rest of the data. Outliers can make it hard to use a set of data effectively, because they significantly alter the line of best fit.

8. Press [2nd] [TRACE] [STAT] [1] [8] [0] [0] on the calculator to estimate the time to collapse for 1800 pounds, adjusting the window as necessary. Is this an accurate estimation? Why might it not be?

9. Edit the scatterplot coordinates and find one point that may be an outlier. Press [DEL] in both columns to remove this point from the data set.

10. Follow steps 4–6 above to find a new equation for a line of best fit.

Math Reasoning

Analyze What is one disadvantage of eliminating possible outliers? Think of one instance where an outlier should not be eliminated?

11. Using this new line, follow step 8 above and estimate the time it will take a bookshelf to collapse under the weight of 1800 pounds. Has the estimate changed? What caused the change? Is the new estimate more accurate than the one obtained above?

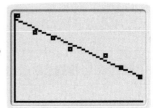

Lab Practice

Using the chart "Bookshelf Test", remove another one of the data points and fit a new line to the data. Does removing one of the other data points cause a significant change in the line of best fit? Can you still make an accurate estimate of the time to collapse for a given weight? In another data set, how can it be determined if a point were an outlier?

Finding Areas of Polygons Using Matrices

Warm Up

1. *(SB 13)* **Vocabulary** The point (0, 0) on the Cartesian plane is called the _____.

2. *(100)* Add the two matrices $\begin{bmatrix} 1 & 0 \\ 0 & 1 \end{bmatrix} + \begin{bmatrix} -1 & 4 \\ 3 & 0 \end{bmatrix}$.

3. *(107)* **Multiple Choice** If the perimeters of these shapes are equal, which shape has the greatest area?

    A square   **B** rectangle

    C rhombus   **D** parallelogram

New Concepts

The area of a polygon on the coordinate plane can be found using the determinant of a matrix.

Determinant of an $m \times 2$ Matrix

Given $A = \begin{bmatrix} x_1 & y_1 \\ x_2 & y_2 \\ x_3 & y_3 \\ \vdots & \vdots \\ x_m & y_m \end{bmatrix}$, the determinant of A is given by the formula:

$\det(A) =$
$(x_1y_2 + x_2y_3 + x_3y_4 + \ldots + x_my_1 - y_1x_2 - y_2x_3 - y_3x_4 - \ldots - y_mx_1)$

Hint

Since the vertices of any polygon can be listed as a two-column matrix, the formula for the determinant given here will work for any polygon.

Example 1   Taking the Determinant of a Matrix

Calculate the determinant of $A = \begin{bmatrix} 6 & 1 \\ 9 & 2 \\ 3 & 1 \end{bmatrix}$.

SOLUTION

Substitute $x_1 = 6$, $y_1 = 1$, $x_2 = 9$, $y_2 = 2$, $x_3 = 3$, and $y_3 = 1$ into the determinant formula.

$\det(A) = x_1y_2 + x_2y_3 + x_3y_1 - y_1x_2 - y_2x_3 - y_3x_1$
$\det(A) = (6)(2) + (9)(1) + (3)(1) - (1)(9) - (2)(3) - (1)(6)$
$\det(A) = 12 + 9 + 3 - 9 - 6 - 6$
$\det(A) = 3$

The determinant of A is 3.

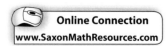

Online Connection
www.SaxonMathResources.com

Matrices can be used to find the area of any convex polygon by putting the polygon's vertices into a vertical matrix and finding the determinant of that matrix divided by 2.

Matrix Method of Computing Area of a Polygon

Given a convex polygon with n vertices, (x_1, y_1), (x_2, y_2), up to (x_n, y_n), the area of the polygon is equal to the absolute value of one-half of the determinant of the matrix listing the coordinates vertically.

$$A = \left| \frac{1}{2} \det \begin{bmatrix} x_1 & y_1 \\ x_2 & y_2 \\ \vdots & \vdots \\ x_n & y_n \end{bmatrix} \right|$$

So, the area of the polygon can be given by the formula.

$$A = \left| \frac{1}{2}(x_1 y_2 + x_2 y_3 + x_3 y_4 + \ldots + x_n y_1 - y_1 x_2 - y_2 x_3 - y_3 x_4 - \ldots - y_n x_1) \right|$$

Example 2 Calculating Area Using Matrices

Calculate the area of the pentagon with vertices $(0, 2)$, $(1, 1)$, $(1, -2)$, $(-2, 0)$, and $(-3, 2)$.

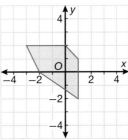

SOLUTION

Use the formula, $\left| A = \frac{1}{2}(x_1 x_2 + x_2 y_3 + x_3 y_4 + x_4 y_5 + x_5 y_1 - y_1 x_2 - y_2 x_3 \right.$
$\left. - y_3 x_4 - y_4 x_5 - y_5 x_1). \right|$

Using the ordered pairs in the order given, $x_1 = 0$, $y_1 = 2$, $x_2 = 1$, $y_2 = 1$, $x_3 = 1$, $y_3 = -2$, $x_4 = -2$, $y_4 = 0$, $x_5 = -3$, and $y_5 = 2$.

Substitute into the formula and solve to find the area of the pentagon.

$$A = \left| \frac{1}{2}[0(1) + 1(-2) + 1(0) + (-2)(2) + (-3)(2) - 1(2) - 1(1) - \right.$$
$$\left. (-2)(-2) - (-3)(0) - 0(2)] \right|$$

$$A = \left| \frac{1}{2}(0 - 2 + 0 - 4 - 6 - 2 - 1 - 4 - 0 - 0) \right|$$

$$A = \left| \frac{1}{2}(-19) \right|$$

$$A = -\frac{19}{2}$$

The area of the pentagon is $\frac{19}{2}$ or 9.5 square units.

Example 3 Application: Fencing

Joshua has a unique fence around his house. The plans for his house are drawn on a grid. The coordinates of the corners of his fence are located at $(-2, 0)$, $(-3, -3)$, $(1, -4)$, and $(3, 4)$. Determine the area of Joshua's land if 1 unit on the grid is 1 yard.

SOLUTION

Joshua's fence is a quadrilateral around the origin. Use the determinant area formula.

$$A = \frac{1}{2}(x_1y_2 + x_2y_3 + x_3y_4 + x_4y_1 - y_1x_2 - y_2x_3 - y_3x_4 - y_4x_1)$$

Using the pairs in the order given, $x_1 = -2$, $y_1 = 0$, $x_2 = -3$, $y_2 = -3$, $x_3 = 1$, $y_3 = -4$, $x_4 = 3$, and $y_4 = 4$. Substitute into the formula to find the area of Joshua's land.

$$A = \left| \frac{1}{2}[(-2)(-3) + (-3)(-4) + (1)(4) + (3)(0) - (-3)(0) - (1)(-3) \right.$$

$$\left. - (3)(-4) - (-2)(4)] \right|$$

$$A = \left| \frac{1}{2}(6 + 12 + 4 + 0 + 0 + 3 + 12 + 8) \right|$$

$$A = \left| \frac{1}{2}(45) \right|$$

$$A = \frac{45}{2}$$

The area of Joshua's land is 22.5 square yards.

Caution

When using this method, you may want to sketch the polygon first. Note that this method can only be applied to convex polygons.

Lesson Practice

a. Calculate the determinant of $\begin{bmatrix} 5 & 6 \\ 4 & 2 \\ 5 & 0 \end{bmatrix}$.
(Ex 1)

b. Calculate the determinant of $\begin{bmatrix} -2 & -4 \\ 1 & 2 \\ 0 & 3 \end{bmatrix}$.
(Ex 1)

c. Calculate the area of the hexagon with vertices $(1, 4)$, $(2, 3)$, $(1, -4)$, $(-1, -4)$, $(-2, 0)$, and $(-1, 4)$.
(Ex 2)

d. **Sports** The gravel for the infield of a baseball diamond forms a pentagon. The vertices of the pentagon are located at $(0, -35)$, $(-70, 35)$, $(-35, 70)$, $(35, 70)$, and $(70, 35)$. If one unit equals one foot, determine the area of the field.
(Ex 3)

Practice Distributed and Integrated

* 1. **Flooring** Bart wants to paint a design on his tiled floor. He determines the vertices of the polygon he is going to paint are located at $(-5, 7)$, $(-1, 15)$, $(5, 2)$, $(0, -8)$, and $(-10, -3)$. What is the area Bart wants to paint?
(118)

* 2. Describe the correlation in the scatterplot.
(117)

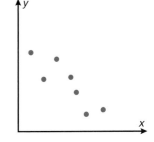

* 3. (**Architecture**) On a grid of the floor plan of a building, the corners of
(118) the building are located at the points (5, 7), (7, 0), (1, −5), and (−6, 2).
If one unit represents one yard, what is the area of the actual floor?

4. Analyze Explain what happens to the graph of cosecant and secant when
(116) $\sin \theta = 0$ and when $\cos \theta = 0$.

5. What is the distance between the point (−9, 4) and the line $y = 3$?
(42)

6. Write Using Theorem 64-1, explain why, if a tangent to a circle has a point of
(64) tangency, T, and $\overline{TA}$ is a diameter of the circle, then $\overline{TA}$ is perpendicular to
the tangent.

7. Find the volume of this right cylinder from which a smaller right cylinder
(115) has been removed. Give your answer to the nearest tenth of a cubic
centimeter.

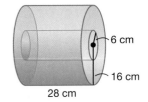

* 8. Calculate the determinant of $\begin{bmatrix} 5 & 4 \\ 4 & 5 \\ 2 & 1 \end{bmatrix}$.
(118)

9. Determine the area of the shaded part of the circle by
(112) determining the area of the segment and subtracting
that from the area of the circle. Write your answer in
terms of π.

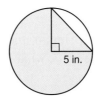

10. Analyze A region is defined by the disjunction of the inequalities $y > -x + 2$ and
(114) $y < x + 2$. Graph the region, then add $y > -2x + 3$. If $y > -2x + 3$ were
included in the disjunction, would the region be larger?

11. Draw front, top, and side views of the solid at right.
(93)

12. Algebra Find the equation of the line of symmetry of
(76) the function $y = |x - 3|$.

*13. Multi-Step Find the circumferences and areas of the inscribed and circumscribed
(106) circles if the regular hexagon has a side length of 10 inches.

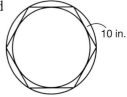

14. Multiple Choice The conjecture, "The bisector of an angle measuring 180° or
(14) less divides the angle into two congruent acute angles," has
A no counterexamples. **B** many counterexamples.
C exactly one counterexample. **D** exactly two counterexamples.

15. Find the line perpendicular to $y = 12x - 4$ that passes through the point (4, 1).
(37)

*16. Calculate the determinant of $\begin{bmatrix} 1 & 2 \\ 0 & 0 \end{bmatrix}$.
(118)

*17. Formulate If the height and radius of a cone are H and R, respectively,
(85) what is a formula for the area of the circular cross section that is
h units from the vertex of the cone?

*18. Write Is the line in the scatterplot a good choice for line of best fit? Explain why or why not.
(117)

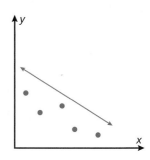

19. An isosceles right triangle has a hypotenuse of 16 feet. What is the perimeter, to the nearest tenth of a foot?
(53)

20. What is the equation of a circle with a radius of 5 and a center point located at $(6, -2)$?
(75)

21. (**Megaphones**) Matteo is making a homemade megaphone out of newspaper for Recycling Awareness Week at his school. He plans to have the megaphone be a right circular cone that is 2 feet long, with an opening that is 8 inches across at one end, and be as thick as three sheets of newspaper. How much newspaper does he need?
(77)

22. In this figure, $\overline{VU}$ is a diameter. What is the measure of $\angle VTU$?
(47)

23. (**Flight**) An airplane is flying at an elevation of 1 mile. Its location is 35 miles east and 140 miles south of the airport. Another airplane is flying at an elevation 2.25 miles, and its location is 300 miles west and 95 miles north of the airport. Find the distance to the nearest tenth of a mile between the planes.
(111)

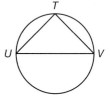

24. $\triangle ABC$ and $\triangle DEF$ are similar and have sides related by the ratio 3:4. If the area of $\triangle ABC$ is 15.3 square units, what is the area of $\triangle DEF$?
(87)

25. Multiple Choice Which of the following values of θ is not in the domain of cot θ?
(116)

A 90°　　　　　　　　　　**B** 0°

C 45°　　　　　　　　　　**D** 270°

26. (**Construction**) A construction company is having cinder blocks specially made for the walls of a building. For structural support, each block must contain approximately 3000 cubic inches of concrete. How tall should each block be to meet the specifications? Give your answer to the nearest whole inch.
(115)

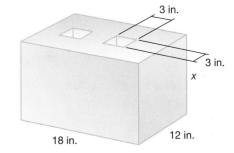

27. Looking down from her balcony, Celeste sees her car at a 25° angle of depression. The balcony is 16 feet above the ground, and Celeste is about 5 feet tall. How far away is she from her car, to the nearest foot?
(73)

*28. Error Analysis Oscar and Reena drew different lines of best fit in a scatterplot. Which line of best fit is more suitable for the data in the scatterplot? Explain.
(117)

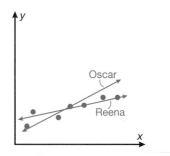

*29. Calculate the area of the pentagon with vertices $(0, 2)$, $(-3, 1)$, $(-2, -1)$, $(1, -1)$, and $(3, 1)$.
(118)

30. Solve the system of inequalities $y > 5x + 1$ and $y \leq 5x - 11$.
(88)

Platonic Solids

1. **Vocabulary** A section of a cone or a pyramid that has
(103) two parallel bases is called a _____. (**base, frustum, vertex**)

2. Draw the cross section created by the plane's intersection
(85) with this triangular pyramid.

3. Find the volume of a cone with an 8-inch height and a
(77) 6-inch diameter in terms of π.

4. Find the radius of a sphere with a volume of 972π cubic centimeters.
(80)

New Concepts

In a regular polygon, all angles are congruent and all sides are congruent. A **regular polyhedron** is a polyhedron in which all faces are congruent, regular polygons, and the same number of faces meet at each vertex. Regular polyhedrons are also called **Platonic solids**. There are five Platonic solids: the tetrahedron, the cube, the octahedron, the dodecahedron, and the icosahedron. These are the only regular polyhedrons. Each one is pictured and defined below.

Tetrahedron: A polyhedron with four faces. A regular tetrahedron has equilateral triangles as faces, with three faces meeting at each vertex.

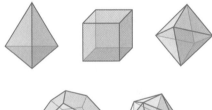

Cube: A polyhedron with six faces. All the sides are squares, with three faces meeting at each vertex.

Octahedron: A polyhedron with eight faces. A regular octahedron has equilateral triangles as faces, with four faces meeting at each vertex.

> **Math Reasoning**
>
> **Generalize** List all the shapes that make up the faces of the Platonic solids.

Dodecahedron: A polyhedron with 12 faces. The faces of a regular dodecahedron are regular pentagons, with three faces meeting at each vertex.

Icosahedron: A polyhedron with 20 faces. A regular icosahedron has equilateral triangles as faces, with five faces meeting at each vertex.

The number of vertices, edges, and faces in each of the platonic solids is shown in the table. The last column demonstrates that the platonic solids follow Euler's formula.

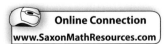
Online Connection
www.SaxonMathResources.com

Properties of Platonic Solids				
Regular Polyhedron	Vertices (V)	Edges (E)	Faces (F)	V − E + F
tetrahedron	4	6	4	2
cube	8	12	6	2
octahedron	6	12	8	2
dodecahedron	20	30	12	2
icosahedron	12	30	20	2

Example 1 Hexagons and Platonic Solids

Why are none of the Platonic solids made with regular hexagons? Explain using angle measures.

SOLUTION

Each angle in a regular hexagon measures 120°. At least three hexagons would need to be placed at each vertex.

This means that each vertex would have a total measure of 120° × 3 = 360°.

But, 360° is the angle of a point on a plane. If the angles added to 360°, the hexagons would make a tessellation, not a polyhedron.

Math Language

A **tessellation** is a repeating pattern of polygons that completely covers a plane with no gaps or overlaps.

Example 2 Angles of Platonic Solids

a. What is the sum of the measures of the angles at a vertex of a regular octahedron?

SOLUTION

Each vertex is the meeting point of four equilateral triangles. The measure of each angle in an equilateral triangle is 60°.

60° × 4 = 240°

The sum of the angles is 240°.

b. What is the sum of the measure of the angles at a vertex of a cube?

SOLUTION

Each vertex is the meeting point of three squares. The measure of each angle in a square is 90°.

90° × 3 = 270°

The sum of the angles is 270°.

c. Make a general statement about the sum of interior angles for a vertex of a Platonic solid.

SOLUTION

The sum of interior angles for a vertex of a Platonic solid must be less than 360°. If the sum of interior angles is 360°, the faces would all lie in a plane and therefore could not be part of a polyhedron.

Because the sum of interior angles of a vertex cannot be greater than 360°, a Platonic solid cannot be constructed with faces that have more than 5 sides. As you saw in Example 1, hexagons cannot be used to construct a Platonic solid, because a vertex where 3 hexagons meet sums to 360°.

Example 3 Proving the Five Platonic Solids

Prove that the five Platonic solids can only be made from equilateral triangles, squares, or regular pentagons. Also, explain why there are no other possibilities for Platonic solids.

SOLUTION

As shown above, the sum of interior angles of a vertex of a Platonic solid must be less than 360°. Also, at least three polygons must meet at each vertex.

The measure of each angle in an equilateral triangle is 60°. Three possibilities exist for the number of equilateral triangles that can meet at a vertex of a Platonic solid.

Three faces: $3 \times 60° = 180°$ tetrahedron
Four faces: $4 \times 60° = 240°$ octahedron
Five faces: $5 \times 60° = 300°$ icosahedron

The measure of each angle in a square is 90°. For squares, only one possibility exists for the number of faces that can meet at a vertex of a Platonic solid.
Three faces: $3 \times 90° = 270°$ cube

The measure of each angle in a regular pentagon is 108°. Only one possibility exists for the number of faces that can meet at a vertex of a Platonic solid.

Three faces: $3 \times 108° = 324°$ dodecahedron

A Platonic solid cannot be constructed with polygons that have more than five sides. Therefore, only five Platonic solids are possible.

Math Reasoning

Justify Explain why it is not possible for only two faces to meet at a vertex of a polygon.

Lesson Practice

a. Explain why regular octagons cannot be used to construct a Platonic solid.
(Ex 1)

b. What is the sum of the interior angles of a vertex in an icosahedron?
(Ex 2)

c. Explain why four squares cannot meet at a vertex of a polyhedron.
(Ex 3)

1. **Error Analysis** Brittney solves $\sin \theta$ in terms of $\csc \theta$ like this: $\sin \theta = \dfrac{\sqrt{\csc^2 \theta - 1}}{\csc \theta}$.
(116)   Write the correct answer and explain Brittney's error.

2. Using Euler's formula, determine how many edges a polyhedron with 10 vertices
(49)   and 7 faces has.

* 3. Which of the Platonic solids can be classified as a pyramid?
(119)

* 4. **Multiple Choice** Which of the following is *not* a net for a tetrahedron?
(119)

A   B

C   D

5. **Write** A rectangular prism has lengths of 7 units, 3 units, and 5 units. What are the
(113)   dimensions of the prism after a rotation of 270°? Explain your answer.

6. **Multi-Step** Find the perimeter of this triangle. Round your answer to the
(98)   nearest tenth.

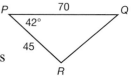

7. The function $y = 8x + 1$ makes an acute angle with the positive x-axis. What is
(82)   the angle measure to the nearest degree?

* 8. **Write** The area of a triangle with vertices $(0, 0)$, (a, b), and (c, d) is equal to $\frac{1}{2}$
(118)   times the determinant of $\begin{bmatrix} a & b \\ c & d \end{bmatrix}$. Use this to explain the formula for the area of a
   convex polygon about the origin given in Lesson 118.

9. (**Sewing**) A tailor is cutting out the pattern shown to make a dress. Find
(40)   the area of the tailor's pattern to the nearest tenth of a square inch.

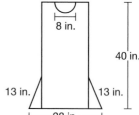

*10. **Write** Use the determinant area formula to explain why the area is dilated
(118)   by a^2 when the side lengths of a convex polygon are dilated by a.

11. Use a trigonometric ratio to find the value of x in the diagram to
(68)   the nearest hundredth.

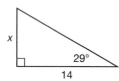

*12. (**Space Walks**) Two astronauts are moving in space on unassisted space walks. The
(108)   first astronaut begins at $(3, -1, 4)$ and moves to $(9, 2, 11)$. The second begins at
   $(-49, -15, -2)$ and moves to $(-38, -11, 4)$. Show that the astronauts' paths will
   cross at the point $(39, 17, 46)$ if they stay on the same linear paths.

13. Multiple Choice In which scatterplot is there a strong positive correlation?
(117)

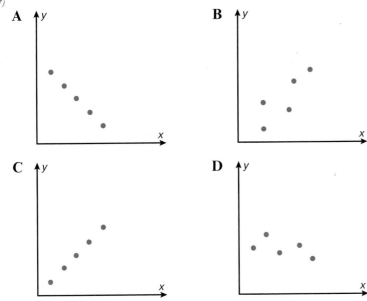

A

B

C

D

*14. How many axes of rotation does a tetrahedron have?
(119)

15. In the circle shown, chords $\overline{AB}$ and $\overline{CD}$ intersect at E. Find the value of x.
(86)

*16. Analyze Calculate the area of the triangle with vertices (2, 3), (4, 5), and (1, 6).
(118)

17. (**Storage**) Alastair's attic is in the shape of a rectangular pyramid.
(70) If the floor's dimensions are 10 feet by 20 feet, and the height of the attic is 6 feet, what volume can theoretically be stored in Alastair's attic?

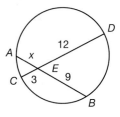

18. Analyze What geometric figure will result from dilating a circle centered about the
(95) origin with a 5-unit radius by a scale factor of 0?

19. Graph the solution of the conjunction of $2x + 3y > 1$ and the disjunction of
(114) $x - 3y \geq 1$ and $5y - 4x \geq 10$.

20. Find the value of x. Give your answer in simplified radical form.
(29)

21. (**Design**) The logo for a communications corporation is a regular pentagon
(112) inscribed in a circle with a radius of 3 inches. Determine the combined area of the five circle segments in square inches, to the nearest tenth.

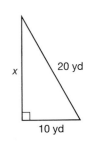

22. Algebra The point $(x, -3)$ is translated 3 units right. The image has coordinates
(71) $(3, -3)$. Find the value of x.

23. (**Running**) Beth runs 150 meters at 43° north of west. Which is larger, her
(89) northward displacement or her westward displacement, and by how much? Round your answer to the nearest tenth.

24. Is the following statement always, sometimes, or never true?
(52)

A square is a rhombus.

25. (**Products**) A jar of peanut butter with a 3.5-inch diameter has had a 1-inch-radius
(115) hemispherical scoop taken out of it. What volume of peanut butter remains in the
jar, to the nearest tenth?

26. **Algebra** Find the side length of an equilateral triangle in terms of r, if it is
(106) inscribed in a circle with a radius, r. What is the side length if the triangle
circumscribes the circle instead?

27. Use trigonometric identities to find the tangent of an angle when the
(116) secant of that angle is $\frac{3}{2}$. Express your answer in simplified radical form.

28. **Multi-Step** Find the perimeter of the triangle. Round your answer to the nearest
(94) whole number.

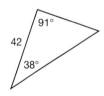

*29. **Analyze** Which Platonic solids can produce a cross section that is a pentagon?
(119)

30. Find the value of a in this scalene triangle.
(94)

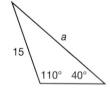

LESSON 120

Topology

Warm Up

1. **Vocabulary** A transformation about a point, P, such that each point and its image are the same distance from P is called a _____. (*dilation, reflection, rotation, translation*)
(78)

2. What properties are shared by rectangles and squares?
(65)

3. **Multiple Choice** The set of points in space that are a fixed distance from a given center point is called which of the following?
(80)

   A cone   **B** cube

   C cylinder   **D** sphere

New Concepts

An extension of geometry that deals with the nature of space is a topic called **topology**. This branch of geometry deals with the properties of a geometric figure that do not change under stress.

As an elementary example, a square and a circle are topologically the same in that they both have a boundary that isolates an area within a two-dimensional space. You can think of starting with a circle and then pulling at four equally spaced points along its circumference to form a square.

> **Hint**
>
> The term *under stress* means that if a force is applied to a shape, the force could cause the shape to change. In topology, there must also be a *restorative force* that can undo changes to restore the shape back to its original form.

A circle and the number 8, however, are not topologically the same. A circle divides a two-dimensional space into two distinct regions, while the number 8 divides the space into three regions. The diagram shows a circle being distorted. No matter how it is altered, the shape still only divides the space into two regions.

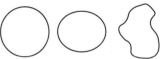

A circle under stress

Shapes in topology are considered infinitely pliable. That is, they can be stretched, compressed, and put under any other kind of stress. However, they cannot be cut, torn, or glued to form other topological shapes.

Example 1 Topological Transformations

Explain why a torus, which is a doughnut shape, can be considered topologically the same as a coffee cup.

SOLUTION
A pliable torus can be reshaped to form a coffee cup by thinning out one side of the torus where the handle will be, and decreasing the size of the hole while moving it closer to the side with the handle. Then, a depression that will eventually form the hollowed portion of the coffee cup can be created. This transformation is shown in the diagram.

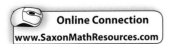

Online Connection
www.SaxonMathResources.com

If two shapes or solids are topologically equivalent, it means they share the same topological properties in the way they divide space.

Example 2 Topological Equivalence

Which object is topologically equivalent to the given object?

A   B   C   D

SOLUTION

The given object does not separate the two-dimensional plane into different regions. From the choices, the only object that does not separate the two-dimensional plane into different regions is choice **D**. Therefore, the object in choice **D** is topologically equivalent to the given object. If the object were simply straightened out, it would be identical to **D**.

To classify a set of objects, use a rule that will divide them into topological classes. For example, the lowercase letters of the alphabet can be categorized according to the number of regions into which they divide the plane.

Letters without a hole do not divide the plane. Letters with one hole divide the plane into two regions—the outside and the inside.

The letters without a hole are {c, f, h, k, l, m, n, r, s, t, u, v, w, x, y, z}.
The letters with one hole are {a, b, d, e, o, p, q}.
The letter with two holes is {g}.
The letters with two parts are {i, j}.

Math Reasoning

Analyze Why might the grouping of lowercase letters be altered for someone handwriting these letters?

Example 3 Topological Classes

Classify the uppercase letters into topologically equivalent classes.

SOLUTION

The letters can be grouped into letters with no holes, letters with one hole, and letters with two holes.
The letters with no holes are {C, E, F, G, H, I, J, K, L, M, N, S, T, U, V, W, X, Y, Z}.
The letters with one hole are {A, D, O, P, Q, R}.
The letter with two holes is {B}.

Exploration   **Observing Properties of Möbius Strips**

In this exploration, you will create a Möbius strip and investigate some of its properties.

1. Cut a strip of paper approximately 1 inch wide and as long as possible.

2. Give one end of the strip a half twist and tape this end to the other end of the strip, making a loop with a half twist in it. This loop is referred to as a Möbius strip.

3. Starting at the middle of the strip on the outside, draw a line along the middle of the strip's surface until you have gone completely around the strip. Explain what you notice. What conclusions can you draw based on this observation?

4. Cut along the line you drew in step 3. What is the result? How does this support the conclusion you made in step 3?

Math Reasoning

Connect Construct a second Möbius strip. Cut the strip along a line that is one-third the distance from a side. Continue to cut along this line until you reach the point where you started cutting. What is the result?

Lesson Practice

a. Is the symbol for infinity, ∞, topologically the same as the capital letter
(Ex 1) B? Explain.

b. Which object is topologically equivalent to the given object?
(Ex 2)

A   **B**   **C**   **D**

c. Classify the ten digits (1, 2, 3, 4, 5, 6, 7, 8, 9, 0) into topological classes
(Ex 3) by the number of regions they divide the plane into.

Practice   Distributed and Integrated

1. The triangle ABC is translated 4 units right and 3 units down. Write a point
(100) matrix for $\triangle ABC$ and find the matrix that transforms $\triangle ABC$ to $\triangle A'B'C'$.

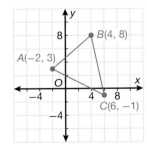

2. Given a fixed area, which shape will have the least perimeter?
(107)

3. Multiple Choice Which region represents the conjunction of $y \geq 3x+1$
(88) and $y \leq -x$?

A *A*

B *B*

C *C*

D *D*

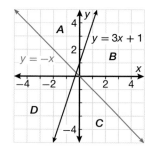

4. Verify Use the determinant area formula to verify that the area of
(118) a square with a side length of s has an area of s^2.

5. Find x and m$\widehat{KJ}$, if $\overline{JL} \cong \overline{JM}$, $\overline{JM} \cong \overline{KN}$, and m = 120°.
(104)

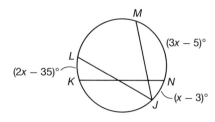

6. A composite transformation reflects a figure over the x-axis, rotates it 120°, and
(90) then flips it across the y-axis. Will the resulting image be the same, regardless of
which order these transformations are performed? Explain.

7. What is the measure of $\angle A$?
(79)

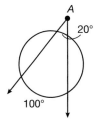

8. Multi-Step Is $\sin \theta = \dfrac{\sqrt{1 + \cot^2\theta}}{\cot \theta}$ true? Use trigonometric identities to prove it or use
(116) a counterexample to disprove it.

9. Find *FD*.
(50)

10. (**Games**) During a board game, Bhavna has pieces at points $(2, 5)$, $(-3, 1)$, and
(118) $(1, -6)$ on the game board. Imagine the pieces are the vertices of a polygon. What
is the area of the polygon?

11. (**Lighthouses**) A lighthouse flashes its light to warn ships that are closer than
(73) 1500 meters away. If the lighthouse keeper knows his lighthouse is 50 meters tall,
at what angle of depression should he begin flashing the light at the ship? Round
your answer to the nearest hundredth.

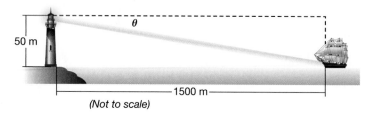

*12. A square pyramid has vertices $R(1, 1, 2)$, $S(3, 1, 2)$, $T(3, -1, 2)$, and $U(1, -1, 2)$.
(111) If the vertex of the pyramid is $V(2, 0, 6)$ what is the height of the pyramid?

 *13. Write The faces of a soccer ball are regular polygons. Why is a soccer ball
(119) not an example of a Platonic solid?

*14. Multiple Choice Which letters are topologically the same whether they are
(120) uppercase or lowercase?

   A a and P   **B** b and D

   C e and D   **D** J and k

15. Algebra Find x if $AB = x + 6$.
(55)

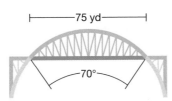

*16. Draw a net for an octahedron.
(119)

17. To the nearest whole square centimeter, find the surface area of a sphere with
(80) a 15-centimeter radius.

18. (**Civil Engineering**) Determine the area of the circle segment of this bridge, to
(112) the nearest square yard.

$$\longmapsto \text{75 yd} \longmapsto$$

70°

19. Multi-Step A triangle has vertices $X(-2, 3)$, $Y(1, 1)$, and $Z(-1, -3)$.
(98) Find all the measures in the triangle. Round angles to the nearest
degree and sides to the nearest tenth.

*20. (**Physics**) When an object rests on a surface, a force vector called the *normal*
(89) *force* holds the object up. This vector is always perpendicular to the surface.
Find the horizontal and vertical components F_V and F_H of the normal force,
vector F_N in the diagram, if the normal force is 12 pounds. Round your
answers to the nearest hundredth of a pound.

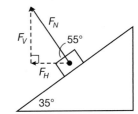

21. Formulate The point x is the midpoint of a and b on the number line. The
(11) point y is the midpoint of c and d on the number line. Determine an
expression for the midpoint of x and y. Interpret your result.

*22. Analyze Is a knotted loop of string topologically equivalent to an
(120) unknotted loop of string? Explain.

23. Formulate Determine a formula for the area of a segment with a central angle
(112) measure of $0° < x° < 180°$, in a circle with a radius, r.

24. Write a two-column proof to show that if two parallelograms both have an area of A and a base of b, corresponding to heights h_1 and h_2 respectively, then $h_1 = h_2$.
₍₂₇₎

*25. Error Analysis Joseph conjectures, "A closed soup can is topologically distinct from an open soup can." Identify and correct his error.
₍₁₂₀₎

26. Find the distance between $P(2, 5, 6)$ and $H(-2, 7, 1)$.
₍₁₁₁₎

27. Describe the correlation shown in the scatterplot as negative or positive, and strong or weak.
₍₁₁₇₎

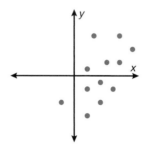

*28. Write Could a fork and a spoon be considered topologically equivalent? Explain.
₍₁₂₀₎

29. (**Basketball**) Shirley is playing basketball and wants to pass the ball, but is not sure to which teammate to throw the ball. She decides to pass the ball to whomever is closest. To whom should she pass the ball, and why?
₍₅₁₎

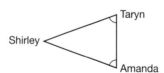

30. The drawing shows a part of a solid and its center of rotation. Draw the complete solid with order of rotation 3.
₍₁₁₃₎

Polar Coordinates

A **polar coordinate system** is a system in which a point in a plane is located by its distance, r, from a point called the pole, and by the measure of a central angle θ.
The **pole** corresponds to the origin of the coordinate plane. The **polar axis** is the horizontal ray with the pole as its endpoint that extends infinitely to the right.

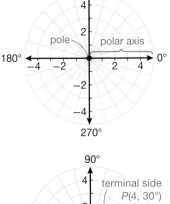

The angle θ is formed from two rays, the initial side and the terminal side. The **initial side** is the ray that lies on the polar axis, and the **terminal side** is the ray that is rotated relative to the polar axis.

The angle is measured counterclockwise from the initial side.

The graph shows the location of point P given by $P(4, 30°)$, where $r = 4$ and $\theta = 30°$.

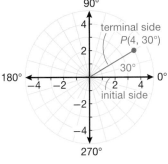

To convert from polar to Cartesian coordinates, use a trigonometric ratio.

The initial side, the terminal side, and the line segment that is perpendicular to initial side from Point A form a right triangle.

1. Use a trigonometric ratio to write and solve an equation that describes CB. Round the answer to the nearest hundredth.

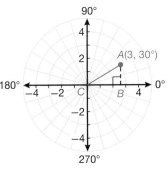

2. **Formulate** The length of $\overline{CB}$ is the horizontal coordinate of point A in the Cartesian coordinate system. Write a general equation for finding the x-coordinate of a point in terms of r and θ.

3. **Predict** Use step 1 and 2 to write a general equation for finding the y-coordinate of a point in terms of r and θ.

You can also convert from Cartesian coordinates to polar coordinates by looking at the right triangle made by the x-axis and a perpendicular line from the x-axis to the point.

4. Use the Pythagorean Theorem to write an equation to find the value of r for the point $(-8, 6)$, shown here.

5. Use a trigonometric ratio to write and solve an equation that describes θ_2.

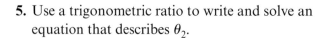

6. Formulate Use steps 4 and 5 to write a general equation for finding the distance r and angle θ_2 of a point in terms of x and y.

7. Since a polar coordinate must be measured counterclockwise from the positive x-axis, what is the value of θ? What is the polar coordinate for $(-8, 6)$?

8. Formulate Write an equation for finding θ in terms of θ_2 in each of Quadrants II, III, and IV.

A **polar function** is a function that has θ as its independent variable and r as its dependent variable.

Polar functions can be graphed in polar coordinates. For example, to graph the function $r = 1 + \sin \theta$, use a table of values with multiples of 30°.

<table>
<tr><td>

θ	r
0°	1
30°	1.5
60°	1.87
90°	2
120°	1.87
150°	1.5
180°	1

</td><td>

θ	r
210°	0.5
240°	0.13
270°	0
300°	0.13
330°	0.5
360°	1

</td></tr>
</table>

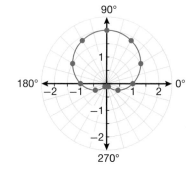

The graph of $r = 1 + \sin \theta$ is an example of a cardioid curve.

9. Plot the graph of the polar equation $r = 2.5$. Describe the shape of the graph.

10. Generalize Using your graph from step 9, write an equation that will graph a circle with a radius, t in polar coordinates.

11. Create a table of values from 0° to 360° by multiples of 15° for the polar function $r = \cos 2\theta$.

12. Plot the points from your table of values. When r is negative, take the absolute value of r.

Math Reasoning

Analyze Which point in the table coincides with the pole? Would it be correct to give the coordinates of this point as (0, 0°)?

Math Language

The graph of $y = \cos(2\theta)$ is an example of a **quadrifolium**, meaning "four leaves."

Hint

When θ has a value greater than 360° this indicates more than one rotation around the circle. 480°, for example, is identical to 120°, because it is one full circle of 360° and then another 120°.

13. Plot the graph of the polar equation $r = \frac{\theta}{360°}$. Use multiples of 45° for θ.

14. Extend your plot to cover the range $0 \le \theta \le 1080°$. Describe the shape of the graph.

Investigation Practice

a. Convert these polar coordinates to Cartesian coordinates.

  1. $(4, 180°)$

  2. $(1, 270°)$

  3. $(3, 360°)$

b. Convert these Cartesian coordinates to polar coordinates.

  1. $(-3, -3)$

  2. $(0, 2)$

  3. $(3, -2)$

c. Plot the graph of the polar equation $r = 1 + \cos \theta$. Describe the shape of the graph, and its relationship to the graph of $r = 1 + \sin \theta$.

d. **Analyze** Plot the graph of the polar equation $r = 1 - \sin \theta$. Describe the shape of the graph, and its relationship to the graph of $r = 1 + \sin \theta$.

Skills Bank

Order of Operations and Absolute Value

Skills Bank 1

When simplifying expressions, it is important to follow the Order of Operations:
1. Evaluate expressions in parentheses or other grouping symbols first.
2. Evaluate exponents and roots.
3. Multiply and divide from left to right.
4. Add and subtract from left to right.

The order of operations can be abbreviated as PEMDAS: Parentheses, Exponents, Multiplication, Division, Addition, and Subtraction.

Example 1 Using Order of Operations

Evaluate $\left[9 \div 3 - (5 \times 3)^2\right] + 2$.

SOLUTION

When more than one pair of grouping symbols are present, work from the inside out. The expression in parentheses must be evaluated first: $5 \times 3 = 15$. Next, evaluate the exponent: $15^2 = 225$. Then, divide: $9 \div 3 = 3$. To complete the operations in the brackets, subtract: $3 - 225 = -222$. Now that there are no further parentheses, the last step is to add: $-222 + 2 = -220$.

$$\left[9 \div 3 - (5 \times 3)^2\right] + 2$$
$$\left[9 \div 3 - (15)^2\right] + 2$$
$$\left[9 \div 3 - 225\right] + 2$$
$$\left[3 - 225\right] + 2$$
$$\left[-222\right] + 2$$
$$-220$$

The **absolute value** of a number is the distance from that number to zero on a number line. The notation for absolute value of a number x is $|x|$.

Example 2 Using Absolute Value

Find $|2|, |-2|, |5|, |-5|$, and $|0|$.

SOLUTION

When the numbers are plotted on a number line, finding the distance of each from zero is simple. The number 2 is 2 units from zero, so $|2| = 2$. The number -2 is also 2 units from zero, but in the opposite direction, so $|-2| = 2$. The numbers 5 and -5 are each 5 units from zero, so $|5|$ and $|-5|$ are both 5. Zero is 0 units from itself, so $|0| = 0$.

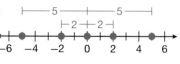

Skills Bank Practice

Find each absolute value.

  a. $|8|$ **b.** $|-157|$

Evaluate each expression.

  c. $(2 + 3)^2$ **d.** $2 + 3^2$ **e.** $2 - |-4|$ **f.** $|2 - (-4)|$

Properties of Arithmetic

Skills Bank 2

The basic properties of addition and multiplication are outlined in the table below. These properties only apply when a, b, and c are real numbers.

If a and b are real numbers, then…

Properties of Addition		Properties of Multiplication	
Closure	$a + b$ is a real number	Closure	ab is a real number
Commutative	$a + b = b + a$	Commutative	$ab = ba$
Associative	$(a + b) + c = a + (b + c)$	Associative	$(ab)c = a(bc)$
Identity	$a + 0 = a$	Identity	$(a)(1) = a$
		Multiplication Property of Zero	$(a)(0) = 0$
Other Properties of Arithmetic			
Distributive Property	$a(b + c) = ab + ac$	Transitive Property of Equality	If $a = b$ and $b = c$, then $a = c$

Example 1   Identifying Properties of Arithmetic

What property of addition or multiplication is being shown in each problem?

a. $3 + -8 = -8 + 3$

SOLUTION

Commutative Property of Addition

b. $(5)(2 - 2) = 0$

SOLUTION

Multiplication Property of Zero

Skills Bank Practice

What property of addition or multiplication is being shown in each problem? Assume c is a real number.

a. $(3 + 2) + 5 = 3 + (2 + 5)$

b. $5(3 \times 4) = (5 \times 3) \times 4$

c. $6(4 + 1) = (6)(4) + (6)(1)$

d. $(5)(2) = (2)(5)$

e. $(1)(3c) = 3c$

f. $5c$ is a real number

Classifying Real Numbers

Skills Bank 3

Numbers can be organized into sets based on their properties.

The **natural numbers** are the counting numbers.	$\{1, 2, 3, 4, ...\}$	
The **whole numbers** are the natural numbers and 0.	$\{0, 1, 2, 3, ...\}$	
The **integers** are the whole numbers and their opposites.	$\{..., -2, -1, 0, 1, 2, ...\}$	
The **rational numbers** are numbers that can be written as the ratio of two integers.	Examples: $\left\{5.7, -\dfrac{1}{2}, \dfrac{4}{3}, 0, 2\right\}$	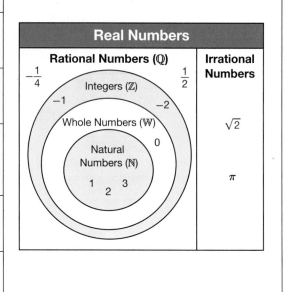
Irrational numbers cannot be written as the ratio of two integers.	Examples: $\left\{\pi, \sqrt{2}, -\sqrt{33}\right\}$	
The **real numbers** are the rational numbers and the irrational numbers.	Examples: $\left\{-\dfrac{3}{41}, 0, \sqrt{5}\right\}$	

Example 1   **Classifying Real Numbers**

List all the sets of numbers to which -25 and $\sqrt{7}$ belong.

SOLUTION

Negative twenty-five is not a natural number or a whole number since it is negative, but it is an integer since it is the opposite of a natural number. Since it is an integer, it is also a rational number and a real number.

The square root of seven is not a natural number, a whole number, or an integer, since 7 is not the square of any integer. It is also not a rational number, since 7 is not the square of any rational number. However, since it is a real number—it lies between 2 and 3 on the number line—it must be irrational. So, $\sqrt{7}$ is both irrational and real.

Skills Bank Practice

List the sets to which each number belongs.

    a. -156 **b.** $0.\overline{3}$

    c. $150,000$

    d. Are any numbers both rational and irrational?

Ratios, Proportions, and Percents

Skills Bank 4

A **ratio** compares two quantities and is often written as a fraction. For example, if you exchange 4 nickels for 2 dimes, the ratio of nickels to dimes is $\frac{4}{2}$. A ratio shows that two quantities are in direct variation, meaning they have a linear relationship that can be written in the form $y = kx$, where k is a nonzero constant. In the example above, the number of nickels (y) and the number of dimes (x) are in direct variation and $k = 2$.

Example 1 | Writing Ratios

Janine picked fruit from her garden. She picked 5 tomatoes and 3 peaches. What is the ratio of tomatoes to peaches that Janine picked?

SOLUTION

The ratio of tomatoes to peaches is $\frac{5}{3}$.

A **proportion** is used to show that two ratios are equivalent. Proportions always have an equal sign. If, in the example above, Janine had picked 10 tomatoes, and she wanted to pick enough peaches to ensure the same ratio, she could use a proportion. Since the number of peaches is unknown, it is represented by a variable: $\frac{5}{3} = \frac{10}{x}$. Solving the equation shows that Janine needs to pick 6 peaches to ensure a proportionate ratio.

Example 2 | Using Proportions

Dylan is using a recipe to make cookies. The recipe calls for 2 cups of flour, but Dylan mistakenly measures out 2.5 cups of flour. The recipe also calls for 1 cup of sugar. How many cups of sugar should Dylan use to make sure he has the same ratio of flour to sugar as in the recipe?

SOLUTION

Write two ratios. The first should show the ratio in the original recipe. The recipe calls for 2 cups of flour and 1 cup of sugar, so the ratio is $\frac{2}{1}$. The second ratio reflects Dylan's change to the recipe. His dough has 2.5 cups of flour and an unknown amount of sugar, so the ratio is $\frac{2.5}{x}$. To make a proportion, these two ratios must be equal, so $\frac{2}{1} = \frac{2.5}{x}$. Solving the proportion for x shows that Dylan should use 1.25, or $1\frac{1}{4}$ cups of sugar.

A **percentage** shows what part of the whole is a given quantity. In the example above with the coins, the dimes are two coins out of six total coins. This can be converted to a percentage: $\frac{2}{6} = 0.\overline{3} = 33\frac{1}{3}\%$. A percentage can be represented as a decimal. If it is known that there are 6 coins and $33\frac{1}{3}\%$ of them are dimes, then the actual number of dimes can be found by multiplying the total number of coins by the decimal representation of the percent that are dimes: $6 \times 0.\overline{3} = 2$ dimes.

Example 3   Using Percentages

If a shirt that normally costs $50 is being sold at 30% off, how much does it cost?

SOLUTION

To find the discount, multiply the price by the percentage: ($50)(0.30) = $15. Since the discount is $15, the shirt is being sold for $50 − $15 = $35.

Skills Bank Practice

Use a proportion to find the missing quantity.

a. If $4.50 will buy 3 bus tickets, how many bus tickets can be bought with $18?

b. It takes about 3 pounds of lye to make 100 bars of soap. How many pounds of lye would it take to make 70 bars of soap?

Use percentages to find the indicated quantity.

c. A manufacturer makes widgets. Each widget costs $12.50 to make. The wholesale price of each widget is 50% greater than the cost to make it. What is the wholesale price of each widget?

d. A pair of shoes costs $120. A store reduces the price by 40%. What is the new price of the shoes? The store then tells customers they can save another 10% off the reduced price if they trade in an old pair of shoes. What is the price of the shoes after the 40% reduction and the 10% discount?

Estimation

Skills Bank 5

Often the exact value of a number is unknown or difficult to work with. In these cases, the number can be estimated by rounding. To round a number to a given place value, refer to the digit on the right of that place value. If the digit is 4 or less, round down; if it is 5 or greater, round up.

Example 1   Rounding to a Place Value

Find the value of π to the nearest hundred-thousandth.

SOLUTION

The decimal representation of π begins: 3.14159265.... The digit in the hundred-thousandths place is 9. The digit to the right of the hundred-thousandths place is 2. Since 2 is less than 5, the number rounds down: 3.14159.

Skills Bank Practice

Round each number to the given place value.

a. 1.45974763 to the nearest tenth

b. 32,897 to the nearest hundred

c. 516,201.721 to the nearest whole

d. 0.0099909 to the nearest hundredth

Exponents and Roots

Skills Bank 6

There are two parts to any exponential expression. In the expression a^n, a is called the **base** and n is called the **exponent**. In the statement below, 3 is the base and 5 is the exponent.

$$3^5 = 3 \times 3 \times 3 \times 3 \times 3$$

When an expression contains a negative exponent n, it is equal to the reciprocal of the base with an exponent of $|n|$. For example:

$$3^{-5} = \left(\frac{1}{3}\right)^5 = \frac{1}{3^5}$$

A **root** is the inverse of an exponent. In mathematics, a root is denoted by a **radical symbol**. The radical symbol has a radicand and an index. If the index is 2, it is not necessary to show it. In the root below, x is the radicand and y is the index.

$$\sqrt[y]{x}$$

Taking the nth root of a number finds a factor that, when raised to the nth power, results in the original number. For example:

$$\sqrt[4]{16} = 2 \qquad 2^4 = 16$$

Any root can be rewritten as an exponential expression by using the radicand as the base and the reciprocal of the index as the exponent. For example:

$$\sqrt[4]{16} = 16^{\frac{1}{4}}$$

A square root can also be simplified by factoring the radicand into perfect squares, which can then be evaluated and pulled outside the root. This is called simplified radical form. For example:

$$\sqrt{80} = \sqrt{16 \times 5} = \sqrt{16} \times \sqrt{5} = 4\sqrt{5}$$

Example 1 | Expanding and Evaluating Exponents

Evaluate each expression.

a. 4^3

SOLUTION

$4^3 = 4 \times 4 \times 4 = 64$

b. $\sqrt[3]{64}$

SOLUTION

$\sqrt[3]{64} = 64^{\frac{1}{3}} = 4$

Skills Bank Practice

Evaluate each expression.

a. 2^5

b. 3^{-2}

c. $\sqrt{16}$

d. $\sqrt{27}$

e. 5^{-3}

f. $\sqrt[3]{8}$

Scientific Notation and Significant Figures

Skills Bank 7

When numbers are very large or very small, it can often be easier to express them in **scientific notation**. A number in scientific notation is written as a number whose absolute value is greater than 0 and less than 10, multiplied by a power of ten. To determine which power of ten applies to a given number, count the number of spaces that the decimal must be shifted for the given number to be between 0 and 10. When shifting the decimal left, the power is positive; when shifting right, the power is negative.

Example 1   Expressing in Scientific Notation

Express $-3,410,000$ in scientific notation.

SOLUTION

If the decimal point were shifted six spaces to the left, the absolute value of the number would be 3.41, between 0 and 10. Therefore $-3,410,000$ is equal to -3.41×10^6.

The digits in a number that are known to be exact are called significant digits, or **significant figures**. The number of significant digits in a number can be determined by applying the following rules:

- Any nonzero digit is a significant figure.
- Any zero or string of zeroes between two nonzero digits is a significant figure.
- To the right of the decimal point, any zeroes after the rightmost nonzero digit are significant figures.

Example 2   Counting Significant Digits

Find the number of significant digits in 0.00300, 0.003, and 3200.

SOLUTION

By applying the rules above, 0.00300 has three significant digits: the 3 and the two zeroes after it. In contrast, 0.003 has only one significant digit: 3. The number 3200 has two significant digits: the two nonzero digits, 3 and 2.

When expressing a number in scientific notation, use all significant digits. For example, the scientific notation of 0.00300 is 3.00×10^{-3}, whereas the scientific notation of 0.003 is 3×10^{-3}.

Skills Bank Practice

Use significant digits to express each number in scientific notation.

  a. 78,000   **b.** 78,000.1   **c.** 78,000.10   **d.** 0.078001

Units of Measure

Skills Bank 8

The table below gives conversion factors for some commonly used measures of distance.

Metric System	U.S. Customary System
100 centimeters = 1 meter	12 inches = 1 foot
1000 meters = 1 kilometer	3 feet = 1 yard
	1760 yards = 1 mile

To convert to a larger unit, divide by the conversion factor. To convert to a smaller unit, multiply by the conversion factor. For example:

How many inches is 320 yards?

First, convert to feet, then to inches. Since the problem asks us to start with a larger unit and convert to a smaller one, multiplication will be used.

$$(320 \text{ yd})\left(\frac{3 \text{ ft}}{1 \text{ yd}}\right) = 960 \text{ ft}$$

$$(960 \text{ ft})\left(\frac{12 \text{ in.}}{1 \text{ ft}}\right) = 11{,}520 \text{ in.}$$

Example 1 Converting Units

a. Convert 200 centimeters into kilometers.

SOLUTION

$$(200 \text{ cm})\left(\frac{1 \text{ m}}{100 \text{ cm}}\right) = 2 \text{ m}$$

$$(2 \text{ m})\left(\frac{1 \text{ km}}{1000 \text{ m}}\right) = 0.002 \text{ km}$$

b. Convert 1 mile into feet.

SOLUTION

$$(1 \text{ mi})\left(\frac{1760 \text{ yd}}{1 \text{ mi}}\right) = 1760 \text{ yd}$$

$$(1760 \text{ yd})\left(\frac{3 \text{ ft}}{1 \text{ yd}}\right) = 5280 \text{ ft}$$

Skills Bank Practice

Perform each conversion. When necessary, round answers to the nearest hundredth.

a. Convert 10 yards into inches.

b. Convert 24 centimeters into meters.

c. Convert 1240 inches into miles.

d. Convert 120 miles into yards.

e. Convert 400 meters into kilometers.

f. Convert 2 kilometers into centimeters.

Converting Units and Rates

Skills Bank 9

To convert a measurement in one system to another system, use a conversion factor. A conversion factor is a ratio equal to 1, which can be multiplied by a unit of measurement to give the same measurement expressed in another unit.

Common Conversion Factors

	Length	Capacity	Mass/Weight
Metric to Customary	1 cm ≈ 0.394 in. 1 m ≈ 3.281 ft 1 km ≈ 0.621 mi	1 L ≈ 1.057 qt 1 mL ≈ 0.034 oz	1 g ≈ 0.0353 oz 1 kg ≈ 2.205 lb
Customary to Metric	1 in. ≈ 2.54 cm 1 ft ≈ 0.305 m 1 mi ≈ 1.609 km	1 qt ≈ 0.946 L 1 oz ≈ 29.57 mL	1 oz ≈ 28.350 g 1 lb ≈ 0.454 kg

Example 1 Converting Units

Convert 3.5 kilograms to pounds.

SOLUTION

$$3.5 \text{ kg} \times \frac{2.205 \text{ lb}}{1 \text{ kg}} = 7.7175 \text{ lb}$$

To convert a given rate to a rate with different units, use unit multipliers which are equal to 1 and allow for converting units to divide out, leaving the desired units in the converted rate. For conversions within the same system of measure, refer to the Properties and Formulas chart at the end of this book.

Example 2

Convert 60 miles per hour to feet per second.

SOLUTION

$$\frac{60 \text{ mi}}{1 \text{ hr}} \times \frac{1 \text{ hr}}{60 \text{ min}} \times \frac{1 \text{ min}}{60 \text{ sec}} \times \frac{5280 \text{ ft}}{1 \text{ mi}} = \frac{88 \text{ ft}}{\text{sec}}$$

Skills Bank Practice

Convert each measurement into the given units. Round to the nearest tenth.

   a. 15 in. to cm   **b.** 1.89 qt to mL   **c.** 2.02 mi to km

   d. 2.714 g to oz   **e.** 3.14159 kg to oz   **f.** 508 mL to oz

Convert the rate as indicated. Round to the nearest tenth.

   g. 9.8 m/sec to ft/min   **h.** 42 L/min to mL/sec

   i. 50 km/hr to mi/hr   **j.** 8.3 lb/gal to g/L

Calculating Percent Error

Skills Bank 10

Measuring an object imprecisely can lead to an incorrect result. This is commonly expressed as a percentage of the correct measurement and is called **percent error**.

> ### Example 1 Calculating Percent Error
>
> Fernando measures the volume of a pitcher and decides that it is 6.47 liters. The actual volume of the pitcher is 6.5 liters. Find the percent error in Fernando's measurement.
>
> **SOLUTION**
>
> Fernando's measurement is 0.03 liters off. As a percentage: $\dfrac{0.03}{6.5} = 0.0046$
> $= 0.46\%$

Skills Bank Practice

a. Jarrell measures the height of his classroom to be 7.9 feet. The actual height of the classroom is 8 feet. Find the percent error in Jarrell's measurement.

b. Susanne estimates the weight of a brick at 1.4 pounds using a balancing scale. The actual weight is shown to be 1.3 pounds on a digital scale. Find the percent error in Susanne's measurement to the nearest hundredth.

c. Mae uses a measuring cup to scoop 2 cups of flour. She measures the flour again in a larger measuring cup and finds that the original amount she had was only 1.9 cups of flour. What is Mae's percent error?

Measures of Central Tendency

Skills Bank 11

Measures of central tendency refer to different "centers" of a set of data. The three most common such measures are **mean, median,** and **mode**. The mean of a data set is calculated by adding the values of all the data points and then dividing by the total number of data points in the set. The median is calculated by listing all the data points sequentially and then finding the data point in the middle. If there are an even number of data points, there will not be a middle point. In these cases, the median is the mean of the two middle terms. The mode is simply the most common value of a data set. There can be one, several, or no modes of a set of data.

> ### Example 1 Finding Measures of Central Tendency
>
> Find the mean, median, and mode of the following set: {1, 1, 3, 5, 6, 2, 2, 4, 2, 14, 4}.
>
> **SOLUTION**
>
> The mean is $\dfrac{(1+1+3+5+6+2+2+4+2+14+4)}{11}$, which is 4.
>
> To find the median, list the numbers in order: {1, 1, 2, 2, 2, 3, 4, 4, 5, 6, 14}. The median is the value in the middle of the list, which is 3.
>
> The mode is the most common value, which is 2.

Skills Bank Practice

a. Find the median of the following set: $\left\{8, \frac{1}{2}, 4, 0.4, 200, \pi, 0, 2, 3\right\}$.

b. Find the mean of the following set: $\{12, 2, 7, 15, 9, 1, 8, 2\}$.

c. Find the mode of the following set: {square, circle, cube, square, cylinder, cone, square}.

d. Find the mean, median, and mode of the following set: $\{7, 3, 6, 1, 2, 3, 7, 9\}$.

Probability

Skills Bank 12

Probability is the chance that something will happen. For a given event, the set of all the possible outcomes is called the sample space. Probability can be expressed as a fraction, by taking the total number of ways a desired outcome could occur and dividing by the number of possible outcomes in the sample space. When probability is determined in this way, it is called theoretical probability.

If two events happen independently, the probability of each event can be multiplied to find the probability of both events happening together.

Example 1 **Determining Simple Probability**

What is the probability of getting heads twice when flipping two coins?

SOLUTION

There are 4 possibilities: Heads-Heads, Heads-Tails, Tails-Heads, and Tails-Tails. There is only one way to get two heads. So the probability of getting heads twice is $\frac{1}{4}$.

When events interact with each other, they are said to be dependent. In these cases, the second event's probability must be calculated with the knowledge of the first event in mind.

Example 2 **Determining Complex Probability**

A drawer has 5 black socks and 3 red socks in it. What is the probability of randomly drawing a black sock from the drawer followed by a red sock?

SOLUTION

There are 8 possible socks for the first draw, of which 5 are black socks. The probability that the first sock to be drawn would be black is $\frac{5}{8}$. After the first draw, there are only 7 socks left in the drawer, and 3 of them are red, so the probability of drawing a red sock *after drawing a black sock* is $\frac{3}{7}$. To find the probability of both of these events happening together, multiply these two probabilities: $\left(\frac{5}{8}\right)\left(\frac{3}{7}\right) = \frac{15}{56}$.

Another way to determine probability is to set up an experiment. In the example above, an experiment would entail drawing from the sock drawer many times and recording the results of each trial. If probability is determined by experimentation, it is called experimental probability.

Example 3   Experimental Probability

Vicente and Nick flip a coin. If the coin comes up heads, Vicente wins. If the coin comes up tails, Nick wins. After flipping the coin 20 times, Vicente has won 11 times and Nick has won 9 times. What is the experimental probability of getting heads? What is the theoretical probability of getting tails?

SOLUTION

Heads has come up 11 out of the 20 times the coin has been tossed, so the experimental probability of getting heads is $\frac{11}{20}$. Since there are two possible outcomes from a coin flip, and one is tails, the theoretical probability of getting tails on a coin flip is simply $\frac{1}{2}$.

Sometimes it is useful to determine the probability that an event will not happen. This is called the complement of an event. The probability of an event's complement can be found by first determining the probability that the event will happen, and then subtracting that probability from 1.

Example 4   Probability of the Complement of an Event

Joseph rolls two number cubes. What is the probability that he will roll a total less than 12?

SOLUTION

First, find the probability that Joseph will roll a 12. Both cubes would have to land on six in order for Joseph to roll a 12, so there is only one way to roll a 12. Since there are 36 possible rolls in all, the probability is $\frac{1}{36}$. To determine the probability that he will not roll a 12, find the complement of this event. It is $1 - \frac{1}{36} = \frac{35}{36}$.

Skills Bank Practice

a. A bag contains 12 marbles. Half of the marbles are black and half are yellow. If two marbles are taken out of the bag, what is the probability that neither marble is black?

b. A number cube is rolled twice. What is the probability that 3 is rolled both times?

c. A spinner is divided into eight equal regions, numbered 1 through 8. Shelby spins twice. What is the probability that she does not get a 4 on either spin?

The Coordinate Plane

Skills Bank 13

The **coordinate plane** is formed by two perpendicular number lines, the *x*-axis and the *y*-axis, which meet at a point called the origin. These axes divide the plane into four quadrants, numbered as shown.

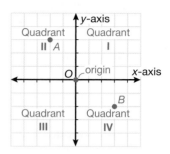

Points on the coordinate plane are identified relative to their position on the *x*-axis and the *y*-axis, giving them a unique pair of numbers called **coordinates**. The pair of numbers in a point's coordinates is called an **ordered pair**. An ordered pair gives the *x*-coordinate first, then the *y*-coordinate. For example, in the diagram, point *A* has the coordinates $(-2, 3)$, while point *B* has the coordinates $(3, -2)$.

Example 1   **Graphing on the Coordinate Plane**

Graph the point $(4, -4)$. In which quadrant is the point located?

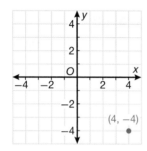

SOLUTION

Find 4 on the *x*-axis. Find -4 on the *y*-axis. Plot the point at the intersection of these coordinates. The point is located in Quadrant IV.

Quadrant I: both *x* and *y* are positive
Quadrant II: *x* is negative and *y* is positive
Quadrant III: both *x* and *y* are negative
Quadrant IV: *x* is positive and *y* is negative

Points that lie on either the *x*- or *y*-axis are not in any quadrant.

Skills Bank Practice

Graph each point and state which quadrant the point is in.

a. $(3, 6)$

b. $(2, -4)$

c. $(-3, -1)$

d. $(-4, 7)$

e. What are the coordinates of the origin?

Evaluating Expressions

Skills Bank 14

An expression that has a variable is called an algebraic expression. To evaluate an algebraic expression, substitute the given value into the expression, replacing the variable.

Example 1 | **Evaluating an Algebraic Expression**

a. Evaluate the expression $5x + 3$ for $x = 4$.

SOLUTION

$5x + 3$	
$5(4) + 3$	Substitute.
$20 + 3$	Simplify.
23	Simplify.

b. Evaluate the expression $\dfrac{3x - 6}{4x + 2}$ for $x = 3$.

SOLUTION

$\dfrac{3x - 6}{4x + 2}$	
$\dfrac{3(3) - 6}{4(3) + 2}$	Substitute.
$\dfrac{9 - 6}{12 + 2}$	Simplify.
$\dfrac{3}{14}$	Simplify.

Skills Bank Practice

Evaluate each algebraic expression for the given value of the variables.

a. $4x^2 - 3$, for $x = 2$

b. $\dfrac{9}{2y + 3}$, for $y = -2$

c. $3(a + 2b)$, for $a = -2$ and $b = 5$

d. $4r - 3$, for $r = 1.5$

e. $\dfrac{3p + p^2}{p}$, for $p = -3$

f. $mn + 3m$, for $m = 2$ and $n = 6$

Solving Linear Equations and Inequalities

Skills Bank 15

To solve linear equations in one variable, the following properties may be used:

Property	Algebra
Addition Property of Equality	$a = b$ $a + c = b + c$
Subtraction Property of Equality	$a = b$ $a - c = b - c$
Multiplication Property of Equality	$a = b$ $ac = bc$
Division Property of Equality	$a = b, c \neq 0$ $\dfrac{a}{c} = \dfrac{b}{c}$

Example 1 Solving Linear Equations

Solve $4x + 12 = 56$ for x.

SOLUTION

$$4x + 12 = 56$$
$$4x + 12 - 12 = 56 - 12 \qquad \text{Substraction Property of Equality}$$
$$4x = 44 \qquad \text{Simplify.}$$
$$\frac{4x}{4} = \frac{44}{4} \qquad \text{Division Property of Equality}$$
$$x = 11 \qquad \text{Simplify.}$$

When solving a linear **inequality**, the addition and subtraction properties of equality can still be used. However, linear inequalities reverse direction when multiplying or dividing by a negative number. The following example illustrates this property.

Value for c:	Multiplication Property of Inequality:	Division Property of Inequality:
positive	If $a < b$ and $c > 0$, then $ac < bc$	If $a < b$ and $c > 0$, then $\dfrac{a}{c} < \dfrac{b}{c}$
negative	If $a < b$ and $c < 0$, then $ac > bc$	If $a < b$ and $c < 0$, then $\dfrac{a}{c} > \dfrac{b}{c}$

Example 2 Solving Linear Inequalities

Solve $-3x - 2 < 7$ for x.

SOLUTION

$$-3x - 2 < 7$$
$$-3x - 2 + 2 < 7 + 2 \qquad \text{Addition Property of Inequality}$$
$$-3x < 9 \qquad \text{Simplify.}$$
$$\frac{-3x}{-3} < \frac{9}{-3} \qquad \text{Division Property of Inequality}$$
$$x > -3 \qquad \text{Simplify.}$$

One other property of inequality is the Comparison Property of Inequality, which states that if $a + b = c$, and $b > 0$, then $a < c$.

Skills Bank Practice

a. Solve for x.

$\frac{2-x}{2} = x$

b. Solve for h.

$46 - 3h = 7$

c. Solve for y.

$\frac{-y}{2} < 14$

d. Solve for p.

$2p - 5 > 3p$

Transforming Formulas

Skills Bank 16

Sometimes, it is necessary to rearrange formulas to make them easier to use. For example, the formula for converting from degrees Fahrenheit to degrees Celsius is $C = \frac{5}{9}(F - 32)$. This formula is straightforward if the temperature in Fahrenheit is known. It is less straightforward, however, when only the temperature in Celsius is known. In this case, it would be useful to rewrite the formula by solving for F.

Example 1 **Transforming Formulas**

Transform the formula $C = \frac{5}{9}(F - 32)$ by solving for F.

SOLUTION

The goal is to have F alone on one side of the equation.

$$C = \frac{5}{9}(F - 32)$$

$$9C = (9)\left(\frac{5}{9}\right)(F - 32) \qquad \text{Multiply both sides by 9.}$$

$$9C = 5(F - 32) \qquad \text{Simplify.}$$

$$\frac{9C}{5} = \frac{5(F - 32)}{5} \qquad \text{Divide both sides by 5.}$$

$$\frac{9}{5}C = F - 32 \qquad \text{Simplify.}$$

$$\frac{9}{5}C + 32 = F - 32 + 32 \qquad \text{Add 32 to both sides.}$$

$$F = \frac{9}{5}C + 32 \qquad \text{Simplify.}$$

Now the rewritten formula makes it easier to determine the temperature in Fahrenheit if the temperature in Celsius is known.

Skills Bank Practice

a. Solve $A = bh$ for h.

b. Solve $\frac{20}{3x} = y$ for x.

c. Solve $A = \frac{1}{2}bh$ for b.

d. Solve $\frac{2}{3}(3p + 8) = q$ for p.

Functions

Skills Bank 17

A relation is a rule that relates two sets of numbers or values to each other. When a relation relates exactly one value in the second set (called the **range**) to each value in the first set (called the **domain**), the relation is called a **function**.

By convention, x is used to represent the domain and y represents the range of a relation.

> ### Example 1 Identifying Functions
>
> Is $y = x^2$ a function? Is $y = |x|$ a function? Is $|y| = x$ a function?
>
>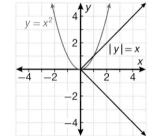
>
> **SOLUTION**
>
> The first two relations are functions because in each, for every x-value, there is only one possible y-value that fits the relation. The third relation is not a function, because for any nonzero x-value, there are two possible values of y that fit the relation. For example, when $x = 2$, y could be either 2 or -2.

Notice that a function may relate more than one number from the domain to the same number in the range. The second relation in the example above is a function even though multiple x-values correspond to the same y-value, because each x-value corresponds to only one possible y-value.

A visual way to test whether a relation is a function, called the **vertical line test**, is to examine the graph of the relation on the coordinate plane. If every vertical line that intersects the graph does so at only one point, then the graphed relation is a function.

A given relation also has an inverse relation, which simply relates each point in the range of the original function to points in the domain of the original relation. If the inverse relation is a function, it is called the **inverse function**.

Many functions can be grouped into families of closely related functions. The simplest function with the defining characteristics of the family is called the parent function. For example, $y = x^2$ is the parent function of $y = (x + 3)^2$, $y = x^2 - 15$, and $y = 5(x - 1)^2 + 1$, since each function in this family shares the value x^2.

Skills Bank Practice

a. Show that $y^2 = x$ is not a function.

b. Give an example of a linear relation that is not a function.

c. Graph $y = x^3$. Is this a function?

d. What is the parent function of $y = 4|x - 3|$?

Binomial Products and Factoring

Skills Bank 18

A binomial is an expression with two terms consisting of variables and constants. To multiply two binomials, use the FOIL method. FOIL stands for "First terms, Outside terms, Inside terms, and Last terms." FOIL refers to the order in which you should multiply the parts of the two binomials. For example:

Multiply the binomials: $(x + 3)(x - 2)$.

The first term of each of these binomials is x. The outside term of the first binomial is also x, and the outside term of the second binomial is -2. The inside term of the first binomial is 3, and the inside term of the second binomial is x. Finally, the last term of the first binomial is 3, and the last term of the second binomial is -2. Multiply each of these pairs together in the FOIL order, and then add them.

$(x)(x) + (x)(-2) + (3)(x) + (3)(-2)$
$x^2 - 2x + 3x - 6$
$x^2 + x - 6$

Example 1 | Multiplying Binomials

Simplify the expression $(2x + 2)(4 - x)$.

SOLUTION

Identify the first terms: $2x$ and 4. The outside terms are $2x$ and $-x$. The inside terms are 2 and 4. Finally, the last terms are 2 and $-x$.

$(2x)(4) + (2x)(-x) + (2)(4) + (2)(-x)$
$8x - 2x^2 + 8 - 2x$
$-2x^2 + 6x + 8$

The result of multiplying the two binomials above is a trinomial. Factoring a trinomial can be a useful way to solve for the value of a variable. To factor a trinomial, reverse the process of FOIL. You will need to work backwards to factor the trinomial down to the two binomials that were combined to make the trinomial.

Example 2 | Factoring Trinomials

Factor the expression $x^2 + 7x + 10$.

SOLUTION

The answer will be in the form of two binomials. Begin with parentheses representing the solution:

$(\quad)(\quad)$

We also know that when the first term of these two binomials are multiplied together, the result is x^2, so the first term in each binomial must be x.

$(x \quad)(x \quad)$

Finally, we know that when the outside terms of these binomials are multiplied together, the result is 10. The last terms must either be 5 and 2, or 10 and 1. Since these terms must also add to get the middle term, $7x$, in the trinomial, the only possibility is 5 and 2.

$(x \ 5)(x \ 2)$

The last step of the problem is to determine the sign in the middle of each binomial. You may have to guess and check your answer by multiplying the binomials. In this case, we know we want the last terms of each binomial to add to 7. This will require that both terms be positive.

$(x + 5)(x + 2)$

Verify this answer by multiplying the binomials.

Skills Bank Practice

Multiply the binomials.

a. $(x - 5)(x + 3)$

b. $(2x - 5)(1 - x)$

Factor the trinomials.

c. $x^2 + 4x + 3$

d. $x^2 - 3x - 18$

Lines in Point-Slope and Standard Forms

Skills Bank 19

There are many ways to write a linear equation. Two of the most common are **point-slope form** and **standard form**.

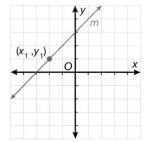

Point-slope form is used to write the equation of a line when a point and the slope are known. A line in point-slope form is given by $y - y_1 = m(x - x_1)$ where (x_1, y_1) is the known point and m is the slope. To graph a line in point-slope form, plot the known point and then draw a line through the point with slope m.

A linear equation in standard form is given by $Ax + By = C$, where A, B, and C are (if possible) integers. Traditionally, the A term is positive, and A, B, and C have no common factors other than 1. Standard form is useful because it can express vertical lines, which cannot be expressed in point-slope form. Standard form is also convenient for solving systems of linear equations.

Example 1 **Transforming Equations**

Express the linear equation $\frac{3}{2}y = \frac{2}{3}x + 1$ in point-slope form and in standard form.

SOLUTION

Point-slope form:		Standard form:	
$\frac{3}{2}y = \frac{2}{3}x + 1$		$\frac{3}{2}y = \frac{2}{3}x + 1$	
$\frac{2}{3}\left(\frac{3}{2}y\right) = \frac{2}{3}\left(\frac{2}{3}x + 1\right)$	Multiply to solve for y.	$6\left(\frac{3}{2}y\right) = 6\left(\frac{2}{3}x + 1\right)$	Multiply both sides by 6.
$y = \frac{4}{9}x + \frac{2}{3}$	Simplify.	$9y = 4x + 6$	Simplify.
		$-4x + 9y = 4x + 6 - 4x$	Subtraction Property of Equality
$y = \frac{4}{9}\left(x + \frac{3}{2}\right)$	Factor out $\frac{4}{9}$.	$-4x + 9y = 6$	Simplify.
		$-1(-4x + 9y) = -1(6)$	Multiply both sides by -1.
		$4x - 9y = -6$	Simplify.

Skills Bank Practice

a. Express $4x + 5y = 13$ in point-slope form.

b. Express $y = \frac{3}{4}x + 6$ in standard form.

c. Express $2x + 2y = 6$ in point-slope form and graph it.

Quadratic Equations and Functions

Skills Bank 20

Quadratic equations are equations in the form $ax^2 + bx + c = 0$, with $a \neq 0$. Two methods that can be used to solve for x in a quadratic equation are factoring and using the quadratic formula.

Example 1   Factoring a Quadratic Equation

Factor $x^2 + 5x - 24 = 0$ and solve for x.

SOLUTION

First, factor the equation.
$$x^2 + 5x - 24 = 0$$
$$(x + 8)(x - 3) = 0$$
To make this expression equal to 0, either $(x + 8)$ or $(x - 3)$ must be equal to 0.
$$x + 8 = 0 \qquad x - 3 = 0$$
$$x = -8 \qquad x = 3$$
In this case, there are two solutions. Every quadratic equation has one, two, or no solutions.

Another way to solve a quadratic equation is to use the quadratic formula. The quadratic formula can be used on any quadratic equation. The solutions for x are:

$$x_1 = \frac{-b + \sqrt{b^2 - 4ac}}{2a} \quad \text{and} \quad x_2 = \frac{-b - \sqrt{b^2 - 4ac}}{2a}$$

Example 2   Using the Quadratic Formula

Use the quadratic formula to solve for x: $2x^2 - 5x - 12 = 0$.

SOLUTION

From the equation, $a = 2$, $b = -5$, and $c = -12$.

$$x = \frac{-b \pm \sqrt{b^2 - 4ac}}{2a}$$

$$x = \frac{5 \pm \sqrt{(-5)^2 - 4(2)(-12)}}{2(2)}$$

$$x = \frac{5 \pm \sqrt{121}}{4}$$

$$x = \frac{5 \pm 11}{4}$$

$$x = \frac{5 + 11}{4} = 4 \qquad \frac{5 - 11}{4} = -1.5$$

The graph of a quadratic function forms a 'U' shape called a parabola. Quadratic functions can be graphed by finding the axis of symmetry and the vertex of the parabola.

For any quadratic function ($a \neq 0$) $y = ax^2 + bx + c$:

The axis of symmetry is the vertical line, $x = \dfrac{-b}{2a}$.

The vertex of the parabola is the point $\left(\dfrac{-b}{2a}, y\right)$.

If $a > 0$, the parabola opens upward.
If $a < 0$, the parabola opens downward.

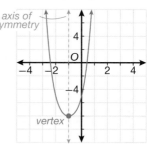

Example 3 | **Graphing a Quadratic Function**

Sketch the graph of $y = x^2 - 5x - 14$. Indicate the axis of symmetry and the vertex.

SOLUTION

The axis of symmetry is the line $x = -\dfrac{(-5)}{2(1)} = 2.5$.

The vertex is the point $\left(-\dfrac{(-5)}{2(1)}, y\right) = (2.5, y)$. To find the value of y, use the original equation and substitute in 2.5 for x: $y = (2.5)^2 - 5(2.5) - 14 = -20.25$. So, the vertex is $(2.5, -20.25)$.

The parabola opens upwards, since $a > 0$.

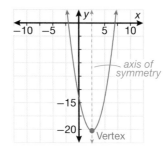

Skills Bank Practice

Solve each quadratic equation using any method.

   a. $x^2 - 2x - 15 = 0$ **b.** $3x^2 + 3x - 6 = 0$

   c. Sketch the graph of the quadratic function $y = x^2 + 2x - 8$. Indicate the graph's axis of symmetry and vertex.

Displaying Data

Skills Bank 21

Different kinds of data are best displayed in different ways. Data that are understood to be parts of a whole are usually best represented in a circle graph (also called a pie chart).

Example 1 | **Making a Circle Graph**

Favorite Flower	Students
Rose	23
Tulip	15
Dandelion	33
Lily	19

Construct a circle graph to display the information about students' favorite flowers from the chart.

SOLUTION

To determine how much area of the circle to allocate to each flower, first find the percentage of the total respondents associated with each. There were 90 total respondents, of which 23 chose 'rose', so $\frac{23}{90} = 25.\overline{5}\%$ of respondents chose 'rose' as their favorite flower. Therefore, $25.\overline{5}\%$ of the circle, or about 92 degrees, should be given to 'rose.' Similarly, $\frac{15}{90} = 16.\overline{6}\%$ chose 'tulip,' so about 60 degrees should be given to 'tulip;' $\frac{33}{90} = 36.\overline{6}\%$ chose 'daffodil,' so about 132 degrees should be given to 'daffodil;' and $\frac{19}{90} = 21.\overline{1}\%$ chose 'lily,' so about 76 degrees should be given to 'lily.'

Data that are disjointed, or not in categories that are directly related to one another, are often best displayed in a bar graph.

Example 2 | Making a Bar Graph

Student	Number of States Traveled
Josie	5
Sebastian	18
Dominique	7
Rob	13

Construct a graph to display the information about students' travels from the chart.

SOLUTION

The data do not represent parts of a whole or meaningful total, so a circle graph would be a poor choice; instead, use a bar graph. Place a number line on the y-axis of the graph to represent the number of states to which each student has traveled, and list the students' names on the x-axis. Draw a bar to the appropriate height for each student.

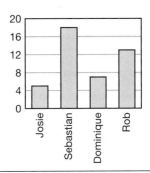

A histogram is a type of bar graph in which the data are organized into intervals. In a histogram, the width of the bars is determined by the category of intervals into which the data fall, so because of this, the bars are shown touching side-by-side. Notice here, the x-axis displays intervals of weight, and the y-axis shows the data being organized into intervals of quantity.	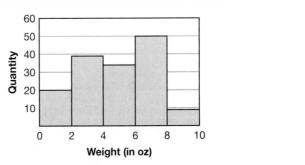
When data can be grouped into sets that have defined relationships, a Venn diagram can be used to graphically show these relationships.	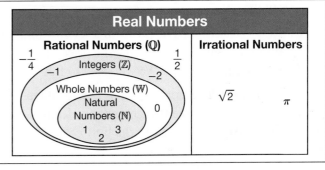

Skills Bank Practice

a. What type of graph would be best to display the data about what students in the class had for breakfast? Why?

b. Display the breakfast data from the table in a graph.

Breakfast	Number of Students
Cereal (hot)	7
Cereal (cold)	9
Pancakes or Waffles	5

c. Construct a histogram from the data on collecting food for a canned food drive.

Cans of Food	Number of People
0–5	4
6–10	16
11–15	7
16–20	3
21–25	1

Problem-Solving Process and Strategies

Skills Bank 22

This four-step process for solving problems is helpful when dealing with word problems.

Step 1: Understand the problem. Read the problem carefully and make sure you know exactly what the problem is asking. List some things that you know and some things that you do not know but need to find out in order to solve the problem.

Step 2: Plan a method. Decide how you will solve the problem. You might be able to solve the problem by drawing a picture, using a coordinate plane, looking for a pattern, writing an equation, or making a chart.

Step 3: Solve the problem. Execute the plan you made above.

Step 4: Check your answer. First, does your answer seem right? Is it reasonable? If possible, plug your answer into the original problem and make sure that it works. You should also make sure you have labeled your answer with the correct units in this step, if there are units being used in the problem.

Example 1   Applying the Four-Step Process

Manuela has $19.50 that she plans to use to rent movies from the video store. Each video costs $3.20 to rent. There is also a 5% tax on each rental. How many movies could Manuela rent?

SOLUTION

Step 1: Understand The problem asks how many rentals can be made with a set amount of money. We know the price of each rental before tax, but not the price of each rental after tax. We will need to find this out.

Step 2: Plan The tax is given as a percentage. To find out what 5% of $3.20 is, we have to convert the percentage to a decimal and multiply by the base cost of the rental. Adding the tax to the base cost of $3.20 will give us the total cost of a rental.

Once we know the total price of a rental, we can simply divide the amount of money Manuela has by the cost per rental.

Step 3: Solve The tax is given by: $0.05 \times \$3.20 = \$0.16$.

Next, we add the tax to the price of the rental: $\$3.20 + \$0.16 = \$3.36$.

Finally, we divide the $19.50 that Manuela has by this new cost: $\frac{\$19.50}{\$3.36} \approx 5.80$.

Manuela can only buy whole rentals, not parts of a rental, so the maximum number of videos she can rent is 5.

Step 4: Check Does the answer make sense? One good way to verify this is to see how much it would cost to rent 6 videos. $(6)(\$3.20) + (6)(0.05)(\$3.20) = \$20.16$, so 6 rentals would have been too many. The answer, 5 *rentals*, seems right.

Skills Bank Practice

Use the four-step process to solve each problem.

a. Alicia is 5 feet tall. In a photograph of Alicia and her younger brother, she is pictured 2 inches tall, and her brother is 1.8 inches tall. Using this ratio, how tall is Alicia's brother?

b. Hoon paid $16 for a meal at a restaurant, and tipped his server exactly 15%. How much change did Hoon receive if he paid with a $20 bill?

c. If you spin a spinner twice, and it has 9 equal sections on it, each numbered 1 through 9, what is the probability that you will get an odd number on both spins?

Properties and Formulas

Properties

Addition Property of Equality
(SB)

For every real number a, b, and c, if $a = b$ then $a + c = b + c$.

Addition Property of Inequality
(SB)

For every real number a, b, and c, if $a < b$ then $a + c < b + c$.

Also holds true for $>$, $\leq$, $\geq$, and $\neq$.

Associative Property of Addition
(SB)

For every real number a, b, and c,
$(a + b) + c = a + (b + c)$.

Associative Property of Multiplication
(SB)

For every real number a, b, and c,
$(ab)c = a(bc)$.

Closure Property of Addition
(SB)

For every real number a and b, $a + b$ is a real number.

Closure Property of Multiplication
(SB)

For every real number a and b, ab is a real number.

Commutative Property of Addition
(SB)

For every real number a and b, $a + b = b + a$.

Commutative Property of Multiplication
(SB)

For every real number a and b, $ab = ba$.

Comparison Property of Inequality
(SB)

For every real number a, b, and c, if $a + b = c$, and $b > 0$, then $a < c$.

Cross Products Property
(41)

For every real number a, b, c, and d, where $b \neq 0$ and $d \neq 0$, if $\dfrac{a}{b} = \dfrac{c}{d}$ then $ad = bc$.

Distributive Property
(SB)

For all real numbers a, b, and c,
$a(b + c) = ab + ac$ and $(b + c)a = ba + ca$.
$a(b - c) = ab - ac$ and $(b - c)a = ba - ca$.

Division Property of Equality
(SB)

For every real number a, b, and c, where $c \neq 0$, if $a = b$, then $\dfrac{a}{c} = \dfrac{b}{c}$.

Division Property of Inequality
(SB)

For every real number a, b, and c, where $c > 0$, if $a < b$, then $\dfrac{a}{c} < \dfrac{b}{c}$.

For every real number a, b, and c, where $c < 0$, if $a < b$, then $\dfrac{a}{c} > \dfrac{b}{c}$.

Also holds true for $>$, $\leq$, $\geq$, and $\neq$.

Identity Property of Addition
(SB)

For every real number a, $a + 0 = a$.

Identity Property of Multiplication
(SB)

For every real number a, $(a)(1) = a$.

Law of Cosines
(98)

In any triangle ABC:

$a^2 = b^2 + c^2 - 2bc \cos A$

$b^2 = a^2 + c^2 - 2ac \cos B$

$c^2 = a^2 + b^2 - 2ab \cos C$.

Law of Detachment
(21)

For two statements p and q, when "If p, then q" is a true statement and p is true, then q is true.

Law of Sines
(94)

In any triangle ABC:
$$\frac{a}{\sin A} = \frac{b}{\sin B} = \frac{c}{\sin C}.$$

Law of Syllogism
(21)

When "If p, then q" and "If q, then r" are true statements, then "If p, then r" is a true statement.

Multiplication Property of Equality
(SB)

For every real number a, b, and c, and $a = b$, then $ac = bc$.

Multiplication Property of Inequality
(SB)

For every real number a, b, and c, where $c > 0$, if $a < b$, then $ac < bc$.

For every real number a, b, and c, where $c < 0$, if $a < b$, then $ac > bc$.

Also holds true for $>$, $\leq$, $\geq$, and $\neq$.

Multiplication Property of Radicals
(SB)

If m and n are real numbers, and $m \neq 0, n \neq 0$ then, $\sqrt{m}\,\sqrt{n} = \sqrt{mn}$ and $\sqrt{mn} = \sqrt{m}\,\sqrt{n}$.

Multiplication Property of Zero
(SB)

For every real number a, $(a)(0) = 0$.

Negative Exponent
(SB)

For any nonzero real number x and integer n,
$$x^{-n} = \frac{1}{x^n} \text{ and } \frac{1}{x^{-n}} = x^n.$$

Order of Operations
(SB)

To simplify expressions:
1. Evaluate expressions in parentheses or other grouping symbols first.
2. Evaluate exponents and roots.
3. Multiply and divide from left to right.
4. Add and subtract from left to right.

Properties of Isometries and Transformations
(67, 71, 74, 78, 84, 90)

- An isometry maps a figure to a congruent figure.
- A translation is an isometry.
- A reflection is an isometry.
- A rotation is an isometry.
- A dilation maps a figure to a similar figure.
- The composition of two isometries is an isometry.
- The composition of two reflections across two parallel lines is a translation.
- The composition of two reflections across two intersecting lines is a rotation about the point of intersection. The angle of rotation is twice the angle formed by the intersecting lines.

Properties of Parallelograms
(34, 52)

- Opposite sides of a parallelogram are congruent.
- Opposite angles of a parallelogram are congruent.
- Consecutive angles of a parallelogram are supplementary.
- Diagonals of a parallelogram bisect each other.
- The diagonals of a rectangle are congruent.
- The diagonals of a rhombus are perpendicular.
- Each diagonal of a rhombus bisects opposite angles.

Properties of Special Right Triangles
(53, 56)

- In a 45°-45°-90° right triangle, both legs are congruent and the length of the hypotenuse is the length of a leg multiplied by $\sqrt{2}$.
- In a 30°-60°-90° triangle, the length of the hypotenuse is twice the length of the short leg, and the length of the longer leg is the length of the shorter leg times $\sqrt{3}$.

Properties of Trapezoids and Kites
(69)

- The midsegment of a trapezoid is parallel to both bases and has a length that is equal to half the sum of the bases.
- Base angles of an isosceles trapezoid are congruent.
- The diagonals of an isosceles trapezoid are congruent.
- The diagonals of a kite are perpendicular to each other.

Reflexive Property of Similarity
(41)

$\triangle ABC \sim \triangle ABC$

Reflexive Property of Congruence

$\triangle ABC \cong \triangle ABC$

Reflexive Property of Equality

$a = a$

Scientific Notation
(SB)

A number written as the product of two factors in the form $a \times 10^n$, where $1 \leq |a| < 10$ and n is an integer.

Subtraction Property of Equality
(SB)

For every real number a, b, and c and $a = b$, then $a - c = b - c$.

Subtraction Property of Inequality
(SB)

For every real number a, b, and c, if $a < b$, then $a - c < b - c$.

Also holds true for $>$, $\leq$, $\geq$, and $\neq$.

Symmetric Property of Similarity
(41)

If $\triangle ABC \sim \triangle DEF$, then $\triangle DEF \sim \triangle ABC$.

Transitive Property of Equality
(SB)

For every real number a, b, and c, if $a = b$ and $b = c$, then $a = c$.

Transitive Property of Similarity
(41)

If $\triangle ABC \sim \triangle DEF$ and $\triangle ABC \sim \triangle XYZ$, then $\triangle ABC \sim \triangle XYZ$.

Zero Property of Exponents
(SB)

For every nonzero number x, $x^0 = 1$.

Zero Product Property
(SB)

For every real number a and b and $ab = 0$, then $a = 0$ or $b = 0$.

Formulas

Perimeter

Rectangle	$P = 2b + 2h$ or $P = 2(b + h)$
Square	$P = 4s$

Area

Rectangle	$A = bh$
Square	$A = s^2$
Rhombus	$A = \frac{1}{2}d_1 d_2$
Parallelogram	$A = bh$
Triangle	$A = \frac{1}{2}bh$
Trapezoid	$A = \frac{1}{2}(b_1 + b_2)h$
Kite	$A = \frac{1}{2}d_1 d_2$
Regular polygon	$A = \frac{1}{2}aP$

Circles

Circumference	$C = 2\pi r$ or $C = \pi d$
Arc length	$L = 2\pi r \left(\dfrac{m°}{360°} \right)$
Area	$A = \pi r^2$
Area of a sector	$A = \pi r^2 \left(\dfrac{m°}{360°} \right)$
Area of a circle segment	$A_{segment} = A_{sector} - A_{triangle}$
Equation of a circle	$(x - h)^2 + (y - k)^2 = r^2$

Lateral Area

Prism	$L = Ph$
Cylinder	$L = 2\pi rh$
Pyramid	$L = \frac{1}{2}Pl$
Cone	$L = \pi rl$

Surface Area

Cube	$S = 6s^2$
Cylinder	$S = 2\pi r^2 + 2\pi rh$
Cone	$S = \pi r^2 + \pi rl$
Prism	$S = L + 2B$
Pyramid	$S = L + B$
Sphere	$S = 4\pi r^2$

Volume

Prism	$V = Bh$
Cylinder	$V = \pi r^2 h$
Pyramid or cone	$V = \frac{1}{3}Bh$
Sphere	$V = \frac{4}{3}\pi r^3$

Frustum of a cone

$$V = \frac{1}{3}h(B_1 + \sqrt{B_1 B_2} + B_2)$$

Linear Equations

Slope formula	$m = \dfrac{y_2 - y_1}{x_2 - x_1}$
Slope-intercept form	$y = mx + b$
Point-slope form	$y - y_1 = m(x - x_1)$
Standard form	$Ax + By = C$

Trigonometric Ratios

$$\text{sine of } \angle A = \frac{\text{length of leg opposite to } \angle A}{\text{length of hypotenuse}}$$

$$\text{cosine of } \angle A = \frac{\text{length of leg adjacent to } \angle A}{\text{length of hypotenuse}}$$

$$\text{tangent of } \angle A = \frac{\text{length of leg opposite } \angle A}{\text{length of leg adjacent } \angle A}$$

Percents

$$\text{Percent of change} = \frac{\text{amount of change}}{\text{original amount}}$$

Probability

$$P(\text{event}) = \frac{\text{number of desired outcomes}}{\text{number of possible outcomes}}$$

$P(A)$ probability of event A

Probability of complement

$$P(\text{not event}) = 1 - P(\text{event})$$

Probability of independent events

$$P(A \text{ and } B) = P(A) \cdot P(B)$$

Probability of dependent events

$$P(A \text{ then } B) = P(A) \cdot P(B \text{ after } A)$$

Distance

Distance formula

$$d = \sqrt{(x_2 - x_1)^2 + (y_2 - y_1)^2}$$

Midpoint of a segment

$$M = \left(\frac{x_1 + x_2}{2}, \frac{y_1 + y_2}{2}\right)$$

Additional Formulas

Angle sum of an n-gon	$(n - 2)180°$
Euler's Formula	$V - E + F = 2$

Symbols

Geometric

≅	is congruent to
~	is similar to
°	degree(s)
∠ABC	angle ABC
m∠ABC	the measure of angle ABC
△ABC	triangle ABC
⊙A	circle A
$\overleftrightarrow{AB}$	line AB
$\overline{AB}$	segment AB
$\overrightarrow{AB}$	ray AB
AB	the length of $\overline{AB}$
$\overarc{AC}$	minor arc AC
$\overarc{ABC}$	major arc ABC
∟	right angle
⊥	is perpendicular to
∥	is parallel to

Comparison Symbols

<	less than
>	greater than
≤	less than or equal to
≥	greater than or equal to
≠	not equal to
≈	approximately equal to

Real Numbers

$\mathbb{R}$	the set of real numbers
$\mathbb{Q}$	the set of rational numbers
$\mathbb{Z}$	the set of integers
$\mathbb{W}$	the set of whole numbers
$\mathbb{N}$	the set of natural numbers

Additional Symbols

±	plus or minus
$a \cdot b$, ab or $a(b)$	a times b
$|-5|$	the absolute value of -5
%	percent
π	pi, $\pi \approx 3.14$, or $\pi \approx \frac{22}{7}$
$f(x)$	function notation: f of x
a^n	a to nth power
a_n	nth term of a sequence
$\sqrt{x}$	square root of x
(x, y)	ordered pair
$x{:}y$	ratio of x to y, or $\frac{x}{y}$
{ }	set braces

Metric Table of Measures

Length

1 kilometer (km) = 1000 meters (m)

1 meter = 100 centimeters (cm)

1 centimeter = 10 millimeters (mm)

Capacity and Volume

1 liter (L) = 1000 milliliters (mL)

Mass

1 kilogram (kg) = 1000 grams (g)

1 gram = 1000 milligrams (mg)

Customary Table of Measures

Length

1 mile (mi) = 5280 feet (ft)

1 mile = 1760 yards (yd)

1 yard = 3 feet

1 yard = 36 inches (in.)

1 foot = 12 inches

Capacity and Volume

1 gallon (gal) = 4 quarts (qt)

1 quart = 2 pints (pt)

1 pint = 2 cups (c)

1 cup = 8 fluid ounces (fl oz)

Weight

1 ton = 2000 pounds (lb)

1 pound = 16 ounces (oz)

Customary and Metric Measures

1 inch = 2.54 centimeters

1 yard ≈ 0.9 meters

1 mile ≈ 1.6 kilometers

Time

1 year = 365 days

1 year = 12 months

1 month ≈ 4 weeks

1 year ≈ 52 weeks

1 week = 7 days

1 day = 24 hours

1 hour (hr) = 60 minutes (min)

1 minute = 60 seconds (s)

Postulates and Theorems

Postulate 1: Ruler Postulate
(2)

The points on a line can be paired in a one to one correspondence with the real numbers such that:

1. any two given points can have coordinates 0 and 1.
2. the distance between two points is the absolute value of the difference of their coordinates.

Postulate 2: Segment Addition Postulate
(2)

If B is between A and C, then $AB + BC = AC$.

Postulate 3: Protractor Postulate
(3)

Given a point, X, on $\overleftrightarrow{PR}$, consider $\overrightarrow{XP}$ and $\overrightarrow{XR}$, as well as all the other rays that can be drawn, with X as an endpoint, on one side of $\overleftrightarrow{PR}$. These rays can be paired with the real numbers from 0 to 180 such that:
1. $\overrightarrow{XP}$ is paired with 0, and $\overrightarrow{XR}$ is paired with 180.
2. If $\overrightarrow{XA}$ is paired with a number, c, and $\overrightarrow{XB}$ is paired with a number, d, then $m\angle AXB = |c - d|$.

Postulate 4: Angle Addition Postulate
(3)

If point D is in the interior of $\angle ABC$, then $m\angle ABD + m\angle DBC = m\angle ABC$.

Postulate 5
(4)

Through any two points there exists exactly one line.

Postulate 6
(4)

Through any three noncollinear points there exists exactly one plane.

Postulate 7
(4)

If two planes intersect, then their intersection is a line.

Postulate 8
(4)

If two points lie in a plane, then the line containing the points lies in the plane.

Postulate 9
(4)

A line contains at least 2 points. A plane contains at least 3 noncollinear points. Space contains at least 4 noncoplanar points.

Postulate 10: Parallel Postulate
(5)

Through a point not on a line, there exists exactly one line through the point parallel to the line.

Postulate 11: Corresponding Angles Postulate
(Inv 1)

If two parallel lines are cut by a transversal, then the corresponding angles formed are congruent.

Postulate 12: Converse of the Corresponding Angles Postulate
(12)

If two lines are cut by a transversal and the corresponding angles formed are congruent, then the lines are parallel.

Postulate 13: Side-Side-Side (SSS) Triangle Congruence Postulate
(25)

If three sides of one triangle are congruent to three sides of another triangle, then the triangles are congruent.

Postulate 14: Arc Addition Postulate
(26)

The measure of an arc formed by two adjacent arcs is the sum of the measures of the two arcs.

Postulate 15: Side-Angle-Side (SAS) Triangle Congruence Postulate
(28)

If two sides and the included angle of one triangle are congruent to two sides and the included angle of another triangle, then the triangles are congruent.

Postulate 16: Angle-Side-Angle (ASA) Triangle Congruence Postulate
(30)

If two angles and the included side of one triangle are congruent to two angles and the included side of another triangle, then the triangles are congruent.

Postulate 17: Parallel Lines Postulate
(37)

If two lines are parallel, then they have the same slope. All vertical lines are parallel to each other.

Postulate 18: Perpendicular Lines Postulate
(37)

If two nonvertical lines are perpendicular, then the product of their slopes is −1. Vertical and horizontal lines are perpendicular to each other.

Postulate 19: Area Congruence Postulate
(40)

If two polygons are congruent, then they have the same area.

Postulate 20: Area Addition Postulate
(40)

The area of a region is equal to the sum of the areas of its non-overlapping parts.

Postulate 21: Angle-Angle (AA) Triangle Similarity Postulate
(46)

If two angles of one triangle are congruent to two angles of another triangle, then the triangles are similar.

Theorems

Theorem 4-1

If two lines intersect, then they intersect at exactly one point.
Proved in Lesson 48, Example 1

Theorem 4-2

If there is a line and a point not on the line, then exactly one plane contains them.
Proved in Lesson 27, Example 1

Theorem 4-3

If two lines intersect, then there exists exactly one plane that contains them.
Proved in Lesson 48, Practice

Theorem 5-1

If two parallel planes are cut by a third plane, then the lines of intersection are parallel.
Proved in Lesson 29, Practice

Theorem 5-2

If two lines in a plane are perpendicular to the same line, then they are parallel to each other.
Proved in Lesson 28, Practice

Theorem 5-3

In a plane, if a line is perpendicular to one of two parallel lines, then it is perpendicular to the other one.
Proved in Lesson 27, Example 4

Theorem 5-4

If two lines are perpendicular, then they form congruent adjacent angles.
Proved in Lesson 31, Lesson Practice

Theorem 5-5

If two lines form congruent adjacent angles, then they are perpendicular.
Proved in Lesson 27, Lesson Practice

Theorem 5-6: Right Angles Theorem

All right angles are congruent.
Proved in Lesson 27, Practice

Theorem 5-7: Transitive Property of Parallel Lines

If two lines are parallel to the same line, then they are parallel to each other.

Proved in Lesson 27, Practice

Theorem 6-1: Congruent Complements Theorem

If two angles are complementary to the same angle or to congruent angles, then they are congruent.

Proved in Lesson 60, Example 3

Theorem 6-2: Congruent Supplements Theorem

If two angles are supplementary to the same angle or to congruent angles, then they are congruent.

Proved in Lesson 31, Example 1

Theorem 6-3: Linear Pair Theorem

If two angles form a linear pair, then they are supplementary.

Proved in Lesson 28, Practice

Theorem 6-4: Vertical Angles Theorem

If two angles are vertical angles, then they are congruent.

Proved in Lesson 27, Example 3

Theorem 6-5

If a point lies on the perpendicular bisector of a segment, then the point is equidistant from the endpoints of the segment.

Proved in Lesson 29, Practice

Theorem 6-6

If a point is equidistant from the endpoints of a segment, then the point lies on the perpendicular bisector of the segment.

Proved in Lesson 36, Practice

Theorem 6-7

If a point lies on the bisector of an angle, then the point is equidistant from the sides of the angle.

Proved in Lesson 30, Practice

Theorem 6-8

If a point is equidistant from the sides of an angle, then the point lies on the bisector of the angle.

Proved in Lesson 36, Practice

Theorem 8-1: Pythagorean Theorem

The sum of the squares of the legs, a and b, of a right triangle is equal to the square of the hypotenuse, c, and is written $a^2 + b^2 = c^2$.

Proved in Investigation 8

Theorem 10-1: Alternate Interior Angles Theorem

If two parallel lines are cut by a transversal, then the alternate interior angles are congruent.

Proved in Lesson 31, Example 4

Theorem 10-2: Alternate Exterior Angles Theorem

If two parallel lines are cut by a transversal, then the alternate exterior angles are congruent.

Proved in Lesson 27, Practice

Theorem 10-3: Same-Side Interior Angles Theorem

If two parallel lines are cut by a transversal, then the same-side interior angles are supplementary.

Proved in Lesson 29, Practice

Theorem 12-1: Converse of the Alternate Interior Angles Theorem

If two lines are cut by a transversal and the alternate interior angles are congruent, then the lines are parallel.

Proved in Lesson 27, Practice

Theorem 12-2: Converse of the Alternate Exterior Angles Theorem

If two lines are cut by a transversal and the alternate exterior angles are congruent, then the lines are parallel.

Proved in Lesson 70, Practice

Theorem 12-3: Converse of the Same-Side Interior Angles Theorem

If two lines are cut by a transversal and the same-side interior angles are supplementary, then the lines are parallel.

Proved in Lesson 82, Practice

Theorem 18-1: Triangle Angle Sum Theorem

The sum of the measures of the angles of a triangle is equal to 180°.

Proved in Lesson 31, Example 2

Corollary 18-1-1: Third Angle Theorem

If two angles of one triangle are congruent to two angles of another triangle, then the third angles are congruent.

Corollary 18-1-2

The acute angles of a right triangle are complementary.

Proved in Lesson 27, Lesson Practice

Corollary 18-1-3

The measure of each angle of an equiangular triangle is 60°.

Corollary 18-1-4

A triangle can have at most one right or one obtuse angle.

Theorem 18-2: Exterior Angle Theorem

The measure of each exterior angle of a triangle is equal to the sum of the measures of its two remote interior angles.

Proved in Lesson 27, Lesson Practice

Theorem 26-1

In the same or congruent circles, congruent arcs have congruent central angles.

Proved in Lesson 35, Practice

Theorem 30-1: Angle-Angle-Side (AAS) Triangle Congruence Theorem

If two angles and a non-included side of one triangle are congruent to two angles and the corresponding non-included side of another triangle, then the triangles are congruent.

Proved in Lesson 30, Teacher's Manual Challenge Problem

Theorem 32-1: Centroid Theorem

The centroid of a triangle is located $\frac{2}{3}$ the distance from each vertex to the midpoint of the opposite side.

Proof will be developed in other Saxon High School Math courses

Theorem 33-1: Converse of the Pythagorean Theorem

If the sum of the squares of the two shorter sides of a triangle is equal to the square of the longest side of the triangle, then the triangle is a right triangle.

Proved in Lesson 33, Example 2

Theorem 33-2: Pythagorean Inequality Theorem

In a triangle, let a and b be the lengths of the two shorter sides and let c be the length of the longest side. If $a^2 + b^2 < c^2$, then the triangle is obtuse. If $a^2 + b^2 > c^2$, then the triangle is acute.

Proved in Lesson 41, Practice

Theorem 36-1: Leg-Angle (LA) Right Triangle Congruence Theorem

If a leg and an acute angle of one right triangle are congruent to a leg and an acute angle of another right triangle, then the triangles are congruent.

Proved in Lesson 36, Practice

Theorem 36-2: Hypotenuse-Angle (HA) Right Triangle Congruence Theorem

If the hypotenuse and an acute angle of one right triangle are congruent to the hypotenuse and an acute angle of another right triangle, then the triangles are congruent.

Proved in Lesson 36, Example 2

Theorem 36-3: Leg-Leg (LL) Right Triangle Congruence Theorem

If the two legs of one right triangle are congruent to the two legs of another right triangle, then the triangles are congruent.

Proved in Lesson 37, Practice

Theorem 36-4: Hypotenuse-Leg (HL) Right Triangle Congruence Theorem

If the hypotenuse and a leg of one right triangle are congruent to the hypotenuse and a leg of another right triangle, then the triangles are congruent.

Proved in Lesson 36, Example 4

Theorem 38-1: Triangle Angle Bisector Theorem

If a line bisects an angle of a triangle, then it divides the opposite side proportionally to the other two sides of the triangle.

Proved in Lesson 60, Practice

Theorem 39-1

If one side of a triangle is longer than another side, then the angle opposite the first side is larger than the angle opposite the second side.

Proved in Lesson 52, Practice

Theorem 39-2

If one angle of a triangle is larger than another angle, then the side opposite the first angle is longer than the side opposite the second angle.

Proved in Lesson 48, Example 2

Theorem 39-3: Exterior Angle Inequality Theorem

The measure of an exterior angle of a triangle is greater than the measure of either remote interior angle.

Proved in Lesson 39, Example 2

Theorem 39-4: Triangle Inequality Theorem

The sum of the lengths of any two sides of a triangle must be greater than the length of the third side.

Proved in Lesson 55, Practice

Theorem 40-1: Hinge Theorem

If two sides of one triangle are congruent to two sides of another triangle and the included angle of the first triangle is greater than the included angle of the second triangle, then the third side of the first triangle is longer than the third side of the second triangle.

Proved in Lesson 42, Practice

Theorem 40-2: Converse of the Hinge Theorem

If two sides of one triangle are congruent to two sides of another triangle and the third side of the first triangle is longer than the third side of the second triangle, then the measure of the angle opposite the third side of the first triangle is greater than the measure of the angle opposite the third side of the second triangle.

Proved in Lesson 48, Practice

Theorem 42-1

Through a line and a point not on the line, there exists exactly one perpendicular line to the given line.

Proved in Lesson 48, Practice

Theorem 42-2

The perpendicular segment from a point to a line is the shortest segment from the point to the line.

Proved in Lesson 48, Example 4

Theorem 42-3

The perpendicular segment from a point to a plane is the shortest segment from the point to the plane.

Proved in Lesson 48, Practice

Theorem 42-4

If two lines are parallel, then all points on one line are equidistant from the other line.

Proved in Lesson 42, Example 4a

Theorem 43-1

If a diameter is perpendicular to a chord, then it bisects the chord and its arcs.

Proved in Lesson 43, Lesson Practice

Theorem 43-2

If a diameter bisects a chord other than another diameter, then it is perpendicular to the chord.

Proven in Lesson 43, Practice

Theorem 43-3

The perpendicular bisector of a chord contains the center of the circle.

Proved in Lesson 43, Example 3

Theorem 43-4

In a circle or congruent circles (1) chords equidistant from the center are congruent (2) congruent chords are equidistant from the center of the circle.

Proven in Lesson 57, Practice

Theorem 44-1

If two polygons are similar, then the ratio of their perimeters is equal to the ratio of their corresponding sides.

Proven in Lesson 44, Example 3

Theorem 46-1: Side-Side-Side (SSS) Similarity Theorem

If the lengths of the sides of a triangle are proportional to the lengths of the sides of another triangle, then the triangles are similar.

Proved in Lesson 46, Example 3

Theorem 46-2: Side-Angle-Side (SAS) Similarity Theorem

If two sides of one triangle are proportional to two sides of another triangle and the included angles are congruent, then the triangles are similar.

Proved in Lesson 46, Practice

Theorem 47-1

The measure of an inscribed angle is equal to half the measure of its intercepted arc.

Proved in Lesson 59, Practice

Theorem 47-2

If an inscribed angle intercepts a semicircle, then it is a right angle.

Proved in Lesson 47, Example 1

Theorem 47-3

If two inscribed angles intercept the same arc, then they are congruent.

Proved in Lesson 47, Lesson Practice

Theorem 47-4

If a quadrilateral is inscribed in a circle, then it has supplementary opposite angles.

Proved in Lesson 47, Practice

Theorem 50-1

If the altitude is drawn to the hypotenuse of a right triangle, then the two triangles formed are similar to each other and to the original triangle.

Proved in Lesson 50, Example 1

Corollary 50-1-1

If the altitude is drawn to the hypotenuse of a right triangle, then the length of the altitude is the geometric mean between the segments of the hypotenuse.

Corollary 50-1-2

If the altitude is drawn to the hypotenuse of a right triangle, then the length of a leg is the geometric mean between the hypotenuse and the segment of the hypotenuse that is closer to that leg.

Theorem 51-1: Isosceles Triangle Theorem

If a triangle is isosceles, then its base angles are congruent. *Proved in Lesson 51, Example 1*

Corollary 51-1-1

If a triangle is equilateral, then it is equiangular.

Theorem 51-2: Converse of the Isosceles Triangle Theorem

If two angles of a triangle are congruent, then the sides opposite those angles are also congruent.

Proved in Lesson 51, Practice

Corollary 51-2-1

If a triangle is equiangular, then it is equilateral.

Theorem 51-3

If a line bisects the vertex angle in an isosceles triangle, then it is the perpendicular bisector of the base.

Proved in Lesson 51, Example 4

Theorem 51-4

If a line is the perpendicular bisector of the base of an isosceles triangle, then it bisects the vertex angle.

Proved in Lesson 51, Example 4

Theorem 55-1: Triangle Midsegment Theorem

The segment joining the midpoints of two sides of a triangle is parallel to, and half the length of, the third side.

Proved in Lesson 55, Example 2

Theorem 55-2

If a line is parallel to one side of a triangle and it contains the midpoint of another side, then it passes through the midpoint of the third side.

Proved in Lesson 55, Lesson Practice

Theorem 58-1

If a line is tangent to a circle, then the line is perpendicular to a radius drawn to the point of tangency.

Proved in Lesson 59, Practice

Theorem 58-2

If a line in the plane of a circle is perpendicular to a radius at its endpoint on the circle, then the line is tangent to the circle.

Proved in Lesson 58, Example 3

Theorem 58-3

If two tangent segments are drawn to a circle from the same exterior point, then they are congruent.

Proved in Lesson 58, Lesson Practice

Theorem 60-1: Triangle Proportionality Theorem

If a line parallel to one side of a triangle intersects the other two sides, then it divides those sides proportionally.

Proved in Lesson 60, Lesson Practice

Theorem 60-2: Converse of the Triangle Proportionality Theorem

If a line divides two sides of a triangle proportionally, then it is parallel to the third side.

Proved in Lesson 60, Practice

Theorem 60-3

If parallel lines intersect transversals, then they divide the transversals proportionally.

Proved in Lesson 60, Practice

Theorem 60-4

If parallel lines cut congruent segments on one transversal, then they cut congruent segments on all transversals.

Proved in Lesson 60, Example 3

Theorem 64-1

The measure of an angle formed by a tangent and a chord is equal to half the measure of the intercepted arc.

Proved in Lesson 64, Lesson Practice

Theorem 64-2

The measure of an angle formed by two chords intersecting in a circle is equal to half the sum of the intercepted arcs.

Proved in Lesson 64, Example 2

Theorem 79-1

The measure of an angle whose vertex is outside a circle is equal to half the difference of the intercepted arcs.

Proved in Lesson 79, Example 1

Theorem 85-1: Cavalieri's Principle

If two solids have the same height and the same cross-sectional area at each level, then they have the same volume.

Proof will be developed in other Saxon High School Math courses

Theorem 86-1

If two chords intersect in a circle, then the products of the chord segments are equal.

Proved in Lesson 86, Example 1

Theorem 87-1

If two similar figures have a scale factor of $a:b$, then the ratio of their perimeters is $a:b$, and the ratio of their areas is $a^2:b^2$.

Proved in Lesson 87, Example 1

Theorem 99-1

If two similar solids have a scale factor of $a:b$, then the ratio of the perimeters of their corresponding faces is $a:b$, the ratio of the areas of their corresponding faces is $a^2:b^2$, and the ratio of their volumes is $a^3:b^3$.

Theorem 101-1

If two secant segments are drawn through a circle from an external point, then the product of one secant segment and its external segment is equal to the product of the other secant segment and its external segment.

Proved in Lesson 101, Example 2

Theorem 101-2

If a secant segment and a tangent segment are drawn through a circle from an external point, then the product of the lengths of the secant segment and its external segment is equal to the length of the tangent segment squared.

Proved in Lesson 101, Lesson Practice

Theorem 104-1

In the same or congruent circles (1) congruent arcs have congruent chords (2) congruent chords have congruent arcs.

Proved in Lesson 104, Example 2 and Lesson Practice

English/Spanish Glossary

English	Example	Spanish

A

absolute value
(2)

The distance from zero on a number line.

$$|3| = |-3| = 3$$

valor absoluto
(2)

La distancia desde cero.

---

acute angle
(3)

An angle that measures greater than 0° and less than 90°.

ángulo agudo
(3)

Ángulo que mide más de 0° y menos de 90°.

---

acute triangle
(13)

A triangle with three acute angles.

triángulo acutángulo
(13)

Triángulo con tres ángulos agudos.

---

adjacent angles
(6)

Two angles in the same plane with a common vertex and a common side, but no common interior points.

∠1 and ∠2 are adjacent angles.

ángulos adyacentes
(6)

Dos ángulos en el mismo plano que tienen un vértice y un lado común pero no comparten puntos internos.

---

adjacent arcs
(26)

Two arcs of the same circle that intersect at exactly one point.

$\overarc{RS}$ and $\overarc{ST}$ are adjacent arcs.

arcos adyacentes
(26)

Dos arcos del mismo círculo que se cruzan exactamente en un punto.

---

alternate exterior angles
(Inv 1)

For two lines intersected by a transversal, a pair of angles that lie on opposite sides of the transversal and outside the other two lines.

∠4 and ∠5 are alternate exterior angles.

ángulos alternos externos
(Inv 1)

Dadas dos rectas cortadas por una transversal, un par de ángulos no adyacentes ubicados en los lados opuestos de la transversal y fuera de las otras dos rectas.

---

alternate interior angles
(Inv 1)

For two lines intersected by a transversal, a pair of nonadjacent angles that lie on opposite sides of the transversal and between the other two lines.

∠3 and ∠6 are alternate interior angles.

ángulos alternos internos
(Inv 1)

Dadas dos rectas cortadas por una transversal, un par de ángulos no adyacentes ubicados en los lados opuestos de la transversal y entre las otras dos rectas.

English	Example	Spanish

A

altitude of a cone
(77)

The perpendicular segment from the vertex to the plane containing the base.

altura de un cono
(77)

Segmento que se extiende desde el vértice hasta el plano de la base y es perpendicular al plano de la base.

---

altitude of a cylinder
(62)

A segment that is perpendicular to, and has its endpoints on, the planes of the bases.

 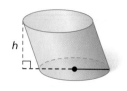

altura de un cilindro
(62)

Segmento con sus extremos en los planos de las bases que es perpendicular a los planos de las bases.

---

altitude of a prism
(59)

A segment that is perpendicular to, and has its endpoints on, the planes of the bases.

 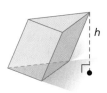

altura de un prisma
(59)

Segmento con sus extremos en los planos de las bases que es perpendicular a los planos de las bases.

---

altitude of a pyramid
(70)

The perpendicular segment from the vertex to the plane containing the base.

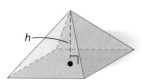

altura de una pirámide
(70)

Segmento que se extiende desde el vértice hasta el plano de la base y es perpendicular al plano de la base.

---

altitude of a triangle
(32)

The perpendicular segment from a vertex to the line containing the opposite side.

altura de un triángulo
(32)

Segmento perpendicular que se extiende desde un vértice hasta la recta que forma el lado opuesto.

---

angle
(3)

A figure formed by two rays with a common endpoint.

ángulo
(3)

Figura formada por dos rayos con un extremo común.

---

angle bisector
(3)

A ray that divides an angle into two congruent angles.

$\overrightarrow{JK}$ is the angle bisector of $\angle LJM$.

bisectriz de un ángulo
(3)

Rayo que divide un ángulo en dos ángulos congruentes.

English	Example	Spanish

angle of depression
(73)

The angle formed by a horizontal line and a line of sight to a point below.

ángulo de depresión
(73)

Ángulo formado por una recta horizontal y una línea visual a un punto inferior.

---

angle of elevation
(73)

The angle formed by a horizontal line and a line of sight to a point above.

ángulo de elevación
(73)

Ángulo formado por una recta horizontal y una línea visual a un punto superior.

---

angle of rotation
(Inv 12)

An angle formed by a rotating ray, called the terminal side, and a stationary reference ray, called the initial side.

terminal side, 45°, 135°, *y*, *x*, 0 initial side
The angle of rotation is 135°.

ángulo de rotación
(Inv 12)

Ángulo formado por un rayo rotativo, denominado lado terminal, y un rayo de referencia estático, denominado lado inicial.

---

angle of rotational symmetry
(76)

The smallest angle through which a figure with rotational symmetry can be rotated to coincide with itself.

90° 90° 90° 90°

ángulo de simetría de rotación
(76)

El ángulo más pequeño alrededor del cual se puede rotar una figura con simetría de rotación para que coincida consigo misma.

---

annulus
(97)

The region between two concentric circles.

corona circular
(97)

Región comprendida entre dos círculos concéntricos.

---

apothem
(66)

The perpendicular distance from the center of a regular polygon to one of its sides.

a

apotema
(66)

Distancia perpendicular desde el centro de un polígono regular hasta un lado del polígono.

---

arc
(26)

An unbroken part of a circle consisting of two points on the circle, called the endpoints, and all the points on the circle between them.

R *S*

arco
(26)

Parte continua de un círculo formada por dos puntos del círculo denominados extremos y todos los puntos del círculo comprendidos entre éstos.

English	Example	Spanish

A

arc length
(35)

The distance along an arc measured in linear units.

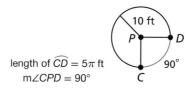

length of $\overset{\frown}{CD} = 5\pi$ ft
m∠CPD = 90°

longitud de arco
(35)

Distancia a lo largo de un arco medida en unidades lineales.

---

arc marks
(3)

Marks used on a figure to indicate congruent angles.

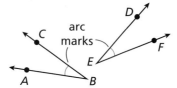

marcas de arco
(3)

Marcas utilizadas en una figura para indicar ángulos congruentes.

---

area
(8)

The number of non-overlapping unit squares of a given size that will exactly cover the interior of a plane figure.

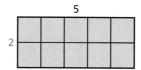

The area is 10 square units.

área
(8)

Cantidad de cuadrados unitarios de un determinado tamaño no superpuestos que cubren exactamente el interior de una figura plana.

---

auxiliary line
(31)

A line drawn in a figure to aid in a proof.

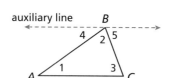

recta auxiliar
(31)

Recta dibujada en una figura como ayuda en una demostración.

---

axis of symmetry
(113)

The line that a figure can be rotated about so that the image corresponds with the preimage.

axis of symmetry

eje de simetría
(113)

Línea que divide una figura plana o una gráfica en dos mitades reflejadas congruentes.

B

base angle of an isosceles triangle
(51)

One of the two angles that have the base of the triangle as a side.

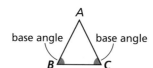

ángulo base de un triángulo isósceles
(51)

Uno de los dos ángulos que tienen como lado la base del triángulo.

---

base angle of a trapezoid
(69)

One of a pair of consecutive angles whose common side is a base of the trapezoid.

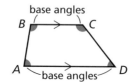

ángulo base de un trapecio
(69)

Uno de los dos ángulos consecutivos cuyo lado en común es la base del trapecio.

---

base of a cone
(77)

The flat circular surface of the cone.

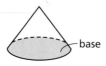

base de un cono
(77)

Cara circular del cono.

English	Example	Spanish
B		

base of a cylinder
(62)

One of the two flat circular surfaces of the cylinder.

base de un cilindro
(62)

Una de las dos caras circulares del cilindro.

---

base of an isosceles triangle
(51)

The side opposite the vertex angle.

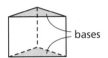

base de un triángulo isósceles
(51)

Lado opuesto al ángulo del vértice.

---

base of a prism
(49)

One of the two congruent parallel faces of the prism.

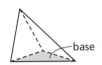

base de un prisma
(49)

Una de las dos caras paralelas y congruentes del prisma.

---

base of a pyramid
(70)

The face of the pyramid that is opposite the vertex.

base de una pirámide
(70)

Cara de la pirámide opuesta al vértice.

---

base of a trapezoid
(69)

One of the two parallel sides of the trapezoid.

base de un trapecio
(69)

Uno de los dos lados paralelos del trapecio.

---

base of a triangle
(13)

Any side of a triangle.

base de un triángulo
(13)

Cualquier lado de un triángulo.

---

between
(2)

On the same line as and in the space separating two points.

A   B   C

entre
(2)

Sobre la misma línea y en el espacio que separa a dos puntos.

---

biconditional statement
(20)

A statement that can be written in the form "p if and only if q."

A figure is a triangle if and only if it is a three-sided polygon.

enunciado bicondicional
(20)

Enunciado que puede expresarse en la forma "p si y sólo si q".

---

bisect
(3)

To divide into two congruent parts.

$\overrightarrow{JK}$ bisects $\angle LJM$.

trazar una bisectriz
(3)

Dividir en dos partes congruentes.

English	Example	Spanish
center of a circle *(23)* The point inside a circle that is equidistant from every point on the circle.		**centro de un círculo** *(23)* Punto dentro de un círculo que se encuentra a la misma distancia de todos los puntos del círculo.
center of a regular polygon *(Inv 3)* The point that is equidistant from all vertices of a regular polygon.		**centro de un polígono regular** *(Inv 3)* Punto equidistante de todos los vértices del polígono regular.
center of a sphere *(80)* The point inside a sphere that is equidistant from every point on the sphere.		**centro de una esfera** *(80)* Punto dentro de una esfera que está a la misma distancia de cualquier punto de la esfera.
center of dilation *(84)* The intersection of the lines that connect each point of the image with the corresponding point of the preimage.		**centro de dilatación** *(84)* Intersección de las líneas que conectan cada punto de la imagen con el punto correspondiente de la imagen original.
center of rotation *(78)* The point around which a figure is rotated.		**centro de rotación** *(78)* Punto alrededor del cual rota una figura.
central angle of a circle *(26)* An angle whose vertex is the center of a circle.		**ángulo central de un círculo** *(26)* Ángulo cuyo vértice es el centro de un círculo.
central angle of a regular polygon *(Inv 3)* An angle whose vertex is the center of a regular polygon and whose sides pass through consecutive vertices.		**ángulo central de un polígono regular** *(Inv 3)* Ángulo cuyo vértice es el centro del polígono regular y cuyos lados pasan por vértices consecutivos.

English	Example	Spanish

C

centroid of a triangle
(32)

The point of concurrency of the three medians of a triangle. Also known as the *center of gravity.*

The centroid is *P.*

centroide de un triángulo
(32)

Punto donde se encuentran las tres medianas de un triángulo. También conocido como *centro de gravedad.*

---

chord
(43)

A segment whose endpoints lie on a circle.

cuerda
(43)

Segmento cuyos extremos se encuentran en un círculo.

---

circle
(23)

The set of points in a plane that are a fixed distance from a given point called the center of the circle.

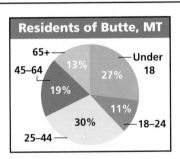

círculo
(23)

Conjunto de puntos en un plano que se encuentran a una distancia fija de un punto determinado denominado centro del círculo.

---

circle graph
(SB21)

A way to display data by using a circle divided into non-overlapping sectors.

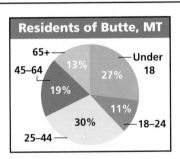

Residents of Butte, MT

65+ 13%
45–64 19%
25–44 30%
Under 18 27%
18–24 11%

gráfica circular
(SB21)

Forma de mostrar datos mediante un círculo dividido en sectores no superpuestos.

---

circumcenter of a triangle
(38)

The point of concurrency of the three perpendicular bisectors of a triangle.

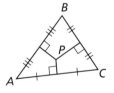

The circumcenter is *P.*

circuncentro de un triángulo
(38)

Punto donde se cortan las tres mediatrices de un triángulo.

---

circumference
(23)

The distance around a circle.

circunferencia
(23)

Distancia alrededor de un círculo.

---

circumscribed circle
(38)

A circle that contains all vertices of a polygon.

círculo circunscrito
(38)

Un círculo que contiene a todos los vértices de un polígono.

English	Example	Spanish

C

circumscribed polygon
(106)

A polygon whose sides are all tangent to the same circle.

polígono circunscrito
(106)

Un polígono con todos sus lados tangentes a un mismo círculo.

---

collinear
(1)

On the same line.

K, L, and M are collinear points.

colineal
(1)

Puntos que se encuentran sobre la misma línea.

---

common tangent
(72)

A line that is tangent to two circles.

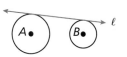 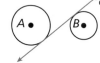

tangente común
(72)

Recta que es tangente a dos círculos.

---

compass
(Lab 1)

A tool used to draw arcs.

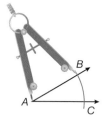

compás
(Lab 1)

Un instrumento que se utiliza para dibujar arcos.

---

complement of an angle
(6)

An angle whose measure equals 90° less the measure of the other angle.

The complement of a 53° angle is a 37° angle.

complemento de un ángulo
(6)

La suma de las medidas de un ángulo y su complemento es 90°.

---

complement of an event
(SB12)

All outcomes in the sample space that are not in an event E, denoted $\overline{E}$.

In the experiment of rolling a number cube, the complement of rolling a 3 is rolling a 1, 2, 4, 5, or 6.

complemento de un suceso
(SB12)

Todos los resultados en el espacio muestral que no están en el suceso E y se expresan $\overline{E}$.

---

complementary angles
(6)

Two angles whose measures have a sum of 90°.

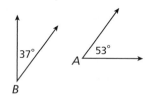

ángulos complementarios
(6)

Dos ángulos cuyas medidas suman 90°.

---

component form
(63)

The form of a vector that lists the vertical and horizontal change from the initial point to the terminal point.

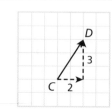

The component form of $\overrightarrow{CD}$ is $\langle 2, 3 \rangle$.

forma de componente
(63)

Forma de un vector que muestra el cambio horizontal y vertical desde el punto inicial hasta el punto terminal.

English	Example	Spanish

composite figure
(40)

A plane figure made up of simple shapes, or a three-dimensional figure made up of simple three-dimensional figures.

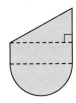

figura compuesta
(40)

Figura plana compuesta por figuras simples, o figura tridimensional compuesta por tridimensionales simples.

---

composite transformation
(90)

A combination of transformations.

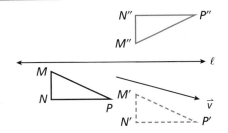

composición de transformaciones
(90)

Una combinación de transformaciones.

---

compound statement
(20)

Two statements that are connected by the word *and* or *or*.

The sky is blue and the grass is green.
I will drive to school or I will take the bus.

enunciado compuesto
(20)

Dos enunciados unidos por la palabra *y* u *o*.

---

concave polygon
(15)

A polygon in which a diagonal can be drawn such that part of the diagonal contains points in the exterior of the polygon.

concave
quadrilateral

polígono cóncavo
(15)

Polígono en el cual se puede trazar una diagonal tal que parte de la diagonal contiene puntos ubicados fuera del polígono.

---

concentric circles
(97)

Coplanar circles with the same center.

círculos concéntricos
(97)

Círculos coplanares que comparten el mismo centro.

---

conclusion
(10)

The part of a conditional statement following the word *then*.

If $x + 1 = 5$, then $\underbrace{x = 4}_{\text{conclusion}}$.

conclusión
(10)

Parte de un enunciado condicional que sigue a la palabra *entonces*.

---

concurrent lines
(32)

Three or more lines that intersect at one point.

líneas concurrentes
(32)

Tres o más líneas rectas que se intersecan en un punto.

conditional statement (10) A statement that can be written in the form "if p, then q," where p is the hypothesis and q is the conclusion.	 If $x + 1 = 5$, then $x = 4$. hypothesis conclusion	**enunciado condicional** (10) Enunciado que se puede expresar como "si p, entonces q", donde p es la hipótesis y q es la conclusión.
cone (49) A three-dimensional figure with a circular base and a curved lateral surface that connects the base to a point called the vertex.	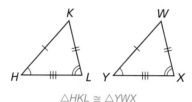	**cono** (49) Figura tridimensional con una base circular y una superficie lateral curva que conecta la base con un punto denominado vértice.
congruence statement (2) A statement that indicates that two polygons are congruent by listing the vertices in the order of correspondence.	 $\triangle HKL \cong \triangle YWX$	**enunciado de congruencia** (2) Enunciado que indica que dos polígonos son congruentes enumerando los vértices en orden de correspondencia.
congruence transformation (67) See *isometry*.		**transformación de congruencia** (67) Ver *isometría*.
congruent (2) Having the same size and shape, denoted by $\cong$.	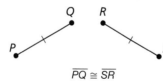 $\overline{PQ} \cong \overline{SR}$	**congruente** (2) Que tiene el mismo tamaño y la misma forma, expresado por $\cong$.
congruent angles (3) Angles that have the same measure.	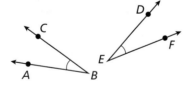 $\angle ABC \cong \angle DEF$	**ángulos congruentes** (3) Ángulos que tienen la misma medida.
congruent arcs (26) Arcs that are in the same or congruent circles and have the same measure.		**arcos congruentes** (26) Dos arcos que se encuentran en el mismo círculo o en círculos congruentes y que tienen la misma medida.

C

English	Example	Spanish
congruent circles *(23)* Circles that have congruent radii.		**círculos congruentes** *(23)* Dos círculos que tienen radios congruentes.
congruent polygons *(15)* Polygons whose corresponding sides and angles are congruent.		**polígonos congruentes** *(15)* Dos polígonos cuyos lados y ángulos correspondientes son congruentes.
congruent segments *(2)* Segments that have the same length.	 $\overline{PQ} \cong \overline{SR}$	**segmentos congruentes** *(2)* Dos segmentos que tienen la misma longitud.
congruent triangles *(25)* Triangles whose corresponding sides and angles are congruent.	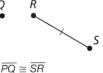	**triángulos congruentes** *(25)* Triángulos cuyos lados y ángulos correspondientes son congruentes.
conjecture *(7)* A statement that is believed to be true.	A sequence begins with the terms 2, 4, 6, 8, 10. A reasonable conjecture is that the next term in the sequence is 12.	**conjetura** *(7)* Enunciado que se supone verdadero.
conjunction *(20)* A compound statement that uses the word *and*.	3 is less than 5 AND greater than 0.	**conjunción** *(20)* Enunciado compuesto que contiene la palabra *y*.
consecutive interior angles *(Inv 1)* See *same-side interior angles*.		**ángulos internos consecutivos** *(Inv 1)* Ver *ángulos internos del mismo lado*.
construction *(Lab 1)* A method of creating a figure that is considered to be mathematically precise.	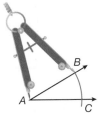	**construcción** *(Lab 1)* Método para crear una figura que es considerado matemáticamente preciso.
contraction *(84)* See *reduction*.		**contracción** *(84)* Ver *reducción*.

C

---

contrapositive
(17)

The statement formed by both exchanging and negating the hypothesis and conclusion of a conditional statement.

Statement: If $n + 1 = 3$, then $n = 2$

Contrapositive: If $n \neq 2$, then $n + 1 \neq 3$

contrapuesto
(17)

Enunciado que se forma al intercambiar y negar la hipótesis y la conclusión de un enunciado condicional.

---

converse
(10)

The statement formed by exchanging the hypothesis and conclusion of a conditional statement.

Statement: If $n + 1 = 3$, then $n = 2$

Converse: If $n = 2$, then $n + 1 = 3$

expresión recíproca
(10)

Enunciado que se forma intercambiando la hipótesis y la conclusión de un enunciado condicional.

---

convex polygon
(15)

A polygon in which no diagonal contains points in the exterior of the polygon.

convex quadrilateral

polígono convexo
(15)

Polígono en el cual ninguna diagonal contiene puntos fuera del polígono.

---

coordinate
(SB13)

A number used to identify the location of a point.

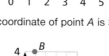

The coordinate of point A is 3.

The coordinates of point B are $(1, 4)$.

coordenada
(SB13)

Número utilizado para identificar la ubicación de un punto.

---

coordinate plane
(SB13)

A plane that is divided into four regions by a horizontal line called the x-axis and a vertical line called the y-axis.

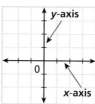

plano cartesiano
(SB13)

Plano dividido en cuatro regiones por una línea horizontal denominada eje x y una línea vertical denominada eje y.

---

coordinate proof
(45)

A style of proof that uses coordinate geometry and algebra.

prueba de coordenadas
(45)

Tipo de demostración que utiliza geometría de coordenadas y álgebra.

---

coplanar
(1)

In the same plane.

coplanar
(1)

En el mismo plano.

---

corollary
(18)

A theorem whose proof follows directly from another theorem.

corolario
(18)

Teorema cuya demostración proviene directamente de otro teorema.

---

corresponding angles of lines intersected by a transversal
(Inv 1)

For two lines intersected by a transversal, a pair of angles that are on the same side of the transversal and on the same sides of the other two lines.

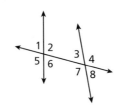

∠1 and ∠3 are corresponding.

ángulos correspondientes de líneas cortadas por una transversal
(Inv 1)

Dadas dos rectas cortadas por una transversal, el par de ángulos ubicados en el mismo lado de la transversal y en los mismos lados de las otras dos rectas.

---

corresponding angles of polygons
(25)

Angles in the same relative position in two different polygons with the same number of angles.

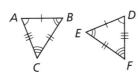

∠A and ∠D are corresponding angles.

ángulos correspondientes de los polígonos
(25)

Ángulos que tienen la misma posición en dos polígonos diferentes que tienen el mismo número de ángulos.

---

corresponding sides of polygons
(25)

Sides in the same relative position in two different polygons with the same number of sides.

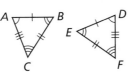

$\overline{AB}$ and $\overline{DE}$ are corresponding sides.

lados correspondientes de los polígonos
(25)

Lados que tienen la misma posición en dos polígonos diferentes que tienen el mismo número de lados.

---

cosecant of an angle
(116)

In a right triangle, the ratio of the length of the hypotenuse to the length of the side opposite the angle.

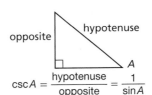

$$\csc A = \frac{\text{hypotenuse}}{\text{opposite}} = \frac{1}{\sin A}$$

cosecante
(116)

En un triángulo rectángulo, la razón de la longitud de la hipotenusa a la longitud del cateto opuesto.

---

cosine of an angle
(68)

In a right triangle, the ratio of the length of the leg adjacent to the angle to the length of the hypotenuse.

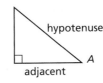

$$\cos A = \frac{\text{adjacent}}{\text{hypotenuse}} = \frac{1}{\sec A}$$

coseno
(68)

En un triángulo rectángulo, la razón de la longitud del cateto adyacente a la longitud de la hipotenusa.

GLOSSARY/ GLOSARIO

cotangent of an angle
(116)

In a right triangle, the ratio of the length of the side adjacent to the angle to the length of the side opposite the angle.

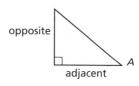

$$\cot A = \frac{\text{adjacent}}{\text{opposite}} = \frac{1}{\tan A}$$

cotangente
(116)

En un triángulo rectángulo, la razón de la longitud del cateto adyacente a la longitud del cateto opuesto.

---

counterexample
(14)

An example that proves that a conjecture or statement is false.

contraejemplo
(14)

Ejemplo que demuestra que una conjetura o enunciado es falso.

CPCTC
(25)

An abbreviation for "Corresponding Parts of Congruent Triangles are Congruent," which can be used as a justification in a proof after two triangles are proven congruent.

PCTCC
(25)

Abreviatura que significa "Las partes correspondientes de los triángulos congruentes son congruentes", que se puede utilizar para justificar una demostración después de demostrar que dos triángulos son congruentes (CPCTC, por sus siglas en inglés).

---

cross products
(41)

In a proportion, the product of the means and the product of the extremes.

$$\frac{1}{2} = \frac{3}{6}$$

Product of means: $2 \cdot 3 = 6$
Product of extremes: $1 \cdot 6 = 6$

productos cruzados
(41)

En una proporción, el producto de los valores medios y el producto de los valores extremos.

---

cross section
(85)

The intersection of a three-dimensional figure and a plane.

sección transversal
(85)

Intersección de una figura tridimensional y un plano.

---

cube
(49)

A prism with six square faces.

cubo
(49)

Prisma con seis caras cuadradas.

---

cylinder
(49)

A three-dimensional figure with two parallel congruent circular bases and a curved lateral surface that connects the bases.

cilindro
(49)

Figura tridimensional con dos bases circulares congruentes y paralelas y una superficie lateral curva que conecta las bases.

English	Example	Spanish

D

decagon
(15)

A ten-sided polygon.

decágono
(15)

Polígono de diez lados.

---

deductive reasoning
(21)

The process of using logic to draw conclusions.

razonamiento deductivo
(21)

Proceso en el que se utiliza la lógica para sacar conclusiones.

---

definition
(1)

A statement that describes a mathematical object and can be written as a true biconditional statement.

definición
(1)

Enunciado que describe un objeto matemático y se puede expresar como un enunciado bicondicional verdadero.

---

degree
(3)

A unit of angle measure, equal to $\frac{1}{360}$ of a circle.

grado
(3)

Unidad de medida de ángulos; un grado es $\frac{1}{360}$ de un círculo.

---

denominator
(SB)

The bottom number of a fraction, which tells how many equal parts are in the whole.

The denominator of $\frac{3}{7}$ is 7.

denominador
(SB)

El número inferior de una fracción, que indica la cantidad de partes iguales que hay en un entero.

---

diagonal of a polygon
(15)

A segment connecting two nonconsecutive vertices of a polygon.

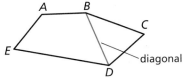

diagonal de un polígono
(15)

Segmento que conecta dos vértices no consecutivos de un polígono.

---

diagonal of a polyhedron
(49)

A segment whose endpoints are vertices of two different faces of a polyhedron.

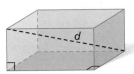

diagonal de un poliedro
(49)

Segmento cuyos extremos son vértices de dos caras diferentes de un poliedro.

---

diameter
(23)

A segment that passes through the center of a circle and has endpoints on the circle; the length of such a segment.

diámetro
(23)

Segmento que atraviesa el centro de un círculo y cuyos extremos están sobre el círculo; longitud de dicho segmento.

English	Example	Spanish
D		

dilation
(84)

A transformation that changes the size of a figure but not its shape.

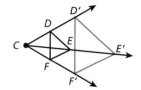

dilatación
(84)

Una transformación que cambia el tamaño de una figura sin cambiar su forma.

---

direct reasoning
(48)

The process of reasoning that begins with a true hypothesis and builds a logical argument to show that a conclusion is true.

razonamiento directo
(48)

Proceso de razonamiento que comienza con una hipótesis verdadera y elabora un argumento lógico para demostrar que una conclusión es verdadera.

---

direct variation
(SB4)

A linear relationship between two variables, x and y, that can be written in the form $y = kx$, where k is a nonzero constant.

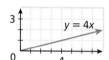

variación directa
(SB4)

Relación lineal entre dos variables, x e y, que puede expresarse en la forma $y = kx$, donde k es una constante distinta de cero.

---

direction of a vector
(63)

The orientation of a vector, which is determined by the angle the vector makes with a horizontal line.

dirección de un vector
(63)

Orientación de un vector, determinada por el ángulo que forma el vector con una recta horizontal.

---

disjunction
(20)

A compound statement that uses the word *or*.

John will walk to work or he will stay home.

disyunción
(20)

Enunciado compuesto que contiene la palabra *o*.

---

distance between two points
(2)

The length of the segment connecting two points.

$$AB = |a - b| = |b - a|$$

distancia entre dos puntos
(2)

La longitud del segmento que conecta los dos puntos.

---

distance from a point to a line
(42)

The length of the perpendicular segment from the point to the line.

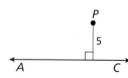

The distance from P to $\overleftrightarrow{AC}$ is 5 units.

distancia desde un punto hasta una línea
(42)

Longitud del segmento perpendicular desde el punto hasta la línea.

English	Example	Spanish

D

dodecagon
(15)

A twelve-sided polygon.

dodecágono
(15)

Polígono de 12 lados.

dodecahedron
(119)

A polyhedron with 12 faces.

dodecaedro
(119)

Poliedro con 12 caras.

E

edge of a three-dimensional figure
(49)

A segment that is the intersection of two faces of the figure.

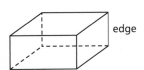
edge

arista de una figura tridimensional
(49)

Segmento que constituye la intersección de dos caras de la figura.

endpoint
(2)

A point at an end of a segment or the starting point of a ray.

extremo
(2)

Punto en el final de un segmento o punto de inicio de un rayo.

enlargement
(84)

A dilation that produces an image larger than its preimage.

agrandamiento
(84)

Una dilatación que produce una imagen más grande que la imagen original.

equal vectors
(63)

Two vectors that have the same magnitude and the same direction.

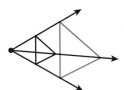
$|\vec{u}| = |\vec{v}| = 2\sqrt{5}$

vectores iguales
(63)

Dos vectores de la misma magnitud y con la misma dirección.

equiangular polygon
(15)

A polygon in which all angles are congruent.

polígono equiangular
(15)

Polígono cuyos ángulos son todos congruentes.

equiangular triangle
(13)

A triangle with three congruent angles.

triángulo equiangular
(13)

Triángulo con tres ángulos congruentes.

equidistant
(11)

The same distance from two or more objects.

X is equidistant from *A* and *B*.

equidistante
(11)

Igual distancia de dos o más objetos.

equilateral polygon (15)		**polígono equilátero** (15)
A polygon in which all sides are congruent.		Polígono cuyos lados son todos congruentes.

equilateral triangle (13)		**triángulo equilátero** (13)
A triangle with three congruent sides.		Triángulo con tres lados congruentes.

Euclidean geometry (109)		**geometría euclidiana** (109)
The system of geometry described by Euclid. In particular, this system of geometry satisfies the Parallel Postulate.		El sistema de geometría descrito por Euclides. En particular, el sistema de geometría que satisface el Postulado de líneas paralelas.

Euler line (38)		**recta de Euler** (38)
The line containing the circumcenter, centroid, and orthocenter of a triangle.		Recta que contiene el circuncentro (U), el centroide (C) y el ortocentro (O) de un triángulo.

event (SB12)	In the experiment of rolling a number cube, the event "an odd number" consists of the outcomes 1, 3, 5.	**suceso** (SB12)
An outcome or set of outcomes in a probability experiment.		Resultado o conjunto de resultados en un experimento de probabilidad.

expansion (84)		**expansión** (84)
See *enlargement*.		Ver *agrandamiento*.

experiment (SB)	Tossing a coin 10 times and noting the number of heads.	**experimento** (SB)
An operation, process, or activity in which outcomes can be used to estimate probability.		Una operación, proceso o actividad en la que se usan los resultados para estimar una probabilidad.

exterior angle of a polygon (15)	∠4 is an exterior angle.	**ángulo externo de un polígono** (15)
An angle formed by one side of a polygon and the extension of an adjacent side.		Ángulo formado por un lado de un polígono y la prolongación del lado adyacente.

English	Example	Spanish

E

exterior of an angle
(3)

The set of all points outside an angle.

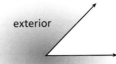
exterior

exterior de un ángulo
(3)

Conjunto de todos los puntos que se encuentran fuera de un ángulo.

---

exterior of a circle
(79)

The set of all points outside a circle.

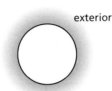
exterior

exterior de un círculo
(79)

Conjunto de todos los puntos que se encuentran fuera de un círculo.

---

exterior of a polygon
(15)

The set of all points outside a polygon.

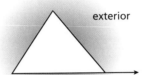
exterior

exterior de un polígono
(15)

Conjunto de todos los puntos que se encuentran fuera de un polígono.

---

external secant segment
(101)

A segment of a secant that lies in the exterior of a circle with one endpoint on the circle.

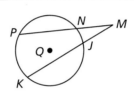
$\overline{NM}$ is an external secant segment.

segmento secante externo
(101)

Segmento de una secante que se encuentra en el exterior del círculo y tiene un extremo sobre el círculo.

---

extremes of a proportion
(41)

In a proportion such that $\frac{a}{b} = \frac{c}{d}$, a and d are the extremes.

valores extremos de una proporción
(41)

En la proporción $\frac{a}{b} = \frac{c}{d}$, a y d son los valores extremos.

F

face of a polyhedron
(49)

A flat surface of a polyhedron.

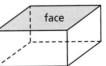
face

cara de un poliedro
(49)

Superficie plana de un poliedro.

---

fair
(SB12)

When all outcomes of an experiment are equally likely.

When tossing a fair coin, heads and tails are equally likely. Each has a probability of $\frac{1}{2}$.

justo
(SB12)

Cuando todos los resultados de un experimento son igualmente probables.

Fibonacci sequence
(Inv 11)

The infinite sequence of numbers beginning 1, 1, … such that each term is the sum of the two previous terms.

1, 1, 2, 3, 5, 8, 13, 21, …

sucesión de Fibonacci
(Inv 11)

Sucesión infinita de números que comienza con 1, 1, … de forma tal que cada término es la suma de los dos términos anteriores.

---

flip
(67)

See *reflection*.

inversión
(67)

Ver *reflexión*.

---

flowchart proof
(31)

A style of proof that uses boxes and arrows to show the structure of the proof.

$\angle 1 \cong \angle 2$
Given

$\angle 1$ and $\angle 2$ are supplementary.
Lin. Pair Thm.

$\angle 1$ and $\angle 2$ are right angles.
$\cong \angle$ supp. → rt. $\angle$

demostración con diagrama de flujo
(31)

Tipo de demostración que se vale de cuadros y flechas para mostrar la estructura de la prueba.

---

formula
(8)

A mathematical relationship expressed with symbols.

fórmula
(8)

Una relación matemática expresada con símbolos.

---

fractal
(Inv 10)

A figure that is generated by iteration.

fractal
(Inv 10)

Figura generada por iteración.

---

frieze pattern
(Inv 8)

A pattern that has translation symmetry along a line.

patrón de friso
(Inv 8)

Patrón con simetría de traslación a lo largo de una línea.

---

frustum of a cone
(103)

A part of a cone with two circular parallel bases.

tronco de cono
(103)

Parte de un cono con dos bases paralelas circulares.

---

frustum of a pyramid
(103)

A part of a pyramid with two square parallel bases.

b_1

b_2

tronco de pirámide
(103)

Parte de una pirámide con dos bases paralelas.

---

function
(SB17)

A relation in which every input is paired with exactly one output.

function: $\{(0, 5), (1, 3), (2, 1), (3, 3)\}$

not a function: $\{(0, 1), (0, 3), (2, 1), (2, 3)\}$

función
(SB17)

Una relación en la que cada entrada corresponde exactamente a una salida.

G

geometric mean
(50)

For positive numbers a and b, the positive number x such that $\frac{a}{x} = \frac{x}{b}$.

$$\frac{a}{x} = \frac{x}{b}$$
$$x^2 = ab$$
$$x = \sqrt{ab}$$

media geométrica
(50)

Dados los números positivos a y b, el número positivo x tal que $\frac{a}{x} = \frac{x}{b}$.

---

geometric probability
(Inv 6)

A form of theoretical probability determined by a ratio of geometric measures such as lengths, areas, or volumes.

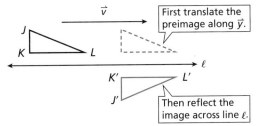

The probability of the pointer landing on blue is $\frac{2}{9}$.

probabilidad geométrica
(Inv 6)

Método para calcular probabilidades basado en una medida geométrica como la longitud o el área.

---

glide reflection
(90)

A composition of a translation and a reflection across a line parallel to the translation vector.

First translate the preimage along $\vec{y}$.

Then reflect the image across line ℓ.

deslizamiento con inversión
(90)

Composición de una traslación y una reflexión sobre una línea paralela al vector de traslación.

---

glide reflection symmetry
(90)

A type of symmetry that describes a pattern that coincides with its image after a glide reflection.

simetría de deslizamiento con inversión
(90)

Un patrón tiene simetría de deslizamiento con inversión si coincide con su imagen después de un deslizamiento con inversión.

---

golden ratio
(Inv 11)

The ratio of a quantity to a larger quantity that is equal to the ratio of the larger quantity to the sum of the two quantities; the numerical value is $\frac{1+\sqrt{5}}{2}$.

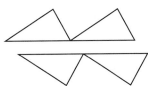

$$\frac{BC}{AB} = \frac{AB}{AC} = \frac{1+\sqrt{5}}{2}$$

razón áurea
(Inv 11)

Si se divide un segmento en dos partes de forma tal que la razón entre la longitud de todo el segmento y la de la parte más larga sea igual a la razón entre la longitud de la parte más larga y la de la parte más corta, entonces dicha razón se denomina razón áurea. La razón áurea es igual a $\frac{1+\sqrt{5}}{2}$.

English	Example	Spanish

G

golden rectangle
(Inv 11)

A rectangle in which the ratio of the length of the longer side to the length of the shorter side is the golden ratio.

rectángulo áureo
(Inv 11)

Rectángulo en el cual la razón entre la longitud del lado más largo y la longitud del lado más corto es la razón áurea.

---

great circle
(80)

A circle on a sphere that divides the sphere into two hemispheres.

great circle

círculo máximo
(80)

En una esfera, círculo que divide la esfera en dos hemisferios.

H

head-to-tail method
(83)

A method of adding two vectors by placing the tail of the second vector on the head of the first vector; the sum is the vector drawn from the tail of the first vector to the head of the second vector.

método de cola a punta
(83)

Método para sumar dos vectores colocando la cola del segundo vector en la punta del primer vector. La suma es el vector trazado desde la cola del primer vector hasta la punta del segundo vector.

---

height of a triangle
(13)

The segment from a vertex that forms a right angle with the line containing the base.

altura de un triángulo
(13)

Segmento que se extiende desde el vértice y forma un ángulo recto con la línea de la base.

---

hemisphere
(80)

Half of a sphere.

hemisferio
(80)

Mitad de una esfera.

---

heptagon
(15)

A seven-sided polygon.

heptágono
(15)

Polígono de siete lados.

---

hexagon
(15)

A six-sided polygon.

hexágono
(15)

Polígono de seis lados.

English	Example	Spanish

H

horizon
(54)

The horizontal line in a perspective drawing that contains the vanishing point(s).

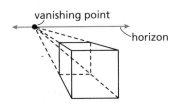

horizonte
(54)

Línea horizontal en un dibujo en perspectiva que contiene el punto de fuga o los puntos de fuga.

---

hypotenuse
(Inv 2)

The side opposite the right angle in a right triangle.

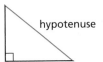

hipotenusa
(Inv 2)

Lado opuesto al ángulo recto de un triángulo rectángulo.

---

hypothesis
(10)

The part of a conditional statement following the word *if*.

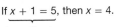

If $x + 1 = 5$, then $x = 4$.
hypothesis

hipótesis
(10)

La parte de un enunciado condicional que sigue a la palabra *si*.

I

icosahedron
(119)

A polyhedron with 20 faces.

icosaedro
(119)

Poliedro con 20 caras. Las caras de un icosaedro regular son triángulos equiláteros y cada vértice es compartido por 5 caras.

---

identity
(91)

An equation that is true for all values of the variable.

$3 = 3$
$2(x - 1) = 2x - 2$

identidad
(91)

Ecuación verdadera para todos los valores de las variables.

---

image
(67)

A shape that results from a transformation of a figure known as the preimage.

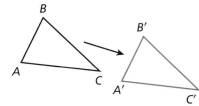

imagen
(67)

Forma resultante de la transformación de una figura conocida como imagen original.

---

incenter of a triangle
(38)

The point of concurrency of the three angle bisectors of a triangle.

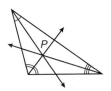

P is the incenter.

incentro de un triángulo
(38)

Punto donde se encuentran las tres bisectrices de los ángulos de un triángulo.

---

included angle
(28)

The angle formed by two adjacent sides of a polygon.

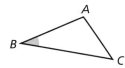

$\angle B$ is the included angle between $\overline{AB}$ and $\overline{BC}$.

ángulo incluido
(28)

Ángulo formado por dos lados adyacentes de un polígono.

English	Example	Spanish
included side (28) The common side of two consecutive angles of a polygon.	$\overline{PQ}$ is the included side between $\angle P$ and $\angle Q$.	**lado incluido** (28) Lado común de dos ángulos consecutivos de un polígono.
indirect measurement (8) A method of measurement that uses formulas, similar figures, and/or proportions.		**medición indirecta** (8) Método para medir objetos mediante fórmulas, figuras semejantes y/o proporciones.
indirect proof (48) A proof in which the statement to be proven is assumed to be false and a contradiction is shown.		**demostración indirecta** (48) Prueba en la que se supone que el enunciado a demostrar es falso y se muestra una contradicción.
inductive reasoning (7) The process of reasoning that a rule or statement is true because specific cases are true.		**razonamiento inductivo** (7) Proceso de razonamiento por el que se determina que una regla o enunciado son verdaderos porque ciertos casos específicos son verdaderos.
inequality (SB15) A statement that compares two expressions by using one of the following signs: $<$, $>$, $\le$, $\ge$, or $\ne$.	$x > 2$	**desigualdad** (SB15) Enunciado que compara dos expresiones utilizando uno de los siguientes signos: $<$, $>$, $\le$, $\ge$, o $\ne$.
initial point of a vector (63) The starting point of a vector.	initial point	**punto inicial de un vector** (63) Punto donde comienza un vector.
initial side (Inv 12) The ray that lies on the polar axis when an angle is drawn in standard position.	135° 45° initial side	**lado inicial** (Inv 12) Rayo que se encuentra sobre el eje x positivo cuando se traza un ángulo en posición estándar.

English	Example	Spanish

I

inscribed angle
(47)

An angle whose vertex is on a circle and whose sides contain chords of the circle.

ángulo inscrito
(47)

Ángulo cuyo vértice se encuentra sobre un círculo y cuyos lados contienen cuerdas del círculo.

---

inscribed circle
(106)

A circle that is tangent to each side of a polygon.

círculo inscrito
(106)

Círculo en el que cada lado del polígono es tangente al círculo.

---

inscribed polygon
(38)

A polygon for which every vertex lies on the same circle.

polígono inscrito
(38)

Polígono cuyos vértices se encuentran sobre el círculo.

---

integer
(SB3)

A member of the set of whole numbers and their opposites.

$\{\ldots -3, -2, -1, 0, 1, 2, 3, \ldots\}$

entero
(SB3)

Miembro del conjunto de números cabales y sus opuestos.

---

intercepted arc
(47)

The arc formed by an inscribed angle.

$\overset{\frown}{DF}$ is the intercepted arc.

arco abarcado
(47)

El arco formado por un ángulo inscrito.

---

interior angle
(15)

An angle formed by two sides of a polygon with a common vertex.

∠1 is an interior angle.

ángulo interno
(15)

Ángulo formado por dos lados de un polígono con un vértice común.

---

interior of a circle
(23)

The set of all points inside a circle.

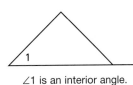

interior de un círculo
(23)

Conjunto de todos los puntos que se encuentran dentro de un círculo.

---

interior of an angle
(3)

The set of all points between the sides of an angle.

interior de un ángulo
(3)

Conjunto de todos los puntos entre los lados de un ángulo.

---

interior of a polygon
(15)

The set of all points inside a polygon.

interior de un polígono
(15)

Conjunto de todos los puntos que se encuentran dentro de un polígono.

English	Example	Spanish
intersection (1) The point or set of points at which two figures meet.	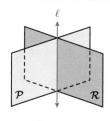	**intersección** (1) El punto o conjunto de puntos donde una figura se une con otra.
inverse (17) The statement formed by negating the hypothesis and conclusion of a conditional statement.	Statement: If $n + 1 = 3$, then $n = 2$ Inverse: If $n + 1 \neq 3$, then $n \neq 2$	**inverso** (17) El enunciado que se forma al negar la hipótesis y la conclusión de un enunciado condicional.
inverse cosine (82) The measure of an angle whose cosine ratio is known.	If $\cos A = x$, then $\cos^{-1} x = m\angle A$.	**coseno inverso** (82) La medida de un ángulo cuya razón coseno es conocida.
inverse function (SB17) The function that results from exchanging the input and output values of a one-to-one function.	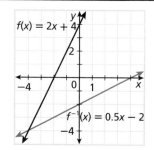 The function $f^{-1}(x) = \frac{1}{2}x - 2$ is the inverse of the function $f(x) = 2x + 4$.	**función inversa** (SB17) Función que resulta de intercambiar los valores de entrada y salida de una función uno a uno. La función inversa de $f(x)$ se indica $f^{-1}(x)$.
inverse sine (82) The measure of an angle whose sine ratio is known.	If $\sin A = x$, then $\sin^{-1} x = m\angle A$.	**seno inverso** (82) Medida de un ángulo cuya razón seno es conocida.
inverse tangent (82) The measure of an angle whose tangent ratio is known.	If $\tan A = x$, then $\tan^{-1} x = m\angle A$.	**tangente inversa** (82) Medida de un ángulo cuya razón tangente es conocida.
irrational number (SB3) A real number that cannot be expressed as a ratio of integers.	$\sqrt{2}$, π, e	**número irracional** (SB3) Número real que no se puede expresar como una razón de dos enteros.
irregular polygon (15) A polygon that is not regular.		**polígono irregular** (15) Polígono que no es regular.

English	Example	Spanish

L

isometric drawing
(54)

A way of drawing three-dimensional figures using *isometric dot paper*, which has equally spaced dots in a repeating triangular pattern.

dibujo isométrico
(54)

Forma de dibujar figuras tridimensionales utilizando *papel punteado isométrico*, que tiene puntos espaciados uniformemente en un patrón triangular que se repite.

---

isometry
(67)

A transformation that does not change the size or shape of a figure.

Reflections, translations, and rotations are all examples of isometries.

isometría
(67)

Transformación que no cambia el tamaño ni la forma de una figura.

---

isosceles trapezoid
(69)

A trapezoid in which the legs are congruent.

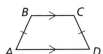

trapecio isósceles
(69)

Trapecio cuyos lados no paralelos son congruentes.

---

isosceles triangle
(13)

A triangle with at least two congruent sides.

triángulo isósceles
(13)

Triángulo que tiene al menos dos lados congruentes.

---

iteration
(Inv 10)

The repeated application of the same rule.

iteración
(Inv 10)

Aplicación repetitiva de la misma regla.

K

kite
(19)

A quadrilateral with exactly two pairs of congruent consecutive sides.

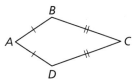

Kite *ABCD*

cometa o papalote
(19)

Cuadrilátero con exactamente dos pares de lados congruentes consecutivos.

Koch snowflake
(Inv 10)

A fractal formed from a triangle by replacing the middle third of each segment with two segments that form a 60° angle.

copo de nieve de Koch
(Inv 10)

Fractal formado a partir de un triángulo sustituyendo el tercio central de cada segmento por dos segmentos que forman un ángulo de 60°.

English	Example	Spanish

L

lateral area
(59)

The sum of the areas of the lateral faces of a prism or pyramid, or the area of the lateral surface of a cylinder or cone.

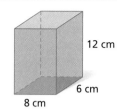

12 cm
6 cm
8 cm

lateral area = (28)(12) = 336 cm²

área lateral
(59)

Suma de las áreas de las caras laterales de un prisma o pirámide, o área de la superficie lateral de un cilindro o cono.

---

lateral edge
(59)

An edge of a prism or pyramid that is not an edge of a base.

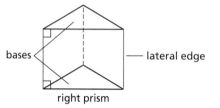

bases — lateral edge
right prism

borde lateral
(59)

Borde de un prisma o pirámide que no es el borde de una base.

---

lateral face
(49)

A face of a prism or a pyramid that is not a base.

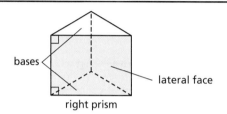

bases — lateral face
right prism

cara lateral
(49)

Cara de un prisma o pirámide que no es la base.

---

lateral surface of a cone
(77)

The curved surface of a cone.

superficie lateral de un cono
(77)

La superficie curva de un cono.

---

lateral surface of a cylinder
(62)

The curved surface of a cylinder.

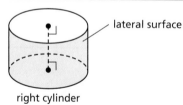

lateral surface
right cylinder

superficie lateral
(62)

Superficie curva de un cilindro o cono.

---

leg of a right triangle
(Inv 2)

One of the two sides of the right triangle that form the right angle.

leg
leg

cateto de un triángulo rectángulo
(Inv 2)

Uno de los dos lados de un triángulo rectángulo que forman el ángulo recto.

---

leg of a trapezoid
(69)

One of the two nonparallel sides of the trapezoid.

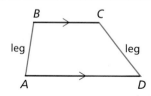

B   C
leg   leg
A   D

cateto de un trapecio
(69)

Uno de los dos lados no paralelos del trapecio.

English	Example	Spanish
leg of an isosceles triangle *(51)* One of the two congruent sides of an isosceles triangle.		**cateto de un triángulo isósceles** *(51)* Uno de los dos lados congruentes del triángulo isósceles.
length *(2)* The measure of the distance between the two endpoints of a segment.	 $AB = \lvert a - b \rvert = \lvert b - a \rvert$	**longitud** *(2)* Distancia entre los dos extremos de un segmento.
line *(1)* An undefined term in geometry; a straight path that has no thickness and extends infinitely.	$\longleftrightarrow \;\ell$	**línea** *(1)* Término indefinido en geometría; una línea es un trazo recto que no tiene grosor y se extiende infinitamente.
line of best fit *(117)* The line that comes closest to all of the points in a data set.	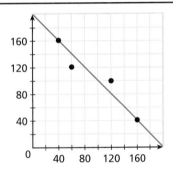	**línea de mejor ajuste** *(117)* Línea que más se acerca a todos los puntos de un conjunto de datos.
line of symmetry *(76)* A line that divides a plane figure into two congruent reflected halves.		**eje de simetría** *(76)* Línea que divide una figura plana en dos mitades reflejas congruentes.
line symmetry *(76)* A type of symmetry that describes a figure that can be reflected across a line so that the image coincides with the preimage.		**simetría axial** *(76)* Figura que puede reflejarse sobre una línea de forma tal que la imagen coincida con la imagen original.
linear equation *(16)* An equation whose graph is a line.	 $y = 3x + 5$ is a linear equation.	**ecuación lineal** *(16)* Una ecuación cuya gráfica es una línea.

GLOSSARY/ GLOSARIO

English	Example	Spanish

L

linear pair
(6)

A pair of adjacent angles whose noncommon sides are opposite rays.

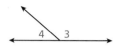
∠3 and ∠4 form a linear pair.

par lineal
(6)

Par de ángulos adyacentes cuyos lados no comunes son rayos opuestos.

---

locus
(Lab 3)

A set of points that satisfies a given condition.

lugar geométrico
(Lab 3)

Conjunto de puntos que cumple con una condición determinada.

---

logically equivalent statements
(17)

Related conditional statements that have the same truth value.

enunciados lógicamente equivalentes
(17)

Enunciados que tienen el mismo valor de verdad.

M

magnitude
(63)

The length of a vector, written $\left|\overrightarrow{AB}\right|$ or $\left|\vec{v}\right|$.

$\left|\vec{u}\right| = 5$

magnitud
(63)

Longitud de un vector, que se expresa $\left|\overrightarrow{AB}\right|$ o $\left|\vec{v}\right|$.

---

major arc
(26)

An arc of a circle whose points are on or in the exterior of a central angle.

$\widehat{ADC}$ is a major arc of the circle.

arco mayor
(26)

Arco de un círculo cuyos puntos están sobre un ángulo central o en su exterior.

---

matrix
(100)

An ordered rectangular arrangement of numbers.

$$\begin{bmatrix} 1 & 0 & 3 \\ -2 & 2 & -5 \\ 7 & -6 & 3 \end{bmatrix}$$

matriz
(100)

Arreglo rectangular ordenado de números.

---

means of a proportion
(41)

In a proportion such that $\frac{a}{b} = \frac{c}{d}$, b and c are the means.

valores medios de una proporción
(41)

En la proporción $\frac{a}{b} = \frac{c}{d}$, b y c son los valores medios.

---

measure of an angle
(3)

The number of degrees in the interior of an angle.

m∠M = 26.8°

medida de un ángulo
(3)

Los ángulos se miden en grados. Un grado es $\frac{1}{360}$ de un círculo completo.

---

measure of a major arc
(26)

The difference of 360° and the measure of the associated minor arc.

$m\widehat{ADC} = 360° - x°$

medida de un arco mayor
(26)

Diferencia entre 360° y la medida del arco menor asociado.

---

measure of a minor arc
(26)

The measure of its central angle.

$m\widehat{AC} = x°$

medida de un arco menor
(26)

Medida de su ángulo central.

---

median of a triangle
(32)

A segment whose endpoints are a vertex of the triangle and the midpoint of the opposite side.

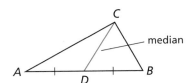

mediana de un triángulo
(32)

Segmento cuyos extremos son un vértice del triángulo y el punto medio del lado opuesto.

---

midpoint
(2)

The point that divides a segment into two congruent segments.

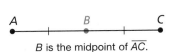

B is the midpoint of $\overline{AC}$.

punto medio
(2)

Punto que divide un segmento en dos segmentos congruentes.

---

midsegment of a trapezoid
(69)

The segment whose endpoints are the midpoints of the legs of the trapezoid.

segmento medio de un trapecio
(69)

Segmento cuyos extremos son los puntos medios de los catetos del trapecio.

---

midsegment of a triangle
(55)

A segment that joins the midpoints of two sides of the triangle.

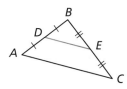

segmento medio de un triángulo
(55)

Segmento que une los puntos medios de dos lados del triángulo.

---

midsegment triangle
(55)

The triangle formed by the three midsegments of a triangle.

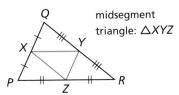

midsegment triangle: △XYZ

triángulo de segmentos medios
(55)

Triángulo formado por los tres segmentos medios de un triángulo.

---

minor arc
(26)

An arc of a circle whose points are on or in the interior of a central angle.

$\widehat{AC}$ is a minor arc of the circle.

arco menor
(26)

Arco de un círculo cuyos puntos están sobre un ángulo central o en su interior.

English	Example	Spanish

N

natural number
(SB)

A counting number.

1, 2, 3, 4, 5, 6, …

número natural
(SB)

Número de conteo.

---

negation
(17)

The opposite of a statement.

Statement: Erin is here.
Negation: Erin is not here.

negación
(17)

La opuesto de un enunciado.

---

net
(Inv 5)

A diagram of the faces of a three-dimensional figure arranged in such a way that the diagram can be folded to form the three-dimensional figure.

10 m
6 m
10 m
6 m

plantilla
(Inv 5)

Diagrama de las caras y superficies de una figura tridimensional que se puede plegar para formar la figura tridimensional.

---

n-gon
(15)

An *n*-sided polygon.

n-ágono
(15)

Polígono de *n* lados.

---

nonagon
(15)

A nine-sided polygon.

nonágono
(15)

Polígono de nueve lados.

---

noncollinear
(1)

Not on the same line.

A   B
• D

Points *A*, *B*, and *D* are non collinear.

no colineal
(1)

Puntos que no se encuentran sobre la misma línea.

---

noncoplanar
(1)

Not in the same plane.

S
T   U   V
$\mathcal{RR}$

T, *U*, *V*, and *S* are non coplanar.

no coplanar
(1)

Puntos que no se encuentran en el mismo plano.

---

non-Euclidean geometry
(109)

A system of geometry in which the Parallel Postulate does not hold.

In spherical geometry, there are no parallel lines. The sum of the angles in a triangle is always greater than 180°.

geometría no euclidiana
(109)

Un sistema de geometría en el cual el Postulado de líneas paralelas no es válido.

---

numerator
(SB)

The top number of a fraction, which tells how many parts of a whole are being considered.

The numerator of $\frac{3}{7}$ is 3.

numerador
(SB)

El número superior de una fracción, que indica la cantidad de partes de un entero que se consideran.

O

English	Example	Spanish
oblique cone (77) A cone where the altitude does not intersect the center of the base.		**cono oblicuo** (77) Un cono cuya altura no interseca al centro de la base.
oblique cylinder (62) A cylinder with bases that are not aligned directly above one another.		**cilindro oblicuo** (62) Un cilindro cuyas bases moesta'n alineadas una directaments arriba de la otra.
oblique prism (59) A prism that has at least one nonrectangular lateral face.		**prisma oblicuo** (59) Prisma que tiene por lo menos una cara lateral no rectangular.
obtuse angle (3) An angle that measures greater than 90° and less than 180°.		**ángulo obtuso** (3) Ángulo que mide más de 90° y menos de 180°.
obtuse triangle (13) A triangle with one obtuse angle.		**triángulo obtusángulo** (13) Triángulo con un ángulo obtuso.
octagon (15) An eight-sided polygon.		**octágono** (15) Polígono de ocho lados.
octahedron (119) A polyhedron with eight faces.		**octaedro** (119) Poliedro con ocho caras.
one-point perspective (54) A perspective drawing with one vanishing point.	vanishing point	**perspectiva de un punto** (54) Dibujo en perspectiva con un punto de fuga.
opposite rays (3) Two rays that have a common endpoint and form a line.	F　E　G $\overrightarrow{EF}$ and $\overrightarrow{EG}$ are opposite rays.	**rayos opuestos** (3) Dos rayos que tienen un extremo común y forman una recta.

English	Example	Spanish
opposite reciprocal *(37)* The opposite of the reciprocal of a number.	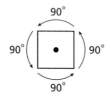 The opposite reciprocal of $\frac{2}{3}$ is $\frac{-3}{2}$.	**recíproco opuesto** *(37)* Opuesto del recíproco de un número.
opposite vectors *(63)* Vectors that have the same magnitude and opposite directions.	The opposite vector of $\langle 3, -2 \rangle$ is $\langle -3, 2 \rangle$.	**vectores opuestos** *(63)* Vectores que tienen la misma magnitud y direcciones opuestas.
order of rotational symmetry *(76)* The number of times a figure with rotational symmetry coincides with itself as it rotates 360°.	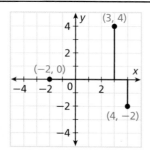 order of rotational symmetry: 4	**orden de simetría de rotación** *(76)* Cantidad de veces que una figura con simetría de rotación coincide consigo misma cuando rota 360°.
ordered pair *(SB)* A pair of numbers (x, y) that can be used to locate a point on a coordinate plane.	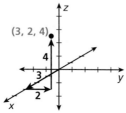	**par ordenado** *(SB)* Un par de números (x, y) que pueden ser utlizados para localizar un punto en un plano coordenado.
ordered triple *(108)* A set of three numbers that can be used to locate a point (x, y, z) in a three-dimensional coordinate system.	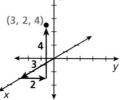	**tripleta ordenada** *(108)* Conjunto de tres números que se pueden utilizar para ubicar un punto (x, y, z) en un sistema de coordenadas tridimensional.
origin *(SB13)* The intersection of the x- and y-axes in a coordinate plane. Its coordinates are $(0, 0)$.		**origen** *(SB13)* Intersección de los ejes x e y en un plano coordenado. Sus coordenadas son $(0, 0)$.
orthocenter of a triangle *(32)* The point of concurrency of the three altitudes of a triangle.	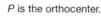 P is the orthocenter. 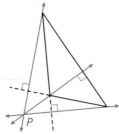	**ortocentro de un triángulo** *(32)* Punto de intersección de las tres alturas de un triángulo.

English	Example	Spanish

O

orthographic drawing
(93)

A drawing that shows a three-dimensional object in which the line of sight for each view is perpendicular to the plane of the picture.

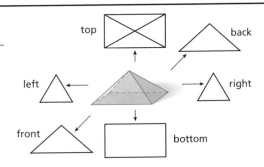

dibujo ortográfico
(93)

Dibujo que muestra un objeto tridimensional en el que la línea visual para cada vista es perpendicular al plano de la imagen.

---

outcome
(SB12)

A possible result of a probability experiment.

In the experiment of rolling a number cube, the possible outcomes are 1, 2, 3, 4, 5, and 6.

resultado
(SB12)

Resultado posible de un experimento de probabilidad.

P

paragraph proof
(31)

A style of proof in which the statements and reasons are presented in paragraph form.

demostración con párrafos
(31)

Tipo de demostración en la cual los enunciados y las razones se presentan en forma de párrafo.

---

parallel lines
(5)

Lines in the same plane that do not intersect.

$r \parallel s$

líneas paralelas
(5)

Líneas rectas en el mismo plano que no se cruzan.

---

parallel planes
(5)

Planes that do not intersect.

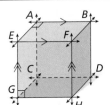

Plane *AEF* and plane *CGH* are parallel planes.

planos paralelos
(5)

Planos que no se cruzan.

---

parallelogram
(19)

A quadrilateral with two pairs of parallel sides.

paralelogramo
(19)

Cuadrilátero con dos pares de lados paralelos.

---

parallelogram method
(83)

A method of adding two vectors by drawing a parallelogram using the vectors as two of the consecutive sides; the sum is a vector along the diagonal of the parallelogram.

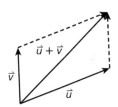

método del paralelogramo
(83)

Método mediante el cual se suman dos vectores dibujando un paralelogramo, utilizando los vectores como dos de los lados consecutivos; el resultado de la suma es un vector a lo largo de la diagonal del paralelogramo.

English	Example	Spanish

P

parent function
(SB17)

The simplest function with the defining characteristics of the family.

$f(x) = x^2$ is the parent function for
$g(x) = x^2 + 4$ and
$h(x) = 5(x + 2)^2 - 3$

función madre
(SB17)

La función más básica que tiene las características distintivas de una familia.

Pascal's triangle
(Inv 11)

A triangular arrangement of numbers in which every row starts and ends with 1 and each other number is the sum of the two numbers above it.

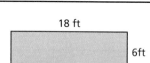

triángulo de Pascal
(Inv 11)

Arreglo triangular de números en el cual cada fila comienza y termina con 1 y cada uno de los otros números es la suma de los dos números que están encima de él.

---

pentagon
(15)

A five-sided polygon.

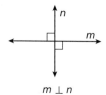

pentágono
(15)

Polígono de cinco lados.

---

perimeter
(8)

The sum of the side lengths of a closed plane figure.

18 ft
6ft
Perimeter = 18 + 6 + 18 + 6 = 48 ft

perímetro
(8)

Suma de las longitudes de los lados de una figura plana cerrada.

---

perpendicular
(5)

Intersecting to form 90° angles, denoted by ⊥.

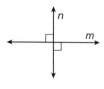
n
m
$m \perp n$

perpendicular
(5)

Que se cruza para formar ángulos de 90°, expresado por ⊥.

---

perpendicular bisector of a segment
(Lab 3)

A line, ray, or segment that is perpendicular to a segment and divides the segment into two congruent parts.

ℓ
A B

ℓ is the perpendicular bisector of $\overline{AB}$.

mediatriz de un segmento
(Lab 3)

Línea perpendicular a un segmento en el punto medio del segmento.

---

perpendicular lines
(5)

Lines that intersect at 90° angles.

n
m

$m \perp n$

líneas perpendiculares
(5)

Líneas que se cruzan en ángulos de 90°.

English	Example	Spanish
P		

perspective drawing
(54)

A drawing in which nonvertical parallel lines appear to meet at one or two points called *vanishing points*.

vanishing point

dibujo en perspectiva
(54)

Dibujo en el cual las líneas paralelas no verticales parecen encontrarse en uno o dos puntos llamados *puntos de fuga*.

---

pi
(23)

The ratio of the circumference of a circle to its diameter, denoted by the Greek letter π.

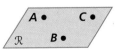
If a circle has a diameter of 5 inches and a circumference of C inches, then $\frac{C}{5} = \pi$, or $C = 5\pi$ inches, or about 15.7 inches.

pi
(23)

Razón entre la circunferencia de un círculo y su diámetro, expresado por la letra griega π (pi).

---

plane
(1)

An undefined term in geometry; a flat surface that has no thickness and extends infinitely.

plane R or plane ABC

plano
(1)

Término indefinido en geometría; un plano es una superficie plana que no tiene grosor y se extiende infinitamente.

---

plane of symmetry
(113)

A plane that divides a three-dimensional figure into two congruent reflected halves.

Plane $\mathcal{N}$ is the plane of symmetry for this figure.

plano de simetría
(113)

Un plano que divide a figuras tridimensionales en dos mitades reflejadas congruentes.

---

plane symmetry
(113)

A type of symmetry that describes a three-dimensional figure that can be divided into two congruent reflected halves by a plane.

plane symmetry

simetría de plano
(113)

Una figura tridimensional que se puede dividir en dos mitades congruentes reflejadas por un plano tiene simetría de plano.

Platonic solid
(119)

One of the five regular polyhedra: a tetrahedron, a cube, an octahedron, a dodecahedron, or an icosahedron.

sólido platónico
(119)

Uno de los cinco poliedros regulares: tetraedro, cubo, octaedro, dodecaedro o icosaedro.

---

point
(1)

An undefined term in geometry; names a location and has no size.

$P \bullet$
point P

punto
(1)

Término indefinido de la geometría que denomina una ubicación y no tiene tamaño.

English	Example	Spanish
P		

point matrix
(100)

A matrix that represents the coordinates of the vertices of a polygon.

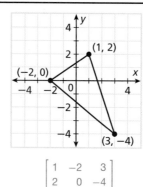

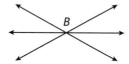

matriz de puntos
(100)

Matriz que representa las coordenadas de los vértices de un polígono.

---

point of concurrency
(32)

A point where three or more lines intersect.

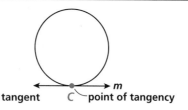

punto de concurrencia
(32)

Punto donde se cruzan tres o más líneas.

---

point of tangency
(43)

The point of intersection of a circle or sphere with a tangent line or plane.

punto de tangencia
(43)

Punto de intersección de un círculo o esfera con una línea o plano tangente.

---

point symmetry
(76)

Rotational symmetry of order 2.

simetría puntual
(76)

Simetría rotacional de orden 2.

---

point-slope form
(SB)

$y - y_1 = m(x - x_1)$, where m is the slope and (x_1, y_1) is a point on the line.

forma de punto y pendiente
(SB)

$(y - y_1) = m(x - x_1)$, donde m es la pendiente y (x_1, y_1) es un punto en la l ínea.

---

polar axis
(Inv 12)

In a polar coordinate system, the horizontal ray with the pole as its endpoint that extends infinitely to the right.

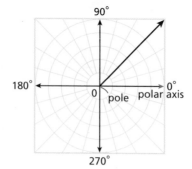

eje polar
(Inv 12)

En un sistema de coordenadas polares, el rayo horizontal, cuyo extremo es el polo, que se encuentra a lo largo del eje x positivo.

English	Example	Spanish

P

---

polar coordinate system
(Inv 12)

A system in which a point in a plane is located by its distance *r* from a point called the pole, and by the measure of a central angle θ.

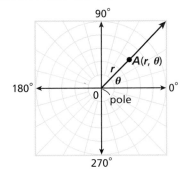

sistema de coordenadas polares
(Inv 12)

Sistema en el cual un punto en un plano se ubica por su distancia *r* de un punto denominado polo y por la medida de un ángulo central θ.

---

pole
(Inv 12)

The point from which distances are measured in a polar coordinate system.

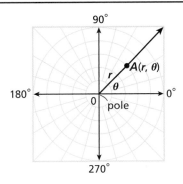

polo
(Inv 12)

Punto desde el que se miden las distancias en un sistema de coordenadas polares.

---

polygon
(15)

A closed plane figure formed by three or more segments such that each segment intersects exactly two other segments only at their endpoints and no two segments with a common endpoint are collinear.

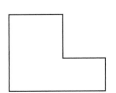

polígono
(15)

Figura plana cerrada formada por tres o más segmentos tal que cada segmento se cruza únicamente con otros dos segmentos sólo en sus extremos y ningún segmento con un extremo común a otro es colineal con éste.

---

polyhedron
(49)

A closed three-dimensional figure formed by four or more polygons that intersect only at their edges.

poliedro
(49)

Figura tridimensional cerrada formada por cuatro o más polígonos que se cruzan sólo en sus aristas.

---

postulate
(2)

A statement that is accepted as true without proof. Also called an *axiom*.

postulado
(2)

Enunciado que se acepta como verdadero sin demostración. También denominado *axioma*.

English	Example	Spanish
preimage *(67)* The original figure in a transformation.	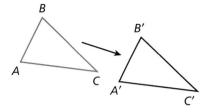	**imagen original** *(67)* Figura original en una transformación.
primes *(67)* Symbols used to label the image in a transformation.	$A'B'C'$	**apóstrofos** *(67)* Símbolos utilizados para identificar la imagen en una transformación.
prism *(49)* A polyhedron formed by two parallel congruent polygonal bases connected by lateral faces that are parallelograms.		**prisma** *(49)* Poliedro formado por dos bases poligonales congruentes y paralelas conectadas por caras laterales que son paralelogramos.
probability *(SB12)* A number from 0 to 1 (or 0% to 100%) that is the measure of how likely an event is to occur.	A bag contains 3 red marbles and 4 blue marbles. The probability of randomly choosing a red marble is $\frac{3}{7}$.	**probabilidad** *(SB12)* Número entre 0 y 1 (o entre 0% y 100%) que describe cuán probable es que ocurra un suceso.
proof *(24)* An argument that uses logic to show that a conclusion is true.		**demostración** *(24)* Argumento que se vale de la lógica para probar que una conclusión es verdadera.
proof by contradiction *(48)* See *indirect proof*.		**demostración por contradicción** *(48)* Ver *demostración indirecta*.
proportion *(41)* A statement that two ratios are equal.	$\frac{2}{3} = \frac{4}{6}$	**proporción** *(41)* Ecuación que establece que dos razones son iguales.
protractor *(3)* A tool used to measure angles.	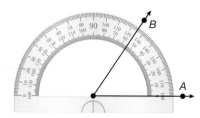	**transportador** *(3)* Un instrumento utilizado para medir ángulos.

English	Example	Spanish

P

pyramid
(49)

A polyhedron formed by a polygonal base and triangular lateral faces that meet at a common vertex.

pirámide
(49)

Poliedro formado por una base poligonal y caras laterales triangulares que se encuentran en un vértice común.

Pythagorean triple
(29)

A set of three nonzero whole numbers a, b, and c such that $a^2 + b^2 = c^2$.

$\{3, 4, 5\}$ $3^2 + 4^2 = 5^2$

Tripleta de Pitágoras
(29)

Conjunto de tres números cabales distintos de cero a, b y c tal que $a^2 + b^2 = c^2$.

Q

quadrant
(SB13)

One of the four regions into which the x- and y-axes divide the coordinate plane.

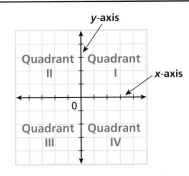

cuadrante
(SB13)

Una de las cuatro regiones en las que los ejes x e y dividen el plano coordenado.

---

quadrilateral
(19)

A four-sided polygon.

cuadrilátero
(19)

Polígono de cuatro lados.

R

radical expression
(29)

An expression that indicates a root of a quantity.

$\sqrt{4}$
$\sqrt{x + 3}$

expresión radical
(29)

Una expresión que indica la raíz de una cantidad.

---

radical symbol
(SB)

The symbol $\sqrt{}$ used to denote a root. The symbol is used alone to indicate a square root or with an index, $\sqrt[n]{}$, to indicate the nth root.

$\sqrt{36} = 6$
$\sqrt[3]{27} = 3$

símbolo de radical
(SB)

Símbolo $\sqrt{}$ que se utiliza para expresar una raíz. Puede utilizarse solo para indicar una raíz cuadrada, o con un índice, $\sqrt[n]{}$, para indicar la *enésima* raíz.

---

radicand
(SB)

The number or expression under a radical sign.

expression: $\sqrt{x + 3}$
radicand: $x + 3$

radicando
(SB)

Número o expresión debajo del signo de radical.

English	Example	Spanish
radius of a circle (23) A segment whose endpoints are the center of a circle and a point on the circle; also the length of that segment.	 radius	**radio de un círculo** (23) Segmento cuyos extremos son el centro y un punto de un círculo; también la longitud de ese segmento.
radius of a cone (77) The distance from the center of the base of the cone to any point on the edge of the base.	 r	**radio de un cono** (77) Distancia desde el centro de la base del cono hasta un punto cualquiera de la base.
radius of a cylinder (62) The distance from the center of a base of the cylinder to any point on the edge of that base.	 r	**radio de un cilindro** (62) Distancia desde el centro de la base del cilindro hasta un punto cualquiera de la base.
radius of a sphere (80) A segment whose endpoints are the center of the sphere and any point on the sphere.	 r	**radio de una esfera** (80) Segmento cuyos extremos son el centro de una esfera y cualquier punto sobre la esfera.
rate of change (16) A ratio that compares the amount of change in a dependent variable to the amount of change in an independent variable.	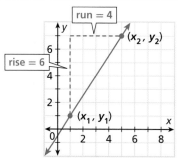 $$\text{rate of change} = \frac{\text{change in } y}{\text{change in } x} = \frac{6}{4} = \frac{3}{2}$$	**tasa de cambio** (16) Razón que compara la cantidad de cambio de la variable dependiente con la cantidad de cambio de la variable independiente.
ratio (41) A comparison of two quantities by division.		**razón** (41) Comparación de dos cantidades mediante una división.
rational number (SB) A number that can be written in the form $\frac{a}{b}$, where a and b are integers and $b \neq 0$.	$3, 1.75, 0.\overline{3}, -\frac{2}{3}, 0$	**número racional** (SB) Número que se puede expresar como $\frac{a}{b}$, donde a y b son números enteros y $b \neq 0$.

English	Example	Spanish

R

ray
(3)

A part of a line that starts at an endpoint and extends infinitely in one direction.

rayo
(3)

Parte de una recta que comienza en un extremo y se extiende infinitamente en una dirección.

---

rectangle
(19)

A quadrilateral with four right angles.

rectángulo
(19)

Cuadrilátero con cuatro ángulos rectos.

---

reduction
(84)

A dilation that creates an image smaller than its preimage.

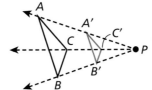

reducción
(84)

Dilatación que crea imagen que es más pequeña que la imagen original.

---

reference angle
(Inv 12)

For an angle in standard position, the positive acute angle formed by the terminal side of the angle and the horizontal axes.

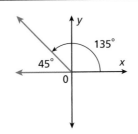

ángulo de referencia
(Inv 12)

Dado un ángulo en posición estándar, el ángulo de referencia es el ángulo agudo positivo formado por el lado terminal del ángulo y el eje *x*.

---

reflection
(67)

A transformation across a line, called the line of reflection, such that the line is the perpendicular bisector of each segment joining each point and its image.

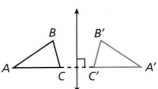

reflexión
(67)

Transformación sobre una línea, denominada la línea de reflexión. La línea de reflexión es la mediatriz de cada segmento que une un punto con su imagen.

---

reflection symmetry
(76)

See *line symmetry*.

simetría de reflexión
(76)

Ver *simetría axial*.

---

regular polygon
(15)

A polygon that is both equilateral and equiangular.

polígono regular
(15)

Polígono equilátero de ángulos iguales.

English	Example	Spanish
R		

regular polyhedron
(119)

A polyhedron in which all faces are congruent regular polygons and the same number of faces meet at each vertex. See also *Platonic solid*.

poliedro regular
(119)

Poliedro cuyas caras son todas polígonos regulares congruentes y en el que el mismo número de caras se encuentran en cada vértice. Ver también *sólido platónico*.

---

regular pyramid
(70)

A pyramid whose base is a regular polygon and whose lateral faces are congruent isosceles triangles.

pirámide regular
(70)

Pirámide cuya base es un polígono regular y cuyas caras laterales son triángulos isósceles congruentes.

---

regular tessellation
(Inv 9)

A repeating pattern of congruent regular polygons that completely covers a plane with no gaps or overlaps.

teselado regular
(Inv 9)

Patrón que se repite formado por polígonos regulares congruentes que cubren completamente un plano sin dejar espacios y sin superponerse.

---

relation
(SB)

A set of ordered pairs.

$\{(0, 5), (0, 4), (2, 3), (4, 0)\}$

relación
(SB)

Conjunto de pares ordenados.

---

remote interior angle
(18)

An interior angle of a polygon that is not adjacent to the exterior angle.

The remote interior angles of ∠4 are ∠1 and ∠2.

ángulo interno remoto
(18)

Ángulo interno de un polígono que no es adyacente al ángulo externo.

---

resultant vector
(63)

The vector that represents the sum of two given vectors.

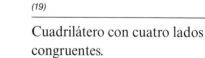

vector resultante
(63)

Vector que representa la suma de dos vectores dados.

---

rhombus
(19)

A quadrilateral with four congruent sides.

rombo
(19)

Cuadrilátero con cuatro lados congruentes.

---

right angle
(3)

An angle that measures 90°.

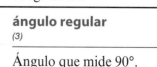

ángulo regular
(3)

Ángulo que mide 90°.

English	Example	Spanish

R

---

right cone
(77)

A cone where the altitude intersects the center of the base.

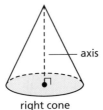

right cone

cono recto
(77)

Un cono cuya altitud interseca en centro de la base.

---

right cylinder
(62)

A cylinder with bases that are aligned directly above one another.

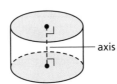

axis

cilindro recto
(62)

Un cilindro con bases que están alienadas una directamente arriba de la otra.

---

right prism
(59)

A prism whose lateral faces are all rectangles.

prisma regular
(59)

Prisma cuyas caras laterales son todas rectángulos.

---

right triangle
(13)

A triangle with one right angle.

triángulo rectángulo
(13)

Triángulo con un ángulo recto.

---

rigid transformation
(67)

See *isometry*.

transformación rígida
(67)

Ver *isometría*.

---

rise
(16)

The difference in the *y*-values of two points on a line.

For the points $(3, -1)$ and $(6, 5)$, the rise is $5 - (-1) = 6$.

distancia vertical
(16)

Diferencia entre los valores de *y* de dos puntos de una línea.

---

rotation
(67)

A transformation about a point *P*, also known as the center of rotation, such that each point and its image are the same distance from *P*.

rotación
(67)

Transformación sobre un punto *P*, también conocido como el centro de rotación, tal que cada punto y su imagen estén a la misma distancia de *P*.

---

run
(16)

The difference in the *x*-values of two points on a line.

For the points $(3, -1)$ and $(6, 5)$, the run is $6 - 3 = 3$.

distancia horizontal
(16)

Diferencia entre los valores de *x* de dos puntos de una línea.

English	Example	Spanish
same-side interior angles *(Inv 1)* For two lines intersected by a transversal, a pair of angles that are on the same side of the transversal and between the two lines.	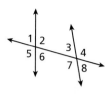 ∠2 and ∠3 are same-side interior angles.	**ángulos internos del mismo lado** *(Inv 1)* Dadas dos rectas cortadas por una transversal, el par de ángulos ubicados en el mismo lado de la transversal y entre las dos rectas.
sample space *(SB)* The set of all possible outcomes of a probability experiment.	In the experiment of rolling a number cube, the sample space is {1, 2, 3, 4, 5, 6}.	**espacio muestral** *(SB)* Conjunto de todos los resultados posibles de un experimento de probabilidad.
scalar *(63)* A quantity that is completely expressed by its magnitude and has no direction.		**escalar** *(63)* Una cantidad que se expresa de manera completa por su magnitud y que no tiene dirección.
scalar multiplication of a vector *(63)* The process of multiplying a vector by a constant.	$3\langle -8, 1\rangle = \langle -24, 3\rangle$	**multiplicación escalar de un vector** *(63)* Proceso por el cual se multiplica un vector por una constante.
scale *(110)* The ratio between two corresponding measurements.	1 cm : 5 mi	**escala** *(110)* Razón entre dos medidas correspondientes.
scale drawing *(110)* A drawing that uses a scale to represent an object as smaller or larger than the actual object.	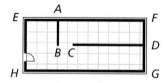 A blueprint is an example of a scale drawing.	**dibujo a escala** *(110)* Dibujo que utiliza una escala para representar un objeto como más pequeño o más grande que el objeto original.
scale factor *(84)* The multiplier used on each dimension of a figure to change it into a similar figure.	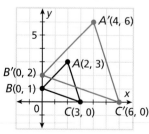 scale factor: 2	**factor de escala** *(84)* El multiplicador utilizado en cada dimensión para transformar una figura en una figura semejante.

English	Example	Spanish
scale model *(110)* A three-dimensional model that uses a scale to represent an object as smaller or larger than the actual object.		**modelo a escala** *(110)* Modelo tridimensional que utiliza una escala para representar un objeto como más pequeño o más grande que el objeto real.
scalene triangle *(13)* A triangle with no congruent sides.		**triángulo escaleno** *(13)* Triángulo sin lados congruentes.
scatter plot *(117)* A graph with points plotted to show a possible relationship between two sets of data.	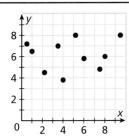	**diagrama de dispersión** *(117)* Gráfica con puntos dispersos para demostrar una relación posible entre dos conjuntos de datos.
secant of an angle *(116)* In a right triangle, the ratio of the length of the hypotenuse to the length of the side adjacent to the angle.	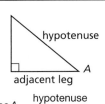 $\sec A = \dfrac{\text{hypotenuse}}{\text{adjacent}} = \dfrac{1}{\cos A}$	**secante de un ángulo** *(116)* En un triángulo rectángulo, la razón de la longitud de la hipotenusa a la longitud del cateto adyacente al ángulo.
secant of a circle *(43)* A line that intersects a circle at two points.	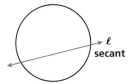	**secante de un círculo** *(43)* Línea que corta un círculo en dos puntos.
secant segment *(101)* A segment of a secant with at least one endpoint on the circle.	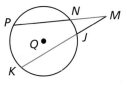 $\overline{NM}$ is an external secant segment. $\overline{JK}$ is an internal secant segment.	**segmento secante** *(101)* Segmento de una secante que tiene al menos un extremo sobre el círculo.
sector of a circle *(35)* A region inside a circle bounded by two radii of the circle and their intercepted arc.		**sector de un círculo** *(35)* Región dentro de un círculo delimitado por dos radios del círculo y por su arco abarcado.

GLOSSARY/ GLOSARIO

English	Example	Spanish
segment bisector *(Lab 3)* A line, ray, or segment that divides a segment into two congruent parts.	 $\overleftrightarrow{CD}$ is the segment bisector of $\overline{AB}$.	**bisectriz de un segmento** *(Lab 3)* Línea, rayo o segmento que divide un segmento en dos segmentos congruentes.
segment of a circle *(112)* A region inside a circle bounded by a chord and an arc.		**segmento de un círculo** *(112)* Región dentro de un círculo delimitada por una cuerda y un arco.
segment of a line *(2)* A part of a line consisting of two endpoints and all points between them.		**segmento de una línea** *(2)* Parte de una línea que consiste en dos extremos y todos los puntos entre éstos.
self-similar figure *(Inv 10)* A figure that can be divided into parts so that each part is similar to the entire figure.		**autosemejante** *(Inv 10)* Figura que se puede dividir en partes, cada una de las cuales es semejante a la figura entera.
semicircle *(26)* An arc of a circle whose endpoints lie on a diameter.		**semicírculo** *(26)* Arco de un círculo cuyos extremos se encuentran sobre un diámetro.
semiregular tessellation *(Inv 9)* A repeating pattern formed by two or more regular polygons in which the same number of each polygon occurs in the same order at every vertex and completely covers a plane with no gaps or overlaps.	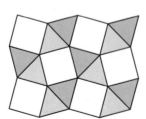	**teselado semirregular** *(Inv 9)* Patrón que se repite formado por dos o más polígonos regulares en el que el mismo número de cada polígono se presenta en el mismo orden en cada vértice y cubren un plano completamente sin dejar espacios vacíos ni superponerse.
side of a polygon *(15)* One of the segments that form a polygon.	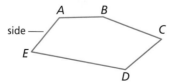	**lado de un polígono** *(15)* Uno de los segmentos que forman un polígono.

right cone
(77)

A cone where the altitude intersects the center of the base.

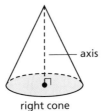

right cone

cono recto
(77)

Un cono cuya altitud interseca en centro de la base.

---

right cylinder
(62)

A cylinder with bases that are aligned directly above one another.

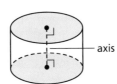

cilindro recto
(62)

Un cilindro con bases que están alienadas una directamente arriba de la otra.

---

right prism
(59)

A prism whose lateral faces are all rectangles.

prisma regular
(59)

Prisma cuyas caras laterales son todas rectángulos.

---

right triangle
(13)

A triangle with one right angle.

triángulo rectángulo
(13)

Triángulo con un ángulo recto.

---

rigid transformation
(67)

See *isometry*.

transformación rígida
(67)

Ver *isometría*.

---

rise
(16)

The difference in the *y*-values of two points on a line.

For the points $(3, -1)$ and $(6, 5)$, the rise is $5 - (-1) = 6$.

distancia vertical
(16)

Diferencia entre los valores de *y* de dos puntos de una línea.

---

rotation
(67)

A transformation about a point *P*, also known as the center of rotation, such that each point and its image are the same distance from *P*.

rotación
(67)

Transformación sobre un punto *P*, también conocido como el centro de rotación, tal que cada punto y su imagen estén a la misma distancia de *P*.

---

run
(16)

The difference in the *x*-values of two points on a line.

For the points $(3, -1)$ and $(6, 5)$, the run is $6 - 3 = 3$.

distancia horizontal
(16)

Diferencia entre los valores de *x* de dos puntos de una línea.

S

English	Example	Spanish
same-side interior angles *(Inv 1)* For two lines intersected by a transversal, a pair of angles that are on the same side of the transversal and between the two lines.	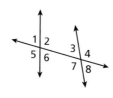 ∠2 and ∠3 are same-side interior angles.	**ángulos internos del mismo lado** *(Inv 1)* Dadas dos rectas cortadas por una transversal, el par de ángulos ubicados en el mismo lado de la transversal y entre las dos rectas.
sample space *(SB)* The set of all possible outcomes of a probability experiment.	In the experiment of rolling a number cube, the sample space is {1, 2, 3, 4, 5, 6}.	**espacio muestral** *(SB)* Conjunto de todos los resultados posibles de un experimento de probabilidad.
scalar *(63)* A quantity that is completely expressed by its magnitude and has no direction.		**escalar** *(63)* Una cantidad que se expresa de manera completa por su magnitud y que no tiene dirección.
scalar multiplication of a vector *(63)* The process of multiplying a vector by a constant.	$3\langle-8, 1\rangle = \langle-24, 3\rangle$	**multiplicación escalar de un vector** *(63)* Proceso por el cual se multiplica un vector por una constante.
scale *(110)* The ratio between two corresponding measurements.	1 cm : 5 mi	**escala** *(110)* Razón entre dos medidas correspondientes.
scale drawing *(110)* A drawing that uses a scale to represent an object as smaller or larger than the actual object.	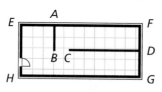 A blueprint is an example of a scale drawing.	**dibujo a escala** *(110)* Dibujo que utiliza una escala para representar un objeto como más pequeño o más grande que el objeto original.
scale factor *(84)* The multiplier used on each dimension of a figure to change it into a similar figure.	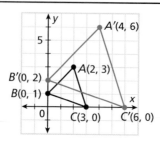 scale factor: 2	**factor de escala** *(84)* El multiplicador utilizado en cada dimensión para transformar una figura en una figura semejante.

English	Example	Spanish

S

---

side of an angle
(3)

One of the two rays that form an angle.

$\overrightarrow{AC}$ and $\overrightarrow{AB}$ are sides of $\angle CAB$.

lado de un ángulo
(3)

Uno de los dos rayos que forman un ángulo.

Sierpinski triangle
(Inv 10)

A fractal formed from a triangle by removing triangles with vertices at the midpoints of the sides of each remaining triangle.

triángulo de Sierpinski
(Inv 10)

Fractal formado a partir de un triángulo al cual se le recortan triángulos cuyos vértices se encuentran en los puntos medios de los lados de cada triángulo restante.

---

similar
(41)

Having the same shape but not necessarily the same size.

semejantes
(41)

Dos figuras con la misma forma pero no necesariamente del mismo tamaño.

---

similar polygons
(41)

Polygons whose corresponding angles are congruent and whose corresponding sides are proportional.

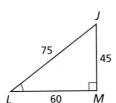

polígonos semejantes
(41)

Polígonos cuyos ángulos correspondientes son congruentes y cuyos lados correspondientes son proporcionales.

---

similarity ratio
(41)

The ratio of two corresponding linear measurements in a pair of similar figures.

Similarity ratio: $\dfrac{3.5}{2.1} = \dfrac{5}{3}$

razón de semejanza
(41)

Razón de dos medidas lineales correspondientes en un par de figuras semejantes.

---

similarity statement
(41)

A statement that indicates that two polygons are similar by listing the vertices in the order of correspondence.

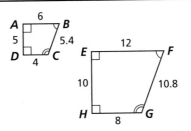

quadrilateral $ABCD \sim$ quadrilateral $EFGH$

enunciado de semejanza
(41)

Enunciado que indica que dos polígonos son semejantes enumerando los vértices en orden de correspondencia.

---

sine of an angle
(68)

In a right triangle, the ratio of the length of the leg opposite the angle to the length of the hypotenuse.

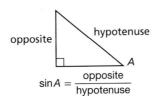

$\sin A = \dfrac{\text{opposite}}{\text{hypotenuse}}$

seno
(68)

En un triángulo rectángulo, la razón de la longitud del cateto opuesto a la longitud de la hipotenusa.

English	Example	Spanish

skew lines
(5)

Lines that are not coplanar.

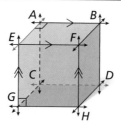

$\overleftrightarrow{AE}$ and $\overleftrightarrow{CD}$ are skew lines.

líneas oblicuas
(5)

Líneas que no son coplanares.

---

slant height of a regular pyramid
(70)

The distance from the vertex of a regular pyramid to the midpoint of an edge of the base.

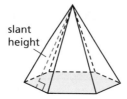

slant height

altura inclinada de una pirámide regular
(70)

Distancia desde el vértice de una pirámide regular hasta el punto medio de una arista de la base.

---

slant height of a right cone
(77)

The distance from the vertex of a right cone to a point on the edge of the base.

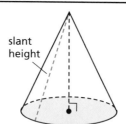

slant height

altura inclinada de un cono regular
(77)

Distancia desde el vértice de un cono regular hasta un punto en el borde de la base.

---

slide
(67)

See *translation*.

deslizamiento
(67)

Ver *traslación*.

---

slope
(16)

A measure of the steepness of a line. The ratio of the vertical change between two points on the line to the horizontal change.

pendiente
(16)

Una medida de la inclinación de una línea. La razón del cambio vertical entre dos puntos en la línea al cambio horizontal.

---

slope-intercept form
(16)

A form of writing a linear equation using the slope (m) and y-intercept (b) of the line; $y = mx + b$.

$y = -2x + 4$
The slope is -2.
The y-intercept is 4.

forma de pendiente-intersección
(16)

La forma de pendiente-intersección de una ecuación lineal es $y = mx + b$, donde m es la pendiente y b es la intersección con el eje y.

---

solid
(49)

A three-dimensional figure.

cuerpo geométrico
(49)

Figura tridimensional.

S

English	Example	Spanish
space *(1)* The set of all points.		**espacio** *(1)* Conjunto de todos los puntos en tres dimensiones.
special right triangle *(53)* A 45°-45°-90° triangle or a 30°-60°-90° triangle.	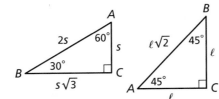	**triángulo rectángulo especial** *(53)* Triángulo de 45°-45°-90° o triángulo de 30°-60°-90°.
sphere *(49)* The set of points in space that are a fixed distance from a given point called the center of the sphere.		**esfera** *(49)* Conjunto de puntos en el espacio que se encuentran a una distancia fija de un punto determinado denominado centro de la esfera.
spherical geometry *(109)* A system of geometry defined on a sphere.		**geometría esférica** *(109)* Sistema de geometría definido sobre una esfera.
square *(19)* A quadrilateral with four congruent sides and four right angles.		**cuadrado** *(19)* Cuadrilátero con cuatro lados congruentes y cuatro ángulos rectos.
standard position *(Inv 12)* An angle that has its vertex at the origin and its initial side on the polar axis.	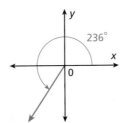	**posición estándar** *(Inv 12)* Ángulo cuyo vértice se encuentra en el origen y cuyo lado inicial se encuentra sobre el eje *x* positivo.
straight angle *(3)* An angle formed by opposite rays that measures 180°.		**ángulo llano** *(3)* Ángulo que mide 180°.
straightedge *(Lab 1)* A tool used to draw straight line segments.		**regla** *(Lab 1)* Un instrumento que se utiliza para dibujar segmentos de líneas rectas.

English	Example	Spanish

S

subtend
(64)

A segment or arc subtends an angle if the endpoints of the segment or arc lie on the sides of the angle.

If *D* and *F* are the endpoints of an arc or chord, and *E* is a point not on $\overline{DF}$, then $\overset{\frown}{DF}$ or $\overline{DF}$ is said to subtend ∠*DEF*.

subtender
(64)

Un segmento o arco subtiende un ángulo si los puntos extremos del segmento o arco yacen sobre los lados del ángulo.

---

supplementary angles
(6)

Two angles whose measures have a sum of 180°.

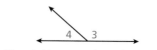

∠3 and ∠4 are supplementary angles.

ángulos suplementarios
(6)

Dos ángulos cuyas medidas suman 180°.

---

surface area
(59)

The total area of all faces and curved surfaces of a three-dimensional figure.

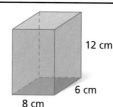

12 cm
6 cm
8 cm

surface area = 2(8)(12) + 2(8)(6) + 2(12)(6) = 432 cm²

área total
(59)

Área total de todas las caras y superficies curvas de una figura tridimensional.

---

symmetry
(76)

A property illustrated when the image of a transformation of a figure coincides with the preimage.

simetría
(76)

Una propiedad ilustrada cuando la imagen de una transformación de una figura coincide con su preimagen.

---

symmetry about an axis
(113)

A type of symmetry that describes a figure for which there is a line it can be rotated about so that the image coincides with the preimage.

simetría axial
(113)

Un tipo de simetría que describe a una figura donde hay una línea alrededor de la cual la figura al ser rotada coincide con su preimagen.

---

system of equations
(81)

A set of two or more equations that have two or more variables.

$2x + 3y = -1$
$3x - 3y = 4$

sistema de ecuaciones
(81)

Conjunto de dos o más ecuaciones que contienen dos o más variables.

T

tangent circles
(72)

Two coplanar circles that intersect at exactly one point.

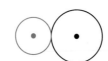

círculos tangentes
(72)

Dos círculos coplanares que se cruzan únicamente en un punto.

English	Example	Spanish
T		
tangent of an angle *(68)* In a right triangle, the ratio of the length of the leg opposite the angle to the length of the leg adjacent to the angle.	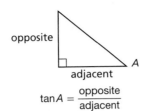	**tangente de un ángulo** *(68)* En un triángulo rectángulo, la razón de la longitud del cateto opuesto al ángulo a la longitud del cateto adyacente al ángulo.
tangent of a circle *(43)* A line that is in the same plane as a circle and intersects the circle at exactly one point.	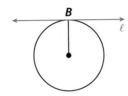	**tangente de un círculo** *(43)* Línea que se encuentra en el mismo plano que un círculo y lo cruza únicamente en un punto.
tangent of a sphere *(80)* A line or plane that intersects the sphere at exactly one point.	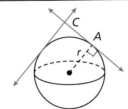	**tangente de una esfera** *(80)* Línea que toca la esfera únicamente en un punto.
tangent segment *(101)* A segment of a tangent line with one point on the circle.	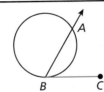 $\overline{BC}$ is a tangent segment.	**segmento tangente** *(101)* Segmento de una tangente con un extremo en el círculo.
terminal point of a vector *(63)* The endpoint of a vector.	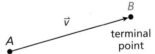	**punto terminal de un vector** *(63)* Extremo de un vector.
terminal side *(Inv 12)* For an angle in standard position, the ray that is rotated relative to the polar axis.	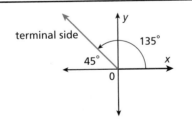	**lado terminal** *(Inv 12)* Para un ángulo en posición estándar, el rayo que se rota en relación con el eje x positivo.
tessellation *(Inv 9)* A repeating pattern of plane figures that completely covers a plane with no gaps or overlaps.	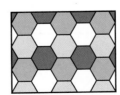	**teselado** *(Inv 9)* Patrón que se repite formado por figuras planas que cubren completamente un plano sin dejar espacios libres y sin superponerse.

English	Example	Spanish

T

tetrahedron
(119)

A polyhedron with four faces.

tetraedro
(119)

Poliedro con cuatro caras.

---

theorem
(4)

A statement that has been proven.

teorema
(4)

Enunciado que ha sido demostrado.

---

theoretical probability
(SB12)

The ratio of the number of equally likely outcomes in an event to the total number of possible outcomes.

In the experiment of rolling a number cube, the theoretical probability of rolling an odd number is $\frac{3}{6} = \frac{1}{2}$.

probabilidad teórica
(SB12)

Razón entre el número de resultados igualmente probables de un suceso y el número total de resultados posibles.

---

three-dimensional coordinate system
(108)

A space that is divided into eight regions by an *x*-axis, a *y*-axis, and a *z*-axis. The locations, or coordinates, of points are given by ordered triples.

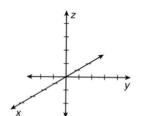

sistema de coordenadas tridimensional
(108)

Espacio dividido en ocho regiones por un eje *x*, un eje *y* un eje *z*. Las ubicaciones, o coordenadas, de los puntos son dadas por tripletas ordenadas.

---

tick marks
(2)

Marks used on a figure to indicate congruent segments.

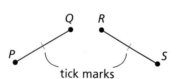

tick marks

marcas "|"
(2)

Marcas utilizadas en una figura para indicar segmentos congruentes.

---

tiling
(Inv 9)

See *tessellation*.

teselación
(Inv 9)

Ver *teselado*.

---

topology
(120)

The study of the properties of a geometric figure that do not change under stress.

topología
(120)

El estudio de las propiedades de una figura geométrica que no cambia al ser sometida a fuerzas.

---

transformation
(67)

A change in the position, size, or shape of a figure or graph.

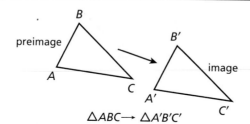

$\triangle ABC \rightarrow \triangle A'B'C'$

transformación
(67)

Cambio en la posición, tamaño o forma de una figura o gráfica.

English	Example	Spanish

T

translation
(67)

A transformation that shifts or slides every point of a figure or graph the same distance in the same direction.

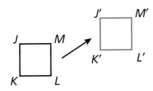

traslación
(67)

Transformación en la que todos los puntos de una figura o gráfica se mueven la misma distancia en la misma dirección.

---

translation symmetry
(Inv 8)

A type of symmetry that describes a figure that can be translated along a vector so that the image coincides with the preimage.

simetría de traslación
(Inv 8)

Una figura tiene simetría de traslación si se puede trasladar a lo largo de un vector de forma tal que la imagen coincida con la imagen original.

---

transversal
(Inv 1)

A line that intersects two coplanar lines at two different points.

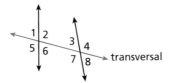

transversal
(Inv 1)

Línea que corta dos líneas coplanares en dos puntos diferentes.

---

trapezium
(19)

A quadrilateral with no parallel sides.

trapezoide
(19)

Un cuadrilátero que no tiene lados paralelos.

---

trapezoid
(19)

A quadrilateral with exactly one pair of parallel sides.

trapecio
(19)

Cuadrilátero con sólo un par de lados paralelos.

---

triangle
(13)

A three-sided polygon.

triángulo
(13)

Polígono de tres lados.

---

triangle rigidity
(25)

A property of triangles that states that if the side lengths of a triangle are fixed, the triangle can have only one size and shape.

rigidez del triángulo
(25)

Propiedad de los triángulos que establece que, si las longitudes de los lados de un triángulo son fijas, el triángulo puede tener sólo una forma.

---

trigonometric ratio
(68)

A ratio of two sides of a right triangle.

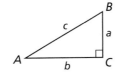

$$\sin A = \frac{a}{c}; \cos A = \frac{b}{c}; \tan A = \frac{a}{b}$$

razón trigonométrica
(68)

Razón entre dos lados de un triángulo rectángulo.

English	Example	Spanish

T

trigonometry
(68)

The study of the relationship between sides and angles of triangles.

trigonometría
(68)

El estudio de la relación entre los lados y los ángulos de triángulos.

---

truth table
(20)

A table that lists all possible combinations of truth values for a statement and its components.

tabla de verdad
(20)

Tabla en la que se enumeran todas las combinaciones posibles de valores de verdad para un enunciado y sus componentes.

---

truth value
(10)

True (T) or false (F); denotes whether or not a statement is true.

valor de verdad
(10)

Verdadero (V) o falso (F); denota si un enunciado es verdadero o falso.

---

turn
(67)

See *rotation*.

giro
(67)

Ver *rotación*.

---

two-column proof
(27)

A style of proof in which the statements are written in the left-hand column and the reasons are written in the right-hand column.

demostración de dos columnas
(27)

Estilo de demostración en la que los enunciados se escriben en la columna de la izquierda y las razones en la columna de la derecha.

---

two-point perspective
(54)

A perspective drawing with two vanishing points.

vanishing points

perspectiva de dos puntos
(54)

Dibujo en perspectiva con dos puntos de fuga.

U

undefined term
(1)

A basic figure that is not defined in terms of other figures. In geometry, they are *point*, *line*, and *plane*.

término indefinido
(1)

Figura básica que no está definida en función de otras figuras. En geometría son el *punto*, la *línea* y el *plano*.

English	Example	Spanish

unit circle
(75)

A circle with a radius of 1, centered at the origin.

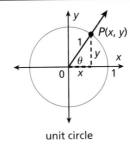

unit circle

círculo unitario
(75)

Círculo con un radio de 1, centrado en el origen.

vanishing point
(54)

In a perspective drawing, a point on the horizon where parallel lines appear to meet.

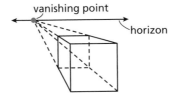

vanishing point
horizon

punto de fuga
(54)

En un dibujo en perspectiva, punto en el horizonte donde todas las líneas paralelas parecen encontrarse.

vector
(63)

A quantity that has both magnitude and direction.

$\vec{u}$

vector
(63)

Cantidad que tiene magnitud y dirección.

Venn diagram
(SB)

A diagram used to show relationships between sets.

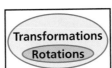

Transformations
Rotations

diagrama de Venn
(SB)

Diagrama utilizado para mostrar la relación entre conjuntos.

vertex angle of an isosceles triangle
(51)

The angle formed by the legs of an isosceles triangle.

vertex angle

E   F

ángulo del vértice de un triángulo isósceles
(51)

Ángulo formado por los catetos de un triángulo isósceles.

vertex of a cone
(77)

The point opposite the base of the cone.

vertex

vértice de un cono
(77)

Punto opuesto a la base del cono.

vertex of a polygon
(15)

The intersection of two sides of the polygon.

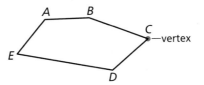

A   B
C
vertex
E
D

A, B, C, D, and E are vertices of the polygon.

vértice de un polígono
(15)

La intersección de dos lados del polígono.

English	Example	Spanish
V		

vertex of a pyramid
(70)

The common vertex of a pyramid's lateral faces.

vértice de una pirámide
(70)

El vértice común a las caras laterales de una pirámide.

---

vertex of a three-dimensional figure
(49)

A point that is the intersection of three or more faces of the figure.

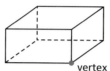

vértice de una figura tridimensional
(49)

Punto que representa la intersección de tres o más caras de la figura.

---

vertex of a triangle
(13)

The intersection of two sides of the triangle.

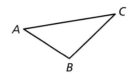

A, B, and C are vertices of $\triangle ABC$.

vértice de un triángulo
(13)

Intersección de dos lados del triángulo.

---

vertex of an angle
(3)

The common endpoint of the sides of the angle.

A is the vertex of $\angle CAB$.

vértice de un ángulo
(3)

Extremo común de los lados del ángulo.

---

vertical angles
(6)

The nonadjacent angles formed by two intersecting lines.

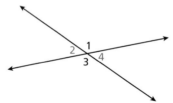

$\angle 1$ and $\angle 3$ are vertical angles.
$\angle 2$ and $\angle 4$ are vertical angles.

ángulos opuestos por el vértice
(6)

Ángulos no adyacentes formados por dos rectas que se cruzan.

---

volume
(59)

The number of nonoverlapping unit cubes of a given size that will exactly fill the interior of a three-dimensional figure.

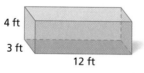

Volume = $(3)(4)(12) = 144$ ft³

volumen
(59)

Cantidad de cubos unitarios no superpuestos de un determinado tamaño que llenan exactamente el interior de una figura tridimensional.

| **W** | | |

whole number
(SB)

The set of natural numbers and zero.

$\{0, 1, 2, 3, 4, 5, ...\}$

número cabal
(SB)

Conjunto de los números naturales y cero.

English	Example	Spanish

X

x-axis
(SB13)

The horizontal axis in a coordinate plane.

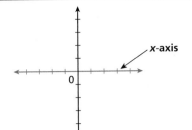

eje x
(SB13)

Eje horizontal en un plano coordenado.

Y

y-axis
(SB13)

The vertical axis in a coordinate plane.

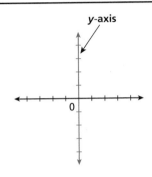

eje y
(SB13)

Eje vertical en un plano coordenado.

Z

z-axis
(108)

The third axis in a three-dimensional coordinate system.

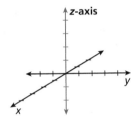

eje z
(108)

Tercer eje en un sistema de coordenadas tridimensional.

Index

265, 270–272, 277–279, 285–287, 292–294, 299–300, 305–307, 312–314, 317–319, 323–326, 331–333, 340–342, 346–348, 352–354, 358–360, 365–367, 371–373, 378–380, 384–386, 393–395, 400–402, 409–411, 415–417, 421–423, 427–429, 433–435, 440–442, 449–450, 454–456, 461–463, 467–469, 474–476, 479–482, 487–489, 492–494, 497–499, 502–505, 509–510, 513–515, 520–523, 527–528, 535–537, 540–542, 545–547, 550–552, 556–559, 562–564, 570–573, 576–578, 582–584, 589–591, 597–599, 603–605, 610–612, 615–617, 621–623, 627–629, 633–635, 639–641, 645–647, 650–652, 659–661, 665–667, 671–673, 677–679, 683–685, 688–690, 693, 695, 701–703, 706, 708, 711–713, 718–720, 723, 725, 728–730, 733–735, 739–741, 744–746, 750, 752–753, 758–760, 764–766, 769–772

Arc(s)
  adjacent, 163
  of circle, measure of, 163
  congruent, 164, 674
  endpoints of, 163
  intercepted, 308
  length, 224
  major, 163
  measures, finding, 310
  minor, 163
  semicircle, measure of, 163
  Theorem 104-1, 674–675

Arc Addition Postulate (Postulate 14), 164–165

Arc measure
  adjacent arcs, 163
  Arc Addition Postulate (Postulate 14), 164–165
  arc of circle, 163
  arcs and angles, identifying, 164
  central angle, 163
  circle, 163
  congruent arcs, 164
  endpoints of arc, 163
  endpoints of semicircle, 163
  finding, 164
  major arc, 163
  minor arc, 163
  semicircle, 163
  semicircle, measure of, 163

Area(s)
  application: land surveying, 439
  circle, 146–147
  composite figure, 258
  coordinates, calculating with, 375
  coordinates, estimating with, 376–377
  cross sections of solids, 554
  differing, 691
  equilateral triangle, 438
  figure, 48
  formulas using geometry, 48
  irregular polygon, 376
  lateral, 389
  maximizing, 692–695
  parallelogram, 137–138
  polygon, 625, 757
  quadrilaterals, 137–140
  ratios, 567
  rectangle, 117, 138
  regular hexagon, 438
  regular polygon, 437
  rhombus, 139–140
  square, 117, 138
  surface, of solid, 389
  trapezoid, 138–139
  triangle, 80

ASA Triangle Congruence Postulate, 189–191

Associative Property of Multiplication, 151

Auxiliary line, 199

Axis
  reflection across, 490
  symmetry about, 726

B _____

Base
  cone, 506
  cylinder, 412
  isosceles triangle, 336
  parallelogram, 137
  pyramid, 464
  trapezoid, 457

Base angles
  isosceles trapezoid, 458
  isosceles triangle, 336
  trapezoid, 457

Biconditional statement, 121

Bisect, 15

Bisector of angle, 15, 40

C _____

Calculator use, 538

Cartesian coordinates, 773

Cavalieri's Principle, 555

Center
  circle, 145
  dilation, 548
  regular polygon, 436
  sphere, 524

Central angle
  arc measure, 163
  defined, 308
  formula, 197
  hexagon, 196
  hexagon, sum of angles, 196
  pentagon, 196
  pentagon, sum of angles, 196
  quadrilateral, 196
  quadrilateral, sum of angles, 196
  triangle, 196
  triangle, sum of angles, 196
  unknown measure, solving, 226

Centroid of triangle
  application: balancing objects, 207
  centroid, finding on coordinate plane, 206
  coordinate plane, finding on, 206
  defined, 205
  segment lengths, find using centroid, 205
  Theorem 32-1: Centroid Theorem, 205
  triangle, altitude of, 206
  triangle, orthocenter of, 206
  Centroid Theorem (Theorem 32-1), 205

Check for Understanding, 4, 10, 17, 24, 31, 38, 44, 50, 55, 61, 69, 74, 81, 87, 93, 100, 105, 112, 118, 124, 133, 141, 148, 154, 160, 166, 172, 177, 184, 191, 201, 207, 215, 221, 234, 241, 247, 253, 260, 269, 277, 285, 291, 299, 305, 312, 317, 323, 331, 340, 346, 352, 357, 365, 371, 378, 384, 392, 399, 409, 415, 421, 426, 433, 439, 448, 454, 461, 467, 474, 479, 486, 492, 497, 502, 509, 513, 520, 526, 534, 540, 545, 550, 556, 562, 570, 576, 582, 588, 597, 603, 609, 615, 620, 627, 632, 638,

644, 650, 665, 676, 683, 688, 693,
701, 706, 711, 718, 723, 728, 733,
739, 744, 750, 758, 763, 768–769

Chord(s)
angle measure, 424
congruent, 674
congruent properties, applying, 284
defined, 281, 560
endpoints of, 674
intersecting, 560
lengths, 561
perpendicular bisector of, 283
Theorem 104-1, 674
Theorem 104-1, proving, 675

Chord-diameter relationships, 282

Circle(s)
angles exterior to, 518
application: astronomy, 497
application: eclipses, 519
application: emergency planning, 620
arc measure, 163
arc of, 163
area, 146–147
center, 145
centered at origin, dilating, 619
centered at the origin, analyzing, 495
chord-diameter relationships, 282
circumference, 145–146
concentric, 630–35
congruent, 145
congruent radii, 145
on coordinate plane, 495
diameter of, 145, 281
dimensions of, altering, 626
equation of, 496, 618–623
equation of, writing from a graph, 496
exterior of, 518
graphing from equation, 496
great, 525
inscribed angles, 308–314
interior, 145
intersecting lines, identifying, 281
intersecting segments, identifying, 281
noncollinear points through three, 250
not centered at origin, dilating, 619

perimeter, 145
perpendicular bisector of chord containing the center, 283
point of tangency, 477
radius of, 145–46, 281
secant of, 281, 518
segment of, 721
tangent of, 518
Theorem 43-1, 282
Theorem 43-2, 282
Theorem 43-3, 283
Theorem 43-4, 283
Theorem 79-1, 518–519
translating, 618–19
unit, 495
unknown measure, solving, 225

Circular cross section, 553

Circumcenter
in coordinate plane, 246
of triangle, 245

Circumference of circle, 145–146

Circumscribed figures, 686–690

Classes, topological, 768–769

Classifying angle(s), 15

Collinear points
defined, 2
finding, 700
lines and, 2

Common tangent
defined, 477
external, 477
internal, 477
solving problems with, 477

Complement, 34

Complementary angles, 34–35, 170

Composite figure
application: architecture, 259
application: volume of gas tank, 738
finding area of, 258
defined, 257
finding perimeter of, 257
finding surface area of, 736–737
finding volumes of, 738

Composite solid
volume of, 736–741

Composite transformations
application: design, 588
composite reflection, rotating by, 586
composition, analyzing order of, 587

glide reflection, performing, 585
two reflections across two parallel lines, 586

Composition of two isometries, 585

Concave polygon, 91

Concentric circles
annulus of, 631
application: targets, 632
equations of, 631
determining two circles are concentric, when 630

Conclusion,
defined, 58
valid, 131

Concurrency, point of, 205

Conditional statement
conclusion, identifying, 58
examining contrapositive of, 105
converse of, 60, 103
defined, 58
hypothesis, 58
inverse of, 104
Law of Detachment, 131–132
Law of Syllogism, 131–132
logically equivalent statements, 104
negation of, 104
truth value, 58, 103
using, 58–62

Cone
altitude of, 506
application: cooking, 508
base of, 506
defined, 320
lateral surface area of, 506–507
oblique, 506
radius, 506
right, 506
slant height of, 506
surface area of, 507
vertex of, 506
volume of, 508
volume of frustum, 669

Congruence
proving with coordinates, 601
equality and, 8
Reflexive Property of, 7–8
segments, congruent, 7
statement, 7, 159
Symmetric Property of, 7–8
Transitive Property of, 7–8

lateral area of, 413
radius of, 412
surface area of, 413
volume of, 414

D

Decagon, 90

Deductive reasoning
given statement, 132
Law of Detachment, 130–132
Law of Syllogism, 131–132
in proof, 169
using, 130

Degrees of angle(s), 14

Depression, angle of, 483–484

Diagonal(s)
as angle bisectors, 344
congruent, 343
of parallelograms, 343
of polygons, 91
of quadrilaterals, 195
of rectangles, 343
of regular polygons, 194
of rhombuses, 344
perpendicular, 344

Diameter
defined, 145
of a circle, 281

Dilation
application: photocopiers, 549
center of, 548
of a circle centered at origin, 619
contracting by, 549
in coordinate plane, 663–667
on coordinate plane, 549
defined, 548
enlarging by, 548

Direct reasoning, 315

Disjunction, 731–732

Distance
formula, 53–57, 374–375
Law of Cosines, 636
Law of Sines, finding with, 613–614
from a point to a line, 273–277
in three dimensions, 716–717

Distributive Property of Equality, 152–153

Division Property of Equality, 151–153

Dodecagon, 90

Dodecahedron, 761–762

Drawing
application: drafting, 357
isometric, 356
one-point perspective, 355
two-point perspective, 356

E

Elevation
angle of, 483, 485

Endpoints
of a arc, 163
of a semicircle, 163

Enlargement, 548

Equality, 8

Equation. *See also* **formula**
of circles, 618
of concentric circles, 631
conic section, 495
of lines, 96–101
of parallel lines, 237–243
of perpendicular lines, 237–243
quadratic, 85

Equiangular polygon, 89

Equilateral polygon, 89

Equilateral triangle, 79, 438

Equivalence, topological, 768

Error Analysis, 6, 11, 17, 31, 38, 51, 56, 62, 69–70, 76, 82, 87, 101, 114, 120, 125, 133, 143, 149, 154, 161, 167, 172, 178, 186–187, 193, 203, 209, 216, 222, 228, 234, 242, 248, 254, 260–261, 270, 285, 293, 300, 306, 313, 325, 333, 347, 360, 365, 372, 380, 386, 393, 394, 400, 415, 422, 428, 433, 441–442, 450, 455, 469, 479, 489, 493, 498, 503, 513, 521, 537, 541, 547, 558, 564, 571, 578, 584, 590, 599, 611, 615, 622, 628, 633, 640, 647, 652, 667, 671, 690, 694, 702, 708, 711, 718, 723, 729, 734, 740, 745, 750, 760, 764, 772

Estimate. *See* **Math Reasoning**

Euclidean geometry, 704

Euler's Formula, 322

Euler's Line, 245

Expansion, 548

Exterior
of angle(s), 13
point, 383
of a polygon, 91

Exterior angle
in regular polygons, 195

Exterior Angle Theorem, 111

External
common tangent, 477
secant segment, 656
tangent circles, 478

Extremes, 266

F

Flowchart proof
defined, 198
interpreting, 198–199
writing, 199

Formula. *See also* **equation**
area of a circle, 146
area of a parallelogram, 137–138
area of a rectangle, 138
area of a regular polygon, 437
area of a rhombus, 139
area of a sector, 225
area of a segment, 721
area of a spherical triangle, 705
area of a square, 138
area of a trapezoid, 138–139
area of a triangle, 80, 83
central angle measure of a
regular polygon, 197
circumference of circle, 145–146
defined, 47
distance, 53
Euler's, 322
exterior angle measure of a
regular polygon, 196
geometry, using, 47–50
golden ratio, 714
Heron's, 83
interior angle measure of a
regular polygon, 195
lateral area of a cylinder, 413
lateral area of a prism, 389
lateral area of a regular
pyramid, 464
lateral area of a cone, 506–7
length of an arc, 224
pi (π), 146
point-slope, 238
Pythagorean Theorem, 211
Pythagorean triples, 181
radius of a circle, 146
slant height of a cone, 506
slope, 96
slope-intercept, 96

Inscribed

    angle theorems, 308–310

    circle, 687

    circle in regular polygon, 687

    figures, 686–690

    polygon, 686

    quadrilaterals, 311

    triangles, 309

Inscribed angles

    application: air traffic control, 311

    circles, 308–314

    finding, 310

    Theorem 47-1, 308

    Theorem 47-2, 308

    Theorem 47-3, 310

    Theorem 47-4, 310

Intercepted arc, 308

Interior

    of angle(s), 13

    of circle, 145

Interior angle

    formula, 195

    hexagon, 194

    pentagon, 194

    polygon, 92

    quadrilateral, 194

    sum of angles, hexagon, 194

    sum of angles, pentagon, 194

    sum of angles, quadrilateral, 194

    sum of angles, triangle, 194

    triangle, 194

Interior of polygon, 91

Internal common tangent, 477

Internal tangent circles, 478

Intersecting chords, 561

Intersecting lines

    identifying, 281

    segment lengths, finding, 398

Intersecting segments, 281

Intersection, 3

Inverse

    of conditional statements, 104

    cosine, 538–539

    sine, 538

    tangent, 538–539

Investigate practice

    analyze, 65

    generalize, 63

    maps, 64

    multi-step, 63–65

    write, 64

Investigation

    angle relationships and transversals, 63–65

    angles of polygons, 194–197

    fractals, 653–655

    geometric probability, 403–405

    golden ratio, 714–715

    inequalities, 263–265

    inequalities of two triangles, 263–265

    nets, 334–335

    patterns, 529–531

    polar coordinates, 773–775

    proving the Pythagorean theorem, 127–129

    Pythagorean Theorem, 127–129

    tessellations, 592–593

    transversals and angle relationships, 63–65

    trigonometric functions, 470–471

Irregular polygon, 89, 376

Isometric drawing, 357

Isometry, 472, 490

Isometry map, 445

Isosceles trapezoid

    base angles, 458

    defined, 458

    diagonals of, 459

    properties of, 458

Isosceles triangle

    application: engineering, 339

    base angle, 336

    base of, 336

    Corollary 51–1–1, 336

    Corollary 51–2–1, 337

    defined, 79

    leg of, 336

    proving, 336–337

    Theorem 51-1: Isosceles Triangle Theorem, 336–337

    Theorem 51-2: Converse of the Isosceles Triangle Theorem, 337

    Theorem 51-3, 338–339

    Theorem 51-4, 338–339

    vertex angle, 336

Iteration, 653

J _____

Justify. *See* **Math Reasoning**

Justifying statements, 169–172

K _____

Kite, 115, 459

Koch snowflake, 654

L _____

Labs

    chords and tangents in Sketchpad, 565–566

    circle through three noncollinear points, 250

    congruent segments and angles, 19–20

    congruent triangles, 180

    distinguishing types of quadrilaterals, 606

    lines of best fit, 754–755

    maximizing area, 696–697

    parallel line through a point, 77

    perpendicular bisectors and angle bisectors, 40–41

    perpendicular line through a point on a line, 33

    perpendicular through a point not on line, 280

    quadrilaterals, 606

    regular polygons, 443–444

    secant segments, 662

    tangents to a circle, 387–388

    transformations, 516–517

Language. *See* **math language**

Lateral area

    cone, 506

    cylinder, 413

    defined, 389

    prism, 389

    regular pyramid, 464

Lateral edge, 321

Lateral face, 321

Law of Cosines. *See also* **cosine**

    angle, finding missing, 637–638

    application: surveying, 638

    distance, find with, 636–637

    right triangle, 637

Law of Detachment, 130–132

Law of Sines. *See also* **sine**

    angles, finding missing, 614

    application: surveying, 614

    distance, finding with, 613–14

    formula, 613

Law of Syllogism, 131–132

622, 690, 701, 707, 719, 724, 752, 763

model, 3, 22, 44, 61–62, 78, 85, 93, 102, 116–117, 157, 161–162, 179, 193, 220, 227, 254, 263, 287, 319, 334, 339, 344, 356, 371, 383, 394, 407, 409, 416, 469, 475, 480, 482, 487, 489, 502, 509, 514, 529, 536, 555, 557, 584, 592, 607, 610, 611–613, 616–617, 623, 628, 647, 672, 685, 690, 712, 751

predict, 31, 95, 144, 164, 247, 251, 256, 270, 348, 386, 403, 427, 454, 466, 473, 550, 565, 592–593, 638, 645, 649, 682, 722, 744

verify, 18, 20, 51–52, 54, 56, 69, 94, 100–101, 104–105, 108, 112–113, 119, 134, 147, 152–153, 156, 158–159, 171, 173, 177, 181, 195, 204, 209, 215, 229, 236, 248, 270, 273, 288, 292, 317, 323, 325, 350, 379, 401, 435, 461, 469, 488, 491, 504, 521–522, 534, 551, 557, 563, 567, 581, 585, 595, 597, 604, 612, 617, 621, 629, 666, 708, 715, 725, 730, 735, 741, 753, 770

write, 5, 11, 18, 31–32, 38–39, 44–45, 47, 51, 56, 61, 70, 74–76, 79, 82–84, 87–88, 93–94, 98, 100–102, 107, 114, 118–120, 125–126, 128, 134, 143, 145, 148–149, 154–155, 161–162, 167–168, 172–173, 178–179, 184, 192, 196–197, 216–217, 222–223, 227, 235–236, 240, 255, 260, 265–266, 272, 274, 278, 283–284, 287, 292–293, 301, 304, 306–307, 315, 319–320, 325, 328, 331, 334–335, 340, 346, 348, 352, 354, 358, 367, 371, 380, 382, 385–386, 393, 399, 401, 411, 417, 425, 427, 435, 440, 455, 461, 468, 475, 481, 494, 499, 505, 510, 514–515, 519, 521, 523, 527–528, 536, 539, 540, 546–547, 550, 551, 564, 571–573, 576, 592, 597, 605, 610, 628, 634–635, 639, 652–655, 667, 672–673, 678–679, 685, 692, 695, 701–702, 707,

709, 711–713, 720, 724, 734, 740, 745, 752, 759–760, 764, 771–772

Math-to-Math

algebra, 6, 11, 17, 18, 26, 31–32, 38–39, 45, 51, 56–57, 62, 65, 69, 75, 82–83, 87, 95, 101, 107–108, 112–113, 118, 126, 135, 142, 148–149, 155, 161–162, 166–168, 173–174, 178–179, 185, 202–204, 215–217, 223, 227, 241, 272, 278–279, 285–286, 294, 300, 306–307, 313–314, 318, 324–326, 331–333, 340, 347–348, 352–353, 358–359, 366–367, 371–373, 379, 385, 394, 400, 411, 414–417, 421–423, 428, 433, 435, 440–442, 449, 454, 463, 468–469, 475, 480, 482, 487, 489, 493–494, 499, 504–505, 509–510, 515, 521, 523, 527–528, 536, 540, 542, 546–547, 550–552, 557–558, 563–564, 571–572, 576, 578, 583, 590, 598–599, 605, 610–612, 616–617, 621, 627–628, 635, 639, 646–647, 651–652, 659–661, 665, 667, 671–672, 679, 685, 689–690, 695, 702–703, 708, 711–713, 718–720, 725, 728, 734, 741, 745–746, 751, 759, 765–766, 771

coordinate geometry, 6, 50, 56, 173, 243, 248, 299, 469, 514, 537, 546, 597, 685

data analysis, 11

probability, 325, 633, 665

Matrix (matrices)

adding, 648

adding to point matrix, 649

application: fencing, 758

application: stained glass, 682–683

area, calculating using, 757

defined, 648

determinant of, taking, 756

multiplication, 680–682

point, 648

reflecting by, 681

rotating by, 682

transformation, 648–650

translating with, 649

Maximizing area, 691–693

Means, 266

Measures of arcs, 310

Measuring angle(s), 15

Median of triangle, 205

Midpoint

application: navigation, 68

defined, 10

finding, 66–68

midpoint on a coordinate plane, 66

midpoint on a number line, 66

of sides, 68

Midsegment of trapezoid, 451

Midsegment of triangle

application: maps, 364

defined, 361

similarity, applying, 363

Theorem 55-1, 361

Theorem 55-2, 362

Minor arc, 163

Model. *See* **Math Reasoning**

Multiple Choice, 5, 11, 17, 31, 39, 46, 51, 57, 61–62, 70, 75–76, 82, 87, 95, 101–102, 107–108, 112, 114, 118, 125, 134, 141, 149, 151, 154–155, 161, 163, 166–167, 174, 178, 185, 192, 203, 209, 216, 222, 227, 235, 242, 249, 254–255, 261, 265, 272, 278, 286, 294, 300, 305, 314, 318, 324, 326, 332–333, 342, 347, 353, 359, 365, 372, 379–380, 385–386, 393, 401, 406, 410–411, 415, 418, 421–422, 427–428, 430, 434–436, 441, 445, 449–451, 454–456, 462–463, 468, 475–476, 480, 483, 487–488, 490, 492, 495, 498–499, 503–505, 509, 513, 518, 520–521, 528, 535–536, 541, 543, 546–547, 550, 553, 557, 560, 563–564, 567, 571, 573, 576, 583, 589, 598, 600, 603, 610–611, 616–618, 622, 624, 628, 634, 641, 647–648, 651, 654–656, 660–661, 663, 666, 673, 677, 684–686, 688, 691, 694, 698, 703, 707–708, 711–713, 720, 724–725, 729, 734, 740, 745, 752–753, 759–760, 764–765, 770, 771

Multiplication matrix, 680–682

Multiplication Properties of Equality, 151–152

Multi-Step, 12, 45, 70, 76, 88, 95, 100–101, 108, 114, 120, 148, 155,

INDEX

trapezium, 115
trapezoid, 115
trapezoids, 601

R _____

Radical expression, 182–183

Radius
    circle, 145–146, 281
    cone, 506
    cylinder, 412
    sphere, 524

Rate of change, 96

Ratio
    defined, 266
    similarity, 267
    writing, 266

Ray(s)
    angles and, naming, 14
    defined, 13
    sides of an angle, 13

Reading math, 8, 115, 121, 127, 213, 238, 267, 301, 418–419, 446, 451, 473, 538, 544, 594, 737

Reason(s) in two-column proofs, 170

Rectangle
    application: architecture, 346
    area of, 117, 138
    defined, 115
    diagonals of, 343
    parallelogram, 431
    perimeter of, 117
    properties, 115
    Properties of a Rectangle
        Congruent Diagonals, 343

Reduction, 548

Reflection
    across a horizontal line, 491
    across a line, 491
    across an axis, 490
    application: visual arts, 492
    defined, 490
    property of, 490

Reflection matrices, 681

Reflections, 445

Reflexive Properties of Equality, 151

Reflexive Property of Congruence, 7

Regular polygon. *See also* **polygon(s)**
    area of, 437
    center of, 436
    central angle measure, 197
    central angles of, 196–197
    defined, 89

exterior angle measure, 196
exterior angles of, 195–196
inscribed circle in, 687
interior angle measure, 195
interior angles of, 194–195
perimeter of, 437
sum of the interior angles, 194

Regular polyhedron, 761

Regular pyramid
    defined, 464
    lateral area formula, 464
    slant height, 464

Remote interior angles, 111, 252

Rhombus
    area of, 139–140
    Diagonals of a Rhombus as
        Angle Bisectors, 344
    diagonals, using properties of, 344
    parallelogram, 430
    Perpendicular Diagonals, 344
    properties, 115

Right angle, 14

Right cone, 506

Right prism
    defined, 391
    volume of, 391

Right triangle
    45°-45°-90° triangle side length,
        properties of, 349
    geometric mean, 327–330
    identifying similar, 328
    Law of Cosines, 636–638
    properties, 78
    special, 349

Right triangle congruence theorems
    application: engineering, 233
    Theorem 36-1: Leg-Angle (LA)
        Right Triangle Congruence
        Theorem, 230
    Theorem 36-2: Hypotenuse-
        Angle (HA) Right Angle
        Congruence Theorem, 231
    Theorem 36-3: Leg-Leg (LL)
        Right Triangle Congruence
        Theorem, 231
    Theorem 36-4: Hypotenuse-
        Leg (HL) Right Triangle
        Congruence Theorem, 232

Root, 182

Rotation
    about the origin, 511
    center of, 511

chosen point, rotating about, 512
definition, 511
transformations, 445–446

Rotation matrices, 682–683

Rotational symmetry
    angle of, 501
    defined, 501
    order of, 501
    in solids, 727

S _____

Same-side angles, 63

SAS Congruence Postulate.
    See **Side-Angle-Side (SAS)**
    Congruence Postulate

Scale, 709

Scale drawing, 709–710

Scale factor, 266, 548

Scale model, 710

Scalene triangle, 79

Scatterplot, 747

Secant
    of an angle, 742
    of circle, 281
    identifying, 743
    identity, tangent and, 743
    intersection by two, 656
    as rays of an exterior angle, 518
    reciprocal of, 742

Secant segment
    defined, 656
    distance to horizon, 658
    external, 656
    intersection of tangent and, 657
    Theorem 101-1, 656–657
    Theorem 101-2, 657

Sector, area of, 225–226

Segment(s)
    application, hiking, 10
    area of, 721
    area using trigonometry, 722
    of circle, 721
    congruence statements, 7
    congruent, 7
    distance, 8
    endpoints, 7
    equality and congruence, 8
    length, 7
    lengths using intersecting
        transversals, 398
    line segment, 7

T

Tangent line
angle measures and, 381
application: glass cutting, 383
defined, 381
exterior point, relationship of
tangents from, 383
identifying, 382
Theorem 58-1, 381
Theorem 58-2, 382
Theorem 58-3, 383

Terminal side, 773

Tessellations, 592–593

Tetrahedron, 760–762

Theorem
Alternate Exterior Angles
Theorem, 64

Alternate Interior Angles Theorem,
64

Angle-Angle-Side (AAS) Triangle
Congruence Theorem, 189

Centroid Theorem, 205

Congruent Complements Theorem,
35

Congruent Supplements Theorem,
35

Converse of Alternate Exterior
Angles Theorem, 72

Converse of the Alternate Interior
Angles Theorem, 71

Converse of the Hinge Theorem,
264

Converse of the Isosceles Triangle
Theorem, 337

Converse of the Pythagorean
Theorem, 212

Converse of the Same-Side Interior
Angles Theorem, 72

Converse of Triangle
Proportionality Theorem, 397

Exterior Angle Inequality
Theorem, 252

Exterior Angle Theorem, 111

Hinge Theorem, 263

Hypotenuse-Angle (HA) Right
Angle Congruence Theorem,
231

Hypotenuse-Leg (HL) Right
Triangle Congruence Theorem,
232–233

Isosceles Triangle Theorem, 336–
337

Leg-Angle (LA) Right Triangle
Congruence Theorem, 230

Leg-Leg (LL) Right Triangle
Congruence Theorem, 231

Linear Pair Theorem, 36

postulates and, 21–22

Pythagorean Theorem, 48

Pythagorean Inequality Theorem,
213

Same-Side Interior Angles
Theorem, 65

SAS Similarity Theorem, 303

SSS Similarity Theorem, 302–303

Transitive Property of Parallel
Lines Theorem, 29

Trapezoid Midsegment Theorem,
457

Triangle Angle Bisector Theorem,
245

Triangle Angle Sum Theorem, 109

Triangle Inequality Theorem,
252–253

Triangle Midsegment Theorem,
361–362

Triangle Proportionality Theorem,
396

two-column proofs, 169–172

Vertical Angle Theorem, 36

30°-60°-90° right triangle
application: engineering, 370
perimeter of, finding unknown
measures, 369
properties, 368
Pythagorean Theorem,
applying, 370
side lengths, finding, 368

Three-dimensional
composite, 736
coordinate system, 698
figure, 321

Three-dimensions
distance formula for, 716
finding distance in, 716–717
midpoint formula in, 717

Tick marks, 7

Topology
classes, topological, 768–69

defined, 767
equivalence, topological, 768
transformations, topological,
767–777

Transformation
application: stained glass
design, 448
defined, 445
identifying, 446
performing, 447
symmetrical figures using, 501
topological, 767–777

Transitive Property of Equality, 151

Transitive Property of Congruence, 7

**Transitive Property of Parallel Lines
(Theorem 5-7),** 29

Transitive Property of Similarity,
267

Translation
defined, 472
matrices, 649
one dimension, 472
as transformation, 445
two dimensional, 473

Transversals
angle relationships, 63–65
angles, alternate exterior, 63
angles, alternate interior, 63
angles, consecutive interior, 63
angles, corresponding, 63
angles, same-sided, 63
defined, 63
Postulate 11: Corresponding
Angles Postulate, 64
Theorem 10-1: Alternate
Interior Angles Theorem, 64
Theorem 10-2: Alternate
Exterior Angles Theorem, 64
Theorem 10-3: Same-Side
Interior Angles Theorem, 65

Trapezium, 115

Trapezoid
application: woodworking, 460
area of, 138–139
base angles of, 457
base of, 457
isosceles, 458
midsegment of, 457
properties, 115
Theorem 69-1: Trapezoid
Midsegment Theorem, 457